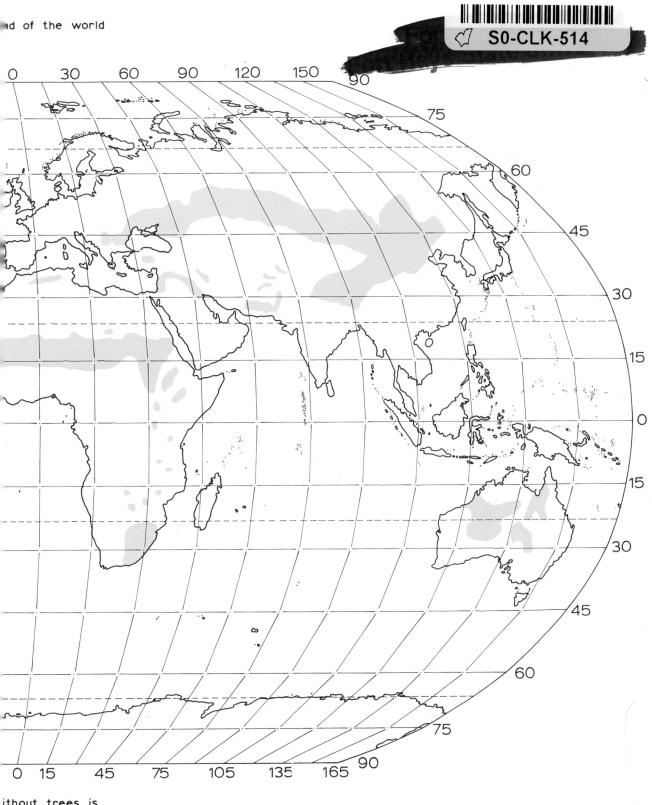

ECOSYSTEMS OF THE WORLD 8B

NATURAL GRASSLANDS

EASTERN HEMISPHERE AND RÉSUMÉ

ECOSYSTEMS OF THE WORLD

Editor in Chief:

David W. Goodall

CSIRO Division of Wildlife and Ecology, Private Bag 4, Midland, W.A. 6056 (Australia)

I. TERRESTRIAL ECOSYSTEMS

    A. Natural Terrestrial Ecosystems
1. Wet Coastal Ecosystems
2. Dry Coastal Ecosystems
3. Polar and Alpine Tundra
4. Mires: Swamp, Bog, Fen and Moor
5. Temperate Deserts and Semi-Deserts
6. Coniferous Forests
7. Temperate Deciduous Forests
8. Natural Grasslands
9. Heathlands and Related Shrublands
10. Temperate Broad-Leaved Evergreen Forests
11. Mediterranean-Type Shrublands
12. Hot Deserts and Arid Shrublands
13. Tropical Savannas
14. Tropical Rain Forest Ecosystems
15. Forested Wetlands
16. Ecosystems of Disturbed Ground

    B. Managed Terrestrial Ecosystems
17. Managed Grasslands
18. Field Crop Ecosystems
19. Tree Crop Ecosystems
20. Greenhouse Ecosystems
21. Bioindustrial Ecosystems

II. AQUATIC ECOSYSTEMS

    A. Inland Aquatic Ecosystems
22. River and Stream Ecosystems
23. Lakes and Reservoirs

    B. Marine Ecosystems
24. Intertidal and Littoral Ecosystems
25. Coral Reefs
26. Estuaries and Enclosed Seas
27. Continental Shelves
28. Ecosystems of the Deep Ocean

    C. Managed Aquatic Ecosystems
29. Managed Aquatic Ecosystems

III. UNDERGROUND ECOSYSTEMS

30. Cave Ecosystems

ECOSYSTEMS OF THE WORLD 8B

NATURAL GRASSLANDS
EASTERN HEMISPHERE AND RÉSUMÉ

Edited by

Robert T. Coupland

Department of Crop Science and Plant Ecology,
University of Saskatchewan
Saskatoon, Sask. S7N 0W0 (Canada)

ELSEVIER
Amsterdam — London — New York — Tokyo 1993

ELSEVIER SCIENCE PUBLISHERS B.V.
P.O. Box 211, 1000 AE Amsterdam, The Netherlands

ISBN 0-444-89557-4 (Vol. 8B)

© 1993 Elsevier Science Publishers B.V. All rights reserved.

No part of this publication may be reproduced, stored in a retrieval system or transmitted in any form or by any means, electronic, mechanical, photocopying, recording or otherwise, without the prior written permission of the publisher, Elsevier Science Publishers B.V., Copyright and Permission Department, P.O. Box 521, 1000 AM Amsterdam, The Netherlands.

Special regulations for readers in the USA — This publication has been registered with the Copyright Clearance Center Inc. (CCC), Salem, MA 01970, USA. Information can be obtained from the CCC about conditions under which photocopies of parts of this publication may be made in the USA. All other copyright questions, including photocopying outside the USA, should be referred to the copyright owner, Elsevier Science Publishers B.V., unless otherwise specified.

No responsibility is assumed by the Publisher for any injury and/or damage to persons or property as a matter of products liability, negligence or otherwise, or from any use or operation of any methods, products, instructions or ideas contained in the material herein.

This book is printed on acid-free paper.

Printed in the Netherlands

PREFACE

This volume is concerned with those parts of the earth's surface where semiarid to subhumid climate determines that the potential (i.e. climax) vegetation is characterized by a more or less continuous layer of grasses and other associated herbs. These grasslands have been modified to a very considerable degree by the activities of man and his domesticated animals. The authors of the various chapters have evaluated the effect of these stresses — cultivation, grazing, browsing, hunting and manipulation of fire — in modifying ecosystem structure and activities of organisms. The extent to which this has been achieved in the various regions varies with the duration and intensity of the stress that was exerted prior to scientific investigation.

The volume commences with an introductory section applicable to natural grassland in general, continues with five sections relating to the respective continental regions, and concludes with a résumé. Because of its length it is published in two parts. The first includes eight synthesis chapters and eleven regional chapters concerning the grasslands of the Western Hemisphere. The Eastern Hemisphere and the résumé compose the second part.

I appreciate the invitation of Dr. David W. Goodall, the Editor-in-Chief of the book series, to contribute this volume. I am also indebted to the 56 contributors from twelve countries who have co-operated in the authorship of this volume. The opportunity to work with them on this project has been very rewarding. Many of the contacts necessary to permit this widespread international effort were the result of activities during the International Biological Programme (IBP). It is largely due to the stimulation provided for multi-disciplinary studies of grassland ecosystems during the IBP that functional processes constitute a considerable part of the discussion in this volume.

Our objective has been to prepare an intensively referenced and indexed synthesis and review of knowledge concerning the natural grasslands of the world in as uniform a style as possible.

My editorial duties have been greatly assisted by the guidance of Dr. Goodall, who made numerous suggestions for the improvement of the various presentations. The contributors have all been very receptive to suggested changes, most of which were directed toward the more uniform treatment of regions and compatibility with other volumes in this series.

ROBERT T. COUPLAND
Professor Emeritus of Plant Ecology
University of Saskatchewan
Saskatoon, Sask., Canada

LIST OF CONTRIBUTORS TO VOLUME 8

DONALD F. ACTON
Agriculture Canada Land Resource Research
  Centre
University of Saskatchewan Campus
Saskatoon, Sask. S7N 0W0 (Canada)

JAMES W. BARTOLOME
Department of Forestry and Resource
  Management
University of California
Berkeley, CA 94720 (U.S.A.)

I.V. BORISOVA
Komarov Botanical Institute
Popova Ulitsa 2
St. Petersburg 197376 (Russia)

MIGUEL A. CAUPHÉPÉ
National Institute of Agricultural Technology
  (INTA)
Balcarce, Prov. Buenos Aires (Argentina)

FRANCIS E. CLARK
Agriculture Research, Western Region
United States Department of Agriculture
Fort Collins, CO 80522 (U.S.A.)

ROBERT T. COUPLAND
Department of Crop Science and Plant Ecology
University of Saskatchewan
Saskatoon, Sask. S7N 0W0 (Canada)

MILO L. COX
School of Renewable Natural Resources
University of Arizona
Tucson, AZ 85721 (U.S.A.)

REXFORD DAUBENMIRE
31150 Interlochen Drive
Mt. Plymouth, FL 32776 (U.S.A.)

VICTOR A DEREGIBUS
Department of Ecology, Faculty of Agronomy
University of Buenos Aires
Buenos Aires (Argentina)

DAVID D. DIAMOND
Texas Natural Heritage Program
Texas Parks and Wildlife Department
4200 Smith School Road
Austin, TX 78744 (U.S.A.)

HERMAN J. DIRSCHL
Environmental Social Advisory Services Inc.
370 Churchill Ave., N.
Ottawa, Ont. K1Z 5C2 (Canada)

JOHN F. DORMAAR
Agriculture Canada Research Station
Lethbridge, Alta. TIJ 4B1 (Canada)

L. CARLOS FIERRO
Rancho Experimental Las Campana
Chihuahua (Mexico)

GEORGE M. FRAME
136 Spruce
Westville, NJ 08093 (U.S.A.)

ANDY N. GILLISON
Tropical Forest Research Centre
CSIRO Division of Wildlife and Ecology
P.O. Box 780
Atherton, Qld. 4883 (Australia)

S.R. GUPTA
Department of Botany
Kurukshetra University
Kurukshetra 132119 (India)

N.P. GURICHEVA
Komarov Botanical Institute
Popova Ulitsa 2
St. Petersburg 197376 (Russia)

C. WAYNE HANSELKA
Texas Agricultural Extension Service
Texas A&M Research and Extension Centre
Corpus Christi, TX (U.S.A.)

HAROLD F. HEADY
1964 Capistrano Ave.
Berkeley, CA 94707 (U.S.A.)

DENNIS J. HERLOCKER
Kenya Range Management Handbook Project
G.T.Z., P.O. Box 47051
Nairobi (Kenya)

ROGER J. HNATIUK
Australian National Botanic Gardens
G.P.O. 1777
Canberra, A.C.T. 2601 (Australia)

Z.V. KARAMYSHEVA
Komarov Botanical Institute
Popova Ulitsa 2
St. Petersburg 197376 (Russia)

JAMES O. KLEMMEDSON
School of Renewable Natural Resources
University of Arizona
Tucson, AZ 85721 (U.S.A.)

JEAN KOECHLIN
1, rue de l'Abreuvoir
17000 La Rochelle (France)

CLAIR L. KUCERA
Division of Biological Sciences
University of Missouri
Columbia, MO 65211 (U.S.A.)

WILLIAM K. LAUENROTH
Department of Range Science
Colorado State University
Fort Collins, CO 80523 (U.S.A.)

RAUL S. LAVADO
Department of Ecology, Faculty of Agronomy
University of Buenos Aires
Buenos Aires (Argentina)

E.M. LAVRENKO (deceased)
Komarov Botanical Institute
Popova Ulitsa 2
St. Petersburg 197376 (Russia)

HENRI NOEL LE HOUÉROU
C.N.R.S. – Centre d'Ecologie Fonctionelle et Evolutive
Centre Louis Emberger
Route de Mende
F-34033 Montpellier Cedex (France)

JORGE H. LEMCOFF
Department of Ecology, Faculty of Agronomy
University of Buenos Aires
Buenos Aires (Argentina)

ROLANDO J.C. LÉON
Department of Ecology, Faculty of Agronomy
University of Buenos Aires
Buenos Aires (Argentina)

ALAN F. MARK
Department of Botany
University of Otago
P.O. Box 56, Dunedin (New Zealand)

DANIEL G. MILCHUNAS
Department of Range Science
Colorado State University
Fort Collins, CO 80523 (U.S.A.)

R. MILTON MOORE
8 Mirning Crescent
Aranda, A.C.T. 2614 (Australia)

R.I. NIKULINA
Komarov Botanical Institute
Popova Ulitsa 2
St. Petersburg 197376 (Russia)

JON J. NORRIS
School of Renewable Natural Resources
University of Arizona
Tucson, AZ 85721 (U.S.A.)

LIST OF CONTRIBUTORS TO VOLUME 8

PHIL R. OGDEN
School of Renewable Natural Resources
University of Arizona
Tucson, AZ 85721 (U.S.A.)

MICHAEL D. PITT
Department of Plant Science
University of British Columbia
Vancouver, B.C. V6T 1W5 (Canada)

T.A. POPOVA
Komarov Botanical Institute
Popova Ulitsa 2
St. Petersburg 197376 (Russia)

ROBERT E. REDMANN
Department of Crop Science and Plant Ecology
University of Saskatchewan
Saskatoon, Sask. S7N 0W0 (Canada)

EARLE A. RIPLEY
Department of Crop Science and Plant Ecology
University of Saskatchewan
Saskatoon, Sask. S7N 0W0 (Canada)

MILENA RYCHNOVSKÁ
Department of Ecology
Palacky University
771 46 Olomouc (Czechoslovakia)

OSVALDO E. SALA
Department of Ecology, Faculty of Agronomy
University of Buenos Aires
Buenos Aires (Argentina)

GLENN D. SAVELLE
P.O. Box, 1330
San Andreas, CA 95249 (U.S.A.)

OSVALDO A. SCAGLIA
Department of Biology
University of Mar del Plata
Mar del Plata, Prov. Buenos Aires (Argentina)

ERVIN M. SCHMUTZ
School of Renewable Natural Resources
University of Arizona
Tucson, AZ 85721 (U.S.A.)

JAI S. SINGH
Department of Botany
Banaras Hindu University
Varanasi 221 005 (India)

FRED E. SMEINS
Department of Rangeland Ecology and
   Management
Texas A&M University
College Station, TX 77843-2126 (U.S.A.)

E. LAMAR SMITH
School of Renewable Natural Resources
University of Arizona
Tucson, AZ 85721 (U.S.A.)

ALBERTO SORIANO
Department of Ecology, Faculty of Agronomy
University of Buenos Aires
Buenos Aires (Argentina)

MICHAEL C. STROUD
P.O. Box 5005
South San Francisco, CA 94080 (U.S.A.)

NEIL M. TAINTON
Department of Grassland Science
University of Natal, Faculty of Agriculture
P.O Box 375
Pietermaritzburg 3200 (Republic of South Africa)

BLANKA ÚLEHLOVÁ
Nerudova 354
Zidlochovice 667 01 (Czechoslovakia)

CARLOS A. VELÁZQUEZ
Museo Municipal de Ciencias Naturales
"Lorenzo R. Scaglia"
C.C. 1207 (7600)
Mar del Plata (Argentina)

BRIAN H. WALKER
CSIRO Division of Wildlife and Ecology
P.O. Box 84
Lyneham, Canberra, A.C.T. 2602 (Australia)

ROBERT G. WOODMANSEE
Natural Resource Ecology
  Laboratory
Colorado State University
Fort Collins, CO 80253 (U.S.A.)

ZHU TING-CHENG
Institute of Grassland Science
Northeast Normal University
Changchun, Jilin, 130024
(People's Republic of China)

CONTENTS OF VOLUME 8A

Section I. The Nature of Grassland

Chapter 1. APPROACH AND GENERALIZATIONS
by R.T. Coupland
Chapter 2. GRASSLAND CLIMATE
by E.A. Ripley
Chapter 3. GRASSLAND SOILS
by D.F. Acton
Chapter 4. WATER FLOW
by E.A. Ripley
Chapter 5. PRIMARY PRODUCTIVITY
by R.E. Redmann
Chapter 6. MICRO-ORGANISMS
by B. Úlehlová
Chapter 7. DECOMPOSITION AS A PROCESS IN NATURAL GRASSLANDS
by J.F. Dormaar
Chapter 8. NUTRIENT CYCLING
by F.E. Clark and R.G. Woodmansee

Section II. Grasslands of North America

Chapter 9. OVERVIEW OF THE GRASSLANDS OF NORTH AMERICA
by R.T. Coupland
Chapter 10. MIXED PRAIRIE
by R.T. Coupland
Chapter 11. SHORT-GRASS STEPPE
by W.K. Lauenroth and D.G. Milchunas
Chapter 12. TALL-GRASS PRAIRIE
by C.L. Kucera

Chapter 13. COASTAL PRAIRIE
by F.E. Smeins, D.D. Diamond and C.W. Hanselka
Chapter 14. FESCUE PRAIRIE
by R.T. Coupland
Chapter 15. PALOUSE PRAIRIE
by R. Daubenmire
Chapter 16. CALIFORNIA PRAIRIE
by H.F. Heady, J.W. Bartolome, M.D. Pitt, G.D. Savelle and M.C. Stroud
Chapter 17. DESERT GRASSLAND
by E.M. Schmutz, E.L. Smith, P.R. Ogden, M.L. Cox, J.O. Klemmedson, J.J. Norris and L.C. Fierro

Section III. Grasslands of South America

Chapter 18. OVERVIEW OF SOUTH AMERICAN GRASSLANDS
by R.T. Coupland
Chapter 19. RÍO DE LA PLATA GRASSLANDS
by A. Soriano with sections by R.J.C. León, O.E. Sala, R.S. Lavado, V.A. Deregibus, M.A. Cauhépé O.A. Scaglia, C.A. Velázquez and J.H. Lemcoff

SYSTEMATIC LIST OF GENERA
AUTHOR INDEX
SYSTEMATIC INDEX
GENERAL INDEX

CONTENTS OF VOLUME 8B[1]

PREFACE V

LIST OF CONTRIBUTORS VII

Section IV. Grasslands of Europe and Asia

Chapter 1. OVERVIEW OF THE GRASSLANDS OF EUROPE AND ASIA
by R.T. Coupland 1

References 2

Chapter 2. STEPPES OF THE FORMER SOVIET UNION AND MONGOLIA
by E.M. Lavrenko and Z.V. Karamysheva
with sections by I.V. Borisova, T.A. Popova, N.P. Guricheva and R.I. Nikulina 3

Introduction 3
Phyto-geographical characteristics of Eurasian steppes 4
The organization and dynamics of steppe ecosystems 17
Comparison of two zonal types of ecosystems 26
The effect of rodents on steppe ecosystems 42
Conclusion 54
References 55

Chapter 3. GRASSLANDS OF CHINA
by Zhu Ting-Cheng 61

Introduction 61
Climate 61
Soils 63
Zonal grassland 64
Azonal vegetation 72
Native fauna 74
Comparison with other Eurasian grasslands 78
Use and management 78
References 80

Chapter 4. GRASSLANDS OF SOUTHERN ASIA
by J.S. Singh and S.R. Gupta 83

Introduction 83
Geographical limits and area 83
Grassland vegetation 88
Standing crop and primary production 99
Stability 103
Wild animals 103
Micro-organisms and their activities 106
Use, management and renovation 108
Acknowledgements 117
References 117

Chapter 5. TEMPERATE SEMI-NATURAL GRASSLANDS OF EURASIA
by M. Rychnovská 125

Introduction 125
Environment 126
Structure and function of primary producers . . . 126
Consumers 146
Decomposers 148
Energy flow and mineral cycles 153
Land use and management 155
Appendix I — Ecological types of semi-natural grasslands in Eurasia 156
References 162

Section V. Grasslands of Africa

Chapter 6. OVERVIEW OF AFRICAN GRASSLANDS
by R.T. Coupland 167

References 169

Chapter 7. GRAZING LANDS OF THE MEDITERRANEAN BASIN
by H.N. Le Houérou 171

Introduction 171
Climate 172
Vegetation 173
Primary production 180
Land use and management 188
Impact of man and livestock 190
Improvement of grazing lands 192
References 194

---

[1] For short contents of Vol. 8A, see p. XI

Chapter 8. **GRASSLANDS OF THE SAHEL**
by H.N. Le Houérou 197

Introduction 197
Climate 197
Seasonality 202
Soils, topography and geomorphology 202
Flora and vegetation 203
Primary production 207
Fauna 209
Land use and management 210
Impact of man and livestock 214
Improvement of grazing lands 216
References 216

Chapter 9. **GRASSLANDS OF EAST AFRICA**
by D.J. Herlocker, H.J. Dirschl and
G. Frame 221

Introduction 221
Pennisetum mid-grass region 222
Pennisetum giant-grass region 227
Panicum–Hyparrhenia tall-grass region 229
Hyparrhenia tall-grass region 232
Themeda mid-grass region 235
Chrysopogon mid-grass region 243
Leptothrium mid-grass region 250
Aristida annual short-grass region 255
Acknowledgements 257
References 257

Chapter 10. **GRASSLANDS OF SOUTHERN AFRICA**
by N.M. Tainton and B.H. Walker . . . 265

Introduction 265
Classification of major communities 265
Contacts with other ecosystems 269
Determinants of the main vegetation types 270
The grassland communities 272
The animal communities 279
Activities of micro-organisms and soil fauna . . . 281
The nature of human impact 282
Current management practices 283
Present and future trends 287
References 288

Chapter 11. **GRASSLANDS OF MADAGASCAR**
by J. Koechlin 291

Introduction 291
Ecological conditions 291
Phytogeographical regions 292
Herbaceous vegetation 292
Fauna 300
Agriculture and stock farming 300
References 301

Section VI. Grasslands of Oceania

Chapter 12. **OVERVIEW OF THE GRASSLANDS OF OCEANIA**
by A.N. Gillison 303

Introduction 303
Origins and affinities 304
Vegetation 305
Concluding remarks 312
References 312

Chapter 13. **GRASSLANDS OF AUSTRALIA**
by R.M. Moore 315

Introduction 315
The grasslands 319
Lands without native grasslands 345
Exotic grasslands (pastures) 345
Grassland community interrelationships 350
True grasslands 352
Management of grasslands 353
References 355

Chapter 14. **INDIGENOUS GRASSLANDS OF NEW ZEALAND**
by A.F. Mark 361

Introduction 361
The grassland environment 365
Vegetation 373
Fauna 393
Micro-organisms 397
Present and future management practices 398
Acknowledgements 404
References 404

Chapter 15. **GRASSLANDS OF THE SUB-ANTARCTIC ISLANDS**
by R.J. Hnatiuk 411

Introduction 411
Climate 414
Vegetation 417
Autecology 421
Seasonal variation in growth 421
Standing crop and primary productivity 421
Nutrient flow 425
Decomposition and soil fauna 426
Impact of native animals on vegetation 427
Man's interference 427
Relationship to tropical high mountain grasslands . . 430
Acknowledgements 431
References 431

Chapter 16. GRASSLANDS OF THE
SOUTH-WEST PACIFIC
by A.N. Gillison 435

Introduction 435
Papua New Guinea 443
Solomon Islands 454
New Caledonia 455
Vanuatu 456
Fiji 457
Micronesia 459
The role of fire and frost in Pacific grasslands . . . 461
Land use 464
Grasslands as a pathway for invasive exotics 466
References 467

Section VII. Résumé

Chapter 17. REVIEW
by R.T. Coupland 471

Introduction 471
Vegetation structure and biomass 471

Plant production 475
Animal biomass 476
Micro-organisms 480
Activity of heterotrophs 481
Turnover of plant biomass 481
Management 482

SYSTEMATIC LIST OF GENERA 483

AUTHOR INDEX 489

SYSTEMATIC INDEX 499

GENERAL INDEX 525

Chapter 1

OVERVIEW OF THE GRASSLANDS OF EUROPE AND ASIA

R.T. COUPLAND

The vast land mass embracing Europe and Asia contains grasslands ranging from temperate to tropical in nature. In his treatment of the grasslands of Asia to the south of the former Soviet Union, Numata (1979) separated them into two categories — monsoonal and arid. He considered the former to be seral communities in which the equilibrium is maintained under the influence of biotic factors; climax grasslands occur only in arid and semiarid regions and in alpine areas.

Sochava (1979) classified the grasslands of the former Soviet Union into the following categories:

(1) Boreal grasslands in the taiga forest region
(2) Nemoral grasslands in the broad-leaved and conifer–broad-leaved forest regions
(3) Herbaceous swamps (in tundra, forest and forest-steppe)
(4) Forb-grass and forb steppes (in forest steppe and northern steppe)
(5) Tuft-grass (bunch-grass) steppes
(6) *Artemisia*-grass steppes (semi-desert)
(7) Subtropical steppes (pseudo-savannas in submontane and mountain areas)
(8) Mountain steppes
(9) Grasslands in the belt of broad-leaved and conifer–broad-leaved forests in mountains
(10) Subalpine meadows
(11) Alpine meadows
(12) Sub-Arctic grasslands of the Pacific type

Types (4), (5) and (6) constituted the major region of grassland in pre-agricultural time. He emphasized the importance of distinguishing between the original forest-free grasslands and those of areas deforested by man.

In his discussion of the distribution of grasslands in Europe, Knapp (1979) indicated that the steppe grasslands originally extended from the Ukraine to the dry plains of Hungary (*puszta*), Romania, Yugoslavia and Bulgaria, but now only small fragments survive. He refers also to "spots" of steppe vegetation in the dry, hot interior valleys of the Alps. Knapp considered only a minor part of the grazing lands along the Mediterranean Sea as natural grassland. Prior to agricultural development, natural grassland also occupied parts of Iceland. However, elsewhere in Europe to the west of the former Soviet Union grassland now occupies only areas that are marshy or deforested, except for small areas of natural grassland on mountain slopes and above the tree line.

The temperate natural and semi-natural grasslands of northern Eurasia have closer affinities with those of North America than with the grasslands of southern Asia (Table 6.1, p. 168). In many instances the important taxa in the grasslands of northern Asia and North America are closely related taxonomically and sometimes even recognized as the same species. As is the situation in the Western Hemisphere, the tribes Agrosteae, Aveneae and Festuceae are much better represented in the grass flora in the temperate parts of the continent, whereas a larger proportion of species of the tribes Andropogoneae, Eragrosteae and Paniceae occur in subtropical and tropical regions (Table 1.1).

The diversity of vegetation types of Europe and Asia is treated in the following four chapters.

In Chapter 2, the semi-arid and desert grasslands of the former Soviet Union are examined, with emphasis on physiological and morphological adaptations which determine how plant species function within the ecosystem. A detailed analytical comparison is made of a steppe grassland in

TABLE 1.1

Percentage of the grass flora contributed by each major tribe in selected localities in northern and southern Asia, compared with the global average [1]

State/Country	Agrosteae	Andropo-goneae	Aveneae	Eragrosteae	Festuceae	Paniceae	Others
NORTHERN ASIA							
Transvolga [2]	17.3	0.0	11.8	5.5	33.1	5.5	26.8
Augara-Sayan [2]	18.1	0.6	12.5	2.5	33.8	3.1	29.4
Ciscaucasia [2].	17.0	2.6	10.4	3.5	37.8	5.7	23.0
Tien Shan [2]	10.2	1.0	8.3	2.5	33.8	2.9	41.3
Kopet-Dag [2]	10.3	3.4	7.8	2.6	35.4	5.2	35.3
Average	14.6	1.5	10.2	3.3	34.8	4.5	31.2
SOUTHERN ASIA							
Palestine and Syria	11.5	6.5	10.0	3.6	30.5	7.5	30.4
North-western Arabia	5.3	7.7	7.7	3.8	30.5	8.4	36.6
Kashmir	12.9	12.9	6.3	7.1	28.0	9.2	23.6
Gujarat, India	0.0	30.3	0.0	17.2	1.0	27.3	24.2
Bombay Deccan, India	0.7	41.9	0.0	16.9	0.0	21.3	19.2
Luzon, Philippine Is.	2.1	33.7	1.1	7.4	2.9	33.7	19.1
Sri Lanka	4.2	27.0	2.5	12.0	1.2	30.3	22.8
Malay Peninsula	1.0	27.4	1.0	7.3	0.5	32.2	30.6
Jawa	2.8	24.7	1.4	9.2	3.2	33.6	25.1
Average	4.5	23.6	3.3	9.4	10.9	22.6	25.7
Normal distribution spectrum (global)	8.2	11.9	6.3	8.1	16.5	24.7	24.3

[1] After Hartley (1950).
[2] Former Soviet Union.

Kazakhstan and a desert grassland in Mongolia. The activity of rodents is discussed in depth, with emphasis on how they modify the plant community.

Chapter 3 is composed of a description of the various types of natural grassland in western and northern China, with emphasis on their floristic composition in relation to varying climatic conditions. The distribution of fauna is also considered.

The grasslands of southern Asia, from Turkey to the Philippines, are considered in Chapter 4. They occur in climates ranging from dry to humid and from subtemperate to tropical in character.

The grasslands within the temperate forest zone across Europe, the former Soviet Union and Japan are treated in Chapter 5. The wide variety of these semi-natural grasslands results partly from the use of deforested lands as hayfields and grazing lands and partly from the exclusion of trees by flooding. An important aspect of this presentation is an original classification of these grasslands.

REFERENCES

Hartley, W., 1950. The global distribution of tribes of the Gramineae in relation to historical and environmental factors. *Aust. J. Agric. Res.,* 1: 355–373.
Knapp, R., 1979. Distribution of grasses and grasslands in Europe. In: M. Numata (Editor), *Ecology of Grasslands and Bamboolands in the World.* VEB Gustav Fischer Verlag, Jena, pp. 111–123.
Numata, M., 1979. Distribution of grasses and grasslands in Asia. In: M. Numata (Editor), *Ecology of Grasslands and Bamboolands in the World.* VEB Gustav Fischer Verlag, Jena, pp. 92–102.
Sochava, V.B., 1979. Distribution of grasslands in the USSR. In: M. Numata (Editor), *Ecology of Grasslands and Bamboolands in the World.* VEB Gustav Fischer Verlag, Jena, pp. 103–110.

Chapter 2

STEPPES OF THE FORMER SOVIET UNION AND MONGOLIA

E.M. LAVRENKO and Z.V. KARAMYSHEVA
With sections by: I.V. BORISOVA, T.A. POPOVA, N.P. GURICHEVA and R.I. NIKULINA

INTRODUCTION*

The present work is concerned with the characteristics of steppe vegetation in the inner part of the Eurasian continent. These steppes constitute a well-defined region occupying a vast area between 48° and 57°N latitude and 27° to 128°E longitude. The great expanse of the steppe region from north to south (800 to 1000 km) and from west to east (about 8000 km) supports an extraordinary diversity of steppe communities.

The steppe landscape is highly affected by human activity. Except in Central Asia, the steppe has been preserved in its natural state only in reservations, or as steppe modified by grazing of domesticated livestock and by tillage. The characteristics of a virgin steppe, as it existed before development and destruction by agricultural pursuits, has been described in many works of the last century and the beginning of the present century (Korzhinskyi, 1888, 1891; Gordyagin, 1900, 1901; Vysotskyi, 1915; Pachoskyi, 1917; Novopokrovskyi, 1921, 1925, 1927, 1931, 1940; Alekhin, 1925, 1934; Reverdatto, 1925; Kleopov, 1928, 1933a, b; Lavrenko and Zoz, 1928; Keller, 1931; Lavrenko and Dokhman, 1933; Lavrenko, 1940, 1954, 1956, 1980a; Krasheninnikov and Kucherovskaya-Rozhanets, 1941).

Subsequently, investigations of steppe were conducted in preserved areas. Important among these are studies by Vandakurova (1950), Yunatov (1950, 1954, 1974), Kuminova (1960), Rescikov (1961), Isachenko and Rachkovskaya (1961), Stepanova (1962), Bilyk (1973), Karamysheva and Rachkovskaya (1973) and Osychnyuk et al. (1976).

Beginning in the late 1950s, complex investigations of various biogeocoenoses (ecosystems), each several years in duration, have been made at field stations in Kazakhstan and in Mongolia, where reference plots of steppe phytocoenoses (plant communities) are reserved for scientific purposes. These studies provide details of the structure of steppe communities and of the biological, morphological, ecological and physiological characteristics of the component species, during fluctuations in environment within years and between years.

The first part of this chapter is concerned with composition, structure and rhythm of development of steppe communities, and with differences resulting from increasing aridity of climate caused by latitudinal increase in solar heat and decrease in precipitation, as well as progressive salinization of soil. Attention is also given to the effects on vegetation of the increasing continentality of climate towards the centre of the continent. In the second part of the chapter, we consider the organization and dynamics of steppe ecosystems. The third part compares the nature of two zonal types of steppe, based on lengthy studies of semi-desert (desertified) steppes in Central Kazakhstan and of desert steppes in Mongolia. The concluding section deals with the influence of burrowing activity of rodents on steppes. This analysis suggests that the effect of rodents on steppes is similar in some ways to that caused by grazing of domesticated ungulates.

---

* By E.M. Lavrenko and Z.V. Karamysheva.

PHYTO-GEOGRAPHICAL CHARACTERISTICS OF EURASIAN STEPPES*

The nature of steppe vegetation

Perennial microtherm xerophilous, and often sclerophyllous, bunch grasses predominate in steppe communities. These grasses are usually species of the genera *Agropyron, Cleistogenes, Festuca, Helictotrichon, Koeleria* and *Stipa*. Dominants also include bunch species of sedges (*Carex*) and, in Central Asia, of species of bunch forbs (*Allium* and *Filifolium*). Chernozems and chestnut soils are typical of many types of steppe.

The synusia[1] of dominant bunch grasses forms the basis of the plant community and provides most of the phytomass. It is usually composed of a combination of relatively tall bunch grasses — mostly species of *Stipa*, of shorter bunch grasses of the genera *Cleistogenes* and *Festuca*, and sometimes of dwarf bunch species of *Carex* or dwarf species of *Stipa*.

Plants in the dominating synusia grow during the whole vegetative period, but during droughty months in summer (July and early August) their rate of development is retarded over much of the steppe region. This period of semi-dormancy occurs throughout the steppes of the European part of the former Soviet Union, as well as in the northern part of Kazakhstan and south-western Siberia. In these regions growth is most rapid during June, the month of greatest precipitation. To the east, in the Daurian steppes[2], maximum precipitation occurs in July and August, and the maximum development of the vegetation is attained later than in western regions.

Other synusiae are of lesser importance than the synusia of bunch grasses and other bunch herbs. A synusia of forbs (herbaceous dicotyledons and non-gramineous monocotyledons, usually of the families Iridaceae and Liliaceae) is often associated with the dominating synusia. The number of species of forbs and their proportional contribution to biomass decreases from north to south within the steppe region, because of increased aridity of climate. These forbs are of various life-forms, especially with respect to the nature of their under-ground parts. Some are tap-rooted (*Centaurea* spp. and *Dianthus* spp.), some are rhizomatous (*Artemisia latifolia, A. pontica, Galium ruthenicum* and *Veronica incana*[3]), and others have root suckers (*Artemisia adamsii, A. austriaca* and *Cymbaria daurica*). In the most arid semi-desert and desert steppes, a synusia of dwarf half-shrubs also occurs. This consists mostly of species of the sub-genera *Seriphidium* and *Artemisia* of the genus *Artemisia*.

The rhizomatous grasses and sedges do not form a clearly defined synusia in zonal steppe communities, but they are abundant in more northerly grasslands. Common members of this group in the Black Sea–Kazakhstan steppes are: *Agrostis vinealis* (= *A. syreistschikowii*), *Bromopsis* (= *Bromus*) *inermis, Bromopsis* (= *Zerna*) *riparia, Elytrigia repens, E. trichophora* and *Poa angustifolia*, whereas *Leymus chinensis* is prominent in the steppes of Transbaykal and Mongolia.

A synusia of sparse shrub thickets often occurs, especially in areas of sandy soils. The genera *Calophaca, Caragana* and *Spiraea* are well represented among these species of shrubs.

The relative vigour of the various species, and of one synusia compared to another, change with fluctuations in precipitation. Therefore, the vegetative cover exhibits considerable variation during each 11-year cycle of solar activity.

Zonal types of steppe

Russian geobotanists traditionally distinguish three or (more often) four zonal types of steppe, which successively replace one another from

---

* By E.M. Lavrenko, Z.V. Karamysheva and R.I. Nikulina.

[1] The term **synusia** is used in this chapter in the sense of Gams (1918). A synusia is a structural part of a plant community, which includes those species that are similar in a biological and ecological sense.

[2] **Dauria** is a geographical–botanical region (Lavrenko, 1970), which includes the basin of the river Selenga, Zabaykalye (Transbaykal — to the south and east of Lake Baykal), the basin of the river Nercha, the steppe east of Hentiyn Nuruu and west of the Lesser Khingan mountains (in Mongolia and China).

[3] Botanical names are according to the *Flora of the USSR* (1934–1964), as modified by Cherepanov (1981).

Fig. 2.1. Meadow steppe in the Black Sea–Kazakhstan subregion [Black Sea (Pontic) steppe province] dominated by *Stipa pennata*. (Mikhaylovskaya virgin-land, near the Psel river, Sumy region, Ukraine). (Photo by V.S. Tkachenko.)

north to south with increasing aridity of climate, as demonstrated by decreasing precipitation, increase of temperature summations, and lengthening of the frost-free period. The vegetative growing season is increasingly interrupted by drought. Our classification of grasslands is as follows:

(1) Meadow steppe (= forest steppe), in semi-humid climate (Fig. 2.1).

(2) True or typical steppe: (a) bunch-grass steppes with many forbs, in semiarid climate (Figs. 2.2 to 2.6); (b) bunch-grass steppes with few forbs, in arid climate (Fig. 2.7).

(3) Desertified bunch-grass and dwarf half-shrub–bunch-grass (semi-desert steppe), in very arid climate (Figs. 2.8 to 2.10).

(4) Desert dwarf half-shrub–bunch-grass steppe in hyperarid climate (Fig. 2.11).

The northern meadow steppes are characterized by the dominance of mesoxerophilous bunch grasses and sedges and by the abundance of xeromesophilous and mesoxerophilous forbs.[1] The typical or true steppes are distinguished by greater aridity of habitat, domination of xerophilous species of bunch grasses, and by a lower abundance of forbs (which are of a more xerophilous character than those in the meadow steppe). The southern semi-desert and desert steppes are more arid than the preceding types, the most xerophilous (euxerophilous and even hyperxerophilous) bunch grasses and dwarf half-shrubs be-

---

[1] We recognize the following gradations of xerophily: **eumesophyte**, **xeromesophyte**, **mesoxerophyte**, **euxerophyte** (= **xerophyte**) and **hyperxerophyte**. Those species which occur throughout the range in aridity of climate from meadow steppe to desert steppe are **euryxerophytes**. Included in this category are: *Festuca valesiaca* (= *F. sulcata*), *Stipa capillata*, *S. krylovii* and *S. lessingiana*. The degree of xerophily of a species is judged on the basis of the prominence of xeromorphic features, and by changes in its ecological position within its area of distribution.

STEPPES OF THE FORMER SOVIET UNION AND MONGOLIA

Fig. 2.4. True (typical) bunch-grass steppe in the Black Sea–Kazakhstan subregion [Black Sea (Pontic) steppe province] dominated by *Stipa pulcherrima* and forbs. (Khomutovskaya steppe, Donetsk region, Ukraine). (Photo by V.S. Tkachenko.)

ing dominant. Few euryxerophilous forbs occur; the co-dominants are xerophilous and hyperxerophilous dwarf half-shrubs of the genus *Artemisia* and sometimes of other genera, particularly *Anabasis* and *Salsola*.

Many other structural features change with increasing aridity. Species diversity is reduced from 40–50 species in a square metre in meadow steppe to 12–15 species in semi-desert and desert steppes. The height of the grass canopy decreases from 80–100 cm in the north to 15–20 cm in the south; and foliage cover decreases from 70–90% to 10–20%, and even less.

Much of the variation in vegetation within each of the zonal types of steppe is related to edaphic characteristics. The zonal characteristics of soil and vegetation are exemplified by the *plakor*[1]. Variations from this which markedly affect the nature of the vegetation are found in sands and loamy sands (psammophytic and hemipsammophytic conditions), in stony and gravelly

---

[1] The term ***plakor*** was introduced by Vysotskyi (1915) to signify a flat, well-drained plain with loamy soils; the groundwater is located too deeply to affect plant growth. The vegetation of *plakor* plains most fully reflects zonal climate.

Fig. 2.2 (top). True (typical) bunch-grass steppe in the Black Sea–Kazakhstan subregion [Black Sea (Pontic) steppe province] dominated by *Stipa lessingiana* and forbs. (Khomutovskaya steppe, Donetsk region, Ukraine). (Photo by V.S. Tkachenko.)

Fig. 2.3 (bottom). True (typical) bunch-grass steppe in the Black Sea–Kazakhstan subregion [Black Sea (Pontic) steppe province] dominated by *Stipa zalesskii* and forbs. (Streltsovskaya steppe, in the valley of the Donets river, Voroshilovgrad region, Ukraine). (Photo by V.S. Tkachenko.)

Fig. 2.5. True (typical) bunch-grass steppe in the Black Sea–Kazakhstan subregion [Black Sea (Pontic) steppe province] dominated by *Stipa lessingiana* and *S. capillata* and forbs. The abundant forb is *Phlomis tuberosa*. (In an outflow (drainage) hollow in Khomutovskaya steppe, Donetsk region, Ukraine.) (Photo by V.S. Tkachenko.)

soils (petrophytic conditions), and in soils with high salt content (halophytic conditions).

Hemi-psammophytic and psammophytic subtypes of steppe occur in sands and loamy sands, in which the principal dominants are specific groupings of bunch-grass species of *Agropyron*, *Festuca*, *Koeleria* and *Stipa*.

Areas of gravelly soils — in *melkosopochniks* (= low hills) and in intermountain plains — are characterized by communities of dwarf half-shrubs of several families, especially Asteraceae, Caryophyllaceae, Lamiaceae and Scrophulariaceae. Species occurring in sparse stands in very stony sites are very specific; examples are *Androsace kozopoljanskii*, *Artemisia hololeuca* and *Hyssopus cretaceus*. Communities of dwarf half-shrubs, with or without a scattering of bunch grasses and sedges, are often formed on rock outcrops almost devoid of soil. The floristic composition of the communities of dwarf half-shrubs varies considerably within the steppe region.

In some places within the steppe zone, shrubs are dominant in diluvial–proluvial loamy, sandy and gravelly deposits. These shrub steppes are widely distributed over the steppe region of Eurasia, especially in Eastern Kazakhstan and in Mongolia. Species of *Caragana* and *Spiraea* are particularly abundant. Thickets of steppe shrubs occur on the slopes of ravines, in gullies formed by water erosion, and in the margin of steppe forests. These are composed of representatives of many genera, including *Amygdalus*, *Calophaca*, *Caragana*, *Cerasus* and *Cotoneaster* (Figs. 2.12 and 2.13).

The vegetative cover of the forest-steppe zone was, in the past, a mosaic of meadow grassland and forest. In the European part of the former Soviet Union the principal tree was *Quercus robur*. These forest stands occurred on well-drained catchments along the slopes of rather large rivers. At present they have been preserved only as small tracts. In the zone of true steppes, patches of trees grow only along the slopes of ravines. In the

Fig. 2.6. True (typical) bunch-grass steppe in the Black Sea–Kazakhstan subregion (Transvolga–Kazakhstan steppe province) dominated by *Stipa zalesskii* and forbs. The shrub is *Caragana frutex*. (In the Karaganda region, Kazakhstan.) (Photo by I.N. Safronova.)

western Siberian forest steppe, as well as in the low mountain ranges of Central Kazakhstan, small tracts of *Betula pendula*, *B. pubescens* and *Populus tremula* are typical around shallow depressions. More or less steppified forests of *Pinus sylvestris* also grow along the flood-plains and terraces of rivers throughout the Eurasian forest steppe and in low mountain ranges in Central and Northern

Fig. 2.7. True (typical) steppe in the Black Sea–Kazakhstan subregion (Transvolga–Kazakhstan steppe province) dominated by *Stipa lessingiana*, *Festuca valesiaca* and *Crinitaria tatarica*. (In the Tselinograd region of Kazakhstan.) (Photo by I.V. Borisova.)

Kazakhstan. *Larix gmelinii* and *L. sibirica* occur in the mountainous forest steppe of Transbaykal and Mongolia, the former only east of the Hentiyn Nuruu (mountains) in Mongolia.

Longitudinal division of the steppe zone

The structure and floristic composition of vegetation of Eurasian steppes change from west to east. These changes are associated with changes in the degree of continentality of the climate, which is expressed particularly in sharp fluctuations in the amount of precipitation and heat in different seasons of the year, as well as between years. The meridional sections can be recognized, as follows: (1) the temperate-continental southern European part of the former Soviet Union and Romania; (2) the continental southern part of Western Siberia and Northern Kazakhstan; (3) the hyper-continental southern part of Eastern Siberia and Central Asia (including the steppes of Transbaykal and Mongolia); and (4) the temperate-continental plain of north-eastern China and adjacent meadow steppe.

The boundary between the continental and hyper-continental meridional zones can be used as a basis for dividing the Eurasian steppe into two subregions: Black Sea–Kazakhstan and Central Asian (Dauro–Mongolian) (Figs. 2.14 and 2.15). To the east of this boundary the climate is

Fig. 2.8. Semi-desert steppe in the Black Sea–Kazakhstan subregion (Transvolga–Kazakhstan steppe province) dominated by *Stipa lessingiana* and *Artemisia gracilescens*. (In the Koksengir mountains, Karaganda region, Kazakhstan.) (Photo by I.N. Safronova.)

highly continental. This is reflected in a considerable extension of the occurrence of permafrost in soils southward into the Mongolian forest steppe and, as patches, even into parts of the typical steppe in Mongolia. The forest-steppe region of north-eastern China (east of Inner Mongolia) lies within the region of influence of the Pacific monsoon, which exerts only a slight influence on the climate of the eastern parts of Mongolia. The "temperate mountain bunch-grass steppes (*Festuca valesiaca* and *Stipa capillata*)" (Khou, 1979) in the southern Altai mountains (within China) constitutes the contact between the two steppe subregions. Lavrenko (1970a, b) presented the following sub-division of the Eurasian steppe zone into two subregions, each consisting of five provinces:

Black Sea–Kazakhstan steppe subregion (see Fig. 2.14)
    Eastern European block
        Balkano–Mesian forest-steppe province
        East European forest-steppe province
        Black Sea (Pontic) steppe province
    West Siberian–Kazakhstan block
        West Siberian forest-steppe province
        Transvolga–Kazakhstan steppe province
Central Asian steppe subregion (Fig. 2.15)
    Dauro–Mongolian block
        Hangay–Daurian mountain forest-steppe province
        Mongolian steppe province
        North-Gobi desert-steppe province
    Manchuro–North-western Chinese block
        Manchurian forest-steppe province
        Shanxi–Gansuian forest-steppe and steppe province

We shall now consider the distinctive environmental and vegetational features of each subregion.

Fig. 2.9. Semi-desert steppe in the Black Sea–Kazakhstan subregion (Transvolga–Kazakhstan steppe province) dominated by *Stipa lessingiana* and *Artemisia sublessingiana*. The shrub is *Caragana pumila*. (In Chingiz-Tau ridge, Semipalatinsk region, Kazakhstan.) (Photo by I.N. Safronova.)

The Black Sea–Kazakhstan subregion

This includes the extensive plains of eastern Europe, the plain in the southern part of the Western Siberian lowland, the plain of Western Kazakhstan to the east and south of the Ural mountains, and the *sopki* (small hills) of the Central-Kazakhstan *melkosopochnik* with separate low mountains, mostly in the eastern part of Kazakhstan. Mountains are situated only in the extreme south-east, including the southern Altai, the Kalbinskiy mountain ridge, and the Tarbagatai and Saur ridges. These exhibit a well-pronounced altitudinal zonality of vegetation.

In the western part of this subregion the climate is temperate-continental. The boundary between the temperate-continental and the continental block of zones passes along the Ural mountains and along the Ural river. June is the month of highest precipitation. Spring (April–May) is relatively warm, with considerable precipitation.

The soils over most of this subregion are deep, typical and southern chernozems. Dark- and light-coloured chestnut soils occur only to the south[1]. Salinized, solonetz and solonchak soils are com-

---

[1] These occur in the extreme southern Ukraine, on the plains near the Black Sea and Sea of Azov (Azovskoye More), on the Yergeni heights, in the southern (elevated) Transvolgian (Zavolzhye) area, the sub-Caspian (Pricaspiy) area (to the north and north-west of the Caspian Sea) and also in Southern Kazakhstan.

Fig. 2.10. Semi-desert steppe in the Black Sea–Kazakhstan subregion (Transvolga–Kazakhstan steppe province) dominated by *Stipa orientalis*. (In Chingiz-Tau ridge, Semipalatinsk region, Kazakhstan.) (Photo by I.N. Safronova.)

Fig. 2.11. Desert steppe in the Central Asian steppe subregion (North-Gobi desert-steppe province) dominated by *Stipa gobica*, *S. glareosa*, *Allium polyrrhizum* and *Iris bungei*. (Photo by I.V. Borisova.)

Fig. 2.12. Bunch-grass steppe in the Black Sea–Kazakhstan subregion dominated by species of *Stipa* and *Festuca* and forbs. The small shrub in the foreground is *Calophaca wolgarica*, whereas the large shrubs in the background are *Prunus spinosa* and *Rhamnus cathartica*. (Khomutovskaya steppe, Donetsk region, Ukraine.) (Photo by V.S. Tkachenko.)

mon in the chestnut soil area and also in the chernozem soils of south-western Siberia.

The dominants include the following species of *Stipa*:

Section *Stipa*
    S. dasyphylla   *S. tirsa*
    S. pennata   *S. ucrainica*
    S. pulcherrima   *S. zalesskii* (= *S. rubens*)
Section *Leiostipa*
    S. capillata   *S. sareptana*
Section *Barbatae*
    S. korshinskyi
    S. orientalis
    S. lessingiana

The species of the Section *Stipa* have a greater role than those of the Section *Leiostipa*. *Stipa orientalis*, a short bunch grass, dominates on stony outcrops in Kazakhstan. *Festuca valesiaca* is another important short bunch grass. Also of consequence are *Koeleria cristata* and species of *Agropyron* and *Poa*. *Cleistogenes squarrosa* is sometimes quite abundant in loamy sands.

Species of *Artemisia* (Section *Seriphidium*) dominating in semi-desert steppes include (from west to east): *A. santonica, A. taurica, A. lerchiana, A. lessingiana, A. gracilescens* and *A. sublessingiana*. Dwarf half-shrubs of the family Chenopodiaceae occurring on strongly solonetzic soils include *Kochia prostrata, Anabasis salsa, Nanophyton erinaceum* and *Atriplex cana*. The relatively mild spring and rather frequent precipitation favour ephemerals, particularly winter annuals. Examples include:

Alyssum turkestanicum   *Erophila verna*
Anisantha tectorum   *Holosteum* spp.
Bromus squarrosus   *Valerianella* spp.
Ceratocephala spp.

Ephemeroids (perennials fruiting in spring and quiescent in summer) are also typical:

Fig. 2.13. Semi-desert steppe in the Black Sea–Kazakhstan subregion dominated by *Stipa lessingiana*, *S. sareptana*, *Festuca valesiaca* and *Ferula soongarica* with thickets of *Spiraea hypericifolia* and *Caragana balchaschensis*. (In the Koksengir Mountains, Karaganda region, Kazakhstan.) (Photo by I.N. Safronova.)

Bellevalia spp.
Bulbocodium spp.
Crocus reticulatus
Geranium tuberosum
Poa spp. (subsection *Bulbosae*)

Rindera tetraspis
Tulipa biebersteiniana
Tulipa biflora
Tulipa schrenkii

Ephemerals and ephemeroids are especially abundant in the southern steppes. Hemi-ephemeroids also occur. Examples are: *Adonis wolgensis*, *Pedicularis dasystachis* and *P. kaufmannii*. Annuals and biennials with long life-cycles, such as *Artemisia scoparia*, *Salsola australis* and *S. tamariscina*, are uncommon and do not have a significant role.

The Central Asian subregion

Prominent features of this subregion are low *sopki* and high mountain ranges [Mongolian and Gobi Altai, the Hangayn Nuruu (Khangai or Changai mountains), Bulnai, Tarbagatai] with well-defined altitudinal differentiation of vegetation. To the east, highly elevated (500–1000 m or more), well-drained plains are situated between the eastern foothills of the Hangayn Nuruu and the western edge of the Greater Khingan Mountains (Da Hinggan Ling).

The climate is distinctly continental, with frequent droughts during the growing season. July and August are the months of maximum precipitation. Spring (April–May) is cold, windy and dry.

The soils of the Mongolian steppe are mostly dark chestnut to light chestnut in colour. They are usually free of salt, predominantly loamy sand or sandy loam in texture, and more or less gravelly. In southern Mongolia, steppe vegetation also occurs in the brown desert-steppe soils, which is not the case in the Black Sea–Kazakhstan subregion.

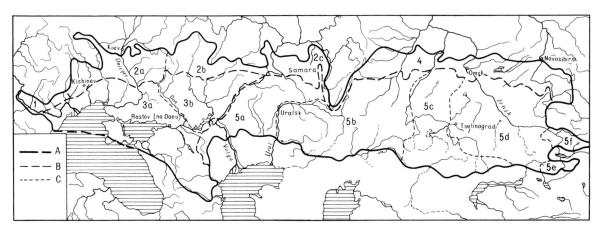

Fig. 2.14. The provincial division of the Black Sea–Kazakhstan subregion of the Eurasian steppe region. Borders: A — of the subregion; B — of provinces; C — of sub-provinces.

1 – Balkano-Mesian forest-steppe province
2 – East-European forest-steppe province
    a – Middle Dnieper sub-province
    b – Middle Russian (Middle Don) sub-province
    c – Transkama-Transvolga sub-province
3 – Black Sea (Pontic) steppe province
    a – Azov–Black Sea sub-province
    b – Middle Don sub-province

4 – West Siberian forest-steppe province
5 – Transvolga–Kazakhstan steppe province
    a – Yergeni–Transvolga sub-province
    b – Western Kazakhstan sub-province
    c – Central Kazakhstan sub-province
    d – Eastern Kazakhstan sub-province
    e – Saur–Tarbagatai (Tacheng) sub-province
    f – Southern Altai steppe province

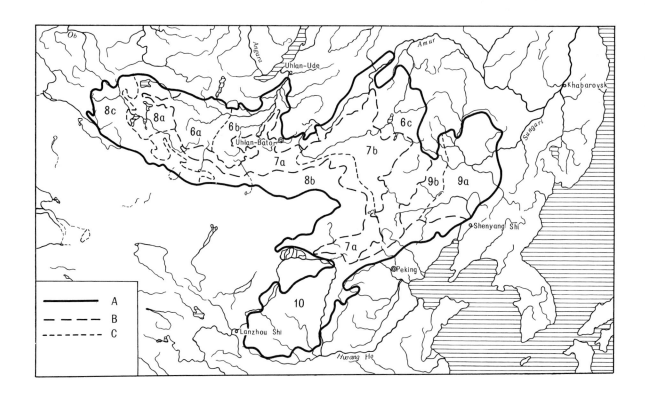

Chernozems occur only in the northern part of the forest steppe in Transbaykal and in the northern part of Mongolia.

The dominant species are as follows:

Agropyron cristatum	*Poa botryoides*
Cleistogenes songorica	*Stipa baicalensis*
Cleistogenes squarrosa	*Stipa glareosa*
Festuca kryloviana	*Stipa gobica*
Festuca lenensis	*Stipa grandis*
Koeleria macrantha	*Stipa klemenzii*
Poa attenuata	*Stipa krylovii*

Stipa capillata and *S. sareptana* are limited to the western part of the subregion. The low-growing species of *Stipa* from the Section *Smirnovia* (*S. klemenzii*, *S. glareosa* and *S. gobica*) dominate in the desert steppe. Three species of the Section *Stipa* (*S. borysthenica*, *S. kirghisorum* and *S. zalesskii*) are found occasionally only in the western part of Mongolia, and do not have a prominent role in the plant cover. *Agropyron cristatum*, the two species of *Cleistogenes*, *Koeleria macrantha*, and the two species of *Poa* are all short bunch grasses. The two species of *Festuca* are found only in mountain steppes.

A very interesting feature of this subregion is the bunch-herb type of steppe composed of *Filifolium sibiricum*, rhizomatous grasses (especially *Leymus chinensis*) and bunch onions (*Allium polyrrhizum*). This type has no analogue in the Black Sea–Kazakhstan subregion. Species of *Caragana* are also very important contributors to the vegetative cover.

Dwarf half-shrubs of semi-desert and desert steppes include:

Ajania fruticulosa	*Artemisia xerophytica*
Anabasis brevifolia	*Asterothamnus* spp.
Artemisia frigida	*Reaumuria songorica*

Artemisia frigida occurs also in more northerly steppes. Ephemerals are absent or rare, those genera typical of the Black Sea–Kazakhstan subregion being absent from Central Asia or limited to its western edge. Among ephemeroids, only *Tulipa uniflora* has been found. Summer–autumn annuals and biennials are abundant in steppes, especially during moist years. These species complete their life-cycles in early autumn. Especially typical are the following annuals:

Artemisia palustris	*Chamaerhodos erecta*
Artemisia pectinata	*Dontostemon integrifolius*
Artemisia scoparia	

The annual grasses in desert steppe are represented by *Aristida heymannii*, *Eragrostis minor* and others.

THE ORGANIZATION AND DYNAMICS OF STEPPE ECOSYSTEMS *[1]

Ecosystems include various levels of organization. Let us discuss some characteristics of each of the three major groups of organisms in steppes.

Plants

The organization of the plant community reflects that of the whole complex of populations of the component species. An understanding of the adaptive ability of these species (especially of the dominants) and of their response to various

---

* By Z.V. Karamysheva and E.M. Lavrenko.

[1] Based on Arnol'di and Yunatov (1969), Yunatov and Lavrenko (1969) and Lavrenko (1981).

Fig. 2.15. The provincial division of the Central Asian subregion of the Eurasian steppe region. Borders: A — of the subregion; B — of provinces; C — of sub-provinces.

6 – Hangay–Daurian mountain forest-steppe province
    a – Western Hangay sub-province
    b – Lower Selenga sub-province
    c – Nerchinsk–Onon sub-province
7 – Mongolian steppe province
    a – Middle Khalkha sub-province
    b – Eastern Mongolian sub-province
8 – North-Gobi desert-steppe province
    a – Big Lakes pan sub-province
    b – North-eastern Gobian sub-province
    c – Mongolo-Altaian mountain-steppe sub-province
9 – Manchurian forest-steppe (meadow-steppe) province
    a – Songhuian sub-province
    b – South Khinganian mountain forest-steppe sub-province
10 – Shanxi–Gansuian forest-steppe and steppe province

environmental factors is of great importance. This reveals the means by which communities maintain their stability even in the presence of catastrophic disturbance. Only by an understanding of morphological, ecological and physiological characteristics of the dominant species is it possible to consider the structural and functional features of a plant community on a holistic basis.

Adaptation

The dominating edificator[1] synusia in steppe communities is of bunch grasses with pronounced xeromorphy of above-ground organs. The main feature of their xeromorphism is the presence of narrow, almost thread-like, more or less folded leaves. The stomata are located on the inner surface of the folded leaves, where they are protected during dry and hot weather. They are able to open at the moment of unfolding of the leaf during moist periods. These characteristics of leaves are very important adaptive features of steppe grasses.

The bunch life-form, which predominates among species of the steppe, evidently has an advantage over the other life-forms in adjusting to a tenuous supply of moisture. This is achieved by burial of the vegetative buds, so that the bases of tillers are positioned deeper in the soil. This protects them better than species of other life-forms from grazing and trampling by wild and domesticated ungulates. The bunches assist in the accumulation of snow and dust, which add to the supply of moisture and nutrients.

The great lability of tiller formation, which is characteristic of steppe bunch grasses, helps them to survive extremely unstable weather conditions (Belostokov, 1957). Usually shoots of monocarpic species do not flower until the third year of development, but if growing conditions are especially favourable in spring they flower in the second or even the first year. The time of tillering is also adjustable, especially in species of *Stipa* (of the Section *Leiostipa*) in Kazakhstan.

The under-ground parts of steppe plants also possess characteristics that facilitate survival under adverse conditions. A fibrous root system is typical of most plants of the steppe, providing a great number of small roots and root hairs which penetrate densely throughout the soil. The surface area of the root system is very great in relation to mass, a characteristic that is particularly important under conditions of low soil moisture. For example, Shalyt (1950, 1952) estimated that the total surface of roots in the dry Black Sea steppe is 0.9 to 2.3 $km^2\ ha^{-1}$. The major amount of underground biomass was found to be in the uppermost 50 cm, in the humus horizon of the soil, whereas in the shallow soil of the southern steppes (Black-Sea steppe and Transvolga–Kazakhstan steppe) the roots are concentrated in the uppermost 30 cm (Shalyt, 1950, 1952; Arnol'di and Yunatov, 1969). In semi-desert steppes, the major portion of the root biomass is located in the top 20 cm of soil. In meadows the root system is also concentrated in the upper 50 cm, but the roots do not penetrate as deeply as in the steppe, where the moisture supply is more limited (Lavrenko, 1941).

The processes of water relations in steppe plants vary widely. Both transpiration rate and osmotic pressure are highly variable during the vegetative period, and even diurnally, indicating an ability to respond rapidly to changes in air humidity and supply of soil moisture. The ranges between maximum and minimum values of transpiration, osmotic pressure and water saturation deficit increase from the northern to the southern steppes and are especially great for bunch grasses growing at the southern limit of their distribution. For example, during the vegetative period the osmotic pressure in *Stipa pennata* in meadow steppes or bunch-grass steppes with many forbs ranged from 22 to 30 atm, whereas in *S. sareptana* in the semi-desert steppe of Kazakhstan the range was from 26 to 57 atm, and in *S. gobica* in the desert steppe of Mongolia the range was from 25 to 56 atm (Sveshnikova, 1979; Lavrenko, 1981). In desert plants the amplitudes of the indices of transpiration rate, osmotic pressure and water potential are essentially less than in steppe plants, although the absolute maximum values of osmotic pressure are higher (Bobrovskaya, 1988).

Steppe plants are able to endure intense desiccation. This is shown from measurements of the sublethal moisture deficiency. Bobrovskaya et al. (1977), who measured this index for *Agropyron cristatum* and *Cleistogenes squarrosa* in the dry

---

[1] An **edificator** is an environment-forming species that is responsible for the floristic composition and physiognomy of the community and the quantity and composition of the phytomass produced (Braun-Blanquet and Paviard, 1922).

steppe of Mongolia, found that the latter species can survive even after the loss of 80% of its moisture. The difference between the lethal and sublethal deficiency of moisture in steppe plants is very small, being much less than in desert species, in which the sublethal period begins at a lesser level of dehydration of tissues. These observations emphasize that plants of steppe and desert adapt in different ways to the lack of moisture.

Most steppe species are both anemophilous and anemochoric. The regimes of flower opening and of pollination are generally controlled by humidity of the air. Grass flowers are chasmogamous (open for pollination), with the exception of species of *Cleistogenes* and several species of *Stipa* (which are cleistogamous). The inflorescences of some species of *Stipa* remain in the closed sheath of the upper leaf of the generative shoot, which dries up and opens only when the seeds are fully ripe. Species of *Cleistogenes* are characterized by permanent axillary cleistogamy. The lateral branches of the inflorescence are formed in the axils of the lower leaves of the reproductive part of the shoot (Bespalova, 1977). The wind, which is almost constant over steppes, is the chief agent of distribution of seeds and fruits. Anemochory is assisted by the structure of the inflorescence (for example, "tumble weed") and of disseminules in some species.

Other adaptive features function at the organismic level. For example, *Stipa gobica* has a rather high photosynthetic ability at temperatures much lower than is optimal for this activity. At 0° and 10°C, it is able to absorb, respectively, 22% and 56% of the amount of carbon dioxide which it fixes at the optimum temperature (which varies from 20° to 25°C). This plasticity is a feature of the adaptation of this species to survival in southern Mongolia, where low temperatures are frequent throughout the vegetative period. Such plasticity is not inherent in desert species.

Many adaptive features operate at the population level. For example, the steppe community is able to compensate for an adverse environment by rare and irregular peaks in rates of germination of seeds and of survival of seedlings. This is especially important in the dry conditions of southern steppes. Even in the meadow steppes of the Black Sea region, where annual precipitation reaches 600 mm, great fluctuations are observed in the time of appearance and abundance of seedlings, as well as in the number of juvenile individuals[1] (Kamenetskaya, 1949). This situation exists even though seeds are regularly produced, both in the meadow steppe and in the more southerly forb–bunch-grass steppes. We found that seedlings of *Stipa* (as many as six different species in a square metre) were frequently abundant in the steppe in the basin of the Donets river, and that juveniles of these species were always present. Juveniles endure droughty years much better than seedlings, so that, when space is freed by death of older plants, they are able to undergo normal development. However, in the desert steppe of Mongolia, where annual precipitation is less than 150 mm, seedlings and juvenile individuals contribute only 1% of the population of the dominant species, *Stipa gobica*. Survival of seedlings and juvenile plants is possible only in rare instances when 2 or 3 or more favourable years occur in succession. Then, seeds germinate which have accumulated in the soil during several previous years. In spite of difficulties in regeneration, populations of the principal species, even in the desert steppe of Mongolia, include a complete range of age groups and are homoeostatic, although individuals of middle age and old age prevail because regeneration is irregular.

Phenological development of steppe species is highly diverse, both in the duration of the vegetative period and in the timing of phenological phases (aspects), their number and their prominence. This variability is in response to variations in temperature and moisture supply. For example, in the steppified meadows[2] of the central chernozem regions (forest steppe) of the European part of the former Soviet Union, where the vegetative period is from April to September, some species are flowering at all times. As many as eleven flowering phases can be identified. In more xerophytic steppes (in the same region), where growth begins in the second half of April and continues until early October, summer is the most colourful period of flowering. During late summer the leaves of grasses dry up somewhat, but some

---

[1] A **juvenile** is a plant which has not yet reached the reproductive stage.

[2] **Steppified meadow** is meadow with steppe species.

forbs continue to flower. In these grasslands no pronounced period of semi-dormancy occurs, as in southern steppe, and seven distinct flowering phases can be detected.

In the bunch-grass dry steppes of the Black Sea region and Kazakhstan, where the vegetative period is still longer (from March to early November), new seedlings of winter ephemerals and new leaves of *Festuca valesiaca* and of *Stipa* spp. begin to appear in autumn. A well-pronounced period of semi-dormancy occurs during summer (late June and early July). Most species flower during spring and early summer, or during late summer and early autumn, depending of the distribution of precipitation.

In the semi-desert steppe of Kazakhstan, where the vegetative period is from 7 to 8 months long, as well as in the dry steppe, the period of semi-dormancy lasts for one to one and a half months. Here the periodicity of phenological development permits utilization of all ecological niches under conditions of irregular supply of moisture. The number of flowering phases (5–6) is similar to that in dry steppes. The changes observed are similar to those in dry steppes.

In the extremely adverse environment in the desert steppe of Mongolia, growth begins between early April and early June, depending on the occurrence of rainfall. Nevertheless, there is a general tendency towards an increase in the number of species growing vegetatively from spring to summer, the time of maximum precipitation. When growth begins early, plant development is delayed by spring frosts. The vegetative season is seldom as long as six months, being repeatedly interrupted by drought, even to the extent of complete cessation of growth. The end of the vegetative season in early October is determined by the repeated occurrence of freezing temperatures. The desert steppe is also characterized by phenological non-uniformity of coenopopulations[4], especially in droughty years. During moist years the phenological phases are lengthened, while in droughty years tillers at various stages of development are present in the same individual plant. Some plants remain dormant for several successive years. Thus, the phenological development of the desert-steppe community in Mongolia exhibits the greatest variability among the various types of Eurasian steppes.

Community structure

Considerable variability also exists in the structure of the above-ground and under-ground parts of communities in different subzones of the steppe. The proportion of the soil surface covered by the vegetative canopy decreases from approximately 90% in the meadow steppe of the Black Sea region to about 10% in the desert steppe of Mongolia. Species diversity decreases along this same gradient from 22–36 species in one square metre to only 7–10 species. Similarly, the height of grasses decreases from 80 cm or more in meadow steppes to only 5–8 cm (10 cm in moist years) in desert steppes (Figs. 2.16 to 2.19).

Although the layer of grasses and dwarf half-shrubs constitutes the basic canopy of steppe vegetation, other strata are usually present. A layer of mosses (most commonly of *Thuidium abietinum*) often is present in meadow steppe, whereas lichens (mainly *Parmelia vagans*) form a rather thick cover on the soil surface in dry and semi-desert steppes of Kazakhstan, especially on sandy and stony soils. The whole Eurasian steppe region is characterized by a well-formed, but rar-

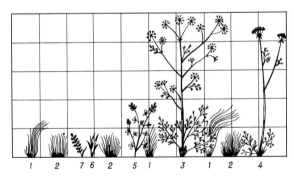

Fig. 2.16. A vertical projection of true (typical) bunch-grass steppe in the Black Sea–Kazakhstan subregion (Transvolga–Kazakhstan steppe province):
1 – *Stipa zalesskii* 5 – *Medicago romanica*
2 – *Festuca valesiaca* 6 – *Jurinea multiflora*
3 – *Peucedanum alsaticum* 7 – *Artemisia latifolia*
4 – *Seseli ledebourii*
(Kokchetav region, Kazakhstan). (Drawing by T.I. Isachenko and E.I. Rachkovskaya.)

---

[1] A **coenopopulation** is the aggregate of individuals in a phytocoenosis (Rabotnov, 1983).

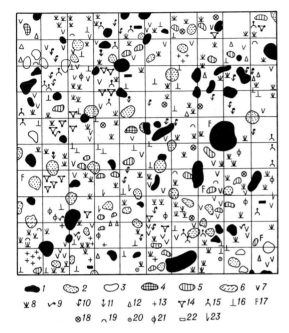
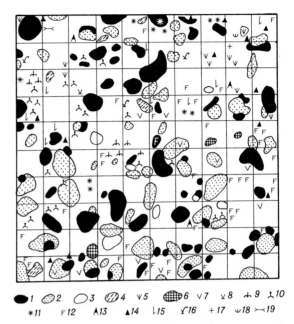

Fig. 2.17. A horizontal projection of *Stipa zalesskii–Festuca valesiaca* steppe in the Black Sea–Kazakhstan subregion (Transvolga–Kazakhstan steppe province), bases of shoots and bunches:

1 – *Stipa zalesskii*	12 – *Galatella angustissima*
2 – *Festuca valesiaca*	13 – *Potentilla humifusa*
3 – *Koeleria cristata*	14 – *Thymus marschallianus*
4 – *Poa stepposa*	15 – *Medicago romanica*
5 – *Phleum phleoides*	16 – *Artemisia sericea*
6 – *Helictotrichon desertorum*	17 – *Artemisia latifolia*
7 – *Bromopsis inermis*	18 – *Senecio integrifolius*
8 – *Pulsatilla multifida*	19 – *Filipendula vulgaris*
P. flavescens	20 – *Achyrophorus maculatus*
9 – *Veronica spicata*	21 – *Stellaria graminea*
10 – *Veronica spuria*	22 – *Onosma transrhimnensis*
11 – *Campanula wolgensis*	23 – *Galium verum*

(Kokchetavskaya hills region, Kazakhstan.) (Drawing by T.I. Isachenko and E.I. Rachkovskaya.)

Fig. 2.18. A horizontal projection of *Stipa zalesskii* steppe in the Black Sea–Kazakhstan subregion (Transvolga–Kazakhstan steppe province), bases of shoots and bunches:

1 – *Stipa zalesskii*	11 – *Seseli ledebourii*
2 – *Festuca valesiaca*	12 – *Artemisia latifolia*
3 – *Koeleria cristata*	13 – *Salvia stepposa*
4 – *Helictotrichon desertorum*	14 – *Jurinea multiflora*
5 – *Bromopsis inermis*	15 – *Galium ruthenicum*
6 – *Poa stepposa*	16 – *Veronica supina*
7 – *Carex supina*	17 – *Potentilla humifusa*
8 – *Carex praecox*	18 – *Potentilla bifurca*
9 – *Peucedanum alsaticum*	19 – *Eryngium planum*
10 – *Medicago romanica*	

(Kokchetav region, Kazakhstan.) (Drawing by T.I. Isachenko and E.I. Rachkovskaya.)

efied shrub layer (particularly of species of *Caragana* and *Spiraea*), which is usually taller than the grasses, reaching a height of 30 to 100 cm. Stratification of the grass–half-shrub layer is not evident in meadow steppes. In the grass canopy of the forb–bunch-grass steppes, and especially of the dry bunch-grass steppes, three sublayers are clearly distinguished: one of tall bunch grasses (species of *Helictotrichon* and *Stipa*) and of dicotyledonous forbs; a second of low bunch grasses and dicotyledonous forbs; and a third stratum of ephemerals, often reaching several centimetres in height. Similar sublayering occurs in semi-desert and desert steppes.

The division of the under-ground parts into sublayers also is evident, especially in the more southerly steppes. In meadow steppes, the uppermost 10 cm is a turf layer consisting of roots and rhizomes, the upper 50 cm of soil contains 89 to 94% of the under-ground biomass, and only solitary roots penetrate deeper than 75 cm. In the forb–bunch-grass steppes, where the stratification of roots is more pronounced than in meadow steppes, the major part of the root mass (76–80%) is contained in the 0- to 30-cm layer, the turf horizon being absent, and the number of rhizomes is considerably reduced. The usual pat-

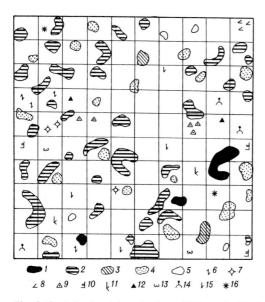

Fig. 2.19. A horizontal projection of *Stipa zalesskii–Festuca valesiaca* bunch-grass steppe in the Black Sea–Kazakhstan subregion (Transvolga–Kazakhstan steppe province), bases of shoots and bunches:

1 –*Stipa zalesskii*	9 –*Galatella divaricata*
2 –*Stipa lessingiana*	10 –*Artemisia austriaca*
3 –*Stipa korshinskyi*	11 –*Astragalus macropus*
4 –*Festuca valesiaca*	12 –*Jurinea multiflora*
5 –*Koeleria cristata*	13 –*Palimbia salsa*
6 –*Agropyron pectinatum*	14 –*Medicago romanica*
7 –*Leymus ramosus*	15 –*Galium ruthenicum*
8 –*Crinitaria tatarica*	16 –*Seseli ledebourii*

(Tselinograd region, Kazakhstan.) (Drawing by T.I. Isachenko and E.I. Rachkovskaya.)

tern of vertical distribution of roots is: a shallow layer of ephemerals; a mid-depth layer of rhizomatous dicotyledons; a deeper stratum of bunch grasses (and a few rhizomatous grasses) and of dicotyledonous forbs; and the deepest layer of tap-rooted dicotyledons. Fine, profusely branched lateral rootlets are typical in the surface layer of dry bunch-grass steppes, where forbs are almost absent and rhizomatous species are rare. Roots are mainly in the uppermost 30 cm of soil. The proportion of the under-ground plant parts that is composed of short-lived roots increases in semi-desert steppes (Shalyt, 1950, 1952).[1]

---

[1] Unfortunately, the root systems of steppe plants have, as yet, been studied insufficiently, especially in Mongolia.

Standing crop

The maximum standing crop of green shoots ranges widely in the steppe region, from 320–420 g m^{-2} in meadow steppes to only 6–30 g m^{-2} in the extremely xerophytic desert steppes in Mongolia. However, the standing crop of underground parts is much more constant. Shalyt (1950, 1952) reported values of from 1520 to 1970 g m^{-2} for meadow steppes on deep chernozem soils in the Black Sea region, 1470 g m^{-2} for more southerly forb–bunch-grass steppes on southern chernozem, from 1490 to 3000 g m^{-2} for dry bunch-grass steppes and from 1700 to 1800 g m^{-2} for semi-desert *Artemisia*–bunch-grass steppes on light-chestnut soils. Thus, the root/shoot ratio ranges from 8 in the north to 30 or more in the south.

Animals

Animals have an important role in the formation, development and functioning of steppe ecosystems. Chernov (1975) and Mordkovich (1982) discussed the quantitative and qualitative differentiation of animal populations throughout the steppe region, as well as their physiological relationships with vegetation. These authors are of the opinion that the very great changes in animal populations between northern and southern steppes are related to increased aridity. According to Chernov, these changes are much more clearly pronounced in the steppe than in other vegetation zones of temperate latitudes, because of the steeper hydrothermic gradient typical of subarid and arid regions.

The forest steppe, owing to an optimal combination of temperature and moisture, provides the most favourable conditions for existence of all biocomponents of the ecosystem and for their highest productivity. The biomass of invertebrates is greatest in communities of forest steppes. This group contributes 95% of faunal biomass in all zonal ecosystems and is the most important in terms of energetic potential. Invertebrate biomass in soils in a dry steppe is only one-fourth of that in a forest steppe. In semi-desert steppes values are from 0.02 to 0.03 g m^{-2} (live weight), and they are even lower in the desert steppe. A similar southward decline in biomass exists with respect to above-ground invertebrates, as well as to vertebrates. For example, the biomass of above-ground insects in the semi-desert steppe is only one-seventh that in the grassland portion of the forest steppe, and the mean biomass (throughout the year) of vertebrates, which in the forest steppe amounts to 1.29 g m^{-2} (live weight), is only 0.6 g m^{-2} in the desert steppe. Earthworms, which have a very important role in the processes of soil formation and especially in improving the structure of soils, decrease markedly in abundance southward within the steppe region — as do other moisture-loving animals. The abundance of saprophages generally decreases and that of herbivores increases from forest steppe to semi-desert steppe. However, some saprophages, notably the darkling beetles (Tenebrionidae) are more abundant to the south.

The effects of these differences in populations and biomass must be considered in relation to the biomass of the plant substrate on which they depend. For example, the ratio of the live-weight biomass of Locustidae to the fresh-weight of above-ground green phytomass changes from 0.001 in the north to 0.1 in the south.

The seasonal rhythm of activity of animals changes from north to south similarly to that of plants. Southwards, periods of activity are shifted towards concentration in spring and autumn, whereas in summer activity is reduced.

Numerous examples were given by Mordkovich (1982) of adaptation of animals to the changeable environment inherent in steppe ecosystems. For example, the development of chitin of specific composition, many-layered in structure, permits conservation of moisture in the bodies of insects. Many insects adapt to the variable conditions in steppes by adjusting their feeding habits. Some combine carnivory with herbivory and even saprophagy.

Mammals, 92 species of which occur in the steppe region (Mordkovich, 1982), also adapt in various ways to the open landscape of the steppe with its low canopy. Most inhabit burrows and depend on the under-ground part of the ecosystem, as a result of which they have developed various morphological features of an adaptive character. Several are gregarious, which is an adaptation to the existence in open spaces. The ability to move rapidly, in search of food and water and for protection, is another important adaptation.

The activities of various groups of animals in steppes have an important influence on the composition and structure of the vegetative cover, on the physical and chemical features of soils, and on the micro-relief of the landscape. The very important roles of rodents will be discussed later in this chapter (pp. 42–55). The effects of insects have received much attention, especially those of herbivores (such as locusts).

The considerable influence of hoofed mammals (Ungulata) in steppe ecosystems is also well known. Large populations of wild ungulates inhabited the steppe in the past. These included *Equus hemionus* (Asiatic wild ass), *E. przewalskii* (wild horse) and *Saiga tatarica* (saiga). These consumed large quantities of canopy phytomass. Grazing by ungulates leads to the disappearance of forbs (Fig. 2.20), followed by decreases in abundance of species of *Stipa* and increases in species of *Festuca* (Fig. 2.21). Very intensive grazing results in dominance of annuals. Investigations of successions (Lavrenko, 1980a) in an absolutely protected steppe (Dokhman, 1956, 1968; Semenova-Tyan-Shanskaya, 1966; Osychnyuk, 1973; Bilyk and Tkachenko, 1976; Osychnyuk et al., 1976; Tkachenko, 1984; Shelyag-Sosonko et al., 1985) have irrefutably proved that the true steppe (with domination of bunch grasses) can exist only under conditions of permanent moderate grazing by ungulates. The absence of ungulates results in accumulation of a large quantity of standing dead shoots and litter, in changes in the air and water regimes in the canopy, and replacement of bunch grasses by rhizomatous species.[1] Also, vegetatively reproducing shrubs spread in some situations in unmowed and ungrazed steppes, as do trees in forest steppes. These observations support the concept that wild ungulates have had a role in the development of steppe ecosystems and that bunch grasses, because of their ability to withstand grazing, have gained their position as edificators in the presence of the selective grazing by wild ungulates.

---

[1] The reservation regime in the meadow steppes of Central Asia does not induce a pronounced increase in the proportion of mesophytes in the vegetation (Guricheva et al., 1984).

Fig. 2.20. A digressive (denuding) clump (bunch form) of *Stipa tirsa* dominating in the Streltsovskaya Steppe Reservation in the Black Sea–Kazakhstan subregion [Black Sea (Pontic) steppe province]. (Voroshilovgrad region, Ukraine.) (Photo by V.S. Tkachenko.)

Micro-organisms

Whereas the above discussion has dealt with zonal changes in the quantitative and qualitative composition of autotrophic higher vascular plants and of the structure of communities formed by them, as well as changes in the ecological–physiological characteristics of their edificators (the bunch grasses), similar (and even more distinct) changes are also typical of the autotrophic algal flora of the surface horizons of the steppe soils and especially of the heterotrophic part of the steppe biocoenoses (Bacteria, Fungi and animals).

The abundance of soil micro-organisms decreases from north to south and fluctuates with variations in content of soil moisture. For example, in steppes in the Central-Kazakhstan subprovince the number of algal cells was estimated to be from 4 to 6 million per gram of dry soil, whereas in semi-desert steppe only 420 000 occurred (Rakhmanina and Arnol'di, 1961). Moisture-loving forms of algae (green filamentous and unicellular yellow-green algae) are more abundant on the soil surface in dry steppes; blue-green species are also important, the accretions of *Nostoc commune* being widely distributed. The microscopic soil fungi are similarly reduced in numbers in more southerly steppes. They are more abundant in the rhizosphere in the dry steppe than in the desert steppe; species of *Aspergillus*, *Cladosporium*, *Geotrichum*, *Penicillium* and *Stenophyllum* are prevalent in the dry steppe, but poorly represented in more southerly steppes. Fungi of the order Mucorales are abundant in desert steppes, but almost absent from dry steppes. In contrast, Bacteria are more abundant in the semi-desert steppe (9–176 million of cells g^{-1} of dry soil) than in the dry steppe (7–8 million g^{-1}). Calculated values for biomass of Bacteria in brown soils of Mongolia are greater than in soils of semi-desert steppes in Kazakhstan.

Fig. 2.21. A denuding variant of *Festuca valesiaca–Stipa ucrainica* steppe in the Black Sea–Kazakhstan subregion [Black Sea (Pontic) steppe province]. (Askaniya Nova Reservation, Kherson region, Ukraine.) (Photo by V.S. Tkachenko.)

Zonal changes also occur in the proportions of the various functional groups of micro-organisms. For example, the oligonitrophiles — species adapted to utilization of extremely small quantities of nitrogen — are abundant in the soil of desert steppes in Mongolia. This group of micro-organisms completes the mineralization of organic substances. In moister situations those groups, such as non-sporulating bacteria, which induce mineralization of more complex organic substances, are prevalent. Southward, the greater abundance of microflora which completely mineralize organic substances, as well as a reduced rate of humification, appear to be the reasons for the lower content of soil humus.

Other factors

Steppe ecosystems possess considerable stability and ability to recover from disturbance caused by heavy grazing, fire and tillage (Gordyagin, 1901; Taliev, 1904; Vysotskyi, 1908, 1909, 1915, 1923; Pachoskyi, 1917; Zalesskyi, 1918; Lavrenko, 1927; Shalyt, 1927, 1938; Reverdatto and Golubintseva, 1930; Gael', 1930, 1932). Only slight changes in composition and structure of steppes occur as a result of fire, and recovery is rapid. Recovery of vegetation in abandoned cropland is also quite rapid, especially in situations where disseminules are available nearby.

The steppe has undergone transformation by the activities of man. A large portion of the Eurasian steppe region has been converted into cropland by ploughing. In the European part of the former Soviet Union and in Western and Northern Kazakhstan, the steppes have been preserved only in reservations, except for areas occupied by halophytic variants of dry and semi-desert steppes (in the Caspian region) and variants in gravelly and stony soil (in Central Kazakhstan).

The steppe has been preserved closest to its natural state in Mongolia, where fire is rare, mowing is uncommon, and grazing is moderate.

COMPARISON OF TWO ZONAL TYPES OF ECOSYSTEMS*

Since the end of the 1950s, complex biogeocoenotic (ecosystem) investigations were carried out in reference plots of steppe coenoses in Kazakhstan and in Mongolia (Arnol'di and Yunatov, 1969; Yunatov and Lavrenko, 1969; Lavrenko and Borisova, 1976; Lavrenko, 1980b, 1981). The results of these investigations provide an insight into the composition and structure, and the morphological, ecological and physiological features of steppe ecosystems and into certain aspects of biological activities during seasonal cycles and cycles of different years.

We have chosen for discussion studies in two of these sites in which the establishment of research facilities facilitated intensive study over several successive years using standardized techniques in similar programs. These are representative of the most southerly steppes and are located about 2500 km apart. Unfortunately, comparable information for more northerly types of steppe (meadow and true steppes) are not yet available.

One of these sites is a complex of two communities in semi-desert steppe dominated by bunch grasses (*Stipa lessingiana–Festuca valesiaca*) and a dwarf half-shrub (*Artemisia gracilescens*) in light-chestnut soils. This complexity is a result of heterogeneity of the soil. The ecosystem is a zonal type typical of the subzone of semi-desert steppes of the Central Kazakhstan subprovince of the Transvolga–Kazakhstan steppe province (of the Black Sea–Kazakhstan subregion) where light-chestnut soils prevail. The 5-ha experimental area in this site was studied for eleven years (1958–1968).

The other site is in desert steppe typical of the North-Gobi desert steppe province (Central Asian steppe subregion). It is dominated by bunch grasses (*Stipa gobica–Cleistogenes songorica*) and by a dwarf half-shrub *Artemisia frigida*, with *Caragana leucophloea* as an associate. The species

---
* By I.V. Borisova and T.A. Popova.

in this type of steppe are adapted to endure a prolonged existence in extremely dry conditions. Desert steppe extends farthest southward (to 45°N latitude) within the steppe region and is characterized by brown desert-steppe soils. Studies were undertaken here for six years (1970–1974, 1976) in an experimental area of 2 ha.

The vegetational structure in the two sites is similar, but intensive study of these steppe communities revealed differences in floristic composition, in the proportions of life-forms, in structure and rhythm of development, and in ecological–physiological and other indices, all of which are related to environmental differences between the two sites. These studies, together with those of ecosystems in other types of steppe, show that all steppe communities are dominated by bunch grasses and possess both a layered (vertical) and complex mosaic (horizontal) structure. An unusual lability of phenological development is also typical. This is observed in the rhythm of tiller formation and the ecology of flowering. It is apparent that, in steppe, only those species can dominate which are able to adjust their ecophysiological indices to a wide range of environmental conditions. Each dominant species is well adapted to cohabiting with the others. In the Kazakhstan site (semi-desert steppe) *Festuca* and *Stipa* develop synchronously and supplement one another, whereas *Artemisia gracilescens* develops later and substitutes for them. In the Mongolian site (desert steppe), *Stipa* develops early, whereas *Artemisia frigida* and *Cleistogenes* develop at the same time and substitute for *Stipa*.

Climate

The climate of the two areas is continental, but the Mongolian site receives only about half as much precipitation as the more westerly one (Table 2.1). Another distinct difference is the distribution of precipitation. Whereas 60% of the precipitation often occurs in June in the Kazakhstan site, almost all (as much as 97%) occurs in July and August in the Mongolian area. Spring is very dry at the latter site.

The evaporative power of the air is much greater at the Mongolian site. Whereas the relative humidity at the Kazakhstan site averages 66% during daylight from June to August, in Mongolia

TABLE 2.1

Comparison of the climatic conditions of the two study sites

Parameter	Kazakhstan semi-desert steppe	Mongolian desert steppe [1]
Air temperature (°C)		
Annual mean	2.1	4.8
Monthly mean		
July	22–25	22
January	−14	−14
December–February	−13 to −16	−13 to −14
Absolute maximum	45	39
Absolute minimum	−37	−36
Annual amplitude	41	37
Daily amplitude	25–31	20–25
Mean annual precipitation (mm)	250	117
Proportion in summer (%)	60–70	85–97
Length of vegetative season	170–190	140–150
Period of active growth (days)[2]	150–180	120–140
Years of observation	1958–1968	1970–1976
Location, N. latitude	48°	44°
E. longitude	71°	101°
Elevation (m)	600	1300

[1] At the Mongolian site the temperature at the soil surface ranges from −42 to 64°C, and the range in annual precipitation is from 60 to 169 mm.

[2] The period of active growth at the Kazakhstan site is that during which the mean daily temperature rises above 10°C (mid to late April until mid to late October); at the Mongolian site it is delayed until the beginning of the rainy season (in May, June or July) and ceases when constant night frosts commence (in late September to mid October).

it is usually 35 to 45% during this time, but falls to 10 to 20% on dry, windy days.

Whereas the mean annual temperature of the Mongolian site (4.8°C) is slightly higher than of the Kazakhstan site (2.1°C), summer is warmer in the latter area. The length of the vegetative period in Kazakhstan is 1.5 to 2 months longer than that of the Mongolian site. In the Mongolian site, the upper soil layers are frozen during winter (to a maximum depth of 1.5–3 m).

Soil

The soil in the Kazakhstan site is light-chestnut in colour, whereas that of the Mongolian site is brown. Both soils are sandy in texture, but the Mongolian soil is the coarser of the two. Its surface is covered by a crust of sand and gravel, and the adsorptive capacity is very low. The Mongolian soil has a lower content of organic matter (0.3 to 0.4%) than that in Kazakhstan (1.6 to 2.2%) and a higher content of carbonates. The Kazakhstan soil is of solonetz character, whereas the Mongolian soil is practically free of salt.

In the Kazakhstan site available soil moisture is most plentiful after snow melt in spring and after significant rainfall. In the Kazakhstan site the sandiness and the presence or absence of gravel on the surface are important factors in affecting the hydrothermal balance in the soil. The moisture content fluctuates widely in the Mongolian soil.

Flora

Although species of *Artemisia* and *Stipa* are among the dominants in both locations, the floristic composition of the two study sites is quite dissimilar. They share no species, and even only 16% of the genera are common to both (Tables 2.2 to 2.4). The Kazakhstan site includes only two-

TABLE 2.2

Relative abundance [1] of the vascular plants present in the various communities studied, classified according to growth form

Species	Kazakhstan semi-desert steppe		Mongolian desert steppe		
	Stipa–Festuca–Artemisia	Artemisia–Stipa–Festuca	Stipa	Stipa–Convolvulus	Caragana–Cleistogenes
SHRUBS					
Caragana leucophloea			R	R	SD
Zygophyllum xanthoxylon			VR		
DWARF SEMI-SHRUBS					
Artemisia austriaca	R	R			
Artemisia frigida			D	A	SD
Artemisia gracilescens	A	D			
Artemisia sublessingiana	R				
Artemisia xerophytica			R		
Ajania fruticulosa			A	C	C
Eurotia ceratoides			R		
Kochia prostrata	R	C			
Oxytropis aciphylla			R		R
Ptilotrichum canescens			A	C	R
Tanacetum santolina	R	R			
PERENNIAL GRASSES AND SEDGES					
Carex stenophylla	R				
Cleistogenes songorica			D	A	D
Cleistogenes squarrosa			VR		
Festuca valesiaca	D	SD			
Koeleria cristata	R	R			
Stipa glareosa			R	R	B
Stipa gobica			D	D	D
Stipa kirghisorum	R				
Stipa lessingiana	D	C			
Stipa sareptana	C	C			
ANNUAL GRASSES					
Aristida heymannii			R	R	R
Eragrostis minor			C	C	C–A
Poa bulbosa	C	C			
Setaria viridis			R		A
PERENNIAL FORBS					
Allium clathratum			VR		
Allium mongolicum			R		C
Allium pallasii	R	R			
Allium polyrrhizum			A	A	A
Asparagus gobicus			R		R
Astragalus grubovii			R		R
Astragalus junatovii			R		R
Astragalus monophyllus			R		R
Astragalus vallestris			C	R	R
Convolvulus ammanii			A	C–D	A
Crinitaria tatarica	R	R			
Crinitaria villosa	VR				
Dontostemon crassifolius			R	R	R
Erysimum leucanthemum	C	C			

(continued)

TABLE 2.2 (continued)

Species	Kazakhstan semi-desert steppe		Mongolian desert steppe		
	Stipa–Festuca–Artemisia	Artemisia–Stipa–Festuca	Stipa	Stipa–Convolvulus	Caragana–Cleistogenes
PERENNIAL FORBS (continued)					
Ferula bungeana					R
Ferula caspica	C	R			
Gagea bulbifera	R	R			
Gagea pusilla	C	R			
Gypsophila desertorum			C	A	C
Haplophyllum davuricum			R		
Heteropappus altaicus			R		
Hymenolyma trichophyllum	R	R			
Iris bungei					R
Iris scariosa	R	R			
Iris tenuifolia			C		R
Jurinea mongolica			R	R	
Lagochilus ilicifolius			R		C
Limonium tenellum			R		
Linaria sp.	R				
Peganum nigellastrum					R
Rheum nanum			R		R
Scorzonera capito			VR		
Scorzonera divaricata			R	R	R
Scorzonera tuberosa	R	R			
Tugarinovia mongolica			VR		
Tulipa patens	C	R			
Valeriana tuberosa	R				
Vincetoxicum sibiricum					R
Zygophyllum rosovii			R		R
ANNUAL FORBS					
Alyssum turkestanicum	C	R			
Androsace maxima	R	R			
Bassia dasyphylla			R		R
Ceratocarpus arenarius		R			
Ceratocephala testiculata	C	C (gr)			
Chamaerhodos erecta					R
Chenopodium acuminatum			R		R
Corispermum mongolicum			R	R	C
Descurainia sophia	R	R			
Euphorbia humifusa			R	R	R
Filago arvensis	R	R			
Meniocus linifolius	R				
Plantago minuta					R
Polygonum patulum		R			
Salsola australis			R	R	R
Salsola collina			R	R	R

[1] D = dominant; SD = subdominant; A = abundant; C = common; R = rare; VR = very rare; (gr) = in groups.

thirds as many species as the Mongolian site and is composed mostly of Eurasian steppe species, in contrast to the main representation of Central-Asian (North Gobi) desert-steppe species in the more easterly site. The projected cover of soil by plant crowns is two to three times as great in the Kazakhstan community, because of a greater number of plants.

The western site is a mosaic of two communities, related to differences in soil structure, especially depth of the carbonate horizon. The major area (70%) of this Kazakhstan site is occupied by a community of *Stipa lessingiana, Festuca valesiaca* and *Artemisia gracilescens* (in order of dominance), together with two lichens (*Parmelia vagans* and *P. ryssolea*) (Table 2.2). In the other community, the principal dominant is *Artemisia gracilescens*, and the most abundant grass is *Festuca valesiaca*.

The community in the Mongolian site is composed of three microcoenoses (sub-communities), differing considerably in floristic composition, both of perennials and annuals. In terms of numbers of plants, the grasses *Cleistogenes songorica* and *Stipa gobica* share dominance throughout the community with the dwarf half-shrub *Artemisia frigida* (Fig. 2.22).

The floristic composition in the Kazakhstan location changed little during the period of study. The contributions of perennial species to the cover remained constant, but the late-spring annuals did not grow at all when spring precipitation was low. In the Mongolian site, variations also occur between years within each sub-community in response to differences in weather. The floristic composition of both annuals and perennials is affected.

Vegetative cover is very sparse. In the Kazakhstan site, foliage cover increases from 40% in spring to 50–70% in summer, whereas the proportion of the soil surface occupied by bases of plants is 10 to 30%. In the Mongolian site, cover is only one-half to one-third as dense.

TABLE 2.3

Numbers of plant species identified in the two study sites, classified into taxonomic groups

Groups of organisms	Kazakhstan	Mongolia
Angiospermae	33	49
Mosses	2	0
Lichens	12	0
Soil algae	26	*
Parasitic (smut and rust) micro-fungi	4	*
Soil microfungi	81	*
Bacteria and Actinomycetes	105	65
Total	263	114

* Not studied.

TABLE 2.4

Comparison of the floristics of vascular plants in two kinds of steppe

Parameter	Kazakhstan semi-desert steppe	Mongolia desert steppe
Number of species	33	49
Number of genera	27	37
Number of families	14	19
Number of species in leading families		
Asteraceae	8	8
Brassicaceae	4	2
Chenopodiaceae	2	6
Fabaceae	0	6
Liliaceae	4	4
Poaceae	6	7
Zygophyllaceae	0	3
Growth-forms (number of species)		
Perennials	25	38
Annuals	8	11
Dominant grasses		
Cleistogenes songorica		×
Festuca valesiaca	×	
Stipa gobica		×
Stipa lessingiana	×	
Dominant dwarf semi-shrubs		
Artemisia frigida		×
Artemisia gracilescens	×	
Other abundant species		
Caragana leucophloea		×
Stipa sareptana	×	

Geographical and community types

True steppe (Black Sea–Kazakhstan) species, which are widely distributed throughout the steppe region, prevail in the vegetation of the western site (Table 2.5). A considerable number of southern (Mediterranean) species also occur, but these do not play any considerable role in the structure of the vegetation. They are mainly ephemerals, ephemeroids and hemi-ephemeroids. The presence of species indicative of sandy, saline and carbonate conditions is also typical.

TABLE 2.5

Percentage representation of phytogeographic, phytocoenotic and ecological types of species in the Kazakhstan study site

	Number of species	Canopy cover
PHYTOGEOGRAPHIC ELEMENTS		
Black Sea–Kazakhstanian and Mongolian	28	35
Palaearctic and western palaearctic	16	39
Zavolzhsko–Kazakhstanian and Dzungaro–Tien-Shanian	16	22
Mediterranean	22	3
Kazakhstan–Middle Asian	9	+[1]
Kazakhstansko–North Turanian	6	+
Holarctic	3	1
PHYTOCOENOTIC TYPES OF SPECIES		
Steppe	40	74
Desert steppe	44	24
Steppe–desert	16	2
ECOLOGICAL TYPES OF SPECIES		
Halophilous–petrophilous-*plakor* steppe	37	17
Hemipsammophilous–halophilous-*plakor* steppe	7	38
Petrophilous–hemicalciphilous-*plakor* steppe	3	33
Petrophilous–halophilous-*plakor* steppe	22	9
Petrophilous-*plakor* steppe	19	1
Petrophilous steppe	6	1
Hemipsammophilous–petrophilous-*plakor* steppe	6	1

[1] + = less than 0.5%.

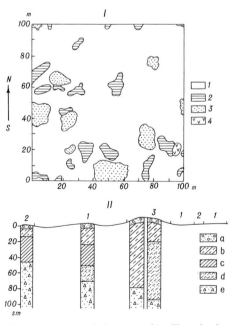

Fig. 2.22. A map of the vegetation (I) and micro-relief and soil structure (II) of various elements of the mosaics of part (1 ha) of the Mongolian study site:

Most species in the Mongolian site have Central-Asian areas of distribution, some (*Cleistogenes songorica* and *Stipa gobica*) being North-Gobian, whereas others (*Allium polyrrhizum*) are of the Gobi–Mongolian type. Only *Artemisia frigida* is holarctic in distribution. The species of desert steppes are most important, both in terms of number of species present (53%) and above-ground phytomass contributed (73%), and steppe species account for most of the remainder, in number (27%) and phytomass (25%). Steppe-desert species account for 16% and desert species for 4% of those present.

1 – the background (*Stipa gobica, Cleistogenes songorica, Artemisia frigida*)
2 – *Stipa gobica, Convolvulus ammanii*
3 – *Caragana leucophloea, Cleistogenes songorica, Stipa gobica, Artemisia frigida*
4 – *Peganum nigellastrum*

a – the sand-cobble armour; b – loamy sands (70%) + sands (30%); c – loams (carbonate horizon); d – sands (70%); e – sandy cobble. (Drawing by T.A. Popova and G.N. Yakunin.)

TABLE 2.6

Contribution of various life-forms of polycarpic herbs to the canopy

Growth-form	Number of species		Canopy biomass (g m^{-2})	
	Kazakhstan site	Mongolian site	Kazakhstan site	Mongolian site
Tap root polycephalous[1]	0	36	0	5.1
Tap root monocephalous	7	4	1	–
Rhizomatous	4	4	0.5	0.1
Caespitose	34	15	80	54.3
Bulbiferous	15	2	0.5	–
Total	60	61	82	60

[1] **Polycephalous** species have a branched caudex, whereas the caudex in monocephalous species is not branched.

Life-forms

The percentage of species of each life-form is very similar in both sites:

	Kazakhstan	Mongolia
Perennial herbs	60	61
Annual herbs	25	23
Dwarf half-shrubs	14	12

The main difference is the presence of shrubs in the Mongolian location. The vegetation is composed almost entirely of polycarpic (perennial) herbs and primitive[1] dwarf half-shrubs. Although the number of monocarpic (annual) herbs is considerable, their contribution to the phytomass is negligible (0.4–0.6%).

Perennial herbs are the main contributors to both numbers of species and canopy biomass in both sites (Table 2.6). In the Kazakhstan site the plant cover is mainly of perennial semi-rosette summer–winter-green[2] herbs, with a period of summer semi-dormancy; the biomass of primitive summer–winter-green dwarf half-shrubs with tap roots is also considerable. Participation of hemi-ephemeroids, ephemeroids and ephemerals is insignificant (Tables 2.7 and 2.8); ephemerals are not present in the Mongolian site (Table 2.9).

Species which are pollinated and disseminated by wind are common, which is generally typical in steppes (Levina, 1967). Many species have adapted to variable weather by combining different means of pollination and distribution of diaspores. Morning and afternoon anemophily prevails in the Kazakhstan site. The flowering spectrum of the Mongolian site (Tables 2.9 and 2.10) shows entomophilous species to be most numerous (43%), but the anemophilous species are of chief phytocoenotic significance, in terms of both number of species and biomass. About one-fourth of the species have been adapted to self-pollination along with the possibility of pollination by another means. Of particular note is the axillary cleistogamy of *Cleistogenes songorica*. Distribution of disseminules is principally by wind, with self dissemination occupying a secondary role.

Euxerophiles and euryxerophiles are dominant (Table 2.10).

Phenology

Differences between the two sites in distribution of precipitation affects phenological development of plants. In the Kazakhstan site the early-summer seasonal synusia is prominent, whereas in the dry-spring environment of Mongolia the late-summer element is more important. The primary dominant grass (a species of *Stipa* in both instances) grows in late spring and early summer in Kazakhstan and in early summer in Mongolia. In

---

[1] *Primitive* half-shrubs are those in which more than half of the annual growth of shoots dies back each year; whereas in *true* half-shrubs the loss is less than half. The nature of the shoots of primitive half-shrubs is similar to that of herbs.

[2] **Summer–winter-green** species are those which have green leaves throughout the year, but the leaves are replaced periodically (two or three generations per year).

both areas, the dominant half-shrub (a species of *Artemisia* in both instances) makes most of its growth and flowers in late summer. However, the secondary grass (*Festuca valesiaca*) in Kazakhstan makes most growth in late spring to early summer, whereas that of the Mongolian site (*Cleistogenes songorica*) develops in late summer.

In Kazakhstan a period of semi-dormancy occurs in summer. This is well defined, except in very moist years, and usually lasts for 1 to 1.5 months. However, in the Mongolian site periods of active growth alternate with periods of suppression. A pronounced period of dormancy or semi-dormancy is absent. Summer–winter-green species prevail in both locations, but the diversity

TABLE 2.7

Percentage representation of species in the Kazakhstan study site according to life-form and growth-form

	Number of species	Canopy cover
LIFE-FORMS		
Long-lived perennials	71	99
Short-lived perennials (2–6 yr)	4	0.5
Annuals	25	0.5
Polycarpic	68	99
Monocarpic	28	0.5
Oligocarpic	4	0.5
Herbs	84	81
Dwarf half-shrubs	16	19
Bunch herbs	25	79
Primitive dwarf half-shrubs	12	17
Short-lived perennials and annuals, bunch- and cluster-rooted	25	0.5
Few- and one-headed (monocephalous), bunch- and cluster-rooted	19	1
Rhizomatous	16	0.5
True dwarf half-shrubs	3	2
Fibrous-rooted	31	79
Tap-rooted	44	19
Cluster-rooted	25	2
GROWTH-FORMS OF HERBS		
Half-rosette [1]	59	99.6
Non-rosette	22	0.4
Rosette herbs	18	+[2]

[1] **Half-rosette** species have a rosette of leaves at the base, with the rest of the leafy stem elongated.
[2] + = less than 0.5%.

TABLE 2.8

Percentage representation of species in the Kazakhstan study site according to phenological and hibernating types

	Number of species	Canopy cover
PHENOLOGICAL TYPES		
Summer–winter-green with summer semi-dormancy	25	78
Summer–winter-green	16	20
Hemi-ephemeroids, ephemeroids and summer ephemerals	25	2
Autumn–winter–spring-green with summer dormancy	22	+[1]
Spring–summer–autumn-green with winter dormancy	12	+
SEASON OF FLOWERING		
Early spring	13	0.4
Mid spring	12	0.3
Late spring	12	0.3
Early summer	25	73
Mid summer	19	6
Late summer	16	5
Early autumn	3	15
HIBERNATING TYPES		
Hemicryptophytes	31	80
Chamaephytes	13	19
Cryptophytes (geophytes)	28	1
Therophytes and therophyto-hemicryptophytes	28	+

[1] + = less than 0.5%.

of phenotypes is greater in the Kazakhstan site, where ephemeroids and ephemerals are present. However, the Mongolian site includes a specific phenotype of periodically vegetating plants (*Allium* spp.) which reflects the irregularity of phenological development under conditions of intermittent drought. The maximum number of species grow actively when soil moisture from precipitation is most readily available. In Kazakhstan this takes place in late spring, whereas in the Mongolian site it occurs in mid to late summer. The period of maximum flowering is in late spring to early summer in the Kazakhstan site and in late summer to early autumn in Mongolia.

In Kazakhstan the composition and number of actively growing species change during the growing season (Fig 2.23). Year-to-year differences also occur in the time required to complete each stage of development. Seasonal development in the Mongolian site is characterized by alterna-

TABLE 2.9

Percentage representation of species in the Mongolian study site according to life-form, flowering type, and diaspore type

	Number of species	Canopy biomass
LIFE-FORM		
Polycarpic herbs	61	60
Monocarpic herbs	23	+[1]
Dwarf half-shrubs	12	35
Shrubs	4	5
FLOWERING TYPE		
Anemophilous	33	80
Entomophilous	43	16
Entomophilous–autogamous	20	3
Entomophilous–anemophilous–autogamous	2	1
Entomophilous–anemophilous	2	+
DIASPORE TYPE[2]		
Anemochores	47	50
Autochores	29	43
Ballistochores	18	7
Ballistochores–anemochores	4	+
Zoochores	2	+

[1] + = less than 0.5%.
[2] Plants are distinguished in relation to the means of distribution of fruits, diaspores and seeds, as follows: **Anemochores** distributed by wind; **Autochores** possess self-dispersing mechanisms; **Ballistochores** are also self-dispersing, but with different missile mechanisms; **Zoochores** are distributed by animals.

TABLE 2.10

Percentage representation of species in the Kazakhstan study site according to method of pollination and time of flowering, ways of distribution of diaspores, and moisture relations

	Number of species	Canopy cover
FLOWERING		
Morning and late-morning anemophilous	13	42
Noon and afternoon anemophilous	12	38
Day anemophilous	16	18
Entomophilous	28	1
Day entomophilous–autogamous	16	1
Autogamous	12	+[1]
Day anemophilous–entomophilous	3	+
DIASPORY		
Anemochores	22	34
Hemianemochores	10	41
Barochores	22	18
Ballistochores	28	0.5
Zoochores and hemianemochores–zoochores	9	6
Autochores	9	0.5
MOISTURE RELATIONS		
Euxerophiles	47	58
Euryxerophiles	13	40
Xeromesophiles	34	2
Mesoxerophiles	6	+

[1] + = less than 0.5%.

tion of periods of active growth and suppression (Table 2.11). In this eastern site, a second phenotype, summer-green, has been recognized, in addition to the prevailing summer–winter-green type. These species have 1 to 3 peaks of flowering activity during late summer and early autumn (Table 2.11). Many of the early-flowering species do not flower every year. Some species flower over a prolonged period, which is interrupted, once or more often, in dry years. The proportion of individuals that flower changes markedly from year to year.

Community structure

The vegetation in both sites is a mosaic of subtypes. In the Kazakhstan site the mosaic is a complex of two communities, but in the Mongolian site three sub-communities all have the same chief

TABLE 2.11

Changes in the number of species flowering and not in flower during the course of five growing seasons in the Mongolian study site

	May	June	July	August	September
NON-FLOWERING[1]					
1970	–	–	17	18	19
1971	19	19	23	24	22
1972	17	24	23	22	20
1973	20	22	26	26	–
1974	–	22	23	22	22
FLOWERING					
1970	–	–	1	3	3
1971	2	2	5	6	10
1972	2	5	3	1	–
1973	4	9	8	9	–
1974	–	4	4	6	12

[1] "Non-flowering" = in the vegetative stage only; – indicates that no data are available.

edificator, *Stipa gobica*; the main edificator synusia is of perennial bunch herbs; and the co-edificator synusia is formed of primitive dwarf half-shrubs. The edificators in the Kazakhstan site are the seasonal synusia of the early-summer bunch grasses in the *Stipa–Festuca–Artemisia* community and the seasonal synusia of late summer–early autumn dwarf half-shrubs in the *Artemisia–Stipa–Festuca* community.

The distribution of plants lacks uniformity. The plants are arranged in small groups, with seedlings of perennials located close to and protected by the adult plants.

Species are combined ecologically into synusiae both in space and time. Space synusiae constitute vertical layers. For example, in the Kazakhstan site there are four above-ground layers of higher plants: (1) the uppermost stratum of *Stipa lessingiana*; (2) the *Artemisia gracilescens* stratum; (3) the *Festuca valesiaca* stratum; and (4) a stratum of ephemerals and ephemeroids. Below these are two strata of soil organisms. In the Mongolian site all dominants occur in the uppermost stratum, whereas the lower strata consist of minor species (Table 2.12). Stratification is fully developed in

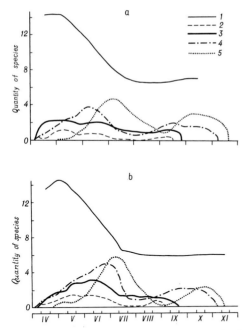

Fig. 2.23. Phenological development of species in the Kazakhstan study site in the *Stipa–Festuca–Artemisia* community (a) and in the *Artemisia* community (b): 1 – non-flowering; 2 – flower budding; 3 – flowering; 4 – fruiting; 5 – seeding. (Drawing by I.V. Borisova, from means of several years.)

TABLE 2.12

Seasonal changes in community structure in the Mongolian study

	1970		1971					1972		1973	
	July	Sept.	June	July	Aug.	Sept.	Oct.	June	Aug.	June	Aug.
MEAN HEIGHT OF PLANTS (cm)											
Stipa gobica	3.2	6.0	8.0	9.5	9.5	9.5	10.0	9.2	9.2	8.0	10.8
Cleistogenes songorica	1.5	2.2	3.0	4.0	4.0	4.5	5.5	3.5	3.5	3.8	4.2
Artemisia frigida	1.5	1.8	2.0	6.5	7.8	9.0	10.5	2.8	7.0	3.2	9.5
Allium polyrrhizum	0.7	4.5	7.5	8.8	6.5	0.6	0.5	5.8	1.5	8.0	11.5
Convolvulus ammanii	–	–	2.7	3.5	3.0	3.0	3.8	–	–	–	–
Ptilotrichum canescens	–	–	3.0	4.7	4.3	8.5	5.7	–	–	–	–
Heteropappus altaicus	–	–	5.7	10.0	8.0	8.0	9.5	–	–	–	–
MEAN PROJECTED COVER OF CROWNS (%)											
Stipa gobica	4.5	5.0	8.0	10.0	10.0	10.0	10.0	8.0	8.2	7.0	10.0
Cleistogenes songorica	3.0	4.0	6.0	8.0	8.0	8.0	8.0	6.0	6.0	6.0	8.0
Artemisia frigida	2.0	3.0	4.0	5.0	5.0	5.0	5.0	4.0	5.0	4.0	5.0
Allium polyrrhizum	0.5	1.0	0.7	2.0	2.0	0.7	0.8	1.5	1.0	1.0	3.0
Convolvulus ammanii	–	–	1.0	1.2	1.5	1.5	1.7	–	–	–	–
Ptilotrichum canescens	–	–	0.1	0.5	0.4	0.5	0.7	–	–	–	–
Heteropappus altaicus	–	–	0.2	0.8	0.8	0.8	0.8	–	–	–	–

– indicates that no data are available.

TABLE 2.13

Seasonal changes in mean height (cm) of the dominants in the *Stipa–Festuca–Artemisia* community of the Kazakhstan study site during 1959 (rather dry) and 1960 (moist)

	Stipa lessingiana				*Festuca valesiaca*				*Artemisia gracilescens*			
	N-F[1]		F		N-F		F		N-F		F	
	1959	1960	1959	1960	1959	1960	1959	1960	1959	1960	1959	1960
April	–	6	0	0	–	5	0	7	0	7	0	0
May	28	20	0	6	14	15	20	31	10	7	10	7
June	30	28	0	35	18	21	35	33	10	9	24	22
July	37	28	0	42	19	21	35	33	10	11	35	23
August	37	28	0	42	19	21	35	33	10	12	35	23
September	37	28	0	42	19	21	35	33	10	12	35	23
October	37	28	0	42	19	21	35	33	10	12	35	23

[1] N-F = non-flowering shoots; F = flowering shoots;
0 indicates that no plants were flowering at this time; – indicates that no data are available.

the Kazakhstan site by late June or early July (summer); however, it is not reached in Mongolia until September (early autumn). The bulk of the shoot biomass is concentrated 0 to 10 cm above the soil surface.

A synusia of bunch grasses, as well as one of primitive dwarf half-shrubs, occurs in the communities of both sites. The temporary synusiae are more pronounced and more varied in the Kazakhstan site, which is richer in phenotypes. For example, the spring synusia of ephemeral species is present there, whereas it is absent from the desert steppe of Mongolia.

Vertical distribution of canopy components in the Kazakhstan site changes as the growing season progresses, and differs from year to year (Table 2.13). Differentiation into strata takes place in late spring. By midsummer the structure becomes more complex because of the appearance of flowering shoots, but by the end of summer it is again simplified. The height and distinctiveness of the strata change from year to year with variations in moisture supply.

The under-ground parts are distributed in three layers, but are not analogous to those of the canopy. The main mass of roots is contained in the uppermost 20 cm of soil in the Kazakhstan site (Fig. 2.24) and in the 5- to 20-cm (sometimes to 30 cm) layer in the Mongolian site. In the latter situation the uppermost 5 cm of soil is very dry and unstable because of disturbance by wind.

Coenopopulations

Each of the dominant species of both locations exhibits a normal (= homoeostatic according to Rabotnov, 1980) type of population; all age groups are represented, with a maximum number of individuals at the reproductive stage of development. Large middle-aged and ageing individuals predominate in populations of the dominant species. However, interspecific differences occur in the proportion of plants of the various age groups (Table 2.14). Death of plants occurs at any stage, but mortality is greatest among young plants.

Seed production and the establishment of new seedlings do not take place annually (Tables 2.15 and 2.16). Sequences of two, three or more years with conditions which permit the development and survival of juvenile plants are a rare occurrence. However, this situation is more frequent in the moister environment of the Kazakhstan site than in Mongolia.

The seed reserve in the soil is considerable in the Kazakhstan site, being 8000 to 9000 seeds m^{-2}. About 56% of these are viable and all of the viable seeds are in the uppermost 5 cm of soil. In the Mongolian site, where severe conditions reduce seed production, the seed reserve in the soil is much lower. There are only 1000 to 2500 seeds m^{-2}, and only 40% of these are viable. However, in the latter site, *Cleistogenes songorica* produces seeds annually because of its cleistogamy.

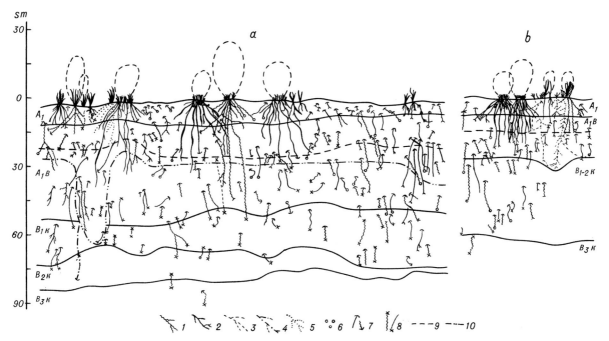

Fig. 2.24. Distribution of roots in the soil in the Kazakhstan study site in the *Stipa–Festuca–Artemisia* community (a) and in the *Artemisia* community (b):

1 – *Stipa lessingiana* and *S. sareptana*
2 – *Festuca valesiaca*
3 – *Artemisia gracilescens*
4 – *Kochia prostrata*
5 – *Poa bulbosa*
6 – roots emerging from the exposed soil profile
7 – roots entering the exposed soil profile
8 – apical part of the root is dead
9 – upper limit of weak effervescence from 10% HCl
10 – upper limit of strong effervescence from 10% HCl

Solid lines are the borders of soil horizons. (Drawing by S. Masyutina.)

TABLE 2.14

Distribution (%) of age groups[1] in the populations of the dominant species of the Mongolian desert-steppe community

Species	Seedlings	Juvenile plants	Generative group				Senile plants
			I	II	III	IV	
Stipa gobica	0.7	0.3	2	14	43	37	3
Cleistogenes songorica	8	1	9	30	27	19	6
Artemisia frigida	5	0	6	29	40	20	0

[1] The categories are defined as follows. **Seedlings** are young plants which still retain a seed lobe (1–3 months of age). **Juvenile plants** are young plants which have not yet reached the reproductive stage (1–5 years of age). **Mature plants** are those which have reached the reproductive stage: I — young, with few dead shoots; II — mature, most shoots living. III — mature, with equal numbers of live and dead shoots; IV — old, with most shoots dead. **Senile plants** are those which have passed the generative stage and are dying.

Biomass

The seasonal dynamics of accumulation of above-ground phytomass differs markedly between sites. In the Kazakhstan site maximum standing crop is attained at the beginning of summer (early July) (Table 2.17) and in the Mongolian location in late summer to early autumn. The

TABLE 2.15

Annual variations in seed yield of various species in the Kazakhstan site (in numbers of seeds m^{-2})

Species	1958	1959	1960	1961	1968
Stipa lessingiana	1683	0	50	0	210
Festuca valesiaca	1925	580	1344	78	75
Artemisia gracilescens	–	4	30	0	–
Kochia prostrata	1134	34	100	126	–
Ferula caspica	–	16	128	–	35
Tulipa patens	–	–	6	–	181
Erysimum leucanthemum	–	–	37	70	–

TABLE 2.16

Annual variations in appearance of new seedlings in the Kazakhstan site (numbers m^{-2})

Species	August		May		June
	1958	1959	1960	1961	1968
Stipa lessingiana	2	120	25	0	0
Festuca valesiaca	15	36	48	0	0
Artemisia gracilescens	5	19	28	0	12
Kochia prostrata	3	1	5	0	15
Ferula caspica	–	–	3	–	–
Tulipa patens	–	–	6	–	–
Total	25	176	115	0	27

date is somewhat variable, depending on weather. Yield of above-ground biomass is low and fluctuates markedly from year to year. For example, the following values were recorded in the Kazakhstan site:

| 1959 | 154 g m^{-2} | 1965 | 62 g m^{-2} |
| 1961 | 180 g m^{-2} | 1968 | 63 g m^{-2} |

This fluctuation is principally due to variations in the amount and distribution of precipitation, as the following data from the Mongolian site demonstrate:

Year	Annual precipitation (mm)	Maximum standing crop (g m^{-2})
1970	60	6
1971	118	18
1972	72	12
1973	169	29
1974	129	15
1975	89	16
1976	116	20

The relative proportion contributed to canopy biomass by each dominant species changes considerably during the growing season and from one year to the next (Tables 2.17 to 2.19). Yield in subsequent years is not reduced by clipping of the grasses early in the growing season (before

TABLE 2.17

Seasonal dynamics of above-ground biomass (g m^{-2}) during 1961 in the Kazakhstan site [1]

Species	May 19	May 24	June 3	June 23	July 5	July 15	July 26	August 4
Festuca valesiaca	58	60	88	85	79	67	66	75
Stipa lessingiana	69	37	38	52	64	27	50	73
Artemisia gracilescens	22	32	22	6	37	57	26	26
Kochia prostrata	5	5					8	
Total canopy biomass	154	134	148	143	180	151	150	174
Litter fall (% of canopy biomass)	25	11	20	25	46	57	25	29
Canopy biomass (% of maximum canopy biomass)	88	74	82	79	100	85	83	96

[1] Each value represents a mean of 5 plots, each 1 m^2 in area, clipped at the soil surface. Statistical analysis revealed that the sampling error was no more than 10%. Only shoots of the current growth season of *Artemisia* and *Kochia* are included. Other species contributed so little biomass that they are ignored.

TABLE 2.18

Seasonal dynamics of above-ground biomass (in g m^{-2}) during 1971 in the Mongolian site[1]

Species	June 10	June 20	July 1	July 15	July 25	August 4
Stipa gobica, S. glareosa	2.2	3.6	5.2	3.6	3.6	3.2
Cleistogenes songorica	0.8	1.1	1.9	2.4	3.6	2.7
Artemisia frigida	0.3	0.5	1.8	2.7	4.2	4.5
Allium polyrrhizum	0.2	0.4	0.3	0.3	0.4	0.2
Other species	2.2	6.0	6.4	5.2	6.1	5.6
Total canopy biomass	5.7	11.6	15.6	14.2	17.9	16.2

[1] The sampling procedure was as described in Table 2.17.

TABLE 2.19

Relative contribution to canopy biomass by each of the principal species in the Mongolian site, 1971–1978[1]

Species	1971	1972	1973	1974	1975	1976	1977	1978
Stipa gobica, S. glareosa	29	36	21	30	40	56	38	50
Cleistogenes songorica	20	18	23	23	17	11	8	8
Artemisia frigida	25	28	28	23	17	13	9	9
Allium polyrrhizum	2	3	24	21	4	2	3	3
Other species	24	15	4	3	22	18	42	30
Total	100	100	100	100	100	100	100	100
Total canopy biomass (g m^{-2})	17.9	11.9	29.4	15.7	15.4	20.4	28.8	18.2

[1] Sampling was done at the time of maximum standing crop, by the procedure described in Table 2.17.

the heading stage), but late cutting (particularly close to the soil surface) does restrict subsequent growth.

The biomass of under-ground parts is much greater than that of the canopy. The difference is particularly great in the Mongolian community, where 97% of the biomass is located under ground; about half of this is in the uppermost 10 cm of soil.

Phytocoenotypes

In the *Stipa–Festuca–Artemisia* community (Kazakhstan site), *Festuca valesiaca* is more plastic than *Stipa lessingiana*, so that it is the more prominent of the two grasses in dry years when growth of *Stipa* is suppressed. *Artemisia gracilescens* is the third species in order of importance, being the co-edificator of the grasses. However, in the *Artemisia–Stipa–Festuca* community, where the bunch grasses are co-dominant, none of the three dominants can be considered as strong edificators as their environment-forming role is approximately the same. All other species of these communities are typical assectators[1].

The *Stipa–Cleistogenes–Artemisia* community in the Mongolian site is also poly-dominant. *Cleistogenes songorica* is the more hardy and plastic (adaptable) of the two grasses. *Stipa gobica* is an edificator, but not a strong one. The abundance of *Artemisia frigida* is a result of grazing by livestock.

Ash content

The ash content of the grasses increases with stage of development (Table 2.20). Silicon is the most abundant element present and is responsible for more than half of the ash content at all

[1] A **co-edificator** is of much lower rank than an **edificator**; an **assectator** is a species which occurs throughout the community, but is of low abundance.

TABLE 2.20

Arrangement of elements in the ash of the dominant species of the Kazakhstan site in order from the highest to lowest relative content

Species and stage of development	Elements									Total ash content (%)
Stipa lessingiana										
Vegetative	Si	K	Ca	P	S	Mg	Na	Fe	Al	7.6
Sheath to flowering	Si	K	P	Ca	S	Mg	Na	Fe	Al	8.8
Fruiting	Si	Ca	K	P	Mg	S	Na	Fe	Al	9.3
Festuca valesiaca										
Vegetative	Si	K	Ca	P	S	Mg	Fe	Na	Al	7.4
Sheath to flowering	Si	K	Ca	P	Mg	S	Fe	Al	Na	8.6
Fruiting	Si	Ca	K	P	S	Mg	Fe	Na	Al	9.2
Artemisia gracilescens										
Flowering:										
herbaceous shoots	K	Ca	S	P	Mg	Si	Na	Al	Fe	5.7
woody shoots	K	Ca	Si	S	Mg	P	Al	Fe	Na	4.0
Fruiting:										
herbaceous shoots	K	Ca	S	Mg	P	Si	Na	Al	Fe	–
woody shoots	Ca	K	Si	S	Mg	P	Al	Na	Fe	–

stages of growth. The ash content of *Artemisia gracilescens* is less than that of the grasses, and the proportion contributed by potassium and by calcium is greater than silicon, even in woody shoots.

Fauna

The animal components of the Kazakhstan site include 25 species of mammals. Of these, 16 species are rodents, 7 are carnivores, and 2 are ungulates. The fauna of the Mongolian site include 17 species of mammals and 3 species of reptiles. Rodents are the most numerous mammals. In both sites the families Tenebrionidae and Curculionidae are well represented among the insects. The chief relations with plants are as food and cover. *Stipa lessingiana* in Kazakhstan and *Cleistogenes songorica* in Mongolia have the richest associated insect flora.

Micro-organisms

The microbiocoenoses of the two study sites differ in species composition, group structure and biomass of Bacteria and Actinomycetes, as a result of differences in environment and in the floristic composition of the vascular plants.

In the Kazakhstan site, 105 species and varieties of Bacteria and Actinomycetes have been identified, as against 65 in the Mongolian site. However, more species of Actinomycetes occur in the Mongolian desert steppe (25) than in the Kazakhstan semi-desert steppe (20). Unlike the situation with vascular plants, many species are common to both sites. Of the taxa recorded, 22% of Actinomycetes and 18% of Bacteria occurred at both sites. The genus *Bacillus* is prevalent at both sites, together with *Pseudomonas* in the Mongolian site and *Bacterium* in Kazakhstan.

The numbers of cells of nitrifying, denitrifying and oily-acidic (butyric acid) bacteria are much greater in the Kazakhstan site, whereas free-living nitrogen fixers (*Azotobacter* spp. and *Clostridium pasteurianum*) and cellulose-decomposing bacteria and Actinomycetes (especially in the rhizosphere) are more abundant in Mongolia. The biomass of micro-organisms is considerable, being almost twice as great in the brown soil of the desert steppe (from 15 to 293 g DM m^{-2}) as in the light-chestnut soil of Kazakhstan (from 37 to 151 g DM m^{-2}).

Fungi, Bacteria and Actinomycetes are mainly concentrated in the rhizosphere, whereas algae are in the uppermost 3 cm of soil.

TABLE 2.21

Seasonal fluctuations in the populations (number of cells g^{-1} of soil) of various groups of micro-organisms in the Kazakhstan site [1]

Group	May	June	July	Sept
Ammonifiers	10^9–10^{10}	10^3–10^{11}	10^9–10^{10}	10^8–10^9
Nitrifiers	10^2–10^3	10^3–10^4	10^3–10^4	10–10^2
Denitrifiers	10^3–10^4	10^2–10^4	10^3–10^4	10^5
Butyric (oily) acidic	10^3–10^4	10–10^5	10^3–10^5	10^3–10^4
Clostridium pasteurianum	10^3	10^3–10^4	10^3–10^4	10^3–10^4
Cellulose decomposers	10^2–10^3	10–10^3	10–10^2	10^2
Cryptogams	340–367	264–630	165–370	150–400

[1] Range of values measured in 1959, 1961 and 1968.

The species composition of microbial synusiae changes with the season of the year and from year to year (Table 2.21). Bacteria and Actinomycetes are often antagonists. Poor hydrothermal conditions reduce development of Bacteria, with a resultant increase in numbers of Actinomycetes.

The most prevalent microscopic fungi are species of *Aspergillus*, *Fusarium* and *Penicillium*. Fungi are relatively low in abundance in the rhizosphere, not exceeding several tens of thousands of embryos (growing spores and primary mycelia) per gram of soil. Seasonal changes in abundance are small, but the population density of microscopic fungi varies, being greater in dry years.

More than half of the species of soil algae are blue-green types, but diatoms, green algae and yellow-green algae are also common.

The total biomass of micro-organisms in soil is considerable (Table 2.22), as is their energy content. In the Kazakhstan site the energy content was 4125 kJ m^{-2} for Bacteria and 268 kJ m^{-2} for Actinomycetes and Fungi — a total of 4393 kJ m^{-2}. Development of micro-organisms proceeds more intensively in the rhizosphere than between the roots (Table 2.23). Several species of micro-organisms are specific to the rhizosphere of certain species of higher plants. *Clostridium pasteurianum* and cellulose-decomposing bacteria are found only in the rhizosphere. The greatest numbers of micro-organisms occur at a depth of 2 to 16 cm in the soil. In the upper soil layer and in the rhizosphere Bacteria are more abundant than Actinomycetes (Table 2.23). Calculations illustrate that a considerable amount of the organic matter in soils is composed of microbial protoplasm. In the Kazakhstan site, for example, the wet weight of the microbial mass equalled 1.3% of the dry weight of the organic matter in the soil.

TABLE 2.22

Seasonal fluctuations in biomass of micro-organisms (g of dry matter m^{-2}) in the uppermost 10 cm of soil in the Kazakhstan site [1]

Year	May	June	July	Sept	Mean
1959	132	110	116	145	123
1960	143	145	218	98	151
1961	39	31	101	65	37
1968	–	100	106	45	84
Mean	105	97	135	88	99

[1] The bulk density of the soil is 1.26.

TABLE 2.23

Comparison of the populations of micro-organisms (millions of cells g^{-1} of soil) in the rhizosphere and the rest of the soil in the Mongolian desert steppe

Medium	Taxon	Soil (2–6 cm)	Rhizosphere
Beef-extract agar	Total numbers	1.30	3.92
Starch-amino agar	Bacteria	14.9	20.8
	Actinomycetes	5.9	1.0
Ashby medium	Bacteria	7.8	19.8
	Actinomycetes	4.9	1.3

Biological activity in the soil

The activity of the large population of ammonifying bacteria (Table 2.21) in hydrolysis of protein results in accumulation of considerable amounts of amino acids in the soil. The concentration of alcohol hydrolysates is as high as 0.72 p.p.m., and the easily hydrolyzed nitrogen amounts to 47 p.p.m. The composition of amino acids found in the soil of the Kazakhstan site was as follows (in p.p.m. of oven-dry soil):

Aspartic acid	0.312	Serine	0.058
Alanine	0.107	Leucine	0.082
Methionine	0.023	Isoleucine	0.026
Glutamine	0.194	Lysine	traces

Decomposition of protein to form amino acids is due to an active enzyme, protease. High activity was also observed in other soil enzymes (Table 2.24). The activity of both hydrolytic enzymes (protease and invertase) and of an oxidation enzyme (catalase) in the rhizosphere was greater than in the rest of the soil.

At the Kazakhstan site, carbon dioxide was also released more rapidly in the rhizosphere (41.0–45.7 mg CO_2 kg^{-1} of soil hr^{-1}) than in the rest of the soil (20.7–30.4 mg CO_2 kg^{-1} hr^{-1}).

In the desert-steppe site the rate of loss of carbon dioxide from the soil surface is extremely low, averaging about 6 kg CO_2 ha^{-1} day^{-1}. The minimum rate is as low as 1 kg CO_2 ha^{-1} day^{-1} (4.2 mg CO_2 kg^{-1} hr^{-1}). Biological activity is especially low during summer in extremely dry years; during such a period, the mean value is 3.4 (14.2 mg CO_2 kg^{-1} hr^{-1}) and the minimum 0.5 kg CO_2 ha^{-1} day^{-1} (2.1 mg CO_2 kg^{-1} hr^{-1}).

TABLE 2.24

Measures of the activity of three enzymes in soil at the Kazakhstan site [1]

Enzyme	Soil	Rhizosphere
Protease (mg nitrogen)	1.05	1.98
Invertase (mg invert sugar)	22.4	35.4
Catalase (ml 0.1 H $KMnO_4$)	13.2	18.4

[1] The activity was measured throughout the vegetative period. Values are on the basis of 5 g of oven-dry soil.

This low activity is accounted for primarily by low numbers of soil micro-organisms at such times.

THE EFFECT OF RODENTS ON STEPPE ECOSYSTEMS *

Small herbivorous mammals that dwell in the soil have an important role in determining the nature of vegetation, soil and microrelief in steppes. Both soil scientists and zoologists made early contributions to this subject and noted the importance of burrowing activity by rodents as a significant factor in soil formation (Levakovskyi, 1871; Dokuchaev, 1883; Vernadskyi, 1889; Bogdan, 1900; Pankov, 1921). Pachoskyi (1917) laid the foundation for the study of interrelations between the animals and vegetation. As a result of lengthy investigations in the steppes of the Kherson (Ukraine) region (at Askaniya Nova Reservation), he concluded that the mechanical action of hoofed animals on standing dead plants and litter was a necessary condition for the normal functioning of steppe phytocoenoses. He also studied the effect of consumption of plant material by rodents, but considered this of minor importance (Lavrenko, 1952).

Formozov and his students established the concept that rodents not only affect plant production but also alter the microrelief of the landscape and the complexity of the soil and vegetation (Formozov, 1928, 1929; Formozov and Voronov, 1939; Voronov, 1950, 1964; Formozov et al., 1954; Kucheruk, 1963). More recent contributions to these subjects are those of Rotshil'd (1968), Zlotin (1975), Chernov (1975, 1978) and Dobrinskyi et al. (1983).

The biogeocoenotic role of phytophages (herbivores) in communities of the arid zone has been discussed by Dinesman (1971, 1977), Khodasheva and Zlotin (1972), Zlotin and Khodasheva (1973, 1974), Abaturov (1973, 1979, 1980), and Zimina and Zlotin (1980). Interrelationships between vegetation and animal populations have been studied in the works of Lavrenko (1952) and Lavrenko and Yunatov (1952).

---

* By N.P. Guricheva.

Principal species

The principal small mammals of steppe belong to the orders Lagomorpha and Rodentia. The most abundant are species belonging to the family Cricetidae:

Lagurus lagurus (steppe lemming), which is widely distributed in the European–Kazakhstan steppe and in the northern part of the desert in eastern Kazakhstan.

Lasiopodomys brandtii (Brandt's vole), which is a species of dry steppe, widely distributed in the steppes of Transbaykal and Mongolia.

Microtus arvalis (common vole), which ranges from the forest (in the west) to desert and occupies the steppe region as far east as Mongolia.

M. gregalis (narrow-skulled vole), which is distributed from the plains and mountains of the tundra region southward to the desert steppe and alpine steppe of Central Asia.

M. socialis (social vole), which occurs in the southern part of the steppe region and in the desert.

Two other species of the family Cricetidae feed on roots and are active in affecting the underground parts of plants: *Ellobius talpinus* (mole vole), which is widely distributed from the southern Ukraine to the eastern part of Mongolia, and *Lasiopodomys mandarinus* (Chinese vole), which occurs in the steppes of Transbaykal and northern Mongolia. *Spalax microphtalmus* (mole rat) of the family Spalacidae is also a root-feeder, and occurs in the forest steppe between the Dnieper (Dnepr) and Volga rivers.

The larger rodents of the family Sciuridae are particularly active in modifying steppe. These are: *Citellus pygmaeus* (suslik), which is distributed in the steppified desert and desert steppes of the southern Ukraine and between the Volga and Ural rivers in Kazakhstan; *Marmota bobak* (marmot or baibak) in the European–Kazakhstan steppes; and *M. sibirica* (tarbagan) (Fig. 2.25), which is distributed widely in the steppes of Transbaykal, in the Chuyskaya steppes of the Altai, and in steppes in Mongolia; *Allactaga sibirica* (Siberian jerboa) of the family Dipodidae is active in desert steppes and steppified deserts from the Ural river to the south of Transbaykal and Mongolia.

Ochotona daurica (Daurian pita) (Fig. 2.26), of the order Lagomorpha, has a considerable influence on the structure of steppes of Transbaykal and Mongolia.

Fig. 2.25. *Marmota sibirica* in steppe south-west of Ulaanbaatar. (Photo by G. SchUnzel.)

Fig. 2.26. *Ochotona daurica* in steppe south-west of Ulaanbaatar. (Photo by G. SchUnzel.)

Most of the above species live in colonies, where burrow holes are concentrated. In such places the vegetation is almost entirely destroyed, the surface layers of soil are loosened, and soil chemistry, microrelief and microclimate are all modified. Almost all of these rodents and lagomorphs are polyphagous — except for the stenophagous *Lasiopodomys mandarinus* which feeds only on *Stellera chamaejasme* and rarely on *Sanguisorba officinalis*. They feed on green parts of plants, especially on the juicy bases of leaves and stems where the renewal buds are located, as well as (for root-feeders) on juicy rhizomes, bulbs and tubers.

The non-hibernating, small mouse-like rodents and lagomorphs, in addition to consuming green phytomass in summer, accumulate considerable stores for winter. For example, *Lasiopodomys brandtii* stores as much as 10 kg of forage per hole in the dry steppe (*Cleistogenes squarrosa* and *Festuca lenensis*) of Mongolia (Lavrenko and Yunatov, 1952), and *Ochotona daurica* stores as much as 56 kg ha^{-1} of dry green phytomass in the meadow steppe (*Stipa baicalensis–Festuca lenensis*–forbs) of the eastern Hangay.

The influence of all rodents on the vegetation and soil increases noticeably during that part of the 4- to 5(10)-year population cycle when they are increasing in number. This is especially typical of small mouse-like rodents. Such a periodic burst in numbers of some species can reach such considerable proportions (10 000 to 20 000 holes ha^{-1}) as to cause a natural calamity (Formozov and Kiris-Prosvirina, 1937).

The influence of each species of rodent on steppe vegetation is rather specific because of its particular nutritional requirements, the nature of its colonies, its mobility, and its season of activity. The reaction of vegetation to this influence is not always unambiguous.

Energy relationships

The effect of grazing by these animals on the vegetation depends on their feeding habits and the reaction of plants to the loss of vegetative parts. This activity, generally, does not affect the structural integrity of the plant communities. To some extent, these animals can be considered to occupy a catalytic role in productive and destructive processes. Although their consumption of living phytomass might be expected to retard plant growth, this is compensated for by improvement of growing conditions. Losses of phytomass to ingestion by animals must be considered in relation to the effect on the rate of biological turnover, which in turn affects the rate of production of ecosystems (Zlotin and Khodasheva, 1973).

From the geobotanical viewpoint, these herbivores initiate local dynamic processes in the plant community which are manifested in changes in abundance and distribution of individual species, and alteration of phenology. Phytocoenologically, these processes may be considered as seasonal or yearly mini-fluctuations of zoogenic character.

Most small rodents are probably not competitors with domesticated livestock during years when their populations are near average, since the species eaten by them are often not those sought by ungulates (Nikol'skyi et al., 1984). Evidently such specialization in the choice of food permits these rodents to exist even in areas where rangeland is under severe grazing stress by livestock.

Effect on yield of phytomass

The biomass of small herbivores is very low compared to annual production of the vegetation ($<1\%$). The quantity of phytomass consumed is also small.

In the meadow steppe of the Central Chernozem Reservation [Streletskaya steppe[1] near Kursk (southern Russia)], *Microtus arvalis* constitutes the major part of the vertebrate fauna. According to Zlotin and Khodasheva (1974), this species, with a mean population of 30 individuals ha^{-1}, consumes from 5.4 to 7.2 g m^{-2} yr^{-1}, equal to from 2 to 4% of the above-ground green biomass. Return to litter as excrement ranges from 0.9 to 1.2 g of dry matter m^{-2}. In the same steppe, *Spalax microphtalmus* feeds chiefly on juicy rhizomes, tubers and tap roots, and destroys about 25% of the annual increment of under-ground phytomass. The entire complex of herbivores (including invertebrates) in meadow steppes consumes about 70 g m^{-2} yr^{-1} of plant biomass, which is only 6% of the total standing crop.

In the forb–bunch-grass steppe of the southern Ukraine (Streltsovskaya steppe), *Marmota bobak* occurs over only about 4 to 12% of the area of the steppe. During the active period of the year this species removes from 20 to 40% of the aboveground phytomass within its feeding area, which amounts to only 1–3% for the steppe as a whole (Abaturov et al., 1980).

Citellus pygmaeus consumes an even larger proportion of plant production. In the steppe and desert communities of the Caspian Sea region, this species (the population of which reaches 70 individuals ha^{-1}) removes about 20% of the aboveground phytomass in the desert communities and up to 50% in forb–grass steppes. However, this does not lower the total productivity of the plant cover (Abaturov, 1979).

These observations suggest that, under natural conditions not affected by man, consumption by small herbivores during periods of average population density is not destructive to the vegetative cover. Even during extreme years, consumption by these herbivores does not exceed 60% of herbage yield, which is lower than the threshold level (70–75%) of tolerance of the plant cover (Abaturov, 1980).

Nevertheless, feeding activity of rodents is sometimes increased manyfold during periods of population explosion, especially as a result of anthropogenic disturbance (for example, in heavily trampled pastures). Lavrenko and Yunatov (1952) described the activity of *Lasiopodomys brandtii* in dry bunch-grass steppe (*Cleistogenes squarrosa* and *Stipa krylovii*) of Mongolia which is intensively grazed by cattle throughout the year. During years of high population density, colonies of this species extend for dozens of kilometres. During early spring it feeds mainly on rhizomes, roots and bulbs (of *Allium polyrrhizum*, *A. tenuissimum*, *Carex duriuscula* and *Stipa krylovii*). In summer it consumes green fodder, clipping off to the base shoots of:

Artemisia frigida	*Iris lactea*
Carex duriuscula	*Leymus chinensis*
Cleistogenes squarrosa	*Potentilla bifurca*
Cymbaria daurica	*Stipa krylovii*

In September, *Lasiopodomys brandtii* begins to store fodder for the winter, often digging out entire plants (of *Artemisia* spp., grasses and sedges).

---

[1] The Streletskaya steppe is near Kursk in southern Russia; the Streltsovskaya steppe is north of Voroshilovgrad (in the south-eastern part of the Ukraine).

The weight of the winter reserves for occupants of one burrow reaches 10 kg and more (wilting material, air-dry weight). As a result of this activity the yield of phytomass is lowered from 30–40 g m^{-2} to from 2–5 g m^{-2}. In this way rodents completely destroy the bunch grasses and in some places turn the steppe into a "bare desert". Such a situation was observed by Bannikov (1948) in the steppes of eastern Mongolia, and by Lavrenko and Yunatov (1952) in the steppes of central Mongolia.

Effect on formation and decay of plant litter

Small mammals, especially mouse-like rodents, increase the rate of accumulation of litter and of decomposition of dead material in the litter and canopy. This increased rate of destruction of dead parts of plants is partly due to increases in the rate of transfer of material from the canopy to the litter layer, and partly to conditions more favourable to microbial decomposition. During the eating process, a portion (as much as 20%) of shoots that are bitten off is not consumed and falls to the litter layer. For example, Zlotin and Khodasheva (1974) found that litter biomass, in locations occupied by *Microtus arvalis* in the meadow steppe of the Central Chernozem Reservation, was four times as great as in unoccupied areas. This increase was caused by additions to litter of shoots that were bitten off but not consumed. This process takes place most actively in late spring and early summer. The rate of decomposition of dead shoots and litter is also increased by the activities of rodents, particularly during the vegetative period. From mid-August to mid-September as much as 80% of standing dead material and 60% of litter fall becomes mineralized within the boundaries of rodent colonies, while elsewhere the process of mineralization proceeds very slowly and the accumulation of dead remains exceeds the rate of mineralization. Mechanical action of the animals in loosening and disintegrating the litter makes the material more vulnerable to decay, as does the addition of their excrement. The latter provides nutrients that favour development of saprophytic micro-organisms, and thus speed up the processes of decomposition.

These processes increase the rate of turnover of nutrients. Nitrogen and ash residues are added to the soil at a faster rate, permitting them to recycle more rapidly. Evidently, the additional supply of nutrients, together with much greater reserves of available moisture on mounds, provides for more luxuriant growth in their proximity and increases the length of the vegetative period by as much as two weeks beyond that of the vegetation in areas not occupied by rodents (Zlotin and Khodasheva, 1974).

A similar phenomenon was noted (Guricheva et al., 1984; Guricheva, 1985) also in meadow steppes of the mountainous forest-steppe region of the eastern Hangay, where the activities of *Microtus gregalis* and other rodents resulted in a considerable increase in fall of litter. The total amount of above-ground phytomass (canopy plus litter) in the area affected by the rodents was much greater than beyond the area of their influence. However, the part made up of litter was greater in the affected area (61–67%) than in the unaffected area (20–22%), whereas the biomass of green shoots in the colonies was less (20–26%) than in surrounding areas. Considerable accumulation of litter takes place in this region because decomposition is retarded by low soil temperatures.

Effect on structure of vegetation

Selectivity of herbivores affects the abundance and age composition of populations of those plant species that are eaten. Animals select certain species, but in addition they seek their most nutritious and succulent parts (flower and vegetative buds, bulbs and fruits). This gives rise to structural reorganization of plant communities. This form of zoogenic regulation is especially pronounced in colonies of stenophagous under-ground rodents, such as *Lasiopodomys mandarinus*. In short bunch-grass vegetation of forest steppe in the eastern Hangay, where this rodent feeds on the roots of *Stellera chamaejasme*, it caused a 60% reduction in the population of adult plants in the colony areas, as compared to uninhabited areas. As a result, the age structure of the coenopopulation was modified because of rejuvenation (Guricheva et al., 1983). The patchy distribution of *S. chamaejasme* observed in this region is often a result of the grazing activity of *Lasiopodomys mandarinus* (Dmitriev, 1980; Guricheva et al., 1983).

A similar influence is undoubtedly exerted by other species of root-feeding soil-inhabiting mammals, particularly *Ellobius talpinus* and *Spalax microphtalmus*, which feed for the most part on succulent under-ground parts of rhizomatous, tuberous-rooted and bunch-rooted herbs. In the meadow steppe of the Central Chernozem Reservation, *S. microphtalmus* consumes as much as 25% of the increment of the under-ground phytomass of these types of plants in the uppermost 20 cm of soil. Although this undoubtedly affects the abundance and the population structure of these species, the influence is less direct than that of *Lasiopodomys mandarinus* because of the polyphagous habits of *Spalax microphtalmus* (Zlotin and Khodasheva, 1974).

Grazing activity of small polyphagous rodents also has significant effects on vegetative structure. Formozov and Voronov (1939) reported that in the semi-desert steppes of the Aktyubinsk region (Kazakhstan) the size of the colonies (compound burrows) of *Lagurus lagurus* reaches 5 to 6 m^2. These colonies are clearly distinguishable as spots with highly rarefied cover. The animals harvest almost all of the herbage of *Poa bulbosa* and of species of *Allium*, *Festuca*, *Medicago* and *Stipa* over extensive areas, and especially in the colonies, where they often leave only such species as *Artemisia austriaca*, *Artemisia maritima* and *Kochia prostrata*. The remaining vegetative cover of the settlements is very sparse and the centre of the colony is often without any plant cover.

The diet of *Ochotona daurica* in the mountain forest steppe of the eastern Hangay includes 75 species of plants from 26 families, Asteraceae, Fabaceae and Rosaceae being best represented. Annual storage of plant shoots as winter fodder amounts to 1 to 5.6 g of dry matter m^{-2}. When summer requirements are included, the annual harvest is estimated to be several times as much. Grazing by *O. daurica* markedly reduces the abundance of the species consumed, especially of *Cymbaria daurica*, *Galium verum*, *Potentilla bifurca* and *Thermopsis lanceolata*. In some places these species have disappeared periodically from the vegetative cover.

Effect on microclimate

Zlotin and Khodasheva (1973) reported that, in the meadow steppe of the Central Chernozem Reservation, the area of leaf surfaces within areas occupied by *Microtus arvalis* was only one-third to one-fourth of that in the surrounding grassland, because of defoliation and trampling. This resulted in the intensity of radiation in the lower vegetative layer being 1.5 to 2 times, and at the soil surface 3 to 3.5 times, that of areas not affected by rodents. In this way animal activity increases the diurnal range of temperatures in the air and uppermost layer of soil. This results in decreased day-time relative humidity and increased relative humidity at night and during the early morning.

Reserves of soil water are greater in the proximity of burrows than elsewhere. By the end of the vegetative period, the amount of moisture in the root layer of rodent colonies has been found to be 25% greater than in the neighbouring uninhabited steppe. The resulting change in microclimate permits growth to continue for 2 or 3 weeks after senescence (due to drought) begins in the surrounding grassland. The standing crop of herbage has been found to increase by 25% during this period (Zlotin and Khodasheva, 1973).

Effects of burrowing activity

Burrowing activity of soil-inhabiting rodents in excavating nesting and protective (those without nests) burrows, as well as food passages, causes a more severe disturbance to the ecosystem than does their feeding activity. Changes are caused in vegetative structure, in nano- and micro-relief of the soil surface, in chemical composition of the soil, and in microclimate. These changes are reflected in micro-groupings of plants of various sizes, which are conspicuous within the much more uniform physiognomy of the surrounding unaffected steppe (Figs. 2.27 and 2.28). These are most prominent where burrows have been inhabited by the animals for extended periods, but are still very evident for a considerable time after the burrows are deserted.

Often the whole plant community undergoes zoogenic disturbance. This is especially so on lower slopes around depressions, where continuous colonies often occur.

In other instances, extensive disturbance is evident from large and complex irregularities in the vegetative cover. In these areas the settlements sometimes include several species of rodents within the limits of one ecotope (Fig. 2.28).

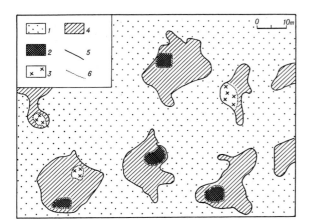

Fig. 2.27. The diffuse type of zoogenic mosaic structure of the short bunch-grass steppes of the eastern Hangay (Mongolia) generated by colonies of *Ochotona daurica* (after Dmitriev and Guricheva, 1983):
1 – background short bunch-grass steppe (*Cleistogenes squarrosa, Koeleria cristata, Stipa krylovii, Carex korshinskyi, Artemisia frigida, Stellera chamaejasme*)
Zoogenic nano- and microgroupings:
2 – grasses and forbs (*Lappula squarrosa, Potentilla bifurca, Leymus chinensis*) in the centres of the colonies
3 – forbs (*Artemisia changaica, Echinops latifolius, Heteropappus hispidus*) in old ecocentres
4 – forbs and grasses (*Leymus chinensis, Agropyron cristatum, Cleistogenes squarrosa, Heteropappus hispidus*) along the periphery of the colony
5 – the borders of zoogenic microgroupings
6 – the borders of zoogenic nanogroupings

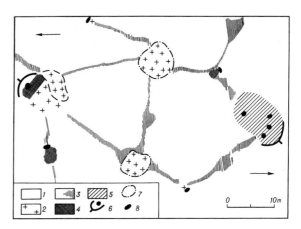

Fig. 2.28. The net-like type of zoogenic mosaic structure in the *Festuca lenensis*–petrophyte–forb steppe of the eastern Hangay (Mongolia) on areas with combined colonies of *Marmota sibirica, Ochotona daurica* and *Microtus gregalis* (after Dmitriev and Guricheva, 1983):
1 – background *Festuca*–petrophyte–forb steppe (*Eremogone capillaris, Potentilla sericea, Festuca lenensis*)
Zoogenic nano- and microgroupings:
2 – forbs and *Stipa* (*Stipa krylovii, Scutellaria scordiifolia, Galium pediformis*) on old hills of *Microtus gregalis* and *Ochotona daurica*
3 – forbs and sedge (*Carex pediformis, Festuca sibirica, Polygonum angustifolium*) on animal paths
4 – *Artemisia* and annual forbs (*Axyris hybrida, Chenopodium album, C strictum, Artemisia changaica*) on excavated soil around burrow openings of *Marmota sibirica*
5 – xeromesophilous forbs (*Phlomis tuberosa, Sanguisorba officinalis*) and *Pentaphylloides fruticosa* along the slopes and in the craters of old *butany*
6 – hills (*butany*) of *Marmota sibirica*
7 – the borders of old colonies of *Ochotona daurica*
8 – protective burrows (without nests) of *Marmota sibirica*

The degree of effect on vegetation caused by these animals can be divided into: temporary, reversible disturbance (Table 2.25, 2); temporary

TABLE 2.25

Comparison of the population density and activity of four species of rodents in the forb–bunch-grass steppe of the eastern Hangay [1]

Parameter	*Lasiopodomys mandarinus*	*Lasiopodomys brandtii*	*Ochotona daurica*	*Marmota sibirica*
Area of disturbance (m^2)	5–2000	3–25	20–200	5–900
Depth of excavations (cm)	5–15	20–50	30–100	200–500
Height of mounds (cm)	0–10	2–20	5–30	10–150
Amount of material in mound (kg)	up to 600	10–30	50–500	50–2000
Number of animals per ha	5–30	200	40	1–10
Number of burrows per ha	1–5	30–60	2–8	1–3
Maximum period of occupation of burrows (yr)	1–2	10+	10+	10+
Degree of effect on vegetation [2]	1; 2	2	2	3

[1] After Guricheva and Dmitriev (1983).
[2] Intensity of effect: 1 = short-term fluctuations; 2 = reversible disturbance; 3 = irreversible disturbance.

disturbance that is repairable by short-term transformation processes (Table 2.25, 1); and irreversible destruction that can be rectified only by long-term primary succession (Table 2.25, 3).

Temporary types of disturbance

Small rodents, which excavate burrows in the uppermost 30 cm of soil, cause only temporary disturbance to plant communities. Among the animals with such digging habits are:

Ellobius talpinus	*Microtus* spp.
Lagurus lagurus	*Ochotona daurica*
Lasiopodomys spp.	*Spalax microphtalmus*

These species form small mounds which are uniformly distributed all over the steppe. In these situations, the normal plant community is replaced by species of early stages of secondary succession. These include rhizomatous and root-sprouting species and various annuals and biennials. Lavrenko (1956) referred to these disturbed areas as "miniature fallows". The burrows are inhabited for many years. After the departure of the animals, succession results in rapid revegetation by the climax dominants. However, for many years after the cover has been restored, the spotty distribution of certain species bears evidence of the former disturbance.

Characteristics and extent of burrows. Colonies of *Lasiopodomys brandtii* sometimes extend for many dozens of kilometres in the steppes of Transbaykal and in the dry bunch-grass steppes of central and eastern Mongolia. The number of burrows reaches 1000 to 3000 ha^{-1} (Lavrenko, 1952). Lavrenko and Yunatov (1952) reported compound burrows in the form of "medallions" 3 to 10 m in diameter, with mounds 3 to 5 cm high, occupying 10 to 20% of the area of a settlement (a group of colonies). In dry steppes in the eastern Hangay, burrows range in size from 3 to 25 m^2 and occupy 6% of the surface within settlements (Dmitriev and Guricheva, 1978).

The mean number of colonies (compound burrows) of *Lagurus lagurus* is 6.5 ha^{-1} in the dry steppe of Western Kazakhstan (Guricheva, 1961). The compound burrows are distributed uniformly and are from 2 to 2.5 m in diameter. Only one-sixth of the burrows are occupied. In the semi-desert steppes of the Aktyubinsk region (Formozov and Voronov, 1939), the colonies of *L. lagurus* are from 5 to 6 m^2 in area.

Ochotona daurica often forms continuous settlements along the lower parts of slopes of low mountains in the short bunch-grass steppe (*Stipa krylovii* and forbs) in the eastern Hangay of Mongolia (Dmitriev and Guricheva, 1978; Guricheva and Dmitriev, 1983). The colonies are amoeboid in shape (Fig. 2.27) and from 20 to 200 m^2 in size. The centre of the colony (5–10 m^2) is most markedly affected, where as many as 10 burrow openings per square metre may occur. The colonies occupy 8–12% of the total settlement.

Spalax microphtalmus leads an under-ground mode of life and excavates an immense amount of the upper soil layers over extensive areas. Colonies occupy about 30% of the meadow steppe of the Central Chernozem Reservation. During the vegetative period (April–October), the density of new small mounds (averaging 0.2 m^2 in area) is as great as 370 ha^{-1}. These, together with the small mounds of the previous years, occupy 1 to 1.5% of the inhabited area (Zlotin and Khodasheva, 1973).

The mounds of *Ellobius talpinus* occupy as much as 7.5% of the ground area in dry and semi-desert steppes of Southern Kazakhstan (Andrushko, 1948). The number of small hills (each up to 35 cm in diameter) reaches 600 ha^{-1}.

In colonies of *Lasiopodomys mandarinus* in the short-forb-bunch-grass steppes of the eastern Hangay, the number of colonies reaches 10 ha^{-1} (Dmitriev, 1980). Burrows may be up to 70 m in length and 350 m^2 in area. The boundaries of settlements constantly shift, so that eventually the whole surface of steppe populated by this species is affected by digging activity and becomes uniformly covered by mounds (Guricheva and Dmitriev, 1983).

Vegetation on the mounds. In areas occupied by *Lasiopodomys brandtii* in the dry bunch-grass steppes of central Mongolia, the populated mounds are occupied mainly by annuals, especially by *Chenopodium strictum* and *Salsola australis* (Lavrenko and Yunatov, 1952). As the animal population decreases, *Artemisia adamsii*, a root-sprouting species, and the rhizomatous *Leymus chinensis* become prominent members of the community. Similarly, in the steppes of the Hentey and the Hangay, the rhizomatous *Artemisia changaica* and *A. dracunculus* appear early in the revegetation process, along with *Leymus chinensis*. In dry steppes, dominated by *Caragana microphylla* and *Stipa krylovii*, in the eastern Hangay, mounds of soil recently exposed by *Lasiopodomys brandtii* are occupied by the following annuals:

Artemisia macrocephala	*Chenopodium strictum*
Chenopodium album	*Lappula squarrosa* (= *L. echinata*)
Chenopodium aristatum	*Salsola australis*

Artemisia adamsii invades as revegetation continues, and finally, after the burrows have been deserted, the prominent species include *Artemisia changaica*, *Heteropappus hispidus*, *Leymus chinensis* and *Schizonepeta multifida* (Dmitriev and Guricheva, 1978).

Disturbance by *Microtus arvalis* in the forb–*Festuca valesiaca*–*Stipa* steppes in the southern part of the Ukraine was described by Formozov and Kiris-Prosvirina (1937). The following dicotyledonous species are abundant in the various successional stages that take place before the grasses become re-established:

Capsella bursa-pastoris	*Kochia prostrata*
Convolvulus arvensis	*Marrubium vulgare*
Cynoglossum officinale	*Salvia nutans*
Goniolimon tataricum	*Stachys recta*

Rhizomatous grasses (*Bromopsis riparia* and *Elytrigia repens*) are abundant after the burrows have been deserted. The bunch-grass dominants of the virgin steppe are the last to reinvade, with *Festuca valesiaca* appearing before *Stipa capillata* and *S. lessingiana*. In areas disturbed by *Microtus socialis*, in the Kizlyar region of Dagestan, the perennials *Artemisia maritima* s.l. and *A. absinthium* dominate, along with the following annual and biennial weeds:

Alyssum turkestanicum	*Onopordum acanthium*
Atriplex tatarica	*Polygonum patulum*
Lepidium perfoliatum	

In this region the undisturbed steppe is dominated by *Anisantha tectorum*, *Artemisia maritima*, *Bromus squarrosus* and *Poa bulbosa*. Re-establishment proceeds over a period of 3 to 4 years (Formozov and Kiris-Prosvirina, 1937). The tall-weed stage is prolonged in areas disturbed by *Microtus arvalis* and *M. socialis* in old fields (abandoned cropland) in the southern Ukraine (Tyulina, 1930; Formozov and Kiris-Prosvirina, 1937).

The most recently deserted burrows of *Lagurus lagurus*, in the dry steppe of Western Kazakhstan (dominated by *Festuca valesiaca* and *Stipa capillata*), support *Artemisia austriaca*, *Bromopsis inermis* and *Echinops ritro*. Mounds of intermediate age are occupied by *Agropyron pectiniforme*, *Leymus ramosus*, *Phlomis agraria* and *Stipa sareptana*. In the areas deserted longest, a large proportion of the vegetation is formed of *Stipa lessingiana*, *S. sareptana* and *Tanacetum achilleifolium* (Guricheva, 1961). The first stages of revegetation in areas disturbed by *Lagurus lagurus*, in the semi-desert steppes of the Aktyubinsk region (Kazakhstan), include *Carduus uncinatus* and *Echinops meyeri*, as well as abundant seedlings of *Artemisia austriaca* (Formozov and Voronov, 1939).

The mounds of *Ochotona daurica*, in the short bunch-grass steppe (*Stipa krylovii* and forbs) in the eastern Hangay of Mongolia, are occupied by an abundance of rhizomatous perennials and of annuals (Dmitriev and Guricheva, 1978; Guricheva and Dmitriev, 1983). These include:

Androsace septentrionalis	*Leymus chinensis*
Artemisia changaica	*Salsola australis*
Axyris amaranthoides	*Thermopsis lanceolata*

On the old mounds, *Echinops latifolius*, *Galium verum* and *Schizonepeta multifida* are abundant. In the final stages of revegetation, *Stipa krylovii*, *S. sibirica* and other species invade from the surrounding steppe.

The vegetation of the mounds of *Spalax microphtalmus* is considered by Zlotin and Khodasheva (1973) to be more richly developed than virgin meadow steppe. It is distinguished by an abundance of rhizomatous and root-sprouting plants, as well as annuals and ruderal species. Seedlings are 9 times as abundant as in virgin soil. Nearly half (48%) of the 38 species found on mounds are of rhizomatous or root-offshoot habit.

In grazed gravelly steppe (*Festuca lenensis*–*Arctogeron gramineum*–*Chamaerhodos altaica*) of the eastern Hangay, the following species are abundant on mounds of *Lasiopodomys mandarinus* 2 to 3 years old: *Agrostis trinii*, *Chenopodium album*, *Galium verum*, *Leymus chinensis* and *Po-*

tentilla bifurca. Stellera chamaejasme is almost totally destroyed by this rodent.

Alteration of soil. In the semi-desert steppes of the Aktyubinsk region (Formozov and Voronov, 1939), the chemical composition of the uppermost 5 cm of soil is noticeably modified by excrement and urine of *Lagurus lagurus*. The amount of compact organic residue is 3 times, and contents of calcium, chlorine and magnesium are 2 to 4 times, that of undisturbed areas.

The amount of soil deposited on the surface and in passages under the surface in rodent colonies is from 2.2 to 2.3 t ha^{-1} yr^{-1}, which corresponds to 0.1% of the weight of the uppermost 20 cm of soil in which most passages are located (Zlotin and Khodasheva, 1974, p. 138). The humus content of the soil in new mounds is increased by the digging process, and the calcium brought into the root layer is 4 times that in the annual increment of phytomass. Within the area of inhabited burrows, the nitrogen content of the humus increases by 7% (Zlotin and Khodasheva, 1973, p. 110).

Irreversible type of disturbance

Large rodent burrowers exert such destructive effects on steppe ecosystems that the damage can be repaired only by primary micro-successions. These are *Citellus pygmaeus*, *Marmota bobak* and *M. sibirica*. They build permanent burrows which are used for a long time by many generations of animals. This building process is accompanied by a radical change in the whole ecotope. The topsoil is buried and the parent material is exposed.

Marmota hills (*surchiny*, *baibakoviny* and *butany*) and *Citellus* hills (*suslikoviny*) are of considerable size and are preserved for many hundreds of years, creating a unique micro-relief (Dinesman, 1971, 1977). Under some conditions they eventually sink to form micro-depressions (Lavrenko, 1952; Formozov et al., 1954). The changes caused by these animals are so great that their former presence can be detected, as light coloured spots, in arable land long after it is brought under cultivation (Lavrenko, 1952).

Effect of *Marmota* spp. The destruction caused by *Marmota bobak* and *M. sibirica* is especially great. In the zonal steppes the hills occupy an average of 4 to 5% (up to 14%) of the area of colonies. Single hills are as much as 25 to 30 m in diameter and are from 30 to 150 cm high. On the whole, the volume of soil excavated is estimated to range from 0.5 to 150 m^3 ha^{-1}, with a mean of 15 to 20 m^3 ha^{-1}. During its active life each individual of *M. bobak* in the Povolzhye steppe (in the basin of the Volga) is estimated to excavate from 0.06 to 0.36 m^3 of soil, whereas estimates for *M. sibirica* in Transbaykal are 0.1 to 0.2 m^3 (Zimina and Zlotin, 1980).

The soil of the hills is greatly transformed. The soil is mechanically displaced and subsoil is brought to the surface. This material is of high carbonate content from the lower soil strata and from loess-like parent material and from chalk (on southern slopes of ravines). Also large amounts of organic matter and mineral elements of animal origin are added. During its life-time, one animal excretes up to 15 kg of faeces (dry weight) and about 6 to 7 kg of urine. As a result, the structural and chemical characteristics of the soil of the *surchiny* are distinctive (Zimina and Zlotin, 1980).

The geometric form of the hills (*surchiny*), their size, and the complexity of their nanostructure depend on the prevailing physico–geographical conditions and on the age of the *surchiny*. In relatively level areas and on gentle slopes of mountains, they are approximately oval in shape. Several types of micro-ecotope can be detected on old inhabited hills. These differ in soil chemistry, water relations, and in microclimatic conditions. They include: (1) bare areas around the burrow opening with excavated soil of new hillocks and an abundance of zoogenic litter; (2) craters where old burrow openings have filled up; (3) slopes of the hills with various exposures; and (4) near-slope sections. Such structure is especially typical of the old *butany* in the steppe of the eastern Hangay (Guricheva and Dmitriev, 1983) and in the steppe of the inner Tien Shan (Zimina and Zlotin, 1980). In the level zonal steppe, especially in eastern Europe and Kazakhstan, the nanostructure of *surchiny* is less clearly pronounced (Guricheva, 1961; Ismagilov, 1961).

The population of *Marmota bobak* has decreased greatly in the steppe of the European part of the former Soviet Union because of tillage of meadow and forb–bunch-grass steppes. However, traces of their former activity are well preserved.

For example, in the Central Chernozem Reservation (near Kursk), the last individuals disappeared about 200 years ago, but their hills (*baibakoviny*) remain. These number as many as 3.5 ha^{-1} and are up to 70 cm (mean of 45 cm) high and 14 m (mean of 7 m) in diameter (Zimina and Zlotin, 1980). In the forb–*Festuca valesiaca*–*Stipa* steppes of the Sea of Azov region (the Khomutovskaya steppe), the last *Marmota* individuals were eradicated 80 to 100 years ago, but their hills still exist (as many as 8.1 ha^{-1}) (Genov, 1975).

In the forb–bunch-grass (Streltsovskaya) steppe of the south-eastern Ukraine, the virgin sections are almost universally populated by *M. bobak*. Its density reaches 5 to 6 individuals ha^{-1} (Seredneva and Abaturov, 1980).

The mean density of hills (*surchiny*) of various ages in the area surveyed in the dry steppe (*Stipa lessingiana–Festuca valesiaca*) of Central Kazakhstan is reported to be 1.8 ha^{-1}, of which 83% were inhabited. These *surchiny* are oval in shape, with a mean height of 18 to 46 cm, are 10 to 18 m in diameter, and are rather uniformly distributed over the steppe (Ismagilov, 1961).

In the steppes of Barga (south of the Chita region) the hills of *Marmota sibirica* are 2 to 12 m in diameter and 30 to 40 cm high. They occupy 5 to 12% of the area of inhabited steppe (Frish, 1967). In the steppes of the eastern Hangay (Dmitriev and Guricheva, 1978; Guricheva and Dmitriev, 1983), the hills occupy from 0.8 to 5% of the inhabited area and their size varies from 5 to 900 m^2.

The vegetation of the *surchiny* is very diverse and dynamic, consisting of a mosaic of progressive micro-successions in various stages of development. It is more xeric in nature than the surrounding plant communities of zonal sites. Species of the drier steppe region are common in the disturbed areas of meadow steppes, whereas those of drier steppes include desert elements; xeromesophytes occupy the more mesic micro-environment of craters. The first stage of revegetation of exposed soil and subsoil consists of annual and rhizomatous species; nitrophilous species occur around the burrow openings, and calciphiles are more abundant in the disturbed areas than elsewhere in the steppe. Finally, species of the surrounding zonal vegetation gradually invade the slopes of old deserted hills.

The areas in the Khomutovskaya steppe (Sea of Azov region) that were disturbed by *Marmota* before their disappearance (80 to 100 years ago) can still be clearly distinguished by the abundance of *Agropyron pectinatum* and *Bromopsis riparia*, which contrasts with the dominance of *Stipa* spp. in the surrounding grassland (Genov, 1975). *Agropyron pectinatum* is also the principal grass of *surchiny* in the southern Ukraine (Voroshilovgrad region), where the zonal dominant is *Stipa lessingiana*. The associated dicotyledons are also mainly calciphilous species, including: *Ceratocephala testiculata*, *Jurinea multiflora*, *Kochia prostrata* and *Lappula patula*. Ruderal nitrophilous species, *Chenopodium foliosum* and *Hyoscyamus niger*, grow near burrow openings. On old hills, *Caragana frutex* sometimes replaces the annual species (Lavrenko, 1952; Seredneva and Abaturov, 1980).

In the *Festuca–Stipa* steppe of Kazakhstan (dark-chestnut and chestnut soils), *Artemisia pauciflora* grows on newly formed *surchiny*, which contain large quantities of carbonates and readily soluble salts, but not beyond their margins (Vinogradov, 1937). On older hills, where easily soluble salts are absent, but the carbonates are still concentrated, *Stipa sareptana* (typical of the more southerly desert steppe) and *Leymus ramosus* are abundant. On still older hills, with less carbonate, *Stipa lessingiana* (the dominant of the zonal steppe) is abundant.

Around the burrow openings, in dry steppes (*Stipa lessingiana–Festuca valesiaca*) in Central Kazakhstan, *Amaranthus albus*, *Ceratocarpus arenarius* and *Leymus ramosus* are abundant. Around this central area, the vegetation is distributed in belts. The uppermost is of *Stipa sareptana*. *Crinitaria tatarica*, *Festuca valesiaca*, *Jurinea multiflora* and *Phlomis agraria* occur below this. In steppes with more solonetzic soil, the central zone is occupied by *Psathyrostachys* (*Elymus*) *juncea*, around which is a zone of *Stipa sareptana* with ephemerals and ephemeroids (*Ceratocephala testiculata* and *Tulipa patens*) and *Leymus ramosus*. The same species are abundant on deserted hills, but *Festuca valesiaca*, *Stipa lessingiana* and other steppe species are also present.

Artemisia pauciflora and *Psathyrostachys juncea* are abundant on hills in solonetzic soil along the Tersakkan river (Guricheva, 1961; Ismagilov, 1961).

Chenopodium foliosum and *Saussurea salicifolia* form dense thickets on hills in the Chuyskaya steppe of the Altai and in the south of the European part of the former Soviet Union (Kolosov, 1939; Lavrenko, 1952).

The principal plant species of "marmot gardens" (Radde, 1862) in Transbaykal are *Rheum* sp. and *Urtica cannabina*.

The species occupying large old inhabited hills in the Mongolian steppe (75–80 km south of Ulaanbaatar), on the slopes of *sopki* (in steppes dominated by *Koeleria cristata*, *Leymus chinensis*, *Poa botryoides* and *Thermopsis lanceolata*) include *Agropyron cristatum*, *Artemisia adamsii*, *A. dracunculus* and *Leymus chinensis*, with large plants of *Urtica cannabina* around the openings to burrows (Lavrenko, 1952). Elsewhere, old hills support *Carex duriuscula* and *C. pediformis* in the craters and *Potentilla bifurca* and *Thermopsis lanceolata* around entrances to burrows. Annuals on recently formed hills include *Lepidium densiflorum* and *Leptopyrum fumarioides* (Lavrenko, 1952).

Formozov (1929) listed *Axyris amaranthoides*, *Chenopodium aristatum*, *Salsola collina* and *Thermopsis lanceolata* as important on *Marmota* hills in the northern Mongolian steppe.

Butany in meadow steppe (*Festuca lenensis* and *Stipa baicalensis*) and in steppified meadow of the eastern Hangay support *Artemisia changaica*, *A. dracunculus* and *Rheum undulatum* (Dmitriev and Guricheva, 1978). In craters of old *butany*, species of meadow steppes are common, including *Geranium pratense*, *Sanguisorba officinalis* and *Valeriana officinalis*. In mountain steppes, the species typical of *butany* include *Achnatherum splendens*, *Echinops latifolius* and *Rheum undulatum*. Within dry steppes (*Stipa krylovii–Caragana microphylla*), *Achnatherum splendens*, *Caragana microphylla* and *Urtica cannabina* are common on these *Marmota* hills (Fig. 2.29).

Effect of *Citellus pygmaeus*. The digging activity of *Citellus pygmaeus* causes considerable modification to the microrelief of arid landscapes. This is the principal rodent in the zonal dry *Festuca lenensis–Stipa* steppes and semi-desert steppes, as well as in the complex flood-plains (clay semi-deserts) of the Caspian Sea region between the Volga river and Ural mountains (Lavrenko, 1952;

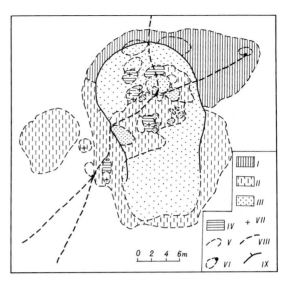

Fig. 2.29. Vegetation of an old inhabited *butan* of *Marmota sibirica* in dry steppe (*Stipa krylovii* with *Caragana microphylla*) in the eastern Hangay (Mongolia) (after Dmitriev et al., 1982):

I – Pre-slope section with microcoenoses dominated by *Stipa sibirica*, *Festuca lenensis*, *Kochia prostrata*, *Artemisia changaica* and *Serratula centauroides*.
II – Microcoenoses dominated by *Achnatherum splendens*
III – Microcoenoses dominated by *Leymus chinensis*, *Agropyron cristatum*, *Carex duriuscula* and *Goniolimon speciosum*
IV – Microcoenoses dominated by *Artemisia dracunculus*, *A. palustris*, *Scutellaria scordiifolia*, *Echinops latifolius* and *Serratula centauroides*
V – Boundary of microcoenoses
VI – Entrance to burrow and bare area of recently excavated soil
VII – Small depression where old burrow entrances have filled up
VIII – *Marmota* paths
IX – Horizontal boundary of main hill of *butan*

Formozov et al., 1954). In this region, *souslikoviny* are an integral part of the relief, especially on the solonetz soils.

These hills are from 3 to 4 m in diameter and 35 to 40 cm high in the steppe of the Don river (Volgograd region); the density of the burrows is 2000 to 3000 ha^{-1} (Formozov and Voronov, 1939; Kirsanov and Fursaev, 1941). The animal population averages from 10 to 15 individuals ha^{-1} and the hills occupy 5 to 10% of the total area in flood-plains (between the small rivers) of the Caspian Sea region (Formozov and Voronov, 1939; Formozov et al., 1954). Near the Dzhanibek station of the Academy of Sciences of the former Soviet Union (in the Ural region), the population of these animals reaches 40 to 66 individuals ha^{-1} in some years, and the number of hills exceeds 1000 ha^{-1} (Abaturov, 1979). In semi-desert steppe in solonetz soils with *Artemisia pauciflora* at the Temir Sonal Experimental Station (Aktyubinsk region), the average area of the hills is 9.8 m^2, their diameter is 4 to 5 m, and their height is 15 to 50 cm (Formozov and Voronov, 1939).

Citellus pygmaeus is commonest in solonetz soils. It transfers as much as 1.5 t of soil and subsoil per hectare annually to the surface from depths of 40 to 200 cm. This material is high in content of easily soluble salts, gypsum, carbonates and aluminium (Abaturov, 1973). Rodent activity also improves the water regime. The soil is moist to a depth of 1.5 m in disturbed areas, but only to 50 cm elsewhere; the reserves of moisture are doubled (190 mm, compared to 90 mm) (Abaturov, 1979). These conditions lead to desalinization over a period of about 20 years (Formozov et al., 1954). However, during increases in population, the process of desalinization is sometimes arrested because of the secondary occupation of the hills by the animals.

The vegetative cover changes as desalinization progresses. In communities with *Artemisia pauciflora* the first seral stage is composed of annuals (*Climacoptera brachiata* and *Lepidium perfoliatum*). Next to appear is *Artemisia pauciflora*, which grows more vigorously on the hills than in undisturbed areas (1.33 t ha^{-1} of phytomass, compared to 0.38 t ha^{-1}). *Artemisia lerchiana* is prominent in the next successional stage, which often coincides with the occupation of *Citellus* hills by *Lagurus lagurus*. Then *Artemisia austriaca* often appears. With further desalinization, *Festuca valesiaca*, *Stipa sareptana* and other species of *Stipa* finally become established. In this way, the activities of *Citellus* result in the formation of small areas of bunch-grass steppe within desert-type communities (Formozov and Voronov, 1939). Ivanov (1950) repeatedly stressed the colossal role of *Citellus* in the processes of desalinization and steppification of semi-deserts.

Old *souslikoviny* in *Festuca valesiaca–Stipa sareptana–S. lessingiana* steppe in the catchment of the Yergeni (Volgograd region) are occupied by solitary plants of *Atriplex sphaeromorpha* and of *Salsola laricina* (a half-shrub) (Lavrenko, 1952). On the eastern slope of the Yergeni, *Anabasis aphylla* (another half-shrub) also occurs. In the upper parts of ravines undisturbed areas are occupied by *Festuca valesiaca* and *Tanacetum* sp., with communities of *Artemisia lerchiana* and *Stipa lessingiana* on light-chestnut coloured soil, and of *Artemisia pauciflora* in flat depressions; *Artemisia pauciflora* is prevalent on recently developed hills, whereas *Agropyron desertorum* (with some *Festuca valesiaca*) occupies the older ones. *Artemisia pauciflora* is also common on *Citellus* hills in the semi-desert steppe at the Temir Sonal Experimental Station (Aktyubinsk region).

In Western Kazakhstan the first stage of succession on recently exposed soil of *Citellus* hills is composed of annuals with *Artemisia pauciflora* and (sometimes) *Camphorosma monspeliaca*. *Artemisia lerchiana*, *Kochia prostrata* and *Tanacetum achilleifolium* are next to appear, and, finally, the species of the zonal steppe (*Festuca valesiaca*, *Koeleria cristata*, *Leymus ramosus* and *Stipa* spp.) dominate (Ivanov, 1950).

CONCLUSION

Bogdan (1900) was the first to recognize that *Citellus pygmaeus*, and other burrowing animals, are "pioneers of draining and demineralization of solonchak steppes". In contrast, in normal steppe they bring salt-bearing materials to the surface and cause the formation of temporary solonetz spots. Species of the family Sciuridae are especially important in the formation of the long-term and very dynamic micro-complexity of soils and vegetation over considerable areas from the forest steppe to the desert steppe and semi-deserts (Lavrenko, 1952).

The dynamic processes of salinization–desalinization in the soil and of denudation–revegetation in the plant cover, which result from the digging activity of rodents, are mostly localized and occur within only a certain part of the geobotanical contour, a part which can be calculated. The form, size and degree of permanence of disturbance caused by rodents in the steppe depend on the species of animal involved and on its biological characteristics. However, many qualitative and quantitative, as well as temporal, parameters of disturbance, especially the character and speed of the processes of denudation and revegetation, are dependent, to a considerable extent, on natural features of the landscape and its environment, as well as on vegetative and edaphic factors.

The eventual effect of the activities of rodents is to complicate the natural mosaic structure of steppe vegetation. In some instances the effect is of short duration, but in others it is long-lasting, being dependent on the rate at which those transformations can occur which are associated with primary succession.

The activity of these animals and the dynamic processes that they initiate are intrinsic features of the steppe ecosystem, and must be regarded as indispensable to the prolonged conservation of zonal steppes in a state of dynamic equilibrium (Formozov, 1929; Lavrenko, 1952; Valter, 1975).

REFERENCES

Abaturov, B.D., 1973. The role of burrowing animals in the movement of chemical matter in soil. In: E.M. Lavrenko and T.A. Rabotnov (Editors), *Problemy Biogeotsenologii*. Nauka, Moscow, pp. 5–11 (in Russian).

Abaturov, B.D., 1979. *Bioproductive Processes in Terrestrial Ecosystems*. Nauka, Moscow, 128 pp. (in Russian).

Abaturov, B.D., 1980. Peculiarities of trophic relations in the type "herbivorous plants" in pasture ecosystems. In: V.E. Sokolov (Editor), *Fitofagi v Rastitel'nykh Soobshchestvakh*. Nauka, Moscow, pp. 31–42 (in Russian).

Abaturov, B.D., Rakova, M.V. and Seredneva, T.A., 1980. The influence of small sousliks on productivity of vegetation in semi-desert. In: V.E. Sokolov (Editor), *Fitofagi v Rastitel'nykh Soobshchestvakh*. Nauka, Moscow, pp. 111–127 (in Russian).

Alekhin, V.V., 1925. *The Vegetative Cover of Steppe in the Central Chernozem Region*. I.vo Sojuza Obsch. i Organ. po Izucheniju Ts. Ch. O. Voronezh, 110 pp. (in Russian).

Alekhin, V.V., 1934. *The Steppe of the Central Chernozem Region*. Communa, Voronezh, 96 pp. (in Russian).

Andrushko, A.M., 1948. On the burrowing activity of some rodents as a factor of soil formation in the folding upland of Kazakhstan. *Vestn. Len. Gos. Univ.*, 9: 44–51 (in Russian).

Arnol'di, L.V. and Yunatov, A.A. (Editors), 1969. *Biocomplex Investigations in Central Kazakhstan, 2. Plant and Animal Communities of the Steppe and Desert of Central Kazakhstan*. Nauka, Leningrad, 496 pp. (in Russian).

Bannikov, A.G., 1948. Materials on Mongolian mammals, 2. Grey voles (*Microtus*), lemmings (*Lagurus*), mole-voles (*Ellphius*). *Byull. Mosk. Ova. Ispyt. Prir.*, 53: 19–36 (in Russian).

Belostokov, G.P., 1957. On tillering of bunch grasses. *Bot. Zh.*, 42: 1267–1277 (in Russian).

Bespalova, Z.G., 1977. On flowering and fruiting peculiarities of *Cleistogenes songorica* (Roshev.) Ohwi and *C. squarrosa* (Trin.) Keng. In: Z.V. Karamysheva (Editor), *Problemy Ekologii, Geobotaniki, Botanicheskoi Geografii i Floristiki*. Nauka, Leningrad, pp. 156–164 (in Russian).

Bilyk, G.I., 1973. Meadow steppe. Typical (true) steppe. Herb–bunch-grass steppe. Desert steppe. Shrub steppe. Semi-savanna steppe. In: A.I. Barbarich (Editor), *Stepi, Kam'yanisti Vidslonennya, Piski*. Naukova Dumka, Kiev, pp. 33–94; 94–170; 229–240; 245–249 (in Ukrainian).

Bilyk, G.I. and Tkachenko, V.S., 1976. The steppe of Tarkhankutsky Peninsula and its protection. *Ukr. Bot. Zh.*, 33: 526–531 (in Ukrainian).

Bobrovskaya, N.I., 1988. Peculiarities of water relations of the dominants. In: R.V. Kamelin (Editor), *Pustyni Zaaltaiskoi Gobi*. Nauka, Leningrad, pp. 107–135 (in Russian).

Bobrovskaya, N.I., Zavadskaya, I.G. and Kobak, K.I., 1977. The influence of different degrees of dehydration on protoplasmic movement, ability of plasmolysis and respiration of *Cleistogenes squarrosa* (Trin.) Keng. and *Agropyron cristatum* (L.) Beauv. In: Z.V. Karamysheva (Editor), *Problemy Ecologii, Geobotaniki, Botanicheskoi Geografii i Floristiki*. Nauka, Leningrad, pp. 201–206 (in Russian).

Bogdan, V.S., 1900. *Report of Valuiskaya Agricultural Experimental Station (Novouzen' District of Samara Region)*. Tipografiya V. Kirshlauma, St. Petersburg, 128 pp. (in Russian).

Braun-Blanquet, J. and Paviard, J., 1922. *Vocabulaire de sociologie végétale*. Roumegous and Dehan, Montpellier, 17 pp.

Cherepanov, S.K., 1981. *The Vascular Plants of the USSR*. Nauka, Leningrad, 509 pp. (in Russian).

Chernov, Yu.I., 1975. *Natural Zonation and Distribution of Animals*. Mysl', Moscow, 222 pp. (in Russian).

Chernov, Yu.I., 1978. *Structure of Animal Populations of the Sub-Arctic*. Nauka, Moscow, 167 pp. (in Russian).

Dinesman, L.G., 1971. Colonization of steppe marmots on Russian plain. *Byull. Mosk. Ova. Ispyt. Prir.*, 76(6): 59–73 (in Russian).

Dinesman, L.G., 1977. *Steppe biogeocoenoses in the Holocene*. Nauka, Moscow, 160 pp. (in Russian).

Dmitriev, P.P., 1980. The ecology of *Lasiopodonis mandarinus* and its role in pasture biocoenoses in northern Mongolia. *Zool. Zh.*, 59: 1852–1861 (in Russian).

Dmitriev, P.P. and Guricheva, N.P., 1978. Small mammals in pasture biocoenoses in eastern Hangay. In: I.A. Bannikova and L.M. Medvedev (Editors), *Geografiya i Dinamika Rastitel'nogo i Zhivotnogo Mira MNR*. Nauka, Moscow, pp. 124–131 (in Russian).

Dmitriev, P.P. and Guricheva, N.P., 1983. The main patterns of the vegetation in settlements of mammals in mountain steppe of eastern Hangay (Mongolia). *Dokl. Akad. Nauk. SSSR*, 271: 250–254 (in Russian).

Dmitriev, P.P., Gugalinskaya, L.A. and Guricheva, N.P., 1982. The role of marmot (*Marmota sibirica* Radde) in the origin of aeolian hills with *Achnatherum splendens* (Trin.) Nevski in Mongolian dry steppe. *Zh. Obshch. Biol.*, 43: 712–718 (in Russian).

Dobrinskyi, L.N., Davydov, V.A., Kryazhimskyi, F.V. and Malafeev, Yu.M., 1983. *Functional Relationships Between*

Animals and Vegetation in Meadow Biocoenoses. Nauka, Moscow, 160 pp. (in Russian).

Dokhman, G.I., 1956. An experience of phytocoenological interpretation of origin of northern steppe. In: V.B. Sochava (Editor), *Akademiku V.N.Sukachevu k 75-Letiyu so Dnya Rozhdeniya.* Nauka, Moscow–Leningrad, pp. 182–208 (in Russian).

Dokhman, G.I., 1968. *The Forest Steppe of the European Part of the USSR: On the Natural Regularity of Forest Steppe.* Nauka, Moscow, 271 pp. (in Russian).

Dokuchaev, V.V., 1883 (1948). *Russian Chernozem* I. vo Deklerona i Evdokimova, St. Petersburg, 1948, 480 pp. (in Russian).

Flora of the USSR, 1934–1964. Volumes I–XXX. Akad. Nauk SSSR, Moscow–Leningrad (in Russian).

Formozov, A.N., 1928. Mammals in the steppe biocoenose. *Ecology,* 9: 449–460. [1978, *Byull. Mosk. Ova. Ispyt. Prir.,* 13(2): 150–156.]

Formozov, A.N., 1929. The mammals of northern Monogolia collected during the expedition of 1926. *Mater. Komiss. po issled. Mongol'sk. i Tannu-Tuvinsk. Narodn. Resp. i Buryat-Mongol'skoi ASSR,* 3. Akad. Nauk SSSR, Leningrad, pp. 1–144 (in Russian).

Formozov, A.N. and Kiris-Prosvirina, I.B., 1937. The activity of rodents in pastures and hay meadows, 2 and 3. *Uchen. Zap. Mosk. Gos. Univ., Zool.,* 8: 39–59 (in Russian).

Formozov, A.N. and Voronov, A.G., 1939. The activity of rodents in pastures and hay meadows of Western Kazakhstan and its economic importance (the biotic relations between rodents and vegetation). *Uchen. Zap. Mosk. Gos. Univ., Zool.,* 20: 3–12 (in Russian).

Formozov, A.N., Khodasheva, K.S. and Golov, B.A., 1954. The influence of rodents on vegetation of pastures and hay meadows of clay semi-desert on the watershed of the Volga and the Ural. In: I.V. Larin (Editor), *Voprosy Uluchsheniya Kormovoi Bazy v Stepnoi, Polupustynnoi i Pustynnoi Zonakh SSSR.* Akad. Nauk SSSR, Moscow–Leningrad, pp. 331–340 (in Russian).

Frish, V.A., 1967. Landscape investigations in Soviet Barga. *Izv. Vses. Geogr. Ova.,* 99(1): 3–18 (in Russian).

Gael', A.G., 1930. *A Guide to Invesigations of Sands.* Sel'khozgiz, Moscow–Leningrad, 135 pp. (in Russian).

Gael', A.G., 1932. The sands of the upper Don. *Izv. Gos. Geofiz. Observ.,* 64: 1–50 (in Russian).

Gams, H., 1918. Prinzipienfragen der Vegetationsforschung. *Vierteljahresschr. Naturforsch. Ges. Zurich,* 63: 293–493 (in German).

Genov, A.P., 1975. The morphological and physical–chemical characteristics of the soils of marmot burrows in Khomutovskaya steppe. In: T.L. Bystritskaya (Editor), *Pochvenno-Biogeotsenologicheskie Issledovaniya v Priazov'e.* Nauka, Moscow, pp. 129–134 (in Russian).

Gordyagin, A.Ya., 1900. The soils and vegetation of Western Siberia. *Tr. Ova. Estestvoispyt. Pri Kazanskom Univ.,* 34: 1-222 (in Russian).

Gordyagin, A.Ya., 1901. The materials on soils and vegetation of Western Siberia. *Tr. Ova. Estestvoispyt. Pri Kazanskom Univ.,* 35: 222–528 (in Russian).

Guricheva, N.P., 1961. The vegetation on burrows of steppe lemming (*Lagurus lagurus* Pall.) and marmot (*Marmota bobac* Müll.) within the dry-steppe station. In: L.V. Arnol'di, B.A. Bykov, A.V. Kalinina and A.A. Yunatov (Editors), *Materialy Kazakhstanskoi Konferentsii po Probleme "Biologicheskie Kompleksy Raionov Novogo Osvoeniya, ikh Ratsional'noe Ispol'zovanie i Obogashchenie".* Akad. Nauk SSSR, Moscow–Leningrad, pp. 92–94 (in Russian).

Guricheva, N.P., 1985. Die Beeinflussung der Steppenvegetation in östlichen Changai durch bodenbewohnende Nagetiere. *Arch. Naturschutz Landschaftsforsch.,* 25: 47–56.

Guricheva, N.P. and Dmitriev, P.P., 1983. The relationship between vegetative cover and animals: the main forms of dynamics and structure of mountain steppe due to the activity of soil-inhabiting mammals. In: E.M. Lavrenko and I.A. Bannikova (Editors), *Gornaya lesostep' Vostochnogo Khangaya (MNR).* Nauka, Moscow, pp. 172–181 (in Russian).

Guricheva, N.P., Izmailova, N.N., Slemnev, N.N. and Lkhagvasuren, S., 1983. *Stellera chamaejasme* (Thymelaeaceae) in the steppe of eastern Hangay. *Bot. Zh.,* 68: 453–463 (in Russian).

Guricheva, N.P., Buevich, Z.G. and Sukhoverko, R.V., 1984. The effect of protection on *Stipa baicalensis* steppe of eastern Hangay (Mongolia). *Bot. Zh.,* 69: 636–647 (in Russian).

Isachenko, T.I. and Rachkovskaya, E.I., 1961. The main zonal types of steppe of Northern Kazakhstan. *Tr. Bot. Inst. Im. V.L. Komarova, Ser. 3, Geobot.,* 13: 133–397 (in Russian).

Ismagilov, M.I., 1961. On types of settlement of steppe marmot (*Marmota bobac* Müll.) and its influence on vegetation in regions of development of virgin lands in Kazakhstan. *Zool. Zh.,* 60: 905–913 (in Russian).

Ivanov, V.V., 1950. Small souslik as a desalinizer of soil. *Izv. Vses.. Geogr. Ova.,* 82: 651–653 (in Russian).

Kamenetskaya, I.V., 1949. The influence of meteorological conditions on seed reproduction of plants in Streletskaya steppe. *Byull. Mosk. Ova. Ispyt. Prir.,* 54(4): 83–89 (in Russian).

Karamysheva, Z.V. and Rachkovskaya, E.I., 1973. *Botanical Geography of the Steppe Part of Central Kazakhstan.* Nauka, Leningrad, 278 pp. (in Russian).

Keller, B.A., 1931. In Khrenovskaya steppe. Fescue and meadow–feather-grass steppe. In: B.A. Keller (Editor), *Stepi Tsentral'no-Chernozemnoi oblasti.* Gos. Sel'.-khoz. i. vo, Moscow–Leningrad, pp. 26–33 (in Russian).

Khodasheva, K.S. and Zlotin, R.I., 1972. A study of the biogeocoenotic role of animals in the landscape of forest steppe. In: D.L. Armand (Editor), *Biogeograficheskoe i landshaftnoe izuchenie lesostepi.* Nauka, Moscow, pp. 180–196 (in Russian).

Khou, S. (Editor), 1979. *A Vegetation Map of China.* Cartographic Publishing House, Peking (in Chinese).

Kirsanov, M.P. and Fursaev, A.D., 1941. Materials for the characteristics of the sub-Don steppe of the Stalingrad region. *Uchen. Zap. Saratovsk. Univ., Biol.,* 15: 82–108 (in Russian).

Kleopov, Yu.D., 1928. The remains of steppe vegetation in the Cherkassk district. In: *Okhorona Pam'yatok Prirodi na Ukraini,* 2. Ukrainskii Komitet okhoroni pam'yatok na Ukraine, Kharkov, pp. 37–49. (in Ukrainian).

Kleopov, Yu.D., 1933a. The remains of steppe vegetation in Kievskaya upland. *Zh. Bio-bot. Tsiklu Vseukr. Akad. Nauk*, 5–6: 135–156. (in Ukrainian).

Kleopov, Yu.D., 1933b. The vegetative cover of the southwestern part of Donetsk Kryazh (upland) (near Stalinsk district). *Visn. Kiivs'k. Bot. Sadu*, 15: 9–162. (in Ukrainian).

Kleopov, Yu.D., 1934a. The geobotanical essay of the left bank of the middle stream of the Dnieper river. *Zhurn. Inst.. Bot. Vseukr. Akad. Nauk*, 2(10): 29–73 (in Ukrainian).

Kleopov, Yu.D., 1934b. Vegetation of the Karlovsky Steppe Reservation of All-Ukrainean Academy of Science. *Visn. Kiivsk. Bot. Sadu*, 17: 41–86 (in Ukrainian).

Kolosov, A.M., 1939. The animals of the south-eastern Altai and of the adjoining part of Mongolia. *Uchen. Zap. Mosk. Gos. Univ., Zool.*, 20: 123–190 (in Russian).

Korzhinskyi, S.I., 1888. The northern border of the chernozem steppe region of the eastern part of European Russia, in relation to botanical geography and soils. t. 1. Kazan', 253 pp. (*Tr. Ova. Estestvoispyt. Pri Kazansk. Univ.*, 18(5) (in Russian).

Korzhinskyi, S.I., 1891. The northern border of the chernozem steppe region of the eastern part of European Russia, in relation to botanical geography and soils. t. 2. Kazan', 201 pp. (*Tr. Ova. Estestvoispyt. Pri Kazansk. Univ.*, 22(6) (in Russian).

Krasheninnikov, I.M. and Kucherovskaya-Rozhanets, S.E., 1941. *The Vegetation of the Bashkir ASSR*. Akad. Nauk SSSR, Moscow–Leningrad, 154 pp. (in Russian).

Kucheruk, V.V., 1963. The influence of herbivorous mammals on the productivity of a stand of steppe grasses and their role in formation of the organic component of steppe soils. In: A.A. Nakhimovich and B.I. Salkin (Editors), *Biologiya, biogeografiya i sistematika mlekopitayushchikh SSSR*. Akad. Nauk SSSR, Moscow, pp. 157–193 (in Russian).

Kuminova, A.V., 1960. *The Vegetative Cover of the Altai*. Sibirsk. Otd. Akad. Nauk SSSR, Novosibirsk, 450 pp. (in Russian).

Lavrenko, E.M., 1927. The digression of pastures on the sands of the lower Dnieper in relation to the work of the Aleshkovskaya Experimental Sand Reclamation Station. *S. Kh. Opyt. Delo*, 3: 50–59 (in Russian).

Lavrenko, E.M., 1940. The steppes of the USSR. In: B.K. Shishkin (Editor), *Rastitel'nost' SSSR*. t. 2. Akad. Nauk SSSR, Moscow–Leningrad, pp. 1–265 (in Russian).

Lavrenko, E.M., 1941. On the relationships between plants and the environment in steppe phytocoenoses. *Pochvovedenie*, 3: 42–58 (in Russian).

Lavrenko, E.M., 1952. Microcomplex and mosaic structure of the vegetative cover of steppe due to activities of animals and plants. *Tr. Bot. Inst. Akad. Nauk SSSR, Ser. 3, Geobot.*, 8: 40–47 (in Russian).

Lavrenko, E.M., 1954. The steppes of the Eurasian steppe region: its geography, dynamics and history. *Voprosy Bot.*, 1: 155–191 (in Russian).

Lavrenko, E.M., 1956. The steppes and agricultural lands on their place. In: E.M. Lavrenko (Editor), *Rastitel'nyi pokrov SSSR*. t. 2. Akad. Nauk SSSR, Moscow–Leningrad, pp. 595–730 (in Russian).

Lavrenko, E.M., 1970a. Division of the Black Sea–Kazakhstan subregion of the Eurasian steppe region. *Bot. Zh.*, 55: 609–625 (in Russian).

Lavrenko, E.M., 1970b. Division of the Central Asian subregion of the steppe region of Eurasia. *Bot. Zh.*, 55: 1734–1741 (in Russian).

Lavrenko, E.M., 1980a. The steppes. In: S.A. Gribova, T.I. Isachenko and E.M. Lavrenko (Editors), *Rastitel'nost' Evropeiskoi chasti SSSR*. Nauka, Leningrad, pp. 203–273 (in Russian).

Lavrenko, E.M. (Editor), 1980b. *The Desert Steppes and Northern Deserts of Mongolia, 1. The Natural Conditions*. Nauka, Leningrad, 184 pp. (in Russian).

Lavrenko, E.M. (Editor), 1981. *The Desert Steppes and Northern Deserts of Mongolia, 2. Investigations on Bulgan Station*. Nauka, Leningrad, 258 pp. (in Russian).

Lavrenko, E.M. (Editor), 1984. *The Dry Steppes of Mongolia, 1. The Natural Conditions*. Nauka, Leningrad, 168 pp. (in Russian).

Lavrenko, E.M. and Bannikova, I.A. (Editors), 1986. *The Steppes of Eastern Hangay*. Nauka, Moscow, 181 pp. (in Russian).

Lavrenko, E.M. and Borisova, I.V. (Editors), 1976. *Biocomplex Investigations in Central Kazakhstan, 3. The Principal Plant Communities of Central Kazakhstan*. Nauka, Leningrad, 292 pp. (in Russian).

Lavrenko, E.M. and Dokhman, G.I., 1933. The vegetation of Starobelsky steppe. *Zh. Bio-bot. Tsiklu Vseukr. Akad. Nauk*, 5–6: 23–133 (in Ukrainian).

Lavrenko, E.M. and Yunatov, A.A., 1952. The regime of fallow land as influenced by field vole (*Marmota brandtii*) in steppe vegetation and soils. *Bot. Zh.*, 37: 128–138 (in Russian).

Lavrenko, E.M. and Zoz, I.G., 1928. The vegetation of virgin land on Mikhailovsky Stud Farm in Sumskoya district. In: O. Fedorovsky and E.M. Lavrenko (Editors), *Okhorona Pam'yatok Prirodi na Ukraini, 2*. Ukrainskii Komitet okhoroni pam'yatok na Ukraine. Kharkov, pp. 23–36 (in Ukrainian).

Levakovskyi, I., 1871. The study of chernozems. *Tr. Ova. Estestvoispyt. Pri Khar'kovskom Univ.*, 4: 15–68.

Levina, R.E., 1967. *Fruits: Their Morphology, Ecology and Practical Importance*. Privolzhskoe Knizhnoe i. vo, Saratov, 216 pp. (in Russian).

Mordkovich, V.G., 1982. *Steppe Ecosystems*. Nauka, Novosibirsk, 204 pp. (in Russian).

Nikol'skyi A.A., Guricheva, N.P. and Dmitriev, P.P., 1984. Winter reserves of *Ochotona daurica* Pall. in grazed steppe. *Byull. Mosk. Ova. Ispyt. Prir.*, 89(6): 9–22 (in Russian).

Novopokrovskyi, I.V., 1921. *Vegetation of the Don Region (Botanical-geographical Outline).*. Rostov-na-Donu Otdelenie Gosizdata, Novocherkassk, 48 pp. (in Russian).

Novopokrovskyi, I.V., 1925. Vegetation of the North Caucasus Region. *Izv. Donsk. Inst. S. Kh. i Melior. 5, Prilozh.*, 1: 1–24 (in Russian).

Novopokrovskyi, I.V., 1927. Some data on steppe and river valley vegetation of the southern Sub-Urals. In: *Trudy Soveshchaniya po Voprosam Lugovedeniya i Opytnogo Lugovodstva, 1*. Izdaniya Lugovogo Inst., Moscow, pp. 87–93 (in Russian).

Novopokrovskyi, I.V., 1931. *On the Vegetation of the Southern*

Sub-Urals. Sel'khozgiz, Moscow–Leningrad, 139 pp. (in Russian).

Novopokrovskyi, I.V., 1940. Vegetation. In: K.Z. Yatsuta (Editor), *Priroda Rostovskoi oblasti*. Rostovskoe oblastnoe knigoizdatel'stvo, Rostov-na-Donu, pp. 111–140 (in Russian).

Osychnyuk, V.V., 1973. Succession of the vegetation in steppe. In: A.I. Barbarich (Editor), *Stepi, Kam'yanisti Vidslonennya, Piski*. Naukova dumka, Kiev, pp. 249–315; 373–398 (in Ukrainian).

Osychnyuk, V.V., Bilyk, G.I., Tkachenko, V.S., Genov, A.P. and Shupranov, N.P., 1976. Vegetative cover. In: T.L. Bystritskaya (Editor), *Pochvenno-Biogeotsenologicheskie Issledovaniya v Priazov'e*, 2: 37–122 (in Russian).

Pachoskyi, I.K., 1917. *Description of the Vegetation in the Kherson Region, 2. The Steppes*. Estestvenno-istorichesky Musey Khersonskogo Gubernskogo zemstva, Kherson, 366 pp. (in Russian).

Pankov, A.M., 1921. Burrowing animals and their role in soil formation. *Vestn. Opytn. Dela Sredne-Chernozemnoi Obl.*, 5–6: 19–58 (in Russian).

Rabotnov, T.A., 1980. Some aspects of a study of autotrophic plants as components of terrestrial biogeocoenoses. *Byull. Mosk. Ova. Ispyt. Prir., Otd. Biol.*, 85(3): 64–80 (in Russian).

Rabotnov, T.A., 1983. *Phytocoenology*. Izd. Mosk. Gos. Univ., Moscow, 296 pp. (in Russian).

Radde, G., 1862. *Reisen im Süden von Ost-Sibirien in den Jahren 1855–1859*. Band 1, Buchdo. der Keiserlichen Univ., St. Petersburg, 328 pp.

Rakhmanina, A.T. and Arnol'di, L.V., 1961. A comparative complex of characteristics of zonal coenoses of steppe and semi-desert in Kazakhstan. In: L.V. Arnol'di, B.A. Bykov, A.V. Kalinina and A.A. Yunatov (Editors), *Materialy Kazakhstanskoi Konferentsii po Probleme: Biologicheskie Kompleksy Raionov Novogo Osvoeniya, ikh Ratsional'noe Ispol'zovanie i Obogashchenie*. Akad. Nauk SSSR, Moscow–Leningrad, pp. 109–120 (in Russian).

Rescikov, M.A., 1961. The steppes of the western Transbaykal (Zabaikal'e). *Tr. Vost.-Sib. Fil. Akad. Nauk SSSR, Ser. Biol.*, 34, 175 pp.

Reverdatto, V.V., 1925. Vegetation of the Abakan steppe region. *Izv. Tomsk. Univ.*, 75, 23 pp. (in Russian).

Reverdatto, V.V. and Golubintseva, V.P., 1930. *Weed Vegetation of Irrigated and Non-Irrigated Areas and Long Fallows in the Southern Siberian Steppe*. Sel'khozgiz, Moscow–Leningrad, 78 pp. (in Russian).

Rotshil'd, E.V., 1968. *Nitrogen-loving Vegetation of Desert in Relation to Animals*. Mosk. Gos. Univ., Moscow, 204 pp. (in Russian).

Semenova-Tyan-Shanskaya, A.M., 1966. *Dynamics of Steppe Vegetation*. Akad. Nauk SSSR, Moscow–Leningrad, 169 pp. (in Russian).

Seredneva, T.A. and Abaturov, B.D., 1980. The influence of steppe marmots on plant production in the Ukrainian steppe. In: V.E. Sokolov (Editor), *Fitofagi v Rastitel'nykh Soobshchestvakh*. Nauka, Moscow, pp. 128–141 (in Russian).

Shalyt, M.S., 1927. The effect of grazing by sheep on the condition of steppe vegetation on Askaniya-Nova Reservation. *Byull. Zootekhn. Opytn. Plem. St. v Goszapov.*, Novaya Derevnya, Moscow, 2: 128–153 (in Russian).

Shalyt, M.S., 1938. The steppe vegetation of Askaniya-Nova Reservation. *Izv. Krymsk. Ped. Inst.*, 7: 45–132 (in Russian).

Shalyt, M.S., 1950. The under-ground parts of some meadow, steppe and desert plants and phytocoenoses, 1. *Tr. Bot. Inst. Akad. Nauk SSSR, Ser. 3, Geobot.*, 6: 205–442 (in Russian).

Shalyt, M.S., 1952. The under-ground parts of some meadow, steppe and desert plants and phytocoenoses, 2. *Tr. Bot. Inst. Akad. Nauk SSSR, Ser., 3, Geobot.*, 8: 71–139 (in Russian).

Shelyag-Sosonko, Yu.R., Andrienko, T.L., Osychnyuk, V.V. and Dubyna, D.V., 1985. Basic trends of anthropogenic changes of Ukrainian vegetation. *Bot. Zh.*, 70: 451–463 (in Russian).

Stepanova, E.F., 1962. *The Vegetation and Flora of Tarbagatai Mountain Range*. Kazakhsk. Akad. Nauk, Alma-Ata, 434 pp. (in Russian).

Sveshnikova, V.M., 1979. *The Dominants of Kazakhstan Steppes: Ecological–Physiological Characteristics*. Nauka, Leningrad, 192 pp. (in Russian).

Taliev, V.I., 1904. Vegetation of Cretaceous outcrops in southern Russia, 1904–1905. *Tr. Ova. Ispyt. Prir. Pri Khar'kovsk. Univ.*, 39: 81–238 (in Russian).

Tkachenko, V.S., 1984. On the nature of the meadow-steppe Mikhailovskaya Tselina Reservation and the prognosis of its development under protection. *Bot. Zh.*, 69: 448–457 (in Russian).

Tyulina, L.N., 1930. A study of long fallows in the Chapli State Reservation (formerly Askaniya-Nova). *Vist. Derzh. Stepovogo Zapovidnika "Chapli" (k. Askaniya-Nova)*, 7: 89–137 (in Russian).

Valter, G., 1975. *Vegetation of the Earth*, 3. Progress, Moscow, 428 pp. (in Russian).

Vandakurova, E.V., 1950. *Vegetation of the Kulunda Steppe*. Knizhnoe Izdatel'stvo, Novosibirsk, 130 pp. (in Russian).

Vernadskyi, V.I., 1889. The travel notes by V.I. Vernadskyi on soils of the basin of the river Chaplinki in Novomoskovsk district, Ekaterinoslav region. *Tr. Imp. Vol'nogo Economich. Ova.*, 3: 22–29 (in Russian).

Vinogradov, B.S., 1937. A study of marmot (*Marmota bobac* Müll.) as an excavator of soil. *Zh. Bio-bot. Tsiklu Vseukr. Akad. Nauk*, 3: 149–157 (in Ukrainian).

Voronov, A.G., 1950. Influence of animals on soils and vegetation in the steppe zone. In: S.A. Zernov and N.Ya. Kuznetsov (Editors), *Zhivotnyi Mir SSSR*, 3. Akad. Nauk SSSR, Moscow–Leningrad, pp. 527–538 (in Russian).

Voronov, A.G., 1964. A study of the effect of terrestrial vertebrates on the vegetation. In: E.M. Lavrenko and A.A. Korchagin (Editors), *Polevaya Geobotanika*, 3. Nauka, Moscow–Leningrad, pp. 451–500 (in Russian).

Vysotskyi, G.N., 1908. *On Conditions for Forest Growth in Samara Region: Soil–Botanical–Forestry Considerations, 1*. Tipografiya Sib. Gradonachal'stva, St. Petersburg, 235 pp. (in Russian).

Vysotskyi, G.N., 1909. *On Conditions for Forest Growth in Samara Region: Soil–Botanical–Forestry Considerations, 2*. Tipografiya Sib. Gradonachal'stva, St. Petersburg, 226 pp. (in Russian).

Vysotskyi, G.N., 1915. The Ergeny peneplains: cultural–

phytological considerations. *Tr. Byuro Po Prikl. Bot.*, 8: 1113–1436 (in Russian).

Vysotskyi, G.N., 1923 (1922–1923). On prospects of field crop cultivation and cattle breeding in the steppe region. *Tr. Byuro Po Prikl. Bot. i Selekts.*, 13: 3–20 (in Russian).

Yunatov, A.A., 1950. The main features of the vegetative cover of Mongolian People's Republic. *Tr. Mongol'skoi Komissii (Akad. Nauk SSSR)*. Moscow–Leningrad, 224 pp. (in Russian).

Yunatov, A.A., 1954. *Fodder Plants of Pastures and Hay Meadows in the Mongolian Steppes*. Akad. Nauk SSSR, Moscow–Leningrad, 352 pp. (in Russian).

Yunatov, A.A., 1974. *Desert Steppes of the Northern Gobi in Mongolian People's Republic*. Nauka, Leningrad, 132 pp. (in Russian).

Yunatov, A.A. and Lavrenko, E.M. (Editors), 1969. Biocomplex characteristics of the main coenose-forming plants in the vegetation of Central Kazakhstan. In: *Biocomplex Investigations in Central Kazakhstan, 2*. Nauka, Leningrad, 336 pp. (in Russian).

Zalesskyi, K.M., 1918. *Vegetation of Long-Fallows and Pastures of Donetsk Region*. Tipografiya A.I. Ter-Abramyana, Rostov-na-Donu, 84 pp. (in Russian).

Zimina, R.P. and Zlotin, R.I., 1980. Effect of marmots on the vegetative cover. In: R.P. Zimina and Yu.A. Isakov (Editors), *Surki. Biotsenoticheskoe i Prakticheskoe Znachenie*. Nauka, Moscow, pp. 81–97 (in Russian).

Zlotin, R.I. (Editor), 1975. *The Role of Animals in the Functioning of Ecosystems*. Nauka, Moscow, 220 pp. (in Russian).

Zlotin, R.I. and Khodasheva, K.S., 1973. Influence of animals on the autotrophic cycle of biological rotation. In: E.M. Lavrenko and T.A. Rabotnov (Editors), *Problemy Biogeotsenologii*. Nauka, Moscow, pp. 105–117 (in Russian).

Zlotin, R.I. and Khodasheva, K.S., 1974. *The Role of Animals in the Biological Rotation of Forest-Steppe Ecosystems*. Nauka, Moscow, 200 pp. (in Russian).

Chapter 3

GRASSLANDS OF CHINA

ZHU TING-CHENG

INTRODUCTION

Natural grassland is an important resource of China which provides grazing for a large proportion of the livestock population. An understanding of the structure and function of the various types of grassland that have developed in the diversified climates of this country is basic to rangeland management under the pressures imposed by an expanding livestock industry. The following characterization of grasslands is based on my own investigations in seven provinces in north-eastern and north-central China and on a survey of the relevant literature concerning the remaining parts of the country.

The natural grasslands in China are temperate in nature, and occupy 2.0 million km^2 — about 21% of the land area of the country (Zhu, 1988). They form a broad belt from the plains of the north-east to the Tibetan plateau (Xizang Gaoyuan) in the south-west, and lie within a latitudinal range from 35° to 50°N (Fig. 3.1). They merge northward with the grasslands of Mongolia and the former Soviet Union which have been discussed in Chapter 2. Westward they are bounded by desert, whereas deciduous and coniferous forests are located to the south, east and north-east. The main body of grassland is located on the plateau of Inner Mongolia (Nei Mongol Zizhiqu, Fig. 3.2). An eastward extension occupies the plain of the three north-eastern provinces (Heilongjiang, Jilin and Liaoning), and a south-western extension involves the loess plateau (Huangtu Gaoyuan) [in parts of Shanxi, Shaanxi and Gansu provinces and the Ningxia Huizu Zizhiqu (autonomous region)]. In the Tibetan plateau, zonal grassland gives way to alpine meadow. In the far north-west, isolated areas of grassland occur in the mountains. Approximately 80% of the area of the grassland zone is occupied by zonal steppe types, and 20% by azonal meadow (Zhu, 1988). Part of the area previously occupied by zonal grassland is now cultivated, particularly in the north-east, where more than half of the Nen Jiang plain is almost completely occupied by cropland and human settlements. About 1% of the total grassland zone is seeded to forage crops (Zheng et al., 1986).

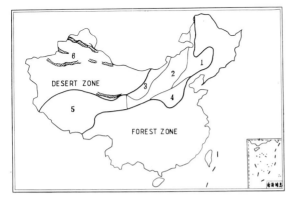

Fig. 3.1. Outline map of China (after Zhu, 1988) showing the grassland subzones: *1* = meadow steppe; *2* = typical steppe; *3* = desert steppe; *4* = shrub steppe; *5* = alpine steppe; *6* = other types.

CLIMATE

The climate of the grassland zone is continental and semiarid to subhumid in character, the value of Walter's (1979) aridity index being 1.5 to 3.5. The winter is long and severe and arid, under

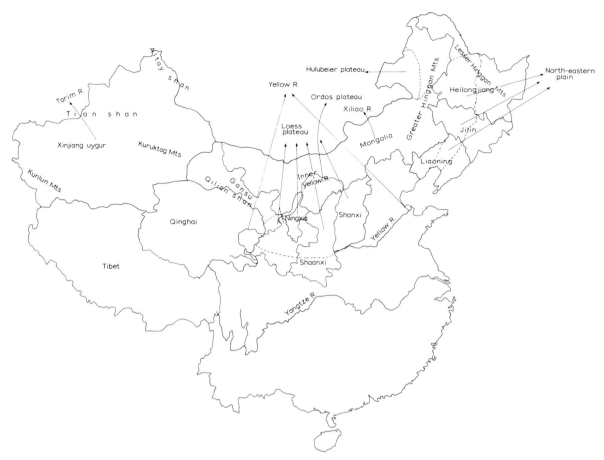

Fig. 3.2. Outline map of China showing the locations of the geographical features mentioned in the text.

the influence of a strong Mongolian atmospheric high-pressure system. Humid oceanic air currents cause the short summer to be warm, with the rainy season coinciding with the warmest season. The amount of precipitation is highly variable; summer temperatures are high and fluctuating.

Because of the location of mountain ranges between the eastern coast and the grasslands, as well as within the grassland zone, the influence of the monsoon gradually weakens from east to west, with a corresponding decrease in annual precipitation from 500 to 200 mm, and an increase in evaporative capacity from 1200 to 3000 mm (Kong, 1976). Correspondingly, the nature of the grassland changes from meadow steppe to typical steppe and then to desert steppe (Fig. 3.3; Table 3.1). Although precipitation is limited, it is concentrated in the growing season and provides excellent conditions for growth and development of plants. From 60 to 90% of the annual precipitation falls during June through September. Hence, herbs flourish during summer and autumn. On the other hand, precipitation is low during winter and spring. Rainfall is scarce during March through May, when temperature is rising rapidly and wind velocity is high. High evaporation and severe spring drought postpone growth until the rainy season begins in June. Consequently, spring ephemerals are not present in the Chinese grasslands.

Radiant energy is high, ranging from 20 000 to 22 000 kJ m^{-2} d^{-1} (Zu and Zhu, 1987) and increasing generally from north-east to south-west with decrease in latitude and reduced cloudiness (Wu, 1980; Kong, 1976). Sunshine occurs during 60 to 70% of daylight, and the number of sunshine

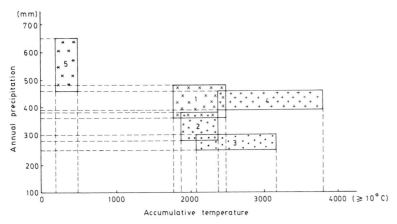

Fig. 3.3. Precipitation and temperature relationships in the various types of Chinese steppe: *1* = meadow steppe (*Stipa baicalensis* association; Zhu et al., 1986); *2* = typical steppe (*Stipa grandis* association; Li, 1962); *3* = desert steppe (*Stipa gobica* association; Wang et al., 1979); *4* = shrub steppe (*Stipa bungeana* association; Wang, 1985); *5* = alpine steppe (*Stipa purpurea* association; Xia, 1981).

hours ranges from 2700 to 3100 annually. Various factors, especially drought and low temperature, reduce the utilization rate of light energy in zonal grassland to less than 0.5%.

Annual air temperature accumulations (above 10°C) range from 500 in alpine steppe to 4000 in shrub steppe (Fig. 3.3, Table 3.1). In winter the Mongolian high-pressure system causes a flow of air from the north-west, which results in extremely low winter temperatures. Since the mountain region to the south stops the flow of cold air in that direction, cold air near the surface stagnates and causes the winter season to be very long (Numata, 1979). Mean temperature during the coldest month (January) ranges from −17° to −27°C.

The wind speed is high, particularly in spring. North-west winds prevail in winter, and south and south-east winds in summer. Average wind velocity is 3 m s^{-1}. Gales of various intensities occur on 20 days or more each year, causing sandstorms on 10 days or more as a result of erosion of the soil surface where it is exposed by heavy grazing or by cultivation. The formation of dunes is common where the soil is sandy.

SOILS

Zonal soil development near the edge of the forest in the north-eastern grasslands has resulted in the formation of a narrow belt of black soils without a lime layer, whereas to the west the zonal soil character is chernozemic, with the colour changing progressively from black to chestnut to brown (Hou and Wang, 1956). Black soils typify meadow steppe. These occur to the east of the Greater Khingan mountain range (Da Hinggan Ling), which separates the plain of the north-east from the Mongolian plateau, and also in the eastern and western foothills of these mountains. The organic content of the surface layer ranges from 3.5 to 6%, and the colour is black to grey-black. Structure is typically granular, and a well-defined lime layer is present. Chestnut soils are characteristic of typical steppe. The uppermost layer is chestnut to grey-brown in colour and contains 1.5 to 4.5% organic matter. Reaction is weakly basic and increases with depth. Brown soils occur in desert steppe, where aridity prevents the soil developing to the extent experienced to the east. As compared to chestnut soils, these are lighter in colour, more shallow, lower in organic-matter content (1.0 to 1.8%), and more basic (pH 9 to 9.5). The surface is often covered by sand, gravel or a crust similar to that found in desert.

The commonest azonal soil types within the grassland zone are meadow soils, saline–alkaline soils, and sandy soils. Due to differences in moisture content, salinity and texture, these soils provide an environment contrasting with that of zonal soil.

TABLE 3.1

A comparison of the characteristics of the zonal grassland regions

Characteristic	Meadow steppe	Typical steppe	Desert steppe	Shrub steppe	Alpine steppe
Precipitation (mm)	350–500	280–400	250–310	380–460	450–700
Accumulated temperatures,[1]	1800–2500	1900–2400	2100–3200	2400–4000	<500
Foliage cover (% of soil surface)	50–80	30–50	15–25	30–60	30–50
Species diversity,[2]	17–23	14–18	8–11	6–10	8–10
Forage yield (t ha^{-1})	1.5–2.5	0.8–1.0	0.2	0.5	0.20–0.35
Grazing capacity (sheep per hectare)	2.0–2.5	1.0–1.25	0.65	0.6	0.4

Important species (in order of decreasing importance) include:

Meadow steppe:	Steppe species:	*Stipa baicalensis, Filifolium sibiricum, Leymus chinensis*
	Meadow species:	*Calamagrostis epigeios, Lathyrus quinquenervius, Hemerocallis minor*
Typical steppe:	Grasses:	*Stipa grandis, S. krylovii, S. breviflora*
	Legumes:	*Caragana microphylla, Astragalus melilotoides, Trigonella ruthenica*
Desert steppe:	Steppe species:	*Stipa glareosa, S. gobica, S. klemenzii*
	Desert species:	*Reaumuria songorica, Calligonum mongolicum, Salsola passerina*
Shrub steppe:	Steppe species:	*Stipa bungeana, Andropogon ischaemum, Themeda triandra*
	Shrubs:	*Vitex negundo, Ziziphus spinosa*
Alpine steppe:	Steppe species:	*Stipa purpurea, Festuca ovina, Poa alpina*
	Cushion plants:	*Androsace tapete, Arenaria musciformis, Thylacospermum caespitosum*

[1] Accumulated temperatures = annual sum of daily mean temperatures above 10°C.
[2] Number in a square metre.

ZONAL GRASSLAND

Lavrenko (1940, 1954, 1973) classified the Eurasian steppe into five zonal types, of which four occur in China — typical steppe, meadow steppe, desert steppe and shrub steppe. In classifying grassland vegetation, one must consider the ecological and biological features of the dominant species, as well as the structure of the vegetation as revealed by density, species diversity, and layering. Variations in these factors reflect differences in zonal climate, as well as local microclimate. This is the basis applied in the following discussion, in which the zonal grasslands of China are divided (albeit somewhat arbitrarily) into five kinds of steppe. A comparison of some of their characteristics is presented in Table 3.1.

The dominants of these grasslands are perennial herbaceous xerophilous species, usually grasses of caespitose or rhizomatous habit, but occasionally including forbs. Many species adapt to adverse growing conditions in a temperate climate by the living parts surviving under ground during winter. The species composing these temperate grasslands are meso-microtherms which are resistant to low temperatures. They are also xerophytic in character, both morphologically and physiologically. These meso-microthermal and xerophytic characteristics of the species identify a zonal vegetation type limited in distribution to a portion of the subhumid to semiarid region of the temperate zone. Where the natural plant cover has been severely disrupted by mismanagement, desertification takes place and desert species invade.

Perennial grasses flourish better in grassland than do forbs for several reasons. They are more resistant to defoliation and trampling by livestock. The density and orientation of leaves permit a high level of interception of sunlight. Grasses also appear to be more able than forbs to endure violent changes in temperature. Fire, which is of frequent occurrence in grassland, is less destructive to grasses (Heady, 1975). Perennial grasses respond better to high levels of soil nitrogen than do perennial forbs (unpublished research by the author).

Meadow steppe

Meadow steppe (Fig. 3.4) is humid grassland dominated by mesoxerophytes with some associated mesophytes (Zhu, 1963; Li, 1986; Zhu et al., 1986). This type is mostly located within the plain of the north-east at an elevation of 120 to 250 m above sea level (Fig. 3.1). However, it extends westward to contact typical steppe in the eastern part of the inner Mongolian plateau. Annual precipitation varies from 350 to 500 mm, annual air temperature accumulations (above 10°C) range from 1800 to 2500, and Walter's aridity index ranges from 0.7 to 1.2 (Fig. 3.3; Table 3.1). The topography is generally fairly flat. The soil is meadow chernozem or dark chestnut. Where drainage is restricted, a complex of sodium–saline soil exists.

Bunch grasses grow densely to a height of 60 cm, so that foliage covers 50 to 80% of the soil surface. Diversity ranges from 17 to 23 species in a square metre, the forb content being very high. Herbage yield is from 150 to 250 g m^{-2} of dry matter, and stocking density is 2 to 2.5 sheep per hectare (Li and Zhu, 1983).

The commonest dominants are *Stipa baicalensis* and *Filifolium* (= *Tanacetum*) *sibiricum*; two rhizomatous species, *Leymus chinensis* (= *Aneurolepidium chinense*) and *Arundinella hirta*, are subdominant (Zhu et al., 1981; Zhu Ting-cheng, 1983). Because of minor fluctuations in topography, meadow steppe often gives way to intermingling patches of components of halophytic meadow, particularly *Polygonum sibiricum*, *Puccinellia tenuiflora* and *Suaeda corniculata*. Common species include the grass *Spodiopogon sibi-*

Fig. 3.4. Meadow steppe in north-eastern China, dominated by *Leymus chinensis*. (Photo by Xu Zong-yao.)

ricus and the legumes *Astragalus adsurgens, Lespedeza daurica, Oxytropis myriophylla* and *Trigonella* (= *Medicago*) *ruthenica*. Forbs of other families are also numerous, the commonest being *Adenophora stenophylla, Cymbaria daurica, Ligularia mongolica, Rhaponticum uniflorum* and *Thalictrum simplex*. Their flowers beautify the landscape.

Several meadow species are common in this steppe (Zhu, 1955, 1958). These include:

Cacalia aconitifolia	*Lathyrus quinquenervius*
Calamagrostis epigeios	*Sanguisorba officinalis*
Carex pediformis	*Stipa effusa*
Delphinium glandiflorum	*Stipa sibirica*
Hemerocallis minor	

Along the eastern boundary of the northeastern plain, a gradual transition occurs to steppe meadow, which, in turn gives way to deciduous forest through an ecotone of forest-steppe. Here islands of sparse woodland occur in valleys and on the lower slopes of the mountains. These are composed of xero-mesophytic broad-leaved trees — such as *Quercus mongolica* and small-leaved species of *Ulmus*. *Ulmus* spp. are also scattered in parts of the meadow steppe.

Typical (dry) steppe

This is a drier type of steppe (Fig. 3.5), consisting of xerophilous herbs, but dwarf half-shrubs and meso-xerophytic herbs also are well represented. It is located centrally within the grassland zone, since it occupies the north-eastern and central parts of the Inner Mongolian plateau and extends southward into the loess plateau (Fig. 3.1) (Li, 1962). Elevation ranges from 800 to 1400 m. Annual precipitation is between 300 and 400 mm, annual air temperature accumulations (above 10°C) range from 1900 to 2400, and the aridity index from 1.2 to 2.0 (Fig. 3.3; Table 3.1). The soil is chestnut in colour.

Fig. 3.5. Typical steppe dominated by *Stipa grandis*, representative of the vegetation of eastern and central Inner Mongolia. The abundant dicotyledon in the foreground is *Artemisia frigida*.

The vegetation is uniformly low in stature, and treeless throughout. The grasses reach 30 to 50 cm in height and foliage covers 30 to 50% of the soil surface. Diversity ranges from 14 to 18 species in a square metre. Herbage yield is about 80 to 100 g m^{-2}, and density of stocking is 1 to 1.25 sheep per hectare. The dominants are xerophytic caespitose species of *Stipa*, which replace one another from region to region. *Stipa grandis* is typical in the Hulunbeier plateau in north-eastern Inner Mongolia along the border with Mongolia, *S. krylovii* in valleys (e.g. of the Xilin Gol river), and *S. breviflora* of the Ordos plateau (the northern part of the loess plateau) (Jiang and Chen, 1985).

Three xerophytic caespitose grasses, *Cleistogenes squarrosa*, *Koeleria gracilis* and *Poa sphondylodes*, are common associates with species of *Stipa*. *Artemisia frigida* is the most abundant dwarf half-shrub. Typical meso-xerophytes include *Agropyron cristatum* and *Leymus chinensis*. Forbs are comparatively scarce; the commonest ones include *Allium mongolicum*, *Aster altaicus*, *Bupleurum scorzonerifolium*, *Filifolium sibiricum*, *Kochia prostrata* and *Potentilla chinensis*. Common legumes include *Astragalus adsurgens*, *A. melilotoides*, *Lespedeza daurica*, *L. hedysaroides*, *Oxytropis myriophylla* and *Trigonella ruthenica*. Where soils are well drained, *Caragana microphylla*, *C. pygmaea* and other dwarf and medium-sized shrubs are scattered throughout the community. This shrubby component of typical steppe is in contrast to the completely herbaceous nature of the ground cover under the more favourable climate of meadow steppe. Similarly, the shrubby character of this semiarid steppe increases westward towards the transition to desert steppe.

Desert steppe

Desert steppe is the most arid type of grassland (Fig. 3.6). The dominants are ultra-xerophytic herbs, and ultra-xerophytic dwarf half-shrubs are common associates. These are the most drought-resistant among grassland species. Desert steppe occupies a belt across Inner Mongolia west of the typical steppe and extends into Ningxia Huizu Zizhiqu and the eastern part of Gansu province (the northern and north-western parts of the loess plateau) (Fig. 3.1) (Zhang Zu-tong, 1982). Annual precipitation ranges from 250 to 310 mm and annual air temperature accumulations (above 10°C) from 2100 to 3200. Snow does not accumulate in winter and the aridity index is between 2.0 and 4.0 (Fig. 3.3; Table 3.1). Soils vary from dark brown to sierozem.

Since this type is transitional to desert, the vegetative cover is low in stature, sparse and of low diversity. The height of the canopy usually does not exceed 20 to 30 cm, foliage cover does not exceed 30%, and diversity is only 8 to 11 species in a square metre. Conspicuous xerophytic characteristics include narrow leaves and extensive development of roots, both vertically and laterally. Herbage yield, which is dependent on moisture supply, is highly variable from year to year, but is always less than 100 g m^{-2}. Stocking density is 0.65 sheep per hectare (Zhao, 1984).

The main components of the vegetation are species of *Stipa* and *Artemisia*. The dominant grasses, *Stipa breviflora*, *S. glareosa*, *S. gobica* and *S. klemenzii*, are densely caespitose in character. Associated shrubs include *Artemisia xerophytica*, *Caragana microphylla* and *C. stenophylla*, as well as some annual forbs (such as *Artemisia pectinata*). In addition, the following xeromorphic shrubs (belonging to the Asteraceae, Chenopodiaccac, Tamaricaceae and Zygophyllaceae) compose a complex layer within the community:

Ajania achilleoides   *Olgaea leucophylla*
Anabasis brevifolia   *Reaumuria songorica*
Calligonum mongolicum   *Salsola passerina*
Eurotia ceratoides

Allium mongolicum and *A. polyrrhizum* also are common.

Shrub steppe

Shrub steppe, as identified here, is different from the shrub steppe of the former Soviet Union and Mongolia, as described by Lavrenko (1954). It is characterized by a grass component of caespitose species and is fragmented in its distribution by its association with gravelly soil. I agree with several other authors (Brown, 1954; Karamysheva, 1961; Zhu Zhi-cheng, 1983; Wang, 1985) that shrub steppe is a separate zonal type, in which secondary thermophilous shrubs occur with the dominant perennial xerophilous and broad-leaved mesophytic herbs.

Fig. 3.6. Desert steppe dominated by *Stipa gobica*, representative of the vegetation of western Inner Mongolia. The prominent shrub is *Caragana microphylla*, which is usually associated with *S. gobica*.

Shrub steppe is transitional between grassland and shrub forest (Fig. 3.7). It is unlike shrub forest in having only islands of shrubs among the dominant herbs rather than shrubs forming a closed canopy. This type is located in the warm-temperate zone, occupying the central and western parts of the loess plateau at elevations between 1000 and 1500 m (Fig. 3.1). Annual precipitation ranges from 380 to 460 mm and annual air temperature accumulations (above 10°C) from 2400 to 4000 (Fig. 3.3; Table 3.1). The soil is dark loess, with a surface dissected by gullies and ravines. The vegetation is low-growing and scattered, with a diversity of 6 to 10 species in a square metre, and the foliage cover averages 30 to 60%. Herbage yield is about 50 g m^{-2}, and density of stocking is 0.6 sheep per hectare (Zhu, 1988). In some places, this vegetation type results from destruction of the original deciduous forest.

The commonest dominant species are *Aneurolepidium* (= *Leymus*) *dasystachys*, *Artemisia giraldii*, *A. gmelinii*, *Peganum harmala*, *Stipa bungeana*, and *Thymus serphyllum*. An important feature of shrub steppe is that mesophytic shrubs adapted to the warmer conditions grow within the herb layer. The families most commonly occurring are Betulaceae, Caprifoliaceae, Oleaceae, Rhamnaceae, Rosaceae and Thymelaceae, species of which seldom occur in other types of steppe. The following thorn-bushes are common:

Armeniaca vulgaris *Rosa hugonis*
Caragana korshinskii *Rosa xanthina*
Hippophae rhamnoides *Syringa oblata*
Lonicera maackii *Vitex negundo*
Ostryopsis davidiana *Wickstroemia chamaedaphne*
Prinsepia uniflora *Ziziphus spinosa*

Fig. 3.7. Shrub steppe occupied by *Stipa bungeana*, and the shrubs *Armeniaca sibirica* and *Vitex negundo*.

Common mesophytic broad-leaved thermophilous grasses include *Andropogon ischaemum*, *Arundinella hirta* and *Themeda japonica*.

Shrub steppe was the first steppic type in China to be converted to cropland (Zou, 1981). During the millennia since farming began, this area in the middle of the Yellow river (Huang He) catchment has been subjected to marked changes. Increased aridity, caused by removal of the forest and other forms of human pressure, impedes succession back towards the original vegetative cover, even where the soil has not been cultivated. Conclusions as to the nature and extent of this type of steppe are based on inference from the present plant cover in refuges of natural vegetation occupying 2 to 3% of the area. These have survived on steep slopes, along ditches, in graveyards, and near temples. Interpretation from these relicts is difficult because of invasions of weeds, such as:

Artemisia capillaris	*Ixeris denticulata*
Chloris virgata	*Panicum miliaceum* var. *ruderale*
Eragrostis pilosa	*Salsola collina*
Ixeris chinensis	*Setaria viridis*

Opinions vary as to whether the original vegetation of this part of the loess plateau should be considered as grassland, and confusion exists concerning the extent to which the secondary communities are indicative of the nature of the original vegetation.

Alpine steppe

Alpine steppe is dominated by xeric perennial microtherms mixed with some alpine cushion species. This type covers extensive areas of the Tibetan plateau of south-western China (Figs. 3.1 and 3.8). Annual precipitation ranges from 450 to 700 mm and annual air temperature accumulations (above 10°C) are from 200 to 500. This is the largest, highest and youngest plateau on earth. Because of low temperatures and shallow soil (often with a permafrost layer), the grass canopy is no taller than 20 to 30 cm. Root systems are interwoven to form a tough turf with a foliage cover of about 50% (Deng, 1983). Species diversity is 8 to 10 in a square metre (Chang, 1983). Herbage yield is about 20 to 35 g m^{-2}, and the stocking density is 0.4 sheep per hectare. The dominant species are *Festuca ovina*, *Poa alpina*, *Stipa purpurea* and such cushion species as *Androsace tapete*, *Arenaria musciformis* and *Thylacospermum caespito-*

Fig. 3.8. Yaks grazing in alpine steppe in the Tibetan plateau, with the Himalaya mountains in the background.

sum, together with some soft-stemmed species of *Artemisia* (Ren, 1984). This vegetation supports yaks (*Bos grunniens*) and Tibetan sheep (a subspecies of *Ovis aries*).

The vegetation of the Tibetan plateau is described by Walter and Box (1983a, pp. 232–235) elsewhere in this book series.

Other types of steppe

Steppe vegetation also occurs in mountains within the desert region of the north-west, contacting desert at lower elevations (900 m) and extending as high as 2000 m (Wang, 1963) (Figs. 3.1 and 3.9). The nature of this grassland is variable, with some of the features of typical steppe and others of desert steppe. It is distributed over broad basins, on plateaus, and on mountain slopes. Variations in slope aspect and gradient cause great complexity. On some mountain slopes, steppe occurs as a narrow belt between desert at lower altitudes and forest at higher altitudes — within an altitudinal range of a few hundred metres. Thus, the species of various vegetation types have invaded the grassland in these situations.

The belts of steppe are mainly oriented in an east–west direction. Altitudinal limits gradually ascend with decreasing latitude and are higher on south-facing than on north-facing slopes. In the far north in the Altai mountains, steppe begins to appear at the 900 m level. Southward, on the northern slopes of the Tien Shan mountains, the lower margin of steppe is at 1000 m, whereas on the southern slopes it contacts desert at 1200 to 1600 m. Still farther southward, on the northern slopes of the Kunlun Shan, desert does not yield to steppe until the altitude reaches 1700 m.

Annual precipitation ranges from 300 to 500 mm, but the vigour of the vegetation is less than expected from this amount of moisture, probably because of the drying influence of the surrounding desert. The height of the canopy is from 20 to 30 cm, and foliage cover is from 20 to 40%.

The important dominants are species of *Festuca* (*F. ovina* and *F. sulcata*) and of *Stipa* (*S. capillata*, *S. kirghisorum*, *S. laxiflora* and *S. orientalis*); species of *Artemisia* (*A. borotalensis*, *A. gracilescens* and *A. transiliensis*) are common associates. Many others make important contributions to the vegetative cover in various regions. In

Fig. 3.9. Steppe in the mountains of Xinjiang Uygur autonomous region, dominated by species of *Stipa* and *Festuca*.

one part of Xinjiang Uygur Zizhiqu (autonomous region), *Andropogon ischaemum*, usually occurring in the warm-temperate zone, is dominant. At the foot of the Tien Shan, *Agropyron cristatum*, *Artemisia frigida*, *Koeleria gracilis* and *Stipa capillata* form a Mongolian–Gobi type of alpine steppe. In the Qilian Shan mountains and the Qinghai plateau, *Stipa aliena* and *S. purpurea* dominate in an alpine steppe which includes a complex of cryptophytes and cushion plants. *Koeleria gracilis* is also common.

A difference of opinion exists regarding the classification of this vegetation (Zhu, 1963). One argument is that it should be divided into subtypes corresponding to zonal steppic types to which the variants of these mountain steppes bear some similarity. However, I suggest that environments of mountain ranges are sufficiently distinct to justify the recognition of a separate type of grassland. The restriction of this vegetation to a part of the country in which the other types of grassland do not occur provides further support for this view. Features of these grasslands which distinguish it from other types relate to difference in the appearance of the vegetation and its floristic composition, atmospheric and edaphic factors, and uses and management. The cold mountain climate favours the occurrence of cryophytes and causes the vegetation to be of low stature and low herbage yield. Certain species (e.g. several species of *Ptilagrostis*) are limited in their distribution to this vegetation type. Because of the preponderance of gravel in mountain soils, lithophytic communities are common. Thus, xerophilous shrubs, especially species of *Cotoneaster*, *Rosa* and *Spiraea* (all Rosaceae) and *Caragana* are more abundant than in non-mountainous grassland. Unlike the continuous grazing practised in other grassland regions, livestock are moved from one elevation level to another with the change of season.

Altitudinal zonation of vegetation in this region is discussed in more detail elsewhere in this book series (Walter and Box, 1983b, pp. 164–179).

TABLE 3.2

A comparison of the density of the canopy and yield of dry matter in four types of meadows in northern China

Vegetation type	Foliage cover (%)	Hay yield (t ha^{-1})	Dominant species
Typical meadow	80–95	2.3–3.7	*Bromus inermis, Sanguisorba officinalis, Vicia amoena*
Marshy meadow	70–90	1.3–2.5	*Carex schmidtii, Kobresia tibetica, Sanguisorba officinalis*
Halophytic meadow	50–80	1.0–2.0	*Achnatherum splendens, Puccinellia tenuiflora, Suaeda corniculata*
Alpine meadow	60–90	0.1–0.3	*Carex moorcroftii, Kobresia pygmaea, Polygonum viviparum*

AZONAL VEGETATION

Meadows are floristically complex, including a wide variety of life forms. The dominants alone include more than 70 species of grasses, sedges, rushes and forbs. The canopy is dense (Table 3.2). Soils are relatively deep and fertile. A distinctive feature of meadow soils is the abundance of anaerobic Bacteria, which play a major role in nutrient cycling. Four types of meadow are recognized.

Typical meadow

The vegetation of a typical meadow is composed of mesophytes well adapted to moderate temperatures and moisture levels. Soils are well drained, black in colour, and with a high content of organic matter. The most prominent species are mesophytic forbs (Fig. 3.10). The luxuriance of these meadows has resulted in their designation as "multi-flower meadows" (*bai-hua cao tang*). They are characteristic mainly of the forest zone in the north-eastern provinces and of mountains in eastern Inner Mongolia (Chang, 1955). Tall grasses form a dense cover, with leaves reaching 40 to 50 cm in height and culms about 1 m. Herbage cover is often as high as 90 to 100%. Herbage yield is high and is relatively stable from year to year. Diversity ranges from 20 to 25 species in a square metre. The dominants are xero-mesophytes. The soil is of meadowy black earth.

A representative community of this type is the *Leymus chinensis*–forb meadow. Leaf height is about 50 cm, foliage cover reaches 70 to 90%, and diversity is 15 to 20 species in a square metre. Forbs associated with the dominant *Leymus chinensis* include *Achyrophorus ciliatus*, *Galium verum*, *Lathyrus quinquenervius*, *Thalictrum simplex* and *Trifolium lupinaster*. This is more suitable for haying than any other natural grassland. About 60% of the herbage is good-quality forage. Other important species include *Cacalia aconitifolia*, *Calamagrostis epigeios*, *Delphinium glandiflorum*, *Lespedeza daurica* and *L. hedysaroides*.

The dominants of the upper layer are members of the Liliaceae (*Hemerocallis minor*, *Lilium amabile* and *Veratrum nigrum*). Other common species in this layer include *Bromus inermis*, *Iris kaempferi*, *Patrinia scabiosaefolia*, *Sanguisorba officinalis* and *Vicia amoena*. Dense tufts of *Carex pediformis* dominate the lower layer.

Marshy meadow

Marshy meadows are vegetated by a few species of hydrophytes, mostly of the Cyperaceae. The habitat consists of low-lying lands with restricted drainage and anaerobic soil conditions. Permafrost is often present. Under these conditions decomposition rates are low and semi-peaty humus is formed.

Extensive areas of marshy meadow, dominated by *Carex schmidtii* and *Sanguisorba parviflora* (Table 3.2), occur in low-lying areas along river valleys in the Greater Khingan, Lesser Khingan (Xiao Hinggan Ling) and Changbai Shan mountains of the north-east. In permafrost conditions and on glei soil adjacent to lakes at the front of alpine glaciers in the Qinghai–Tibetan plateau, the dominants are species of *Kobresia*, which provide herbage of high palatability.

Fig. 3.10. Typical meadow vegetation in the plateau of Inner Mongolia, demonstrating the abundance of dicotyledons (species of *Artemisia*, *Inula*, *Ligularia*, *Rumex*, *Sanguisorba* and *Verbascum*. (Photo by Wang Yi-fung.)

Halophytic (saline) meadow

Halophytic meadows are composed of halo-mesophytic herbs that grow in low-lying areas and along sea-coasts where salts have accumulated in the soil at various concentrations (Fig. 3.11; Table 3.2). Species occupying these habitats have acquired various types of adaptation permitting them to thrive where there is difficulty in extraction of water from the saline soil (Hou, 1983). A means of coping with top-soil of high salt content overlying a less saline subsurface is exemplified by the deep root systems of *Achnatherum splendens* and *Glycyrrhiza uralensis*. Succulent tissue is an adaptive feature of *Kalidium foliatum*, *Polygonum sibiricum* and species of *Suaeda*. Excessive ion concentration in plants is avoided by excretion in *Limonium bicolor* and *Tamarix chinensis*. Among graminoids, *Puccinellia tenuiflora* is subdominant and *Carex schmidtii* is common.

One representative community of this type of meadow, widely distributed throughout the steppe and desert zones of Eurasia, is dominated by *Achnatherum splendens*, a large tussock grass (Fig. 3.11). This community occurs also throughout much of the Chinese steppe in the valleys of dried-up rivers, in old river courses and beside lakes. The dominant species forms a mound-shaped structure 1.5 to 2 m in diameter and 1 to 2 m in height. Forage yield is similar to that of steppe meadow. These meadows provide grazing for all classes of ungulates in spring, and horses use them to some extent in winter.

Alpine meadow

The vegetation of alpine meadow consists of mesophytes which tolerate the moist, cold habi-

Fig. 3.11. Halophytic meadow in which *Achnatherum splendens*, a large caespitose grass, is prominent. The principal species between the bunches are *Artemisia anethifolia* and *Iris ensata*.

tat prevailing at high elevations (Fig. 3.12; Chang, 1983). This type of meadow is typical of the eastern part of the vast south-western plateau from eastern Qinghai province westward across Tibet (Dall and McKell, 1986), being distributed at elevations ranging from 3000 to 5300 m. Elevational limits increase to the south and west. A mixture of shrubs is common near the lower margin of distribution, whereas at the upper margin meadow gives way to sparse cushion-like vegetation or to bare rocky slopes. The climate is characterized by high wind, intense radiation and low temperature, with ground frost occurring even in midsummer. Permafrost persists in the thin soil, so root systems are shallow and distorted (Chang, 1981). A low, carpet-like sod is formed by densely growing, short, rhizomatous species of *Kobresia*, particularly *K. capillifolia*, *K. graminifolia*, *K. humilis* and *K. pygmaea* (Table 3.2) (Wang and Li, 1982). Common companion species include *Gentiana algida*, *Polygonum sphaerostachyum*, *P. viviparum*, *Saussurea superba* and *Stipa aliena* (Xia, 1981). These low-growing species are highly pubescent and are able to propagate in the adverse environment. *Kobresia pygmaea* is the commonest dominant in alpine meadow; it is nutritious and suited to grazing by yaks and sheep.

NATIVE FAUNA

Many of the animal species in the natural temperate grasslands in China are representative of faunal populations of Central Asia. Some species of the temperate-forest fauna mix into the eastern grasslands along the forest margin, whereas desert species invade from the west. The species composition of the grassland fauna is much simpler than

Fig. 3.12. Brown-headed gull (*Larus brunnicephalus*) in alpine meadow. This bird nests in pits on lake-shores in the western parts of Xinjiang Uygur autonomous region and Qinghai province and in the Tibetan plateau.

that of forest. Rodents are abundant in grassland, usually being aggregated into colonies. Their running and burrowing characteristics protect them against carnivorous mammals and birds of prey. Few species of ungulates and resident birds occur. Amphibians and reptiles are also rare. Some species of birds adapt to the adverse grassland environment by nesting in rodent burrows.

A short growing season, long winter, and spring drought over much of the grassland affect hibernation and food-preservation habits and cause reproduction periods to be short. The climate becomes more and more adverse westward, to animals as well as plants (Jiang and Chen, 1985). Extreme diurnal and annual fluctuations in temperature and erratic occurrences of precipitation interfere with animal survival and result in marked fluctuations in populations, especially of rodents. Restricted plant growth often causes ungulates to migrate long distances in search of herbage.

Ungulates

The only ungulates commonly surviving are *Procarpa guffurosa* (Mongolian gazelle) and *Gazella subgutturosa* (goitred gazelle). The area of distribution of the former corresponds closely to the boundaries of typical steppe. Although it is still common, sizeable herds are now rare. *G. subgutturosa* is widespread in the desert steppe of Central Asia, and occurs occasionally in the western part of the typical steppe. Both species migrate in winter towards the south-west in search of grasslands not covered deeply by snow.

Equus przewalskii (Mongolian wild horse) and *Camelus bactrianus* (Bactrian camel) were once distributed throughout the steppe zone, but are now near extinction. *Equus* survives only in remote parts of the desert steppe, where it feeds on narrow-leaved grasses in summer and digs out their withered herbage from under snow in winter. Sometimes the animals aggregate into herds. *Camelus* survives only in small inter-mountain grasslands in the Kuruktag mountains along the lower part of the Tarim river and in the Altai mountains (Zhen, 1959). This species grazes on grasses and forbs, particularly on species of *Allium* and *Salsola*. The animals band into small groups and migrate over distances as great as 600 km.

Most of the species in alpine steppe and alpine meadow of the Tibetan plateau are specific to that region. *Bos grunniens mutus* (wild yak), *Pantholops hodgsoni* (Tibetan antelope) and *Equus hemionus* (wild Tibetan donkey) are common. Herds of a hundred or more are seen occasionally in the central part of the plateau, but these species are rare towards its periphery.

Rodents

The predominant rodents are *Marmota* spp. and genera of the Microtinae, Myospalacinae and Ochotonidae. Regional differences in dominant species are pronounced.

Microtus brandtii (a vole) occurs throughout the meadow and typical steppes. This species is characterized by colonial behaviour, holes in deteriorated grasslands sometimes numbering 4000 to 8000 ha^{-1} (Zhang, 1979). These dense populations compound the disturbance caused by overgrazing of livestock to the point where the existence of the rodents is threatened by severe adversities in weather. Reduced food supply and disease, which follow abnormalities in weather, result in marked reductions in population and cause long-distance migrations. *M. gregalis* (another vole) resides mainly in the moister habitat of meadow steppe, its range overlapping with that of *M. brandtii*. Recent investigations have revealed that deterioration of grasslands through the processes of desertification and salinization provides conditions favouring the increase of *M. brandtii* to the exclusion of *M. gregalis* from its natural haunt.

Myospalax aspalax (zokor), *Citellus dauricus* (ground squirrel) and *Marmota bobak* (marmot) occur in various grasslands, but also favour deteriorated locations. The first species is subterranean in habit, whereas the other two inhabit the uppermost layer of soil. *Marmota bobak* is particularly common in the lower steppe belt of the Tien Shan of the north-west and on gentle slopes of the Greater Khingan mountains of Inner Mongolia. In recent years this species has invaded cut-over woodlands. *Phodopus sungorus* (striped hairy-footed hamster) is typical of dry steppe. Fluctuation in its populations is less pronounced than in the two species of *Microtus* already discussed. *Meriones unguiculatus* (clawed gerbil) and *Phodopus roborouskii* (desert hamster) occur in desert steppe. The first species is also abundant in desertified and salinized portions of the western part of typical steppe and becomes dominant in croplands, where it severely damages crops. *Dipus sagitta* (northern three-toed jerboa) and *Allactaga sibirica* (Mongolian five-toed jerboa), which are common in the arid areas of Central Asia, are not numerous in the steppe zone of China. *Tamias sibiricus* (Siberian chipmunk) occurs in shrubby grasslands, but is not abundant. Common rodents in the Tibetan plateau are: *Microtus oeconomus* (vole), *Myospalax fontanierii* (common Chinese zokor), *Ochotona curzoniae* (mousehare) and *O. rutila* (Turkestan red pika). These are usually light yellow-brown in colour without bright spots and are well adapted to arid environments.

Among the above species, only *Citellus dauricus* and *Marmota bobak* hibernate. The others collect herbage for consumption during winter.

Rodents contribute significantly to degeneration of grasslands in China. Their burrowing activities result in soil deposits on the soil surface in which successional plant communities begin to develop during periods of low rodent populations (see also discussion of their activities in the former Soviet Union to the north and west, pp. 42–55). However, these are soon destroyed as populations again increase. Thus, rodents not only play a part in degrading grasslands, but in interfering with regeneration.

Other groups

The common carnivorous vertebrates are *Mustela altaica* (weasel), *M. putorius* (European polecat), *M. sibirica* (Siberian weasel), *Vormela peregusna* (marbled polecat) and *Vulpes corsac* (corsax fox). These species have a role in controlling rodent populations. However, recent rodent-control programs, involving the application of poison baits, have reduced the populations of carnivores through reduction of their food supply

(Qian, 1980). Evidence of this trend was apparent by a sharp reduction in the number of pelts of *Vulpes* marketed during the year following the initiation of large-scale rodent-control measures.

The variety and numbers of birds in the grasslands are not impressive. The commonest widely distributed resident species are:

Alauda arvensis	sky lark
Eremophila alpestris	horned lark
Melanocorypha mongolica	Mongolian lark
Oenanthe isabellina	isabelline wheatear
Oenanthe oenanthe	wheatear

Migrants are particularly common in the eastern steppe. These include *Apus pacificus* (northern white-rumped snipe), *Emberiza aureola* (yellow-breasted bunting) and *E. spodocephala* (grey-headed black-faced bunting). *Calandrella rufescens* (lesser sand lark) is the most abundant resident species in desertified steppe. *Otis tarda* (great bustard) and *Syrrhaptes paradoxus* (sand grouse) are commonly hunted. They are omnivorous and are speedy runners. *Tadorna tadorna* (shel-duck) makes use of the holes of *Marmota bobak* in which to lay eggs. Birds that prey on rodents include:

Accipiter nisus	sparrow-hawk
Accipiter gentilis	gros-hawk
Aquila chrysaetos	golden eagle
Buteo hemilasius	upland buzzard
Milvus migrans	black kite

The principal birds of the alpine meadow of the Tibetan plateau are: *Columba leuconta* (snow pigeon), *Montifringilla* sp. (finch) and *Tetraogallus tibetanus tibetanus* (Tibetan snowcock); birds of smaller size include (Zhang Xiao'ai, 1982):

Alauda gulgula	small sky lark
Eremophila alpestris	horned lark
Motacilla citreola	grey wagtail
Passer montanus	tree sparrow
Tringa totanus	redshank

The common birds of the Tien Shan of the north-west are: *Emberiza cia* (rock bunting), *Montifringilla nivalis* (snow finch) and *Serinus pusillus* (gold-footed serin).

Many species of waterfowl occur within the steppe zone. In summer, large flocks of *Fulica atra* (coot) and *Larus brunnicephalus* (brown-headed gull) are common in and near bodies of water (Fig. 3.12). Summer residents of water bodies also include (Zhen, 1959):

Acrocephalus arundinaceus	great reed warbler
Anas clypeata	northern shoveller
Anas formosa	Baikal teal
Anthus novaeseelandiae	pipit
Ardea cinerea	grey heron
Botaurus stellaris	Eurasian bittern
Cygnus olor	mute swan
Pelecanus philippensis	spot-billed pelican
Phalacrocorax carbo	great cormorant
Podiceps cristatus	great crested grebe
Vanellus vanellus	lapwing

Among reptiles, the lizards *Eremias argus* and *Phrynocephalus frontalis* are common. The commonest snake is *Elaphe dione*. This species is widely distributed, whereas *Coluber spinalis* is common only in the northern part of the steppe region. Amphibians only occur in the moister portions of the grassland zone, *Bufo raddei* being most abundant and *Rana temporaria asiatica* ranking second.

Insects, especially grasshoppers, are the principal herb-eating consumers of the grasslands (Ma, 1959). The species of grasshoppers present vary with the vegetation type. The common species in meadow steppe are *Epacromia coerulipes* and *Oedaleus infernalis infernalis*, whereas *Locusta migratoria migratoria* is abundant in saline areas. In typical steppe and shrub steppe the characteristic species are *Chorthippus brunneus*, *C. dorsatus*, *C. dubius* and *C. hammarstroemi hammarstroemi*. These are distributed widely in communities of *Artemisia frigida* and *Stipa grandis* (Li et al., 1983). In typical steppe the representative species are:

Angaracris barabensis	*Bryodema tuberculatum dilutum*
Bryodema holdereri holdereri	*Bryodema zaisanicum fallax*
Bryodema luctuosum luctuosum	

Because of the absence of obstacles to migration, several insect species representative of the Central-Asian fauna intermingle in Chinese steppe with Chinese species. These include grasshoppers — *Calliptamus abbreviatus* in typical steppe, *C. barbarus cephalotes* and *C. italicus italicus* of desert steppe, and *Oedaleus asiaticus*, which is distributed throughout desert steppe. In the alpine steppe and alpine meadow of the Tibetan plateau, the following grasshoppers are predominant: *Chorthippus chinensis*, *C. dubius*, *C. fallax* and *Myrmeleotettix palpalis* (Wu and Chin, 1982). When these are plentiful, they occur in swarms with as many as 126 individuals m^{-2} (Cheng and Han, 1974). In such instances forage deficiencies have resulted in the death of livestock. This problem has been reduced considerably in recent years by the application of insecticides from aircraft.

COMPARISON WITH OTHER EURASIAN GRASSLANDS

Certain features of the Chinese grasslands contrast with those to the north and west. Monsoonal influences cause the steppe to occur at a lower latitude in China than farther west. The predominance of caespitose and rhizomatous grasses causes the Chinese steppe to be of low basal cover. The vegetation is less luxuriant because forbs are relatively deficient; they are less able to compete with grasses than in grasslands farther west (Zhu Ting-cheng, 1983). Bryophytes and lichens are rare.

The Chinese steppe differs in structure from that of the former Soviet Union because much of the former is located at a higher elevation, on a plateau rather than in a plain. In addition, the air is drier, and a smaller proportion of the area is affected by the accumulation of salts. Height and density of the canopy and species diversity decrease southwards from Siberia through Mongolia into the Chinese steppe. Another difference is the absence of synusiae of ephemeral species to constitute a vernal aspect, the result of shorter, drier and windier weather in spring. The only species contributing commonly to the spring aspect in most of the steppe are those of *Iris*, *Potentilla* and *Pulsatilla*. The Chinese steppe does not exhibit dormancy in summer, as does that farther to the north. This is because the rainy season, although starting later, extends into autumn. Vigorous growth is concentrated into the period from June to August. The second period of flowering, which is typical of the more northerly grasslands, with their dry late summer and early autumn frosts, does not occur in the Chinese steppe. Another contrasting feature is the abundance of psammophytes in Chinese steppes. These are rhizomatous species of *Agropyron*, *Calamagrostis* and *Leymus* that thrive in sandy soil, so typical in this region. Finally, these steppes differ in the prominence of species of *Stipa* of Section *Capillatae* (e.g. *S. baicalensis*, *S. bungeana* and *S. krylovii*) and of Section *Barbatae* (e.g. *S. glareosa* and *S. gobica*), whereas species of Section *Pennatae* (e.g. *S. joannis* and *S. lessingiana*) dominate in the European and Kazakhstan steppe (cf. p. 14).

USE AND MANAGEMENT

Grazing capacity for livestock decreases with increasing aridity of the grassland. Thus, the area required to support one sheep ranges from 0.4–0.5 ha in meadow steppe to 2.5 ha or more in alpine steppe (Li, 1985).

There is widespread agreement that the natural grasslands in northern China have deteriorated during the last three decades, during which nomadic grazing systems have been replaced by sedentary agriculture. Sparse grass cover and low forage yield suggest that the vegetation is being overgrazed (Fig. 3.13). This is difficult to demonstrate in the absence of fence-line contrasts (in a land where all livestock are herded and wood is scarce). Interpretation of population data (FAO, 1984) indicates that the grazing pressure within the country increased by 20% during the 18-year period ending in 1984. Frequent dust-storms and gully erosion suggest that the vegetation has become so weakened that it no longer can protect the soil adequately. Increased salinization is also common. During the 1980s action has been taken to reduce the rate of deterioration of the natural grassland. In 1985 the national government proclaimed a national grassland law which includes certain stipulations concerning tenure, protection, utilization and management intended to assure more rational land-management procedures. Regional governments have introduced statutes to administer this law. The implementation of the "individual responsibility system" (Gates, 1986) to replace the collective system (with no clear division of responsibility) is also considered to be protective of rangeland.

Nutrient factors are important in determining the potential value of each grassland type as a

Fig. 3.13. Desert steppe in western Inner Mongolia which has deteriorated because of overgrazing. The climax dominants (species of *Agropyron* and *Stipa*) have declined in vigour to leave much of the soil surface exposed. Species of *Artemisia*, *Caragana* and *Kochia*, which are of low palatability to livestock, have increased in abundance.

grazing resource. The nutritive value of all types of Chinese grassland is relatively high (Zhang Zutong, 1982). Crude protein and fat content of the dominant bunch grasses reach 10% and 3%, respectively, in typical steppe during the growing season, whereas values for some of the forbs (e.g. Chenopodiaceae) and dwarf half-shrubs (e.g. *Artemisia*) are even higher. The content of crude protein increases with aridity, whereas carbohydrate levels decrease. In meadow steppe, crude protein is 6 to 8% and nitrogen-free extract is as high as 44 to 49%, compared to maximum values in desert steppe of 14 and 30%, respectively (Table 3.3).

About 40 million hectares of native grassland are not presently utilized (Lu, 1984). This provides a basis for increasing livestock output from grassland by 15%. Further gains in livestock production will be possible by altering management practices so as to reduce the age at which animals are marketed. Plans are being made for a greater degree of co-ordination of livestock and farming enterprises. Considerable increases in production are expected to result from conversion from grass-fed to feedlot finishing of livestock. At present, the area of artificial grassland is only 2 million hectares (about 1% of the area of natural grassland) and, because irrigation and fertilization are limited, the average yield of herbage is only 150 g m^{-2} of dry matter. As a result of encouragement by government policy, expectations are that by the end of this century the area of artificial grassland will increase to 33 million hectares.

Meadows have considerable value as pastures and haylands, but they also have potential as croplands because of the high organic-matter content of the soil. Plant productivity in some types of meadow is much higher than in the surrounding

TABLE 3.3

A comparison of the composition (%) of herbage (at the time of maximum standing crop) in three types of steppe [1]

Vegetation type	Humidity coefficient,[2]	Crude protein	Crude fat	Crude fibre	Nitrogen-free extract	Ash
Meadow steppe	0.6–1.0	6.6	3.3	23.3	49.2	5.0
Typical steppe	0.3–0.6	13.2	2.9	28.7	30.2	5.8
Desert steppe	0.13–0.3	14.8	2.9	28.1	29.5	6.0

[1] After Zhang Zu-tong (1982).
[2] This was calculated as follows:

$$\text{Coefficient of humidity} = \frac{r}{E_o} = \frac{r}{0.0018(25+t)^2(100-f)}$$

where: r = mean annual precipitation (mm); t = mean annual temperature (°C); f = mean annual relative humidity; E_o = mean annual potential evaporation (mm).

steppe. Much of the forage is palatable and provides an important contribution to the support of livestock. Meadows also provide materials for paper making, for knitting and weaving, and for the manufacture of medicines. The vegetation of meadows is very important as a means of retarding soil erosion on slopes and embankments and is a source of "green manure". Meadows are being transformed by various human activities and by mismanagement. Their productive capacity is being decreased in forest regions by deforestation of adjacent land, and in steppe regions by overgrazing and intensive haying. For these reasons, forage quality has declined and poisonous species now flourish. Through application of modern ecological principles to land management, their conversion to highly productive artificial grasslands may be possible. The success of this venture will depend on the extent to which structure and function in meadow ecosystems become understood through intensive research.

REFERENCES

Brown, D., 1954. *Methods of Surveying and Measuring Vegetation*. Bulletin 42, Commonwealth Agricultural Bureaux. Farnham Royal, Bucks., England, 223 pp.
Chang, H.S. David, 1981. The vegetation zonation of the Tibetan Plateau. *Mountain Res. Dev.*, 1(1): 29–48.
Chang, H.S. David, 1983. The Tibetan Plateau in relation to the vegetation of China. *Ann. Missouri Bot. Garden*, 70: 564–570.
Chang Yu-ling, 1955. The plant communities of the Greater Khingan Mountains. *Contribution of Plant Ecology and Geobotany*, No. 1. Science Press, Beijing, pp. 20–35 (in Chinese).
Cheng Tse-ming and Han Rong-kuen, 1974. A study of grasshoppers from Hainan Tibetan Autonomous Prefectre, Qinghai Province. *Acta Entomol. Sin.* (Beijing), 17: 428–440 (in Chinese).
Dall, B.E. and McKell, C.M., 1986. Use and abuse of China's deserts and rangelands. *Rangelands*, 8: 267–271.
Deng Li-you, 1983. The types of grasslands in Tibet and the evaluation of their resources. *Acta Phytoecol. Geobot. Sin.* (Beijing), 7: 136–142 (in Chinese with English summary).
FAO, 1984. FAO Yearbook Tape. Food and Agriculture Organization of the United Nations, Rome.
Gates, D.H., 1986. Rangeland management and livestock production in northeastern China. *Rangelands*, 8: 229–233.
Heady, H.F., 1975. *Rangeland Management*. McGraw-Hill, New York, 460 pp. (also translated into Chinese by Zhan Jing-re and published by Agriculture Press, Beijing).
Hou Hsioh-yu, 1983. Vegetation of China with reference to its geographical distribution. *Ann. Missouri Bot. Garden*, 70: 509–548.
Hou Hsioh-yu and Wang Xian-pu, 1956. *The Vegetation of China with Special Reference to the Main Soil Types*. Science Publishing House, Beijing, 26 pp.
Jiang Shu (Editor) and Chen Zuo-zhong, 1985. *Introduction to Inner Mongolian Grassland Ecosystem Research Station, Bainxile Livestock Farm, Xilinhot City, Inner Mongolian Autonomous Region*. Institute of Botany, Academia Sinica, Beijing, pp. 4–6 (in Chinese).
Karamysheva, Z.V., 1961. Data about the flora of Kazakhstan. *Alma-Ata*, 11: 27–50 (in Russian).
Kong Zhao-chen, 1976. Vegetation and climatic changes of China during the past 100 years. *Acta Phytotaxon. Sin.* (Beijing), 14: 82–89 (in Chinese with English summary).
Lavrenko, E.M., 1940. Steppes of the U.S.S.R. In: E.M. Lavrenko (Editor), *The Vegetation of the U.S.S.R.*, No. 2,

150 pp. (in Russian, translated into Chinese by Zhu Tingcheng and published in 1959 by Science Publishing House, Beijing).
Lavrenko, E.M., 1954. Changes in the Eurasian steppe regions: their geography, dynamics and history. In: E.M. Lavrenko (Editor), *Botanical Problems*, No. 1, pp. 15–25 (in Russian, translated into Chinese by Zhu Tingcheng and published in 1959 by Science Publishing House, Beijing).
Lavrenko, E.M., 1973. Concerning some characteristics of the structure of plant communities in the Central Asian steppes. *Bot. J.* (Moscow), 58: 1603–1607 (in Russian).
Li Bo, 1962. Main types of zonal vegetation in Inner Mongolia, with their ecological distribution. *Inner Mongolia Univ. Bull.*, 4: 41–74 (in Chinese with English summary).
Li Hong-chang, Xi Rui-hua and Chen Yong-lin, 1983. Studies on the feeding behaviour of Acrididae in the typical steppe subzone of the Inner Mongolian Autonomous Region, I. Characteristics of food selection within artificial cages. *Acta Ecol. Sin.* (Beijing), 3: 214–228 (in Chinese with English summary).
Li Jian-dong, 1986. Successional changes in the *Aneurolepidium chinensis* grassland of China. In: P.J. Joss (Editor), *Rangelands: A Resource Under Siege*. Proc. 2nd Int. Rangeland Congr., Australian Academy of Sciences, Canberra, pp. 52–53.
Li Yu-tang, 1985. The new situation of grassland animal husbandry in China. *Grasslands China* (Huhhot), 6: 1–6 (in Chinese).
Li Yue-shu and Zhu Ting-cheng, 1983. The regularities of formation of aboveground biomass of *Leymus chinensis* population. *Acta Phytoecol. Geobot. Sin.* (Beijing), 7: 289–298 (in Chinese with English summary).
Lu Liang-shu, 1984. Recent status and prospects of food and agricultural production in China. *Sci. Agric. Sin.* (Beijing), 1: 1–9 (in Chinese).
Ma Shi-jun, 1959. Geographical Division of Animals in China. In: *Geographical Division of Animals and Insects in China*. Natural Resources Division, Academia Sinica, Science Press, Beijing, pp. 72–88 (in Chinese).
Numata, M., 1979. Climate and soils in Asian grassland areas. In: M. Numata (Editor), *Ecology of Grasslands and Bamboolands in the World*. Junk, The Hague, pp. 35–42.
Qian Guo-zhen, 1980. Population quantity and its dynamics: the role of rainfall and snow in the life of animals. In: Qian Guo-zhen (Editor), *Animal Ecology*, Vol. 1. People's Educational Publishing House, Beijing, pp. 127–162 (in Chinese).
Ren Ji-zhou, 1984. The grassland characteristics and productivity in East Asia. In: W. Siderius (Editor), *Proceedings of the Workshop on Land Evaluation for Extensive Grazing*. International Institute for Land Reclamation and Improvement, Netherlands, pp. 43–53.
Walter, H., 1979. *Vegetation of the Earth and Ecological Systems of the Geo-Biosphere*. Springer–Verlag, New York, 2nd ed., 274 pp.
Walter, H. and Box, E.O., 1983a. The deserts of Central Asia. In: N.E. West (Editor), *Temperate Deserts and Semi-Deserts*. Ecosystems of the World, 5. Elsevier, Amsterdam, pp. 193–236.
Walter, H. and Box, E.O., 1983b. The orobiomes of middle Asia. In: N.E. West (Editor), *Temperate Deserts and Semi-Deserts*. Ecosystems of the World, 5. Elsevier, Amsterdam, pp. 161–191.
Wang Jin-ting and Li Bo-sheng, 1982. Main types and characteristics of high, cold steppe in the Qiantang Plateau of Tibet. *Acta Phytoecol. Geobot. Sin.* (Beijing), 6: 1–13 (in Chinese with English summary).
Wang Yi-fung, 1963. The fundamental characteristics of steppes in the Tianshan Mountains. *Acta Phytoecol. Geobot. Sin.* (Beijing), 1: 110–129 (in Chinese with English summary).
Wang Yi-fung, 1985. *The Vegetation of Inner Mongolia*. Science Publishing House, Beijing, pp. 467–516 (*Stipa grandis* formation) and pp. 547–556 (*Stipa gobica* formation) (in Chinese).
Wang Yi-fung, Yong Shi-peng and Liu Zhong-ling, 1979. Characteristics of the vegetational zones in the Inner Mongolian Autonomous Region. *Acta Bot. Sin.* (Beijing), 21: 274–284 (in Chinese with English summary).
Wu Yan and Chin Cui-xia, 1982. The meadow vegetation and insects. In: Xia Wuping (Editor), *Alpine Meadow Ecosystem*. Contributions from the Haibei Research Station of Alpine Meadow Ecosystem, Northwest Plateau Institute of Biology, Academia Sinica, Vol. 1. Gansu People's Publishing House, pp. 110–116 (in Chinese).
Wu Zheng-yi, 1980. Climate. In: Wu Zheng-yi (Editor), *The Vegetation of China*. Science Publishing House, Beijing, pp. 27–30 (in Chinese).
Xia Wu-ping (Editor), 1981. *Alpine Meadow Ecosystem*. Contributions from the Haibei Research Station of Alpine Meadow Ecosystem, Northwest Plateau Institute of Biology, Academia Sinica, Vol. 1. People's Publishing House, Gansu, 218 pp. (in Chinese).
Zhang Rong-zhu, 1979. Fauna of temperate grasslands. In: Zhang Rong-zhu (Editor), *Animal Geography, Vol. 6. Physical Geography of China*. Science Press, Beijing, pp. 82–98 (in Chinese).
Zhang Xiao'ai, 1982. On the structure of bird communities in an alpine meadow area. In: Xia Wu-ping (Editor), *Alpine Meadow Ecosystem*. Contributions from the Haibei Research Station of Alpine Meadow Ecosystem, Northwest Plateau Institute of Biology, Academia Sinica, Vol. 1. People's Publishing House, Gansu, pp. 117–128 (in Chinese).
Zhang Zu-tong, 1982. Rational utilization of the grassland resources of Inner Mongolia. *Nat. Resour.* (Beijing), 4: 31–37 (in Chinese).
Zhao Song-qiao (Chao Sung chiao), 1984. The sandy deserts and the Gobi in China. In: F.E. Baz (Editor), *Deserts and Arid Lands*. Martinus Nijhoff Publishers, The Hague, pp. 95–113.
Zhen Zuo-xin, 1959. *Geographical Division of Animals in China*. Natural Resources Division, Academia Sinica, Science Press, Beijing, pp. 17–25 (in Chinese).
Zheng Hui-ying, Li Jian-dong and Zhu Ting-cheng, 1986. Classification and ordination of plant communities in

the south of Song-nen Plain. *Acta Phytoecol. Geobot. Sin.* (Beijing), 10: 171–179 (in Chinese with English summary).

Zhu Ting-cheng, 1955. A study of the vegetation of Sartu in Heilongjiang Province. *Acta Bot. Sin.* (Beijing), 4: 117–135 (in Chinese with English summary).

Zhu Ting-cheng, 1958. General comments on the steppes of northeast China. *Northeast Normal Univ. Bull.* (Changchun), 1: 98–116 (in Chinese with English summary).

Zhu Ting-cheng, 1963. On the classification, distribution and basic characteristics of grassland in China. *Northeast Normal Univ. Bull.*, 1: 59–109 (in Chinese with Russian summary).

Zhu Ting-cheng, 1983. A study of the productivity of yancao (*Leymus chinensis*) pastures in the northeast of China. *Proc. Fifth World Conference on Animal Production. Japanese Society of Zootechnical Science*, 2: 629–631.

Zhu Ting-cheng, 1988. Grassland resources and development of grassland agriculture in temperate China. *Rangelands*, 10: 124–127.

Zhu Ting-cheng, Li Jian-dong and Yang Dian-chen, 1981. A study of the ecology of yang-cao (*Leymus chinensis*) grassland in northeastern China. In: J.A. Smith and V.A. Hays (Editors), *Proceedings of the XIV International Grassland Congress*, Westview Press, Boulder, Colo., pp. 429–431.

Zhu Ting-cheng, Li Jian-dong and Zu Yuan-gang, 1986. Grassland resources and future development of grassland farming in temperate China. *Bull. Bot. Res.* (Harbin), 6: 99–115 (in Chinese).

Zhu Zhi-cheng, 1983. The range of forest-steppe zone on the Loess Plateau of the northern part of Shanxi Province. *Acta Phytoecol. Geobot. Sin.* (Beijing), 7: 122–132 (in Chinese with English summary).

Zou Hou-yuan, 1981. A brief description of vegetation in the loess plateau of northern Shanxi and suggestions for development of agriculture and grazing in various vegetative regions. *Acta Phytoecol. Geobot. Sin.* (Beijing), 5: 169–176 (in Chinese with English summary).

Zu Yuan-gang and Zhu Ting-cheng, 1987. The stability analysis and energy flow through *Leymus chinensis* population. *Acta Bot. Sin.* (Beijing), 29: 95-103 (in Chinese with English summary).

Chapter 4

GRASSLANDS OF SOUTHERN ASIA

J.S. SINGH and S.R. GUPTA

INTRODUCTION

This chapter is concerned with the grasslands of Asia south of the Former Soviet Union and China, from Turkey in the west to the Philippines and Indonesia in the east, but excluding the countries of the Levant and the Arabian Peninsula. The majority of tropical and subtropical grasslands in this area owe their origin to destruction of forests and abandonment of cultivation (Bor, 1942; Holmes, 1951, 1956; Johnston and Hussain, 1963; Whyte, 1974a, b; Singh, 1976; Singh et al., 1983). They have become widespread in a variety of habitats as seral communities maintained under the impact of such biotic factors as grazing, cutting and burning. Because of the ever-increasing livestock population, these have been heavily exploited, and often exhibit degraded floristic composition, poor growth, and low carrying capacity for livestock. Marked differences in habitat (climate, soil and topography), age and mode of origin, and intensity of biotic processes result in an array of very diverse grassland communities, which capture energy and use water with varying degrees of efficiency (Singh and Joshi, 1979a).

Parts of this region have been discussed elsewhere in this series by Breckle (1983) and by Orshan (1986).

GEOGRAPHICAL LIMITS AND AREA

The countries of southern Asia lie between the Equator and 42°N, except for Indonesia which is in the Southern Hemisphere (Table 4.1). Although the nature of the climax is questionable over parts of the area, natural grassland apparently occupied the greatest proportion of the landscape in the arid and semiarid western portion of this region. Further eastward the most extensive grasslands are in India, Indonesia and Pakistan. An attempt is made in Table 4.1 to quantify the extent of grassland in each country in the region by referring to data for "permanent meadows and pasture" from the Food and Agriculture Organization of the United Nations (FAO). However, information is not available to adjust these data to reflect: (1) the proportion of the land assigned to this use which was originally forested; (2) the area of cropland that was originally natural grassland; and (3) the fact that the area grazed by livestock in all of these countries includes open forests.

Contacts with other ecosystems

Grasslands in southern Asia are often secondary in nature, having originated from forest clearing and burning by man rather than in response to soil and climate (Sanford and Wangeri, 1985). Grasslands also occupy canopy gaps in open forests and form distinct associations (Champion and Seth, 1968). In arid and semiarid regions, the association of grasslands with thorn forests and shrub vegetation often imparts a savanna-like aspect to the landscape (Singh and Joshi, 1979b; Blasco, 1983). Grasslands are related to agro-ecosystems by providing a rich gene pool and including species which serve as collateral hosts for many plant diseases, insects, and nematodes. Also, most of the agriculture depends on animal power, which derives energy from the grasslands by grazing or by stall-feeding of dried and cured herbage.

TABLE 4.1

Geographical limits of countries in southern Asia, and areas of permanent pasture[1]

Country	Location		Area ($\times 10^3$ km^2)		Area of grass-land as % of area of country
	latitude (°)	longitude (°)	total	grassland	
ARID TO SEMIARID WESTERN REGION					
Afghanistan	29.5–38.6N	60.8– 75.0E	648	300.0	46.33
Iran	25.0–39.8N	43.8– 63.2E	1648	440.0	26.70
Iraq	29.0–37.3N	38.8– 48.5E	435	40.0	9.20
Turkey	25.6–44.8N	36.0– 42.0E	781	90.0	11.53
TROPICAL MONSOON REGION					
Bangladesh	21.1–26.7N	88.1– 92.8E	144	6.0	4.17
Bhutan	26.8–28.0N	89.0– 92.0E	47	2.2	4.64
Burma	9.5–28.6N	92.2–101.2E	677	3.6	0.53
India	6.8–37.0N	68.1– 97.4E	3288	119.0	3.62
Nepal	26.3–30.2N	80.2– 88.2E	141	17.9	12.68
Pakistan	23.7–36.8N	60.9– 75.5E	804	50.0	6.00
Sri Lanka	5.9– 9.8N	79.7– 81.9E	66	4.4	6.69
HUMID SOUTH-EASTERN ASIA					
Indonesia	5.6–11.0S	95.0–141.0E	1905	118.5	6.22
Laos	14.0–22.7N	100.2–107.8E	237	8.0	3.38
Malaysia	1.0– 7.0N	100.0–129.3E	330	0.3	0.08
Philippines	5.8–20.5N	98.2–105.3E	300	11.4	3.80
Thailand	5.7–20.5N	97.3–105.5E	514	3.1	0.60
Vietnam	8.7–23.2N	102.3–109.5E	330	2.7	0.83

[1]Calculated data on "permanent pasture" from the FAO Production Yearbook 1985.

The climates of the Indian subcontinent (except for the alpine region of the Himalaya, above timberline) are suited to the development of forests and deserts; the major portion of the grasslands is of recent origin (Bor, 1960). Large-scale destruction of natural forests, especially over the last 150 years, has resulted in their replacement by grassland communities (Whyte, 1977). Several workers have described the parallelism and association of grasslands and forest types in India (Yadava and Singh, 1977; Saxena and Singh, 1980). Khattak (1976), while reviewing the history of forest management in Pakistan, concluded that, except for the deserts, the country was under forest cover thousands of years ago. Nemati (1985) reported that most of the mountain slopes and uplands in Iran were covered by trees and shrubs as recently as the sixteenth century. In Afghanistan, centuries of overgrazing by sheep, goats, donkeys and camels have replaced the original plant communities and led to the creation of an *Artemisia* steppe in its present form (Neubauer, 1955). The tropical and subtropical region of Nepal was also covered by tropical deciduous and subtropical pine (*Pinus* spp.) and broad-leaved forests until recently. The natural vegetation of Sri Lanka is characterized by tropical rain forests in humid areas at low altitudes, monsoon forests in the dry zone, and thorn scrub in the arid zone. In South-east Asia open forests with a ground layer of grasses are of common occurrence. These have originated due to the change in physiognomy and structure of dense dipterocarp forests (Rollet, 1962, 1972; Legris and Blasco, 1972; Blasco, 1983). The majority of open forests and savanna woodlands represent subclimax communities stabilized by the influence of regular burning (Blasco, 1983).

Physical setting

Climate

The climate exhibits a wide range of conditions, from extremely arid to humid, and is characterized by strong seasonal and regional contrasts. The grassland communities occur under most of these climatic conditions.

The climate of south-western Asia has been described by several workers (Bauer, 1935; Sutcliffe, 1960; Perrin de Brichambaut and Wallen, 1964; Walter and Lieth, 1967). Much of Iran and Turkey have a dry mediterranean-type climate (rainfall during the cool season and a long, dry summer). Turkey is hot and moist near the Black Sea, moderate in the Namara and Aegean coastal regions, and the climate is of the mediterranean type in the south. Except for the northern hills, most of Iraq is arid, with high temperatures during the summer. Iran includes three climatic zones: (1) semiarid to humid region (annual precipitation 400 mm or more); (2) steppe (100 to 400 mm); and (3) arid desert (less than 100 mm) (Nemati, 1985). The semiarid and desert zones compose 87% of the area. In the Caspian zone, annual precipitation ranges from 400 mm in the east to 1700 mm in the west, with a minimum in June and maximum in autumn; mean annual temperatures range between 16° and 19°C (Pabot, 1967).

Afghanistan also lies in the arid zone, with a highly variable climate depending upon elevation. Desert conditions prevail in the south, where maximum temperatures may reach 44°C. The winters in the mountainous regions are extremely cold, with minimum temperatures as low as −26°C. Annual precipitation is highest (250–375 mm) in the valleys of the north-eastern mountains and lowest (50–75 mm) in the south-west.

Aridity is the most important feature of the climate of Pakistan, as a large part of the country is extremely dry and hot in summer and very cold in winter. The annual rainfall over most of the region ranges from 120 to 360 mm, except for the north-east, where it reaches 1250 to 1500 mm. The winter months of December and January are relatively dry. The maximum temperature ranges from 26 to 47°C, and the minimum from 11 to 14°C.

The Thar Desert of Rajasthan in India and the adjoining area of Pakistan are extremely arid; the western coastal region of India and all of Sri Lanka and Bangladesh have a wet tropical climate. The Indo–Gangetic plains (comprising the northern parts of India, most of Pakistan, and Bangladesh) have a tropical to subtropical continental, monsoonal climate, whereas the high Himalayan peaks are permanently covered with ice and snow. The monsoons cause considerable variability in amount and pattern of rainfall. Temperature gradients are also marked in space and time. In India, the rainfall is highest in the Assam hills (in the north-east) (11 490 mm yr^{-1}) and lowest in the north-west (220 mm). In most parts of the country, 75 to 90% of annual rainfall occurs during the monsoon (mid-June to September). In some areas of southern India and in Kashmir, variable amounts of rainfall occur during October and November; and irregular, low pressures associated with air flows from the west cause scanty rainfall during early December and January. Temperatures are lowest (10°–25°C) during winter (November to February) and range from 28° to 43°C during summer (April to mid-June). The rainy season is warm, with moderate temperatures and low diurnal variability.

Sri Lanka has two well-defined climatic zones: wet and dry. The low, hilly south-west constitutes the wet zone. Both south-west and north-east monsoons result in rainfall ranging from 1800 to 6000 mm. The dry zone generally receives 600 to 2500 mm of rainfall from north-east monsoons. Proximity to the Equator results in high temperatures in the lowlands, with little diurnal or monthly variation. (The mean monthly temperature at Colombo ranges from 25°C in December to 28°C in May.) In the hills, temperatures remain low, and the mean daily temperature approximates 16°C.

Nepal falls within the monsoonal system of the Indian subcontinent. Owing to the higher elevations, the average temperature of Nepal is low; for example, the monthly mean temperature in the Kathmandu valley ranges from 27°C (May–June) to 2°C (December–January). The tropical and subtropical region lies between 800 and 1800 m and receives a mean annual rainfall of about 2000 mm in the east and 1000 mm in the west. In eastern Nepal, the mountains are directly exposed to monsoons across the open gaps between the Raj Mahal hills and Shillong plateau. According to Shankar and Shrestha (1985), five climatic belts are distinguishable: (1) hot monsoonal climate in the *terai*[1]; (2) warm temperate monsoonal climate to an elevation of 2130 m, with warm and wet summers and cool and dry winters; (3) cool tem-

---

[1] The **terai** is the plain adjacent to the Himalayan foothills.

perate monsoon climate to 3350 m, with mild wet summers and cold dry winters; (4) alpine climate in the higher mountain ranges to an elevation of about 4870 m, with low temperatures in summer and extremely frosty conditions in winter; and (5) tundra-type climate above the snowline. The climate of Bhutan is temperate.

The climate of Bangladesh is warm tropical. The mean annual rainfall ranges from 1250 to 5000 mm and occurs mostly during the south-eastern monsoon from April to October; two-thirds of the total rainfall is received from June to August. Rainfall is scanty in winter (December–February). The mean monthly temperatures are mild and consistently high, ranging from 32° to 38°C.

In Burma the north-east monsoon season is from November to March and the summer monsoon lasts from June to September (Koteswaram, 1974). April and May are hottest. The temperatures are generally high throughout the year, except in the northern and central regions. Annual rainfall (due to the south-west monsoon) exceeds 2800 mm in the coastal areas and 4000 mm along the western slopes of the hills, whereas parts of the interior receive less than 400 mm. The highest temperatures (over 38°C) occur inland during March to May.

Seasonal distribution of rainfall over South-east Asia indicates July to be the rainiest month in Laos, Thailand and Vietnam, and January to be the rainiest month in Indonesia, Malaysia and the Philippines (Koteswaram, 1974). April and October are wettest in Sumatra (Indonesia). Under humid or perhumid climates in Laos, Malaysia, Thailand and Vietnam, the annual rainfall exceeds 2000 mm, with a short dry season. Under dry climatic conditions, annual rainfall is less than 1000 mm, with a dry season of 6 months or more. In the warm areas, the mean temperature of coldest months (January or February) is higher than 15°C and the absolute minimum temperature never falls below 8°C. The warm climates are generally found in plains and low plateaux below 1000 m elevation.

Soil

Little information is available concerning grassland soils of Turkey, Iran, Iraq and Afghanistan (Kaul and Thalen, 1979). Typical steppe soils in Iran are sierozems which are mostly oropedic, shallow with an A–C profile almost devoid of a humiferous layer, grey in colour and, for the most part, highly calcareous (Dewan and Famouri, 1964; Nemati, 1985). These soils have developed under arid conditions (rainfall less than 250 mm), with cool to cold winters and hot and dry summers. Depending upon topography and lithology these soils vary in depth, texture and content of organic matter (Nemati, 1985). Sierozems, brown soils and loess-like soils occupy vast stretches of steppe in Iraq and Turkey (Zohary, 1973).

The grasslands of India occur on a wide variety of soils. Raychaudhuri et al. (1963) classified Indian soils into eight categories:

alluvial	mountain and hilly
black	arid and desert
red	saline and alkaline
laterites and lateritic	peat.

Alluvial soils cover most of the Gangetic plains; black soils are predominant in central India; red soils in the Deccan plateau; and laterite and lateritic soils on the summits of hills in the Deccan trap (a geological formation in southern India). Arid and desert soils, which are predominantly sandy, occur in the State of Rajasthan and some parts of the States of Haryana and Punjab. Saline and alkaline soils are generally associated with arid and semiarid areas of the country. Peat and other types of organic soils in humid areas result from deposition of large amounts of organic matter. The physical and chemical properties of soil under grassland vegetation range widely: pH from 6.5 to 9.0; organic-matter content from 0.09 to 4.7%; content of nitrogen from 0.001 to 0.126% and of phosphorus from 0.001 to 0.275%; proportion of coarse sand from 0.06 to 42%, of fine sand from 0.47 to 85%, and of clay from 30 to 63% (Yadava and Singh, 1977). These ranges are based on data from sites located in the States of Gujarat (Khirasara), Rajasthan (Jodhpur, Pilani, Udaipur), Haryana (Kurukshetra), Madhya Pradesh (Ambikapur, Ratlam, Sagar, Ujjain), Uttar Pradesh (Varanasi, Jhansi), Orissa (Behrampur, Sambalpur) and Delhi.

The soils of Pakistan are characterized as nutrient-poor, immature, and deficient in organic matter (Zulfiqar, 1962). Those of the Indus plain are on alluvial deposits high in calcium carbonate and low in organic matter, nitrogen and phosphorus. The desert soils are of sandy and strongly saline alluvial type or solonchaks of coastal deltaic origin. In the northern hilly region, lithosols and brown forest soils predominate, whereas in the western hilly region sierozems, regosols and lithosols are commonest.

The soils of Sri Lanka are red-yellow podzolic in the wet zone and hills, and reddish-brown earths in the dry zone. Joachim and Kandiah (1942) reported that the soils of dry *patana*[1] grasslands are acidic and have small amounts of organic matter, nitrogen, exchangeable bases, and calcium. The soils of the *damana* grasslands are low in carbon and nitrogen and have low water-holding capacity and porosity. Those of the coastal region are alluvial and wind-blown, and salt marshes are extensive. Additional information on soils was provided by Perera (1969) and Andrew (1971).

The soils of Bangladesh can be broadly classified as hill soils, old alluvial soils, and recent alluvial soils, the last being the most important group (Islam, 1966). They are rich in organic matter (1.12–2.8%), total nitrogen (0.05–0.83%) and available phosphorus (0.02–0.12%), and have a pH range of 4.5 to 8.5 (Karim, 1966).

The soils of the South-east Asian savannas were described by Blasco (1983). Depending upon the parent material and local topographic circumstances, a great variety of types occur: acid lithosols or skeleton soils, red and yellow podzolic soils, and grey podzolic soils. Acid lithosols are predominantly infertile shallow, eroded soils, found in hilly areas in northern Thailand, southern Cambodia and east of the Mekong (Pendleton and Montrakun, 1960; Crocker, 1963). Grey podzolic soils and red and yellow podzolic soils cover a major part of South-east Asia, including more than half of the total area of Vietnam (Dudal et al., 1974). Hard or hardenable laterite is present at various depths in the subsoil under open forests in Thailand.

Geology and topography

The present land form in south-west Asia is a result of crustal movements that caused folding, fracturing, uplift and submergence (Kaul and Thalen, 1979). This was followed by volcanic activity, erosion and sedimentation. In Turkey, Anatolia forms a vast broken plateau 800 to 1200 m in elevation, bounded on the north, south and east by mountain chains reaching 2000 to 3000 m. The major part of Iraq consists of depressions between the Arabian plateau to the west and the Zagros mountain range to the north and east. The central and southern part of this depression is covered by alluvial material, mostly from the Tigris and Euphrates. There are four main physiographic provinces in Iran and Afghanistan, viz. mountains, plateau, Khuzestān and the low-lying southern coastal plain, and the Caspian sea coast. The grasslands occur mainly in the central plateau. The plateau is dissected and includes mountains and foothills, isolated hills, lake basins and several alluvial plains ranging in elevation from 500 to 2500 m (Dewan and Famouri, 1964).

Geomorphologically, the Indian subcontinent can be divided into four east–west zones: the Himalaya mountains; the Indo–Gangetic plains, extending from Pakistan to Bangladesh in the north; the plateau of peninsular India; and the island of Sri Lanka. India includes the first three of these. Pakistan includes the Indus plains, the Himalaya mountains flanking the north-east, and the plateau in the west and north-west. Sri Lanka consists of a central mass of mountains surrounded by broad coastal plains. Many of the central mountains range from 900 to 2400 m in elevation. In the north, the central plain is almost flat; elsewhere the topography is somewhat irregular. Except for some hilly tracts, Bangladesh is chiefly made up of the flood-plains of the Ganga, Brahmaputra and Meghna rivers. Nepal is bounded in the north by the great Himalaya mountains and in the south mainly by the *terai*.

India, Pakistan, and Nepal are chiefly made up of two crustal blocks of different nature and geological origin, the Himalaya and the Deccan plateau, with the Indo–Gangetic plains of Pleistocene and Recent formation lying between them. The Himalaya comprise long, folded beds of young rocks; the Deccan plateau is a block of ancient crystalline rocks (Wadia, 1961). Sri Lanka is on ancient rocks of Archaean age and is an integral part of the Karnataka gneissic terrain. The flood-plains and the coastal plains are wind- or water-transported sediments of Recent origin. The Bengal Basin, composed of Holocene and Neogene sediments, constitutes a major portion of

---

[1] *Patana* is the montane region; *damana* refers to lowlands in the dry zone.

Bangladesh (M.S. Khan, 1977). Neogene folded sediments occupy the northern, north-eastern, and south-eastern parts. Pleistocene sediments occupy most of Dacca-Mymensingh, the western region, and the Lalmai elevation. The rest of Bangladesh is made up of Recent to sub-Recent alluvial sediments.

Whyte (1974a) emphasized the importance of major tectonic events in relation to the early history of grasses in the western monsoon area of Asia. When peninsular India drifted from the Southern Hemisphere across the Equator to southern Asia, the plate was covered with tropical evergreen forests and deciduous trees with no grasses. Whyte (1974a) was of the view that the uplift of the Himalaya was responsible for changing the climate of the insular Gondwana plate region into the modern semiarid monsoon climate of the subcontinent and for the emergence of various types of grasses.

GRASSLAND VEGETATION

Grassland regions

Descriptions and classifications of the grassland vegetation are available for various regions of southern Asia: for Afghanistan, Iran, Iraq and Turkey (Zohary, 1946, 1973; Springfield, 1954; Whyte, 1968; Al-Ani et al., 1970; Mobayen and Tregubov, 1970); for India (Dabadghao, 1960; Puri, 1960; Dabadghao and Shankarnarayan, 1973); for Pakistan (Johnston and Hussain, 1963); for Sri Lanka (Holmes, 1956; Pemadasa and Amarsinghe, 1982); for Nepal (Numata, 1966); and for South-east Asia (Whyte, 1974a, b; Blasco, 1983; Singh et al., 1985). However, little information exists about the grass cover of Bangladesh (Rahman, 1956; M.S. Khan, 1977). The following discussion is based on the above reports.

Turkey, Iraq, Iran and Afghanistan

Steppe vegetation is characteristic of the high (800–1200 m) plateau of central Anatolia in Turkey (Birand, 1954). Severe overgrazing in this region suppresses *Artemisia fragrans* (an important range resource), which is replaced by communities of less palatable species, including *Eryngium campestre*, *Euphorbia macroclada*, and species of *Ballota*, *Bromus* and *Phlomis*.

Springfield (1954) classified the vegetation of Iraq into six categories: desert, steppe, forest, alpine, marshland and cultivated lands. Native forage plants contribute 87% of total forage available to livestock. Al-Ani et al. (1970) classified the perennial cover types of the southern and western deserts, and related the vegetation cover to land form and soil type.

Grassland vegetation is distributed in Iran from the coastal regions to the dry alpine zones (Rol, 1956; Pabot, 1960). Whyte (1968) gave an account of the grasses in the plains and hills of Khuzestān, where the grass flora includes:

Aeluropus littoralis   *Cymbopogon laniger*
Aeluropus repens   *Hyparrhenia hirta*
Aristida coerulescens   *Pappophorum persicum*
Aristida plumosa   *Stipa barbata*
Cenchrus ciliaris

The species in the Zagros mountains include:

Andropogon ischaemum   *Oryzopsis holciformis*
Bromus tomentellus   *Secale montanum*
Festuca valesiaca   *Stipa barbata*
Melica sp.

Aristida coerulescens and *Pennisetum orientale* are found in southern Iran. *Cymbopogon schoenanthus*, *Lasiurus hirsutus* and *Tricholaena teneriffae* occur in savanna and savanna-like vegetation. In the arid and semiarid zones of Iran, steppe vegetation of *Artemisia herba-alba*, *Astragalus* spp. and grasses (e.g. *Stipa* spp.) is predominant (Mobayen and Tregubov, 1970). The steppe flora in the vegetation zone with cold winters is predominantly an *Artemisia herba-alba* community. *Noaea mucronata* is widespread and *Poa bulbosa* is abundant on non-eroded soils, along with *Carex stenophylla*. *Aristida plumosa* is the dominant grass in areas of sandy soils.

According to Neubauer (1955), the vegetation of Afghanistan exists in two markedly distinct forms: the xeromorphous steppe which has a more or less desertic character and covers most of the country; and the hydrophilous form found on the banks of streams, rivers and irrigation channels. *Artemisia* steppe is characteristic of large areas in the Irano–Turanian zone. Pabot (1964) described seven vegetation regions of Afghanistan. In the northern plains at low altitude (1000 m), the natural flora is abnormally poor and the dominants are *Aristida plumosa*, *Poa bulbosa* and species of *Aeluropus*, *Agropyron* and *Stipa*. *Bromus* spp., *Poa* spp., *Stipa barbata* and *Trigonella cachemiriana* occur at elevations from 1000 to 2000 m. Above 2000 m, the vegetation becomes richer in composition and abundance in response to the shorter dry season. Species of the genera *Astragalus*, *Bromus*, *Elymus*, *Festuca*, *Koeleria*, *Oryzopsis*, *Poa*, *Stipa* and *Trigonella* are predominant. In loamy soils of the southern deserts the principal species are *Aristida plumosa*, *Artemisia herba-alba*, *Astragalus* sp., *Carex stenophylla* and *Poa bulbosa*. On sandy dunes, some of the predominant species are *Aristida pennata*, *Aeluropus mucronatus*, *Cyperus laevigatus*, *Heliotropium* sp., *Pennisetum dichotomum* and *Stellera* sp. *Artemisia herba-alba* is rare on the eroded hillsides to the north and west of Herat. Of the palatable species, only a few survive, such as *Hulthemia berberifolia*, *Iris songarica* and *Peganum hermala*. To the south, the vegetation is floristically richer and denser, and the predominant species is *Artemisia herba-alba*. The winter rangelands of south-western Afghanistan are important for the livestock industry (Hassanyar, 1988). *Haloxylon persicum* covers a large area of mobile sand dunes of the Registan desert, forming a mixed community with *Aristida pennata* and *Calligonum molle*. In the saline soils of the extreme south, *Haloxylon salicornicum* is the principal dominant. *Reaumuria stocksii* occurs commonly on rocky soils. The shrub community is dominated by *Calligonum* spp. and is mixed with *Aristida pennata*, *Artemisia herba-alba* and *Haloxylon persicum*. *Artemisia herba-alba* forms a secondary community which appears after *Haloxylon persicum* is destroyed by overgrazing. In this half-shrub community, the dominants are:

Anabasis setifera	*Salsola arbuscula*
Artemisia herba-alba	*Seidlitzia rosmarinus*
Ephedra scoparia	*Zygophyllum fabago*
Euphorbia cheirolepis	

Pakistan

Johnston and Hussain (1963) identified five grassland regions in Pakistan (Fig. 4.1).

Dichanthium–Cenchrus–Lasiurus **region.** This includes most of the alluvial basin of the Indus river and extends from the former North-West Frontier Province through the Punjab to Sind and as far as Baluchistan. The grass cover is composed of one of the following types: *Aristida–Eragrostis*, *Cenchrus*, *Cynodon*, *Eleusine* and *Lasiurus*. On low and stable sand dunes in Cholistan and Dera Ghazi Khan, the important grasses are *Cenchrus ciliaris*, *Eleusine compressa*, *Lasiurus hirsutus* and *Panicum antidotale*. When protected from grazing, *Cenchrus ciliaris* and *Lasiurus hirsutus* re-establish dominance in deteriorated grassland.

Chrysopogon **region.** This covers most of the Baluchistan area, where rainfall occurs only during winter. The topography is variable, ranging from relatively level to hilly. In the arid southern portion, *Chrysopogon aucheri* is the characteristic grass; whereas in the moist northern half *Chrysopogon fulvus* is abundant. At Hazarganji, Khan (1962) reported that *Cymbopogon schoenanthus* was dominant on hilltops and *Artemisia maritima* on the lower grazed areas. Partial protection favours more palatable species, such as *Chrysopogon aucheri*, *Enneapogon persicus*, *Oryzopsis aequiglumis* and *Stipa szowitsiana*. In the Isplinji valley, *Cenchrus*, *Cymbopogon* and *Stipa* occur on coarse alluvial soils, whereas *Aristida* spp., *Cenchrus ciliaris*, *Chrysopogon aucheri*, *Pennisetum orientale* and *Tetrapogon villosus* inhabit gravel fans. In low-lying areas near Quetta, annual species of *Bromus* and *Poa* are important. Hussain (1962) reported that in Kohistan, the grassland is of the *Chrysopogon* and *Dichanthium–Cenchrus–Lasiurus* types.

Themeda–Arundinella **region.** This includes the hilly range areas of Murree, Hazara and Swat, and is in a degraded form. *Heteropogon contortus* is dominant in much of the area because of intensive summer grazing and frequent burning. In the Murree hills, Khan and Bhatti (1956) found *Chrysopogon fulvus*, *Heteropogon contortus*, *Panicum orientale* and *Themeda anathera* to be important species. Chaudhri (1960) reported that the important grasses of the valleys are *Chrysopogon serrulatus*, *Heteropogon contortus*, *Oryzopsis munroi* and *Themeda anathera*.

Saccharum **azonal community.** This type is characteristic of young alluvial soils along river courses with frequent alternation between erosion and deposition of soil. Champion (1936) reported *Sporobolus marginatus* on saline soils and *Cyn-*

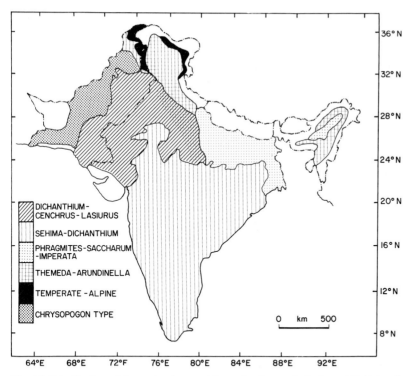

Fig. 4.1. Distribution of the regional potential grassland types in India and Pakistan (based on Dabadghao and Shankarnarayan, 1973, and Johnston and Hussain, 1963). Much of the area of each region is occupied by forest or is under agriculture.

odon dactylon on inundated soils. Khan (1962) also noted *Cynodon dactylon, Desmostachya bipinnata* and *Phragmites* spp. along with *Saccharum bengalense* and *S. spontaneum*.

Alpine grasslands. Chaudhri (1960) reported that alpine meadows are composed entirely of perennial herbs and indicated the presence of various temperate species of such genera as *Bromus, Festuca, Poa* and *Potentilla*. In Pakistan, the vast areas of Chagai–Kharan and the Makran coast have saline depressions, where *Haloxylon ammodendron* (a species of moderate palatability) is widespread, but *Diplachne fusca* has been reported in waterlogged saline areas (M. Khan, 1966).

India

Dabadghao and Shankarnarayan (1973) recognized five major grass cover types in India. The area shown for each type in Fig. 4.1 represents the potential area, because a major portion of this area for each cover type is now either cultivated or under forest. The nature of the grass cover is primarily governed by climatic factors and especially by latitude (Whyte, 1968). Thus, *Sehima–Dichanthium* grassland occurs in tropical regions, whereas *Dichanthium–Cenchrus–Lasiurus, Phragmites–Saccharum–Imperata* and *Themeda–Arundinella* grasslands are subtropical in distribution. Altitude separates *Themeda–Arundinella* cover (restricted to the northern hills) from *Phragmites–Saccharum–Imperata* (in moist to wet habitats) and *Dichanthium–Cenchrus–Lasiurus* (in relatively dry habitats) types, which occur in the plains. The grasslands of more elevated areas northward (temperate–alpine region) exist under temperate climate. The savannas of India have been discussed elsewhere in this series (Misra, 1983).

Sehima–Dichanthium region. This includes peninsular India, south-west Bengal, southern Bihar, and the southern hilly portions of the States of Uttar Pradesh and Rajasthan. The topography is undulating to hilly, rainfall ranges from

300 to 6550 mm (optimal range, 500–900) and the colour of the soil is pale brown to dark grey. The grassland vegetation includes 24 important perennial grasses, several annual grasses, and 129 forbs, including 56 legumes. The foliage cover of a well-developed *Sehima nervosum* community is as high as 89%, whereas the *Dichanthium annulatum* community attains a maximum cover of 80% in the plains. Important associated grasses are:

Aristida setacea	*Dichanthium caricosum*
Bothriochloa pertusa	*Eulalia trispicata*
Chrysopogon fulvus	*Heteropogon contortus*
Cymbopogon spp.	*Themeda triandra*
Cynodon dactylon	

Because of varying physiography, physicochemical characteristics of the soil and degree of use, several distinct communities occur within this region. For example, based on the moisture content of the soil, Burns and Kulkarni (1927) classified the grass flora of the Deccan plateau (in southern India) into three categories: wet, semi-arid and xeric. In the Western Ghats, Bharucha and Shankarnarayan (1958) recognized six major grassland communities, each representing a disclimax stage. Pandeya (1964) identified eight grassland types at Sagar and V.P. Singh et al. (1979) reported nine at Ujjain, both in the State of Madhya Pradesh.

Dichanthium–Cenchrus–Lasiurus region. This includes the northern part of Gujarat, the whole of Rajasthan (except the Aravalli Range in the south), western Uttar Pradesh, Delhi, and Punjab. The topography is broken by the spurs of the southern hills and in western Rajasthan by sand dunes. The grassland vegetation includes 11 important perennial grasses and 45 important forbs, including 19 legumes. *Dichanthium annulatum* dominates in areas of high soil moisture, *Cenchrus setigerus* and *C. ciliaris* under moderate moisture conditions, and *Lasiurus hirsutus* in dry habitats (90–300 mm of rainfall) with loose sandy soils. Saline and alkaline areas within the region, known locally as "usar" land, are dominated by annuals — especially species of *Chloris*, *Eragrostis*, *Eriochloa* and *Sporobolus*. On the other hand, the dominant species in waterlogged loamy soils are *Desmostachya bipinnata*, *Ischaemum rugosum* and *Iseilema wightii*. The semiarid and arid grasslands have been studied extensively in Rajasthan. Based on habitat conditions, soil moisture gradient and the intensity of biotic pressure, several communities have been recognized. For example, Gupta and Saxena (1972) described seven stable types: (1) *Sehima nervosum* on hills and in piedmont areas; (2) *Dichanthium annulatum* on old alluvium; (3) *Cenchrus* spp. on well-drained sandy alluvial soil; (4) *Lasiurus hirsutus* on loose sandy soil; (5) *Desmostachya bipinnata* on young alluvium; (6) *Sporobolus marginatus–Dichanthium annulatum* on low-lying, heavy saline soil; and (7) *Panicum turgidum* on sand dunes.

Phragmites–Saccharum–Imperata region. This includes the Gangetic plains and the Brahmaputra valley, and extends west into the plains of Punjab, including areas both of low and high rainfall. The topography is level, characterized by low-lying, poorly drained lands with a high water table. The grass cover is composed of 19 principal perennial grasses and 56 forbs, including 16 legumes. In the transitional zone between this and the *Sehima–Dichanthium* region, *Bothriochloa pertusa*, *Cynodon dactylon* and *Dichanthium annulatum* form communities. In swampy areas of Sundarbans, tall grassy patches of *Bothriochloa intermedia*, *Imperata cylindrica*, *Phragmites karka* and *Saccharum spontaneum* are associated with some low forests. Other important grasses are *Desmostachya bipinnata*, *Saccharum arundinaceum*, *Sporobolus indicus* and *Vetiveria zizanioides*.

Themeda–Arundinella region. This includes most of the northern and north-western montane tracts (at elevations of 350–2100 m) in the States of Assam, Manipur, West Bengal, Uttar Pradesh, Punjab, Himachal Pradesh and Jammu and Kashmir. Rainfall ranges between 1000 and 2000 mm. There are 16 characteristic perennial grasses and 34 forbs, including 9 legumes. *Themeda anathera* dominates when the grassland is subjected to light grazing. Important species include *Arundinella bengalensis*, *A. nepalensis*, *Bothriochloa intermedia*, *Chrysopogon fulvus*, *C. gryllus*, *Cymbopogon* spp., and *Heteropogon contortus*.

Temperate–alpine region. This includes the high hills of the northern montane belt above 2100 m in the west and above 1500 m in the east. The Nilgiris high-altitude grasslands of southern India may also be placed in this category. There are 25 characteristic perennial grasses and 68 forbs, including 9 legumes. Based on an altitude gradient, several grassland types have been identified (Singh and Saxena, 1980). Gupta and Nanda (1970) recognized four communities (*Chrysopogon gryllus*, *Koeleria cristata*, *Poa pratensis* and *Themeda anathera*) in the temperate zone and four (*Agrostis canina*, *Danthonia cachemyriana*, *Phleum alpinum* and *Puccinellia kashmiriana*) in the cooler subalpine–alpine zone (Figs. 4.2 and 4.3).

Sri Lanka

Holmes (1956) classified the grasslands of Sri Lanka into those of: (1) the wet zone, including *talawa* grasslands, the dry *patana* grasslands, savannas, and wet black *patana* grasslands; and (2) the dry zone, consisting of dry *damana* grasslands and wet *villu* grasslands.

***Talawa* grasslands.** These grasslands occupy: the wet, low landscape of south-western Sri Lanka; the foothills of the central mountain mass; and the slopes of the low, subcoastal hills, between the remnants of tropical evergreen forests and in the coconut plantations. The important species are *Chrysopogon* spp., *Cynodon dactylon*, *Dimeria lehmannii*, *Ischaemum indicum* and *Themeda tremula*. Occasional patches of *Cymbopogon nardus* var. *confertiflorus*, *Imperata cylindrica* and *Themeda tremula* occur on the fertile soils of the lower slopes, and a thin cover of *Massia triseta* occurs in association with sedges on the poor soils of the upper slopes. Important features of these grasslands are the presence of *Cassytha filiformis*, a parasite of grasses and forbs of this area, and of *Nepenthes distillatoria*.

***Patana* grasslands.** These are montane grasslands covering the largest area of Sri Lanka. Based on climate and altitude, De Rasayro (1945) and Holmes (1951) recognized dry and wet *patana* grasslands. Andrew (1971) identified three main communities (*Chrysopogon zeylanicus*, *Cymbopogon nardus* var. *confertiflorus* and *Themeda tremula*) and categorized *patanas* as fire subclimaxes. Based on variation in floral composition and fungal populations, Mueller-Dombois and Perera (1971) suggested further subdivisions of these grasslands, as follows: (1) humid zone dry *patana* dominated by *Cymbopogon nardus* and

Fig. 4.2. A central Himalayan alpine grassland under heavy use by sheep during the summer. The major species are *Cyananthus lobatus*, *Danthonia cachemyriana*, *Geum elatum* and *Polygonum amphibium*.

Fig. 4.3. Alpine grassland just above timberline in the central Himalaya. The dominant grasses are species of *Danthonia*; the shrub is *Rhododendron campanulatum*; and the tree in the background is *Abies pindrow*.

Themeda tremula; (2) dry *patana* of the summer-dry zone, with the important perennial grasses being *Brachiaria distachya*, *Chrysopogon aciculatus*, *Digitaria longiflora*, *Heteropogon contortus* and *Sehima nervosum*; (3) intermediate *patana* on steep slopes; (4) lower wet *patana* occurring below 2000 m, most of it now being under seeded pasture of *Pennisetum clandestinum*; and (5) upper wet *patanas* occurring between 2000 and 2300 m. The characteristic species in the wet *patanas* are *Arundinella villosa*, *Chrysopogon zeylanicus*, *Cymbopogon nardus* var. *confertiflorus*, *Eulalia phaeothrix* and *Ischaemum indicum*. Mueller-Dombois and Perera (1971) reported a floristic similarity of 82% between the two wet *patanas*, based on Sørensen's (1948) index. Ordination analysis based on higher plants and soil fungi affirmed the ecological distinctness of major zones within *patanas* (Pemadasa and Mueller-Dombois, 1979). *Andropogon lividus* and *Arundinella villosa* are the most abundant species in the wet and intermediate *patanas*, and are rare or absent in dry *patanas*. *Chrysopogon zeylanicus*, *Eulalia phaeothrix* and *Fimbristylis pentaptera* occur primarily in the wet *patanas*. *Imperata cylindrica* and *Ischaemum indicum* (of the higher elevations) extend into lower wet *patanas*. *Cymbopogon nardus*, *Dimeria gracilis* and *Themeda tremula* occur predominantly in the dry *patanas* and intermediate *patanas*. *Desmodium triflorum* is confined to both dry *patanas*, whereas *Pycnospora lutescens* occurs mainly in the humid-zone dry *patana*.

Savannas. Savannas occur in a narrow belt between the dry *patana* grasslands of the south-central mountains and the dry evergreen forests of the surrounding plain. Four important coarse grasses are present in the savannas, namely, *Cymbopogon nardus* var. *confertiflorus*, *Imperata cylindrica*, *Themeda arguens* and *Sorghum nitidum*.

Dry *damana* grasslands. These grasslands occur in the lowlands of the dry zone, in patches of various sizes and interspersed with dry, mixed evergreen forests. In these grasslands, tufted or tussock-forming grasses are dominant; the important ones include:

Aristida setacea	*Dactyloctenium aegyptium*
Chloris barbata	*Digitaria adscendens*
Chrysopogon aciculatus	*Imperata cylindrica*
Cymbopogon nardus var. *confertiflorus*	

Wet *villu* grasslands. These grasslands also occur in lowlands, mainly of permanently moist areas. Most of the *villus* occur in the Tamakaduwa district, on either of the reaches of Mahaweli Ganga. These are marshy grasslands with few trees and a rich growth of succulent fodder grasses. *Brachiaria mutica* is invariably the dominant grass in most *villus*. Semi-floating grasses, such as *Oryza rufipogon* and *Paspalidium geminatum*, occur along the water edge. Under semi-moist conditions the important grasses are *Cynodon dactylon*, *Iseilema laxum* and *Paspalidium flavidum*.

Bangladesh

Ibrahim (1967) reported that 6% of the total area of Bangladesh is potential range. According to Rahman (1956), grasslands cover a substantial area (8040 ha) of Sylhet district, the important grasses being *Erianthus ravennae*, *Phragmites karka* and *Saccharum spontaneum*. McClure (1956) reported stands of bamboo along the Karan Phuli river composed of *Bambusa latispiculata*, *Dendrocalamus longipathus*, *Melocanna baccifera*, *Oxystenanthera auriculata* and *Teinostachyum dullooa*. M.S. Khan (1977) reported that, in marshy habitats of the Indo–Gangetic plain of Bangladesh, the important grasses are species of *Arundo*, *Erianthus*, *Imperata*, *Phragmites* and *Saccharum*. On the Chittagong hills, grasslands develop where trees have been removed, as well as along the banks of rivers and swamps, where *Imperata cylindrica* and *Saccharum spontaneum* are important species.

Nepal

The elevated vegetation zones of central Nepal were described by Kitamura (1955). Numata (1966) provided information on parallelism of forest types and grass covers in the subtropical zone of eastern Nepal. He reported that many grasses (such as *Bothriochloa pertusa*, *Cenchrus ciliaris*, *Chrysopogon* spp., *Dichanthium annulatum*, *Heteropogon contortus*, *Phalaris minor* and *Saccharum* spp.) grow well during the rainy season in climax forests of *Bombax*, *Ficus* and *Shorea* in areas of lateritic soils. The vegetation type dominated by *Cynodon dactylon* and *Imperata cylindrica* is subtropical to warm temperate in distribution and disappears at elevations above 2500 m. Large areas of dense stands of *Pennisetum purpureum* (elephant grass) occur in the *terai*, often reaching 4.6 m in height. In catchment areas of the Kulekhani, Daraundi, Myagdi and Mustang rivers, the grasslands/ranges in various altitude zones are dominated by the following species [Agricultural Products Research Centre (APROSC), 1979]:

Tropical zone (800–1000 m)
    Bothriochloa pertusa   *Dichanthium annulatum*
    Chrysopogon aciculatus   *Imperata cylindrica*
    Cynodon dactylon   *Paspalum distichum*

Subtropical zone (1000–1800 m)
    Cynodon dactylon   *Saccharum spontaneum*
    Eragrostis spp.   *Setaria pallidefusca*
    Imperata cylindrica   *Themeda* spp.

Warm temperate zone (1800–2450 m)
    Andropogon spp.   *Ischaemum angustifolium*
    Chrysopogon aciculatus   *Saccharum* spp.
    Dichanthium annulatum   *Setaria glauca*
    Heteropogon spp.   *Themeda* spp.
    Imperata cylindrica

Cool temperate zone (2450–3050 m)
    Arundinella hookeri   *Poa alpigena*
    Festuca spp.   *Poa annua*
    Phleum alpinum   *Themeda hookeri*

Subalpine and alpine zones (3050–4600 m)
    Agropyron spp.   *Phleum* spp.
    Festuca spp.   *Poa* spp.

Bhutan

Numata (1987) reported that rangeland in Bhutan constitutes 3% of the total land area. He described the floristic composition of grassland between Thimbu and Nikkachu. The ground cover of a roadside fragment of vegetation in Thimbu at 2400 m altitude is composed of:

   Chrysopogon aciculatus   *Lepidium virginicum*
   Cosmos bipinnatus   *Mazus delevayi*
   Cynodon dactylon   *Oxalis corniculata*
   Galinsoga parviflora   *Potentilla griffithii*

Rangeland at 3050 m altitude was dominated by *Sagina japonica*, *Senecio chrysanthemoides* and *Iris clarkei*, and other species included:

   Arundinaria falconeri   *Poa annua*
   Carex nubigena   *Potentilla griffithii*
   Galium asperifolium   *Rumex nepalensis*
   Plantago erosa

An overgrazed range on calcareous sandstone soil was dominated by *Carex nubigena*, *Potentilla griffithii*, *Pteridium aquilinum* and *Senecio chrysanthemoides*.

South-east Asia

Elsewhere in this series Blasco (1983) has described the savannas of South-east Asia. Other accounts of the grasslands of this area are those of Schmid (1958, 1974), Whyte (1974a, b), UNESCO (1979) and Singh et al. (1985). The grass savanna is of two main types: climax grassland and secondary savanna.

Climax grasslands develop on narrow low alluvial plains constantly subjected to excessive soil moisture; they are dominated by hygrophilous grasses, including *Echinochloa stagnina*, *Phragmites karka* and *Saccharum narenga*. Trees are unable to thrive in highly hydromorphic soils.

The secondary savannas develop on deep basaltic soil, and are dominated by *Imperata cylindrica* between 300 and 700 m and by *Arundo madagascariensis* above 900 m, whereas on shallower soils *Arundinella* spp., *Cymbopogon nardus* var. *confertiflorus* and *Hyparrhenia eberhardtii* are dominant. These savannas are generally shrub savannas having extremely variable grassy components. Floristically and physiognomically, these grasslands are open woodlands in which the major grasses are *Arundinella setosa*, *Imperata cylindrica*, *Sorghum serratum* and *Themeda gigantea*. In short discontinuous grasslands with and without trees, *Aristida cumingiana*, *Arundinella setosa*, *Schizachyrium brevifolium* and *Themeda* sp. are common. In hilly regions, most of the secondary herbaceous vegetation types are considered to be derived from several forest types due to shifting cultivation. These are composed mainly of *Eupatorium odoratum*, *Imperata cylindrica* and *Pteridium aquilinum*. The important species of trees and shrubs in the secondary savannas are:

Albizia procera   *Grewia* spp.
Careya spp.   *Phyllanthus emblica*
Clerodendrum serratum   *Ziziphus rugosa*
Dillenia ovata

Saccharum spontaneum and species of *Panicum*, *Sorghum* and *Vetiveria* dominate the savanna in northern and north-eastern Thailand, whereas in the more humid south-eastern parts species of *Andropogon*, *Brachiaria*, *Echinochloa* and *Pennisetum* are important.

Imperata cylindrica is a component of native ranges over wide areas in South-east Asia covering an area estimated at 200 million ha (R.S. Martoatmodjo, unpublished). Soerjani (1976) reported that, in Indonesia, 16 million ha are dominated by this species and that the area was increasing at the rate of 1.25% per annum. In Malaysia *I. cylindrica* is stated to dominate an area of 1.6 million ha; but this is largely restricted to the understorey of rubber plantations (Ivens, 1975). *Imperata cylindrica* is also the predominant species in rangeland in the highlands of northern Thailand, where it constitutes a fire climax created by the burning associated with shifting cultivation, covering some 2500 km^2 (Gibson and Van Diepen, 1977). According to Eussen (1976), *I. cylindrica* often occurs in a pure sward, but in the Lesser Sunda Islands (Indonesia) it is associated with other species, including *Andropogon* spp., *Saccharum spontaneum* and *Themeda arundinaceae*. In Jawa (Java), the associated species are (Eussen, 1976):

Digitaria adscendens   *Mimosa invisa*
Eupatorium inulaefolium   *Panicum repens*
Eupatorium odoratum   *Paspalum conjugatum*
Melastoma offine   *Saccharum spontaneum*
Mikania cordata

Savannas, open secondary forests and grassland cover an area of about 30 million ha in Indonesia (Soerianegara, 1970). Savannas are found in the dry region of eastern Jawa (Java), at the northeastern tip of Bali, and in Sumba and Timor. The distribution of grasslands and savannas in Indonesia was mapped by Hannibal (1950), and the floristic composition was described by Soerianegara (1970) as summarized below.

(1) Pine savanna and grasslands (Northern Sumatra, Atjeh):

Andropogon aciculatus   *Paspalum scrobiculatum*
Andropogon amboinicus   *Pogonatherum paniceum*
Andropogon parviflorus   *Rhynchospora rubra*
Digitaria ropalotricha   *Saccharum spontaneum*
Imperata cylindrica   *Sporobolus berterianus*
Manisuris granularis   *Sporobolus diander*
Panicum lutescens   *Themeda gigantea*

The major tree species is *Pinus merkusii*.

(2) *Acacia* and palm savanna, *Eucalyptus* and *Casuarina* savanna and open grassland (Eastern Jawa, Baluran, Lombok, Sumbawa, Sumba, Flores and Timor):

Andropogon sp.   *Cenchrus brownei*
Axonopus compressus   *Chloris barbata*
Bothriochloa compressa   *Chrysopogon aciculatus*
Bothriochloa glabra   *Chrysopogon* sp.
Capillipedium parviflorum   *Cynodon arcuatus*

Cynodon dactylon	*Perotis indica*
Dactyloctenium aegyptium	*Polytoca bracteata*
Dichanthium caricosum	*Polytrias amaura*
Eleusine indica	*Rottboellia exaltata*
Eragrostis amabilis	*Saccharum spontaneum*
Eriochloa punctata	*Setaria geniculata*
Heteropogon contortus	*Setaria verticillata*
Imperata cylindrica	*Sorghum nitidum*
Isachne globosa	*Sorghum plumosum*
Leersia hexandra	*Sporobolus virginicus*
Panicum perakense	*Sporobolus* sp.
Panicum repens	*Themeda gigantea*
Paspalum notatum	*Themeda quadrivalvis*
Paspalum sp.	

The major tree species are *Borassus flabellifer* (in palm savanna), *Eucalyptus alba* (in *Eucalyptus* savanna), *Acacia leucophloea* (in *Acacia* savanna) and *Casuarina junghuhniana* (in *Casuarina* savanna).

Sajise et al. (1976) have recognized the following four types of open grasslands in the Philippines.

(1) The *Imperata cylindrica* community with *Capillipedium parviflorum, Chrysopogon aciculatus, Heteropogon contortus, Manisuris clarkei* and *Paspalum dilatatum* as the main associates.

(2) The *Themeda triandra* community with *Capillipedium parviflorum, Chrysopogon aciculatus, Coelorhachis glandulosa, Imperata cylindrica* and *Paspalum dilatatum* as the major associates.

(3) The *Chrysopogon aciculatus* community with *Alysicarpus vaginalis, Fimbristylis monostachya, Imperata cylindrica, Paspalum dilatatum* and *Themeda triandra* as important species.

(4) The *Capillipedium parviflorum* community with *Chrysopogon aciculatus, Fimbristylis monostachya, Heteropogon contortus, Imperata cylindrica* and *Paspalum dilatatum* as the main associated species.

The first three communities occur over a wide range of soil acidity (pH 4.8–6.2), in contrast to the fourth which occurs over a narrow range of soil pH (5.4–6.2). The latter community is associated with overgrazing, whereas the other three have a much wider ecological amplitude with respect to resistance to the effects of grazing.

In Cambodia, Laos, Vietnam and Malaysia, shifting cultivation produces a rough grass cover in which *Imperata cylindrica*, among others, is dominant (Ogilvie, 1954). Grass savannas cover appreciable areas in Vietnam, on the Darlac plateau, particularly near Pleiku and Ban Me Thuot, at elevations from 300 to 700 m (Blasco, 1983). Schmid (1958) stated that, in this region of Indo-China, grasses other than species of the Bambuseae are of little importance in the climax formation. Grasses play a much more important role in the secondary formations, particularly in open forests and savannas. The first grasses to appear after clearing of the forest are of the tribe Paniceae — *Cyrtococcum patens, C. trigonum, Panicum sarmentosum* and *Paspalum conjugatum*. *Digitaria marginata* and various species of *Paspalum* occur in cultivated land, whereas in the shade of plantations the swards are dominated by *Axonopus compressus, Centosteca latifolia, Cyrtococcum trigonum* and *Paspalum conjugatum*. After several years of shifting cultivation, the sward gradually deteriorates to a savanna of *Imperata cylindrica*. The influence of soil characteristics on the distribution of common grasses appears to be more marked than that of climate.

There are extensive areas of unforested lands in Cambodia which are called *veal* (Delvert, 1961). The herbaceous cover of these grassland communities is composed of *Cynodon dactylon, Imperata cylindrica* (rare), *Pseudopogonatherum* spp., of various species of the families Rosaceae and Cyperaceae, and small bamboos. The plain of Veal Trea is a treeless community, up to 2 m high, of members of the Commelinaceae and Cyperaceae, of *Fimbristylis* spp., and especially of the grasses *Echinochloa crusgalli, Panicum luzonense* and *Themeda* sp. These grasses are burned each year to facilitate hunting, and become green again and flower with the first rains in May.

Verboom (1968) described the grass successions and associations within various forest types in Pahang in western Malaysia, as follows.

(1) Coastal dunes: *Chrysopogon, Eriachne, Heteropogon, Ischaemum, Panicum, Perotis, Spinifex, Thuarea* and *Zoysia*.

(2) Seasonal inland swamp forest (fallow rice paddies): *Eragrostis, Hymenachne, Isachne, Leersia, Oryza* and *Panicum*.

(3) Foothill forest (*Dipterocarpus* and *Shorea*): *Axonopus, Chrysopogon, Desmodium, Imperata* and *Paspalum*.

(4) Forest-fringe grassland: *Acroceras, Brachiaria, Centosteca, Lophatherum, Panicum* and *Setaria*.

(5) Forest-clearing pioneer associations: *Cynodon*, *Digitaria*, *Echinochloa*, *Eleusine*, *Leptochloa* and *Paspalum*.

(6) Fire-climax association (fast growing secondary jungle trees invade as soon as the fire factor is removed): *Imperata cylindrica* with the sedge *Scleria laevis* and species of legumes (*Desmodium* and *Uraria*).

(7) Village clearing/dry *padang*: *Axonopus*, *Chrysopogon*, *Cynodon*, *Paspalum* and *Sporobolus*.

(8) Village clearing/wet *padang*: *Axonopus*, *Chrysopogon*, *Cynodon*, *Eragrostis* and *Paspalum*.

Soil–plant relationships

Little information is available about soil–plant relationships in these grasslands.

Pandeya (1964) compared edaphic conditions with vegetative cover in eight grassland communities in **India** (Table 4.2). He found that *Heteropogon–Andropogon* and *Coix–Ischaemum* communities grow on calcium-poor soils, whereas *Themeda* and *Cymbopogon–Eulalia* types occupy calcium-rich soils. He also concluded that an *Aristida–Melanocenchris* community occurs on coarse soil with low moisture status, whereas the *Themeda*, *Cynodon–Bothriochloa* *Dichanthium* and *Coix–Ischaemum* communities occur in areas of high soil moisture.

For a semiarid grassland, Mankand (1974) reported that total phytomass is dependent on several inter-related factors, especially total amounts of exchangeable sodium, potassium, calcium and phosphorus and of total nitrogen in the soil. Jayan (1970) considered various physical and chemical characters of soil to distinguish eleven ecotypes of *Cenchrus ciliaris* in natural grazing lands of Ahmedabad. Soil characteristics, which were significant variables distinguishing the grazing lands dominated by each of these ecotypes, included: texture, water-holding capacity, pH, organic carbon, total nitrogen, calcium, sodium and phosphorus. Exchangeable potassium and electrical conductivity were not significant variables.

Also on the basis of physico-chemical properties of soil, Singh (1969) recognized nine grassland communities at Ujjain in Madhya Pradesh. The soil under the *Iseilema–Indigofera* community was poorer in nitrogen (0.001–0.486%) and organic carbon (0.061–2.60%) compared to that under the *Sehima–Chrysopogon* community (nitrogen 0.0215–0.653%, and organic carbon 0.13–4.90%).

On the Aravalli Ranges (Rajasthan, India), Kumar and Shankar (1985, 1987) identified seven grassland communities within the desert thorn-forest formation in relation to variations in topography, soil depth, and soil texture. These are:

Aristida spp. type
Cenchrus ciliaris–Cenchrus setigerus
Chrysopogon fulvus–Apluda mutica
Dactyloctenium scindicum–Eleusine compressa
Dichanthium annulatum–Desmostachya bipinnata–
    Cynodon dactylon
Oropetium thomaeum–Eragrostis ciliaris
Sporobolus marginatus–Chloris virgata

Seasonality in vegetation expression and canopy architecture

The strong periodicity in climatic elements causes marked seasonal variations in species composition, phenology, density and biomass. The

TABLE 4.2

Edaphic communities at Sagar (India), based on phytosociological characteristics (from Pandeya, 1964)

Grassland association	Predominant soil type
Themeda quadrivalvis	Stabilized black cotton soil
Cymbopogon–Eulalia	Drifted soil on lime bed
Sehima–Chrysopogon–Tripogon	Thin, coarse, lime-rich soil
Bothriochloa–Dichanthium	Drifted soil not on lime bed
Heteropogon–Andropogon	Laterite soil
Aristida–Melanocenchris	Freshly weathered soil
Coix–Ischaemum	Drifted soil not on lime bed, but very moist
Cynodon–Bothriochloa–Dichanthium	Drifted soil not on lime bed, but moist

number of plant species with green leaves is at a maximum during the rainy season in **India** and is minimum during summer. The number of species with green leaves during the rainy season ranges from 19 (semiarid grassland) to 46 (dry subhumid grassland), whereas the number of species thriving during summer ranges from 28 (subhumid grassland) to 0 (arid and semiarid grasslands) (Singh and Krishnamurthy, 1981). This points to the preponderance of annuals, which complete their life cycle within the rainy season and endure the climatic extremes of winter and summer as seeds. Cryptophytes are the dominant perennials, at all seasons, as these plants are adapted (by subsurface perennating buds) to withstand intensive grazing (Singh and Joshi, 1979a).

Information available on phenology and density of vegetative cover in **Indian** tropical grasslands has been reviewed by several authors (Yadava and Singh, 1977; Singh and Joshi, 1979a; Singh and Krishnamurthy, 1981). With the advent of the monsoon during the last part of June, a large number of new plants and sprouts appear. Continued seed germination and sprouting increases the vegetation density, which peaks late in the rainy season. At that time, a majority of species flower, fruit, and produce mature seeds. Density then decreases rapidly because of mortality of rainy-season annuals and of seasonal tillers of perennial species. Except for the very dry grasslands, this is usually followed by a second or third peak in density because of sprouting of perennials caused by late-winter rainfall and a rise in temperature in March. Although activation of flowering and fruiting in several perennial species occurs during winter, that of other species is favoured by the longer photoperiods and higher temperatures of summer. Thus, some species attain maximum density during the rainy season, others during winter or summer, and some reach two peaks each year (Singh and Yadava, 1974). A similar characterization of species is possible, based on the seasonality of their biomass development (Singh, 1968). This variable temporal pattern of growth reflects differential adaptations and strategies of species in response to seasonal changes in growing conditions.

Depending on growth form, the pattern of vertical distribution of biomass in grassland species is variable. Integration of species with divergent patterns of biomass distribution and different temporal growth patterns into a community results in a multi-layered, dynamic canopy in which each quantum of solar radiation has a greater probability of being intercepted and used than in a single-layered canopy (Singh and Joshi, 1979a). The maximum height of the canopy is limited to about 75 cm in semiarid grasslands (Pandeya et al., 1977), but reaches 200 to 300 cm or more in more humid grasslands (Singh and Yadava, 1974; V.P. Singh et al., 1979). The various layers of vegetation are dominated in different months by different species, according to their light- and moisture-adaptation characteristics, permitting the community to maintain year-long productivity. Furthermore, the height of the canopy adjusts itself to variations in growing conditions, to permit the best use of the opportunities offered by the environment. For example, canopy height is at a maximum during the rainy season, when water supply is abundant and when both species diversity and density of plants are maximal (Singh and Yadava, 1974). Efficiency of energy capture is positively related to chlorophyll content and leaf-area index of the canopy (Mall et al., 1974).

Community-level phenology of a grassland dominated by *Danthonia cachemyriana* above the tree line in the central Himalaya (Fig. 4.3) was examined by Ram et al. (1988). Initiation of growth synchronized with the rise in spring temperature and the resultant snow melt. Of the 142 species examined, 80% were forbs, 17% grasses, and the remaining 3% were evergreen shrubs and semi-shrubs. About 5% of the species completed their growth cycle (from the initiation of growth to the beginning of senescence) within a 2-month period, 54% within from 2 to 4 months, and in 41% of the species the growth cycle extended to more than 4 months. The peaks of the various phenophases — sprouting of propagules and germination of seeds, vegetative growth, flowering, fruiting, seed formation, and senescence — succeeded one after another over a short period of about 4 months, from early June to early October. There was a tendency for species of short growth forms to attain peak growth earlier than the taller forms. However, as the season progressed into the late snow-free period, all species tended to complete their growth cycle in the remaining short time available to them. Consequently in all growth forms, peaks in flowering and seed formation occurred over a relatively short period of time, and senescence and seed setting coincided in several species.

Temporal changes in community biomass also reflect seasonality of climate. Data from a large number of sites in the **Indian** subcontinent indicate that, following the onset of the monsoon in June, the biomass of green shoots actively increases and peaks between July and October. The temporal variation exhibited in attaining peak biomass in different grasslands is related to variation in quantity and frequency of rainfall. In the semiarid grassland at Khirasara, where annual rainfall is limited to about 325 mm, peak biomass occurs during July (Krishnamurthy, 1976), whereas in the more humid grasslands (such as at Sagar where annual rainfall is 1250 mm) peak biomass occurs during October (Jain, 1980). Following this peak and after maturity of rainy-season herbage, the biomass of green shoots declines because of its conversion to dead shoots and litter. This period of decline may be followed by a second or third limited wave of growth during winter or summer, depending on the quantity of residual soil water and winter rains. The standing crop of dead shoots and litter increases in the post-monsoon period. The peak in under-ground plant biomass precedes, synchronizes with, or succeeds the peak growth of green shoots, occurring in either the monsoon or winter season (Singh, 1973). At the time of its peak, more than 75% of the total under-ground mass is concentrated in the upper 10 cm of soil, which causes these grasslands to function as a pulsed system with quick response to rainfall events (Singh and Krishnamurthy, 1981).

STANDING CROP AND PRIMARY PRODUCTION

Standing crop

Information about standing crop and the rate of primary production in tropical grasslands of **India** has been reviewed by several authors (Singh, 1976; Yadava and Singh, 1977; Singh and Joshi, 1979a, b; Singh and Krishnamurthy, 1981; Melkania and Singh, 1989). Maximum values of plant biomass of various primary producer compartments in several grasslands are summarized in Table 4.3.

Maximum green-shoot biomass ranges from 46 g m^{-2} in the semiarid grassland at Khirasara to about 1974 g m^{-2} in dry subhumid grassland at Kurukshetra. In the subhumid grassland at Varanasi (dominated by *Eragrostis*) the peak canopy biomass is reported to be 3296 g m^{-2}. In general, the grasslands in arid and semiarid regions support relatively lower biomass because of the limited availability of soil water. In alpine grassland of the central Himalaya (Figs. 4.2 and 4.3) the peak live-shoot biomass, which was attained in August, ranged from 382 to 409 g m^{-2} (Ram et al., 1989). The peak biomass of dead shoots ranges from 18 to 1268 g m^{-2} and that of litter from 19 to 433 g m^{-2}. The maximum under-ground plant biomass ranges from 51 to 2368 g m^{-2}.

Biomass levels for canopy and under-ground parts in *damana* grazing-lands (1800–2500 mm annual rainfall) in **Sri Lanka**, are presented in Table 4.4.

The relationship between the availability of soil moisture and the proportion of under-ground biomass compared to green canopy biomass in various **Indian** grasslands was examined by Singh and Krishnamurthy (1981). Grasslands in areas with annual rainfall less than 800 mm showed a curvilinear decrease, with increasing rainfall, in percentage of total biomass that occurs under ground (Fig. 4.4). Apparently, this is because low rainfall coupled with high temperature retards growth. In arid areas, even a slight increase in rainfall promotes shoot growth and lateral spread of annuals. Under favourable growing conditions in more humid areas, annuals are replaced by perennials which have their storage organs under ground. This is reflected in an increased proportion of root material as compared to green canopy biomass. However, the proportion of root material does not increase when annual rainfall is more than 1200 mm, because a more even balance is attained between growth of shoots and under-ground parts. This reflects the strategy of vegetation in responding to availability of soil water. In areas of low rainfall, the photosynthetic canopy increases when enough soil water is available; the root system is shallow and absorbs water rapidly from upper layers of soil, following a rain-

TABLE 4.3

Maximum values of biomass of various primary producer compartments and productivity of certain grasslands of India (all values in g m^{-2})

Location	Grassland type	Above-ground biomass			Under-ground biomass	Production yr^{-1}	
		green shoots	dead shoots	litter		above-ground[1]	under-ground
Ambikapur	Mixed grass	423[2]	–	–	–	436	563
Behrampur	*Cynodon dactylon*	1172[2]	–	–	1992	571	1361
Delhi	*Heteropogon*	771[2]	–	–	–	798	–
Jhansi	*Sehima–Heteropogon*	1408[2]	–	226	333	1019	497
Jodhpur	Mixed grass	164[2]	–	–	780	164	570
Khirasara	Plains, ungrazed	201	178	40	205	201	155
Khirasara	Plains, grazed	90	57	40	160	98	247
Khirasara	Hilltops, grazed	62	31	21	52	83	96
Khirasara	Hill bases, grazed	52	18	19	67	92	137
Kurukshetra	Mixed grass	1974	1268	300	1167	2407	1131
Kurukshetra	*Desmostachya bipinnata*	838	740	227	2868	862	1592
Kurukshetra	Mixed grass	424	306	231	1040	617	785
Kurukshetra	*Sesbania bispinosa*	1921	1440	331	900	2143	998
Pilani	Mixed grass	76	27	31	86	217	61
Ratlam	*Sehima nervosum*	363	316	275	873	433	399
Sagar	*Heteropogon–Apluda–Cymbopogon*	572	518	433	1381	914	937
Sambalpur	*Andropogon muricatus–Saccharum spontaneum*	572[2]	–	–	2368	458	1972
Udaipur	*Apluda*	256[2]	–	–	410	237	270
Udaipur	*Apluda*	225[2]	–	–	379	211	227
Udaipur	*Apluda*	394[2]	–	–	495	375	331
Udaipur	*Apluda*	207[2]	–	–	346	201	192
Ujjain	Burned grassland	838	346	185	1063	–	–
Ujjain	Control grassland	643	316	115	920	–	–
Ujjain	*Dichanthium annulatum*	457	422	423	925	520	464
Varanasi	*Desmostachya bipinnata* upland	2360[2]	–	145	788	2218	1377
Varanasi	*Eragrostis nutans* lowland	3296[2]	–	152	1282	3396	1161

[1] Above-ground production was estimated by trough-peak analysis (see text).
[2] Includes biomass of dead shoots. The data sources are identified in Singh and Krishnamurthy (1981). In addition, Gupta and Singh (1982a), Vyas and Vyas (1978), and Mall and Mehta (1978) are cited, respectively, for Kurukshetra, Udaipur, and burned and control grassland at Ujjain.

fall event, before it is evaporated. In areas of moderate rainfall, competition for soil water by under-ground parts is more intense, because of a general stimulation of growth; therefore, increases in rainfall result in an increase in root material as compared to canopy biomass. Apportioning a large proportion of energy to canopy development results in increased competition for light. In areas of high rainfall, a balance is attained between competition for soil water and light, resulting in an approach of the proportion of under-ground plant parts to an asymptote as rainfall increases.

The relationship between under-ground plant biomass and green-shoot biomass is shown in Fig. 4.5. The grasslands responsible for the lower part of the curve are in dry areas and many of the plants are annuals. The curve shows that an increase in shoot biomass beyond a certain value does not influence the quantity of root biomass; evidently a considerable amount of net production is allocated to the development of structural tissue in the enlarged canopy. The grasslands in

TABLE 4.4

Maximum standing crop in four communities of grazed "damana" grassland (1800–2500 mm annual rainfall) in Sri Lanka[1]

Plant community	Biomass (kg ha^{-1})	
	shoots	roots
Perennial grass cover:		
Sporobolus maderaspatanus	230	57
Brachiaria–Digitaria	375	125
Mixed perennial–annual cover:		
Brachiaria–Eragrostis	155	100
Annual herb cover:		
Zornia–Desmodium	130	70

[1] From Mueller-Dombois and Cooray (1968).

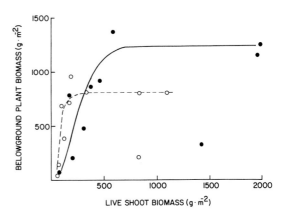

Fig. 4.5. Relationship between under-ground plant biomass and green-shoot biomass in ten grassland communities of India: dots are peak values; circles are season-long, time-weighted mean values.

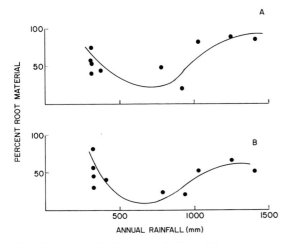

Fig. 4.4. Relationship between rainfall and the percentage of total plant biomass that is composed of under-ground parts in 10 grassland communities of India. A. Season-long, time-weighted mean values. B. Values at the time of maximum green-shoot biomass.

the vicinity of the plateau of the curve are more exploitable in terms of herbage, while the others are more sensitive to grazing pressure due to disturbance of the under-ground plant compartment.

Net production

The annual above-ground net primary production[1] (ANP) of biomass ranges from 83 to 3396 g m^{-2} among grasslands in **India** (Table 4.3). The values for the semiarid grasslands are lower than those for more humid grasslands.

The maximum rate of production invariably occurs during the rainy season, as this period coincides with the major growth phase of annuals and of seasonal tillers in perennials. Rainfall pattern and species composition determine the periodicity of production (Singh and Krishnamurthy, 1981). It is limited to the rainy season in some grasslands; in others it occurs both during the monsoon and winter seasons; and in still others production occurs in all three seasons. In some situations, light grazing may increase productivity (Singh, 1968; Singh and Misra, 1969; Kumar and Joshi, 1972), whereas excessive grazing, particularly in semiarid grasslands, reduces it considerably (Krishnamurthy, 1976). In semiarid grasslands, ANP is considerably affected by fluctuations in rainfall. For example, ANP in Khirasara grasslands was greater in years of high rainfall: 828 g m^{-2} in 1970–71 with rainfall of 1150 mm (Mankand, 1974); 400 g m^{-2} in 1971–72 with rainfall of 445 mm (Jain, 1976); and 201 g m^{-2} in 1972–73 with 325 mm of rainfall (Krishnamurthy, 1976).

Considerable variation (61–1972 g m^{-2}) occurs also in the rate of net under-ground production

---

[1] In most of the studies cited, net above-ground production was estimated as the sum of peak biomass attained by each major species and group of minor species. In a few instances it was calculated by "trough–peak" analysis — increases in green biomass plus increases in dead shoots during periods when green biomass also increased. Simultaneous losses (i.e. by decomposition) were not taken into account.

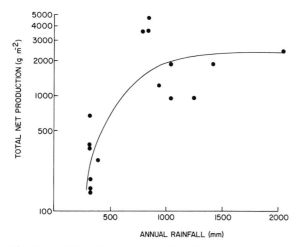

Fig. 4.6. Relationship between total net biomass production of above-ground and under-ground parts and annual rainfall in Indian grasslands.

among grasslands in different climatic regions (Table 4.3).

Singh and Krishnamurthy (1981) demonstrated a functional relationship between total net production (TNP) and annual rainfall for 16 Indian grasslands on a semilog plot (Fig. 4.6). Net dry-matter output exponentially increases with increasing rainfall in semiarid to dry subhumid regions, where water is the factor limiting growth. The vegetation acts opportunistically by taking full advantage of increased soil water for growth. TNP values become asymptotic with further increase in rainfall; some other factor — perhaps light — evidently becomes limiting (as canopy biomass increases, mutual shading also increases). Another factor influencing net production — often more important than rainfall for a given locality — is species composition (Singh and Joshi, 1979a). Gupta and Singh (1982a) reported that, among three floristically different stands in the same grassland and in the same year, TNP ranged from 1403 to 3141 g m^{-2}.

J.S. Singh et al. (1979) calculated the proportion of total incident short-wave solar radiation fixed by photosynthesis in grasslands of various ecoclimatic zones to be 0.005% in semiarid regions, 0.5% in subhumid regions, 0.2% in moist subhumid regions, and 0.26% in humid regions.

The maximum net yield of herbage in monospecific reseeded stands of grasses and forbs in grasslands in the arid zone of **Iran** occurred at flowering time in spring (Nemati, 1985). Most species (harvested at ground level for hay) produced more than 1 t ha^{-1} (100 g m^{-2}) of air-dry herbage; for some species the values exceeded 4 t ha^{-1} (400 g m^{-2}). However, for the 17 species studied, the average air-dry forage yield was approximately 600 kg ha^{-1} (60 g m^{-2}).

In the **Philippines** (Bukidnon province), the above-ground fresh-weight yield of *Imperata cylindrica* ranged from 300 to 500 g m^{-2}, compared to 400 to 2500 g m^{-2} for pasture seeded to *Pennisetum purpureum* (Zablan, 1970). In the *Imperata–Themeda* native range at Carranglan, Nueva Ecija, the net above-ground yield was 0.44 g m^{-2} day^{-1} (total of 80 g m^{-2} for 6 months) when fertilized with 22 kg of phosphorus per hectare (Sajise et al., 1974). Net production increased to 2.54 g m^{-2} day^{-1} (total 480 g m^{-2} for 6 months) when the rangeland was fertilized with 150 kg of nitrogen ha^{-1} (with no phosphorus). Under field conditions, when light and nutrients were not limiting, Sajise (1972) reported that, during a 6-month period, four rhizome buds of *Imperata cylindrica* set out in 1 m^2 plots produced an average rhizome biomass of 495 g of dry matter; the total dry matter accumulated, including shoots, was approximately 112 g m^{-2} month^{-1}.

In the highlands of northern **Thailand**, the peak mean dry-matter yield of a 5-yr-old sward was 4192 kg ha^{-1} (419 g m^{-2}) (Falvey et al., 1979b). The yield of plots harvested for the second time varied from 461 to 743 kg ha^{-1} (46–74 g m^{-2}) during the 126-day regrowth period. The rate of dry-matter accumulation of *Imperata cylindrica* was 4.2 g of dry matter m^{-2} week^{-1} at a cutting frequency of 12 weeks.

Turnover of shoots and under-ground parts

Calculations suggest that annual turnover rates (ratio of ANP to maximum canopy biomass) of shoots in **Indian** grasslands range from 0.49 to 1.55, averaging about 0.80 (Singh and Krishnamurthy, 1981). Turnover rates for under-ground parts are also high, ranging from 0.44 to 2.06. These high rates of turnover evidently result from the occurrence of more than one flush of growth each year, from limited longevity of shoots and roots, and from the pulsed nature of the system. Turnover rates in relatively dry regions and in grazed grasslands are more rapid than in moist and ungrazed grasslands.

Nutrient uptake and restitution

Marked variations occur in the rate of uptake of nitrogen and phosphorus among grasslands in various climatic zones (J.S. Singh et al., 1979). Nitrogen uptake ranges from 2.9 g m^{-2} yr^{-1} (in semiarid areas) to 25.6 g m^{-2} yr^{-1} (in dry sub-humid areas); phosphorus uptake ranges from 0.5 to 5.0 g m^{-2} yr^{-1}. Of the amount absorbed by the vegetation, from 59% (dry subhumid) to 78% (semiarid) of the nitrogen and from 15% (humid) to 98% (semiarid) of the phosphorus is returned to the soil. The grasslands of dry regions appear to need less nitrogen to support each unit of energy flow than do those in humid regions.

STABILITY

A considerable amount of surplus dry matter accumulates under protection in these successional tropical and subtropical grasslands (Singh, 1976; Singh and Joshi, 1979a). A comparison of rates of carbon output in soil respiration and of decomposition of organic matter in tropical grassland suggested that about 35% of decomposing plant organic matter is conserved as new soil organic matter (Gupta and Singh, 1982b). This indicates an accumulation of energy, which provides a basis, under protection from grazing, for succession towards woodland. On the other hand, in the alpine grassland of the central Himalaya, the total amount of dry matter that disappeared almost balanced the total net production of organic matter, indicating that the system is in equilibrium (Ram et al., 1989).

The ratio of TNP to total time-weighted biomass has been used as an index of stability (Singh and Krishnamurthy, 1981). This stability index proportionally reflects the "frequency of fluctuations" in the system and is inversely related to stability (Margalef, 1968). Grasslands in areas of low rainfall show an exponential increase in stability with increasing rainfall (Fig. 4.7). In dry areas, the frequency of fluctuations in biomass is greater because it depends on fluctuations in rainfall. Biomass fluctuations are fewer and stability

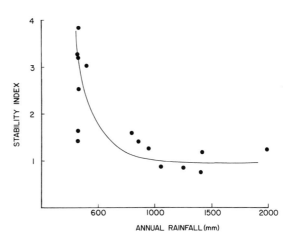

Fig. 4.7. Relationship between stability (ratio of total net biomass production to maximum standing crop of both above-ground and under-ground parts) and annual rainfall in Indian grasslands. The lowest values indicate greatest stability.

is greater in high-rainfall regions, where rainfall oscillations are less pronounced.

WILD ANIMALS

Information about the identity and activities of vertebrates and invertebrates occurring in the grasslands of southern Asia is scanty.

Vertebrates

Prakash (1974) listed 54 species of mammals, belonging to the orders Artiodactyla, Carnivora, Chiroptera, Insectivora, Lagomorpha, Pholidota, Primates and Rodentia, which occur in the desert grasslands of **India**.

Gerbillus nanus indus (gerbil), *Meriones hurrianae hurrianae* (desert gerbil) and *Tatera indica indica* (Indian gerbil) are the commonest rodents in the grass–scrub vegetation at Jodhpur (Prakash and Jain, 1971). The Artiodactyla depend on grasses and crops grown in the desert, and *Gazella gazella bennetti* (Indian gazelle) also feeds on various desert grasses in drought periods (Prakash, 1974). *Meriones hurrianae hurrianae* and *Tatera indica indica* feed mostly on seeds during winter, on rhizomes and stems in summer, and on leaves and flowers during the rainy and post-monsoon seasons (Prakash, 1962). They feed on insects in periods of scarcity.

Prakash (1969) reported that at Maulasar (in Rajasthan) *Meriones hurrianae hurrianae* feeds mainly on grasses in the monsoon season and that the annual food requirements of a population of 477 of these rodents in a fenced area of 1 ha was 1044 kg of grasses. The mean monthly body weight of adults of this species fluctuates between 60 and 70 g, and the pattern of body weight varies according to availability of food. The maximum weight is reached during the monsoon season (when green food is abundant) and is at a minimum during summer (when food is acutely scarce) (Prakash, 1972). However, rodent biomass seems to be independent of the density (standing crop) of the vegetation (Prakash, 1975).

The hare *Lepus nigricollis dayanus* is very common in arid rangelands. This species litters throughout the year, and maximum production occurs during the monsoon when green forage is abundant (Prakash and Taneja, 1969).

According to Rodgers (1988), the regional distribution of ungulates in Indian grasslands is as follows:

Alluvial grasslands:
  Axis porcinus — hog deer
  Bubalus bubalis (Fig. 4.8) — wild water buffalo
  Cervus duvauceli duvauceli — swamp deer or "bara singha"
  Rhinoceros unicornis — Indian rhinoceros

Arid grasslands:
  Antilope cervicapra — black buck
  Equus hemionus khur — wild ass
  Gazella gazella — chinkara

High-altitude grassland:
  Capra ibex — ibex
  Equus hemionus kiang — Tibetan wild ass
  Hemitragus jemlahicus — tahr
  Nemorhaedus goral — goral
  Ovis ammon — argali
  Ovis orientalis — urial
  Pantholops hodgsoni — chiru
  Procapra picticaudata — Tibetan gazelle
  Pseudois nayaur — bharal

Secondary grasslands:
  Antilope cervicapra — black buck
  Axis axis (Fig. 4.9) — chital
  Bos gaurus — gaur
  Gazella gazella — chinkara

In protected areas, the density of grazing ungulates varies widely (Rodgers, 1988), as follows:

Species	Mean live weight (kg)	No. km^{-2}
Antilope cervicapra	80	75–120
Axis axis	37	28–30
Axis porcinus	40	18–20
Cervus duvauceli duvauceli	180	2–40

Aleem (1978) reported on the re-introduction of *Antilope cervicapra*, which was once a native of the Cholistan Desert in **Pakistan**. Three males and seven females were placed in an enclosed area in 1970. By 1978, the population increased to 22. Other native species were introduced in the same area, including *Axis porcinus*, *Boselaphus tragocamelus* (nilgai) and *Gazella gazella*. Game birds occurring in this area include *Francolinus francolinus* (black partridge), *F. pordicerianus* (grey partridge) and *Pterocles* spp. (sand grouse); predators include *Felis caracal* (carcal cat), *F. libyca* (desert cat) and *Viverricula indica* (small Indian civet). Other mammals in the area are *Canis aureus* (jackal), *Hystrix indica* (porcupine), *Lepus nigricollis* and *Vulpes bengalensis* (Indian fox).

Some important animals in the grasslands of **Sri Lanka** are *Bubalus bubalis*, *Cervus* spp., *Elephas maximus* (Asian elephant) and *Panthera pardus* (leopard). *Cervus* spp. are commonly found in wet *patana* grasslands. The savannas include *Elephas maximus*, *Bubalus bubalis* and *Panthera pardus*. The dry *damana* grassland area includes one of the important national parks (Ruhuna) of Sri Lanka, where large numbers of *Bubalus bubalis* occur, and *Axis* sp. (axis deer) is also present. During the wet season, *Elephas maximus* comes to these areas to feed. Several other herbivores thriving in these dry *damana* grasslands include *Lepus* sp., *Panthera pardus*, *Sus scrofa cristatus* (wild boar), several species of deer, and many small mammals. A wide variety of snakes are present in the savannas.

Mueller-Dombois and Cooray (1968) reported that the feeding activity of *Elephas maximus* on short-grass cover of dry *damana* grasslands results in patchy scalping[1] and disappearance of grasses. McKay (1973) found that the preferred plants

---

[1] **Scalping** refers to the pulling out of the soil of whole plants with shallow root systems.

GRASSLANDS OF SOUTHERN ASIA

Fig. 4.8. *Sehima–Heteropogon* grassland used by buffaloes (at Jhansi, India). The dominant species are *Heteropogon contortus* and *Sehima nervosum*.

Fig. 4.9. A herd of *Axis axis* (chital) in grassland in the Kanha National Park (Madhya Pradesh, India).

grazed by *E. maximus* in the Gal Oya area are *Carex* sp., *Cymbopogon nardus* var. *confertiflorus*, *Cynodon dactylon*, *Digitaria* sp., *Eleusine indica* and *Eragrostis* sp. In an area of 1792 km^2 of Gal Oya National Park, the population of *Elephas maximus* was 300.

In **South-east** Asia, the grassy and swampy areas are the favourite habitat for several species of large mammals, including *Axis porcinus*, *Bubalus bubalis*, *Cervus duvauceli duvauceli*, elephant and rhinoceros (Pfeffer, 1974). The elephant (*Elephas maximus*) is prevalent from Assam to Sumatra. The three Asian species of rhinoceros are *Didermocerus sumatrensis* (Sumatra rhinoceros), *Rhinoceros sondaicus* (Java rhinoceros) and *R. unicornis* (Indian rhinoceros). *Didermocerus sumatrensis* and *Rhinoceros sondaicus* are on the verge of extinction, but a few individuals are found in the forest reserves. *Rhinoceros unicornis* occurs in two reserves in Assam (Kaziranga) and west Bengal (Jaldapara), and remains in the Nepati valley in Nepal. *Bubalus bubalis* has almost disappeared in its wild state. The most typical of the various species of deer in the humid grassy zones are *Axis porcinus* and *Cervus duvauceli*.

Invertebrates

Several studies were made of population density and biomass of above-ground insects in a grassland at Kurukshetra (**India**) (Vats and Singh, 1978; Kaushal, 1979; Gupta, 1980). In a stand dominated by *Desmostachya bipinnata*, Vats and Singh (1978) found that herbivores constituted 84% of the total population and predators accounted for 11%. Kaushal (1980) reported that the maximum population density of the total insect fauna was 40.5 m^{-2} and the maximum biomass was 5880 mg m^{-2}. Orthoptera were the predominant order, with a peak density of 38.5 m^{-2} and peak biomass of 5816 mg m^{-2}. The total density and biomass of Orthoptera were highest in the rainy season and lowest during winter (Gupta, 1980). The activity of the termite *Odontotermes gurdaspurensis* was greatest during early winter, and the probability of attack on litter and dead roots (that is, the proportion affected) ranged between 0.44 and 0.97 (44 and 97%) (Gupta et al., 1981).

In a moist grassland at Behrampur (Orissa), the biomass of enchytraeids ranged from 21 mg wet weight m^{-2} (April) to 240 mg (December); the total oligochaete population ranged from 560 to 7800 m^{-2} (Thambi and Dash, 1973). Dash et al. (1974) estimated that the annual secondary production of Megascolecidae and Ocnerodrilidae in this grassland was 140 g wet weight m^{-2}. In relatively humid rangeland at Sambalpur (Orissa), Senapati and Das (1981) reported the presence of five species of earthworms: *Drawidia calabi*, *D. willsii*, *Lampito mauritii*, *Ocnerodrilus occidentalis* and *Octochaetona surensis*. These oligochaetes were the dominant soil invertebrates, contributing about 80% of total invertebrate biomass. The average oligochaete live biomass amounted to 56 g m^{-2} (11 g dry weight) and 41 g m^{-2} (8 g dry weight) in the ungrazed and grazed plots, respectively. Secondary production of oligochaetes was 26.3 to 31.2 g dry weight m^{-2} yr^{-1} (510–600 kJ m^{-2} yr^{-1}). The turnover of oligochaete biomass (ratio of secondary production to average monthly biomass) varied from 2.4 to 3.9, and secondary production accounted for 3.2 to 4.6% of the total net primary production in ungrazed and grazed grassland. The energetics of earthworm populations of *Lampito mauritii* (in lowland range) and *Octochaetona surensis* (in upland range) was studied by Senapati and Das (1983). They found that the ratio of energy utilized in respiration to secondary production was higher in upland range. Dash and Patra (1977) and Senapati and Dash (1981) suggested that earthworms utilize about 15% of the total energy input into the decomposer system of tropical grasslands in Orissa.

MICRO-ORGANISMS AND THEIR ACTIVITIES

The population of microfungi in soil and their qualitative distribution among Ascomycetes, Deuteromycetes and Phycomycetes were studied by Dwivedi (1965) in six grass communities at Varanasi (**India**). The population of microfungi was highest in the *Dichanthium–Cynodon* community (27×10^3 to 76×10^3 g^{-1} of dry soil). The population was usually greatest in the 0- to 15-cm layer. Mishra (1966) isolated 48, 46, 45 and 38 fungal species, respectively, from the soil of stands of *Desmostachya bipinnata*, *Vetiveria zizanioides*, *Saccharum spontaneum* and *Dichanthium annulatum*. The fungal population increased with the onset of rains and was highest in October. In a semiarid grassland, the bacterial population in the 0- to 9-cm soil layer ranged from 6.23×10^7 to 8.37×10^7 g^{-1} (Pandeya and Jain, 1979). The fungal population increased with the onset of rains and was highest in October. In a semiarid grassland, the bacterial population in the 0- to 9-cm soil layer ranged from 6.23×10^7 to 8.37×10^7 g^{-1} (Pandeya and Jain, 1979).

Pandeya et al. (1986) made monthly enumerations of bacteria in soil of a grazed rangeland at Rajkot (India). In the surface layer, the number of bacteria was lowest in winter. Galleries built on litter and standing-dead shoots by the termite *Odontotermes gurdaspurensis* were studied by Gupta et al. (1981). Populations of micro-organisms (bacteria, fungi and Actinomycetes) were higher in termite-modified soil from these galleries than in the surrounding soil. Total populations of micro-organisms (colonies $\times 10^4$ g^{-1} of dry soil) varied from 68.4 to 95.7 in unmodified soil and from 100.8 to 151.0 in termite-modified soil.

In a tropical grassland at Kurukshetra, the succession of fungi on plant material of *Desmostachya bipinnata*, mixed grass species, and *Chenopodium album* was studied by Mehrotra and Aneja (1979) and Aneja and Mehrotra (1979, 1980a, b). On the decomposing leaf litter of *Desmostachya bipinnata* (kept on the soil surface, or buried at 5 cm depth), the predominant species of fungi were *Acrophialophora fusispora*, *Aspergillus* spp., *Curvularia* spp., *Penicillium* spp. and *Periconia minuissima*. Peak populations of bacteria and fungi occurred during rapid phases of decomposition. Microbial populations declined towards the final stages of decomposition. On *Chenopodium* leaf litter, the dominant fungi were *Aspergillus* spp., *Curvularia* spp., *Fusarium* spp., *Myrothecium roridum* and *Penicillium* (Mehrotra and Aneja, 1979). The succession of fungi on decomposing litter showed that initial colonizers on senescent tissues were species belonging to Deuteromycotina; with the progress of decomposition, the species of Ascomycotina were co-dominant with those of Deuteromycotina.

Sixteen species of microfungi were isolated by Dash et al. (1979) from soil and earthworm tissue in a tropical rangeland. *Thielavia terricola* was isolated from the senescent tissue of *Octochaetona surensis*; *Rhizopus nigricans* and *Syncephalastrum racemosum* were isolated from the bodies of diapaused earthworms; and *Aspergillus* spp., *Fusarium* spp. and *Rhizopus nigricans* were isolated from dead cocoons and empty cocoon cases.

Pemadasa and Mueller-Dombois (1979) studied soil fungi in montane grasslands (*patanas*) in **Sri Lanka**. They affirmed the ecological distinctiveness of five major zones within *patanas* (as described on pp. 92–93) based on ordination of soil fungi. *Circinella umbellata* and *Mucor genevensis* were confined mainly to wet *patanas*, whereas *Circinella* sp., *Cunninghamella* sp., *Helicostylum pyriforme* and *Penicillium javanicum* extended from wet *patanas* to the humid-zone dry *patana* and the intermediate *patana*. *Phoma* sp. and *Syncephalastrum racemosum* were more restricted in distribution, occurring only from the upper wet to the humid-zone dry *patana*. *Mucor microsporus* and *Thielavia* sp. were common in the dry *patanas*. *Penicillium wortmanii* and *Rhizopus nigricans* were confined primarily to the humid-zone dry *patana*.

In a grassland at Kurukshetra (**India**), microbial activity (as estimated by output of carbon dioxide from soil) was highest in the rainy season, moderate in summer, and least during winter (Gupta and Singh, 1981a). Decomposition of above-ground litter and roots was also highest during the rainy season and lowest in winter (Gupta and Singh, 1981b). Total output of carbon dioxide from soil in this grassland amounted to 470 g C m^{-2} yr^{-1}, and the rate of disappearance of above-ground plus under-ground biomass was equivalent to 724 g C m^{-2} yr^{-1}. Thus, the carbon dioxide flux from soil was less than the organic-matter input (779 g C m^{-2} yr^{-1}) and actual decomposition of above-ground and under-ground litter. This carbon balance indicated about 35% conservation of plant debris as new soil organic matter (Gupta and Singh, 1982b). In the grassland soils of Orissa, Dash et al. (1985) reported that total microbial biomass (estimated by the fumigation technique) varied from 170 to 310 mg C kg^{-1} dry soil.

Decomposition rates of the roots of five species at depths of 5, 15, 25 and 35 cm under ground in a dry subhumid grassland indicated highest loss of mass at a depth of 5 cm, due to higher activity of micro-organisms and soil fauna in surface layers. At greater depths, the decomposition rates were comparatively low. A multiple correlation between mean relative decomposition rates and weather variables (temperature, rain per day, and percentage of days with rainfall) suggested that from 13 to 39% of the variability in decomposition of litter materials and from 22 to 52% of the variability in that of root materials was due to the combined effect of temperature and rainfall (Gupta and Singh, 1981b). For the grazing lands at Khirasara (near Rajkot, India), Pandeya et al. (1986) reported that 33% of the roots disappeared during October and November (after the monsoon), and 27% in December and January; there was little root decomposition during spring months.

The contribution of root respiration to total soil metabolic activity has been reported by Gupta and Singh (1981a) and Pati et al. (1983). For the grassland system at Kurukshetra (Haryana), the root respiration accounted for 42% of total carbon-dioxide evolution. For the tropical grassland of Orissa (Pati et al., 1983), 62% of the total carbon-dioxide evolution from soil was attributed to the microbial component and 38% to the root component. Pati et al. (1983) found the ratio of fungal biomass to that of bacteria to be 12, indicating the dominance of fungal activity in the soil.

USE, MANAGEMENT AND RENOVATION

Grazing intensity and carrying capacity

The grazing load of domesticated animals in southern Asia is composed approximately 46% of cattle, 32% of water buffaloes, and 20% of goats and sheep (Table 4.5). The intensity of grazing is higher in the **Indian** subcontinent than elsewhere in this region.

TABLE 4.5

Relative grazing load [thousands of cattle units (CU)][1] of livestock relative to the area (000 ha) of permanent meadows and pastures in countries of southern Asia, 1985[1]

Country	Cattle	Buffaloes	Goats	Sheep	Asses	Horses	Camels	Mules	Total	Area (ha)	ha CU^{-1}
Afghanistan	3,750	0	1,500	5,000	1,250	410	270	30	12,210	30,000	2.46
Iran	8,350	460	6,800	8,625	1,800	316	27	123	26,501	44,000	1.66
Iraq	1,500	290	1,175	2,125	450	50	55	28	5,673	4,000	0.71
Turkey	17,300	1,088	6,550	10,098	1,226	623	3	213	37,101	9,000	0.24
South-west	30,900	1,838	16,025	25,848	4,726	1,399	355	394	81,485	87,000	1.07
Bangladesh	36,500	3,600	5,250	278	0	44	0	0	45,672	600	0.01
Bhutan	320	58	23	11	18	16	0	9	455	218	0.48
India	182,410	129,000	40,650	10,325	1,000	910	1,100	135	365,530	11,900	0.03
Nepal	7,050	9,000	1,325	638	0	0	0	0	18,013	1,786	0.10
Pakistan	16,549	26,140	14,863	6,259	2,750	451	900	64	67,976	5,000	0.07
Sri lanka	1,750	1,980	318	7	0	1	0	0	4,056	439	0.11
South-central	244,579	169,778	62,429	17,518	3,768	1,422	2,000	208	501,701	19,943	0.04
Burma	9,550	4,200	550	105	0	117	0	9	14,531	361	0.02
Indonesia	6,859	4,848	5,587	1,240	0	527	0	0	19,060	11,850	0.62
Laos	615	2,400	33	0	0	40	0	0	3,088	800	0.26
Malaysia	570	520	173	16	0	5	0	0	1,284	27	0.02
Philippines	1,900	8,650	965	8	0	300	0	0	11,823	1,140	0.10
Thailand	4,800	12,500	37	11	0	20	0	0	17,368	308	0.02
Vietnam	2,150	5,600	134	5	0	130	0	0	8,019	272	0.03
South-east	26,444	38,718	7,478	1,384	0	1,139	0	9	75,172	14,758	0.20
Total	301,923	210,334	85,931	44,750	8,494	3,960	2,355	611	658,358	121,701	0.18

[1] Livestock numbers and pasture areas from the *FAO Production Yearbook 1985* are converted arbitrarily into cattle units (CU) as follows: 1 head of cattle = 1 CU; 1 buffalo = 2 CU; 1 goat = 0.5 CU; 1 sheep = 0.25 CU; 1 ass = 1 CU; 1 horse = 1 CU; 1 camel = 1 CU; 1 mule = 1 CU. FAO defines permanent meadows and pastures as "land used permanently (5 years or more) for herbaceous forage crops, either cultivated or growing wild (wild prairie or grazing land)". Probably a significant proportion of the grazing load is supported by cropland and woodland.

Gupta (1975) estimated that the mean wet (fresh) biomass of large herbivores in the State of Rajasthan (**India**) is 124 kg ha^{-1}, of which 62% is accounted for by cattle, 16% by camels, 18% by sheep and goats, and 4% by donkeys, other ungulates and rodents. The gain in body weight of large herbivore grazers is highest during the monsoon season because of the greater variety, quantity and quality of herbage, whereas in summer body weight either is only maintained or actually declines (Ahuja, 1966; Ahuja et al., 1968; Shankarnarayan et al., 1972; Shankarnarayan et al., 1975).

Because of this large population, most of the grasslands remain under severe grazing pressure throughout the year. Many areas, especially the high-altitude rangeland and arid and semiarid grasslands, suffer from severe seasonal grazing stress caused by migrating livestock. Frequent drought and famine conditions in western Rajasthan lead to a mass-scale migration of the human and livestock population to Haryana, Punjab, Madhya Pradesh and Gujarat. For example, about 70% of the cattle and 90% of the sheep migrate for about 6 to 7 months, between January and July, from western Rajasthan to nearby areas of fair grass cover (Prajapati, 1970). Similarly, a large number of animals migrate from Gujarat to Madhya Pradesh. There are 30 nomadic and seminomadic pastoral communities in India (Singh, 1988). These are located particularly in the northern and western parts of the country in the States of Jammu and Kashmir, Himachal Pradesh, Uttar Pradesh, Haryana, Rajasthan and Gujarat. In southern India, their concentration is lower. The pastoral nomads of Himalayan states practice horizontal and vertical migration due to snow-cover at higher elevations, which necessitates a move to lower locations for the winter (Fig. 4.2). The pattern of livestock grazing in the Himalayan States of Jammu and Kashmir is summarized in Table 4.6.

Chakravarty (1971) estimated that an area of about 1 million ha in the arid and semiarid regions of India was subjected to grazing by 7.4 million cattle, 5.7 million sheep, 3.4 million goats, and 0.5 million of other kinds of livestock. The year-long carrying capacity, however, is estimated at only 2.5 sheep ha^{-1} or 1 cow for 2.4 to 4 ha. In sown pastures of *Cenchrus ciliaris* (Fig. 4.10), the year-long carrying capacity ranges from 0.6 to 6.9 cattle units ha^{-1} (165 mm to 515 mm annual rainfall), whereas for *Lasiurus hirsutus* rangeland it is from 0.29 to 7.10 cattle units ha^{-1} at a 60% utilization rate (Gupta et al., 1972). Bhimaya and Ahuja (1969) recommended stocking densities of 3 to 4, 5, 6 and 8 ha per adult cattle unit (0.12, 0.17, 0.2, and 0.25 to 0.33 adult cattle unit per hectare), respectively, in excellent, good, fair and poor arid rangelands, with corresponding air-dry forage yields of 1500, 1000, 750, and 500 kg ha^{-1} (150–50 g m^{-2}).

For the arid and semiarid regions of Rajasthan, B.C. Patnayak (unpublished) reported that a dry-matter yield of 0.3 to 1.1 t ha^{-1} is available from the rangelands, with a recommended stocking density of 0.5 to 3.0 sheep or goats ha^{-1}. A dry-matter yield of 2 to 3 t ha^{-1} in seeded pasture can support 5 to 7 sheep ha^{-1}.

Patil and Pathak (1978) reported the carrying capacity for well-developed rangelands as follows (sheep unit ha^{-1}):

Cenchrus setigerus	10.5	*Dichanthium*	
Cenchrus ciliaris	7.0	*annulatum*	6.2
Heteropogon contortus	7.0	*Chrysopogon fulvus*	4.1

They indicated that *Cenchrus setigerus* could support up to 2.1 cattle units ha^{-1}.

The livestock population is also high in **Pakistan** (Said, 1961). In addition to this resident animal population, large numbers of animals (3.74 million) are moved by nomadic grazers from Afghanistan during winter, mainly to the Quetta–Kalat region. In **Bangladesh**, the livestock population in 1956 was reported to be 16 million head (Winters, 1956), but increased to 50 million by 1985 (Table 4.5).

In **Pakistan**, Johnston and Hussain (1963) estimated that the range area (ha) required for one "cattle unit" was:

Range condition	Poor	Fair	Good
Dichanthium–Cenchrus–Lasiurus	32	16–20	10
Chrysopogon	73	20	12

In **Nepal**, the area required per animal (cattle) unit for year-long grazing is estimated as 2.5 ha (0.4 animal unit per hectare) in subtropical and warm temperate rangeland, and 1.7 ha in cool temperate and subalpine grassland (APROSC, 1979).

TABLE 4.6

Pattern of livestock grazing in various ecoclimatic zones of Jammu and Kashmir [1]

Cold arid zone
  Sheep: Grazing in subalpine ranges, supplemented by dry fodder from cropland.
  Goats
    Summer: Grazing in alpine meadows.
    Winter: Stall feeding of hay from alpine meadows.
  Cattle
    Summer: Grazing in mid-altitude subalpine meadows and stream-bank grazing lands.
    Winter: Subsistence-level stall feeding of meadow hay and of forage and residues from cropland.

Temperate zone (Kashmir Valley)
  Sheep
    Summer: Grazing in alpine and subalpine meadows and on middle slopes.
    Winter: Subsistence-level stall feeding of forage from cropland and rangeland.
  Goats
    Summer: Grazing in alpine and subalpine meadows.
    Winter: Stall feeding of forage from cropland and rangeland.
  Cattle
    Summer: Grazing on lower and middle slopes and in village rangelands.
    Winter: Subsistence-level stall feeding of forage from cropland and rangeland.

Intermediate zone
  Sheep
    Summer: Grazing in alpine and subalpine meadows of temperate Himalaya.
    Winter: Grazing on lower and middle slopes of lesser Himalaya and slopes of Shiwalik hills.
  Goats
    Summer: Grazing partly in alpine and subalpine meadows of the temperate zone and partly on slopes and high-elevation areas of outer and middle mountains.
    Winter: Grazing on lower and middle slopes of lesser Himalaya and slopes of Shiwalik hills.
  Cattle
    Summer: Grazing on middle and upper slopes of lower and middle mountains.
    Winter: Grazing on lower and middle slopes of lesser Himalaya and slopes of Shiwalik hills.

Subtropical zone
  Sheep: Year-around grazing on lower fringes of middle mountains and slopes Shiwalik hills.
  Goats: Year-around grazing on lower fringes of middle mountains and slopes of Shiwalik hills and adjacent areas.
  Cattle: Year-around grazing on lower fringes of middle mountains and slopes of Shiwalik hills, supplemented by output from cropland.

[1] After Ahmad and Tahir (1987).

Grazing by domesticated animals is very limited in the humid *patana* of **Sri Lanka**, whereas, in the summer-dry *patana* zone, grazing is continuous. The principal livestock species are cattle, buffaloes, and goats. The *villu* grasslands support a cattle population of 10,000 head on 2020 ha of land.

In the rangelands of **Iran**, the forage required by the existing number of livestock exceeds the forage-producing capacity "by at least five-fold" (Nemati, 1985). For the winter grazing areas of Mazandaran province (north-east Iran), the carrying capacity at 50% utilization rate varies from 0.9 to 5 sheep units per hectare (Mesdaghi, 1988). In rangeland of central Anatolia in **Turkey**, the actual density of stocking was 0.26 animal (cattle) unit per hectare (Erkun, 1964). In this grassland the plant cover was 27% (grasses 15%, legumes

Fig. 4.10. A seeded *Cenchrus ciliaris–Stylosanthes hamata* pasture used by heifers under rotational grazing (in India).

2%, and other forbs 10%).

Some workers have discussed the nutritive value and animal production potential of *Imperata cylindrica* in **South-east Asia** (Magadan et al., 1974; Soewardi et al. 1975; Holmes et al., 1976; Soewardi and Sastradipradja, 1976). The actual stocking density of grasslands dominated by this species has been reported to be 0.59 to 1.25 animal (cattle) units per hectare in the Philippines (Magadan et al., 1974), 0.2 to 1.0 animal unit per hectare in Indonesia (Soewardi and Sastradipradja, 1976) and 0.04 to 0.14 animal unit per hectare in the highlands of Thailand (Falvey, 1977). The rate of intake of *Imperata* may be restricted by the low digestibility of herbage after several weeks' growth.

Holmes et al. (1976) studied the digestibility of *Imperata cylindrica* using the nylon-bag technique. They found that herbage of this species was lower in relative digestibility than that of *Cenchrus ciliaris*, *Pennisetum purpureum* and *Setaria anceps* of the same age. They concluded that animal production from *Imperata cylindrica* would be expected to be lower than that obtained from other tropical grasses. Soewardi and Sastradipradja (1976) recorded a live-weight gain of 0.14 kg day^{-1} in heifers being fed herbage of *I. cylindrica* of 60.9% dry-matter digestibility. When this was supplemented with urea, carbohydrate sources, salt, bone meal and trace elements, the mean live-weight gain was increased to 0.21 kg day^{-1}. In the **Philippines**, Magadan et al. (1974) recorded live-weight gains of cattle grazing *I. cylindrica*, at a stocking rate of 1 animal ha^{-1}, as 100 kg year^{-1} (0.27 kg day^{-1}), whereas on seeded pastures of *Brachiaria mutica* and *Centrosema pubescens* the rate of production was more than three times this level.

In northern **Thailand**, cattle grazing mainly on *Imperata cylindrica* in native highland ranges gained an average of only 0.04 kg (live-weight) day^{-1} (Falvey et al., 1979a). This low level of animal production, in an environment with appar-

TABLE 4.7

Maximum standing crop of herbage and carrying capacity of grasslands[1] in Timor (Indonesia)[2]

Location	Puluthi	Konotoef	Bena I	Bena II	Bena III
Height of canopy (cm)	4.5–10.5	29–43	13–18.5	8.6–17.2	17.5–28.5
Mean basal area (cm^2 m^{-2})	116	120	72	5	140
Dry matter yield (kg ha^{-1})	212	1076	293	154	367
Required dry matter (kg (animal unit[3])$^{-1}$ month^{-1})	189	189	189	189	189
Carrying capacity (animal units ha^{-1})	0.34	1.47	0.42	0.22	0.52

[1] Floristic composition (in decreasing order of abundance of species):
  Puluthi: *Themeda sp., Brachiaria reptans, Chrysopogon* sp.
  Konotoef: *Paspalum sp., Dactyloctenium aegyptium, Bothriochloa pertusa, Digitaria* sp.
  Bena I: *Brachiaria reptans, Echinochloa colona, Brachyachne sp.*
  Bena II: *Dichanthium sp., Panicum* sp.
  Bena III: *Digitaria sp., Chloris barbata, Themeda sp.*
[2] Based on Susetyo et al. (1969).
[3] Animal unit = cattle unit = mature head of cattle.

ently higher potential, can be attributed to deficiency of nutrients, particularly nitrogen. Falvey et al. (1979a) suggested that *I. cylindrica* can support cattle at only slow rates of gain, but that these rates will probably be higher at low stocking intensities.

The estimated carrying capacities of rangelands in Timor (Indonesia) are shown in Table 4.7.

Effects of grazing

Grazing influences the structure and function of grasslands in many ways, depending on vegetation type, rainfall, and on season and intensity of grazing. Most grasslands in India, Pakistan and Sri Lanka undergo severe grazing throughout the year. Heavy grazing results in open canopies and extensive areas of exposed soil. This results in invasion by inferior annual and perennial grasses and the initiation of soil erosion. Overgrazing in moist climates (such as that of Assam) reduces the tall-grass cover to tufted-grass types of *Chrysopogon aciculatus* and *Imperata cylindrica*, whereas in Bengal overgrazing leads to replacement of grasses by *Careya herbacea*, an unpalatable shrub (Bor, 1960).

The effect of grazing on various types of grass cover in **India** was studied by Dabadghao and Shankarnarayan (1973). In *Sehima–Dichanthium* grassland, intensive grazing results in replacing the perennial communities by annuals, especially species of *Aristida, Eragrostis* and *Melanocenchris*, which are nutritionally poor. In *Dichanthium–Cenchrus–Lasiurus* vegetation in moist habitats, heavy grazing leads to deterioration of perennial cover and invasion by annual grass species, whereas in drier habitats (such as that of Rajasthan) grazing and trampling loosen the soil, allowing wind erosion and dune formation. The *Phragmites–Saccharum–Imperata* cover type is subjected to cutting, resulting in the appearance of *Phragmites karka* in marshy habitats, *Desmostachya bipinnata* and *Vetiveria zizanioides* in dry habitats, and *Sporobolus indicus* in humid areas. When heavily grazed, *Themeda anathera* is slowly replaced by *Arundinella bengalensis* and *A. nepalensis*.

In **Sri Lanka**, overgrazing is frequent during the summer in the dry-zone *patanas* and in dry *damana* grasslands. The short-grass cover is the result of excessive grazing pressure. Extensive areas of exposed soil occur in dry-zone *patanas* that are heavily grazed. Such barren spots are invaded by coarse annuals and perennials.

In **Pakistan**, animals are grazed on village commons and uncultivated lands, regardless of the number of livestock and the availability of forage (Said, 1961). The initial impact of intensive grazing is reduction in cover of palatable grasses, which are then replaced by less desirable grasses. Ultimately, the grass is grazed so short that growth ceases, the plants die, and the soil surface becomes bare and compact and is washed away by rain-water.

In addition to excessive numbers of livestock, the factors responsible for deterioration of the rangelands in **Iran** include grazing too early in the spring, grazing for too long a season (often year-long), excessive fuel-gathering of range plants needed for forage and ground cover, and widespread ploughing and cultivation of land unsuitable for sustained crop production (Nemati, 1985).

Singh et al. (1985) reported that in the **Philippines**, overstocking has led in many places to the replacement of native grasses by unwanted broadleaved species such as *Amaranthus* spp., *Chromolaena odorata* (= *Eupatorium odoratum*), *Hyptis suaveolens*, *Lantana camara*, *Pseudoelephantophus spicatus* and *Sida acuta*. Stocking of *Imperata cylindrica* rangeland at one animal unit ha^{-1} increased foliage cover of broad-leaved weeds from 7 to 65%. Soil compaction also results in replacement by species which can survive poorly drained and eroded conditions (such as *Aristida cumingiana*, *Brachiaria subquadripara*, *Chrysopogon aciculatus*, *Eragrostis zeylanica* and *Panicum walense*).

Most of the rangelands in **Indonesia** are in a critical condition due to overgrazing, degradation of vegetative cover, and erosion (Soerianegara, 1970). There is need for land-use zoning followed by rational grassland management, establishment of ranches or mixed farms, fencing, fodder crop production and hay making, forest planting to increase protective forests, weed control in grassland, and introduction of more valuable range and pasture grasses for the improvement of degraded rangeland.

In the highland region of northern **Thailand**, Falvey and Hengmichai (1979) related the disappearance of *Imperata cylindrica* through invasion by *Eupatorium* spp. (low woody perennials) to grazing intensity. Under shifting cultivation, heavy grazing of *I. cylindrica* promoted invasion by *Eupatorium* spp., which in turn allowed regeneration of forest species, because *Eupatorium*-dominated communities are less fire-prone than *Imperata*-dominated sward.

Management and renovation

Since the major cause of deterioration of grasslands is unplanned and intensive grazing, the first step for improvement is to control and regulate grazing (Figs. 4.11 to 4.13). Mann and Ahuja (1974) recommended that continuous grazing at controlled rates, based on carrying capacity, is one of the best ways to maintain ranges for sustained moderate productivity in the arid zone of **India**. Prajapati (1970) found that early deferment for four months (August to November) is beneficial. Upadhayay et al. (1971) found deferred-rotational grazing to be better for maintaining range productivity than continuous grazing at Jhansi. Bhimaya et al. (1967) observed that a 2-yr protection and controlled grazing system in the arid zone resulted in an increase in herbage yield of 148, 92, and 116%, respectively, on poor-, fair-, and good-condition ranges. Results from a simulation–optimization model showed that a combination of moderate grazing with an occasional light-grazing year (to allow the legumes to recuperate) is the best grazing strategy over a 12-yr grazing plan, for a subhumid grassland (Swartzman and Singh, 1974).

The results of range improvement studies in **Pakistan** suggest that complete protection from grazing for a period of at least five years is required to bring the grassland to the maximum level of production (Said, 1961). Khan (1966, 1970, 1971) suggested that practices for improving rangelands should include closing them to grazing, rehabilitating degraded ranges, reducing the number of livestock to the carrying capacity of the range, and using appropriate grazing intensity in subsequent years. According to Malik and Khan (1966), constructing ponds to collect rain-water, reseeding, and controlled grazing improved the range condition in parts of Cholistan. S.M. Khan (1977) reported a 3-fold increase in forage production in alpine grassland at Payer by protecting it from grazing for 5 to 7 years.

The range improvement and management practices in the arid zones of **Iran** have shown that protection from grazing increased the carrying capacity of protected rangelands (Nemati, 1978, 1985). The non-protected areas were extremely depleted of plant cover, with almost no usable forage. Protection for a minimum of five years, to-

Fig. 4.11. *Sehima nervosum–Heteropogon contortus* grassland in good condition, with a hay-stack in the background for provision of forage in the dry months (at Jhansi, India).

gether with water-conservation practices, resulted in the recovery of the native vegetation. Planting of *Atriplex* spp. for protection and water conservation in the steppe zone (100–200 mm of annual rainfall) increased the carrying capacity from 0.1 to 1.0 sheep unit per hectare within four years. Approximately 4.12 million hectares of sandy land have been stabilized by seeding such species as *Aristida pennata, Haloxylon persicum, Panicum antidotale, Tamarix pallassii* and *T. stricta*.

Basak (1964) suggested several measures for range development in **Afghanistan**, including the prevention of the conversion of rangelands to dryland cropping, reclamation of old and abandoned ranges by seeding to more suitable grasses, and provision of drinking water and shelters. Hassanyar (1988) also emphasized that essential requirements for the improvement of rangelands in Afghanistan are controlled grazing, the establishment of shelter-belts, the stabilization of sand dunes, and the availability of water for livestock. Hassanyar (1988) reported that revegetation of sand dunes can be accomplished by inter-seeding of *Aristida pennata, Calligonum comosum* and *Haloxylon persicum*. Erkun (1964) observed that, to avoid further deterioration in the grazing lands of Turkey, the reduction of livestock numbers is essential.

Reseeding the deteriorated rangeland with higher-producing grasses and legumes (both native and introduced) has also been tried with success in India (Krishnaswamy and Reddi, 1953; Whyte, 1964). Broadcast seeding without tillage gave better results with *Lasiurus hirsutus* than sowing in rows, whereas rows resulted in better establishment of *Cenchrus ciliaris*. Thorough tillage and removal of all shrubs before seeding contributed to better establishment of grasses (Chakravarty et al., 1966; Chakravarty and Verma, 1972). The monsoon season is the best season

Fig. 4.12. *Iseilema laxum* rangeland, with heavy black soil in a low-lying area, under moderate use (in north-western Uttar Pradesh, India).

for reseeding. Before species are introduced they should be screened carefully for yield and resistance to drought and grazing (Senaratna, 1934; Pande and Singh, 1981; Misra and Singh, 1981). M. Khan (1966) recommended *Diplachne fusca* as suitable for degraded grasslands in parts of Pakistan. C.M.A. Khan (1966) discussed artificial reseeding in the Thal ranges of Pakistan.

Most of the area of the lower wet *patana* grassland in Sri Lanka has been converted to seeded pastures. Wijesinghe (1962) found encouraging results in a pasture-cum-shelter-belt practice for stabilizing soils of dry *patanas* against strong winds. *Brachiaria brizantha* was seeded between rows of trees of *Acacia mollissima*, *Cupressus macrocarpa* and *Eucalyptus saligna*. Improved cultivars of pasture grasses, such as *Brachiaria brizantha*, *Panicum maximum* and *Setaria sphacelata*, have also been introduced in the dry *damana* grasslands. Sivasupiramanian et al. (1973) stressed the need for rebuilding the fertility of natural grasslands before converting them into seeded pastures.

In Rajasthan (**India**), Ahuja and Mann (1975) recognized the importance of planting trees with edible leaves, such as *Acacia nilotica* var. *indica*, *Prosopis cineraria* and *P. juliflora*, along the boundaries of arid rangelands. The trees provide nutritionally rich fodder and shade for animals and serve as wind-breaks.

The response of several pastures and rangelands to fertilization by nitrogen, phosphorus and potassium has been tested by various workers (Das and Gupta, 1964; Dabadghao et al., 1965; Chakravarty and Verma, 1970; Dabadghao and Shankarnarayan, 1970; Shankarnarayan and Rai, 1975; Shankarnarayan et al., 1976). Nitrogen was found to be especially beneficial in the majority of cases. However, the economics of fertilization vis-à-vis herbage yield, and the alternative of increasing soil nitrogen by introducing improved

Fig. 4.13. Heavily grazed *Chrysopogon fulvus–Heteropogon contortus* grassland (in Sonbhadra, Uttar Pradesh, India). *Bothriochloa pertusa* is subdominant.

legumes, have not been considered thoroughly. Experiments to control weeds in pastures and rangelands with herbicides have shown promising results (Singh et al., 1970).

The use of fire to improve grazing lands has been examined. According to Trivedi et al. (1972), controlled burning promoted *Heteropogon contortus* but not *Sehima nervosum*. Burning in *Dichanthium annulatum* grassland increased herbage yield by about 150% (Pandey, 1974). Burning of *patanas* in **Sri Lanka** is a frequent practice. Senaratna (1942) recommended burning of the *Cymbopogon* stage in wet *patanas* to favour the *Themeda* stage, which provides better forage.

Seavoy (1975) discussed the origin and succession of grasslands in Kalimantan, **Indonesia**. *Imperata cylindrica* grassland is purposely created by repeated burning of abandoned croplands created by slash and burn. Maintenance of this grassland requires repeated burning to avoid re-establishment of trees. Under controlled grazing, *I. cylindrica* is replaced by *Axonopus compressus* and *Chrysopogon aciculatus*. These grasses form a dense sod which prevents growth of *I. cylindrica*. If the grassland is then released from grazing, the grasses grow tall and the ground mat is destroyed, the fuel load increases and attracts fire, and *I. cylindrica* becomes re-established.

Thus, several means are available to restore these solar-powered, pulsed, frequently water-limited systems of high productive potential but often degraded condition (because of excessive biotic stress). These include: deferred-rotation or continuous grazing at stocking rates limited to the carrying capacity; control of shrubs and seeding of forage species; and reservation of certain areas for producing hay. Various procedures are helpful in reducing the grazing pressure on existing grasslands. These include: enhancement of the fodder supply by seeding and fertilizing pastures; provision of silva-pastoral top feed; production of fodders (maize, cow-peas and berseem) on marginal agricultural land; and rotation of short-term leys of grass/legume pasture with cash crops on arable land.

ACKNOWLEDGEMENTS

A major portion of this synthesis was accomplished and the manuscript prepared during the stay of the senior author at the Natural Resource Ecology Laboratory (NREL), Colorado State University, with the support of the Council for International Exchange of Scholars, Washington, D.C., under the Indo–United States exchange program. Support from the NREL and the Ministry of Environment and Forests (New Delhi) for manuscript preparation is gratefully acknowledged. Thanks are due to A.K. Jha, C.B. Pandey and R.S. Singh for assistance. Thanks are also due to Janice Hill and Susan Taylor for meticulously processing the manuscript.

REFERENCES

Ahmad, A. and Tahir, H.M., 1987. An appraisal of the livestock and rangelands in Jammu and Kashmir. In: P. Singh (Editor), *Silver Jubilee Souvenir*. Indian Grassland and Fodder Research Institute, Jhansi, pp. 31–36.

Ahuja, L.D., 1966. Growth of ram of Marwari breed on "fair" rangelands in semi-arid zone. *Ann. Arid Zone*, 5: 229–237.

Ahuja, L.D. and Mann, H.S., 1975. Rangeland development and management in western Rajasthan. *Ann. Arid Zone*, 14: 29–44.

Ahuja, L.D., Bhimaya, C.P. and Samraj, P., 1968. Preliminary studies on the effect of different intensities of grazing on the growth of heifers in rangelands of west Rajasthan. In: R. Misra and B. Gopal (Editors), *Recent Advances in Tropical Ecology*. International Society for Tropical Ecology, Varanasi, pp. 612–619.

Al-Ani, T.A., Al-Mufti, M.M., Ouda, N.A., Kaul, R.N. and Thalen, D.C.P., 1970. Range resources of Iraq, I. Range cover types of western and southern desert. *Technical Report No. 16, Institute of Applied Research and Natural Resources,* Abu-Ghraib, Iraq (mimeo.).

Aleem, A., 1978. Re-introduction of blackbuck in Pakistan. *Pak. J. For.*, 28: 1–6.

Andrew, W.D., 1971. *Report to the Government of Ceylon on Grassland Improvement in the Montane Zone*. FAO/UNDP, Rome, 43 pp.

Aneja, K.R. and Mehrotra, R.S., 1979. Qualitative and quantitative changes in the microflora on *Desmostachya bipinnata* litter buried in soil. *Bot. Prog.*, 2: 50–54.

Aneja, K.R. and Mehrotra, R.S., 1980a. Studies on microorganisms decomposing aboveground parts of 'the grass' (*Desmostachya bipinnata*). *Proc. Natl. Acad. Sci., India*, 50: 12–20.

Aneja, K.R. and Mehrotra, R.S., 1980b. Microbial degradation of mixed litter of grasses on the soil surface. *Proc. Natl. Acad. Sci., India*, 50: 163–170.

APROSC, 1979. *Range and Pasture Production*. Agricultural Projects Services Centre, Kathmandu, Nepal, 88 pp. (mimeo.).

Basak, K.C., 1964. *Report to the Government of Afghanistan on Planning of Agricultural Development*. FAO/EPTA Report No. 190, FAO, Rome, 35 pp.

Bauer, G., 1935. Luftzirkulation und Niederschlagverhältnisse in Vorder-Asien. *Gerlands Beitr. Geophys.*, 45: 38.

Bharucha, F.R. and Shankarnarayan, K.A., 1958. Studies on the grasslands of the western Ghats, India. *J. Ecol.*, 46: 681–705.

Bhimaya, C.P. and Ahuja, L.D., 1969. Criteria for determining condition class of rangelands in western Rajasthan. *Ann. Arid Zone*, 8: 73–79.

Bhimaya, C.P., Kaul, R.N. and Ahuja, L.D., 1967. Forage production and utilization in arid and semiarid rangelands of Rajasthan. In: *Proceedings of a Joint Symposium on India's Food Problem*. National Institute of Science, India, and ICAR, New Delhi, pp. 215–223.

Birand, H., 1954. La végétation de l'Anatolie centrale et son equilibre hydrologique. In: *Actes du Colloque d'Ankara sur l'Hydrologie de la Zone Aride*. UNESCO, Paris, 8 pp.

Blasco, F., 1983. The transition from open forest to savanna in continental southeast Asia. In: F. Bourlière (Editor), *Tropical Savannas*. Ecosystems of the World, 13, Elsevier, Amsterdam, pp. 167–181.

Bor, N.L., 1942. Ecology: Theory and practice, Presidential address to the Botany Section. In: *Proceedings of the 29th Indian Science Congress*, Baroda. Indian Science Congress Association, Calcutta, pp. 145–179.

Bor, N.L., 1960. *Grasses of Burma, Ceylon, India, and Pakistan*. Pergamon Press, London, 767 pp.

Breckle, S.W., 1983. Temperate deserts and semi-deserts of Afghanistan and Iran. In: N.E. West (Editor), *Temperate Deserts and Semi-Deserts*. Ecosystems of the World, 5, Elsevier, Amsterdam, pp. 271–319.

Burns, W. and Kulkarni, L.B., 1927. A line survey of grasslands with reference mainly to rainfall. *J. Indian Bot. Soc.*, 16: 103–108.

Chakravarty, A.K., 1971. Pasture production and its use in the arid and semiarid areas of Rajasthan (India) and Kazakhstan (USSR). *Ann. Arid Zone*, 10: 251–256.

Chakravarty, A.K. and Verma, C.M., 1970. Study on the pasture establishment techniques, V. Effect of reseeding of natural pastures with *Cenchrus ciliaris* by different soil working methods and fertilizers on pasture production. *Ann. Arid Zone*, 9: 236–243.

Chakravarty, A.K. and Verma, C.M., 1972. Effect of different spacing and weeding on the establishment and forage production of *Cenchrus ciliaris*, *Lasiurus sindicus* and *Panicum antidotale* under arid conditions. *Ann. Arid Zone*, 11: 61–66.

Chakravarty, A.K., Debroy, R., Verma, C.M. and Das, R.B., 1966. Studies on pasture establishment technique, I. Effect of seed rates, method of sowing and seed germination on seedling emergence of *Cenchrus ciliaris* and *Lasiurus sindicus*. *Ann. Arid Zone*, 5: 145–148.

Champion, H.G., 1936. A preliminary survey of the forest types of India and Burma. *Indian For. Rec.*, 1: 1–280.

Champion, H.G. and Seth, S.K., 1968. *A Revised Survey of the Forest Types of India*. Government of India, Delhi, 404 pp.

Chaudhri, I.I., 1960. The vegetation of the Kaghan valley. *Pak. J. For.*, 10: 285–294.

Crocker, C.D., 1963. *Carte générale des sols du Cambodge à 1/1.000.000*. Royaume du Cambodge, Secrétariat d'Elat à l'Agriculture, Phnom Pehn.

Dabadghao, P.M., 1960. Types of grass cover in India and their management. In: *VIII Int. Grassland Congr. Proc.*, pp. 226–230.

Dabadghao, P.M. and Shankarnarayan, K.A., 1970. Studies of *Iseilema*, *Sehima* and *Heteropogon* communities of the *Sehima-Dichanthium* zone. In: *XI Int. Grassland Congr. Proc.*, University of Queensland Press, pp. 36–38.

Dabadghao, P.M. and Shankarnarayan, K.A., 1973. *The Grass Cover of India*. Indian Council of Agricultural Research, New Delhi, 713 pp.

Dabadghao, P.M., Chakravarty, A.K., Das, R.B., Debroy, R. and Marwaha, S.P., 1965. Response of some promising desert grasses to fertilizer treatments. *Ann. Arid Zone*, 4: 120–135.

Das, R.B. and Gupta, B.S., 1964. A note on the effect of different fertilizers on protein status of some grasses. *Ann. Arid Zone*, 2: 185–187.

Dash, M.C. and Patra, U.C., 1977. Density, biomass and energy budget of a tropical earthworm population from a grassland site in Orissa, India. *Rev. Ecol. Biol. Soc.*, 14: 461–471.

Dash, M.C., Patra, U.C. and Thambi, A.V., 1974. Comparison of primary production of plant material and secondary production of oligochaetes in a tropical grassland of southern Orissa. *Trop. Ecol.*, 15: 16–21.

Dash, M.C., Senapati, B.K. and Behera, B.N., 1979. Microfungi associated with decomposition of earthworm tissue in a pasture soil. *Biol. Bull.*, India, 1: 21–23.

Dash, M.C., Behara, N., Satpathy, B. and Pati, D.P., 1985. Comparison of two methods for microbial biomass estimation in some tropical soils. *Rev. Ecol. Biol. Soc.*, 22: 13–20.

Delvert, J., 1961. *Le Paysan Cambodgien*. Mouton, Paris and The Hague, 740 pp.

De Rasayro, R.A., 1945. The montane grasslands (*patanas*) of Ceylon, Part I. *Trop. Agric. Ceylon*, 101: 206–216.

Dewan, M.L. and Famouri, J., 1964. *The Soils of Iran*. FAO, Rome, 319 pp.

Dudal, R., Moormann, F. and Riquier, J., 1974. Soils of humid tropical Asia. In: Natural Resources of Humid Tropical Asia. *Nat. Resour. Res.*, 12: 159–178.

Dwivedi, R.S., 1965. Ecology of soil fungi on some grasslands of Varanasi, I. Edaphic factors and fungi. *Proc. Indian Natl. Acad. Sci.*, 35B: 255–274.

Erkun, V., 1964. Balallcesi Mer'alari Uzerinde arastirmalar (Investigation of Bala country Ranges) (unpublished, typed).

Eussen, J.H.H., 1976. Biological and ecological aspects of alang-alang (*Imperata cylindrica* L. Beauv.). In: *Proceedings of BIOTROP Workshop on Alang-alang*, July 27–29, 1976. Bogor, Indonesia. BIOTROP Special Publication No. 5, Seameo Regional Centre for Tropical Biology, Bogor, pp. 15–22.

Falvey, L., 1977. *Ruminants in the Highlands of Northern Thailand — An Agrosociological Study*. Australian Development Assistance Bureau, Tippanetre Press, Chiang Mai, Thailand, 124 pp.

Falvey, L. and Hengmichai, P., 1979. Invasion of *Imperata cylindrica* (L.) Beauv. by *Eupatorium* species in northern Thailand. *J. Range Manage.*, 32: 340–344.

Falvey, L., Hengmichai, P. and Hoare, P., 1979a. Productivity of cattle grazing native highland pastures. *Thai Agric. Sci.*, 12: 61–69.

Falvey, L., Hengmichai, P. and Pongpiachan, P., 1979b. *The Productivity and Nutritive Value of Imperata cylindrica (L.) Beauv. in the Thai Highlands*. Report of the Thai Australian Highland Agricultural Project, Chiang Mai University, Thailand, 9 pp.

Gibson, T.A. and Van Diepen, D., 1977. *The Area Under Grassland in Northern Thailand*. Report of the U.N. Programme in Drug Abuse Control, Chiang Mai, Thailand (mimeo.).

Gupta, A., 1980. *Study of Seasonal Variation in Population Density and Biomass of Orthoptera in a Grassland and Energetics of Phlaeoba infumata Brunn*. Ph.D. dissertation, Kurukshetra University, 239 pp.

Gupta, R.K., 1975. Range ecology and development. In: R.K. Gupta and I. Prakash (Editors), *Environmental Analysis of the Thar Desert*. English Book Depot, Dehradun, pp. 322–360.

Gupta, R.K. and Nanda, P.C., 1970. Grassland types and their succession in the western Himalayas. In: *XI Int. Grassland Congr. Proc.*, University of Queensland Press, pp. 10–13.

Gupta, R.K. and Saxena, S.K., 1972. Potential grassland types and their ecological succession in Rajasthan desert. *Ann. Arid Zone*, 11: 198–211.

Gupta, R.K., Saxena, S.K. and Sharma, S.K., 1972. Aboveground productivity of three promising desert grasses of Jodhpur under different rainfall conditions. In: *Proc. Int. Symp. Ecophysiological Foundation of Ecosystem Productivity in the Arid Zone*. USSR Academy of Science, Moscow, pp. 134–138.

Gupta, S.R. and Singh, J.S., 1981a. Soil respiration in a tropical grassland. *Soil Biol. Biochem.*, 13: 261–268.

Gupta, S.R. and Singh, J.S., 1981b. The effect of plant species, weather variables, and chemical composition of plant material on decomposition in a tropical grassland. *Plant Soil*, 59: 99–118.

Gupta, S.R. and Singh, J.S., 1982a. Influence of floristic composition on the net primary production and dry matter turnover in a tropical grassland. *Aust. J. Ecol.*, 7: 363–374.

Gupta, S.R. and Singh, J.S., 1982b. Carbon balance of a tropical successional grassland. *Oecol. Generalis*, 3: 459–467.

Gupta, S.R., Rajvanshi, R. and Singh, J.S., 1981. The role of the termite *Odontotermes gurdaspurensis* (Isoptera: Termitidae) in plant decomposition in a tropical grassland.

Pedobiologia, 22: 254–261.
Hannibal, L.W., 1950. *Vegetation Map of Indonesia*. Scale 1:2,500,000, Directorate of Forest Inventory and Planning, Bogor.
Hassanyar, Amir S., 1988. Revegetation of the Registan desert in the South Western Afghanistan. In: P. Singh, V. Shankar and A.K. Srivastava (Editors), *Abstracts Volume I, 3rd International Rangeland Congress*, New Delhi, pp. 194–196.
Holmes, C.H., 1951. *The Grass, Fern and Savannah Lands of Ceylon, their Nature and Ecological Significance*. Institute Paper No. 28, Imperial Forestry Institute, Univ. Oxford, 95 pp.
Holmes, C.H., 1956. The broad pattern of climate and vegetational distribution in Ceylon. In: *Proceedings of the Kandy Symposium, Kandy, Ceylon*. UNESCO, Paris, pp. 99–113.
Holmes, J.H.G., Lemerle, C. and Schottler, J.H., 1976. The use of *Imperata cylindrica* (L.) Beauv. by grazing cattle in Papua New Guinea. In: *Proceedings of BIOTROP Workshop on Alang-alang*, July 27–29, 1976. Bogor, Indonesia. BIOTROP Special Publication No. 5, Seameo Regional Centre for Tropical Biology, Bogor, pp. 179–192.
Hussain, T., 1962. Dabeji range management plan. West Pakistan Forest Department. Unpublished. (Quoted by Johnston and Hussain, 1963.)
Ibrahim, K.M., 1967. The pasture, range and fodder crop situation in the near east. In: *Summary Report and Bibliography*. FAO, Rome, pp. 76–90.
Islam, M.A., 1966. Soils of east Pakistan. In: *Scientific Problems of Humid Tropical Zone Deltas and Their Implications*. Proc. Dacca Symp., UNESCO, Paris, pp. 83–87.
Ivens, G.W., 1975. Studies on *Imperata cylindrica* (L.) Beauv. and *Eupatorium odoratum* L. *Tech. Rep. No. 37, Weed Research Organization, Agricultural Research Council*, England, 27 pp.
Jain, H.K., 1976. *Ecosystem Analysis of Grasslands Near Rajkot*. Ph.D. dissertation, Saurashtra University, Rajkot, 218 pp.
Jain, S.K., 1980. Total phytomass, net community productivity and system transfer functions in subhumid grasslands at Sagar (M.P.) India. *Flora*, 170: 251–260.
Jayan, P.K., 1970. *Ecology of Cenchrus ciliaris complex*. Ph.D. dissertation, Saurashtra University, Rajkot, 213 pp.
Joachim, A.W.R. and Kandiah, S., 1942. Studies in Ceylon soils. *Trop. Agric. Ceylon*, 48: 15–29.
Johnston, A. and Hussain, E., 1963. Grass cover types of West Pakistan. *Pak. J. For.*, 13: 239–247.
Karim, A.G.M.B., 1966. Nutrient status of different soil tracts of east Pakistan. In: *Scientific Problems of Humid Tropical Zone Deltas and Their Implications*. Proc. Dacca Symp., UNESCO, Paris, pp. 97–101.
Kaul, R.N. and Thalen, D.C.P., 1979. South-west Asia. In: D.W. Goodall and R.A. Perry (Editors), *International Biological Programme 16, Arid Land Ecosystem: Structure, Functioning and Management*. Cambridge Univ. Press, London, pp. 213–271.
Kaushal, B.R., 1979. *Studies on Seasonal Variations in Insect Population Density and Biomass in a Tropical Grassland and Energetics of Two Insects*. Ph.D. dissertation, Kurukshetra University, 242 pp.

Khan, A. and Bhatti, A.G., 1956. Effects of closure on growth of grasses. *Pak. J. For.*, 6: 187–190.
Khan, A.H., 1962. Ecological assessment of the effects of closure in Quetta Division forest. *Pak. J. For.*, 13: 167–193.
Khan, C.M.A., 1966. Artificial reseeding in Thal ranges. *Pak. J. For.*, 16: 28–42.
Khan, C.M.A., 1970. Range management — a challenge in West Pakistan. *Pak. J. For.*, 20: 329–350.
Khan, C.M.A., 1971. Range management strategy in Pakistan. *Pak. J. For.*, 21: 313–324.
Khan, M., 1966. Kallar grass (a suitable grass for saline lands). *Pak. Agric.*, 17: 375.
Khan, M.S. (Editor), 1977. *Flora of Bangladesh*. Bangladesh Agricultural Research Council, Dacca, No. 4, pp. 4–11.
Khan, S.M., 1977. Ecological changes in the alpine pasture vegetation at Paya (Kaghan) due to complete protection from grazing. *Pak. J. For.*, 27: 139–146.
Khattak, G.M., 1976. History of forest management in Pakistan, Part 1. *Pak. J. For.*, 26: 105–116.
Kitamura, S., 1955. Flowering plants and ferns. In: H. Kihara (Editor), *Fauna and Flora of Nepal Himalaya, Scientific Results of the Japanese Expedition to Nepal Himalaya*. Kyoto University, Kyoto, I: 73–290.
Koteswaram, P., 1974. Climate and meteorology of humid tropical Asia. In: UNESCO, Natural Resources of Humid Tropical Asia. *Nat. Resour. Res.*, 12: 27–85.
Krishnamurthy, L., 1976. *Systems Analysis and Optimization of Land Use System with Reference to a Grazingland*. Ph.D. dissertation. Saurashtra University, Rajkot, 122 pp.
Krishnaswamy, V.S. and Reddi, T.V., 1953. *A Short Report on the Fodder, Milk Supply and Land Utilization Problems of the State of Madras*. Government Press, Madras, 131 pp.
Kumar, A. and Joshi, M.C., 1972. The effects of grazing on the structure and productivity of vegetation near Pilani, Rajasthan, India. *J. Ecol.*, 60: 665–675.
Kumar, S. and Shankar, V., 1985. Vegetation ecology of the Guhiya catchment in the Upper Luni Basin-India. *Trop. Ecol.*, 26: 1–11.
Kumar, S. and Shankar, V., 1987. Vegetation ecology of the Bandi catchment in the Upper Luni Basin, Western Rajasthan. *Trop. Ecol.*, 28: 246–258.
Legris, P. and Blasco, F., 1972. Carte internationale du Tapis végétal, "Cambodge", 1/1.000.000, notice. *Trav. Sec. Sci. Tech. Inst. Fr. Pondicherry, H.S.*, No. 11, 240 pp.
Magadan, P.B., Javier, E.Q. and Madamba, J.C., 1974. Beef production and native (*Imperata cylindrica* (L.) Beauv.) and Para grass (*Brachiaria mutica* (Forsk.) Stapf) pastures in the Philippines. *Proc. 12th Int. Grassland Congr., Grass Utilization, 1*. Moscow, pp. 370–378.
Malik, M.N. and Khan, A., 1966. Development of rangelands in West Pakistan. Physico-chemical characteristics of soil and nutritive value of forage grasses of Cholistan. *Pak. J. For.*, 16: 261–273.
Mall, L.P. and Mehta, S.C., 1978. The grassland fire: An ecological study. In: J.S. Singh and B. Gopal (Editors), *Glimpses of Ecology (Professor R. Misra commemoration volume)*. International Scientific Publications, Jaipur, pp. 291–297.

Mall, L.P., Billore, S.K. and Misra, C.M., 1974. Relation of ecological efficiency with photosynthetic structure in some herbaceous stands. *Trop. Ecol.*, 15: 39–42.

Mankand, N.R., 1974. *Net Primary Production Relations of Some Grasslands in Rajkot District.* Ph.D. dissertation, Saurashtra University, Rajkot, 164 pp.

Mann, H.S. and Ahuja, L.D., 1974. *Primary and Secondary Production in Arid Zone Rangelands of Western Rajasthan.* Central Arid Zone Research Institute, Jodhpur, 23 pp. (mimeo.).

Margalef, R., 1968. *Perspectives in Ecological Theory.* University of Chicago Press, Chicago, 111 pp.

McClure, F.A., 1956. Report on bamboo utilization in eastern Pakistan. *Pak. J. For.*, 6: 182–186.

McKay, G.H., 1973. *Behaviour and Ecology of the Asiatic Elephant in Southeastern Ceylon.* Smithsonian Institution, Washington, D.C., 113 pp.

Mehrotra, R.S. and Aneja, K.R., 1979. Microbial decomposition of *Chenopodium album* litter, I. Succession of decomposers. *J. Indian Bot. Soc.*, 58: 189–195.

Melkania, N.P. and Singh, J.S., 1989. Ecology of Indian grasslands. In: J.S. Singh and B. Gopal (Editors), *Perspectives in Ecology.* Jagmander Book Agency, New Delhi, pp. 67–103.

Mesdaghi, M., 1988. Inventory of winter ranges in the northeast of Iran. In: P. Singh, V. Shankar and A.K. Srivastava (Editors), *Abstracts Volume I, 3rd International Rangeland Congress.* New Delhi, pp. 24–26.

Mishra, R.R., 1966. Seasonal variation in fungal flora of grasslands of Varanasi. *Trop. Ecol.*, 7: 100–112.

Misra, G. and Singh, K.P., 1981. Total nonstructural carbohydrates of one temperate and two tropical grasses under varying clipping and soil moisture regimes. *Agroecosystems*, 7: 213–223.

Misra, R., 1983. Indian savannas. In: F. Bourlière (Editor), *Tropical Savannas.* Ecosystems of the World, 13, Elsevier, Amsterdam, pp. 151–166.

Mobayen, S. and Tregubov, V., 1970. *Vegetation Map of Iran.* Scale 1: 2,500,000 and explanatory guide, UNDP/FAO Project IRA 7, Bull. No. 14.

Mueller-Dombois, D. and Cooray, R., 1968. Effect of elephant feeding on shortgrass cover in Ruhuna National Park. In: *Proceedings of the Annual General Meeting of CAAS.* Ceylonese Association of Agricultural Scientists, Paradeniya, Sect. D, Rep. No. 11, pp. 1–12.

Mueller-Dombois, D. and Perera, M., 1971. Ecological differentiation and soil fungal distribution in the montane grassland of Ceylon. *Ceylon J. Sci.*, 9: 1–41.

Nemati, N., 1978. Range improvement practices in Iran. In: *Proc. 1st Int. Rangeland Congr.*, Denver, pp. 631–632.

Nemati, N., 1985. *Range Improvement and Management in Arid zones of Iran. FAO expert consultation on rangeland rehabilitation and development in the Near East.* AGP, 810, 14 p.

Neubauer, H.F., 1955. Versuch einer Kemreich der Vegetationsverhältnisse Afghanistan. *Ann. Naturalist Mus. Wien,* 60: 77–113.

Numata, M., 1966. Vegetation and conservation in eastern Nepal. *J. Coll. Arts Sci., Chiba Univ.*, 4: 559–569.

Numata, M., 1987. Observations of farmlands and pastures in central Bhutan. In: M. Oshawa (Editor), *Life Zone Ecology of the Bhutan Himalaya.* Chiba University, Japan, pp. 117–132.

Ogilvie, C.S., 1954. The behaviour of seladang (*Bibos gaurus*). *Malay. Nature J.*, 9: 1–10.

Orshan, G., 1986. The deserts of the Middle East. In: M. Evenari, I. Noy-Meir and D.W. Goodall (Editors), *Hot Deserts and Arid Shrublands.* Ecosystems of the World, 12B, Elsevier, Amsterdam, pp. 1–28.

Pabot, H., 1960. *Les grandes régions phytogéographiques et écologiques d'Iran.* FAO Report/Teheran, 16 pp. (mimeo.).

Pabot, H., 1964. Phytogeographical and ecological regions. In: M.L. Dewan and J. Famouri (Editors), *The Soils of Iran.* FAO, Rome, pp. 30–40.

Pabot, H., 1967. *Report to Government of Iran on Pasture Development and Range Improvement Through Botanical and Ecological Studies.* FAO Report No. TA 2311, 129 pp. (mimeo.).

Pande, H. and Singh, J.S., 1981. Comparative biomass and water status of four range grasses grown under two soil water conditions. *J. Range Manage.*, 34: 480–484.

Pandey, A.N., 1974. Short term effect of seasonal burning on the composition of *Dichanthium annulatum* stands at Varanasi. *Proc. Indian Natl. Sci. Acad.*, 40B: 107–112.

Pandeya, S.C., 1964. Ecology of grasslands of Sagar, Madhya Pradesh, II. Composition of the fenced grassland associations. *J. Indian Bot. Soc.*, 43: 557–605.

Pandeya, S.C. and Jain, H.K., 1979. Description and functioning of arid to semiarid grazing land ecosystem at Khirasara, near Rajkot (Gujarat). In: *Tropical Grazing Land Ecosystems: A State of Knowledge Report.* UNESCO/UNEP/FAO, UNESCO, Paris, pp. 630–655.

Pandeya, S.C., Sharma, S.C., Jain, H.K., Pathak, S.J., Paliwal, K.C. and Bhanot, V.M., 1977. *The Environment and Cenchrus Grazing Lands in Western India. An Ecological Assessment.* Saurashtra University, Rajkot, 453 pp. (mimeo.).

Pandeya, S.C., Paliwal, K.C., Pandey, A.N., Sharma, R., Mathur, P.K., Bhatt, S.C., Hirani, O.H. and Roa, Y.N., 1986. *Impact of Human Activities on the Organic Matter Productivity of Seminatural Grazing Lands: from Arid to Dry Subhumid Areas of Western India.* MAB Project 3 Report, Saurashtra University, 500 pp.

Pati, D.P., Behara, N. and Dash, M.C., 1983. Microbial and root contribution to total soil metabolism in a tropical grassland soil from Orissa, India. *Rev. Ecol. Biol. Sol*, 20: 183–190.

Patil, B.D. and Pathak, P.S., 1978. *Possible Facets of Agro-Forestry.* Indian Grassland and Fodder Research Institute, Jhansi, 36 pp. (mimeo.).

Pemadasa, M.A. and Amarsinghe, L., 1982. The ecology of a montane grassland in Sri Lanka, I. Quantitative description of the vegetation. *J. Ecol.*, 70: 1–15.

Pemadasa, M.A. and Mueller-Dombois, D., 1979. An ordination study of montane grasslands of Sri Lanka. *J. Ecol.*, 67: 1009–1023.

Pendleton, R.L. and Montrakun, S., 1960. The soils of Thailand. In: *Proc. 9th Pac. Sci. Congr.*, 18: 12–32 (soil map).

Perera, K.S.O., 1969. Soils of Ceylon. *Ceylon Coconut Planters Rev.*, 5: 147–156.

Perrin de Brichambaut, G. and Wallen, C.C., 1964. A study of agroclimatology in semi-arid and arid zones of the Near East. *Tech. Note No. 56, World Meteorological Organisation*, Geneva, 64 pp.

Pfeffer, P., 1974. Fauna of humid tropical Asia. In: UNESCO, Natural Resources of Humid Tropical Asia. *Nat. Resour. Res.*, 12: 295–306.

Prajapati, M.C., 1970. Range improvement in west Rajasthan — need to control migratory grazing. *Indian For.*, 20: 34–37.

Prakash, I., 1962. Ecology of gerbils of the Rajasthan desert, India. *Mammalia*, 26: 311–331.

Prakash, I., 1969. Ecotoxicology and control of Indian gerbil, *Meriones hurrianae* (Jerdon). *J. Bombay Nat. Hist. Soc.*, 65: 581–589.

Prakash, I., 1972. Ecotoxicology and control of the Indian desert gerbil, *Meriones hurrianae* (Jerdon), VIII. Body weight, sex ratio and age structure in the population. *J. Bombay Nat. Hist. Soc.*, 68: 717–725.

Prakash, I., 1974. The ecology of vertebrates of the Indian desert. In: M.S. Mani (Editor), *Biogeography and Ecology of India*. Junk, The Hague, pp. 369–420.

Prakash, I., 1975. The ecology and zoogeography of mammals. In: R.K. Gupta and I. Prakash (Editors), *Environmental Analysis of the Thar Desert*. English Book Depot, Dehradun, pp. 448–467.

Prakash, I. and Jain, A.P., 1971. Some observations on Wagner's gerbil, *Gerbillus nanus indus* (Thomas) in the Indian desert. *Extrait de Mammalia*, 35: 614–628.

Prakash, I. and Taneja, G.C., 1969. Reproduction biology of the Indian desert hare *Lepus nigricollis dayanus* Blanford. *Extrait de Mammalia*, 33: 102–117.

Puri, G.S., 1960. *Indian Forest Ecology*, Vol. I. Oxford Book and Stationary Company, New Delhi, 318 pp.

Rahman, K., 1956. Grasses as raw material for pulp with special reference to its availability in Sylhet District, East Pakistan. *Pak. J. For.*, 6: 2–9.

Ram, J., Singh, S.P. and Singh, J.S., 1988. Community level phenology of grassland above treeline in Central Himalaya. *Arct. Alp. Res.*, 20: 325–332.

Ram, J., Singh, J.S. and Singh, S.P., 1989. Plant biomass, species diversity and net primary production in a central Himalayan high altitude grassland. *J. Ecol.*, 77: 456–468.

Raychaudhuri, S.P., Agrawal, R.R., Datta Biswas, N.R., Gupta, S.P. and Thomas, P.K., 1963. *Soils of India*. Indian Council of Agricultural Research, New Delhi, 496 pp.

Rodgers, W.A., 1988. The wild grazing ungulates of India: an ecological review. In: P. Singh and P.S. Pathak (Editors), *Rangelands: Resource and Management*. Range Management Society of India, Jhansi, pp. 404–420.

Rol, R., 1956. *Rapport au gouvernement de l'Iran sur les études écologiques et systématiques sur le flore ligneuse de la région caspienne*. FAO/ETAP Rep., 520, FAO, Rome, 68 pp.

Rollet, B., 1962. *Inventaire forestier de l'Est Mékong*. Rapport FAO, No. 1500, Rome 184 pp.

Rollet, B., 1972. La végétation du Cambodge. *Bois For. Trop.*, 145: 23–28; 146: 3–20.

Said, M., 1961. Grazing in West Pakistan. *Pak. J. For.*, 11: 176–190.

Sajise, P.E., 1972. *Evaluation of Cogon (Imperata cylindrica (L.) Beauv.) as a Seral Stage in Philippine Vegetational Succession, II. Autecological Studies on Cogon*. Ph.D. Thesis, Cornell University, Ithaca, N.Y., 152 pp.

Sajise, P.E., Orlido, N.M., Castillo, L.C. and Atabay, R., 1974. Phytosociological studies on Philippine grasslands, I. Floristic composition and community dynamics. *Natl. Sci. Dev. Board, Annu. Rep., 1973*, pp. 4–36.

Sajise, P.E., Orlido, N.M., Lales, J.S., Castillo, L.C. and Atabay, R., 1976. The ecology of Philippine grasslands, I. Floristic composition and community dynamics. *Philipp. Agric.*, 59: 317–334.

Sanford, W. and Wangeri, E., 1985. Tropical grasslands: dynamics and utilization. *Nat. Resour.*, 21: 12–27.

Saxena, A.K. and Singh, J.S., 1980. Analysis of forest-grazingland vegetation in parts of Kumaun Himalaya. *Indian J. Range Manage.*, 1: 13–32.

Schmid, M., 1958. Flora agrostologique de l'Indochine. *Agron. Trop.*, 13: 1–103.

Schmid, M., 1974. Végétation du Vietnam. Le massif sud-Annamitique et les régions limitrophes. *Mémoires, Office de la Recherche Scientifique et Technique* Outre-Mer (ORSTOM, Paris), 74: 1–243.

Seavoy, R., 1975. The origin of tropical grasslands in Kalimantan, Indonesia. *J. Trop. Geogr.*, 40: 48–52.

Senapati, B.K. and Dash, M.C., 1981. Effect of grazing on the elements of production in vegetation and oligochaete components of a tropical pasture. *Rev. Ecol. Biol. Sol*, 18: 487–505.

Senapati, B.K. and Dash, M.C., 1983. Energetics of earthworm population in tropical pastures of India. *Proc. Indian Acad. Sci.*, 92: 315–321.

Senaratna, J.E., 1934. Pasture trials at Peradeniya, I. Some notes on the grasses under trial. *Trop. Agric.*, 82: 204–211.

Senaratna, J.E., 1942. Patana burning with particular reference to pasturage and wet patanas. *Trop. Agric.*, 98: 3–16.

Shankar, K. and Shrestha, P.B., 1985. Climate. In: T.C. Majupuria (Editor), *Nepal Natures Paradise*. White Lotus Co. Ltd. Bangkok, pp. 39–44.

Shankarnarayan, K.A. and Rai, P., 1975. Effect of fertilizer application on dry matter production in *Iseilema laxum* Hack grasslands in Jhansi. *Forage Res.*, 1: 155–157.

Shankarnarayan, K.A., Dabadghao, A.K. and Dabadghao, P.M., 1972. Relative grazing values of principal grass species of Sehima–Dichanthium cover. In: *Annu. Rep., Indian Grassland and Fodder Res. Inst.*, Jhansi, pp. 69–70.

Shankarnarayan, K.A., Dabadghao, P.M., Upadhayay, V.S. and Rai, P., 1975. A note on the utilization of speargrass pasture: monthly variation in yield, chemical composition and performance of wethers. *Indian J. Anim. Sci.*, 45: 92–95.

Shankarnarayan, K.A., Dabadghao, P.M., Upadhayay, V.S. and Rai, P., 1976. Effect of fertilizers (N and P) on the net aboveground community productivity in *Heteropogon, Iseilema* and *Sehima* communities. *Trop. Ecol.*, 17: 110–112.

Singh, J.S., 1968. Net aboveground community productivity in the grasslands at Varanasi. In: R. Misra and B. Gopal (Editors), *Recent Advances in Tropical Ecology*. International Society for Tropical Ecology, Varanasi, pp. 631–654.

Singh, J.S., 1973. A compartment model of herbage dynamics for Indian tropical grasslands. *Oikos*, 24: 367–372.

Singh, J.S., 1976. Structure and function of tropical grassland vegetation of India. *Pol. Ecol. Stud.*, 2: 17–34.

Singh, J.S. and Joshi, M.C., 1979a. Primary production. In: R.T. Coupland (Editor), *Grassland Ecosystems of the World: Analysis of Grasslands and Their Uses*. International Biological Programme 18, Cambridge University Press, Cambridge, pp. 197–218.

Singh, J.S. and Joshi, M.C., 1979b. Ecology of the semi-arid regions of India with emphasis on land-use. In: B.H. Walker (Editor), *Management of Semiarid Ecosystems*. Elsevier, Amsterdam, pp. 243–275.

Singh, J.S. and Krishnamurthy, L., 1981. Analysis of structure and function of tropical grassland vegetation of India. *Indian Rev. Life Sci.*, 1: 225–270.

Singh, J.S. and Misra, R., 1969. Diversity, dominance, stability and net production in the grasslands at Varanasi. *Can. J. Bot.*, 47: 425–427.

Singh, J.S. and Saxena, A.K., 1980. The grass cover in the Himalayan region. In: *Proceedings, National Seminar on Resources, Development and Environment in the Himalayan Region*. Department of Science and Technology, Government of India, New Delhi, pp. 164–203.

Singh, J.S. and Yadava, P.S., 1974. Seasonal variation in composition, plant biomass, and net primary productivity of a tropical grassland at Kurukshetra, India. *Ecol. Monogr.*, 44: 351–375.

Singh, J.S., Singh, K.P. and Yadava, P.S., 1979. Ecosystem synthesis. In: R.T. Coupland (Editor), *Grassland Ecosystems of the World: Analysis of Grasslands and Their Uses*. International Biological Programme 18, Cambridge University Press, Cambridge, pp. 231–239.

Singh, J.S., Lauenroth, W.K. and Milchunas, D.G., 1983. Geography of grassland ecosystems. *Prog. Phys. Geogr.*, 7: 46–80.

Singh, J.S., Hanxi, Y. and Sajise, P.E., 1985. Structural and functional aspects of Indian and Southeast Asian Savanna ecosystems. In: J.C. Tothill and J.J. Mott (Editors), *Ecology and Management of the World's Savannas*. The Australian Academy of Sciences, Canberra and Commonwealth Agricultural Bureaux, pp. 34–51.

Singh, M., Pandey, R.K. and Shankarnarayan, K.A., 1970. Problems of grassland weeds and their control in India. In: *XI Int. Grasslands Congr. Proc.*, University of Queensland Press, pp. 71–74.

Singh, P., 1988. Indian rangelands status and improvement. Plenary address. *3rd International Rangeland Congress*, November 7–11, New Delhi. Indian Grassland and Fodder Research Institute, Jhansi, p. 40.

Singh, V.P., 1969. *Ecology of Grasslands of Ujjain*. Ph.D dissertation, Vikram University, Ujjain, India, 291 pp.

Singh, V.P., Dagar, J.C. and Upadhayaya, S.D., 1979. Analysis of structure, production dynamics and successional trends of tropical grassland communities at Ujjain (India). *Sylvatrop. Philipp. For. Res. J.*, 4: 231–254.

Sivasupiramanian, S., Sithamparanathan, J. and Appadurai, R.R., 1973. Studies on the pasture improvement in the hill country dry zone patana of Sri Lanka, I. Effect of pioneer cropping on pasture management; II. Performance of improved pasture on re-conditioned patanas. *Trop. Agric.*, 129: 85–101.

Soerianegara, I., 1970. The status of range research and management in Indonesia. *Rimba Indones.*, 25: 99–115.

Soerjani, M., 1976. Symposium on the prevention and rehabilitation of critical land in an area development (a summary). In: *Proceedings of BIOTROP Workshop on Alang-alang*, July 27–29, 1976, Bogor, Indonesia. BIOTROP Spec. Publ., 5, Seameo Regional Centre for Tropical Biology, Bogor, pp. 9–12.

Soewardi, B. and Sastradipradja, D., 1976. Alang-alang (*Imperata cylindrica*) and animal husbandry. In: *Proceedings of BIOTROP Workshop on Alang-alang*, July 27–29, 1976, Bogor, Indonesia. BIOTROP Spec. Publ., 5, Seameo Regional Centre for Tropical Biology, Bogor, pp. 157–178.

Soewardi, B., Sastradipradja, D., Nasation, A.H. and Hutasoit, J.H., 1975. The influence of corn and cassava meal supplementation on the feeding value of alang-alang (*Imperata cylindrica* (L.) Beauv.) for ongole grad heifers. *Malay. J. Agric. Res.*, 4: 123–130.

Sørensen, T., 1948. A method of establishing groups of equal amplitude in plant sociology based on similarity of species content. *Det Kong. Danske Vidensk. Sels K. Biol. Skr.* (Copenhagen), 5(4): 1–34.

Springfield, H.W., 1954. Natural vegetation in Iraq. *Technical Bulletin of the Ministry of Agriculture*, Baghdad, 23 pp. (mimeo.).

Susetyo, B., Suwardi, I.K., Abdulgani, Kismono and Sudarmadi, 1969. *Carrying Capacities of Some Ranges of Timor (Indonesia)*. Faculty of Animal Husbandry, Bogor Agricultural University, Bogor.

Sutcliffe, R.C., 1960. The Mediterranean in relation to the general circulation. In: *UNESCO/WMO Seminar on Mediterranean Synoptic Meteorology*. Abhandlungen aus dem Institut für Meteorologie und Geophysik der Freie Universität, Berlin.

Swartzman, G.L. and Singh, J.S., 1974. A dynamic programming approach to optimal grazing strategies using a successional model for a tropical grassland. *J. Appl. Ecol.*, 11: 537–548.

Thambi, A.V. and Dash, M.C., 1973. Seasonal variations in numbers and biomass of Enchytraeidae (Oligochaeta) and population in tropical grassland soils from India. *Trop. Ecol.*, 14: 228–237.

Trivedi, B.K., Shankar, V. and Kanodia, K.C., 1972. Effect of frequency of burning with or without grazing on changes in botanical composition in *Sehima* grasslands. In: *Annu. Rep. Indian Grassland and Fodder Res. Inst.*, Jhansi, pp. 74–75.

UNESCO, 1979. *Tropical Grazingland Ecosystems: A State of Knowledge Report*. UNESCO/UNEP/FAO, Paris, 655 pp.

Upadhayay, V.S., Dabadghao, P.M. and Shankarnarayan, K.A., 1971. Continuous versus deferred-rotational grazing system. In: *Annu. Rep. Indian Grassland and Fodder Res. Inst.*, Jhansi, p. 48.

Vats, L.K. and Singh, J.S., 1978. Population, biomass, and secondary net production of aboveground insects in a tropical grassland. *Trop. Ecol.*, 19: 51–64.

Verboom, W.C., 1968. Grassland successions and associations in Pahang, Central Malaya. *Trop. Agric.* (Trinidad), 45: 47–59.

Vyas, L.N. and Vyas, N.L., 1978. Net primary production of *Apluda* community on Aravalli hills of Udaipur. In: J.S. Singh and B. Gopal (Editors), *Glimpses of Ecology (Professor R. Misra Commemoration Volume)*. International Scientific Publications, Jaipur, pp. 111–124.

Wadia, D.N., 1961. *Geology of India*. MacMillan, London, 3rd ed., 536 pp.

Walter, H. and Lieth, H., 1967. *Klimadiagramm — Weltatlas*. Fischer-Verlag, Jena, 245 pp.

Whyte, R.O., 1964. *Grassland and Fodder Resource of India*. Sci. Monogr. No. 22, Indian Council of Agricultural Research, New Delhi, 553 pp.

Whyte, R.O., 1968. *Grasslands of the Monsoon*. Faber and Faber, London, 325 pp.

Whyte, R.O., 1974a. *Tropical Grazinglands: Communities and Constituent Species*. Junk, The Hague, 222 pp.

Whyte, R.O., 1974b. Grasses and grasslands. In: UNESCO, Natural Resources of Humid Tropical Asia. *Nat. Resour. Res.*, 12: 239–262.

Whyte, R.O., 1977. Analysis and ecological management of tropical grazinglands. In: W. Krause (Editor), *Application of Vegetation Science to Grassland Husbandry*. Vol. 13, Junk, The Hague, pp. 1–121.

Wijesinghe, L.C.A., 1962. Some aspects of land use in the dry montane grasslands. *Ceylon For.*, 5: 128–136.

Winters, R.K., 1956. Forestry problem analysis and research programme for East Pakistan. *Pak. J. For.*, 6: 259–288.

Yadava, P.S. and Singh, J.S., 1977. *Grassland Vegetation: Its Structure, Function, Utilization and Management*. Today and Tomorrow's Printers and Publishers, New Delhi, 132 pp.

Zablan, T.A., 1970. Problems in evaluating the productivity and grazing capacity of improved and native pastures. In: *XV National Livestock and Poultry Production Week Celebration*. Bai, Manila, (mimeo.).

Zohary, M., 1946. The flora of Iraq and its phytogeographical sub-division. *Directorate General of Agriculture, Iraq, Bull.*, 31, 201 pp.

Zohary, M., 1973. *Geobotanical Foundation of the Middle East*. Volumes I and II, Fischer-Verlag, Stuttgart.

Zulfiqar, A., 1962. Rangelands of West Pakistan, their problems and potentialities. *Pak. J. For.*, 12: 205–216.

Chapter 5

TEMPERATE SEMI-NATURAL GRASSLANDS OF EURASIA [1]

MILENA RYCHNOVSKÁ

INTRODUCTION

Grasslands of the temperate zone of Europe and Asia are either natural — as in the steppe zone (see Chapters 2 and 3) — or semi-natural — the meadows and pastures in the forest zone. These semi-natural grasslands are located mainly within a belt of coniferous forest, or within the areas of deciduous forest which extend southward in the western and eastern parts of the continent. The latitudinal range over which these meadows and pastures exist extends from 50° to 70°N in the coniferous forest zone and extends from 50°N as far south as 40°N in the deciduous forest zone. The portion of the deciduous forest zone in central and western Europe is under the influence of air masses from the Atlantic Ocean, whereas that of China and Japan is affected by air flows from the Pacific Ocean (Walter, 1968) (Fig. 5.1).

It is not possible to locate exactly the boundary between semi-natural grasslands in the forest zone and natural grasslands of the steppe and semi-desert zones, because even within the latter secondary mesophytic grasslands occur azonally on flood-plains of rivers and in moist valleys (Walter, 1968). In addition, accumulation of water from snow and rain in depressions ("pody") in the steppe results in the formation of mesophytic meadows (of *Alopecurus pratensis*, *Thalictrum minus* and *Trifolium montanum*) (Shennikov, 1941).

Although most of the grasslands within the temperate forest zone of Eurasia originated because of human activities, they display many of the characteristics of natural vegetation. If mowing and grazing were to cease, these grasslands would revert to woodlands by natural succession. Primary natural grasslands are scarce in the forest zone; they are confined to sites above the tree line in the mountains, to peat bogs and swamps, to flood-plains of rivers, and to fragments of forest–steppe communities.

Semi-natural grasslands of the forest zone first appeared during the Neolithic period, and are still forming. The basic requirement is deforestation combined with management that prevents the return of trees by natural means. The origin of such grassland is depicted in Table 5.1. The species composition in these grasslands is spontaneous and develops in harmony with the type of biotope. Human activity affects both the environmental characteristics of the habitat and species selection. However, in spite of this interference, the autochthonous character of the vegetation, which is well balanced with the environment and is highly homoeostatic, is predominant in these grasslands. Their maintenance requires much less external energy than agronomic ecosystems of the forest zone. Although they have been developed mainly for economic reasons, the intensity of their utilization is surprisingly low. According to Alberda (1980), only 25% of the available production of semi-natural grasslands is made use of in Europe, and 4 to 5% in Asia. This suggests that the potential for increased fodder production in these areas is still considerable, even in the densely populated areas.

---

[1] Manuscript submitted in July, 1982; revised in November, 1987. Nomenclature is according to Rothmaler (1970) and Flora of the USSR (1964).

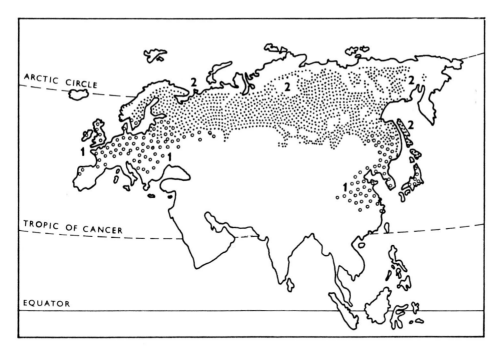

Fig. 5.1. Distribution of semi-natural grasslands in the forest zone of Eurasia: *1* = zone of the deciduous broad-leaved forests; *2* = zone of the coniferous boreal forests.

ENVIRONMENT

Semi-natural grasslands exist under a wide range of climatic conditions. The characteristics of the sites that were studied during the International Biological Programme (IBP) are as follows (Balátová-Tuláčková, 1979): annual mean temperature from about 0° to 10°C, with means in the coldest month (January) averaging between $-20.5°$ and $+3.4°C$, and in the warmest month between 10.0° and 22.5°C; and mean annual precipitation ranges from 430 mm to 2330 mm. Differences in weather between consecutive years are often considerable (Fig. 5.2), even exceeding those between the climates of different vegetation zones. The daily amplitude of temperature sometimes exceeds 30°C.

Soils are predominantly acid, with pH ranging mostly from 5.5 to 6.5, although ecotopes with pH of approximately 4.5 (oligotrophic brown soil) on the one hand, and pH from 8 (rendzina, chernozem) to 9 (solonetz) on the other hand, are not exceptional. Generally, the soils of these grasslands are rich in organic matter (from 5 to 27%) with a wide C/N ratio (from 8 to 29) (Balátová-Tuláčková, 1979). Extremely acid or alkaline soils, or those with high salinity, are occupied by plant communities different from those of semi-natural temperate grasslands.

These grasslands are adapted to hydrological conditions featuring a considerable, but not excessive, quantity of soil moisture and no marked summer drought. In these properties, semi-natural grasslands differ both from natural types of steppe vegetation and from wetland and peat-bog vegetation.

STRUCTURE AND FUNCTION OF PRIMARY PRODUCERS

Types of vegetation

Semi-natural temperate grasslands are secondary plant communities composed of perennial mesophytic grasses and forbs, which are physiologically active throughout the growing season and show no marked depression in summer. A decrease in their rate of growth, or its complete arrest, during winter is usual (Rabotnov,

TABLE 5.1

Succession of species and their percentage share of the above-ground biomass in a meadow in West Germany which developed after deforestation of an *Alnus* forest [1]

Species	Original forest 1937	Deforested, slightly drained and fertilized 1944	Fully fertilized hay meadow and pasture 1964
Scirpus sylvaticus	22		
Juncus effusus	32	6	
Juncus acutiflorus	24	14	
Cirsium palustre	7	3	
Agrostis canina	3	+	
Lotus uliginosus	3	1	
Carex leporina	1	8	
Carex panicea	+	6	
Carex hirta	+	5	
Carex echinata	+	4	
Angelica sylvestris	3	+	+
Holcus lanatus		6	+
Festuca rubra		14	7
Ranunculus acer		2	7
Poa trivialis		2	13
Festuca pratensis		4	21
Poa pratensis		1	30
Ranunculus repens		+	3
Taraxacum officinale			3
Dactylis glomerata			4
Phleum pratense			5
Other species	5	24	7
Total	100	100	100

[1] After Klapp (1965).
+ denotes solitary individuals not contributing significantly to the biomass.

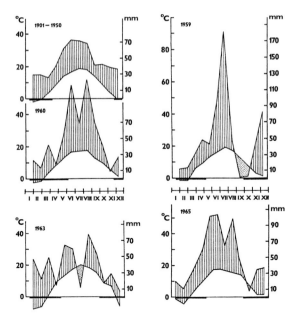

Fig. 5.2. Climatic diagrams (Walter, 1963) showing differences in weather between years at Znojmo, Czechoslovakia, compared to the mean climatic conditions. The upper line = monthly means of precipitation; the lower line = monthly means of temperature; the thick line on the X-axis indicates months with a mean daily minimum temperature below 0°C.

1974). Unlike the situation in pastures, part of the herbage is regularly exported from the hay meadow version of this ecosystem. This results in a predominance of the detritus food chain over the grazing food chain in hay meadows. However, in both situations a great amount of under-ground litter is available to undergo specific decomposition processes which help to bring forth properties typical of grassland soils.

Semi-natural grasslands are a multiform system of herbaceous plants, generally dominated by monocotyledons (mainly Poaceae, Cyperaceae and Juncaceae). The most important families of dicotyledons (in declining order of plant abundance) are Asteraceae, Ranunculaceae, Rosaceae, Fabaceae, Boraginaceae and Apiaceae. Ferns and mosses are less abundant. The prevailing life forms are hemi-cryptophytes and geophytes, whereas few chamaephytes (plants with buds close above the soil surface) and phanerophytes (half-shrubs and shrubs) occur. Annuals, which overwinter as seeds (therophytes), are exceptionally present. An increase in their incidence marks a disturbance of the phytocoenosis (plant community) (Table 5.2). Hemi-cryptophytes are the most important group for fodder production. These include most of the grasses and dicotyledonous dominants, many of which retain some green leaves throughout winter, thus producing a permanent pool of living above-ground biomass. An increase in the proportion of geophytes leads to a reduction of this winter-green pool. In areas with mild winters, semi-natural grasslands are practically evergreen, but even in areas with severe frosts they do not completely lose their above-ground pool of biomass, and produce an active, living layer above the soil surface throughout the year. Grasslands are grazed all year in western Europe, whereas elsewhere they are grazed only during the growing season and most of the sum-

TABLE 5.2

Life-form spectra in meadows in Czechoslovakia

Dominant species	Response to moisture	No. of species	Percentage distribution [1]					
			Ph	Ch	H	G	A	T
Lowland meadows [2]								
Festuca rupicola	meso-xerophytic	83	1	4	73	16	–	6
Alopecurus pratensis	xero-mesophytic	56	–	2	75	18	5	–
Alopecurus pratensis	mesophytic	44	–	2	70	23	5	–
Phalaris arundinacea	hydro-mesophytic	41	–	5	63	12	20	–
Glyceria maxima	hydrophytic	26	–	–	42	8	50	–
Upland meadows [3]								
Festuca rubra	mesophytic	76	3	4	79	10	–	4
Nardus stricta	mesophytic	72	3	4	83	9	–	1

[1] Ph = phanerophytes; Ch = chamaephytes; H = hemi-cryptophytes; G = geophytes; A = hydrophytes; T = therophytes.
[2] Near Lanžhot, along the Morava and Dyje rivers.
[3] Near Kameničky, in the Bohemian–Moravian Uplands.

mer surplus of biomass is preserved for feed in winter either by natural drying in the field, by artificial drying and pelletting, or as silage.

Semi-natural temperate grasslands display considerable uniformity in their ecological functions, and considerable variation in their floristic composition. Selective stress caused by regular mowing or grazing has resulted in the predominance of heliophilous species capable of fast regeneration and possessing a wide ecological amplitude with regard to their requirements for water and mineral nutrients. These features make them suitable for economic exploitation by man. Therefore, most of the cultivated grasses used as fodder for livestock in the temperate zones of the world are of European origin. Grasses are the most important contributors to biomass production in most intensively managed meadows.

The wide ecological amplitude and universality of distribution of the dominant species impedes classification of the vegetation. Because selectivity and indicator values of the dominants of mesophytic semi-natural grasslands are low, they form an unsatisfactory basis for a system of classification founded on principles similar to those used for other natural phytocoenoses. Accordingly, Braun-Blanquet (1951) devised a system of phytocoenological classification based on "characteristic" and "differential" species. This is dependent on either the presence or the absence of species which possess an important indicator value, even though they need not be important participants in the physiognomy of the stand. This approach has been used in classifying the vegetation of western and central Europe and, in part, that of eastern Europe and Japan. However, economic considerations have required that importance be given to the dominant species in determining yield and quality of fodder for livestock. This physiognomical standpoint is the basis of another approach referred to as "grassland typology", and evaluation of grasslands and pastures of eastern Europe and of the temperate zone of Asia is founded upon such a classification. This distinction in approach to classification is related to the degree to which human activity has influenced the nature of vegetation. Because of more intensive management (fertilization and regulation of the water regime), semi-natural grasslands in western Europe have been influenced more than have the semi-natural stands of Asia, which are controlled mainly by natural levels of mineral nutrients and natural water regimes.

An ecological classification of the vegetation of semi-natural meadows and pastures throughout the temperate zone of Eurasia is presented in Appendix I (pp. 156–162). This combines the two approaches discussed above — providing an image of physiognomy as well as of the floristic composition and the habitat; it is based on rich information concerning grasslands in Europe and Japan but less abundant information about the semi-natural meadows and pastures of eastern Europe and the vast territory east of the Urals. The sources of information about the vegetation of Europe and Japan were based on phytocoenological principles according to Braun-Blanquet (Oberdorfer, 1957; Ellenberg, 1963; Klapp, 1965; Suganuma, 1966; Knapp, 1979; Numata, 1979b, c; Speidel, 1979), whereas a little-known exhaustive work by Shennikov (1941)[1] became the basis for the classification of grasslands in the temperate zone of the former Soviet Union. Shennikov's ground-work is based on both physiognomical and ecological characteristics of the vegetation.[2] Drafts of the whole ecological classification were amended by grassland ecologists Denisa Blazkova and T.A. Rabotnov who have studied these types of vegetation intensively.

This classification does not include climax grasslands (such as steppes), wetlands, peat-bogs, and montane grasslands above the tree line. On the other hand, it includes secondary stands in deforested sites, where the reinvasion of trees has been prevented by mowing or pasturing. However, the borderline between these groups is not clearly defined, because of the existence of park forests, forest grasslands, and forest steppes, as well as various communities which appear during the infilling of lakes and on flood-plains. A simplified summary of types of semi-natural grasslands is presented in Table 5.3.

Differences in the anthropogenic influence also determine the nature of grassland types. Thus, a semi-natural grassland may originate from a spontaneous colonization of clearings, or may be a stage of succession on abandoned fields. On the other hand, it may be a crop intensively managed by fertilization and irrigation. When amounts of available nutrients and their turnover are insufficient, grasslands are degraded to such a degree that they become barren stands with half-shrubs and ferns; these may display such high stability that they prevent the development of forest, even though the site has ceased to be a pasture. Such areas have not been regarded as grassland in the present survey.

Apart from the influence of man, there are two main factors which always participate in determining the character of semi-natural grasslands. These are soil fertility and water supply; their level determines the natural floristic composition of forests and of secondary grassland communities in the boreal zone. Southward, in the steppe and desert zones, the water supply is insufficient for development of forests and mesophytic grasslands. However, the flood-plains of large rivers are occupied by azonal forest and grassland vegetation extending far into the steppe and desert zones.

Flooding is the main factor accounting for the presence of azonal vegetation in river valleys. Floods influence the vegetation both directly (by the level of the water-table above or below the soil surface) and indirectly (by periodical enrichment of the soil with nutrients transported from other parts of the catchment). This results in selection of species resistant to both flooding and sedimentation (mostly rhizomatous species), and, in summer, to water stress (in communities on raised terraces). For instance, Shennikov (1941) reported the presence of 1200 species of vascular plants in the catchment of the Volga river, as compared with 300 species in the flood-plain; of these only about 70 to 80 are components of grasslands. Floods are thus responsible for the occurrence of mesophytic species, even in the steppe and desert zones. On the contrary, flooding of streams flowing northward causes eutrophication of grasslands in the floristically poor taiga, thus enabling penetration by southern species which enrich the flood-plains floristically. Therefore, river flood-plains and river terraces are treated separately in classification (Table 5.3).

Numbers and morphological characteristics of species

Semi-natural grasslands are the only agricultural ecosystems possessing natural richness and high species diversity. The presence of as many

---

[1] This book was published just before the siege of Leningrad, during which most of the edition was destroyed. Consequently, the work is unknown to most botanists of the region, even to Russians. The author is much indebted to Professor T.A. Rabotnov of Moscow University for making a copy available for this study.

[2] It is probable that the vegetation now differs from that of Shennikov's time due to human activity. Thus the Appendix reflects the potential grassland types rather than their real extent and present economic value.

TABLE 5.3
A simplified scheme of various types of semi-natural grasslands in the temperate zone of Eurasia [1]

Region	Plains above river valleys		Uplands and mountains	River basins and valleys	
	dry	moist		unflooded terraces	flooded areas
Western and central Europe (10°W –25°E)	Short grasses and forbs *Bromus erectus* *Festuca valesiaca*	Tall-grass meadow *Arrhenatherum elatius* Short-grass pasture *Lolium perenne*	Tall-grass–forb meadow (rich soil) *Trisetum flavescens* Short-grass–forb meadow/pasture (poor soil) *Festuca rubra* Short-grass pasture (poor soil) *Nardus stricta*	Short-grass–forb *Molinia coerulea* *Polygonum bistorta* Tall forbs *Filipendula ulmaria*	Tall grasses and forbs *Alopecurus pratensis* Tall grasses *Glyceria maxima* *Phalaris arundinacea*
Eastern Europe (25°E–60°E)	Short grasses and forbs (in forest zone) *Agrostis tenuis* Tall grasses and forbs (in forest-steppe zone) *Bromus riparius*	Tall-grass (in forest zone) *Deschampsia caespitosa* Tall forbs *Polygonum bistorta*	Tall grasses and forbs (Caucasus) *Bromus variegatus* Short forbs and sedges *Carex humilis* Short-grass–forb pasture (Crimea) *Alchemilla jailae*	Short grasses and forbs *Festuca rubra* *Festuca rupicola* Tall grasses and forbs *Sanguisorba officinalis*	Tall grasses *Alopecurus pratensis* *Agrostis stolonifera* Tall grasses *Phalaris arundinacea*
Western Siberia (60°E–90°E)	Tall grasses (in forest zone) *Festuca pratensis* *Dactylis glomerata* Tall forbs (in forest-steppe zone) *Peucedanum officinale*	Tall grasses (in forest-steppe zone) *Alopecurus ventricosus*	Tall forbs (Tien-shan, Pamir) *Prangos pabularia* *Dactylis glomerata* Tall grasses and forbs (Altai) *Dactylis glomerata* *Hemerocallis flava*	Short grasses *Poa pratensis* *Phleum phleoides* Tall forbs *Festuca pratensis* *Filipendula ulmaria*	Tall grasses and sedges *Carex gracilis* *Phalaris arundinacea*
Eastern Siberia (90°E–160°E)	Short-grass–forb (in forest zone) *Festuca lenensis*	Tall grasses (in forest zone) *Calamagrostis langsdorfii*		Tall-forb–grass meadow *Heracleum dissectum* *Bromus sibiricus* Short grasses and forbs *Hordeum brevisubulatum* *Poa pratensis*	Tall grasses *Bromus sibiricus* *Alopecurus ventricosus*
Far East (160°E–170°W)	Tall grasses and forbs *Arundinella anomala*	Tall grasses and sedges *Carex schmidtii*		Tall forbs *Bromus ciliatus* *Hemerocallis flava*	Tall grasses *Calamagrostis langsdorfii*
Northern Japan	Dwarf bamboo *Sasa nipponica* (meadow) *Poa pratensis* (pasture)				
Middle Japan	Tall-grass meadow *Miscanthus sinensis* Short-grass pasture *Zoysia japonica*				

[1] Compiled after Shennikov (1941), Oberdorfer (1957), Ellenberg (1963), Klapp (1965), Suganuma (1966), Numata (1979a, b) and Speidel (1979). The species listed are representative of many others that are present. For more detail, see Appendix I. Critical revision by T.A. Rabotnov is gratefully acknowledged.

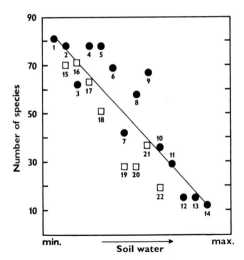

Fig. 5.3. Species diversity in a grassland hydrosere in the Morava and Dyje river lowlands (*1* to *14*, according to Vicherek, 1962; and *15* to *22*, according to Balátová-Tuláčková, 1966), from meso-xerophytic grassland dominated by *Festuca rupicola* and *F. valesiaca* through tall grass–forb meadows dominated by *Alopecurus pratensis* and hygrophytic meadows dominated by *Phalaris arundinacea*, tall sedges and *Glyceria maxima*, to wetland stands of *Phragmites communis*.

as 80 species of vascular plants (and even more) in a stand is not exceptional. Conditions which are suboptimal for the production of biomass tend to result in a richer floristic composition, whereas optimal or supraoptimal conditions result in a reduction in species diversity and in a more marked dominance of competitively strong species. This was confirmed in studies of hydroseres of floodplain meadows in central Europe (Fig. 5.3). Another factor leading to a morphological and functional monotony of semi-natural grasslands is mineral fertilization (Rabotnov, 1973). Halva and Lesák (1979), studying the effect of fertilization on semi-natural stands dominated by *Nardus stricta* in Czechoslovakia, found that four years of fertilization, with annual applications of 100 kg N, 22 kg P, and 41.5 kg K ha^{-1}, reduced the number of species from 47 to 36, and applications of double this rate, to 28 species. However, the suppressed species persist for many years either as seeds or under-ground organs and are always available to replenish a damaged stand when growing conditions permit.

The supply of seeds for regeneration of vegetation is abundant in these grasslands. The reserve of seeds in grassland soil (to a depth of 30 cm) ranges from 17 000 to 38 000 m^{-2} (Harper, 1977). These seeds are either of local origin or imported. Rabotnov (1950) found that a meadow mown in June produced no seeds, that one mown in July produced 2260 seeds m^{-2}, and one mown in August produced 13 400 seeds m^{-2}. Pyatin (1971) recorded as many as 270 seeds of birch (*Betula*) and 300 of willow (*Salix*) in a square metre, transported by the wind from surrounding areas to grassland. Annual changes in seed production differ by one order of magnitude. Jakrlová (1985) recorded the production of 16 770 and 2780 seeds m^{-2} in two consecutive years in unmown grassland dominated by *Nardus stricta*. Most seeds are in a state of dormancy, but they maintain their germinative capacity for decades. The quantity of seeds germinating each year represents a small fraction of the permanent pool. Numata (1979c) in Japan found 2000 and 20 000 seeds m^{-2} in soil to a depth of 10 cm in *Miscanthus* meadows and *Zoysia* pastures, respectively, which suggests that the mode of management (i.e. mowing or pasturing) affects the quantity of seeds stored in the soil.

Greater variations in natural factors of the stand, particularly in moisture conditions, favour particular growth strategies that enable plants either to resist unfavourable conditions or to avoid them. The example in Table 5.4 shows that the diversity of morphological types increases with the amplitude of fluctuation in moisture supply. Over several years, the dominant position passes from one species to another. Various strategies are associated with poorly understood compensation mechanisms which turn semi-natural grasslands into highly homoeostatic systems which are relatively stable. This feature, as well as the possession of a rich pool of genes, plays an important ecological role, which is under-rated, particularly by agriculturalists.

Population structure of stands

Because these grasslands are composed of perennial plants, the age structure of participating coenopopulations (see definition on p. 20) is of considerable importance. Rabotnov (1950), Uranov (1960) and their followers defined the following ontogenetic phases: (1) seeds (SE); (2)

TABLE 5.4

Morphological diversity and ecotopes in a hydrosere in flooded meadows [1]

Ecotope	Dry	Moist	Damp	Wet
Water level during the growing season				
No. of days above the soil surface	2	10	27	41
No. of days deeper than 30 cm in soil	86	66	55	30
Average soil moisture (% of dry weight)	25	54	60	74
Dominant plant species	*Festuca rupicola*	*Alopecurus pratensis*	*Phalaris arundinacea*	*Glyceria maxima*
Proportion of morphological types [2]				
Broad-leaved grasses	17	58	90	100
Narrow-leaved grasses	35	–	–	–
Monocotyledons (type *Colchicum*)	4	1	–	–
Forbs, erect with large leaves	4	23	6	–
Forbs, creepers (type *Vicia*)	9	6	–	–
Low forbs	31	12	4	–
Mean above-ground production (g m^{-2} yr^{-1})	540	721	1041	1153

[1] Near Lanžhot, Czechoslovakia (1966–1972); see also Fig. 5.8.
[2] Percentage of the above-ground standing crop.

seedlings, not yet fully autotrophic (P); (3) juvenile plants (J), frequently showing even morphological differences from other stages; (4) immature plants (IM); (5) virginal, vegetatively mature plants (V); (6) young generative[1] plants (G_1); (7) mature generative plants (G_2); (8) old generative plants (G_3); (9) subsenile plants (SS); and (10) senile plants (S). Developmental stages are not identical with the age of the plant. In semi-natural grasslands, seeds remain viable for decades and seedlings enter the juvenile stage within one year. They may persist in the juvenile stage for many years during which they complete their development and gradually enter the virginal phase. Virginals are found frequently in abundance in plant communities and sometimes they persist in this phase until death. Their transition to the generative phase depends on both external and internal factors. Polycarpic plants in grassland may persist in the generative phase for 30 years or more, but they do not necessarily flower each year. They may survive in the vegetative stage for several years, sometimes even latently in the form of underground organs. This secondary "dormancy" can be interrupted by a change in environment. The life cycle terminates in the loss of reproductive (sexual) organs and the death of the individual. The duration of this stage is short and its participation in grassland populations is small.

Examples of population analysis of grassland ecosystems are shown in Figs. 5.4 and 5.5. The invasive, stable, or regressive character of every population can be derived from its demographic structure. This enables one to forecast the successional development of the stand under the influence of new ecological factors.

Structure of the canopy

The proportion of live and dead material in the canopy depends on the type of stand and on the intensity of its economic exploitation. At the time of maximum standing crop, mown meadows have roughly 15 to 25% of this maximum as standing dead material and unmown grasslands up to 60%. In autumn, the portion of dead biomass increases to about 60 to 90% of the standing crop. In winter, part of the standing dead material changes into litter and decomposes, but, even so, some of it persists until the next season and exceeds the amount of the living herbage in spring (Fig. 5.6).

---

[1] "Generative" refers only to sexual reproduction.

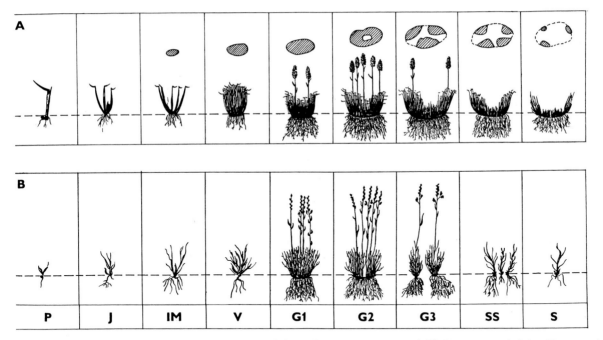

Fig. 5.4. Ontogenetic development of perennial grasses: (A) *Deschampsia caespitosa*; and (B) *Festuca pratensis* (after Uranov and Serebryakova, 1976). P = seedlings, J = juvenile plants, IM = immature plants, V = virginal plants, G_1 to G_3 = generative plants, SS = subsenile plants, S = senile plants.

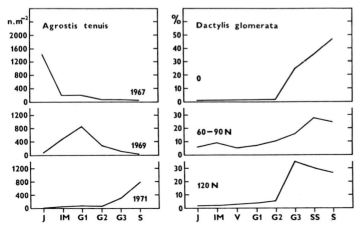

Fig. 5.5. Population analysis of *Agrostis tenuis* during succession on an abandoned field and of *Dactylis glomerata* at three levels of fertilization (kg ha^{-1}) in a meadow (after Uranov and Serebryakova, 1976). Abbreviations are the same as in Fig. 5.4.

Standing dead leaves persist in those grasses with a high proportion of mechanical tissues, even after 2 to 3 years; this leads to the accumulation of dead shoot biomass in unmown grasslands (Fig. 5.7).

A considerable part of the above-ground biomass is formed of mechanical and conductive tissues, even in mesophytic grasslands. In a mown meadow, stems and leaf petioles form about 40% of the total above-ground phytomass, leaf blades about 52% and generative (sexual) organs about 8% (Alekseenko, 1967). Another example is illustrated in Fig. 5.7. Geyger (1964) found that from 40 to 80% of the assimilation surface of grasses and forbs occurred on stems; this means that the assimilation surface cannot be regarded as identical with the leaf area.

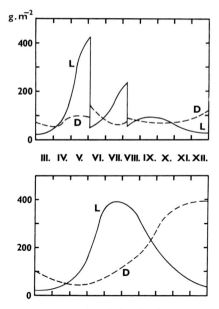

Fig. 5.6. Seasonal variation in above-ground biomass in a mown *Alopecurus pratensis* meadow in lowland in Czechoslovakia (above) and in an unmown mesophytic *Arrhenatherum elatius* meadow in Poland (below): L = living biomass; D = dead biomass (after Jankowska, 1967; Petřík, 1972; Jakrlová, 1975).

The share of various species in the vertical structure of the stand depends on the type of plant community and on environmental conditions. Stands composed mainly of tall grasses display a considerable uniformity in the vertical distribution of biomass. In stands dominated by tufted grasses or forbs, most of the biomass is concentrated in the lower layers and its density diminishes rapidly towards the upper levels of the canopy. Examples of the spatial structure of a semi-natural grassland are given in Fig. 5.8, and the proportional representation of species in Table 5.5.

The radiation profile of the stand depends on the density of foliage. Absorption curves of the photosynthetically active radiation (PhAR) in various layers are shown in Fig. 5.8. These curves indicate that absorption of radiation in all layers of the canopy is greatest in grassland types high both in species diversity and in production; in such stands as little as 3 to 6% of the incoming PhAR reaches the soil surface (Fig. 5.8: B and C). The phyllosphere of a uniform, mostly monospecific wetland stand (Fig. 5.8A) allowed as much as 20% of PhAR to pass through and so remain unutilized, in spite of a high leaf-area index (LAI). Because of the poor above-ground biomass of the dry stand (Fig. 5.8D), 17 to 20% of PhAR penetrated to the soil surface. This suggests that mixed mesophytic stands, with different strategies in the distribution of their radiation receptors, are most effective in intercepting radiation.

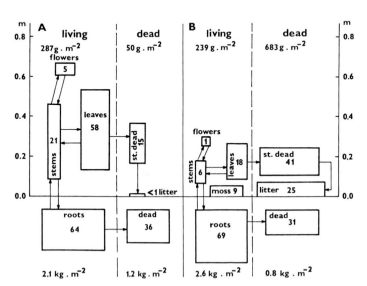

Fig. 5.7. Distribution of biomass in meadows in Bohemian–Moravian upland in Czechoslovakia (at Kameničky): a mown meadow dominated by *Poa pratensis* (A), and an unmown meadow dominated by *Nardus stricta* (B) (after Jakrlová and Kršková, 1982). The maximum standing crop of shoot biomass is indicated.

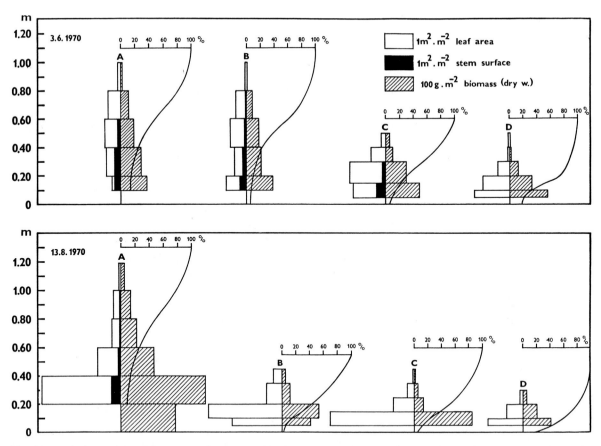

Fig. 5.8. Vertical structure of the canopy (before the first and second cut) in four meadows in lowlands of the Morava river, Czechoslovakia (after Rychnovská et al., 1972a). The radiation profile (in relative units of incoming PhAR) is indicated by curves. A = wet grassland, unmown, dominated by *Glyceria maxima*; B = damp meadow, mown, dominated by *Phalaris arundinacea*; C = moist meadow, mown, dominated by *Alopecurus pratensis*; and D = dry meadow, mown, dominated by *Festuca rupicola*. The meadows are the same as in Table 5.4.

TABLE 5.5

Vertical distribution of above-ground biomass (%) by species in a meadow [1]

Species	Layer above soil surface (cm)						
	0–10	10–20	20–30	30–40	40–50	50–60	60–70
Alchemilla sp.	58	40	2	–	–	–	–
Anthoxanthum odoratum	64	26	4	3	3	–	–
Anthriscus sylvestris	25	48	17	7	3	–	–
Deschampsia caespitosa	47	20	10	8	5	5	5
Equisetum arvense	50	49	1	–	–	–	–
Festuca pratensis	36	23	17	12	6	3	3
Filipendula ulmaria	31	38	26	5	–	–	–
Geranium pratense	20	32	34	14	–	–	–
Phleum pratense	31	19	29	17	3	1	–
Poa pratensis	32	18	13	15	8	14	–
Trifolium medium	23	32	44	1	–	–	–

[1] After Alekseenko (1967).

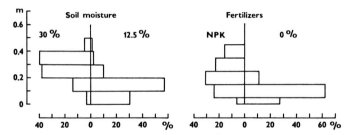

Fig. 5.9. Effect of contrasting levels of moisture and nutrients in soil on the vertical structure of meadows. Left = a meadow dominated by *Phleum pratense* and *Geranium pratense* near Lake Ladozhskoye, Russia; soil moisture (June and July) was 30% in a moist year and 12.5% in a dry year (Alekseenko, 1967). Right = a meadow dominated by *Nardus stricta* and *Polygonum bistorta* in the Bohemian–Moravian uplands of Czechoslovakia; plots fertilized with 100 kg N, 22 kg P and 41.5 kg K ha^{-2} are compared to unfertilized plots (Poslušná, 1975). The X-axis represents percentage of biomass; the Y-axis represents height above the soil surface. The measurements were made at the time of maximum standing crop.

Environmental factors affect the structure of the canopy partly by influencing the representation of various species and partly in the morphological adaptation of these species. The reaction of the same grassland stand to various moisture factors and to different levels of fertilization is shown in Fig. 5.9.

These observations suggest that the structure of the canopy is determined by ecological factors and by indigenous properties of the component species. Optimal development and extent of leaf surface in grassland in the temperate zone are limited neither by light nor by temperature. The two main determining factors are nutrients and moisture. The influence of the combined action of these factors is illustrated in Fig. 5.10 and Table 5.6. Abundant foliage develops only if the amount of available nutrients is sufficient and the supply of water is uninterrupted. If these conditions are not fulfilled, the stands are low and dominated by narrow-leaved grasses (Table 5.6: b, c and d). Leaf area is great in fertile, intensively managed mesophytic meadows. Xero-mesophytic meadows, psammophytic grassland and meso-hydrophytic pastures, if managed, are used for extensive grazing only. However, the latter are very important in protecting the landscape from soil erosion.

Structure of under-ground biomass

Under-ground biomass consists of roots, bulbs, tubers and other reserve organs. No estimate has been made of the share of various species in the under-ground biomass of floristically rich, semi-natural grasslands. However, depths of root penetration of various species have been studied. Kotańska (1970) classified the species in a stand dominated by *Arrhenatherum elatius* according to their depth of penetration as follows:

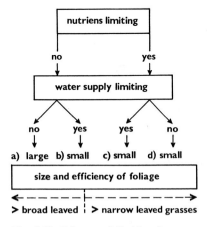

Fig. 5.10. Scheme of limiting factors controlling the area (LAI) and biomass of foliage in semi-natural temperate grasslands (after Rychnovská et al., 1985). The designations *a*, *b*, *c* and *d* correspond to those in Table 5.6.

Shallow-rooted, to 10 cm in depth

Achillea millefolium
Anthoxanthum odoratum
Galium verum
Primula elatior

More deeply rooted, as much as 100 cm

Dactylis glomerata
Festuca pratensis
Plantago lanceolata
Trisetum flavescens

Roots of medium depth, as great as 50 cm

Agrostis vulgaris
Chrysanthemum leucanthemum
Ranunculus acer
Trifolium repens

Most deeply rooted, to depths as great as 170 cm

Alopecurus pratensis
Arrhenatherum elatius
Pimpinella saxifraga
Rumex acetosa

TABLE 5.6

Examples of European grasslands according to different moisture levels[1]

Vegetation type	Dominant species	LAI ($m^2\ m^{-2}$)	Above-ground standing crop ($g\ m^{-2}$)
a. Mesophytic meadow	Arrhenatherum elatius	12.9	672
b. Xero-mesophytic meadow	Festuca rupicola	2.1	195
c. Psammophytic grassland	Corynephorus canescens	0.4	135
d. Meso-hydrophytic pasture	Nardus stricta	1.2	205

[1] This table relates to Fig. 5.10. After Geyger (1964), Berger-Landefeldt and Sukopp (1965), Rychnovská et al. (1972a) and Poslušná (1975).

The age of under-ground organs is associated with the population structure of the stand. The ratio of under-ground to above-ground phytomass changes in the course of ontogenesis. The proportion of biomass under-ground increases from the juvenile to the virginal phase. During the generative phase above-ground biomass may predominate. As the plants grow old, their above-ground parts diminish and finally disappear, whereas their under-ground organs can survive for a prolonged period of time. The under-ground reserve organs lose their contents and decrease in mass density, which ranges from 0.3 to 0.01 g cm^{-3} of tissue. The age of the primary root is identical to that of the plant, whereas that of adventive roots is shorter, being 3 to 6 years for grasses (Persikova, 1959). Root tips survive for shorter periods. Speidel (1976) found, in a meadow dominated by Festuca rubra, that 54% of root tips survived for less than one month, 10 to 15% lived for more than 4 months, and only 0.2% survived for 13 months.

The standing crop of under-ground biomass varies greatly among various vegetation types and management regimes. Values reported for intensively managed mesophytic meadows range from 738 to 946 g m^{-2}, whereas grasslands without management show an amplitude of from 759 to 3498 g m^{-2}. However, for alluvial or peaty sites, values of 1221 to 3840 g m^{-2} were found, even where managed (Table 5.7).

Shalyt (1950) related the surface area to the weight of roots (Table 5.8). His data, based on detailed measurements of volume and diameter of roots, suggest that the surface area of 10 g of roots is approximately 1 m^2. This very extensive biological surface below the soil surface is greater by at least one order of magnitude than that in the phyllosphere. In this study, short-lived root hairs were not considered; their inclusion would increase the biologically active surface of the rhizosphere to at least 5 times the area measured.

Under-ground biomass decreases from the soil surface downward. In mesophytic grasslands, 75 to 80% of under-ground biomass is concentrated in the uppermost 15 to 20 cm of soil. This is the layer of the soil of greatest interaction among plant species, of the deposition and translocation of reserves, of vegetative reproduction, and of the greatest activity of the rhizosphere micro-flora. However, the more sparsely rooted, deeper layers also contain considerable absorptive surface. Shalyt (1950) found a root surface area of 5.5 m^2 per m^2 of soil surface in the 70- to 100-cm layer of a poor Nardus stricta grassland.

Root/shoot biomass ratios usually range between 1 and 2 for regularly fertilized mesophytic grassland, but variations occur during the course of the growing season. Pilát (1969) reported a ratio of 5.1 in spring in mesophytic grassland dominated by Alopecurus pratensis, which changed to 2.1 before the first cut of hay.

The extensive permanent pool of under-ground biomass is one of the most important features distinguishing semi-natural grasslands from croplands. Most of the cycling of geobioelements and carbon occurs in this pool, the biomass enriching the soil with organic substances and improving its physical properties. To a reduced extent, this also takes place in seeded grasslands. This considerable regenerative reserve is made use of in improving soil properties by including perennial grassland in systems of crop rotation.

TABLE 5.7

Typical estimates of net annual production of above-ground and under-ground biomass production in meadows (g m^{-2})[1]

No.	Dominant taxa	Location	Mean under-ground biomass	Production		Ratio of root biomass to shoot production	Turnover time of root biomass (yr)
				shoots	under-ground		
1	*Festuca rupicola*	Czechoslovakia	1556	957	389	1.62	4.0
2	*Alopecurus pratensis*	Czechoslovakia	1221	1016	407	1.20	3.0
3	*Phalaris arundinacea*	Czechoslovakia	1664	1388	554	1.19	3.0
4	*Festuca rubra–Agrostis tenuis*	West Germany					
	no fertilization		946	316	371	2.99	2.5
	NPK fertilization		738	808	381	0.91	1.9
5	*Trisetum flavescens–Alchemilla crinita*	Poland	1791	700	697	2.55	2.5
6	*Brachypodium pinnatum–Agrimonia eupatoria*	Poland	3498	500	1367	6.99	2.5
7	*Arrhenatherum elatius–Poa pratensis*	Poland					
	no fertilization		884	253	336	3.49	2.6
	NPK fertilization		729	993	244	0.73	3.0
8	*Alchemilla monticola–Agrostis tenuis*	former Soviet Union	759	263	399	1.90	1.9
9	*Alchemilla monticola–Alopecurus pratensis*	former Soviet Union	951	347	478	2.74	2.0
10	*Nardus stricta–Deschampsia caespitosa*	Czechoslovakia					
	no fertilization		2580	268	810	9.62	3.1
	NPK fertilization		3840	784	2099	4.89	1.8

[1] Estimates of production are based on the sum of 2 or 3 harvests from the same plots in mown meadows and on maximum standing crop in unmown meadows. Treatments No. 5, 6, 8, 9 and 10 were not mown; all others were mown. Treatments No. 5, 6, 8 and 9 were not fertilized.

References.
No. 1, 2, 3: Rychnovská (1979a).
No. 4: Speidel and Weiss (1972), Speidel (1976).
No. 5, 6: Kotańska (1970).
No. 7: Traczyk et al. (1976), Plewczyńska and Kuraś (1976).
No. 8, 9: Makarevich (1978).
No. 10: Fiala (1979), Halva and Lesák (1979).

Seasonal aspects

The nature of the canopy varies seasonally (Fig. 5.6) with the phenophases of the individual specific components, that is, formation of vegetative shoots, formation of buds and flowering, and the development and ripening of seeds. This last stage is severely limited in mown stands.

An almost synchronous beginning of flowering of various grasses and forbs in early summer is typical of temperate semi-natural grasslands. An investigation of 22 grassland stands in the flood-plain of the Opava river in Czechoslovakia showed that 72% of species came into flower in May and June and only 28% in July and August, regardless of whether the stand had or had not been mown during the current year. The predominant flower colour changed during the season. Yellow prevailed in early May, later giving way to light violet and white. Bluish-violet was a common shade in June; and yellow and purple appeared in late summer (Balátová-Tuláčková, 1971). Colours are less varied in the moister types of grassland dominated by anemophilous species (graminoids) and more varied in drier types, in which a remarkable display of colour is exhibited by flowers of forbs. These colour aspects are apparently associated with the dynamics of insect populations engaged in pollination. The colourful aspects of semi-natural temperate grasslands

TABLE 5.8

A comparison of biomass and surface area of roots in semi-natural grasslands [1]

Dominant species	Biomass (g m^{-2}) total [2]	Biomass (g m^{-2}) roots	Root surface (m^2 m^{-2})
Nardus stricta Achillea millefolium Deschampsia caespitosa	2956	2783	301
Nardus stricta Festuca pratensis Potentilla erecta	1983	1460	123
Festuca rubra Poa trivialis Agrostis alba Vicia cracca	805	679	66
Galium verum Achillea millefolium Phleum pratense	1110	1081	106

[1] After Shalyt (1950).
[2] Includes under-ground parts that are not roots.

represent a characteristic element of the densely inhabited and intensively managed landscape of Europe.

Dynamics of above-ground biomass and leaf development

Quantity and quality of canopy biomass depend partly on environmental conditions and partly on management regimes. Fertilization and regular harvesting are very important in providing for high productivity. The greatest values of standing crop in semi-natural meadows of the temperate zone range from about 1000 to 1100 g m^{-2} (Table 5.9).

Most of the living above-ground biomass functions as an assimilation surface. Foliage (stems, leaf blades and flower-stalks) is, however, also the site of transpiration. Accordingly, optimal biomass production is achieved when radiation receptors occupy the largest possible space while allowing least moisture loss by excessive evapotranspiration. The relationship between extent of leaf area and environment is exemplified in Fig. 5.11. Nutrients were supplied in sufficient quantities either by flooding or fertilizers; thus, moisture supply and temperature were considered as limiting factors (Table 5.10 and Fig. 5.11). In western Germany, the growth rate was greatest in mountain grasslands in early summer, surpassing 8 g m^{-2} d^{-1}, apparently due to the late onset of spring and the long duration of the summer day. The rate of biomass production was slower in lowland stands where the growing season starts as early as March, whereas grassland at higher geographical latitudes showed a similar rate of production to that of mountain meadows. According to Alekseenko and Martynova (1964), seeded stands of Phleum pratense growing in the area of the Gulf of Finland produced, within two summer months, more than 1700 g m^{-2} of above-ground biomass at an average rate of 27 g m^{-2} d^{-1}.

The limiting effect of the water supply results in a relatively reduced leaf cover in lowland (Fig. 5.11). An increase in LAI at higher altitudes appears to be associated with higher soil moisture content and relative humidity of the atmosphere. Leaf areas reported by eastern European authors are less variable than those from western Europe. Values for LAI in a hydrosere of meadows in the flood-plain of the Oka river (Russia), range from 2.68 m^2 m^{-2} in a dry stand of Festuca rupicola and Trifolium montanum to 4.58 m^2 m^{-2} in a moist stand of Phalaris arundinacea. Maximum LAI values at the peak of summer were 2.66 to 6.40 m^2 m^{-2} in mesophytic natural meadows in the lowland in the vicinity of Lake Ladozhskoye (St. Petersburg district), 4.94 to 7.37 m^2 m^{-2} in floodplain meadows in the northern part of the Dvina river region, and 4.33 to 4.44 m^2 m^{-2} in subalpine meadows of the northern Caucasus (Alekseenko, 1967). Estimated LAI values (ranging from 2.1 to 4.3 m^2 m^{-2}) in a grassland hydrosere under subcontinental climate in Czechoslovakia were depressed compared to those in analogous floodplain meadows in various other climatic zones in Europe. Geyger (1964) reported LAI values in flood-plain meadows in the delta of the Elbe river (in a region of oceanic climate) ranging from 4.8 to 16.2 m^2 m^{-2}.

Dynamics of under-ground biomass

The rate of production of under-ground biomass in perennial grassland is difficult to estimate accurately, because under-ground plant

TABLE 5.9

Typical herbage yields (g m^{-2}) in semi-natural grasslands

Dominant species	Location	Unfertilized	Fertilized	Reference
Arrhenatherum elatius *Calamagrostis epigeios*	Elbe river lowland, western Germany	889	–	Geyger (1964)
Cirsium oleraceum *Calamagrostis epigeios*	Elbe river lowland, western Germany	740	–	Geyger (1964)
Festuca rubra *Agrostis tenuis*	Solling hills, western Germany	316	808	Speidel and Weiss (1972)
Festuca rubra *Agrostis tenuis*	Carpathian mountains, Czechoslovakia	209	897	Krajčovič and Regál (1976)
Festuca pratensis *Alopecurus pratensis*	Carpathian mountains, Czechoslovakia	433	1085	Krajčovič and Regál (1976)
Alopecurus pratensis	Morava river lowland, Czechoslovakia	721	–	Rychnovská (1979a)
Phalaris arundinacea	Morava river lowland, Czechoslovakia	1041	–	Rychnovská (1979a)
Nardus stricta *Deschampsia caespitosa*	Bohemian–Moravian Upland, Czechoslovakia	180	731	Halva and Lesák (1979)
Carex fusca *Festuca rubra*	Wisła river lowland, Poland	196	–	Traczyk (1971)
Festuca pratensis *Holcus lanatus*	Wisła river lowland, Poland	585	–	Traczyk (1971)
Arrhenatherum elatius *Poa pratensis*	Wisła river lowland, Poland	278	1046	Traczyk et al. (1976)
Bromus inermis *Agropyron repens*	Oka river lowland, Russia	543	904	Rabotnov (1973)
Agrostis stolonifera *Phleum pratense*	Oka river lowland, Russia	345	708	Rabotnov (1973)
Agrostis tenuis *Alopecurus pratensis*	Lake Ladozhskoye district, Russia	270	–	Makarevich (1978)
Miscanthus sinensis	Honshu, Japan	679	–	Numata (1979b)
Zoysia japonica	Honshu, Japan	284	–	Numata (1979b)

parts display all the phases of their existence at the same time. They grow, die off and decompose uninterruptedly. Estimates of under-ground biomass production are usually based on the difference between the maximum and minimum standing crop during the course of the year, but this method provides only an estimate of net production and underestimates the rate of energy flow in the ecosystem.

Fiala (1979) devised a method of estimating gross under-ground production by using a parallel sampling technique. In one series of soil cores, the shoots are excised at the soil surface and the cores are inserted into nylon nets and replaced in the same holes for a certain interval. The decrease in weight of biomass is a result of decomposition during this interval, which can be used to adjust upwards the estimate of net production obtained by the usual difference method. The sum of both values is an estimate

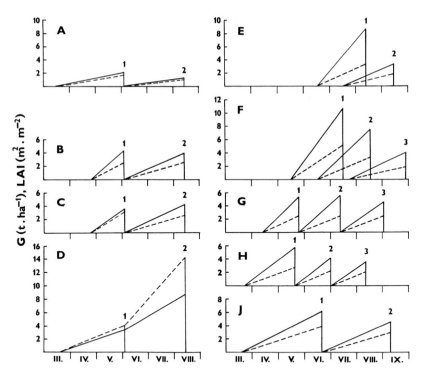

Fig. 5.11. Scheme of accumulation of above-ground biomass and of development of leaf area in successive growth periods after mowing. (A–D) Alluvial grasslands in Czechoslovakia (Rychnovská et al., 1972a). (E–J) Grasslands at various altitudes from north-western to south-western Germany (Geyger, 1977). Dashed line = biomass in t ha^{-1}; solid line = LAI in m^2 m^{-2}. Letters A to J correspond to those in Table 5.10.

TABLE 5.10

Rates of accumulation of canopy biomass (g m^{-2} day^{-1}) in various meadows [1]

Biotope	Altitude (m)	Location	Dominant taxa	First cut	Second cut	Third cut
A. Dry meadow	155	Lanžhot, Czechoslovakia	Festuca rupicola	2.46	1.69	–
B. Moist meadow	155	Lanžhot, Czechoslovakia	Alopecurus pratensis	7.00	3.66	–
C. Damp meadow	155	Lanžhot, Czechoslovakia	Phalaris arundinacea	8.53	3.51	–
D. Wet grassland [2]	155	Lanžhot, Czechoslovakia	Glyceria maxima	4.95	14.35	–
E. Pasture	1500	Feldberg, Germany	Festuca rubra– Agrostis tenuis	5.68	2.45	–
F. Pasture	1000	Menzenschwand, Germany	Festuca rubra– Agrostis tenuis	8.58	5.00	2.76
G. Meadow	450	Solling, Germany	Festuca rubra– Agrostis tenuis– Holcus lanatus	6.04	5.21	4.71
H. Meadow	200	Fulda Valley, Germany	Festuca rubra– Alopecurus pratensis– Poa pratensis	4.74	5.47	4.88
J. Meadow	5	Gose–Elbe lowland, Germany	Filipendula ulmaria– Glyceria maxima	4.20	3.65	–

[1] The German data are from Geyger (1977); the Czechoslovakian data are those of the author. See also Fig. 5.11.
[2] Unmown. All others are mown.

of gross under-ground production. By using this method, Fiala (1979) estimated annual production of under-ground biomass in a poor meadow to be 810 g m^{-2}. This is considerably higher than the values reported in semi-natural grassland by other workers (Table 5.7), but much lower than reported for seeded grassland.

Estimates of net under-ground biomass production are lower than maximum standing crop of herbage in some grasslands and higher in others. The latter situation occurs in stands in which canopy development is suppressed by such factors as water stress (Table 5.7, No. 6) or boggy, waterlogged soil (Table 5.7, No. 10).

Photosynthesis of grassland species

Laboratory studies of photosynthetic rates of the dominant species of the lowland grasslands considered in Table 5.4 were made by Gloser (1976, 1977). Differences between the highly productive species (*Alopecurus pratensis*, *Glyceria maxima* and *Phalaris arundinacea*) were minimal. The maximum rate of photosynthesis (at 20°C in a fully saturated radiation flux) was 20 to 26 mg CO_2 dm^{-2} of leaf surface hr^{-1}. However, a rate as low as 15 mg CO_2 was assessed for the dominant (*Festuca rupicola*) of the dry grassland under optimal conditions. Differences were also found in dark respiration. The output was 1.2 to 1.6 mg CO_2 dm^{-2} hr^{-1} for the three highly productive species, but was 2.4 mg CO_2 for *F. rupicola*, measured under identical conditions. These results suggest that *F. rupicola*, which is adapted to drier biotopes, can never attain the high production capacity of the three species of mesic and hydric sites, even under optimum growing conditions.

Photosynthetic activity decreases in all species when the complex of ecological factors is not optimal. A decrease in the soil-water potential from -8×10^4 Pa to -10×10^4 Pa reduces the photosynthetic activity of the three highly productive species by 30%, but flood-plain grasslands are seldom exposed to such a high deficit of soil water. However, a decrease in atmospheric moisture occurs frequently during hot summer days; consequently, the difference in water vapour tension between the leaf and the atmosphere attains values of more than 120 Pa, and the photosynthetic activity is reduced by about 50%.

Photosynthetic rates of important grasses and forbs of low-producing submontane grasslands under optimum conditions of radiation and temperature were (in CO_2 dm^{-2} hr^{-1}) as follows (Gloser, 1979):

Polygonum bistorta	28 mg
Deschampsia caespitosa	27 mg
Festuca capillata	25 mg
Nardus stricta	25 mg
Sanguisorba officinalis	19 mg

These values are similar to those obtained with lowland grasses, and suggest that the productive capacity of the stand cannot be regarded as a direct reflection of the potential photosynthesis of its components. The actual production level of the stand is determined by limiting factors of the habitat and by the growth strategies of the component species. Under natural conditions, the photosynthetic rate, even of such highly productive species under optimum conditions as *Dactylis glomerata* and *Phleum pratense*, is only about 6 to 7 mg CO_2 dm^{-2} hr^{-1} (Alekseenko, 1962).

Data on root respiration were made available by Tesařová (1979b) and by Gloser and Tesařová (1978). According to their results, the carbon-dioxide output during summer (in g m^{-2} day^{-1}) is about 6 for a dry grassland, about 7.5 for a mesophytic grassland in the flood-plain of the Dyje river, and about 4 to 5 for a poor submontane grassland dominated by *Nardus stricta*. However, even in late autumn, this activity never falls below 1.5 g CO_2 m^{-2} day^{-1}.

Translocation of assimilates

Assimilates are translocated in both directions, from above-ground to under-ground organs and vice versa. A study in a *Alopecurus pratensis* meadow (Rychnovská, 1972a) revealed that deposits of carbonaceous and nitrogenous reserve substances in under-ground biomass exceeded those in above-ground parts. The rate of carbohydrate deposition in shoots was about 1 g m^{-2} day^{-1}, whereas the average rate of the simultaneous translocation of carbohydrates to the roots exceeded 7 g m^{-2} day^{-1}. The rate of protein accumulation was about 0.5 g m^{-2} day^{-1} in shoots, and about 2.5 g m^{-2} day^{-1} in roots. Maximum concentration of carbohydrates was about 150 g

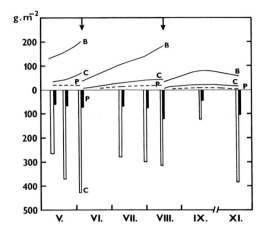

Fig. 5.12. Seasonal variation in hydrolyzable carbohydrates (*C*) and proteins (*P*) in the above-ground and under-ground biomass of a short-grass–forb meadow dominated by *Festuca rupicola* (Lanžhot, Czechoslovakia, after Rychnovská, 1972a). Green above-ground biomass (*B*) was mown twice each year, as indicated by arrows.

m^{-2} in above-ground biomass and 400 g m^{-2} in under-ground biomass. Respective concentration levels of proteins were about 45 and 130 g m^{-2}. During regeneration of foliage after mowing, 35% of carbohydrate reserves, and as much as 60% of protein reserves, passed from roots into shoots. As soon as the new leafage developed, the flow of assimilates was again reversed. The dynamics of hydrolyzable carbohydrates in a dry meadow dominated by *Festuca rupicola* is illustrated in Fig. 5.12.

The transmission of assimilates between shoots and roots was illustrated also by a mathematical model of photosynthetic production of a stand in Czechoslovakia dominated by *Glyceria maxima* (Gloser, 1977). A comparison of calculated values of photosynthetic production with measured accumulations of above-ground biomass indicated that shoots of *Glyceria maxima* were not fully autotrophic in spring until about mid-April, but depended on reserve substances translocated from under-ground organs. However, from the end of June onwards, photosynthetic activity greatly surpassed the actual increment in shoot biomass, the difference being a measure of the quantity of assimilates deposited as under-ground reserves.

Utilization of solar energy

Little information is available concerning efficiency of fixation of solar energy in temperate, semi-natural meadows. Such a study of the dominant species of a hydrosere of flood-plain meadows in Czechoslovakia (Table 5.11) (Rychnovská et al., 1985) indicated that species of mesic or hydric stands (*Alopecurus pratensis*, *Glyceria maxima* and *Phalaris arundinacea*) possess a high potential of utilizing solar energy in photosynthesis (3.4 to 4.4% in laboratory studies), but performed less efficiently (1.1 to 3.3%) under normal growing conditions. The dominant of a drier stand (*Festuca rupicola*) was much less efficient (1.4% in the laboratory and 1.0% under natural conditions). Field measurements approximated laboratory values only for *Glyceria*. The lower values in the field are presumably accounted for by lower photosynthetic activity of the remaining components of the stand and by competition for light and water. The limiting effect of environmental factors is indicated explicitly by values calculated for the entire growing season, showing only 0.7 to 3.3% utilization of global radiation. These actual values decrease with increased degree of water stress in different ecosystems.

These levels of utilization of solar energy are lower than those reported for arable grassland and cropland. For example, the rate of accumulation of biomass (above-ground plus under-ground) in a recently seeded grassland was estimated to be 25 to 35 g m^{-2} day^{-1}, which is equivalent to a use-efficiency of 3.3 to 4.6% of solar energy (Rychnovská et al., 1985), and reaches that of maize (Máthé and Précsényi, 1971). However, in perennial grassland (both arable and semi-natural) a much larger proportion of the assimilates are deposited in under-ground organs, where they contribute to soil enrichment processes.

Water relations

Water is not a limiting factor in most economically important mesophytic to hygrophytic grasslands, either because of adequate precipitation or because of availability of ground- or flood-water. In such meadows, rates of transpiration exceed those of evaporation from an open water surface. The quantity of water fixed directly in canopy

TABLE 5.11

Efficiency of conversion of solar energy in a hydrosere in a flood-plain [1]

Vegetation type:	Dry, low grass–forb	Moist, tall grass–forb	Damp, tall grass–forb	Wet [2] tall grass
Dominant species:	*Festuca rupicola*	*Alopecurus pratensis*	*Phalaris arundinacea*	*Glyceria maxima*
Input of global radiation per growing season (kJ m^{-2}) [3]	2 719 642	2 332 823	2 339 913	1 702 281
Energy content of annual biomass production (kJ m^{-2}):				
Above-ground	12 171	18 443	27 531	41 890
Under-ground	6 284	6 586	8 954	14 226
Total	18 456	25 033	36 480	56 116
Efficiency of conversion (%):				
Calculated from above measurements	0.68	1.07	1.56	3.30
Estimated maximum during period of intensive growth	1.03	2.22	2.00	3.26
Maximum values measured in the laboratory (in leaves of dominants) [4]	1.39	4.41	3.37	3.42

[1] After Rychnovská (1979a). These values are means of the 1965 to 1971 growing seasons in a site near the Morava river near Lanžhot, Czechoslovakia.
[2] Wet grassland was not mown; the others were mown twice each year.
[3] The growing season is considered to be from the beginning of growth to the peak of the cumulative growth curve, which in the three moist to wet sites is after the retreat of flood waters.
[4] After Gloser (1976).

TABLE 5.12

Water relations in meadows near Lanžhot, Czechoslovakia [1]

Parameter	Dry meadow	Mesophytic meadow	Damp meadow
Living shoot biomass [2] (g m^{-2})	222	428	378
Water content of shoot biomass (g m^{-2})	516	1380	1187
Transpiration per unit of biomass (g g^{-1} d^{-1}):			
graminoids (mean)	10.2	6.7	12.6
forbs (mean)	20.2	16.9	25.2
total stand	14.9	11.6	11.2
Transpiration per unit of water content [3] (g g^{-1} d^{-1})	6.4	3.6	3.5
Transpiration per unit area of stand (kg m^{-2} d^{-1})	3.3	4.9	4.2

[1] After Rychnovská (1976). The measurements were made on 25 May 1971.
[2] All biomass values are dry weight.
[3] Corresponds to rate of water turnover.

biomass is 1 to 2 kg m^{-2}, but the daily output of water on a warm summer day amounts to 4 to 10 kg m^{-2}, suggesting that the water contents of the canopy are exchanged 4–5 times daily (Table 5.12) (Rychnovská, 1972b, 1976). Under very long daylight periods in the north, water reserves in the canopy were estimated to be exchanged even more frequently (7–10 times daily) (Alekseenko, 1975).

Considerable variation in transpiration rate occurs among species of the same community. Maximum rates of transpiration of mesophytic grasses, under adequate conditions of soil moisture, reach 10 to 30 mg g^{-1} of dry matter min^{-1}, whereas those of mesophytic forbs are from 26 to 48 mg g^{-1} min^{-1} (Rychnovská, 1976). For hygrophytic forbs these values are as high as 60 mg g^{-1} min^{-1}. However, even a satisfactory water reserve cannot prevent the occurrence, in a grassland stand, of a water saturation deficit as high as 20% on a hot day. The weakest link in the water regime is the flow of water from roots to leaves. Transpiration from grassland under optimum conditions of soil moisture is 3 to 4 times that from wheat fields (Rychnovská et al., 1972b). Transpiration rate is not determined by species composition, but by the amount of canopy biomass and by environmental conditions (Fig. 5.13). Increased transpiration is associated with increases in above-ground biomass resulting from fertilization, which is a useful means employed in biological drainage of waterlogged areas.

Stability of primary production

The high species diversity of semi-natural grasslands provides for better adaptation to fluctuations in weather (from year to year) (Fig. 5.2) than do agricultural monocoenoses. Grasslands rich in species possess a multitude of growth strategies and adaptation mechanisms that enable them to cope with various environmental changes. Thus, a relative stability of primary production is achieved by adjustment of the relative prominence of different species according to their capacity to perform under variable circumstances. The compensating function of various groups of species in a 6-year observation series is shown in Figs. 5.14 and 5.15. Whereas the variation coefficient of the water factor was considerable (60%), that of the above-ground biomass was much less (22 to 25%). On the other hand, the compensation mechanisms of various groups of species assert themselves in full force, with representation of dominants attaining a variation coefficient as high as 77% (forbs up to 55% and grasses to 96%) (Rychnovská, 1979b). Titlyanova (1979) concluded that the stabilizing element in natural grasslands is underground biomass, with stability increasing during the course of succession. Semi-natural grasslands may thus be regarded as a stabilizing element in the intensively exploited and urbanized European landscape.

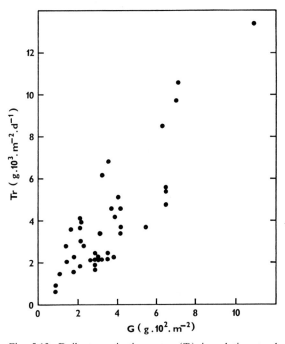

Fig. 5.13. Daily transpiration rates (Tr) in relation to dry weight of shoot biomass (G) in meadows on sunny days in summer. Data from Central Europe (Rychnovská et al., 1972b; Rychnovská, 1976, 1979c).

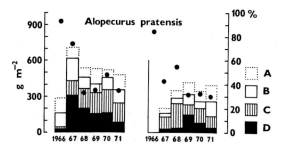

Fig. 5.14. The proportion (g m^{-2}) of various components in the above-ground standing crop in successive years in a moist alluvial *Alopecurus pratensis* meadow at Lanžhot, Czechoslovakia. Left, at time of first cut (May to June); right, second cut (August): A = dead biomass; B to D = green biomass. B = forbs; C = graminoids excluding dominant; D = dominant. Mean soil moisture (% of dry weight of soil) during the growing period is shown by dots.

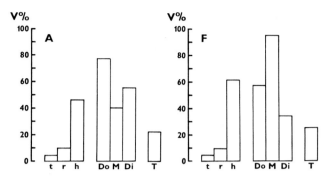

Fig. 5.15. Variation coefficients (in %) for the annual means of the principal environmental factors and maximum above-ground standing crop values during a six-year period in two meadows at Lanžhot, Czechoslovakia. Environment: t = mean air temperature, r = incoming global radiation, h = soil moisture. Biomass: Do = dominant, M = other monocotyledons, Di = dicotyledons, T = total canopy. Left, moist *Alopecurus pratensis* meadow; right, dry *Festuca rupicola* meadow.

CONSUMERS

The number of animal species associated with semi-natural grassland ecosystems is one order of magnitude higher than the number of vascular plant species. This is mainly due to hundreds of invertebrate species. Quantification of the grazing food chain is generally more difficult than that of the detritus food chain, because of taxonomical difficulties and because many animals are both herbivores and carnivores or detritivores. Most of the energy lost from the consumer subsystem is lost by respiration. The rest enters the decomposition chain as urine, faeces and dead tissue, and only a small portion remains as secondary production. Semi-natural grasslands are used partly as hay meadows and partly (at various levels of intensity) as pastures for domesticated animals. Although the grazing food chain is reduced in hay meadows, the effect of animal components is nevertheless considerable.

Hay meadows

Herbivores consume about 10% of above-ground biomass of meadows. In addition, plant tissue is transferred from the canopy to the litter layer as fragments that are bitten off or damaged, but not ingested. Losses from the canopy by this means sometimes exceed the amounts consumed. In Poland, insects directly consumed 11% of total biomass in unfertilized meadows and 5% in fertilized sites. In addition, damage caused by invertebrates to plants (mainly to tillering nodes and roots) while feeding resulted in estimated losses in plant production of 13% and 10% in unfertilized and fertilized sites, respectively (Andrzejewska, 1976).

Rodents have a very important role in these grasslands. Pelikán (1979, 1985) found that the population density of *Microtus arvalis* in semi-natural highland meadows in Czechoslovakia was proportionate to the quantity of food available (Table 5.13). Consumption usually ranged from about 2 to 3% of annual herbage production, but reached 6.8% when the population was highest.

Birds are of little importance in the energy balance, but are important in distributing diaspores. Pelikán (1985) reported that as few as 6 species of birds nested and foraged in meadows, whereas 31 species, which nested in other biotopes, fed in the meadows that he studied. Such migrants remove a significant amount of biomass from one system to another. None of the birds investigated was purely herbivorous; about 38% were granivorous, but even these fed occasionally on insects. The other birds were predators.

The most important predatory vertebrate is *Talpa europaea* (European mole). According to Grulich (1958, 1959), a direct relationship occurs between the quantity of soil fauna and the population density of the mole. This was demonstrated by results of his investigation in flood-plain meadows of the Dyje river basin. Moles were absent in biotopes with 15 g (fresh weight) of soil fauna biomass m^{-2}; in sites with 30 to 40 g m^{-2} of soil fauna, he observed 500 molehills ha^{-1}; and in those with 170 to 320 g m^{-2} of soil fauna, 1500 to

TABLE 5.13

Annual bioenergetic parameters of the population of *Microtus arvalis* in meadows in Czechoslovakia [1]

Condition of meadow: Dominant plant species:	Natural (neither mown nor fertilized) *Nardus stricta*	seeded (mown and fertilized) *Dactylis glomerata* *Phleum pratense*
Biomass values:		
Above-ground primary production (g m^{-2})	180	670
Mean standing crop biomass of *Microtus*:		
number per hectare	29	89
dry weight (g ha^{-1})	194	581
Consumption (g m^{-2})	3.9	11.5
Energy values (MJ ha^{-1}):		
Consumption	684	2045
Net production	15	42
Faeces	128	382
Urine	26	78

[1] Near Kameničky, in the Bohemian–Moravian Uplands. Mean values for 1973 to 1978. After Pelikán (1979).

2000 molehills ha^{-1} were found.

Studies relating to the role of phytophagous insects in energy flow were conducted in Finland (Gyllenberg, 1969), in Poland (Olechowicz, 1976; Breymeyer, 1978), in Czechoslovakia (Lapáček, 1979; Škapec, 1979) and elsewhere (Table 5.14). These reveal that the abundance of herbivorous insects in the canopy ranged from 94 to 5057 specimens m^{-2}, depending on the methodology employed, type of meadow, and year of sampling. Andrzejewska and Gyllenberg (1980) reported fluctuations in the population of Homoptera in consecutive years from 110 to 4 to 26 specimens per m^2. Herbage production was similar during all three years. According to Breymeyer (1978), Orthoptera were the most important contributors to the insect biomass (0.14 to 0.33 g dry matter m^{-2}) in meadows in the Wisła valley. They consumed 24 to 48 g plant biomass m^{-2} yr^{-1} (5 to 9% of herbage production). Homoptera consumed less than 1% of herbage production.

Less information is available concerning phytophagous invertebrates feeding on under-ground plant parts. Ricou (1979) found densities of herbivorous soil macrofauna from 17 to 109 individuals per m^2, with a biomass of 0.13 to 3.5 g m^{-2} (dry weight). Andrzejewska (1976) reported that the biomass of insects inhabiting the rhizosphere was 68 times that of insects in the above-ground plant parts of unfertilized meadows, and 41 times in fertilized sites.

According to Spitzer (1978), in the semi-natural grasslands of southern Bohemia, Lepidoptera play an important indicative role, as judged from his finding that a decrease in number of species of this order (204 of which were found) invariably denotes that the stability of the entire ecosystem is decreasing. Breymeyer (1978) and Pelikán (1985) made detailed studies of the consumer functions in semi-natural meadows.

Pastures

In contrast to hay meadows, where the export of forage accounts for about two-thirds of herbage production, about 85% of herbage is consumed by cattle and sheep in intensively grazed pastures (Breymeyer and Kajak, 1976; Ricou, 1979). Although more above-ground biomass is consumed by grazing than the quantity harvested by mowing, 35% of the energy contained in the biomass that is consumed is returned to the ecosystem as urine and faeces. Rabotnov (1974) estimated that such returns to the soil in a sheep pasture were associated with nutrients which amounted to 78 kg of nitrogen ha^{-1} yr^{-1}, 11 kg of phosphorus, and

TABLE 5.14

Relative abundance and contribution to secondary production of various orders of insects in the foliage of semi-natural grasslands

Parameter:	Proportional distribution of individuals (%)					Secondary production [1] ($mg\ m^{-2}\ yr^{-1}$) Bródno, Poland	
Location:	Solling, Germany	Kameničky, Czechosl.		Bródno, Poland			
Method of sampling:	Photoeclector	Sweeping		Cover cage, 30×30 cm			
Fertilization:	0	0	NPK	0	NPK	0	NPK
Diptera	67	32	45	59	51	184	492
Hymenoptera	16	20	9	4	4	5	12
Homoptera	12	21	12	32	43	69	177
Heteroptera	+	14	5	+	+	8	5
Lepidoptera	+	4	+	+	+	4	8
Thysanoptera	2	4	24	–	–	–	–
Coleoptera	3	3	5	3	2	16	45
Orthoptera	+	+	+	0	+	30	72
Others	–	2	+	2	–	–	–
Total	100	100	100	100	100	316	811
No. of individuals (m^{-2})	5057	94	349	655	2547	–	–
Shoot production ($g\ m^{-2}\ yr^{-1}$)	316	180	670	210	948	–	–
Reference	1	2	3	4	5	6	7

[1] Values for secondary production are in terms of dry matter.
+ = less than 0.5%; – = not estimated. Shoot production estimates are sums of two or three harvests of the same plots each year in mown grassland and maximum standing crop in unmown meadows.

References:
1: Haas (1976); mown meadow, at an altitude of 450 m, dominated by *Festuca rubra* and *Agrostis tenuis*.
2: Škapec (1979); unmown grassland, at an altitude of 620 m, dominated by *Nardus stricta* and *Sanguisorba officinalis*.
3: Škapec (1979); mown, fertilized meadow, at an altitude of 640 m, dominated by *Dactylis glomerata* and *Phleum pratense*.
4, 6: Olechowicz (1976); mown, unfertilized meadow, at an altitude of 69 m, dominated by *Poa pratensis* and *Dactylis glomerata*.
5, 7: Olechowicz (1976); mown, fertilized meadow, at an altitude of 69 m, dominated by *Dactylis glomerata* and *Arrhenatherum elatius*.

104 kg potassium. The results of several studies of the flow of materials through consumers are compared in Table 5.15. Detailed studies on secondary production of consumers on pastures in Europe were made by Breymeyer and Kajak (1976), Heal and Perkins (1978), and Ricou (1979).

Grazing by livestock has both beneficial and detrimental effects. Selection of the more valuable fodder species, particularly by intensive grazing, leads to reduction in number of species. Mechanical pressure of hooves on grass tufts, which for a cow exceeds 4 kg cm^{-2}, is detrimental to several grass species, particularly *Arrhenatherum elatius*, *Molinia coerulea* and *Phalaris arundinacea*. Deposits of urine and faeces during prolonged, intensive grazing have a degrading effect in favouring the invasion of nitrophilous plants, particularly weeds. However, the addition of animal wastes during short-term grazing has a beneficial effect in increasing plant production and herbage quality.

DECOMPOSERS

Although decomposition begins in the canopy as leaves die and their biotic and abiotic disintegration continues in the litter layer on the surface of the soil, most decomposition takes place within the soil. The organisms that are associated with decay processes in the soil include macrofauna and mesofauna (Table 5.16), fungi, algae and micro-organisms. Quantification of the activ-

TABLE 5.15

Consumption of herbage by herbivores [1]

Dominant taxa	Characteristics	Location	Principal herbivore	Herbage production (g m^{-2} yr^{-1})	Percentage of herbage consumed	Consumption (kJ m^{-2} yr^{-1})	Output (% of consumption) excreta and respiration	Output (% of consumption) secondary production
Hay meadows:								
1 Poa pratensis–Dactylis glomerata	Unfertilized, mown	Bródno, Poland	Invertebrates	210	11 (+ 13 damaged)	–	–	–
2 Dactylis glomerata–Arrhenatherum elatius	Fertilized, mown	Bródno, Poland	Invertebrates	948	5 (+ 10 damaged)	–	–	–
3 Nardus stricta–Sanguisorba officinalis	Unfertilized, unmown	Kameničky, Czechoslovakia	Microtus arvalis	180	2 (+ 6 damaged)	68	Excreta 22.5	2.24
4 Dactylis glomerata–Phleum pratense	Fertilized, mown	Kameničky, Czechoslovakia	Microtus arvalis	670	2 (+ 8 damaged)	204	Excreta 22.5	2.05
Pastures:								
5 Lolium perenne–Poa trivialis		Pin-au-Haras, France	Cattle	970	85	14 226	Excr. + Resp. 85	15
6 Agrostis tenuis–Festuca ovina		Snowdonia, Wales	Sheep	271	92	4659	Excr. 50.1 Resp. 38.5	4.4
7 Anthoxanthum alpinum–Poa alpina	Dry grassland	southern Norway	Sheep	95	44	–	–	–
8 Carex nigra–Festuca rubra	Wet grassland	southern Norway	Sheep	90	32	–	–	–

[1] References: 1, 2: Andrzejewska (1976); 3, 4: Pelikán (1979); 5: Ricou (1979); 6: Perkins (1978); 7, 8: Wielgolaski (1976).

TABLE 5.16

Soil mesofauna and macrofauna in lowland meadows [1]

Group	Number of species	Number of organisms (m^{-2} × 10^3)	Biomass (mg m^{-2})	Locality	Reference
Nematodes	74			Bródno, Poland	Wasilewska (1976)
Fungivorous		50–850	4–92 *		
Microbivorous		500–3000	50–500 *		
Plant parasites		100–2000	50–1400 *		
Omnivorous		50–1000	50–1700 *		
Enchytraeidae		3.3–4.3	467–517 *	Bródno, Poland	Makulec (1976)
Lumbricidae	24	0.061–0.23	5600–26300 **	Bródno, Poland	Zajonc (1970), Nowak (1976)
Acarina	22	86	82–372 *	Bródno, Poland	Zyromska-Rudzka (1976)
Apterygota	58 [1]	17.6–69.7		Lanžhot, Czechosl.	Rusek (1984)
Diptera larvae		0.019–0.034		Bródno, Poland	Nowak (1976)
Coleoptera larvae [2] Lepidoptera larvae [2]		0.01–0.073	300–1800 **	Bródno, Poland	Nowak (1976)

[1] Includes species of the orders Collembola, Diplura and Protura.
[2] Includes herbivores only.
* = fresh weight; ** = dry weight.

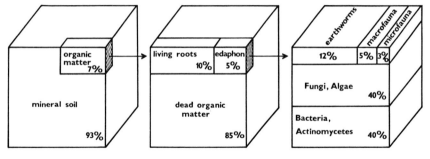

Fig. 5.16. Proportional distribution of living organisms in the soil of semi-natural grasslands (after Pelikán, 1985).

ity of each of these groups is difficult because many of these organisms are polyphagous, some are both consumers and decomposers, and some are even autotrophic. However, their activities all result in the release and cycling of mineral nutrients and contribute to the soil-forming processes typical of grasslands (Fig. 5.16).

Soil macrofauna and mesofauna

The abundance of soil animals increases with declining body size. Pelikán (1985) reported that the numbers of individuals of the various groups of soil fauna per m^2 in a meadow, to a depth of 30 cm, were as follows:

Protozoa	10^9 to 10^{10}	other Arthropoda	10^4
Nematoda	10^6 to 10^{10}	Enchytraeidae	10^5
Acarina	10^5	Lumbricidae	200 to 800
Collembola	10^5		

The soil fauna of grasslands is richer, both in quantity and quality, than that of arable soil. According to Zajonc (1970), the biomass of earthworms in soils of semi-natural grassland is about 3 times as great as in arable soils and the number of species of earthworms in grassland soils is twice that in cropland. In his opinion, an earthworm synusia of grasslands is more similar to that of a forest than to that of a culture of clover several years old. This suggests that even a highly managed semi-natural grassland, with an undisturbed species composition, is more similar in its function to a natural forest ecosystem that to an agrocoenosis.

Animals contribute to decomposition by consuming detritus and degrading it in their digestive tracts. Because assimilation efficiency in this process is low, a large quantity of excrements results. These chemically degraded materials are mixed with inorganic substances and are colonized by micro-organisms. MacFadyen (1963) found that the metabolism of detritophages was 2 to 3 times as great in grassland soils as in forest soils. Pelikán (1985) suggested that earthworms consume about 50% of litter of the above-ground plant biomass, beetles and Diptera together about 6%, Myriapoda 5%, Isopoda 3%, Collembola and Acarina together 1%, the remaining Arthropoda 2%, and snails 1%. This suggests that about 70% of dead plant matter passes through the intestinal tract of soil fauna. As their assimilation efficiency is low, only about 10% of total litter decomposition could be attributed to soil fauna.

Micro-organisms

The detritus and grazing food chains terminate with the decomposing activity of Bacteria, yeasts, Actinomycetes, microscopic Fungi and Protozoa. Their activity leads partly to mineralization and partly to humification. Thus humus is produced that constitutes a pool of energy reserves related to soil fertility. Humification processes dominate in mesophytic and hygrophytic grasslands, rather than mineralization.

Úlehlová (1979) reported the numbers of soil micro-organisms in various temperate grasslands, by direct or plate count, to be as follows (per g of dry soil within the rooting layer):

Bacteria	10^4 to 10^9
Actinomycetes	10^3 to 10^7
Fungi	10^3 to 10^5

Apparently, the numbers of micro-organisms in grassland soils are greater than in forest or arable soils. This is due to the more pronounced rhizosphere effect associated with the large quantity and diversity of under-ground biomass in grassland soils and to its rapid rate of turnover.

Bacterial biomass in soils of these grasslands shows considerable seasonal variation. Estimates range from 3 to 940 g dry matter m^{-2}, to a depth of 10 cm, some of this variation being due to the method employed. Úlehlová (1979) reported calculated values (based on the amount of materials available for decomposition) from 168 to 873 g m^{-2} for potential annual production of bacterial biomass in various types of grasslands. Differences between the potential production and actual amount of bacterial biomass result from their intensive turnover rate and fluctuations in environmental factors which control rates of activity of micro-organisms.

Decomposition processes

The total biomass of soil organisms does not always reflect the degree of their activity. The latter depends on the quality and quantity of plant biomass entering the decomposition chain, on environmental factors, and on the diversity of soil organisms and the enzymes that they produce. The rate of decomposition can be determined by various methods, some of which are based on primary production data, others on direct estimates of rates of disappearance of dead biomass or of pure cellulose. Cellulolytic activity generally increases with the productivity of the stand (Table 5.17), being dependent on soil moisture, temperature and type of soil (Tesařová, 1977; Úlehlová, 1979). The leaves of forbs decompose faster than do their roots, whereas stems decompose even more slowly.

Data on rate of decomposition of under-ground biomass are scarce, because growth of under-ground organs occurs simultaneously and losses in weight of under-ground biomass may be due partly to translocation of materials to shoots. So far it has not been possible to separate these processes *in situ*. By using a paired technique (see p. 140 and Table 5.7), Fiala (1979) observed a considerably higher rate of decomposition of under-ground biomass than reported by other workers (as much as 28 g m^{-2} day^{-1}). However, decomposition may have been over-estimated because no allowance was made for translocation and loss of assimilates by respiration.

An integrating measure of the decomposition activity in a soil system is the production of carbon

TABLE 5.17

Mean rates of decomposition (mg g^{-1} day^{-1}) of plant litter in meadows

Type	Dominants	Location	Material[1]	Summer	Winter	Method[2]	Reference
Lowland, dry	Festuca rupicola	Lanžhot, Czechoslovakia	A A	1.8 6.4	1.0 –	LB PP	Tesařová (1976, 1977) Tesařová (1976)
Alluvial, moist	Alopecurus pratensis	Lanžhot, Czechoslovakia	A A	3.2 10.5	1.7 –	LB PP	Tesařová (1976, 1977)
Alluvial, wet	Glyceria maxima	Lanžhot, Czechoslovakia	A A	3.7 10.7	2.1 –	LB PP	Tesařová (1976, 1977)
Lowland, unfertilized	Poa pratensis– Dactylis glomerata	Bródno, Poland	A A	9.2 13.6	2.5 3.5	LB LS	Jakubczyk (1976)
Lowland, fertilized (NPK)	Dactylis glomerata– Arrhenatherum elatius	Bródno, Poland	A A	11.1 15.5	2.5 3.5	LB LS	Jakubczyk (1976)
Natural, unmown unmanaged	Nardus stricta– Sanguisorba officinalis	Kameničky, Czechoslovakia	A A R	1.4 3.6 12.5	0.3 0.8 –	LB PP PC	Tesařová (1979a) Tesařová (1983) Fiala (1979)
Natural, managed	Nardus stricta– Polygonum bistorta– Deschampsia caespitosa	Kameničky, Czechoslovakia	A A	2.7 7.5	0.6 1.5	LB PP	Tesařová (1988)
Seeded, managed	Dactylis glomerata– Phleum pratense	Kameničky, Czechoslovakia	A A	3.4 8.2	0.7 1.4	LB PP	Tesařová (1988)
Natural	Agrostis tenuis– Alopecurus pratensis– Alchemilla monticola	Otradnoj, former Soviet Union	A U	4.7 2.2	1.0 1.1	LB LB	Miroshnichenko (1978)

[1] The above-ground material used was collected from the soil surface between the tufts of grasses. A = above-ground litter; U = underground parts; R = roots.
[2] LB = litter bags (0.3 mm mesh); LS = loose samples on soil surface (under a net of 2 cm mesh); PP = paired plots, Wiegert and Evans (1964); PC = paired soil cores.

TABLE 5.18

Mean rates of soil and root respiration (in g CO_2 evolved m^{-2} day^{-1}) in meadows

Type	Dominants	Location	Summer	Winter	Reference
Lowland, dry	Festuca rupicola	Lanžhot, Czechoslovakia	9.1	–	Tesařová and Gloser (1976)
Flooded, moist	Alopecurus pratensis	Lanžhot, Czechoslovakia	10.5	–	Tesařová and Gloser (1976)
Lowland, unfertilized	Poa pratensis– Dactylis glomerata	Bródno, Poland	4.9	–	Kubicka (1976)
Lowland, fertilized (NPK)	Dactylis glomerata– Arrhenatherum elatius	Bródno, Poland	6.3	–	Kubicka (1976)
Natural, unmown, unmanaged	Nardus stricta– Sanguisorba officinalis	Kameničky, Czechoslovakia	7.9	1.5	Tesařová (1979b)
Natural, managed	Nardus stricta– Polygonum bistorta– Deschampsia caespitosa	Kameničky, Czechoslovakia	10.0	1.3	Tesařová (1988)
Seeded, managed	Dactylis glomerata– Phleum pratense	Kameničky, Czechoslovakia	10.1	0.8	Tesařová (1988)

dioxide measured directly in the field. Measurements of carbon dioxide vary widely (Table 5.18), partly due to the use of different methods. Temperature is very important; values of $Q_{10}=1.6$ to 2 are given by Drobník (1962) for respiration of grassland soil (within biotic limits). A review of European studies suggests that live roots and their rhizosphere micro-flora account for about 40% of carbon-dioxide evolution, soil fauna about 10%, and micro-organisms the remainder; and in some types of grassland, the litter layer accounts for 20 to 30% of soil respiration (Zlotin, 1974; Tesařová and Gloser, 1976; Gloser and Tesařová, 1978). However, Zlotin (1974) estimated that 30 to 60% of the carbon dioxide released from chernozems rich in carbonates in the Kurdistan steppe was of abiotic origin.

ENERGY FLOW AND MINERAL CYCLES

The energy balance for two meadows at Lanžhot, Czechoslovakia is illustrated in Fig. 5.17. The input of solar energy depends on the latitude, the season, the time of day, and the weather. Altitude also has an effect, as shown by Šmíd's (1972) and Bár's (1979) reported value, at an elevation of 624 m, of 2435×10^3 kJ m^{-2} of net radiation during the growing season, which is 20% less than that of the lowland site (150 km distant) considered in Fig. 5.17, but which is located only 155 m above sea level. Estimates in the Lanžhot sites (Fig. 5.17) indicated that 23 to 51% of the input of solar energy was lost by transpiration, that 0.6 to 1.8% accumulated in plant biomass, and that one-third to one-half (and sometimes even all) of the latter entered the detritus food chain in unmown meadows. Úlehlová and Úlehla (1983) found that 129 to 226 MJ m^{-2} was accumulated in the soil to a depth of 25 cm as humus; this value is about one order of magnitude higher than the energy content of the standing crop of plant biomass.

Carbon and nitrogen cycles are highly specific in grassland ecosystems. They differ from those of both forest and agricultural ecosystems in the abundant and biologically active underground biomass which is supplemented continually with reserve substances from aerial parts of the stand. A considerable quantity of carbonaceous substances is released from the permanent

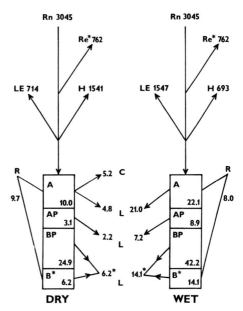

Fig. 5.17. Annual rates of energy flow (MJ m^{-2}) in two meadow ecosystems at Lanžhot, Czechoslovakia: left, dry *Festuca rupicola* meadow; right, wet *Glyceria maxima* meadow. Rn = net radiation, Re = reflection, LE = latent heat flux, H = sensible heat flux, R = respiration, A = above-ground biomass production, AP = above-ground permanent pool, B = under-ground biomass production, BP = under-ground permanent pool, C = crop harvest, L = litter production; * denotes an estimate only. (After Úlehlová and Úlehla, 1983). See additional information in Fig. 5.8.

pool of under-ground biomass, consisting mainly of grass roots uninterruptedly growing and dying, and a large pool of the organic matter accumulates. The carbon cycle of a mesophytic, alluvial flood-plain meadow in the Lanžhot area is illustrated in Fig. 5.18.

The nitrogen cycle is far more complicated than the carbon cycle. Nitrogen enters the ecosystem directly from the atmosphere in abiotic form, and also by numerous additional biotic pathways, such as nitrogen-fixing soil micro-organisms, symbiotic bacteria and others (including fertilizers). Úlehlová (1985) estimated that nitrogen was distributed in highland meadow ecosystems as follows: about 94% is bound in soil organic matter, 1.5% as mineral forms (mostly as ammonium ions), 2 to 3% in living plants (mostly in the roots), 0.5% in soil micro-organisms, and 1 to 2% in dead plant biomass. Plants and micro-organisms compete in the uptake of mineral nitrogen, because

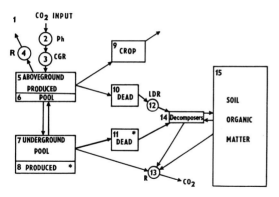

Fig. 5.18. Carbon cycle in two meadows (at Lanžhot, Czechoslovakia (after Tesařová and Gloser, 1976; Rychnovská, 1979a; Rychnovská et al., 1980; Úlehlová and Úlehla, 1983): A = moist alluvial *Alopecurus pratensis* meadow, F = dry *Festuca rupicola* meadow. Values are g of carbon (C) m^{-2}, unless other units are indicated.

	A	F
1 = Annual input from atmosphere	786	690
2 = Photosynthesis of dominant (20°C) in mg dm^{-2} hr^{-1}	5.5–7.0	2.7–4.0
3 = Crop growth rate (mean for growing season) in g m^{-2} day^{-1}	2.4	1.4
4 = Dark respiration of dominant (20°C) in mg dm^{-2} hr^{-1}	0.3–0.4	0.6
5 = Annual accumulation in shoot biomass	481	318
6 = Permanent pool of shoot biomass (over-wintering canopy, including mosses)	66	82
7 = Permanent pool of under-ground biomass	549	700
8 = Annual production of under-ground biomass (estimated only)	183	175
9 = Annual hay harvest	236	131
10 = Annual decomposition of aboveground litter	207	121
11 = Annual decomposition of underground litter (estimated only)	183	175
12 = Litter (above ground) decomposition rate (g m^{-2} day^{-1}) in growing season		
in summer	0.8	0.4
in winter (average)	0.2	0.3
13 = Soil and root respiration (g m^{-2} day^{-1}) in growing season	2.5	2.7
14 = Biomass (mean) of micro-organisms	39	22
15 = Soil organic matter	7910	6610

grassland soils also contain a large quantity of carbon. If the C/N ratio is higher than 20 to 25, the entire mineral nitrogen released by decomposition is immediately taken up by micro-organisms. Soil micro-organisms are always more successful than vascular plants in their competition for nitrogen (Úlehlová, 1985). Figure 5.19 illustrates the nitrogen cycle of fertilized grassland in the Bohemian–Moravian Uplands (in Czechoslovakia). The capacity of absorbing great quantities of nitrogen, by both plants and micro-organisms, reduces its concentration in surface waters carrying dissolved fertilizers from croplands at higher elevations. Thus, the presence of semi-natural grasslands on slopes and in flood-plains reduces eutrophication of surface waters.

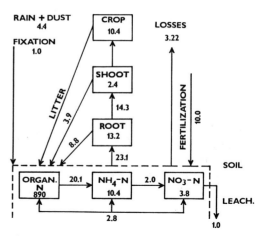

Fig. 5.19. Annual nitrogen budget in a semi-natural fertilized upland meadow dominated by *Nardus stricta* and *Polygonum bistorta* (near Kameničky, Czechoslovakia). Distribution of nitrogen is in g m^{-2} and flow rates are in g m^{-2} yr^{-1} (after Úlehlová et al., 1980).

Characteristics of the cycles of the other geo-bioelements vary considerably in semi-natural grasslands, and they are not yet well understood. However, some aspects are of considerable practical importance. Accumulation of mineral elements differs among plant species. For example, concentration of nitrogen prevails in Fabaceae, calcium and magnesium in forbs, and silicon in graminoids. Specific accumulation of micro-elements is exemplified by data from mesophytic meadows dominated by *Alopecurus pratensis* (in the flood-plain of the Morava river). Bednář (1979) found higher accumulation of aluminium, iron, strontium and zinc in forbs than in graminoids, whereas copper and manganese showed the opposite relationship. The mean concentrations (mg per 100 g dry matter) were as follows:

	Graminoids	Forbs
Aluminium	36.4	52.7
Copper	2.5	1.7
Iron	35.5	40.5
Manganese	16.6	14.5
Strontium	1.3	5.1
Zinc	5.4	6.5

The decreasing order of mobility of various elements, as reflected in the percentage of the amount accumulated in living biomass which is retained in dead biomass, is as follows:

Potassium	9.4	Calcium	39.9
Phosphorus	10.4	Manganese	53.0
Magnesium	17.4	Zinc	59.6
Nitrogen	22.7	Copper	76.6
Strontium	24.2	Iron	79.3
Sodium	32.4		

In unfertilized flooded meadows of the floodplain of the Morava river, Jakrlová (1975) found that the annual export of geobioelements (in harvest of hay) amounted to 109 kg nitrogen, 11 kg phosphorus, 120 kg potassium, 37 kg calcium, and 6 kg magnesium per hectare. The source was mostly run-off waters from the whole catchment area (Table 5.19).

LAND USE AND MANAGEMENT

Semi-natural grasslands are important, both because of their productive and economic role and for various ecological functions related to structure and stability of the agricultural landscape. Because these systems prosper over a wider range of environmental conditions than do arable crops, herbage varies greatly in yield and dietetic quality. The biotopes range from xerophytic through mesophytic to hygrophytic sites, from oligotrophic through mesotrophic to eutrophic habitats — the latter where ruderal weeds prevail. The marginal types (poor oligotrophic sites and those with surplus nutrients) are unsuitable for intensive fodder production, and usually have been tilled and seeded to appropriate field crops.

Forage yield can be improved by management, within the limits imposed by the environmental characteristics of the habitat. As a result, semi-natural grassland exists under a wide range of situations from being nearly in balance with the natural environment to those that involve intensive management. Economic improvement often results in ecological impairment, as intensive management causes a decline in species diversity and deterioration of the habitat. The additional energy input for managing perennial grasslands is always less than that required for annual crop production.

Semi-natural grasslands are acquiring importance as a stabilizing factor in intensively urbanized and agriculturally exploited landscapes. They are especially useful in management of sources of potable water, being used to protect the immediate vicinity of reservoirs and springs, and they serve to reduce soil erosion from slopes. They are also used as sites for recreation, sports activities and tourism. The standing crop of under-ground biomass and a permanent biomass pool aboveground function throughout the year in these grasslands to protect the soil against erosion by wind and water, to reduce the concentration of minerals in flood waters, and to provide a biologically active surface intermediating between the soil and the atmosphere.

Representative stands of semi-natural grasslands must be conserved as future gene pools. They are a source of genetic information deposited in the genotype of individual plant and animal species and of soil micro-organisms. These species possess many properties and adaptation mechanisms which can be used for improvement of crop plants and animal species, as well as being a potential source of substances or biotechnologies for use in human and veterinary medicine and in industry. More than 70% of grassland species contain various pharmaceutically active substances. Rough estimates suggest that about 2 to 4% of species of vascular plants have disappeared from central Europe already and that an additional 50% are endangered by human activity in converting grasslands to arable fields.

Grassland soils are specific in their chemical composition, biological activity, and physical structure. The return of 60 to 80% of plant biomass to the soil through detritus and grazing food chains provides for increases in and maintenance of soil fertility; the fibrous root systems

TABLE 5.19

Content of mineral elements in biomass and the annual nutrient budget of semi-natural grasslands [1]

Vegetation type	Dominant taxa	Nutrient content (kg ha^{-1})				
Biomass measured:		N	P	K	Ca	Mg
Moist alluvial meadow [2]	*Alopecurus pratensis*					
Above-ground		102–218	4–22	60–239	19–74	5–11
Forage exported		51–109	2–11	30–120	10–37	2– 6
Under-ground		66–222	6–25	48–208	4–22	3– 8
Dry alluvial meadow [2]	*Festuca rupicola*					
Above-ground		63–123	6–14	35– 92	19–37	4– 8
Forage exported		21– 41	2– 5	12– 31	6–13	1– 3
Under-ground		96–197	8–20	45–132	4–21	5–10
Lake basin [3]	*Agrostis tenuis–Alopecurus pratensis*					
Above-ground		18– 44	3– 6	33– 80	14–38	7–13
Under-ground		60– 75	7–13	42– 45	48–71	28–53
Montane pasture [4]	*Agrostis tenuis–Festuca ovina*					
Above- plus under-ground		276	24	163	41	70
Entering grazing pathway		59	6	45	13	11
From herbivores to decomposers		43	4	34	9	8
To decomposers from canopy and litter		73	8	27	11	41
To decomposers from all above-ground sources		116	12	61	20	49

[1] Above-ground biomass is the sum of 2 or 3 harvests each year in mown grassland and the maximum standing crop in unmown sites. Under-ground biomass is mean standing crop.
[2] After Jakrlová (1975).
[3] After Druzina (1978).
[4] After Perkins (1978).

of perennial grasses help to maintain soil structure. Such soil-forming conditions gave origin to rich chernozems in the past and provide an opportunity to restore fertility to deteriorated arable soils in the present and future. Recent knowledge suggests that semi-natural grasslands are easily renewable ecosystems which survive to assist, together with wetlands and forests, in stabilizing intensively managed landscapes. They can provide ecological first aid in mountainous areas where forests have disappeared due to air pollution. This economic and ecological benefit of grasslands will be one of the tools contributing to rational land use and stable food production in the future, upon which the continuing well-being of society depends.

APPENDIX I — ECOLOGICAL TYPES OF SEMI-NATURAL GRASSLANDS IN EURASIA

I. Western and central Europe

a. Plains and slopes
aa. On slopes and drier plateaux

– Meso-xerophytic grasslands on rich soils in sites of former oak (*Quercus*) forests are typical of the sub-Mediterranean zone. These short grass–forb communities (*Mesobromion* W. Koch 1926) are floristically very diverse with numerous orchids. Klapp (1965) reported the presence of 121 species of grasses and forbs, the major species including *Bromus erectus*, *Koeleria pyramidata*, *Brachypodium pinnatum*, *Festuca ovina*, *Carex humilis*, *Sanguisorba minor* and *Anthyllis vulneraria*.[5]

[5] In this Appendix species are arranged in decreasing order of their importance for classification, including both dominants and characteristic taxa. Only a few species have been selected from lengthy lists available for each grassland type.

- Sub-xerophytic grasslands (*Festucion valesiacae* Klika 1931) of a similar floristic diversity grow in the subcontinental climate of central and south-eastern Europe. The dominants include *Festuca valesiaca, Melica transsilvanica, Stipa capillata, Astragalus danicus* and *Adonis vernalis*. In western Europe they are represented by *Xerobromion* Br.-Bl. et Moor 1938. The extensive Danubian plain is colonized by *Festucion vaginatae* Soó 1929 with stands of a structure similar to that of the foregoing communities. In the past all types were used for extensive grazing; at present most of the area has been cultivated. However, the remnants are protected as a highly aesthetic element of the landscape owing to the variable seasonal aspect of their noteworthy flowers (species of *Iris, Scabiosa, Centaurea, Inula, Dianthus, Salvia* and others).

ab. On lowlands and moister plains

Mesophytic grasslands on rich and mostly fertilized soils where water is either supplied by rainfall, or there is a sufficient source of under-ground water, but flooding never occurs.

- Tall-grass meadows (*Arrhenatherion* W. Koch 1926) with the grasses including *Arrhenatherum elatius, Alopecurus pratensis, Festuca pratensis, Dactylis glomerata, Trisetum flavescens, Festuca rubra* and *Agrostis tenuis*, whereas the dicotyledons include *Heracleum sphondylium, Crepis biennis, Galium mollugo, Anthriscus sylvestris, Sanguisorba officinalis, Trifolium repens* and *T. pratense*. Klapp (1965) reported 105 species of grasses and forbs participating in the floristic diversity of this type of grasslands (*Arrhenatheretum elatioris* Braun 1915). These are the most productive grassland stands, mown 2 to 3 times a year. In the past they were the basis of the most productive fodder industries in Holland, Denmark, north-western Germany, Belgium, Great Britain, and northern France.
- Short-grass mesophytic pastures (*Cynosurion* Tx. 1947). Species include *Lolium perenne, Cynosurus cristatus, Festuca rubra, Phleum pratense* and *Trifolium repens*. Oberdorfer (1957) recorded 63 species of grasses and forbs participating in this type of intensive pasture (*Lolio-Cynosuretum* Tx. 1937).

b. Uplands, submontane and montane grasslands

- Mesophytic, mown meadows on rich and fertilized soils. These are tall grass–forb communities (*Polygono-Trisetion* Br.-Bl. et Tx. ex Marschall 1947), rich in species with the dominants including *Trisetum flavescens, Festuca rubra, Alopecurus pratensis, Agrostis tenuis, Nardus stricta, Holcus lanatus, Trifolium pratense, T. repens* and *Leucanthemum vulgare*. Klapp (1965) reported about 90 species.
- Mesophytic, mown meadows on poor, acid soils. This is a short-grass–forb community with *Festuca rubra, Agrostis tenuis, Anthoxanthum odoratum, Briza media* and *Alchemilla vulgaris*. Oberdorfer (1957) reported about 60 species, a large number of which are of subcontinental character.
- Pastures on acid, minimally fertilized soils. These are short-grass communities with the dominants including *Festuca rubra, Agrostis tenuis, Cynosurus cristatus, Festuca ovina, Nardus stricta, Anthoxanthum odoratum, Holcus lanatus, Trifolium repens, Taraxacum officinale, Plantago lanceolata* and *Achillea millefolium* (*Festuco-Cynosuretum* Tx. 1940).
- Pastures on poor, acid, unfertilized soils. These are short-grass communities growing on the site of former acidophilous, deciduous and coniferous forests which extend also above the tree-line (*Eu-Nardion* Luquet. 1926; *Violion caninae* Schwick. 1944). The dominants are *Nardus stricta, Festuca ovina, Agrostis tenuis, Festuca rubra* var. *commutata, Meum athamanthicum, Sieglingia decumbens, Calluna vulgaris* and others. These grasslands occurring in the suboceanic zone of Europe are richer in species (68 species in western Europe), whereas the subcontinental variant is poorer.

c. River valleys and basins

The grasslands on alluvium can he regarded as azonal. Duration and depth of flooding and the amount of sedimentation of silt, sand, and nutrients are determining factors. Due to their varying effects across the river valley, various vegetation zones occur on alluvium. The different biotopes that can be distinguished are the river channel, a low flooded terrace, a raised river terrace beyond the reach of flood water, and the river bank proper, continuing to valley slopes and plains above. Generally, the river has a levee made up of coarse sediments tending to desiccate considerably during summer. The first terrace of fine fluvial deposits, with high under-ground water-table, forms a generally moist site rich in nutrients. Nevertheless, even here water stress occurs during summer due to lowering of the under-ground water-table. The second (raised) terrace is the driest biotope of the valley, and provides an avenue for spreading of steppe plants on terraces and particularly on banks of the river extending far into the forest zone. On the other hand, waterlogging at the foot of the second terrace enables northern species to migrate to the forest-steppe and to enter steppe regions. This phenomenon results in the convergence of plant communities along the valleys of large rivers of the entire Eurasian continent. The vegetation of alluvia of small rivers and brooks differs least from that of the zone in which the valley is located and relates to the pertinent zonal vegetation (Shennikov, 1941; Walter, 1968). The following scheme of flood-plain grassland types is based on the availability of water.

- Hygrophytic tall-grass communities on terraces near the river exposed to prolonged flooding (*Phragmition* W. Koch 1926; *Magnocaricion* W. Koch 1926; *Agropyro-Rumicion* Nordh 1940). The species include *Phalaris arundinacea, Glyceria maxima, Carex gracilis, Poa palustris, Agrostis stolonifera, Agropyron repens, Alopecurus geniculatus, Festuca arundinacea* and *Ranunculus repens*. Generally, stands are uniform, poor in species, dominated by one to several grass species, and highly productive (e.g. those situated at the mouth of the Odra river).
- Mesophytic–mesohygrophytic tall-grass–forb communities on rich soils suffering frequently from a retreat of the groundwater level in summer (*Calthion* Tx. 1936). The species include *Alopecurus pratensis, Cirsium oleraceum, Festuca pratensis, Poa pratensis, Festuca rubra, Holcus lanatus, Ranunculus acer, Plantago lanceolata, Bromus racemosus, Angelica sylvestris* and *Lychnis flos-cuculi*. Klapp (1965)

reported 109 species participating in this type of grassland. In climatically drier regions another type develops in which *Sanguisorba officinalis* and *Silaum silaus* are important, in addition to the species already mentioned.
- Mesohygrophytic forb communities typical of submontane alluvia of the upper reaches of rivers, oligotrophic. The species include *Polygonum bistorta*, *Sanguisorba officinalis*, *Deschampsia caespitosa*, *Cirsium palustre*, *Ranunculus auricomus* and *Achillea ptarmica*. Balátová-Tuláčková et al. (1977) reported 56 species.
- Mesohygrophytic short-grass–forb communities on rich, calcareous unfertilized soils, with a sharp decline in the under-ground water during the summer (*Molinion* W. Koch 1926). Although these were well distributed in the past, increased fertilization and more frequent mowing has changed them into more fertile tall-grass communities of cultivated grasses. The important species include *Molinia coerulea*, *Carex panicea*, *Bromus erectus*, *Deschampsia caespitosa*, *Serratula tinctoria* and *Sanguisorba officinalis*. This type is best developed in the more continental part of Europe (Oberdorfer, 1957).
- Another type develops on acidophilous soils mainly in the Atlantic–sub-Atlantic part of Europe. Oberdorfer (1957) reported 69 species (*Junco-Molinietum* Preisg. 1951), among which *Molinia coerulea*, *Holcus lanatus*, *Festuca rubra*, *Juncus effusus* and *Succisa pratensis* are important.
- A subcontinental type was reported by Balátová-Tuláčková (1966) from the Morava river valley, with 61 species. These include *Festuca rupicola*, *F. rubra*, *Anthoxanthum odoratum*, *Galium boreale*, *G. verum*, *Colchicum autumnale*, *Sanguisorba officinalis*, *Centaurea jacea* and *Serratula tinctoria*.

II. Eastern Europe

a. Plains and slopes
aa. On slopes and drier plateaux

- Mesophytic grasslands in the coniferous zone of the European part of the former Soviet Union, growing mostly on trophically poor podzolized soils (southern border: Belorussiya, Moscow and Gorkiy districts, southern Urals). Water is supplied by precipitation, which is frequently deficient. These are short-grass–forb mixed communities without marked dominants, consisting of several tens of species. Among them are *Agrostis tenuis*, *Anthoxanthum odoratum*, *Briza media*, *Poa pratensis*, *Festuca rubra*, *Phleum pratense*, *Deschampsia caespitosa*, *Leucanthemum vulgare*, *Centaurea jacea*, *Pimpinella saxifraga*, *Alchemilla* spp., *Trifolium pratense* and *T. repens*. Fertilization increases the dominance of mesophytic cultivated grasses; mowing without fertilization causes degradation to mossy, waterlogged grasslands with *Nardus stricta*, *Festuca ovina*, *Luzula multiflora*, *Succisa pratensis* and *Potentilla erecta*.
- Mesophytic grasslands in the forest-steppe zone of the European part of the former Soviet Union, "steppe meadows" or "hydrophilous steppes" occurring in the districts of Kursk, Orël, Tula, Ryazan', Tambov (35°–42°E, 52°–55°N) in warm, dry climate on chernozem soils. These are polydominant tall-grass–forb communities with *Bromus riparius*, *Agrostis syreistschikowii*, *Koeleria delavignei*, *Phleum phleoides*, *Avenastrum schellianum* and *Poa angustifolia*, in addition to *Festuca pratensis*, *F. rubra*, *Briza media*, *Alopecurus pratensis* and *Carex humilis* on moister sites and northern slopes. The forbs include *Leucanthemum vulgare*, *Alectorolophus major*, *Filipendula hexapetala*, *Salvia dumetorum*, *Galium verum*, *Myosotis sylvatica*, *Pedicularis comosa*, *Medicago falcata*, *Trifolium montanum* and *Onobrychis arenaria*, as well as *Adonis vernalis* and *Pulsatilla patens*.
- Meso-xerophytic grasslands in depressions of the desert zone, with a sufficient supply of under-ground water (Caspian plain, the plain between the rivers Volga and Ural, Azerbaydzhan). The dominants are *Agropyron cristatum* var. *pectiniforme*, *Calamagrostis epigeios*, *Bromus inermis*, *Poa pratensis*, *Carex stenophylla*, *Medicago falcata* and *Glycyrrhiza glabra*. These grasslands are mown.

ab. On lowlands with a supply of under-ground water, waterlogged sites, at the foot of slopes

- Mesophytic–mesohygrophytic tall-grass communities in the forest zone of the European part of the former Soviet Union, enriched with nutrients transferred from cropland in run-off water or moderately fertilized. The important species include *Deschampsia caespitosa*, *Festuca pratensis*, *Alopecurus pratensis* and *Phleum pratense*. This type ranks among the best grasslands in the forest zone. It has numerous variants. At a lower trophic level, short-grass communities occur with *Deschampsia caespitosa*, *Anthoxanthum odoratum*, *Agrostis tenuis*, *Festuca rubra*, *Ranunculus acer* and *R. auricomus*. At sites with variable moisture conditions, the forbs include *Polygonum bistorta*, *Trollius europaeus* and *Geum rivale*. On peat-bog soils with *Deschampsia caespitosa*, there is a marked incidence of *Agrostis canina*, *Carex nigra*, *Juncus filiformis*, *Eriophorum latifolium* and *Comarum palustre* which form a transition to peat-bog stands with tall and short sedges.
- Mesophytic tall-forb meadows on slopes irrigated by through-flowing under-ground water. The species include *Polygonum bistorta*, *Rumex acetosa*, *Filipendula ulmaria*, *Geum rivale*, *Alchemilla* sp., *Ranunculus acer*, *R. auricomus*, *Lychnis flos-cuculi*, *Lathyrus pratensis*, *Carex caespitosa*, *Eriophorum latifolium* and *E. angustifolium*. The southern border of the distribution of this type of grasslands, with the dominant *Deschampsia caespitosa*, is formed by the northern border of the forest-steppe zone. Owing to chernozem soils in these plains above the river valleys, they are highly productive.
- Mesophytic grasslands in depressions of the eastern European steppe zone (regions of the Black Sea and Kherson) are influenced by fluctuations in the water regime in some years. During moister years, the dominant plant is *Alopecurus pratensis*; during drier years it is *Agropyron pseudocaesium*. Other species present are *Juncus atratus*, *Butomus umbellatus*, *Inula britannica*, *Fritillaria meleagroides* and *Beckmannia eruciformis*.
- Mesophytic tall-grass communities in the desert and semi-desert zones (Caspian lowlands and the Kazakhstan plains) form indistinct depressions among the surrounding semi-

deserts; these depressions are flooded by deluvial (run-off) waters in the spring. These grasslands resemble alluvial grasslands of river valleys. They are greatly influenced by variation in the water level in some years and by the salinization of the soils. *Agropyron pseudocaesium* dominates the lower-situated biotopes in the area bordered by the rivers Volga, Uzen' and Kushum, and in the surroundings of the Kamysh–Samarsk lakes. This species develops stands 1.5 m in height, which are highly productive. *Agropyron cristatum* colonizes higher biotopes. The grasslands of Azerbaydzhan occupy an ecological niche between wetlands with *Phragmites communis* and desert vegetation. They are of the solonchak type with *Atropis distans, Aeluropus litoralis* and *Juncus gerardi*, sometimes accompanied by *Alopecurus ventricosus*.

b. Upland, submontane, and montane grasslands

- Mesophytic tall-grass–forb communities of an anthropogenic origin in the Caucasus region distributed between steppes and subalpine grasslands. The species include *Bromus riparius, Brachypodium pinnatum, Festuca pratensis, Dactylis glomerata, Filipendula hexapetala, Galium verum, Geranium sanguineum, Inula cordata, Onobrychis inermis, Trifolium pratense* and *Trifolium ambiguum*.
- Low forb–sedge communities converging to natural steppe grasslands. Important species include *Carex humilis, Bromus riparius, Phleum phleoides* and *Koeleria gracilis*. According to the local ecotope, they may be enriched by *Bromus variegatus* and/or by *Festuca ovina, F. valesiaca, F. varia* and *Nardus stricta*. The remaining grassland stands in forest clearings of the northern Caucasus are identical to boreal grasslands of the forest zone (*Calamagrostis arundinacea* and *Agrostis tenuis*) enriched with elements of the Caucasian flora (e.g. *Bromus variegatus, Trifolium ambiguum, Onobrychis inermis, Ranunculus caucasicus* and *Trollius caucasicus*).
- Meso-xerophytic low grass–forb pastures in the Crimea mountains (previously Jaila mountains). Owing to the climate and the "karst" character of the area, these grasslands closely resemble steppes and bear evidence of prolonged grazing by sheep. The important species include *Alchemilla jailae, A. leptantha, Carex humilis, Filipendula hexapetala, Festuca pseudovina, Viola oreades, Brachypodium pinnatum, Betonica officinalis, Polygonum bistorta, Helianthemum* sp. and *Teucrium* sp. Similar stands occur in the north-western corner of the Caucasus.

c. River valleys and flood-plains
ca. In the forest zone

The vegetation along the course of rivers in the forest zone of the European part of the former Soviet Union does not differ greatly from conditions in central Europe in its zonation and floristic composition.

- Hygrophytic tall-grass communities in sites which are flooded for the longest period. *Phalaris arundinacea* dominates in some areas, tall sedges (*Carex gracilis* and *C. vulpina*) in others. Stands dominated by *Glyceria maxima* are scarce in north-eastern Europe, being present only in the valley of the upper Volga river and its tributaries.
- Mesohygrophytic tall-grass communities in the alluvial zone, with short-term flooding. The species include *Alopecurus pratensis, Poa palustris, Phleum pratense, Dactylis glomerata, Festuca pratensis, Veronica longifolia, Galium boreale* and *Ranunculus repens*. Uniform tall stands of *Agropyron repens, Bromus inermis, Heracleum sibiricum, Agrostis stolonifera* var. *gigantea* and *Equisetum arvense* grow in the unstable zone of sand aggradation terraces.
- Mesophytic tall-grass–forb communities, also on flooded terraces. This is the most widely distributed and the most important type of grasslands in large river valleys. They originate from the mesohygrophytic type while fluvial deposits are gradually growing and moisture is decreasing. The species include *Agrostis stolonifera* var. *gigantea, Agropyron repens, Alopecurus pratensis, Bromus inermis, Festuca pratensis, Phleum pratense, Poa palustris, P. pratensis, Trifolium pratense, T. repens, Lathyrus pratensis, Vicia cracca, Achillea millefolium, Filipendula ulmaria, Galium boreale, Geranium pratense, Leucanthemum vulgare, Ranunculus acer* and *Heracleum sibiricum*.
- Meso-xerophytic short-forb mixed communities, mostly beyond the reach of flood waters, well drained, displaying some signs of water stress (from Vologda, Onega to Mezen', Sysola and Pechora). Species diversity is high (more than 200 species, predominantly forbs) with a rich representation of Fabaceae. The species include *Festuca rubra, Agrostis stolonifera, A. tenuis, Phleum pratense, Trifolium pratense, T. repens, Thalictrum minus, Leucanthemum vulgare* (determining the physiognomy of the stand), *Campanula glomerata, Achillea millefolium, Ranunculus acer, Equisetum arvense, E. pratense, Galium boreale* and *Rumex haplorhizus*. Extrazonal steppe grasslands in southern parts of the forest zone along the upper reaches of the rivers Dnepr and Volga and in the basin of the Kama river are composed of *Festuca valesiaca, Koeleria delavignei, Agrostis syreistschikowii, Phleum phleoides, Medicago falcata, Galium verum, Trifolium montanum* and others.

cb. In the forest steppe zone

Distinctive in their character are lowland grasslands in the forest-steppe zone of eastern Europe. The soils are richer in inorganic and organic sediments, loess, chernozem with mineral elements. They are neutral or slightly basic owing to more fertile catchments and to the nature of soil-forming processes in the dry and warm climate. Variations in the water regime are greater, from flooding to summer drought. These are mainly in the more continental eastern part (basin of the Volga river). The clearing of forests and ploughing of arable land in the catchment area, together with rapid thawing of the snow, result in a considerable discharge of alluvial sediments. An increased capillary ascent of the water in summer increases the salinity of the soil, so that halophytes are present. Alluvium and river banks are colonized by numerous euxerophytes, ephemerals and ephemeroids (see definitions, pp. 5 and 14). River terraces flooded for a short period provide habitats favourable for the development of a very complicated community with deep-rooted mesophytes of great diversity, a compensation in the floristic composition of the communities enabling the co-existence of mesophytes, hygromesophytes and xeromesophytes. This is brought about by sharp fluctuations in the water regime in certain years, which exclude the dominance of one or other of the strong competing species. Herbage yields are less stable than in the forest zone. In moist years, yields are higher than those from flood-plains of the forest zone; in dry years (without flooding) the biomass production of grasses is low. Growth is not as severely restricted during a dry summer which follows good flooding in spring; but without it, even a rainy summer cannot save the harvest. The steppic character of these communities increases from the north to the south and from the west to the east.

- Tall-grass communities on sand deposits in big rivers (e.g. the central reaches of the Volga and the Don). *Phalaris arundinacea*, growing to a height of more than 2 m, sometimes is accompanied by *Bromus inermis, Carex gracilis, Agropyron repens* and *Beckmannia eruciformis*. Stands of *Glyceria maxima* cover small areas near the water edge. Another type is formed by uniform stands of *Poa palustris–Alopecurus pratensis* growing on the flooded terrace, with an admixture of several halophilic species, such as *Alopecurus ventricosus, Atropis distans* and *Juncus gerardi*.
- Tall-forb communities on flooded terraces adapted to sharp variations in water regime which are responsible for the origin of an ecologically and floristically complicated grassland. *Alopecurus pratensis* is again the co-dominant, with tall forbs. These communities previously covered almost the entire valleys of the rivers Volga, Kama, Belaya, Don and Ural. Accompanying species include *Sanguisorba officinalis, Filipendula ulmaria, Serratula coronata, Asparagus officinalis, Thalictrum minus, Cenolophium fischeri, Eryngium planum, Rumex haplorhizus, R. confertus, R. crispus, Euphorbia virgata, E. palustris* and *Tragopogon brevirostris*.

III. Western Siberia

a. Plains and slopes

This region is clearly influenced by the continental climate — hot, dry summers and cold winters. This is associated with much variation in the water regime leading to a considerable diversity in the grassland vegetation.

- Tall-grass mesophytic meadows and pastures in the southern part of the taiga and dispersed among groves in the birch (*Betula*) zone. The dominants are exacting species which indicate a considerable fertility of the soil. These include *Festuca pratensis, Dactylis glomerata, Bromus inermis, Filipendula ulmaria, Thalictrum* sp., *Libanotis sibirica, Sanguisorba officinalis, Leucanthemum vulgare, Trifolium pratense* and *Galium boreale*, in addition to specific Siberian species — *Trollius asiaticus, Vicia megalotropis, Iris ruthenica, Hemerocallis flava* and *Lilium martagon*. The southern border of the forest zone is entered by species of meadow steppes (e.g. *Filipendula hexapetala* and *Phleum phleoides*).

- Tall-forb mesophytic grasslands of the forest-steppe zone. These are a continuation of the birch(*Betula*)-zone grasslands; in the south they pass into the steppe. The flat plateau of western Siberia is little dissected and has fertile soils, providing satisfactory conditions for the development of richer and more productive grasslands than those of the eastern-European forest steppe, with fewer grass species but rich in tall forbs. The floristic aspect is determined by *Peucedanum officinale*. There is an abundance of *Libanotis montana, Heracleum sibiricum, Pleurospermum uralense, Sanguisorba officinalis, Filipendula hexapetala, F. ulmaria, Thalictrum minus, T. simplex, Serratula coronata, Phlomis tuberosa, Trifolium lupinaster, Onobrychis arenaria, Vicia amoena, Trifolium pratense, Medicago falcata* and *Vicia cracca*.
- In flat waterlogged lowlands, wetlands are occupied by uniform stands of *Alopecurus ventricosus, Hordeum brevisubulatum* and *Atropis distans*. These species are accompanied either by mesophytes (e.g. *Agrostis stolonifera, Agropyron repens, Filipendula ulmaria* and *Triglochin palustris*) or by halophytes (*Carex disticha, C. diluta, Glaux maritima, Juncus gerardi, Plantago maritima*) which pass into solonchak with stands of *Salicornia herbacea* and other halophilic species.
- Short-grass xerophytic grasslands in flat unsalinized depressions of deserts in Central Asia. Stands of *Cynodon dactylon* are typical. A shift of the growing season to the winter and spring months becomes evident in the southern zone of these deserts. The grasslands of submontane, loess plains of Central Asia are composed of the dominants *Carex pachystylis* and *Poa bulbosa* with an admixture of numerous ephemerals. Being dependent on rainwater only, they become green after the autumn rains and attain their maximum biomass in April and May, after which they survive more or less in a state of anabiosis. Owing to the richer soils, if the summers are sufficiently moist the yield of hay from these grasslands is higher than that from meadows of the forest zones.

b. Uplands, submontane and montane grasslands

As a result of the dry climate of Central Asia, submontane areas are covered with steppes (see Chapters 2 and 3) which, at higher elevations, mainly on northern slopes, pass into brush, forest and meadow formations.

- Mesophytic tall-forb grasslands in the piedmonts of the Tien Shan and the Pamir-Altai. These communities are very flowery, impressive, growing to a height of from 80 to 100 cm, with a smaller proportion of grasses (particularly *Dactylis glomerata* and *Bromus inermis*) and rich in forbs (*Ligularia* spp., *Senecio* spp., *Delphinium* spp., *Eremurus* spp., *Phlomis tuberosa, Heracleum dissectum* and others, with *Lathyrus pratensis, Campanula glomerata, Galium verum* and others in the lowest layer). The submontane grasslands in the more southerly parts of these mountain ranges, as well as in Tadzhikistan and Uzbekistan, are remarkable for the gigantic umbelliferous species (*Prangos* spp., *Ferula* spp. and others) accompanied by other common boreal species. The growing season of these grasslands is also shifted towards the winter and the spring.

– Mesophytic tall-grass–forb grasslands in the piedmont of the Altai. These are a continuation of meadow steppes which are secondary communities in the forest zone and are identical with western Siberian grasslands of the forest-steppe zone. The important species include *Dactylis glomerata, Calamagrostis arundinacea, Koeleria gracilis, Avena pubescens, Phleum phleoides, Brachypodium pinnatum, Poa nemoralis, Filipendula hexapetala, Geranium pseudosibiricum, Hemerocallis flava, Iris ruthenica, Hieracium umbellatum, Orobus lathyroides, Sanguisorba officinalis, Trifolium lupinaster, Trollius asiaticus* and numerous other boreal species.

c. River valleys and basins

Lowlands along the rivers Ob and Irtysh situated in the northern part of the forest zone of western Siberia cover a belt 10 to 15 km in width. They remain flooded for more than two months in some years and are covered with uniform stands dominated in the more northerly parts by *Carex gracilis* and *C. aquatilis*, and more to the south by *Phalaris arundinacea* and *Calamagrostis langsdorfii*.

– The southern part of the western Siberian forest zone is colonized by secondary grasslands growing both directly in the flood zone and on alluvial terraces. They are tall-forb meadows with *Festuca pratensis, Phleum pratense, Alopecurus pratensis, Agropyron repens, Bromus inermis, Phalaris arundinacea, Vicia cracca, V. sepium, Trifolium pratense, T. repens, Lathyrus pratensis, L. palustris, Filipendula ulmaria, Sanguisorba officinalis, Rumex haplorhizus, Thalictrum simplex, T. minus, Galium boreale* and others. The stand is rich, interspersed with such southern elements as *Serratula coronata, Phlomis tuberosa,* and typical, specific Siberian species — *Hemerocallis flava, Anemone dichotoma, Aster discoideus, A. impatiens, Vicia megalotropis, Trifolium lupinaster, Trollius asiaticus, Geranium pseudosibiricum, Erythronium sibiricum, Iris ruthenica, Lilium martagon, Heracleum dissectum* and *Pleurospermum uralense*. Meadows in drier ecotopes, with *Poa pratensis, Festuca rubra, Alopecurus pratensis* and *Phleum phleoides* replace the Festucetea rubrae of northern Europe.

IV. Eastern Siberia

a. Plains and slopes

A severe continental climate with cold winters (but little snow), dry summers, and permafrost soil are the factors responsible for grassland stands becoming more xerophytic. Grassland complexes of the Yakutsk region are synanthropic communities replacing larch (*Larix*) forests.

– Xeromesophytic short-grass–forb meadows, partly of steppic character, develop in elevated areas. Species of *Artemisia* dominate together with low grasses (e.g. *Festuca lenensis, Koeleria gracilis* and *Poa attenuata*) and forbs.
– Mesohygrophytic tall-grass stands occurring in concentric belts around the lake depressions on saline soils. The important species include *Scolochloa festucacea, Carex lithophilla, Alopecurus ventricosus* and *Hordeum brevisubulatum*. Another type of grassland occurs in lowlands of shallow rivers, with dominant tufts of *Calamagrostis langsdorfii* (to a height of 1.5 m) and a dense stand of *Carex juncella, C. caespitosa* and *C. schmidtii*.
– Mesohalophytic tall-grass communities in shallow depressions of the steppe and desert regions of Kazakhstan and Central Asia. These are stands with dominant tufts of *Lasiagrostis splendens* (height 1.5 m) or stands with *Carex stenophylla* s.l., *Glycyrrhiza glabra, G. uralensis, Festuca arundinacea, Aeluropus litoralis, Cynodon dactylon* and others.

b. River basins and valleys

Anthropogenic meadows are infrequent; lowlands are either covered with forests or form wetlands. The basin of the Yenisey river belongs to this type.

– The basin of the Lena river south of Aldan (near Yakutsk) is rich in meadows and pastures. The lowest inundated terrace is covered with a tall-grass mesohygrophytic community with *Bromus sibiricus, Alopecurus ventricosus* and *Hordeum brevisubulatum*. The highest flooded terrace is covered by an abundantly distributed mesophytic type of short-grass–forb community with *Hordeum brevisubulatum, Poa pratensis, P. subfastigiata, Equisetum arvense, Vicia cracca, Lathyrus palustris, Allium schoenoprasum, Thalictrum simplex, Sanguisorba officinalis, Cnidium dahuricum, Galium boreale, Senecio jacobaea, Taraxacum ceratophorum* and others. This type is similar to that of eastern-European meadows with the dominant *Agrostis stolonifera*. A mesophytic type with *Festuca rubra, Agrostis borealis* and several forbs, closely resembling an analogous European type, occurs above the flood line.
– Synanthropic grasslands in the basins of the rivers Aldan and Amga in the central and south-eastern Yakutsk area are secondary communities replacing forests with a tendency to rapid succession. Park-like tall-forb–grass meadows consist of *Bromus sibiricus, Agropyron repens, Heracleum dissectum, Anemone dichotoma, Thalictrum minus, T. simplex, Sanguisorba officinalis, Trifolium lupinaster, Vicia amoena, Lilium dauricum, Filipendula palmata, Dianthus chinensis* and *Geranium erianthum*, which are indicators of the flora of the far eastern part of the continent. These stands have been impoverished by a long-term exploitation without fertilization. They are less dense, dominated by *Kobresia filifolia, Pulsatilla angustifolia* and *Bupleurum scorzonerifolium*. Intensive grazing leads to a further deterioration of the stand with the dominants *Kobresia filifolia, Carex rhyzina* and *Artemisia commutata*.

V. Far East

The temperate forest zone extends farther to the south in the coastal region due to the influence of the oceanic climate. The limited information available suggests that human impact on the vegetation has been different from that in the Japanese archipelago.

a. Slopes and plains

Anthropogenic grasslands are infrequent in the Far East and information on them is scarce. The meadows on slopes dotted with oak (*Quercus*), hazel (*Corylus*) and *Lespedeza* attain heights of 1.10 to 1.20 m in the Primorskiy district. The stands consist of *Arundinella anomala, Miscanthus purpurascens, Calamagrostis epigeios, Veronica sibirica, Aster scaber, Vicia cracca, V. unijuga* and *Lathyrus alatus* and differ floristically from those described above. They belong to the Manchurian flora.

- Peat-bog grasslands on plateaux of the Primorskiy and Ussuri regions are secondary communities of birch (*Betula*) and alder (*Alnus*) forests. They form a dense high stand of grasses, sedges and forbs. The important species include *Carex schmidtii, Calamagrostis langsdorfii, Agrostis clavata, Sanguisorba parviflora, Trollius chinensis, Valeriana officinalis, Ranunculus japonicus, Vicia cracca, V. amoena* and *Lathyrus pilosus*.

b. River basins and valleys

The basins of the rivers Amur, Ussuri and Sujfun are rich in grassland stands exploited by mowing and grazing. Among the primary ecological factors are summer monsoon floods, before which hay must be harvested. The standing dead biomass is frequently burned in the spring.

- Mesophytic tall-grass communities on flooded river-bank biotopes with the dominant species *Calamagrostis langsdorfii* and *C. neglecta* and several forbs including *Sanguisorba tenuifolia, Filipendula palmata, Iris kaempferi, Stellaria radians, Trollius chinensis, Vicia amoena, Ranunculus japonicus, R. chinensis* and *Anemone dichotoma*. There is also a considerable incidence of grassland ferns, e.g. *Onoclea sensibilis* and *Aspidium thelypteris*.
- Mesophytic tall-forb communities in drier localities. Inconspicuous grass species include *Bromus ciliatus, Trisetum sibiricum* and *Poa palustris*. The Liliaceae and Iridaceae are more conspicuous, in addition to other forbs. The important species include *Hemerocallis flava, H. minor, H. middendorfii, Lilium dauricum, L. tenuifolium, Polygonatum officinale, Iris setosa, I. kaempferi, Aster scaber, A. tataricus, Cacalia hastata, Ligularia speciosa, Filipendula palmata, Sanguisorba officinalis, Veronica sibirica, Angelica anomala, Vicia unijuga, V. pseudorobus* and *Trifolium lupinaster*. These grasslands occur in the basins of the rivers Amur and Shilka. Another type consisting of forbs several metres high has developed on rich soils which are covered with a thick layer of snow in winter. This type occurs mainly in the Kamchatka peninsula and on Sakhalin island. It includes *Filipendula kamtschatica, Polygonum sachalinense* and *Senecio cannabifolius*.

VI. Japan

The archipelago of Japan includes sub-Arctic, cool temperate, and warm temperate zones, in all of which forest constitutes the climax biome. This accounts for an absence of natural grasslands — apart from some minor exceptions (natural grasslands existing on some biotopes above the tree line). Secondary semi-natural grasslands occupy an area of 1.4 million ha. Of this area, 57% is used for grazing and the remainder is mown for hay (Suganuma, 1966). The first two types (below) are the most widely distributed grasslands in Japan, being secondary communities in the zone of deciduous broad-leaved forests (Numata, 1979c).

- Tall-grass mesophytic communities (Miscanthetalia sinensis), with the dominant tall bunch grass *Miscanthus sinensis* are distributed widely in the southern part of Hokkaido island, and throughout Honshu island. In the south, this type continues to the western shore and occupies the western side of Kyushu island. This important type of stand which is mown, requires a deep humic soil rich in nutrients. Species associated with the dominant include *Lespedeza cyrtobotrya, L. serpens, Osmunda japonica, Isodon japonicus, Imperata cylindrica, Ischaemum antephoroides, Pleioblastus chino* and others (Numata, 1979b). This community responds poorly to grazing or trampling because its regenerative capacity is relatively low. Therefore, stands are mown once a year only (in September) in order to maintain stable yields. More frequent mowing, or earlier harvest in summer (formerly practised), cause *Pteridium aquilinum* to become co-dominant (Speidel, 1979). On the other hand, without mowing the stand becomes overgrown by weedy species which suppress *Miscanthus*. Occasional growth of woody species is generally limited by burning during winter.
- Short-grass mesophytic communities (*Zoysion japonicae*). This is the most widely distributed type of pasture in Japan. The dominant *Zoysia japonica*, with its long rhizomes, does not require such fertile soils as the *Miscanthus* type, from which it may develop under the influence of intensive grazing. It stands up well to trampling and to intensive grazing. Other species of the stand include *Hydrocotyle ramiflora, Carex nervata, Agrimonia pilosa, Ranunculus japonicus, Pteridium aquilinum* and *Lotus corniculatus* among about 70 species recorded by Suganuma (1966).
- Dwarf bamboo types of grasslands with dominant *Sasa* spp. distributed on the island of Hokkaido. This is a secondary community to coniferous forest. Mowing increases stability of this type, whereas grazing has a degrading influence. The dominants (*S. nipponica, S. veitchii* and *S. chokaiensis*) are accompanied by *Ixeris dentata, Carex nervata, Potentilla freyniana, Polygala japonica, Carex lanceolata, Hosta albomarginata, Agrimonia pilosa, Gentiana triflora, Artemisia japonica* and others, in addition to *Pteridium aquilinum*. Because of the intolerance of dwarf bamboo grasslands to grazing, *Poa pratensis, Dactylis glomerata, Trifolium repens* and *Festuca ovina* were introduced as seeded pasture species in this area about 100 years ago. At present, the latter species are the most extensive and most important grassland types of Hokkaido (Numata, 1979c); natural stands of *Sasa* are becoming assigned more to recreational use (Suganuma, 1966).

REFERENCES

Alberda, Th., 1980. Possibilities of dry matter production from forage plants under different climatic conditions. In: E. Wojahn and H. Thöns (Editors), *Proceedings of XIII International Grassland Congress*. Leipzig, GDR, 18–27 May, 1977, Vol. I. Akademie-Verlag, Berlin, pp. 61–69.

Alekseenko, L.N., 1962. Productivity and the rate of photosynthesis of some mesophytic plants in meadows of the Leningrad region. *Nautch. Dokl. Vys. Shkoly, Biol. Nauki*, 1: 140–144 (in Russian).

Alekseenko, L.N., 1967. *Productivity of Meadow Plants in Relation to Environmental Conditions*. Izd. Leningrad University, 168 pp. (in Russian).

Alekseenko, L.N., 1975. Amplitude of variation in transpiration rate in grassland plants of the humid zone. *Byull. Mosk. Ova. Ispyt. Prir., Otd. Biol.*, 80(3): 116–127 (in Russian with English summary).

Alekseenko, L.N. and Martynova, M.F., 1964. Characteristics of development and efficiency of photosynthetic apparatus in seeded plots of meadow grasses. *Fyziol. Rast.*, 11(3): 417–423 (in Russian).

Andrzejewska, L., 1976. The influence of mineral fertilization on the meadow phytophagous fauna. *Pol. Ecol. Stud.*, 2(4): 93–109.

Andrzejewska, L. and Gyllenberg, G., 1980. Small herbivore subsystem. In: A.I. Breymeyer and G.M. Van Dyne (Editors), *Grasslands, Systems Analysis and Man*. International Biological Programme 19, Cambridge University Press, Cambridge, pp. 201–268.

Balátová-Tuláčková, E., 1966. Synecological characteristics of flooded meadows in southern Moravia. *Rozpr. Česk. Akad. Věd.*, 76(1): 1–40 (in German).

Balátová-Tuláčková, E., 1971. Phenological diagrams and their interpretation in meadows of Opava River valley. *Acta Sci. Nat. Acad. Sci. Bohemoslov.* (Brno), 5(6): 1–60 (in German).

Balátová-Tuláčková, E., 1979. Semi-natural temperate meadows and pastures. Introduction. In: R.T. Coupland (Editor), *Grassland Ecosystems of the World: Analysis of Grasslands and Their Uses*. International Biological Programme 18, Cambridge University Press, Cambridge, pp. 115–126.

Balátová-Tuláčková, E., Zelená, V. and Tesařová, M., 1977. Synecological characteristics of important meadow types in the protected area of the Žďárské Vrchy hills. *Rozpr. Česk. Akad. Věd*, 87(5): 1–115 (in German).

Bár, I., 1979. Climatic characteristics of the experimental area at Kameničky. In: M. Rychnovská (Editor), *Function of Grasslands in Spring Region — Kameničky Project*. Institute of Botany, Czechoslovakian Academy of Science, Brno, pp. 35–43.

Bednář, V., 1979. Mineral elements and nitrogen in aboveground biomass of a natural meadow with *Alopecurus pratensis* (*Alopecuretum pratensis*). *Acta Univ. Palacki. Olomuc., Fac. Rerum Nat.*, 63: 59–64 (in Czech).

Berger-Landefeldt, U. and Sukopp, H., 1965. Notes on the synecology of dry sandy grasslands, especially of stands dominated by *Corynephorus canescens*. *Ver. Prov. Brandenburg*, 102: 41–98 (in German).

Braun-Blanquet, J., 1951. *Plant Sociology*. 2nd ed., Springer Verlag, Vienna, 631 pp. (in German).

Breymeyer, A.I., 1978. Analysis of the trophic structure of some grassland ecosystems. *Pol. Ecol. Stud.*, 4(2): 55–128.

Breymeyer, A.I. and Kajak, A., 1976. Drawing models of two grassland ecosystems: A mown meadow and a pasture. *Pol. Ecol. Stud.*, 2(2): 41–49.

Drobník, J., 1962. The effect of temperature on soil respiration. *Folia Microbiol.*, 7: 132–140.

Druzina, V.D., 1978. Variation in content of mineral elements and nitrogen in plant biomass. In: V.M. Ponyatovskaya (Editor), *Productivity of Meadow Communities*. Nauka, Leningrad, pp. 195–217 (in Russian).

Ellenberg, H., 1963. Vegetation of Central Europe including the Alps. In: H. Walter (Editor), *Introduction to Phytology*, Vol. IV/2. Eugen Ulmer Verlag, Stuttgart, 943 pp. (in German).

Fiala, K., 1979. Estimation of annual increment of underground plant biomass in a grassland community (*Polygalo–Nardetum*). *Folia Geobot. Phytotaxon.*, 14(1): 1–10.

Flora of the USSR, 1964. *Flora Unionis Republicarum Socialisticarum Sovieticarum. Indices alphabetici I–XXX*. Nauka, Moscow, 262 pp. (in Russian).

Geyger, E., 1964. Methods of approach to the estimation of assimilation surfaces of meadow. *Ber. Geobot. Inst. ETH, Stiftung Rübel* (Zürich), 35: 41–112 (in German).

Geyger, E., 1977. Leaf area and productivity in grasslands. In: W. Krause (Editor), *Handbook of Vegetation Science XIII. Application of Vegetation Science to Grassland Husbandry*. W. Junk Publishers, The Hague, pp. 499–520.

Gloser, J., 1976. Photosynthesis and respiration of some alluvial meadow grasses: responses to irradiance, temperature and CO_2 concentration. *Acta Sci. Nat.* (Brno), 10(2): 1–39.

Gloser, J., 1977. Photosynthesis and respiration of some alluvial meadow grasses: responses to soil water stress, diurnal and seasonal courses. *Acta Sci. Nat.* (Brno), 11(4): 1–34.

Gloser, J., 1979. Photosynthetic characteristics of several important plant species. In: M. Rychnovská (Editor), *Progress Report on MAB Project No. 91: Function of Grasslands in Spring Region — Kameničky Project*. Institute of Botany, Czechoslovakian Academy of Science, Brno, pp. 147–152.

Gloser, J. and Tesařová, M., 1978. Litter, soil and root respiration measurement. An improved compartmental analysis method. *Pedobiologia*, 18: 76–81.

Grulich, I., 1958. Changes of plant communities in meadows under the impact of digging activity of moles (*Talpa europaea*) under conditions in south Moravia. *Preslia*, 30: 341–356 (in Czech).

Grulich, I., 1959. The consequences of digging activity of moles (*Talpa europaea*) in Czechoslovakia. *Pr. Brněnské Zákl. Česk. Akad. Věd.*, 31(3): 157–212 (in Czech).

Gyllenberg, G., 1969. The energy flow through a *Chorthippus parallelus* (Zett.) (Orthoptera) population on a meadow in Tvärminne, Finland. *Acta Zool. Fenn.*, 123: 1–74.

Haas, H., 1976. Abundance, phenology and production of imagines of Homoptera, Auchenorrhyncha. *Pol. Ecol. Stud.*, 2(2): 153–162.

Halva, E. and Lesák, J., 1979. Fertilization and exploitation effects in natural and man-made grasslands at Kameničky. In: M. Rychnovská (Editor), *Progress Report on MAB Project No. 91: Function of Grasslands in Spring Region — Kameničky Project*. Institute of Botany, Czechoslovakian Academy of Science, Brno, pp. 137–143.

Harper, J.L., 1977. *Population Biology of Plants*. Academic Press, New York, N.Y., 892 pp.

Heal, O.W. and Perkins, D.F. (Editors), 1978. *Production Ecology of British Moors and Montane Grasslands*. Ecological Studies, Vol. 27, Springer-Verlag, Berlin, 426 pp.

Jakrlová, J., 1975. Primary production and plant chemical composition in flood-plain meadows. *Acta Sci. Nat.* (Brno), 9(9): 1–52.

Jakrlová, J., 1985. Seed production of a *Polygalo–Nardetum* grassland, I. Quantitative changes recorded in the course of years. *Ecology* (CSSR), 4(2): 177–183.

Jakrlová, J. and Kršková, J., 1982. Above-ground biomass and its structure in two types of meadows in Bohemian–Moravian upland. *Rep. Kameničky Proj.*, 30: 1–8 (in Czech).

Jakubczyk, H., 1976. The dependence of the rate of plant material decomposition in a meadow upon mineral fertilization and environmental factors. *Pol. Ecol. Stud.*, 2(4): 259–296.

Jankowska, K., 1967. Seasonal variation of vegetation and net primary production in a meadow community *Arrhenatheretum elatioris*. In: A. Medwecka-Kornas (Editor), *Studies of a Beech Forest Ecosystem and a Meadow in Ojców National Park*. Panstwowe Wydawnictwo Naukowe, Kraków, pp. 153–173 (in Polish).

Klapp, E., 1965. *Grassland Vegetation and Its Habitat. Examples from West, Middle and South Germany*. Verlag Paul Parey, Berlin, 384 pp. (in German).

Knapp, R., 1979. Distribution of grasses and grasslands in Europe. In: M. Numata (Editor), *Ecology of Grasslands and Bamboolands in the World*. VEB Gustav Fischer Verlag, Jena, pp. 111–123.

Kotańska, M., 1970. Morphology and biomass of underground organs of plants of meadows in Ojców National Park. *Stud. Naturae, Ser. A*, Nr 4: 1–107 (in Polish).

Krajčovič, V. and Regál, V., 1976. *Biology and Ecology of Grasslands. Synthetic Final Report on the Research Programme, Grassland Research Institute*, Banská Bystrica. 72 pp. (in Czech).

Kubicka, H., 1976. The effect of mineral fertilization on CO_2 evolution in meadow soil, Part I. *Pol. Ecol. Stud.*, 2(4): 323–332.

Lapáček, V., 1979. Food consumption of the grasshopper *Chorthipus biguttulus*. In: M. Rychnovská (Editor), *Function of Grasslands in Spring Region — Kameničky Project*. Institute of Botany, Czechoslovakian Academy of Science, Brno, pp. 179–184.

MacFadyen, A., 1963. The contribution of the microfauna to total soil metabolism. In: J. Doeksen and J. van der Drift (Editors), *Soil Organisms*. Amsterdam, pp. 3–17.

Makarevich, V.N., 1978. Variation of plant biomass with the analysis of its structural units. In: V.M. Ponyatovskaya (Editor), *Productivity of Meadow Communities*. Nauka, Leningrad, pp. 169–194 (in Russian).

Makulec, G., 1976. The effect of NPK fertlizer on a population of Enchytraeid worms. *Pol. Ecol. Stud.*, 2(4): 183–193.

Máthé, I. and Précsényi, I., 1971. Plant biomass production of maize grown on a forest-steppe area. *Acta Agron. Acad. Sci. Hung.*, 20: 378–384.

Miroshnichenko, E.D., 1978. Decomposition of dead plant biomass. In: V.M. Ponyatovskaya (Editor), *Productivity of Meadow Communities*. Nauka, Leningrad, pp. 218–247 (in Russian).

Nowak, E., 1976. The effect of fertilization on earthworms and other soil macrofauna. *Pol. Ecol. Stud.*, 2(4): 195–207.

Numata, M., 1979a. Seminatural temperate meadows and pastures, II. Primary producers in meadows. In: R.T. Coupland (Editor), *Grassland Ecosystems of the World: Analysis of Grasslands and Their Uses*. International Biological Programme 18, Cambridge University Press, Cambridge, pp. 127–138.

Numata, M., 1979b. Distribution of grasses and grasslands in Asia. In: M. Numata (Editor), *Ecology of Grasslands and Bamboolands in the World*. VEB Gustav Fischer Verlag, Jena, pp. 92–102.

Numata, M., 1979c. Temperate forests and grasslands in Japan. In: M. Numata (Editor), *Methodological Studies in Environmental Education*. pp. 31–46.

Oberdorfer, E., 1957. *Plant Communities in South Germany*, Part 1. Plant Sociology, Vol. 10, VEB Gustav Fischer Verlag, Jena, 564 pp. (in German).

Olechowicz, E., 1976. The effect of mineral fertilization on insect community of the herbage in a meadow. *Pol. Ecol. Stud.*, 2(4): 129–136.

Pelikán, J., 1979. The common vole (*Microtus arvalis*) in the Kameničky grassland. In: M. Rychnovská (Editor), *Function of Grasslands in Spring Region — Kameničky Project*. Institute of Botany, Czechoslovakian Academy of Science, Brno, pp. 167–172.

Pelikán, J., 1985. Animals in the structure and functioning of meadow ecosystems. In: M. Rychnovská, E. Balátová-Tuláčková, B. Úlehlová and J. Pelikán, *Ecology of Meadows*. Academia, Prague, pp. 159–181 (in Czech).

Perkins, D.F., 1978. The distribution and transfer of energy and nutrients in the *Agrostis–Festuca* grassland ecosystem. In: O.W. Heal and D.F. Perkins (Editors), *Production Ecology of British Moors and Montane Grassland*. Ecological Studies, Vol. 27, Springer-Verlag, Berlin, pp. 375–395.

Persikova, Z.I., 1959. Development of the tuft and life cycle of *Nardus stricta*. *Byull. Mosk. Ova. Ispyt. Prir., Otd. Biol.*, 5: 61–68.

Petřík, B., 1972. Seasonal changes in plant biomass in four inundated meadow communities. In: M. Rychnovská (Editor), *Ecosystem Study on Grassland Biome in Czechoslovakia*. Czechosl. IBP/PT–PP Report No. 2, Brno, pp. 17–23.

Pilát, A., 1969. Underground dry weight in the grassland communities of *Arrhenatheretum elatioris alopecuretosum pratensis* R.Tx. 1937 and *Mesobrometum erecti stipetosum* Vicherek 1960. *Folia Geobot. Phytotaxon.*, 4(3): 225–234.

Plewczyńska-Kuraś, U., 1976. Estimation of biomass of the underground parts of meadow herbage in the three variants of fertilization. *Pol. Ecol. Stud.*, 2(4): 63–74.

Poslušná, A., 1975. *Development of Leaf Area in a Grassland of Nardetum Type*. Diploma thesis, Faculty of Agronomy, University of Agriculture, Brno, 100 pp. (in Czech).

Pyatin, A.M., 1971. *Viable Seeds in Meadow Soils*. Thesis summary, Moscow State University, 33 pp. (in Russian).

Rabotnov, T.A., 1950. Life cycle of perennial grasses and forbs in meadow communities. *Tr. Bot. Inst. im. V.L. Komarova, AN SSSR, Ser. III Geobot.*, 6: 7–204 (in Russian).

Rabotnov, T.A., 1973. *Impact of Mineral Fertilizers on Meadow Phytocoenoses*. Nauka, Moscow, 178 pp. (in Russian).

Rabotnov, T.A., 1974. *Grassland Science*. Moscow State University, 384 pp. (in Russian).

Ricou, G.A.E., 1979. Seminatural temperate meadows and pastures. Consumers in meadows and pastures. In: R.T. Coupland (Editor), *Grassland Ecosystems of the World: Analysis of Grasslands and Their Uses*. International Biological Programme 18, Cambridge University Press, Cambridge, pp. 147–153.

Rothmaler, W., 1970. *Excursion Guide to Flora of Germany*. Volk and Wissen, Berlin, 622 pp. (in German).

Rusek, J., 1984. Soil fauna in three types of flooded meadows in south Moravia. *Rozpr. Česk. Akad. Ved*, 94(3): 1–126 (in German).

Rychnovská, M., 1972a. Variation of glycid and protein contents in the biomass of meadow grasses and their communities. In: M. Rychnovská (Editor), *Ecosystem Study on Grassland Biome in Czechoslovakia*. Czechosl. IBP/PT–PP Report No. 2, Brno, pp. 33–36.

Rychnovská, M., 1972b. Transpiration of meadow communities. In: M. Rychnovská (Editor), *Ecosystem Study on Grassland Biome in Czechoslovakia*. Czechosl. IBP/PT–PP Report No. 2, Brno, pp. 37–43.

Rychnovská, M., 1976. Transpiration in wet meadows and some other types of grassland. *Folia Geobot. Phytotaxon.*, 11: 427–432.

Rychnovská, M., 1979a. Seminatural temperate meadows and pastures. Ecosystem synthesis of meadows. Energy flow. In: R.T. Coupland (Editor), *Grassland Ecosystems of the World: Analysis of Grasslands and Their Uses*. International Biological Programme 18, Cambridge University Press, Cambridge, pp. 165–170.

Rychnovská, M., 1979b. Sources of stability in grassland ecosystems. In: *Fifth International Symposium on Problems of Landscape Ecological Research*, Nov. 19–23, 1979, Slovak Acad. Sci., Bratislava, pp. 287–294.

Rychnovská, M., 1979c. Transpiration in natural and man-made grasslands at Kameničky. In: M. Rychnovská (Editor), *Function of Grasslands in Spring Region–Kameničky Project*. Institute of Botany, Czechoslovakian Academy of Science, Brno, pp. 153–160.

Rychnovská, M., Gloser, J. and Petřík, B., 1972a. Vertical structure of four inundated meadow stands. In: M. Rychnovská (Editor), *Ecosystem Study on Grassland Biome in Czechoslovakia*. Czechosl. IBP/PT–PP Report No. 2, Brno, pp. 25–28.

Rychnovská, M., Květ, J., Gloser, J. and Jakrlová, J., 1972b. Plant water relations in three zones of grassland. *Acta Sci. Nat.* (Brno), 6(5): 1–38.

Rychnovská, M., Úlehlová, B., Jakrlová, J. and Tesařová, M., 1980. Biomass budget and energy flow in alluvial meadow ecosystem. In: E. Wojahn and H. Thöns (Editors), *XIII International Grassland Congress*, Leipzig, GDR, 18–27 May, 1977, Vol. I. Akademie Verlag, Berlin, pp. 473–475.

Rychnovská, M., Balátová-Tuláčková, E., Úlehlová, B. and Pelikán, J., 1985. *Ecology of Meadows*. Academia, Prague, 291 pp. (in Czech).

Shalyt, M.S., 1950. Under-ground parts of some meadow, steppe and desert plants and phytocoenoses. *Tr. Bot. Inst. Akad. Nauk SSSR, Ser. III Geobot.*, 6: 205–442 (in Russian).

Shennikov, A.P., 1941. *Grassland Science*. State University, Leningrad, 507 pp. (in Russian).

Škapec, L., 1979. Two methods used for quantitative sampling of insects in the investigation of the grassland at Kameničky. In: M. Rychnovská (Editor), *Function of Grasslands in Spring Region–Kameničky Project*. Institute of Botany, Czechoslovakian Academy of Science, Brno, pp. 189–194.

Šmíd, P., 1972. Fundamental climatological and hydrological characteristics of grassland ecosystems in the Lanžhot area. In: M. Rychnovská (Editor), *Ecosystem Study on Grassland Biome in Czechoslovakia*. Czechosl. IBP/PT–PP Report No. 2, Brno, pp. 11–16.

Speidel, B., 1976. Primary production and root activity of a golden oat meadow with different fertilizer treatments. *Pol. Ecol. Stud.*, 2(2): 77–89.

Speidel, B., 1979. Semi-natural temperate meadows and pastures: Use and management of meadows. In: R.T. Coupland (Editor), *Grassland Ecosystems of the World: Analysis of Grasslands and Their Uses*. International Biological Programme 18, Cambridge University Press, Cambridge, pp. 181–185.

Speidel, B. and Weiss, A., 1972. Notes on above-ground and under-ground biomass production in a meadow of *Trisetum flavescens* under different fertilization. *Angew. Bot.*, 46: 75–93 (in German).

Spitzer, K., 1978. Notes on the synecology of butterflies (Lepidoptera) in meadows of southern Bohemia. *Acta Sci. Nat. Mus. Bohem. Merid. České Budějovice*, 18: 37–47 (in Czech).

Suganuma, T., 1966. Phytosociological studies on the seminatural grassland used for grazing in Japan. I. Classification of grazing land. *Jpn. J. Bot.*, 19: 255–276.

Tesařová, M., 1976. Litter production and disappearance in some alluvial meadows (preliminary results). *Folia Geobot. Phytotaxon.*, 11: 63–74.

Tesařová, M., 1977. Factors determining the rate of litter decomposition in some meadow ecosystems. In: J. Szegi (Editor), *Soil Biology and Conservation of the Biosphere*. Akademiai Kiado, Budapest, pp. 329–335.

Tesařová, M., 1979a. Seasonal dynamics of plant litter in the Kameničky grassland dominated by *Nardus stricta* L. In: M. Rychnovská (Editor), *Function of Grasslands in Spring Region — Kameničky Project*. Institute of Botany, Czechoslovakian Academy of Science, Brno, pp. 197–202.

Tesařová, M., 1979b. Soil and root respiration in the Kameničky grassland dominated by *Nardus stricta* L. In: M. Rychnovská (Editor), *Function of Grasslands in Spring Region — Kameničky Project*. Institute of Botany, Czechoslovakian Academy of Science, Brno, pp. 203–209.

Tesařová, M., 1983. Organic matter distribution in some grassland ecosystems. *Ecology* (CSSR), 2(2): 155–172.

Tesařová, M., 1988. Microorganisms and the carbon cycle in terrestrial ecosystems. In: V. Vančura and F. Kunc (Editors), *Soil Microbial Associations, Control of Structures and Functions*. Academia, Prague, pp. 339–405.

Tesařová, M. and Gloser, J., 1976. Total CO_2 output from alluvial soils with two types of grassland communities. *Pedobiologia*, 16: 364–372.

Titlyanova, A.A., 1979. Stability of biological cycles in natural ecosystems. In: *Fifth International Symposium on Problems of Landscape Ecological Research*, Nov. 19–23, 1979, Slovak Acad. Sci., Bratislava, pp. 137–156 (in Russian).

Traczyk, T., 1971. Productivity investigation of two types of meadows in the Vistula Valley. *Ekol. Pol.*, 19(7): 93–106.

Traczyk, T., Traczyk, H. and Pasternak, D., 1976. The influence of intensive mineral fertilization on the yield and floral composition of meadows. *Pol. Ecol. Stud.*, 2(4): 39–47.

Úlehlová, B., 1979. Micro-organisms in meadows. In: R.T. Coupland (Editor), *Grassland Ecosystems of the World: Analysis of Grasslands and Their Uses*. International Biological Programme 18, Cambridge University Press, Cambridge, pp. 155–163.

Úlehlová, B., 1985. Decomposers and decomposition processes in grassland ecosystems. In: M. Rychnovská, E. Balátová-Tuláčková, B. Úlehlová and J. Pelikán, *Ecology of Meadows*. Academia, Prague, pp. 182–218 (in Czech).

Úlehlová, B. and Úlehla, J., 1983. Energy flow in grassland ecosystems. *Ecology* (CSSR), 2: 37–48.

Úlehlová, B., Findejsová, M. and Ružička, V., 1980. Yearly nitrogen budget in natural and fertilized grasslands of a Bohemian–Moravian Highlands district. In: E. Wojahn and H. Thöns (Editors), *XIII International Grassland Congress*, Leipzig, GDR, 18–27 May, 1977, Vol. I. Akademie-Verlag, Berlin, pp. 481–483.

Uranov, A.A., 1960. The age composition of individual species in a plant community. *Byull. Mosk. Ova. Ispyt. Prir., Otd. Biol.*, 64(3): 77–92 (in Russian).

Uranov, A.A. and Serebryakova, T.I. (Editors), 1976. *The Coenopopulations of Plants. Basic Features and Structure.* Nauka, Moscow, 216 pp. (in Russian).

Vicherek, J., 1962. Phytocoenoses in alluvial lowland of the lower Dyje River basin with special regard to meadow communities. *Folia Fac. Sci. Nat. Univ. Purkynianae Brunensis, Biol.*, 3(5): 1–113 (in Czech).

Walter, H., 1963. Climatic diagrams as a means to comprehend the various climatic types for ecological and agricultural purposes. In: A.J. Rutter and F.H. Whitehead (Editors), *The Water Relations of Plants*. Blackwell, Oxford, pp. 3–9.

Walter, H., 1968. *Vegetation of the Earth from an Ecophysiological Point of View, Vol. II. Temperate and Arctic Zones*. VEB Gustav Fischer Verlag, Jena, 1001 pp. (in German).

Wasilewska, L., 1976. The role of nematodes in the ecosystems of a meadow in Warsaw environs. *Pol. Ecol. Stud.*, 2(4): 137–156.

Weigert, R.G. and Evans, F.C., 1964. Primary production and the disappearance of dead vegetation on an old field in southeastern Michigan. *Ecology*, 45: 49–63.

Wielgolaski, F.E., 1976. The effect of herbage intake by sheep on primary production, ratios top–root and dead–live above-ground parts (Hardangervidda, Norway). *Pol. Ecol. Stud.*, 2(2): 67–76.

Zajonc, I., 1970. Synusia of earthworms (Lumbricidae) in meadows in the Carpathian area of Czechoslovakia. *Biol. Práce* (Bratislava), 16(8): 1–99 (in Slovak).

Zlotin, R.I., 1974. The study of decomposition processes in relation to investigation of biological turnover in terrestrial ecosystems. In: *Report October 1971 to June 1972, Zoology and Botany*. Moscow University Publications, Moscow, pp. 55–57 (in Russian).

Zyromska-Rudzka, H., 1976. The effect of mineral fertilization of a meadow on the oribatid mites and other soil mesofauna. *Pol. Ecol. Stud.*, 2(4): 157–182.

Chapter 6

OVERVIEW OF AFRICAN GRASSLANDS

R.T. COUPLAND

Shantz and Marbut (1923) reported that grassland covered 42.3% of the area of the African continent, most of this area including an intermittent layer of trees or shrubs. These authors classified the grasslands and savannas into three major regions. The most extensive of these was the "*Acacia*–tall-grass savanna", throughout which areas of open "tall-grass" vegetation occurred, which was estimated to cover 16.8% of the area of the continent. The "high-grass–low-tree savanna" occupied 11.6%, and the "*Acacia*–desert grass savanna", within which areas of "desert grass" occurred, made up 9.6% of the land surface. Other types covered 4.3%.

The proportion of the African vegetation assigned to the major biomes by various authors has varied considerably. Perhaps this is partly because much of the grass cover in Africa is successional in nature due to the effects of fire and grazing by livestock and to "slash and burn" agriculture. Whereas Shantz and Marbut (1923) assigned only about 5% of the land area to categories that were free of woody layers, their interpretation of a map by Langhams (1906) indicated that he considered that 9.8% of Africa was covered by "open grassland without trees". Similarly, Shantz and Marbut designated 39.3% of the continent as desert, although Langhams's map included only 17.5% in this category.

Savannas are often considered as part of the grassland biome in treatments of African vegetation (Shantz and Marbut, 1923; Bews, 1929). Bews (1929) considered that as far as the grasses themselves are concerned "it is not necessary to distinguish between types with trees and those without, since the composition is largely identical in both cases." Other authors (Rattray, 1960; van Wyk, 1979) have emphasized this concept by classifying types of "grass cover" apart from the presence of woody plants.

Rattray (1960) distinguished 23 grassland associations in Africa, which were mapped by van Wyk (1979). The major types, in order of (decreasing) extent were as follows:

Aristida	in desert steppe and savanna
Hyparrhenia	in woodland savanna and grassland
Andropogon	in savanna
Cenchrus	in savanna
Themeda	in savanna and grassland
Chrysopogon	in desert, steppe and woodland
Eragrostis	in savanna, grassland and steppe

The major regions of savanna and grassland are distributed in a sequence from the edge of the tropical forest (astride the Equator in the west), associated with the declining rainfall gradient and lengthening periods of seasonal drought. Bews (1929) described these belts of vegetation, in sequence outwards from the tropical forest, as follows:

(1) Tropical high-grass savannas. This vegetation is essentially a forest margin type which spread over wide areas mostly because of repeated burning. The most characteristic species are coarse members of the Andropogoneae and Paniceae.

(2) Tropical and subtropical bunch-grass savannas. The high-grass savannas grade into tall bunch-grass savannas and in turn (with decreasing rainfall) into short bunch-grass savannas. The more mesophytic and well-developed communities are dominated by members of the Andropogoneae (species of *Andropogon*, *Hyparrhenia* and *Themeda*).

(3) Semi-desert and desert types. These grasslands are composed of species with various affinities that are adapted to the various adverse habitats of dry regions, including saline and sandy situations.

Bews recognized that this sequence is sometimes interrupted, between the high-grass and tall-grass savannas, by a zone of "dry forest" in which tall grasses occupy the forest floor. He considered that the "tropical alpine grasslands" to the east of the tropical forest were not part of this sequence, being essentially temperate in character. Semi-desert and desert grasslands, as well as some savanna vegetation, also occur beyond the desert in North Africa.

Van Wyk's (1979) map suggests that a belt

TABLE 6.1

Proportion of grass species contributed by each major tribe in selected regions with grasslands in Africa, compared with other regions of the world[1]

Locality/region	Agrosteae	Andropogoneae	Aveneae	Eragrosteae	Festuceae	Paniceae	Others
Morocco	13.9	3.7	13.9	0.0	42.7	2.9	22.9
Mali	0.7	18.4	0.7	13.8	0.7	36.2	29.5
North-western Somalia	2.0	13.1	0.0	13.1	5.0	28.3	38.5
Namibia	0.8	11.6	2.4	28.3	2.0	21.1	33.8
Lesotho	5.4	17.0	10.2	15.7	13.6	16.3	21.8
George District, South Africa	3.8	11.3	17.3	9.8	14.3	15.8	27.7
Average	4.4	12.5	7.4	13.5	13.1	20.1	29.0
REGIONAL AVERAGES							
Continental North America	16.0	4.0	6.6	5.9	25.3	14.9	27.3
Continental South America	6.7	9.3	1.7	8.0	8.4	43.4	22.5
Northern Asia	14.6	1.5	10.2	3.3	34.8	4.5	31.2
Southern Asia	4.5	23.6	3.3	9.4	10.9	22.6	25.7
Oceania	8.4	15.2	10.8	9.0	13.0	19.9	23.7
Normal distribution spectrum (global)	8.2	11.9	6.3	8.1	16.5	24.7	24.3

[1] After Hartley (1950).

TABLE 6.2

The proportion (%) of the land area of each continental region under various land uses[1]

	Total land area[2]	Arable land and permanent crops	Permanent pasture	Forest and woodland	Other land	Total
Eurasia	53,790	15.4	20.6	30.3	33.7	100
Africa	29,664	6.2	26.4	23.3	44.1	100
North and Central America	21,356	12.8	16.7	31.9	38.6	100
South America	17,535	7.9	25.9	53.2	12.9	100
Oceania	8,429	5.7	54.5	18.2	21.6	100
World	130,774	11.3	24.2	31.3	33.3	100

[1] Calculated from estimated areas reported by the Food and Agriculture Organization (FAO, 1984). "Permanent pasture" refers to meadows and pastures used permanently (five years or more) for herbaceous forage crops either cultivated or growing wild (wild prairie or grazing land); "other land" includes desert.
[2] In thousands of square km.

of "steppe" (which includes tress or shrubs) is present both to the north and to the south of the Sahara desert and that it occupies large areas in East Africa and southern Africa. He indicated that "fairly large grasslands" (free of woody species) occur only in South Africa, the Sudan and north-western Africa (Morocco, Algeria and Tunisia).

Hartley's (1950) analysis (see Chapter 1 of Volume 8A) revealed that, in tropical regions, the tribe Paniceae contributes a smaller proportion of grass species in Africa than in South America, whereas for Andropogoneae the reverse situation prevails (Table 6.1). Similarly, species of the tribe Eragrosteae are relatively more numerous and Agrosteae less numerous in Africa than in the Western Hemisphere.

According to F.A.O. estimates, the proportion of the land surface that is arable in Africa is only a little more than half that in the world as a whole (Table 6.2).

The following five chapters are concerned with the grasslands of Africa. Those north of the Sahara desert are discussed in Chapter 7. In this region the vegetation has been so altered by overgrazing that the nature of the potential vegetation is uncertain. However, there are extensive areas of semi-shrub steppe occurring under climatic conditions that would be expected to support grassland in other parts of the world. Submontane grasslands occur at higher elevations. The belts of Sahelian vegetation south of the desert are dealt with in Chapter 8. These range in character from semi-desert grassland to tropical savanna. Chapter 9 is concerned with the grasslands of East Africa, ranging from tropical savannas at lower elevations to open grasslands with temperate characteristics at higher elevations. Chapter 10 considers the grasslands of southern Africa. The grasslands and savannas of Madagascar are treated in Chapter 11.

REFERENCES

Bews, J.W., 1929. *The World's Grasses: Their Differentiation, Distribution, Economics and Ecology.* University of Natal Press, Pietermaritzburg (reissued in 1973 by Russell and Russell, New York, N.Y., 408 pp.)

FAO, 1984. *FAO Production Yearbook, Vol. 37 (1983).* Food and Agriculture Organization of the United Nations, Rome, 320 pp.

Hartley, W., 1950. The global distribution of tribes of the Gramineae in relation to historical and environmental factors. *Aust. J. Agric. Res.,* 1: 355–373.

Langhams, P., 1906. *Wandkarte von Afrika zur Darstellung der Bodenbedeckung.* 1:7,500,000, Gotha. (cited by Wagner, 1912).

Rattray, J.M., 1960. *The Grass Cover of Africa.* F.A.O. Agricultural Studies, No. 49. (cited by van Wyk, 1979).

Shantz, H.L. and Marbut, C.F., 1923. *The Vegetation and Soils of Africa.* Res. Ser., 13, American Geographical Society, National Research Council, New York, N.Y., 263 pp. and map.

Van Wyk, J.J.P., 1979. A general account of the grass cover of Africa. In: M. Numata (Editor), *Ecology of Grasslands and Bamboolands in the World.* VEB Gustav Fischer Verlag, Jena, 299 pp.

Wagner, H., 1912. *Lehrbuch der Geographie.* 9th ed., Hannover.

Chapter 7

GRAZING LANDS OF THE MEDITERRANEAN BASIN [1]

HENRI NOEL LE HOUÉROU

INTRODUCTION

The lands of the Mediterranean region are almost 9.5 million km^2 in extent. Of this, a substantial proportion has a Mediterranean type of climate (winter rain and summer drought), distributed geographically as follows:

The whole of:

Algeria	Israel	Lebanon	Saudi Arabia
Cyprus	Jordan	Libya	Syria
Egypt	Kuwait	Morocco	Tunisia

and:

90% of Iraq	50% of Portugal
80% of Iran	30% of Turkey
60% of Greece	20% of Albania
60% of Spain	15% of France
54% of Italy	10% of Yugoslavia

About 57% of the region is occupied by desert, 31% by various types of steppic vegetation in the arid zone, and 12% by degraded forests and shrublands under semiarid to humid climate (Le Houérou, 1981a). Some 80% of the arid part of the region, where annual precipitation ranges from 100 to 400 mm, is used as rangeland. The proportion of natural grazing land in semiarid to humid regions is about 52%. The remaining area is devoted to production of annual crops and tree crops. This chapter is concerned principally with the arid zone, which is limited in distribution north of the Mediterranean Sea to parts of Spain (100 000 km^2) and Greece (10 000 km^2), but is extensive in all countries of North Africa and Asia Minor. The desert zone and the shrublands and forests of the moister zones have been discussed elsewhere in this series (Di Castri et al., 1981; Le Houérou, 1981a, 1986; Ovington, 1983; Evenari et al., 1986).

No climax grassland occurs in the Mediterranean region except, perhaps, part of the *Stipa tenacissima* (alfa grass) steppes of North Africa and Spain. The dwarf-shrub steppes apparently have resulted from degradation of woodland (rather than of grassland) (Figs. 7.1 and 7.2). However, when these are protected from grazing, they evolve towards grassland dominated by perennial grasses (*Stipa* spp., *Stipagrostis* spp., etc.), forming a sort of neo-climax (Figs. 7.3 and 7.4). Grasslands now comprise only 5 to 10% of the area. Overgrazing has resulted in replacement of communities previously dominated by perennial grasses by ones in which other growth forms, particularly chamaephytic undershrubs, are dominant. The only parts of the region now dominated by grasses are: (1) limited areas of meadows ("*prairies*") in edaphically wet places and some dry grasslands ("*pelouses*") by annual and perennial grasses palatable to livestock; and (2) significant areas in the arid zone of North Africa with relatively pure stands of almost unpalatable perennial grasses. *Stipa tenacissima*, which is used in the paper industry, dominates over about 40 000 km^2 (Figs. 7.5 to 7.7); *Lygeum spartum* (esparto grass), used in traditional handicrafts for making ropes and mats, dominates over about 30 000 km^2 (Fig. 7.8); and *Stipagrostis pungens* (spiny dune three-awn), grazed only by camels, dominates over about 20 000 km^2 (Fig. 7.9).

---

[1] Manuscript submitted in July, 1982; revised in 1986–1989.

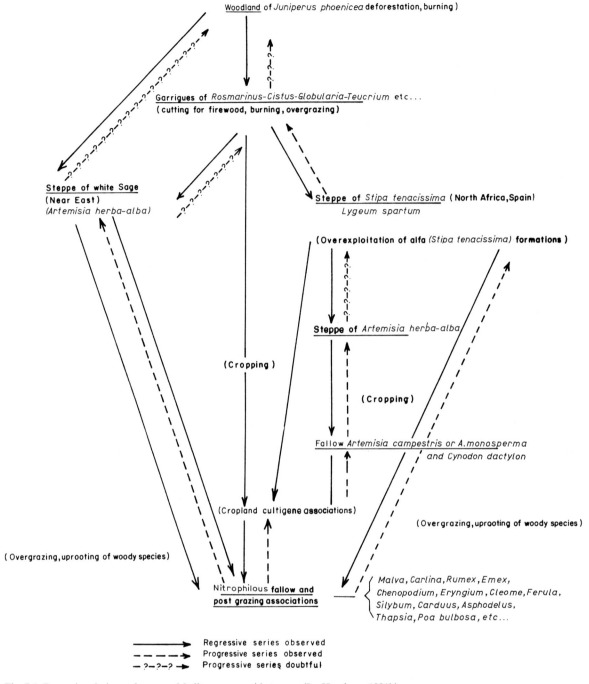

Fig. 7.1. Dynamic relations of western Mediterranean arid steppes (Le Houérou, 1981b).

Almost all of the native vegetation is grazed or browsed (Fig. 7.10) to a greater or lesser degree, especially in the southern and eastern parts of the region.

CLIMATE

The vegetation of the Mediterranean region exists under a wide range of combinations of

Fig. 7.2. Ecotone between open *Juniperus phoenicea* woodland and *Artemisia herba-alba* steppe in central Morocco. Soil shallow on calcrete; elevation 1600 m; annual precipitation 350 mm; mean daily minimum temperature of the coldest month −1°C.

moisture supply and temperature. On the basis of mean annual precipitation the region can be divided into seven climatic zones, as follows:

desert	less than 50 mm
hyperarid	50 to 100 mm
arid	100 to 400 mm
semiarid	400 to 600 mm
subhumid	600 to 800 mm
humid	800 to 1200 mm
hyperhumid	1200 to 2500 mm

Phytogeographic and agronomic justification for these threshold values were presented previously (Le Houérou, 1969b, 1977, 1981a, b; Le Houérou et al., 1975, 1979) and these zones have been subdivided into categories ranging from very warm to cold on the basis of mean daily minimum temperature (during the coldest month), which is negatively correlated with the length of frost-free season. Winter temperature is very important in determining the nature of the vegetation (particularly in respect to species selection) and the length of the winter rest season and of the growing season (Le Houérou, 1981a, pp. 483–485).

VEGETATION

The vegetation of the Mediterranean region ranges from desert to forest, with little grassland. In the desert zone vegetation is limited to temporary water courses and run-in depressions consisting of dwarf and medium-sized shrubs, and sometimes of scattered tall shrubs and small trees (Le Houérou, 1959). In the hyperarid zone, along the northern fringe of the Sahara and Near-Eastern deserts, steppic vegetation is limited to the drainage ways and sandy soils. In the arid zone, steppic vegetation is dominated by dwarf evergreen and deciduous shrubs (0.5 m or less in height) distributed in a diffuse pattern — regularly but sparsely (5 to 50% canopy cover) — over the whole landscape between pediments, hills and mountains. Woodlands and shrublands occupy hills and mountains in the moister part (300 to 400 mm) of this zone. These have a stratum of perennial and annual grasses and forbs (especially legumes) which is very sparse because of centuries of overgrazing by livestock. Shrublands, woodlands and forests also occupy the hilly and

Fig. 7.3. Chamaephytic steppe of *Salsola vermiculata* var. *villosa* and *Artemisia herba-alba* in south-eastern Algeria. Canopy biomass is about 100 g m^{-2}, and annual forage production is about 50 g m^{-2}. Soil is shallow silt on gypsum/calcrete crust; elevation 500 m; annual precipitation 250 mm; mean daily minimum temperature of the coldest month 1.5°C.

Fig. 7.4. The same ecosystem as shown in Fig. 7.3 after total protection for five years. Succession is towards a grass steppe of palatable perennial grasses (*Stipa barbata*, *S. lagascae* and *S. parviflora*).

Fig. 7.5. *Stipa tenacissima* steppe in pristine condition in the central steppic highlands of Algeria. Soil is shallow silt with lime-crust; elevation 1200 m; annual precipitation 280 mm; mean daily minimum temperature of the coldest month −0.5°C.

Fig. 7.6. *Stipa tenacissima* steppe in the early stages of depletion in the highlands of central Morocco. The invading dwarf shrubs are *Artemisia herba-alba* and *A. mesatlantica*. Annual precipitation 250 mm; mean daily minimum temperature of the coldest month −1°C.

montane terrains of the semiarid to humid climatic zones. "Garrigue" shrubland in poorer sites (often on limestone) is dominated by sclerophyllous shrubs 0.5 to 1.5 m in height with a canopy cover of 10 to 50%. Deeper (somewhat acid) soils in humid climate support maquis shrubland composed of shrubs about 2.5 m in height in dense stands (canopy cover 80% or more). Herbaceous

Fig. 7.7. Depleted steppe of *Stipa tenacissima* in the region of Kasserine-Ferina, Tunisia. Similar steppe in pristine condition is shown in Fig. 7.5 and in fair condition in Fig. 7.6. The vegetation in the background is secondary steppe of *Artemisia herba-alba* in a silty depression; elevation 800 m; annual precipitation about 300 mm; mean daily minimum temperature in January 2°C.

species are always present, but with the exception of limited areas of *Ampelodesmos mauritanicum* seldom dominant.

The vegetation of the arid zone is usually referred to as "steppic" — that is, dominated by sparse short perennial grasses or undershrubs (chamaephytes) with various proportions of barren ground in between the perennials during the dry season, and no trees. However, open woodlands, degraded forests or bushland (mixed shrubs and trees), and shrubland (tall shrubs without trees) cover sizeable areas in hills and mountains on skeletal soils (Fig. 7.2). Within the arid zone these are mainly between the 300 mm and 400 mm isohyets, but also occasionally in the 200 to 300 mm zone. Their occurrence depends also on the density of the human population. The hills and mountains near towns and villages are subject to heavy pressure from livestock and from fuel gathering, so their vegetation is often degraded to a steppic stage or sometimes still further to a mineral landscape devoid of all woody species and with a high proportion of bare rock. Relatively unspoiled communities have a complex structure including sparse trees, tall shrubs, low shrubs, forbs and grasses, and sometimes lichens and mosses. As degradation progresses, trees, then tall shrubs, and then low shrubs are successively eliminated until finally only forbs and grasses remain in the cracks in the rocks.

Steppic vegetation is of gramineous, chamaephytic, tragacanthic, succulent and crassulescent types:

(1) Perennial tussock grasses dominate in **gramineous steppe**.
(2) **Chamaephytic steppe**, which is dominated by undershrubs 20 to 50 cm high, has been shown to result from the depletion of gramineous steppe through mismanagement (overgrazing and fire) (Figs. 7.3 and 7.4) (Le Houérou, 1969a, b).

Fig. 7.8. *Lygeum spartum* steppe in southern Tunisia. Soil gypseous sand; elevation 20 m; annual precipitation 150 mm; mean daily minimum temperature of the coldest month 6°C.

(3) **Tragacanthic steppe** is made up of spiny cushion-like xerophytes at altitudes of 2500 to 3000 m. Grasses, graminoids and forbs grow among the xerophytes.
(4) **Succulent steppe** is dominated by glycophytic[1] cactoid species; it occupies limited areas along the Atlantic coasts.
(5) **Crassulescent steppe** is dominated by fleshy halophytes belonging mainly to the Chenopodiaceae.

Pseudo-steppic nanophanerophytic vegetation is dominated by shrubs of medium height on deep non-saline soils on terraces of wadis and in depressions where run-off water adds to the content of soil moisture. This type of vegetation also occurs on sand dunes.

---

[1] A glycophytic species is one which grows in soils with low salt content.

The vegetation in areas heavily trampled by livestock — around human settlements and watering places and along roads — is described as "nitrophilous *erme*" and is dominated by unpalatable species of the "piosphere" (Lange, 1969; Le Houérou, 1986).

Azonal vegetation in the arid zone is represented by moist grasslands ("*prairies*"), salt marshes, and high mountain pastures. Meadows occur adjacent to marshes and swamps in closed, wet depressions where impeded drainage assures a continuous supply of soil moisture. Soils are of medium to fine texture and often somewhat saline (electrical conductivity as high as 10 to 20 mS cm^{-1}). These meadows have a rather homogeneous floristic composition (Fig. 7.11). Although they are limited in extent, they are important in providing forage for livestock in summer. Dry grasslands differ from moist grasslands in that they dry up in summer. The dominant species of salt marshes are specialized halophytes, many

Fig. 7.9. Pristine arid savanna, under limited grazing but protected from fuel collection, in the same district as the arid grassland shown in Fig. 7.10. The tree is *Acacia raddiana*, and the grasses are *Cenchrus ciliaris*, *Digitaria commutata* subsp. *nodosa* and *Stipagrostis plumosa*.

Fig. 7.10. Pristine, but intensively grazed, arid grassland of *Cenchrus ciliaris*, *Digitaria commutata* subsp. *nodosa* and *Sporobolus ioclados* in southern Tunisia. Soil is sandy alluvium on boulders; elevation 200 m; annual precipitation 150 mm; mean daily minimum temperature of the coldest month 5°C.

Fig. 7.11. Halophytic steppe of *Atriplex halimus* in west-central Tunisia. Soil is saline (conductivity 20–30 mS cm^{-1}) alluvial loam, with deep water table; elevation 650 m; annual precipitation 250 mm; mean daily minimum temperature of the coldest month 2°C.

of them being members of the Chenopodiaceae. High mountain pastures are characterized by a cover of spiny, cushion-like xerophytes that are adapted to high altitudes.

Heavy grazing results in a continuing decrease in number, density and cover of the best forage species. They survive only in ecological niches where some protection from overgrazing exists by rugged land surfaces and by spiny shrubs. Invaders that take their place include thistles of the following genera:

Atractylis	*Centaurea*	*Gundelia*
Carduncellus	*Cnicus*	*Onopordum*
Carduus	*Cousinia*	*Scolymus*
Carlina	*Cynara*	*Silybum*
Carthamus	*Galactites*	*Thevenotia*

Other invading groups are: members of the Apiaceae, Euphorbiaceae, Lamiaceae and Liliaceae; and several spiny shrubs (e.g. *Calycotome villosa, Lycium intricatum, Rhus tripartitum, Sarcopoterium spinosum* and *Ziziphus lotus*).

Among the important perennial grasses are:

Agropyron elongatum	*Melica ciliata*
Arrhenatherum elatius	*Melica minuta*
Brachypodium phoenicoides	*Oryzopsis coerulescens*
Brachypodium ramosum	*Oryzopsis holciformis*
Bromus spp.	*Oryzopsis miliacea*
Cenchrus ciliaris	*Oryzopsis paradoxa*
Dactylis glomerata	*Oryzopsis thomasii*
Eragrostis spp.	*Phalaris arundinacea*
Festuca elatior	*Phalaris bulbosa*
subsp. *arundinacea*	*Phalaris coerulescens*
Koeleria spp.	*Phalaris truncata*
Lolium spp.	*Triseteria flavescens*

The commonest annual grasses belong to the following genera:

Aegilops	*Eragrostis*	*Poa*
Avena	*Hordeum*	*Setaria*
Bromus	*Koeleria*	*Trachynia*
Cutandia	*Phalaris*	*Vulpia*
Cynosurus		

Common legumes of the arid zone include about 20 annual species of *Medicago*, of which some are cultivated, and a similar number of annual species of *Trifolium*, some of which provide excellent forage whether cultivated or not. Other common legumes belong to the following genera:

Colutea	Hedysarum	Onobrychis
Coronilla	Lotus	Trigonella
Cytisus	Melilotus	Vicia

Details of the floristic characteristics of the major vegetation types in the hyperarid to semiarid climatic zones are provided in the species distributions in Table 7.1.

PRIMARY PRODUCTION

On the basis of official statistics for animal numbers and estimated stocking rates and assuming an annual intake of 600 kg of dry matter per sheep-equivalent (S.E.)[1] (3.5% of live-weight daily), annual forage yield of dry matter in the arid steppic zone is estimated to be 240 kg DM ha^{-1}. Studies over extensive areas in the arid zone (100 to 400 mm of rainfall) of North Africa have suggested that average values for forage yield range from 219 to 335 kg ha^{-1} (22–34 g m^{-2}), with great variation between range types (Table 7.2, Fig. 7.12).

Other studies in the arid zone have related yield of dry matter to precipitation. For example, in a 15-year study over 120000 km^2 of various vegetation types in the arid zone of Tunisia, Le Houérou (1959, 1962, 1969b) found the average annual yield of forage (including edible portions of woody species) to be 2.8 kg ha^{-1} for each millimetre of annual rainfall (this ratio is termed "Rain Use Efficiency", or R.U.E.), with extreme values of 1 and 10. Le Houérou and his colleagues (Le Houérou and Hoste, 1977; Le Houérou 1982, 1984; Le Houérou et al., 1988) synthesized the data published in some 20 large-scale studies in 12 Mediterranean countries and found the relationship between forage yield and rainfall (R.U.E.) to average 3.0 kg ha^{-1} mm^{-1} over the whole region (both steppe and shrubland).

---

[1] Sheep-equivalent (S.E.) refers to a mature sheep (40 kg) or 0.2 cattle or 0.1 camel or 1.2 goat or 0.15 horse or 0.3 donkey, on the basis of ratios of metabolic weight.

About 25% of the above-ground standing-crop biomass is consumable by herbivores in Mediterranean steppes. In these dry steppes an average of about 80% of forage yield is from dwarf shrubs and 15% from annual grasses and forbs, but the proportions vary greatly from year to year. Loiseau and Sebillotte (1972) reported that in *Artemisia herba-alba* (white sage) steppes of eastern Morocco there is a constant ratio of the annual production of consumable biomass on perennial shrubs (*A. herba-alba*) (y, in kg DM ha^{-1}) to their cover (x, in %):

$$y = 31x$$

Le Houérou (1987c) showed that, for a given community in North Africa, this relationship is fairly constant from year to year and from place to place, the overall average value of the coefficient being 40 ± 5. In other words, on average, each 1% of perennial canopy cover corresponds with 40 ± 5 kg ha^{-1} of perennial standing crop.

Variability in yield of forage is related to variability in precipitation, but 50 to 100% greater (Le Houérou, 1982, 1985a; Le Houérou et al., 1988). The magnitude of the ratio of maximum to minimum annual yield is reported to be 1.85 times that of maximum to minimum rainfall in southern Tunisia (Floret and Pontanier, 1982), compared to a value of 1.88 times in dry rangeland in the U.S.A. (Cook and Sims, 1975) and to 1.57 in the Sahel (Bille, 1977; Grouzis, 1988). Since the coefficient of variation of annual rainfall ranges from 15–20% in the humid zone to 60–80% at the fringe of the desert, the coefficient of variability of forage yield is expected to be of the order of 25 to 40% in the humid region and 90 to 140% in the dry steppes. Series of experimental data for this region to support these estimates include those of Tadmor et al. (1972), Seligman and Gutman (1978), Benjamin et al. (1982), Floret and Pontanier (1982), Naveh (1982) and Aïdoud (1987). In addition, they have been well documented in other arid and semiarid zones (Le Houérou et al., 1984)

Forage yield is directly related to soil characteristics which facilitate infiltration and storage of water. The best and most regular production is in areas of deep silty soils with a sandy uppermost layer; yield is low in shallow soils with calcrete or a gypsum crust. Soils of fine texture

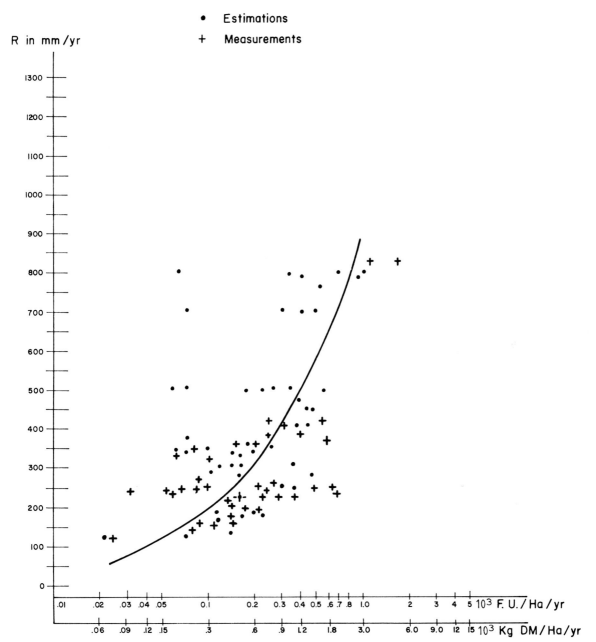

Fig. 7.12. Consumable production in respect to rainfall in the rangelands of North Africa. F.U. = Scandinavian Feed Unit (equivalent to the energy content of 1 kg of barley grain). After Le Houérou (1975).

(loams and clays) favour high production rates during exceptionally rainy years, but their production is negligible during drought. Growth is fair in sandy soils even when rains are somewhat deficient. Moisture enters sandy soils much more easily than fine-textured soils and is more readily available to plants. Relevant observations by Floret and Pontanier (1974) in Tunisia, in the *Rhanterium suaveolens–Stipa lagascae* association, sub-association of *Hammada schmittiana*, have already been reported in this series (Le Houérou, 1986, p. 123). In range in good condition with the sandy top soil intact over silty subsoil, 85% of rainfall entered the soil; in range in fair con-

TABLE 7.1

Distribution, according to climate and habitat, of some important and locally dominant plant species of arid to subhumid parts of the Mediterranean region[1], classified as to growth form

Species	Growth form[2]	Hyperarid zone	Arid steppe	Semiarid to subhumid woodlands, shrublands and *pelouses*	Azonal vegetation		
					meadows	salt marshes	high mountain pastures
PERENNIAL GRASSES AND GRAMINOIDS							
Aeluropus litoralis	Hc/G					×	
Agropyron elongatum	Hc		×		×		
Agropyropsis lolium	Hc				×		
Agrostis stolonifera	G				×		
Alopecurus bulbosus	G				×		
Alopecurus pratensis	Hc				×		
Anthoxanthum odoratum	Hc						×
Avena bromoides	Hc/G			×			
Avena montana	Hc						×
Brachypodium phoenicoides	Hc			×			
Brachypodium ramosum (= *B. retusum*)	Hc			×			
Bromus erectus	Hc						×
Bromus squarrosus	Hc						×
Bromus tomentellus	Hc			×			
Carex fusca	T						×
Cenchrus ciliaris	Hc		×				
Chrysopogon gryllus	Hc			×			
Cynodon dactylon	G		×	×	×		
Cyperus conglomeratus	Hc	×					
Dactylis hispanica	Hc		×	×			
Dactylis glomerata	Hc						×
Digitaria commutata ssp. *nodosa*	Hc		×				
Eleocharis palustris	G				×		
Festuca elatior ssp. *arundinacea*	Hc				×		
Festuca mairei	Hc						×
Festuca ovina	Hc						×
Festuca rubra	Hc						×
Holcus lanatus	Hc			×			×
Hordeum bulbosum	Hc/G			×			
Hyparrhenia hirta	Hc		×	×			
Koeleria splendens	Hc						×
Koeleria vallesiaca	Hc						×
Lasiurus hirsutus (= *L. scindicus*)	Hc	×					
Lolium perenne	Hc			×			
Lygeum spartum	Hc/G		×	×		×	
Melica ciliata	Hc			×			
Melica cupani	Hc			×			
Melica minuta	Hc			×			
Nardus stricta	Hc						×
Oryzopsis coerulescens	Hc			×			
Oryzopsis holciformis	Hc			×			
Oryzopsis miliacea	Hc		×	×			
Oryzopsis paradoxa	T			×			
Panicum turgidum	Hc	×					
Pennisetum asperifolium	Hc		×				
Pennisetum dichotomum (= *P. divisum*)	Hc	×	×				

(*continued*)

TABLE 7.1 (*continued*)

Species	Growth form [2]	Hyperarid zone	Arid to steppe	Semiarid subhumid woodlands, shrublands and *pelouses*	Azonal vegetation		
					meadows	salt marshes	high mountain pastures
PERENNIAL GRASSES AND GRAMINOIDS (*continued*)							
Phalaris arundinacea	Hc				×		
Phalaris coerulescens	Hc			×	×		
Phalaris truncata	Hc		×	×			
Phleum phleoides	Hc				×		
Phleum pratense	Hc				×		
Phragmites communis	G				×		
Puccinellia distans	Hc				×	×	
Schoenus nigricanus	Hc				×	×	
Scirpus holoschoenus	G				×		
Secale montanum	Hc			×			
Sporobolus ioclados	Hc		×	×			
Stipa barbata	Hc		×	×			
Stipa lagascae	Hc	×	×				
Stipa parviflora	Hc		×				
Stipa tenacissima	Hc		×				
Stipagrostis ciliata	Hc		×				
Stipagrostis obtusa	Hc	×	×				
Stipagrostis pennata	Hc	×	×				
Stipagrostis plumosa	Hc	×	×				
Stipagrostis pungens	Hc/G	×	×				
Triseteria flavescens	Hc			×			
ANNUAL GRASSES							
Aristida adscensionis	T		×	×			
Avena barbata	T		×	×			
Bromus rubens	T		×	×			
Cynosurus coloratus	T			×			
Elymus delilaenus	T		×	×	×		
Hordeum maritimum	T			×	×	×	
Hordeum murinum	T		×	×	×		
Koeleria phleoides	T			×			
Koeleria pubescens	T		×	×			
Lepturus incurvus	T					×	
Lolium multiflorum	T				×		
Lolium rigidum	T	×	×	×			
Lolium temulentum	T				×		
Phalaris canariensis	T		×	×			
Phalaris minor	T		×	×			
Polypogon maritimus	T					×	
Sphenopus divaricatus	T					×	
Stipa capensis	T		×	×			
Vulpiella stipoides	T		×	×			
PERENNIAL LEGUMES							
Argyrolobium linnaeanum	Ch			×			
Coronilla glauca	Ch			×			
Coronilla minima	Ch			×			
Hedysarum coronarium	Hc			×			
Lotus corniculatus	Hc			×	×	×	
Lotus creticus ssp. *creticus*	Hc		×	×			

(*continued*)

TABLE 7.1 (*continued*)

Species	Growth form[2]	Hyperarid zone	Arid steppe	Semiarid to subhumid woodlands, shrublands and *pelouses*	Azonal vegetation		
					meadows	salt marshes	high mountain pastures
PERENNIAL LEGUMES (*continued*)							
Lotus creticus ssp. *collinus*	Hc		×	×			
Lotus uliginosus	Hc				×		
Medicago falcata	Hc/G			×			
Medicago gaetula	Hc/G		×	×	×		
Medicago sativa	Hc			×			
Onobrychis argentea	Hc		×	×			
Onobrychis gaubae	Hc			×			
Onobrychis viciaefolia	Hc			×			
Tetragonolobus siliquosus	Hc/G				×		
Trifolium fragiferum	Hc/G				×		
ANNUAL LEGUMES							
Hedysarum capitatum	T			×			
Hedysarum carnosum	T		×			×	
Hedysarum flexuosum	T			×			
Medicago ciliaris	T				×		
Medicago hispida	T		×	×			
Medicago intertexta	T				×		
Medicago littoralis	T		×	×			
Medicago orbicularis	T			×			
Medicago rugosa	T						
Medicago scutellata	T			×			
Medicago truncatula	T		×	×			
Melilotus alba	T			×	×		
Melilotus italica	T			×	×		
Trifolium pratense	T				×		
Trifolium repens	T			×	×		
Trifolium resupinatum	T				×	×	
Trifolium squarrosum	T			×			
Trifolium stellatum	T			×			
Trifolium subterraneum							
ssp. *subterraneum*	T			×	×		
ssp. *brachycalycinum*	T			×			
ssp. *yanninicum*	T				×	×	
Trifolium tomentosum	T			×			
Trigonella arabica	T			×			
FLESHY FORBS							
Plantago crassifolia	Hc					×	
Halopeplis amplexicaulis	T					×	
Spergularia marginata	Hc					×	
Spergularia salina	Hc					×	
WEEDY INVADERS (FORBS)							
Carthamus lanatus	T		×	×			
Cleome arabica	T/Hc	×	×				
Diplotaxis harra	T		×				
Eruca vesicaria	T		×	×			
Euphorbia calyptrata	Ch/Hc	×					
Euphorbia guyoniana	Ch/Hc	×					

(*continued*)

TABLE 7.1 (*continued*)

Species	Growth form [2]	Hyperarid zone	Arid steppe	Semiarid to subhumid woodlands, shrublands and *pelouses*	Azonal vegetation		
					meadows	salt marshes	high mountain pastures
WEEDY INVADERS (FORBS) (*continued*)							
Euphorbia retusa	Ch/Hc	×					
Peganum harmala	Hc		×				
Pergularia tomentosa	Ch/Hc	×	×				
Thapsia garganica	G		×	×			
Urginea maritima	G		×	×	×		
OTHER FORBS							
Aphodelus microcarpus	G		×	×			
Armeria plantaginea	Hc						×
Asphodeline lutea	G						×
Bonjeania recta	Ch				×		
Cardamine pratensis	T						×
Cerastium dichotomum	T						×
Plantago albicans	Hc	×	×	×			
Plantago coronopus	T					×	
Sanguisorba minor	Hc			×			
DWARF SHRUBS							
Agathophora alopecuroides	Ch	×	×				
Alyssum spinosum	Ch						×
Anabasis articulata	Ch	×	×				
Anabasis oropediorum	Ch		×	×			
Anabasis setifera	Ch	×	×				
Anarrhinum brevifolium	Ch		×				
Anthyllis henoniana	Ch	×					
Argyrolobium uniflorum	Ch	×	×				
Artemisia campestris	Ch		×	×			
Artemisia herba-alba	Ch	×	×				
Artemisia judaica	Ch	×					
Artemisia monosperma	Ch	×	×				
Artemisia scoparia	Ch		×				
Astragalus armatus	Ch		×	×			
Atriplex glauca	Ch		×			×	
Atriplex leucoclada	Ch		×			×	
Bupleurum spinosum	Ch						×
Echiochilon fruticosum	Ch	×	×				
Erinacea anthyllis	Ch						×
Farsetia aegyptiaca	Ch	×	×				
Frankenia thymifolia	Ch					×	
Fredolia aretioides	Ch	×	×				
Genista microcephala	Ch			×			
Gymnocarpos decander	Ch	×	×				
Hammada schmittiana	Ch	×	×				
Hammada scoparia	Ch	×	×				
Helianthemum kahiricum	Ch	×	×				
Helianthemum lippii ssp. *sessiliflorum*	Ch	×	×				
Hippocrepis scabra	Ch		×	×			
Marrubium deserti	Ch		×				
Noaea mucronata	Ch		×				

(*continued*)

TABLE 7.1 (continued)

Species	Growth form [2]	Hyperarid zone	Arid steppe	Semiarid to subhumid woodlands, shrublands and *pelouses*	Azonal vegetation		
					meadows	salt marshes	high mountain pastures
DWARF SHRUBS (*continued*)							
Polygonum equisetiforme	Ch		×				
Pseudocytisus mairei	Ch						×
Rhanterium epapposum	Ch	×	×				
Rhanterium suaveolens	Ch	×	×				
Sarcopoterium spinosum	Ch		×	×			
Thymelaea hirsuta	Ch	×	×	×			
Zilla spinosa	Ch	×					
FLESHY DWARF SHRUBS							
Arthrocnemum indicum	Ch	×				×	
Atriplex malvana	Ch		×			×	
Cornulaca monacantha	Ch	×					
Crithmum maritimum	Ch					×	
Halimione portulaccoides	Ch					×	
Halocnemum strobilaceum	Ch	×				×	
Hedysarum carnosum	T/Hc		×				
Inula crithmoides	Ch					×	
Limoniastrum guyonianum	Ch	×				×	
Moltkia ciliata	Ch	×					
Nucularia perrini	Ch	×					
Oudneya africana	Ch	×	×				
Salicornia arabica	Ch					×	
Salicornia fruticosa	Ch					×	
Salicornia herbacea	Ch					×	
Salsola baryosma	Ch	×				×	
Salsola tetrandra	Ch	×				×	
Salsola vermiculata	Ch	×	×			×	
Suaeda fruticosa	Ch					×	
Traganum nudatum	Ch	×	×				
Zygophyllum album	Ch	×	×			×	
Zygophyllum dumosum	Ch	×	×				
SHRUBS OF MEDIUM HEIGHT							
Amygdalus scoparia	Np			×			
Atriplex halimus	Ch/Np		×			×	
Calligonum comosum	Ch/Np	×					
Helianthemum confertum	Np	×					
Lycium intricatum	Np		×	×			
Medicago arborea	Np			×			
Nitraria retusa	Np	×	×			×	
Rhamnus lycioides	Np		×	×			
Rhus pentaphyllum	Np			×			
Rhus tripartitum	Np			×			
Rosmarinus officinalis	Ch			×			
Ziziphus lotus	Np		×	×			
TALL SHRUBS							
Arundo donax	G				×		
Fraxinus xanthoxyloides	Np						×
Juniperus phoenicea	P/Np			×			

(*continued*)

TABLE 7.1 (continued)

Species	Growth form[2]	Hyperarid zone	Arid steppe	Semiarid to subhumid woodlands, shrublands and *pelouses*	Azonal vegetation		
					meadows	salt marshes	high mountain pastures
TALL SHRUBS (*continued*)							
Lygos raetam	Np	×	×				
Pistacia khinjuk	Np	×	×	×			
Pistacia lentiscus	Np			×			
TREES							
Acacia tortilis ssp. *raddiana*	P	×	×				
Argania sideroxylon	P		×	×			
Ceratonia siliqua	P			×			
Olea europaea	P/Np			×			
Pinus halepensis	P			×			
Pistacia atlantica	P	×	×	×			
Tetraclinis articulata	P/Np			×			

[1] These species are all more or less abundant, common or dominant to a greater or lesser extent over large to small areas. Rare and uncommon species are not included.
[2] Raunkaier (1934) growth forms: P = phanerophyte; Np = nano-phanerophyte; Ch = chamaephyte; Hc = hemi-cryptophyte; G = geophyte; T = therophyte.

TABLE 7.2

Summary of results of four surveys of forage resources over extensive areas in the arid zone of North Africa

Location	Chellala region of Algeria[1]	Central and southern Tunisia[2]	Eastern Morocco[3]	Hodna Basin of Algeria[4]
Area surveyed (km^2)	8,000	128,000	49,000	25,000
Overall consumable biomass (kg DM ha^{-1} yr^{-1})	229	295	219	335
Consumable production of individual range types (mean) (kg DM ha^{-1} yr^{-1})	50–800	40–600	50–500	60–900
No. of range types inventoried and mapped	90	130	105	80

[1] Rodin et al. (1970).
[2] Le Houérou (1969a, b).
[3] Loiseau and Sebillotte (1972).
[4] Le Houérou et al. (1975).

dition in which part of the sandy layer had been removed (by wind erosion), 50% of the rainfall infiltrated; and in range in poor condition, where the sandy surface layer had been totally removed, percolation was only 32% of rainfall. The respective annual forage yields were 1069, 614 and 416 kg ha^{-1}, whereas respective values for consumable production were 820, 493 and 293 kg ha^{-1}. Yields of forage were (R.U.E.) 2.61, 1.57 and 0.78 kg ha^{-1} per millimetre of rainfall, respectively (Le Houérou, 1984; Le Houérou et al., 1988). These measurements were made during a year in which rainfall was about twice the long-term mean. The differences among the sites would have been much greater in a year of normal rainfall, and greater still in a dry year.

Forage production takes place within the rainy season, which may extend from October to April. However, this growth period is sometimes shortened by low temperature at elevations above 600 m and at distances greater than 100 km from the sea. The occasional summer storms usually have no significant effect on growth, due to dormancy induced by long photoperiods or high temperatures or a combination of these factors, and also to high potential evapo-transpiration. Many species do not grow in summer, even when irrigated (local populations of *Festuca arundinacea* for instance) (Le Houérou, 1965, 1969a, 1985b). These include most legumes and grasses belonging to the tribes Agrostideae, Aveneae, Festuceae and Poeae. Among those species that do not rest in summer when moisture is sufficient for growth are *Cenchrus ciliaris*, *Cynodon dactylon*, *Digitaria commutata* subsp. *nodosa*, *Medicago sativa*, *Oryzopsis miliacea* and *Sporobolus ioclados*. Usually, species of grasses in tribes with a tropical affinity and possessing the C_4 carboxylation pathway (Andropogoneae, Chlorideae, Eragrostideae, Paniceae, Sporoboleae) exhibit summer growth in moist situations. Germination is strongly affected by photoperiod in most species. Since rest in winter is related to temperature, it is affected by latitude, altitude and continentality.

LAND USE AND MANAGEMENT

Management practices closely integrate natural grazing land with cropland. Almost all of the land is grazed, except that supporting tree crops and irrigated areas. Stubble on cropland is grazed, mainly by sheep, from July to September, whereas fallow is used in winter and spring by sheep and cattle. In the modern agricultural sector of the semiarid zone, the grazing value of fallow is sometimes improved by seeding annual medics (*Medicago hispida*, *M. littoralis*, *M. scutellata* and *M. truncatula*) and by fertilization with phosphate fertilizers.

Recent increases in the human population have resulted in a sharp increase in the proportion of the land that is cultivated, whereas livestock populations have increased (at a similar rate to that of humans) on the reduced area of rangeland. For example, in North Africa, between 1900 A.D. and 1975 A.D. the human population increased from 13 to 82 million and is projected to increase to 154 million by 2000 A.D. The rate of increase is slightly faster in the arid zone than in the region as a whole.

The proportion of the area that is used as rangeland increases with aridity of climate. In North Africa, under an annual precipitation of 300 to 400 mm, the proportion is only 20 to 30%; this figure increases to about 50% under 200 to 300 mm of precipitation, and is over 80% in the 100- to 200-mm rainfall zone. This trend corresponds to increasing restriction of cropland to deep soils and depressions and to the use of run-off water, in addition to the increasing risk of crop failure as mean annual rainfall diminishes.

The proportion between cereals and fallow varies greatly from year to year according to the occurrence of autumn and early-winter rains. In the arid zone of North Africa, for example, 7% of cultivated land is in tree crops (mostly olives), with an average of 31% in annual crops, and 62% in fallow. Some land in the drier parts of the arid zone is cultivated periodically, with a span of several years between successive crops. Therefore, the boundaries between fallow and rangeland are not always distinct; some rangelands are old fallows once tilled and then abandoned and reclaimed by steppic range species, such as *Artemisia herba-alba*. The dynamic stages between cropland, fallow and steppe were described by Long (1954), Le Houérou (1969b, 1981b) and Le Houérou et al. (1975).

Continuous grazing is the prevailing management system in rangeland. There are no fences; since the lands are usually not privately controlled, they are permanently exploited to the maximum of their potential (and beyond) with no compensatory input. Shrubland is grazed all the year by goats, but grazing by sheep and cattle is limited to summer and winter. In the arid steppic zone, year-long grazing of sheep, goats and camels is practised, but camels alone make use of halophytic vegetation (in summer) and of sand-dune types (in winter). About 10% of the zone between the 50- and 100-mm isohyets is used occasionally during the rainy winters. The proportion of forage consumed in steppic vegetation probably averages 50% of phytomass yield, and about 25% of total above-ground perennial standing biomass is utilized.

Traditional practices in the management of herds survive for socio-cultural reasons, but nomadism is decreasing very rapidly in the southern and eastern parts of the Mediterranean region.

Traditional nomadic mentality dictates that the animal must look after its own nutrient needs, since it is considered to be below man's dignity to feed animals; the herder's part is limited to protecting livestock against predators and leading them to appropriate sources of water and pasture. But this way of thinking has changed drastically during the last decade; at present, more than half of the livestock diet is grain and concentrate (Le Houérou and Aly, 1982; Le Houérou, 1985c). Overuse of rangeland is aggravated because possession of a large number of animals has long been regarded as a status symbol, even among individuals who are not, or are no longer, pastoralists. Non-nomads and city dwellers contribute to the overgrazing problem by forming associations to invest in herds of livestock (particularly sheep and goats) (Le Houérou, 1981a).

In the North African highlands, flocks graze in the fringe of the desert in winter, and then they are moved gradually northwards along assigned strips through the highlands to the stubble and fallows of the semiarid zone, which they reach by mid-summer. In the lowlands the desert fringe (50–100 mm) is also grazed in winter when the flocks need little or no water, and various types of range along the migration route are used during spring and autumn.

The rangelands in the arid zone are in an alarming state of degradation due to heavy, continuous overgrazing and the harvesting for firewood of woody species, many of which are important sources of forage during the dry season. After considering that some cattle, horses and sheep are supported on farmland (including cultivated land in fallow and stubble), some of which is irrigated, the overall actual stocking density of rangeland in the arid steppes is about 0.5 S.E.[1] ha^{-1} and often as much as 2 S.E. ha^{-1} with the use of concentrated feed supplements. But the overall stocking density in the arid zone is double the level warranted by the estimated carrying capacity of 0.25 S.E. ha^{-1} without supplemental feed concentrates. The carrying capacity varies with range type from 1 S.E. ha^{-1} in good ranges down to 0.1 S.E. ha^{-1} in poor, depleted ranges. For instance, in the northern coastal strip of Egypt (100–200 mm of rainfall), the carrying capacity was estimated to be 0.14 S.E. ha^{-1} (FAO, 1970). Ranges with a shallow water table or receiving supplemental water from run-off, especially swards of *Cynodon dactylon* and saline shrubland of *Atriplex halimus* (Fig. 7.11), have stocking densities as high as 3 S.E. ha^{-1} (Franclet and Le Houérou, 1971).

Fire plays an essential role in traditional management of Mediterranean shrublands (see Le Houérou, 1981a, pp. 486–498), but not in arid steppes, where biomass is too low to carry a fire over substantial areas. Shrubland is burned when it becomes unsuited for grazing because of the paucity of accessible forage and the development of a dense woody canopy. The direct result of fire is a strong regrowth of new sprouts of shrubs, some of which are relished by goats, and the temporary renewal of the herb layer, which is favoured by mineralization of nutrients from burnt organic matter. In the 1960s about 2000 km^2 of forest and shrubland in the Mediterranean region were affected by fire each year (Le Houérou, 1973). The area affected annually in this way increased to 6500 km^2 in the 1980s (Le Houérou, 1987b). Many of these are wildfires, but others are purposely set by shepherds, particularly in North Africa, Corsica, Crete, Sardinia and Sicily (Le Houérou, 1977; Joffre, 1982, 1987). The extremely degraded condition of the Mediterranean vegetation results from the combined effects of overstocking and repeated burning.

Periodicity and intensity of grazing have a marked effect on forage yield. Yield differences between contiguous areas of the same community with different grazing impact often vary by a factor of 3 to 5, and sometimes as much as 10 (Rodin et al., 1970; Le Houérou, 1971b). For example, Delhaye et al. (1974) have given the following production values for two plant associations in the Hodna Basin in Algeria (in kg dry matter ha^{-1} yr^{-1}).

Artemisia herba-alba–Noaea mucronata association in fairly good condition:

Continuous grazing, 0.67 S.E. ha^{-1}	1562
Rotational grazing, 0.5 S.E. ha^{-1}	2125
Rotational grazing, 0.25 S.E. ha^{-1}	2350
Rotational grazing, 0.17 S.E. ha^{-1}	2575
Protected from grazing for 2 years	2845

---

[1] S.E. = sheep-equivalent; see footnote p. 180.

Anabasis oropediorum–Salsola vermiculata var. *villosa* in poor condition:

Continuous grazing, 0.33 S.E. ha^{-1}	432
Rotational grazing, 0.25 S.E. ha^{-1}	493
Rotational grazing, 0.17 S.E. ha^{-1}	638
Rotational grazing, 0.13 S.E. ha^{-1}	1064
Protected from grazing for 6 years	2266

The proportion of the various species of livestock in the Mediterranean region, in terms of sheep-equivalents, is estimated to be:

cattle	46%	buffalo	3.9%
sheep	31%	camels	1.6%
goats	9.1%	mules	1.5%
donkeys	4.2%	horses	2.4%

Moisture is supplemented in some dryland crops by run-off or flooding by diversion of wadis (temporary streams). This is especially so for cereals in southern Morocco and in Algeria, and for tree crops in Tunisia and Libya. These run-off farming systems are similar to those in the Negev desert of Israel (Evenari et al., 1971–82) and in eastern Iran. They are very ancient, those in North Africa probably dating from before Roman occupation (i.e. before 200 B.C.).

Game, especially *Gazella dorcas* (dorcas gazelle), was present in significant numbers in the arid zone of North Africa until a few decades ago, but has now almost disappeared and game is of no economic importance. Rodents and insects, especially locusts and grasshoppers, levy an important toll on range production under certain circumstances. Consumption by rodents (*Psammomys obesus*) and locusts (*Locusta migratoria*) equivalent to that of several sheep per hectare has been recorded over limited areas in certain years (Franclet and Le Houérou, 1971).

IMPACT OF MAN AND LIVESTOCK

Mediterranean vegetation has been strongly influenced by man since cultivation of cereals and pulses and domestication of sheep and goats commenced some 8000 to 10 000 years ago (Zeuner, 1963; Reed, 1969; Epstein, 1971; Zohary, 1973; Zohary and Spegel-Roy, 1975). The intensity and continuity of human activity during such a long period is comparable only to that of parts of eastern and southern Asia, when arid and semiarid lands are concerned. The explanation of this concentration of human activity in this region probably relates to its geographical position at the junction of three continents and to the existence of a mild climate attractive to man. Invasions, both aggressive and peaceful, have taken place from the time of the barbarians and continue now with tourists. The Mediterranean shores attracted nearly 100 million visitors in 1980 and the annual rate of increase is 5.6%. At this rate, 200 million tourists can be expected in 1993. The number visiting Spain alone was 40 million in 1987. The history of human influence in the region has been covered elsewhere in this series (Le Houérou, 1981a, pp. 479–482).

In North Africa, range exploitation before about 200 B.C. seems to have been limited only by outbreaks of epizootic diseases and by predators [*Acinonyx jubatus* (cheetah), *Hyaena hyaena* (hyaena), *Panthera leo* (lion) and *P. pardus* (leopard) were common until the last century]. Agricultural development took place from about 200 B.C. to 650 A.D., during Phoenician and Roman times. Large tracts of arid lands were cleared for cereal production and for tree crops (olives, figs and grapes). Vestiges of this widespread agriculture remain down to the very edge of the desert in the form of terraced fields and hydraulic structures, such as canals, dams and aqueducts. After the Arab conquest of 650 A.D. most of the cultivated land returned to desert and range, and the rangelands, particularly in the arid zone, were exploited by nomadic herders. The management system was transhumance and nomadism, subject to agreements between tribes or ethnic groups. These groups were in a constant state of strife over ownership of pasture, water resources, grain fields and oases. At the same time endemic and epizootic diseases and periodic droughts kept livestock and population numbers in balance with land resources. There was also some sort of range management policy, at least within some tribes and ethnic groups. Deferred grazing was practised by reserving some pastures for emergency periods, thus permitting better regeneration. Decisions were made each year by the elders' council. This system ("ahmia") developed in the Arabian peninsula as early as the 6th century (Draz, 1969, 1978) and also in North Africa ("g'dal") (Le

Houérou, 1962). Another type of management involved several livestock owners entrusting their animals to a manager who organized migrations and hired pastures for 10 to 20 herds (1000–2000 sheep). This older system, which was in dynamic equilibrium with the environment, broke down during the present century with the introduction of peace and order, control of livestock diseases, and better health services and living conditions for humans. These changes resulted in an explosion of human and livestock populations, which grew at a similar pace of 2.5 to 3.0% annually, and a sharp increase in the area cultivated. The result is general continuous overgrazing and extensive cultivation which causes several hundreds of square kilometres to be added to the desert every year (Le Houérou, 1959, 1968). Migration routes are shortening and sedentarization is increasing, a situation favoured by governments. Also some grazing schemes have been developed based on rotational grazing, control of animal numbers, use of complementary fodder reserves (planted species of *Acacia* and *Atriplex*, and *Opuntia ficus-indica*) and concentrates (bran, barley), and improved herd management (nutrition, breeding and health). In the arid zone 70 to 80% of the people made most of their living directly or indirectly from livestock until recently. This proportion steadily decreased over the last two decades to no more than 10 to 20%.

Concentration of human activity has resulted in severe exploitation and degradation of the vegetal and soil resources. Cultivation, deforestation, disorderly grazing, persistent overstocking, and wildfires have degraded most of the natural vegetation to poor shrublands vulnerable to erosive forces. The effect of grazing has been increased because of year-long exposure made possible by the absence of snow cover. Also, the lack of a need to provide stored fodder has favoured increases in populations of livestock. The presence of a high proportion of goats has caused additional impact on the land resources.

Because of its feeding behaviour and the anatomy and physiology of its digestive tract, the goat is able to ingest larger quantities of low-quality roughage than any other domestic ruminant. This ability is related to a larger abdominal capacity in relation to total body volume, normal values for which are 12.5% for cattle, 15% for sheep, and 33% for goats (McKenzie, 1970). Exceptional values reach 20% for cattle and 55% for goats. In addition, goats seem to be slightly better able to digest very coarse, fibrous material (Wilson, 1977). Consequently, voluntary daily intake of coarse roughage (in terms of dry matter) by goats browsing on woody range routinely reaches 6% of body weight; some authors have reported values of 8% and even 11% (McKenzie, 1970; El Aouini and Sarson, 1976; Carew et al., 1980). Blanchemain (1964) reported that a minimum of 30% of the total dry-matter intake in the form of coarse roughage is a requirement for keeping milk goats in good condition. However, it is a well-known fact that, where only ligneous forage from shrubs is available, only goats and camels can thrive. For example, during the severe droughts in West and East Africa during the early 1970s (when little, if any, herbage was available) the numbers of goats and camels were hardly affected, whereas losses of 30 to 70% were experienced for sheep and cattle. A similar experience during 1946 to 1948 in North Africa was described by Le Houérou (1962).

The feeding behaviour of the goat is also conditioned by its relative independence of water. Range goats are able to adapt to watering every second to fourth day, as their rate of turnover of water is very low. In adapting to drought, they come second to camels among domestic animals (McFarlane et al., 1972; Shkolnik et al., 1972, 1980; Yagil, 1985). Consequently, goats can range much farther (15 to 20 km) from water in hot, dry periods than can cattle (4 to 5 km) and sheep (6 to 8 km). Thus, goats and camels reach rangelands (hills, mountains and steppes) inaccessible to other livestock. Large tracts of Mediterranean rangelands are thus subject to overbrowsing which would otherwise be protected by the scarcity of water during the dry season (Le Houérou, 1981c).

The physical and mental agility of the goat is also a factor in its feeding behaviour. Its climbing ability, into trees and on steep slopes, gives it access to forage inaccessible to other classes of livestock. The goat is a skilful, shrewd and flexible animal in its quest for feed. It can adapt to almost any type of feedstuff (including paper) and gather it in almost any kind of situation.

Degradation of vegetation and soil has been caused by the combined effects of overbrowsing by goats, wildfires (some of which are ignited by shepherds), wood-cutting (lopping for browse, for making charcoal, distillation and for firewood), and clearing for temporary (shifting) cultivation. Browsing affects the vegetation directly through removal of leaves and stems, destruction of seedlings, and consumption of fruits. When browsing pressure is very high, the woody cover is progressively reduced and the proportion of bare soil increases. This results in loss of organic matter, increased run-off, erosion, reduced rate of water intake into the soil, and reduced plant vigour. Shepherds assist this degradation process by lopping inaccessible branches so that their livestock can use them.

Overbrowsing of palatable species results in an increase of species, and of biotypes within a species, which are avoided by goats because of lower palatability, spininess, or the presence of repellent chemicals (Liaccos and Moulopoulos, 1967; Papanastasis and Liaccos, 1980). "Goat resistant" vegetation types develop, such as "garrigues" of *Genista* spp., *Quercus coccifera* and *Ulex parvifolius*. When the canopy of these types of shrubland becomes so dense that it is impenetrable to livestock, shepherds burn the vegetation to permit the development of herbs and young shoots of shrubs to provide better browsing and grazing. Without the intervention of man in this way, the goat would probably be eliminated from such situations and the vegetation would be able to recover by succession towards the climax. Repeated burning often fails to achieve the desired increase in forage cover because it favours the development of stands of *Cistus* spp. (*C. albidus*, *C. monspeliensis* and *C. villosus*) and of *Pteridium aquilinum* and many other species of pyrophytes, which are very low in palatability or even poisonous.

Encroachment of shrubs into grassland is often the result of overgrazing by cattle and sheep (Le Houérou, 1981b, c). Failure to recognize this cause of increased shrub cover has resulted in stockmen increasing the proportion of browsers in their herds, thereby aggravating the problem. In this context, a high proportion of goats is not the cause of rangeland degradation, but the result of it. The land manager's role is to adjust livestock numbers and to design grazing systems compatible with conservation and improvement of the vegetation and soil resources. The time has passed when removal of the goat from Mediterranean rangeland is feasible. It now seems to be necessary in maintaining the balance of the natural ecosystems of which it has been an important part for some 10 000 years.

Soil erosion in the arid zone is severe due to overgrazing, uprooting of woody species, and tillage. Interpretation of data on the silt load of streams and of the rate of silting of reservoirs indicates that water erosion removes an average of 1 to 1.5 mm (15 to 23 t ha^{-1}) of top-soil yearly (Le Houérou, 1969b). Water from flooding wadis carries an average silt load of 50 kg m^{-3} and sometimes as much as 200 kg m^{-3}. Wind erosion at the rate of 1 mm per month has been reported (Le Houérou, 1962, 1985c; Floret and Pontanier, 1974; Khatteli, 1983.

Deposits by wind sometimes exceed 150 m^3 ha^{-1} (225 t ha^{-1}) annually (1.25 mm per month) around tilled fields in sandy steppes. This process results from low cover of perennial plants, primarily due to tillage and secondarily to overstocking. When the ground cover of perennials exceeds 20 to 25%, there is practically no wind erosion (Le Houérou, 1959, 1968, 1987c). Erosion in the desert is mostly, but not only, the result of wind.

IMPROVEMENT OF GRAZING LANDS

During the last four decades much research has been conducted into means of increasing the carrying capacity of Mediterranean grazing lands (Le Houérou, 1981a, pp. 507–509). The methods that have given results in shrublands in subhumid to semiarid climate are, in decreasing order of intensity of management (and consequently of return): (1) replacing the natural communities with stands of seeded forage crops; (2) conversion of marginal cropland into seeded grassland; and (3) regeneration of stands of native forage plants by mechanical means, by fire, and by reduction in stocking rate. The more intensive strategies, (1) and (2), are not applicable in the arid zone of North Africa and of the Near East because of climatic limitations (erratic rainfall), so reduced intensity of grazing offers the only practical means of range improvement.

Because of the high cost of destroying natural vegetation and replacing it with seeded grasslands, this approach can be justified only in level to gently sloping (less than 15%) areas with deep soil where annual precipitation exceeds 500 mm. Spectacular results have been obtained under these conditions in parts of Tunisia, Morocco, Spain, Portugal and southern France. The procedure involves clearing the maquis, ploughing, seeding, and applying chemical fertilizers. The resulting arable grassland is used for grazing or is cut for hay or silage. The species most commonly seeded are: *Festuca elatior* subsp. *arundinacea*, on poorly drained clay soils under semiarid to humid climate; *Hedysarum coronarium* and *Phalaris bulbosa* on marly soils; *Dactylis glomerata* and *Lolium multiflorum* on well-drained soils; and *Trifolium subterraneum* on sandy soils. Annual yield of fodder (in terms of dry matter) reaches 6 to 12 t ha^{-1}, which is 10 to 20 times that of the natural maquis (Thiault, 1955, 1958, 1963; Le Houérou, 1965, 1969a, 1977; Crespo, 1970; Jaritz and Wadsack, 1971; Maignan, 1971, 1978; Gachet and Jaritz, 1972; Etienne, 1977, 1978; Joffre, 1982; Lapeyronie, 1982; Muslera Pardo and Ratera Garcia, 1984; Joffre et al., 1987).

The potential for converting marginal cropland to seeded pasture is high. About half of the 400 000 km² that is allotted to cereal production in the Mediterranean region provides annual grain yields less than 1000 to 1200 kg ha^{-1}, which is a return inferior to the cost of production. These marginal croplands are mostly in the semiarid zone, in areas of shallow red soils with calcareous crusts, and in halomorphic and sandy areas. They also occur in the subhumid zone in areas of brown calcareous clay soils on slopes exceeding 10%. The principal species seeded are:

Fabaceae:
  Hedysarum coronarium
  Lotus corniculatus
  Lotus creticus
  Medicago gaetula
  Medicago sativa
  Medicago truncatula
  Onobrychis sativa
  Trifolium fragiferum
  Trifolium subterraneum

Poaceae:
  Dactylis glomerata
  Festuca elatior
     subsp. *arundinacea*
  Lolium perenne
  Oryzopsis miliacea
  Phalaris bulbosa
  Phalaris truncata
Rosaceae:
  Sanguisorba minor

Annual yield of consumable dry matter is in the order of 3 to 4 t ha^{-1} on shallow soils, but sometimes reaches 10 to 12 t ha^{-1} in moist depressions (Le Houérou, 1965, 1969a, 1971a, b; Maignan, 1971, 1978). Although yield generally is less than in arable grassland converted directly from shrubland, the cost of conversion is also much less.

Seeding and overseeding of forage species (Le Houérou, 1981a) are rarely successful in the arid zone. However, the results of small-scale experiments suggest that some benefit might result from use of such native undershrubs as *Artemisia herba-alba* and *Atriplex glauca*. Plantations of fodder shrubs have been established as a drought-evading strategy. For example, spineless cactus (*Opuntia ficus-indica* var. *inermis*) now occupies about 1000 km² in Tunisia, and an area of some 500 km² in Libya has been planted to species of *Acacia* and *Atriplex* (Le Houérou, 1972, 1987a). Approximately 500 000 ha have been planted to fodder shrubs in North Africa and a few thousand hectares in the Near East. The principal species used are (in decreasing order of importance) *Opuntia ficus-indica*, *Acacia cyanophylla*, *Atriplex nummularia* and *A. halimus*.

Destocking, deferred grazing and reduction of stocking rates are also effective means of regenerating these rangelands. But these approaches require considerable knowledge of the dynamics of the vegetation which is to be transformed. They cannot be applied to vegetation where reduced grazing intensity results in the development of dense stands of shrubs devoid of grazing value. However, the combination of high stocking rates for short periods and periods of rest at certain phenological stages, together with judicious choice of species of livestock, permits restoration of the pastoral value of some types of shrubland, as well as of grasslands and meadows (Thiault, 1955, 1958, 1963). Reduced stocking intensity and deferred grazing have been demonstrated to be effective ways of improvement of degraded rangeland in the arid steppic zone of North Africa and the Near East. Forage yields 3 to 5 times those of areas of uncontrolled grazing have been achieved by application of these methods for periods of 3 to 5 years. In instances of extreme depletion, improvement by as much as ten times were reported (Le Houérou, 1971b).

Cultural practices have not been used extensively in rangeland improvement. Trials with scarifying, subsoiling, pitting, contour furrowing, and contour terracing have not given consistent results, particularly in terms of economic return. In pastures dominated by *Cynodon dactylon* periodic discing every 2 to 4 years is very beneficial in rejuvenating the sward by cutting and multiplying the rhizomes.

REFERENCES

Aïdoud, A., 1987. Les écosystèmes à armoise blanche (*Artemisia herba-alba* Asso), II Phytomasse et productivité primaire. *Biocenose*, 4: 1–2.

Benjamin, R., Eyal, E., Noy-Neir, I. and Seligman, N.G., 1982. Intensive agro-pastoral system in Migda Experimental Farm, northern Negev. *Hassadeh*, 62: 2022–2026 (in Hebrew, with English summary).

Bille, J.C., 1977. Étude de la production primaire nette d'un écosystème sahélien. *Trav. Doc. ORSTOM*, No. 65, Paris, 82 pp.

Blanchemain, A., 1964. Conditions et possibilités d'amélioration de l'élevage de la chèvre. In: *Politiques d'élevage de la chèvre dans la région méditerranéenne et le Proche-Orient*. FAO, PEAT, Rapp. No. 1929, Rome, 25 pp.

Carew, B.A.R., Mosi, A.K., MBa, A.U. and Egbunike, G.N., 1980. The potential of browse plants in the nutrition of small ruminants in the humid forest and derived savanna zones of Nigeria. In: H.N. Le Houérou (Editor), *Browse in Africa*. Intern. Livestock Centre for Africa, Addis Ababa, pp. 307–312.

Cook, C.W. and Sims, P.L., 1975. Drought and its relationships to dynamics of primary productivity and production of grazing animals. In: *Proceedings, Seminar on Evaluation and Mapping of Tropical African Rangelands*. Int. Livestock Centre for Africa, Addis Ababa, pp. 163–170.

Crespo, D., 1970. Some agronomic aspects of selecting subterranean clover (*Trifolium subterraneum* L.) from Portuguese ecotypes. *Proc. XI Int. Grassland Congress*, pp. 207–210.

Delhaye, R.E., Le Houérou, H.N. and Sarson, M., 1974. Améliorations des pâturages et de l'élevage. In: *Études des ressources naturelles et expérimentation et démonstration agricoles dans la région du Hodna*. Rapp. tech. 2, AGS:DP/ALG/66/509, FAO, Rome, 115 pp.

Di Castri, F., Goodall, D.W. and Specht, R.L. (Editors), 1981. *Mediterranean-Type Shrublands*. Ecosystems of the World, 11, Elsevier, Amsterdam, 463 pp.

Draz, O., 1969. *The Hema System of Range Reserves in the Arabian Peninsula: its Possibilities in Range Improvement and Conservation Projects in the Near East*. Mimeo., PL:PFC/13, FAO, Rome, 11 pp.

Draz, O., 1978. Revival of the Hema System of range reserves as a basis for the Syrian range development programme. *Proc. 1st Int. Rangelands Congress*, Denver, Colo., pp. 100–103.

El Aouini, M. and Sarson, M., 1976. *Production primaire et valeur pastorale de certains types de maquis des Mogods non calcaires*. Note Rech. No. 12, Inst. Nat. Rech. Foret, Tunis, 10 pp.

Epstein, H., 1971. *The Origin of the Domestic Animals of Africa*. Two vols., Afr. Publ. Corp., New York, N.Y., 573 and 719 pp.

Etienne, M., 1977. *Bases phytoécologiques du développement des ressources pastorales en Corse*. Thèse, Université des Sciences et Technologie du Languedoc, Montpellier, 210 pp.

Etienne, M., 1978. Amélioration des conditions de parcours et des pâturages naturelles pour la production de proteines animales en zone montagneuse. *8th World Forestry Congress*, Jakarta, pp. 1–15.

Evenari, M., Shanan, L. and Tadmor, N.H., 1971–1982. *The Negev, the Challenge of the Desert*. Harvard University Press, Cambridge, Mass., 345 pp. (1971), 437 pp. (1982).

Evenari, M., Noy-Meir, I. and Goodall, D.W. (Editors), 1985–1986. *Hot Deserts and Arid Shrublands*. Ecosystems of the World, 12A, B, Elsevier, Amsterdam, 365 + 451 pp.

FAO, 1970. *Preinvestment Survey of the Northwestern Coastal Region of the United Arab Republic. Agriculture*. ESE:SP/UAR 49, Tech. Rep. 3, FAO, Rome, 335 pp.

Floret, C. and Pontanier, R., 1974. *Étude de trois formations végétales naturelles du Sud Tunisien: production, bilan hydrique des sols*. Mimeo., Inst. Natl. Rech. Agron., Tunis, 44 pp.

Floret, C. and Pontanier, R., 1982. *L'aridité en Tunisie Présaharienne: climat, sol, végétation et aménagement*. Doct. Sci. Thèse, Université des Sciences et Techniques du Languedoc, Montpellier, 580 pp.

Franclet, A. and Le Houérou, H.N., 1971. *Les Atriplex en Tunisie et en Afrique du Nord*. FAO/UNDP/SF/TUN 11, Rapp. No. 7. FAO, Rome, 249 pp.

Gachet, J.P. and Jaritz, G., 1972. Situation et perspectives de la production fourragère en culture sèche en Tunisie septentrionale. *Fourrages*, 49: 3–15.

Grouzis, M., 1988. *Structure, productivité et dynamique des systèmes écologiques sahéliens (Mare d'Oursi, Burkina Faso)*. Études et Thèses, ORSTOM, Paris, 336 pp.

Jaritz, G. and Wadsack, J., 1971. Mise au point d'un système de production fourragère dans la région des modods. *Séminar sur l'élevage et la production de la viande*. Mimeo., Minist. Agric., Tunis, 15 pp.

Joffre, R., 1982. *Réflexions sur le feu pastoral en Corse*. Multigr., Parc national de la Corse, A.I.D.A., Paris, 29 pp.

Joffre, R., 1987. *Contraints du milieu et réponses de la végétation herbacée dans les Dehesas de la Sierra Norte (Andalousie Espangne)*. Thèse Doct., Univ. Sci. et Techn. du Languedoc, Montpellier, 186 pp. et annexes.

Joffre, R., Vacher, J., De Los Llanos, C. and Long, G.A., 1987. The Dehesa: an agrosilvopastoral system of the Mediterranean region with special reference to the Sierra Morena area of Spain. *Agrofor. Syst.*, 6: 71–96.

Khatteli, M., 1983. *Contribution à l'érosion éolienne dans la Jeffara Tunisienne*. Multigr., Note Techn. 3, Instit. des Régions Arides, Medenine, Tunisia, 38 pp.

Lange, R.T., 1969. The "Piosphere": sheep tracks and dung patterns. *J. Range Manage.*, 22: 396–400.

Lapeyronie, A., 1982. *Les productions fourragères méditerranéennes*. Maisonneuve et Laroset, Paris, 425 pp.

Le Houérou, H.N., 1959. *Recherches écologiques et floristiques sur la végétation de la Tunisie méridionale*. Institut du Recherches Sahariennes, University of Algiers, Algiers, 510 pp.

Le Houérou, H.N., 1962. *Les pâturages naturels de la Tunisie aride et désertique*. Inst. Sci. Écol. Appl., Paris, 106 pp.

Le Houérou, H.N., 1965. *Les cultures fourragères en Tunisie*. Doc. Techn. 13, Inst. Natl. Rech. Agron. de Tunisie, Tunis, 81 pp.

Le Houérou, H.N., 1968. La désertisation du Sahara septentrional et des steppes limitrophes (Algérie, Tunisie, Libye). *Ann. Algér. Geogr.*, 6: 2–27.

Le Houérou, H.N., 1969a. *Principes, méthodes et techniques d'amélioration fourragère et pastorale*. FAO, Rome, 291 pp.

Le Houérou, H.N., 1969b. *La végétation de la Tunisie steppique (avec référence aux végétations analogues d'Algérie, Libye et du Maroc)*. Carte coul. H.T. 1/500,000, Ann. Inst. Natl. Rech. Agron., 42, 5, Tunis, 624 pp.

Le Houérou, H.N., 1971a. An assessment of the primary and secondary production of the arid grazing lands of North Africa. In: L.E. Rodin (Editor), *The Ecophysiological Foundations of Ecosystem Productivity in Arid Zones*. Nauka, Leningrad, pp. 168–173.

Le Houérou, H.N., 1971b. *Les bases écologiques de la production fourragère et pastorale en Algérie*. Div. Product. Plantes, FAO, Rome, 60 pp.

Le Houérou, H.N., 1972. The useful shrubs of the Mediterranean Basin and of the sahelian belt south of the Sahara. In: *Wildland Shrubs, Their Biology and Use*. Intermountain Forest and Exp. Sta., U.S. Forest Service, Ogden, Utah, pp. 26–36.

Le Houérou, H.N., 1973. Fire and vegetation in the Mediterranean basin. *Proceedings, Fire Ecology Conference*. Tall Timbers Research Station, Tallahassee, Fla., 13: 237–277.

Le Houérou, H.N., 1975. The rangelands of North Africa: typology, yields, productivity, and development. *Proceedings, Seminar on Evaluation and Mapping of Tropical African Rangelands*. Int. Livestock Centre for Africa, Addis Ababa, pp. 42–55.

Le Houérou, H.N., 1977. Plant sociology and ecology applied to grazing lands research survey and management in the Mediterranean Basin. In: W. Krause (Editor), *Handbook of Vegetation Science*, Vol. XIII. Junk, The Hague, pp. 231–274.

Le Houérou, H.N., 1981a. Impact of man and his animals on Mediterranean vegetation. In: F. Di Castri, D.W. Goodall and R.L. Specht (Editors), *Mediterranean-Type Shrublands*. Ecosystems of the World, 11, Elsevier, Amsterdam, pp. 479–521.

Le Houérou, H.N., 1981b. Long-term dynamics in arid lands vegetation and ecosystems of North Africa. In: D.W. Goodall and R.A. Perry (Editors), *Arid Land Ecosystems: Their Functioning and Management*. Vol. 2, International Biological Programme 17, Cambridge University Press, Cambridge, pp. 357–384.

Le Houérou, H.N., 1981c. The impact of the goat on the Mediterranean vegetation. *Proceedings, 32th Annual Meeting, European Association of Animal Production*, Zagreb, pp. 1–10.

Le Houérou, H.N., 1982. Prediction of range production from weather records in Africa. *Proceedings, Technical Conference on Climate in Africa*. WMO, Geneva, pp. 286–298.

Le Houérou, H.N., 1984. Rain-use efficiency: a unifying concept in arid land ecology. *J. Arid Environ.*, 7: 213–247.

Le Houérou, H.N., 1985a. The impact of climate on pastoralism. In: R.W. Gates, J.H. Ausubel and M. Berberian (Editors), *Climate Impact Assessments*. SCOPE No. 27, Wiley, New York, N.Y., pp. 155–185.

Le Houérou, H.N., 1985b. Forage and fuel plants in the arid zone of North Africa, the Near and Middle East. In: G.E. Wickens, J.R. Goodin and D.V. Field (Editors), *Plants for Arid Lands*. Proc. Kew Int. Conf. Economic Plants for Arid Lands, George Allen and Unwin, London, pp. 117–141.

Le Houérou, H.N., 1985c. *La Régénération des Steppes Algériennes*. Inst. Nat. Rech. Agric., Rel. Int., Paris, IDOVI, Algér., 45 pp.

Le Houérou, H.N., 1986. The desert and arid zones of northern Africa. In: M. Evenari, I. Noy-Meir and D.W. Goodall (Editors), *Hot Deserts and Arid Shrublands*. Ecosystems of the World, 12B, Elsevier, Amsterdam, pp. 101–147.

Le Houérou, H.N., 1987a. *Les plantations sylvo-pastorales dans la zone aride de Tunisie*. Note Technique du MAB, No. 18, UNESCO, Paris, 81 pp.

Le Houérou, H.N., 1987b. Vegetation wildfires in the Mediterranean basin: evolution and trends. Proceedings, Workshop on the Ecological Consequences of Forest Fires in the Mediterranean Basin. Europ. Sce Foundation and Forest Ecosyst. Res. Network, Communauté Économique Européenne, Brussels. *Oecologia Mediterranea*, 4: 13–24.

Le Houérou, H.N., 1987c. *Aspects météorologiques de la croissance et du développement végétal dans les déserts et les zones menacées de désertisation*. WMO, Geneva, 51 pp.

Le Houérou, H.N. and Aly, I.M., 1982. *Perspective and Evaluation Study on Agricultural Development in Libya: the Rangeland Sector*. UNTF/LIBO18, FAO, Tripoli–Rome, 77 pp.

Le Houérou, H.N. and Hoste, C., 1977. Rangeland production and annual rainfall relations in the Mediterranean basin and in the African Sahelian and Sudanian zones. *J. Range Manage.*, 30: 181–189.

Le Houérou, H.N., Claudin, J., Haywood, M. and Donnadieu, J., 1975. *Étude phyto-écologique du Hodna*. AGS:DP/ALG/66/509, FAO, Rome, 154 pp.

Le Houérou, H.N., Claudin, J. and Pouget, M., 1979. Étude bioclimatique des steppes algériennes. *Soc. Hist. Nat. Afr. Nord*, 68: 33–74.

Le Houérou, H.N., Bingham, R.E. and Skerbek, W., 1984. Towards a probabilistic approach to rangeland development planning. *Proc. 2nd Int. Rangeland Congress*, Adelaide, Australia, 30 pp.

Le Houérou, H.N., Bingham, R.E. and Skerbek, W., 1988. Relationship between the variability of primary production and the variability of annual precipitation in world arid lands. *J. Arid. Environ.*, 15: 1–18.

Liaccos, L.G. and Moulopoulos, C., 1967. *Contribution to the identification of some range types of Quercus coccifera*. University of Thessaloniki, 54 pp.

Loiseau, P. and Sebillotte, M., 1972. *Étude et cartographie des pâturages du Maroc Oriental*. Ministr. Agric., Rabat, 540 pp.

Long, G.A., 1954. Contribution à l'étude de la végétation de la Tunisie Centrale. *Ann. Serv. Bot. Agron.* (Tunis), 27: 1–338.

Maignan, F., 1971. *Essais de culture fourragère à Silakdar (Algérie)*. AGP: SF/Aly., 16, Rapp. Techn. No. 1, FAO, Rome, 121 pp.

Maignan, F., 1978. Productivity of *Lolium rigidum* in a forest of oak trees (*Quercus suber*). *Proc. 1st Int. Rangelands Congress*, Denver, Colo., pp. 239–341.

McFarlane, V.M., Howard, B., Maloty, G.O.M. and Hopcraft, D.D., 1972. Tritiated water in field studies of ruminants metabolism in Africa. In: *Isotope Studies on the Physiology of Domestic Animals*. Int. Atomic Agency, Vienna, pp. 83–94.

McKenzie, D., 1970. *Goat Husbandry*. Faber and Faber, London, 366 pp.

Muslera Pardo, E. and Ratera Garcia, C., 1984. *Prederas y forrajes: producion y aprovechamiento*. Mundi-Prensa Publ., Madrid, 702 pp.

Naveh, Z., 1982. The dependence of the productivity of a semi-arid Mediterranean hill pasture ecosystem on climatic fluctuation. *Agric. Environ.*, 7: 47–61.

Ovington, J.D. (Editor), 1983. *Temperate Evergreen Broad-Leaved Forests*. Ecosystems of the World, 10, Elsevier, Amsterdam, 241 pp.

Papanastasis, V.P. and Liaccos, L.G., 1980. Productivity and management of Kermes oak brushlands for goats. In: H.N. Le Houérou (Editor), *Browse in Africa*. Int. Livestock Centre for Africa, Addis Ababa, pp. 375–384.

Raunkaier, C., 1934. *The Life Form of Plants and Statistical Plant Geography*. Clarendon Press, Oxford, 632 pp., 189 figs.

Reed, C.A., 1969. The pattern of animal domestication in the prehistoric Near East. In: P.J. Ucko and G.W. Dimbley (Editors), *The Domestication and Exploitation of Plants and Animals*. Duckworth, London, pp. 261–380.

Rodin, L.E., Vinogradov, B.V., Microchnitchenko, Y., Pelt, M., Kalenov, H. and Botschantsev, V., 1970. *Étude géobotanique des pâturages du secteur ouest du département de Medea (Algérie)*. Nauka, Leningrad, 124 pp.

Seligman, N.G. and Gutman, M., 1978. Cattle and vegetation response to management of Mediterranean rangelands in Israel. *Proc. 1st Int. Rangelands Congress*, Denver, Colo., pp. 616–618.

Shkolnik, A., Borut, A. and Chosniak, I., 1972. Water economy of the Bedouin goat. *Symp. Zool. Soc. London*, 31: 229–242.

Shkolnik, A., Maltz, E. and Chosniak, I., 1980. The role of ruminant's digestive tract as a water reservoir. In: Y. Ruckebusch and P. Thivend (Editors), *Digestive Physiology and Metabolism in Ruminants*. Medical Technical Press, Lancaster, pp. 731–742.

Tadmor, N.H., Eyal, E. and Benjamin, R., 1972. Primary and secondary production of arid grassland. In: L.E. Rodin (Editor), *The Ecophysiological Foundations of Ecosystem Productivity in Arid Zones*. Nauka, Leningrad, pp. 173–177.

Thiault, M., 1955. L'évolution des pâturages en Tunisie en fonction du mode d'exploitation. *Ann. Serv. Bot. Agron.* (Tunis), 11: 181–208.

Thiault, M., 1958. Les perspectives ouvertes à l'élevage par les nouvelles méthodes de culture de l'herbe. *La Tunisie Agricole*, Avril–Mai, 22 pp.

Thiault, M., 1963. *Rapport au gouvernement Tunisien sur l'amélioration des pâturages et de la production fourragère*. Rapp. PEAT No. 1689, FAO, Rome, 62 pp.

Wilson, A.D., 1977. The digestibility and voluntary intake of the leaves of trees and shrubs by sheep and goats. *Aust. J. Agric. Res.*, 28: 501–508.

Yagil, R., 1985. *The Desert Camel: Comparative Physiological Adaptation*. Karger, Basel, 163 pp.

Zeuner, F.E., 1963. *A History of Domesticated Animals*. Hutchinson, London, 560 pp.

Zohary, D.D., 1973. The origin of cultivated cereals and pulses in the Near East. In: J. Wharam and K.R. Lewis (Editors), *Chromosomes To-Day*. Wiley, New York, N.Y., pp. 307–320.

Zohary, D.D. and Spegel-Roy, P., 1975. Beginning of fruit growing in the old world. *Science*, 187: 319–327.

Chapter 8

GRASSLANDS OF THE SAHEL[1]

HENRI NOEL LE HOUÉROU

INTRODUCTION

The boundaries of the Sahel have been defined in various ways to include more or less territory than is considered in this chapter. In this discussion the Sahel is considered to occupy about 3 million km^2 in a belt from 400 to 600 km wide and nearly 6000 km long extending across the African continent to the south of the Sahara desert (Table 8.1, Fig. 8.1). The northern margin is about the 100-mm isohyet, and the southern limit is about the 600-mm isohyet. This Sahelian zone is subdivided into three ecoclimatic subzones on the basis of the nature of the flora and of the vegetation, the distribution of livestock and wildlife, and land-use patterns (Table 8.2). The Sahelian subzone proper lies between the 200- and 400-mm isohyets. This is bordered northward by the Saharo–Sahelian subzone, which is transitional to desert, and southward by the Soudano–Sahelian subzone, which is transitional to woodland. The Sahel has been discussed elsewhere in this series by Monod (1986) in Volume 12B (Hot Deserts and Arid Shrublands), but with less emphasis on the herbaceous components of the vegetation. It has also been included among other regions in some of the chapters in Volume 13 (Tropical Savannas) (Bourlière, 1983).

CLIMATE

The climate of the Sahel is typically tropical with a unimodal precipitation pattern, rainfall occurring in the short summer season (Fig. 8.2). The rains are provoked by the monsoon of the Gulf of Guinea and the northward movement of the inter-tropical convergence zone.

The rainy season is from June to September, and reaches a peak in August, the rest of the year being virtually without rain. The heaviest annual shower represents 5 to 15% of the average long-term annual precipitation. Of the total amount of precipitation, 50 to 60% falls in daily amounts of less than 10 mm. As in arid zones elsewhere, this proportion increases with aridity.

From agronomic and ecological viewpoints, the growing season corresponds with the period when monthly precipitation in millimetres is equal to or greater than twice the mean temperature in degrees Celsius (Bagnouls and Gaussen, 1953; Walter and Lieth, 1967). This threshold corresponds in most cases to a value of 0.35 of the potential evapo-transpiration using Penman's (1948) formula. As the mean daily temperature is about 28°C at the beginning of the rainy season (Cochemé and Franquin, 1967) and the potential evapo-transpiration is about 6 mm day^{-1} at that time (Davy et al., 1976), the growing season usually starts when the precipitation reaches about 2 mm day^{-1} (0.35 of the potential evapo-transpiration) (Le Houérou and Popov, 1981). In identifying the end of the growing season by this means, the amount of accumulated soil water must be taken into account. When defined in this way, the rainy season lasts only 2 to 4 weeks in the Saharo–Sahelian subzone, 1 to 3 months in the Sahelian subzone, and 3 to 4 months in the Soudano–Sahelian subzone (Table 8.2). The beginning and end of the rainy season occur at fairly regular times each year. The mean annual

---

[1] Manuscript submitted in August, 1985; revised during 1986 and 1988.

TABLE 8.1

Distribution (km$^2 \times 10^3$) of ecoclimatic zones within country boundaries in tropical northern Africa[1]

Country	Saharan		Sahelian		Soudanian		Guinean		Total	
	area	%	area	%	area	%	area	%	area	%
Burkina-Faso[2]	0	–	25	9	249	91	0	–	274	3
Chad	630	49	366	29	288	22	0	–	1284	15
Gambia	0	–	0		11	100	0	–	11	+
Mali	550	44	300	24	390	32	0	–	1240	14
Mauritania	808	78	223	22	0	-	0	–	1031	12
Niger	540	43	723	57	+	+	0	–	1263	14
Nigeria	0	–	50	5	674	73	200	22	924	11
Senegal	0	–	90	44	111	56	0	–	201	2
Sudan	800	32	800	32	480	19	420	17	2500	28
Cameroons, Cape Verde and Ethiopia	0	–	67						67	1
Total	328	38	2644	30	2203	25	620	7	8795	100

[1] From Le Houérou and Popov (1981).
[2] Formerly Upper Volta.

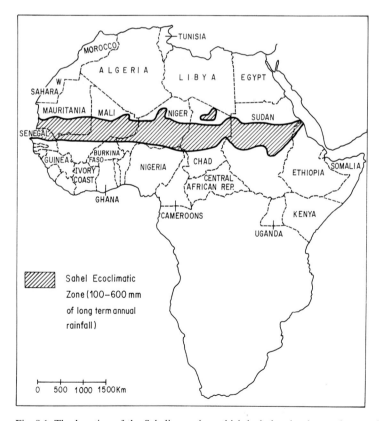

Fig. 8.1. The location of the Sahelian region, which includes the three subzones described in this chapter.

TABLE 8.2

Characteristics of the three ecoclimatic zones of the Sahel[1]

	Saharo–Sahelian	Sahelian	Soudano–Sahelian
Climate	Very arid	Arid	Semi-arid
Mean annual rainfall (mm)	100–200	200–400	400–600
Length of growing season (days)	15–30	30–90	90–120
Vegetation	Contracted and diffuse steppes of perennial grasses	Mimosaceae savanna with annual grass layer	Combretaceae savanna with annual grass layer
Dominant plant species:			
grasses	*Panicum turgidum*	*Aristida mutabilis* *Eragrostis tremula*	*Cenchrus biflorus* *Schoenefeldia gracilis*
trees and shrubs	*Acacia ehrenbergia* *Acacia raddiana*	*Acacia raddiana* *Acacia senegal* *Balanites aegyptiaca* *Commiphora africana*	*Acacia seyal* *Combretum micranthum* *Combretum* spp. *Sclerocarya birrea*
Land use patterns	Nomadic and transhumant pastoralism, no rain-fed cropping	Transhumant and nomadic pastoralism, some flood cropping to millet	Settled and nomadic pastoralism, rain-fed cropping to millet and sorghum

[1] After Le Houérou (1976b, 1977, 1980b), Davy et al. (1976), Le Houérou and Popov (1981), and Boudet (1984).

north–south precipitation gradient is about 1 mm km^{-1} (Le Houérou, 1976b).

The variability in annual precipitation increases with aridity. The coefficient of variation of annual rainfall increases from about 25% in the Soudano–Sahelian subzone to about 45% in the Saharo–Sahelian subzone. In one year out of ten, the annual rainfall is about 135 to 150% of the long-term mean, and is about 50 to 65% of the mean or less with the same frequency. However, as in other dry regions, years of above-average and below-average precipitation tend to aggregate. The 1970 to 1985 mean averaged perhaps 30 to 40% below the long-term mean and locally was even less (Dancette, 1985; Le Houérou and Gillet, 1986). Nevertheless, this level of variability in precipitation is lower than that along equivalent isohyets in the Mediterranean arid zone (Le Houérou, 1982/84), in the North American subtropics, and in northeastern Brazil (Le Houérou and Norwine, 1987). Rainfall is least unreliable at the peak of the rainy season in August (Le Houérou and Popov, 1981). There seems to be no long-term trend in amount of precipitation during the last 3000 years (Le Houérou, 1976a), although Monod (1986) suggests that the region was less arid during some former periods.

The mean annual temperature varies from 25 to 30°C, the mean minimum being 18 to 20°C and mean maximum 35 to 38°C. The mean monthly minimum drops to 10–15°C in January and the mean monthly maximum rises to 38–44°C in May/June. The absolute minimum is 0°C or slightly less in the northern subzone, whereas maxima of 50°C or higher may occur. The Sahel is probably the hottest broad ecological region on earth, even hotter than the Sahara desert when measured in terms of mean annual temperature. The eastern half of the region is hotter than the western half. A strip 50-km wide along the Atlantic Ocean is milder because of the influence of up-welling in this area (the Canaries Current).

Maximum windiness (mean monthly maximum of 4–5 m s^{-1}) occurs in January to May during the latter part of the dry season, when the continental trade winds blow from the northeast and atmospheric humidity is very low (5–15%) (Cochemé and Franquin, 1967). The least windy period is from August to November (latter part of the rainy season and early dry season). The winds during

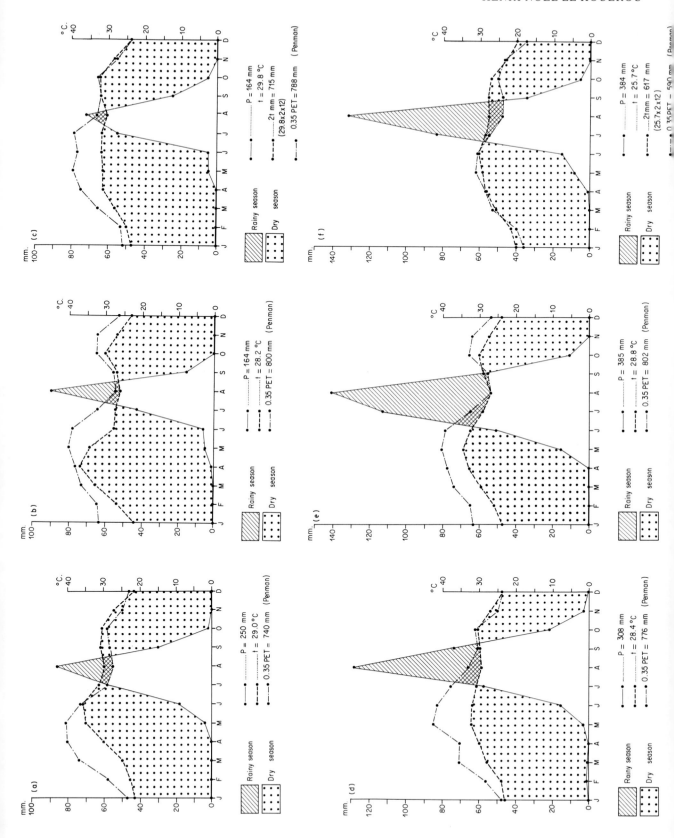

GRASSLANDS OF THE SAHEL

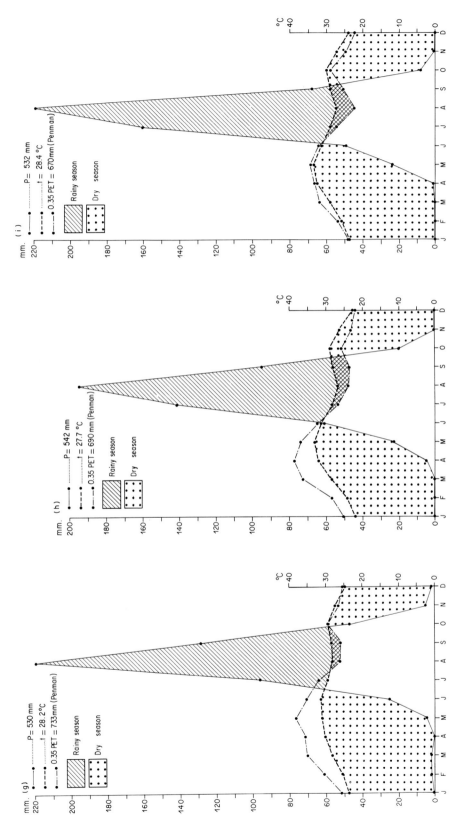

Fig. 8.2. Ombrothermic diagrams of water balance in representative sites within each subzone of the Sahel [using the format of Bagnouls and Gaussen (1953) and Le Houérou and Popov (1981)]. (a) Tombouctou (Mali), Saharo–Sahelian, ecoclimatic subzone. (b) Agadès (Niger), Saharo–Sahelian, ecoclimatic subzone. (c) Khartoum (Sudan), Saharo–Sahelian, ecoclimatic subzone. (d) Podor (Senegal), Sahelian "sensu stricto", ecoclimatic subzone. (e) Tahoua (Niger), Sahelian "sensu stricto", ecoclimatic subzone. (f) El Fasher (Sudan), Sahelian "sensu stricto", ecoclimatic subzone. (g) Linguère (Senegal), Soudano–Sahelian, transition subzone. (h) Mopti (Mali), Soudano–Sahelian, ecoclimatic subzone. (i) Zinder (Niger), Soudano–Sahelian, ecoclimatic subzone.

the rainy season are from the south-west with the monsoon from the Gulf of Guinea.

Evaporation and evapo-transpiration rates are very high (1800–2300 mm yr^{-1}) (Riou, 1975; Davy et al., 1976; Dancette, 1979). The vapour pressure deficit reaches 30 to 45 HPa (equal to millibars) during the day in the dry season.

The region has been exposed recently (1970–1984) to a drought of unprecedented length, as far as instrumental records are available from the middle of the 19th century [for the cities of St-Louis (since 1855) and Dakar (since 1887)]. The mean annual rainfall during this 15-year drought was about 60% of that of the period from 1900 to 1969; during extreme droughts in 1972–1973 and 1983–1984, the rains averaged only 40 to 50% of the long-term mean (Le Houérou and Gillet, 1986). Rivers dried up in 1985 (the Niger for the first time in recorded history; the Chari for the second time, having also disappeared in 1914).

SEASONALITY

The year can be appropriately divided into five seasons: (1) the rainy season, from mid-June to mid-September; (2) the "deferred season", from September to November, when the air is relatively humid and soils and most watering places dry up (nomads and transhumants return to their winter quarters); (3) the cool season, from November to February; (4) the hot, dry season, from March to May; and (5) the May–June pre-rainy season, when air moisture gradually increases with the advancing monsoon, heralded by aborted rains (which evaporate before reaching the land surface) from thunderstorms.

Conditions are extremely harsh for man and herbivores during the hot, dry season. Temperatures exceed 40°C every day and do not drop below 35°C at night. Diurnal fluctuation in relative humidity is from less than 10% to 20–30%. The landscape is parched by intense radiation and scorched by fires that sweep across 15 to 20% of the terrain each year.

New leaves appear in many species of trees and shrubs from 15 to 60 days before the first rains. This early foliation is important to the well-being of domestic herbivores and wildlife, since green herbage has not been available for 8 to 10 months and the dry herbage is devoid of vitamins, carotene, protein, and is very low in phosphorus content (Le Houérou, 1979c, 1980b; Clanet and Gillet, 1980; Lawton, 1980; Walker, 1980). This flush of new growth provides for a balanced diet and recovery to good health by the time of onset of the rainy season. This early foliation is apparently triggered by the steady increase in atmospheric humidity which precedes the northward advance of the monsoon and provides a sharp contrast with the very dry and hot conditions that occur while the north-east trade winds prevail. Development of these new leaves does not appear to involve absorption of water and nutrients by plants; water, proteins and carbohydrates are translocated from roots and stems to twigs, and transpiration is prevented by a thick waxy cuticle which disappears in the early part of the rainy season (Walker, 1980). This early foliation coincides with the late stage of pregnancy in most herbivorous mammals, and it is also the time when the herds start to move towards the northern summer range.

The rainy season is a time of bounty and festivity, when water and milk are plentiful, except during general drought.

SOILS, TOPOGRAPHY AND GEOMORPHOLOGY

Probably more than half of the Sahel is occupied by sandy soils which are slightly acidic (pH 6.0–6.5) and low in content of organic matter, nitrogen and phosphorus. In the FAO classification they are recognized as cambic and luvic arenosols, corresponding to the psamments and aridisols of United States soil taxonomy and to brown subarid soils in the French classification (Hunting Techn. Serv., 1964; FAO/UNESCO, 1973; Fauck, 1973). These are the zonal soils of the northern part of the Sahel (Saharo–Sahelian and Sahelian ecoclimatic subzones). To the south (Soudano–Sahelian subzone) the zonal soils are more acidic (pH 5–6) and evolved towards ferric luvisols, tropical ferruginous soils in French soil taxonomy. Saline and alkaline soils are of very limited extent, unlike the situation in many arid regions elsewhere.

Sand dune formations cover huge areas in the northern part of the Sahel, but are less extensive in the southern part. Drifting sands ("live dunes") are common north of the 150-mm isohyet (Mainguet et al., 1980). Recently, this limit has shifted southwards to the 200- to 300-mm isohyet due to desertization processes resulting from the combination of drought and land abuse. The geomorphology of the remainder of the terrain consists of pediplains, plateaux, clay plains and alluvial flood plains, fluvio-lacustrine deposits, and fossil valleys and wadis (Hunting Techn. Serv., 1964; Pias, 1970; Mainguet, 1972, 1976, 1984; Michel, 1973). Monod (1986) has provided a more lengthy discussion of the soils of this region elsewhere in this series.

The landscape is rather flat to gently rolling and low in elevation (200–600 m). Most of the mountains (up to 3400 m in height) are located along the border between the Sahara desert and the Sahel.

The rivers that traverse the Sahel arise in mountains outside of the region, and the Sahelian area that drains into them is very limited in extent. During the rainy season run-off water fills numerous temporary ponds, the area of each catchment ranging from a few hectares up to a few square kilometres. Ground water is scarce, except in northern Senegal and north-western Niger where bore holes discharge many litres of water per second. Otherwise, deep water is known to exist only in very small local aquifers which discharge only a few cubic metres per hour. This is enough for man and livestock, but not for irrigation (Rodier, 1964, 1975; Roose, 1977).

Based on measurements in several areas, run-off into temporary ponds probably averages no more than 5% of the annual rainfall in the sandy grasslands, but in areas of medium- to fine-textured soils the amount is probably between 30 and 50%. In some areas these ponds number up to 20–30 km^{-2}, and occupy as much as 5% of the land surface (Poulet, 1982). Run-off is considerably increased by wild fires, particularly those in the late part of the dry season (Roose, 1981; Roose and Piot, 1984).

On moderately sloping sandy soils permanently covered with natural vegetation, loss of soil particles by water erosion is probably only about 200 kg ha^{-1} yr^{-1}. However, in cultivated land of various textures losses of up to 18 t ha^{-1} yr^{-1} have been measured (Delwaulle, 1973a, b; Heusch, 1975; Piot and Milogo, 1980).

FLORA AND VEGETATION

The flora of the Sahel is palaeo-tropical in character and low in diversity, with a total of only about 1500 species occurring in this extensive area. Families contributing the greatest numbers of species are:

Acanthaceae   Cyperaceae   Poaceae
Asclepidaceae   Euphorbiaceae   Solanaceae
Capparidaceae   Fabaceae   Zygophyllaceae
Convolvulaceae   Malvaceae

Only about 40 species (3% of the flora) are endemic (White, 1983). Many species are shared with the Soudanian zone, being restricted in the Sahel to water courses, margins of ponds, and other areas subject to flooding. There is a gradient in floristic composition from north to south corresponding to the rainfall gradient, and there are also differences between the eastern and western parts of the region (Le Houérou, 1981).

Saharo–Sahelian subzone

The primeval vegetation of the Saharo–Sahelian subzone was a steppe dominated by perennial grasses. This was probably similar to the diffuse perennial-grass steppe, dominated by highly xeromorphic tussock grasses and graminoids, which still occurs in sandy soil in remote locations far from a permanent supply of water. The plants are distributed sparsely, but regularly, with a canopy cover of 5 to 20%. Although the botanical composition of the vegetation may have been substantially different from the present, the following species, which are now rare and restricted to rocky montane sites, may have played a much more important role before the impact of man became severe (Le Houérou, 1988c).

Andropogon schirensis   *Enneapogon scaber*
Anthephora hochstetteri   *Eremopogon foveolatus*
Bothriochloa pertusa   *Heteropogon contortus*
Cenchrus ciliaris   *Hyparrhenia hirta*
Chrysopogon plumulosus   *Lasiurus hirsutus*
Dichanthium annulatum   *Sporobolus ioclados*
Enneapogon brachystachyus   *Themeda triandra*

Woody vegetation has probably never occurred beyond depressions and other habitats benefitting from run-off water or having a shallow water-table.

At present, stony and gravelly plains and pediments typically support only ephemerals that cover up to 50% of the soil surface during the short (1 to 2 months) rainy season. The silt and clay soils of valleys are covered by a variety of vegetation types composed of short annual grasses (e.g. *Aristida hordacea*) or tall annual grasses (e.g. *Sorghum aethiopicum*), dominated locally by a sparse shrub layer (0.5 to 2.0 m high) and a still sparser layer of small trees (2 to 5 m high). The canopy of the woody species covers up to 10% of the soil surface and consists of as many as 200 stems ha^{-1}. Denser stands of phreatophytic trees and shrubs occur in depressions and along wadis having a deep water-table (subsurface flow).

Sahelian subzone *sensu stricto)*

The vegetation of the Sahelian zone proper is "Mimosaceae thornscrub" which is composed of three layers: herbs, shrubs and small trees (Le Houérou, 1989) (Figs. 8.3 to 8.5). The herbaceous layer consists of annual grasses and forbs with a canopy cover of 20 to 80%. The dominant grasses, which form an upper stratum and reach a height of from 40 to 80 cm, include: *Aristida funiculata*, *A. mutabilis*, *Cenchrus biflorus* and *Schoenefeldia gracilis*. A lower stratum, no more than 40 cm in height, is composed of annual grasses and annual forbs. Common dominant forbs are: *Blepharis linearifolia*, *Borreria (Spermacoce) radiata*, *Gisekia pharnaceoides*, *Limeum viscosum*, *Mollugo nudicaulis* and *Zornia glochidiata*. The shrub layer is made up of spiny species, usually of the genera *Acacia*, *Balanites* and *Ziziphus*, 0.5 to 3 m high with a canopy cover as great as 20%. Some non-spiny species (such as representatives of *Boscia*, *Cordia*, *Euphorbia* and *Guiera*) are also present. The tree layer is of sparse, small trees reaching heights of 3 to 6 m and providing a canopy cover of 5% or less. The dominants in this layer are usually species of *Acacia*, *Balanites*, *Commiphora* and *Grewia*. The number of stems of shrubs and trees sometimes reaches 500 ha^{-1}. Sciaphilous (shade-loving) annual grasses and forbs develop under and near the canopies of the woody species. The most common perennial grasses are *Aristida longiflora* and *Cymbopogon proximus* (in sandy soils).

Fig. 8.3. Annual grassland of *Schoenefeldia gracilis* in the Sahelian subzone of northern Burkina Faso (Upper Volta). The maximum standing-crop biomass of current year growth is about 120 g m^{-2}. Soil is deep silt on pediplain; annual precipitation 300 mm. The woody species in the background are *Acacia senegal*, *A. seyal* and *Pterocarpus lucens*.

Fig. 8.4. Annual grassland of *Cenchrus biflorus* in the Sahelian subzone of northern Senegal. The maximum standing-crop biomass of current year growth is about 350 g m^{-2}. The trees are *Balanites aegyptiaca*. Annual precipitation is about 300 mm.

Fig. 8.5. Annual grassland of *Aristida mutabilis* in the Sahelian subzone of northern Burkina Faso. The maximum standing-crop biomass of current year growth, during the season of poor growth shown, is about 60 g m^{-2}. The woody species are *Acacia tortilis* and *Salvadora persica*. Soil is deep sand on a Pleistocene dune system; mean annual precipitation 280 mm.

Whereas perennial grasses prevail in the drier climate of the Saharo–Sahelian subzone, the present herbaceous layer in the Sahelian subzone is composed almost entirely of annual species. There is apparently no climatic reason why several species of perennial grasses could not thrive in this subzone. The explanation for their absence apparently is the effect of fire. Whereas the amount of fuel (200–500 kg DM ha^{-1}) in the drier subzone to the north is insufficient, and the tussocks of the perennial grasses are too sparsely distributed, to permit widespread burning, biomass is much more plentiful (800–3000 kg DM ha^{-1}) in the Sahelian subzone and fires occur over extensive areas. Probably the perennial grasses have been excluded by the combined effects of fire and the long, severe dry season (Le Houérou and Naegelé, 1972). Studies in various parts of this subzone in Mali, Senegal and the Republic of Sudan indicate that an average of 10 to 35% of the area burns each year and that fire has prevailed at similar rates in the past (Baasher, 1961; Shepherd and Baasher, 1966; Shepherd, 1968; Naegelé, 1971; Le Houérou, 1980b, 1988a, b, 1989; Vanpraet, 1985; G.E. Wickens, pers. commun., 1985).

The minimum average biomass required to permit burning over a substantial area is about 1000 kg DM ha^{-1} (Gillet, 1967). Fires are not as destructive as in the past, because of reduction in biomass due to range depletion resulting from overgrazing. However, they are often very harmful in years when rainfall is close to or above average and dead foliage of grasses forms an almost continuous mat during the dry season.

Soudano–Sahelian subzone

The primeval vegetation of the Soudano–Sahelian subzone was probably an open woodland in which the herbaceous layer was dominated by perennial grasses including: *Andropogon gayanus*, *Andropogon tectorum*, *Aristida longiflora*, *Diheteropogon amplectens*, *Hyperthelia dissoluta*. *Andropogon gayanus* was still common in protected areas prior to the drought of 1969–1973 (Boudet and Leclercq, 1970; M. Grouzis, pers. commun., 1985), but this species succumbed to drought and (with the exception of the areas of Louga and Dahra in Senegal) has not reappeared, even in areas where there has been little or no grazing by livestock, near Niono, Mali, for instance.

The tree and shrub layers are of greater importance than in the Sahelian subzone *sensu stricto* (Fig. 8.6). Their canopies cover 20 to 35% of the soil surface and density ranges between 500 and 1500 stems ha^{-1}, with broad-leaved species being dominant, rather than the spiny, microphyllous species found in the other two subzones. Perennial grasses (60–120 cm in height) occur occasionally, particularly in depressions, in the more continuous herbaceous layer of this subzone. The vestigial perennial grasses are:

Andropogon gayanus	*Dichanthium annulatum*
Andropogon schirensis	*Enneapogon* spp.
Anthephora hochstetteri	*Eremopogon foveolatus*
Aristida longiflora	*Heteropogon contortus*
Bothriochloa pertusa	*Hyparrhenia hirta*
Cenchrus ciliaris	*Hyperthelia dissoluta*
Chrysopogon plumulosus	*Panicum anabaptistum*
Cymbopogon commutatus	*Sporobolus helvolus*
Cymbopogon giganteus	*Sporobolus ioclados*
Cymbopogon proximus	*Themeda triandra*
Cymbopogon schoenanthus	*Vetiveria nigritana*
Cynodon dactylon	

The herbaceous layer is dominated by the following annual grasses:

Andropogon pseudapricus	*Elionurus elegans*
Aristida mutabilis	*Eragrostis tremula*
Cenchrus biflorus	*Loudetia togoensis*
Ctenium elegans	*Pennisetum pedicellatum*
Diheteropogon hagerupii	*Schoenefeldia gracilis*

The perennial grasses in flooded areas include: *Echinochloa pyramidalis*, *Echinochloa stagnina*, *Eragrostis barteri*, *Oryza barthii* (= *O. longistaminata*).

Seasonal dynamics

Because of the short rainy season, annual growth is achieved in from 1 to 3 months. Daily growth increments are very high, particularly at the peak of the growing season in August. In a normal rainy season, water and green forage are plentiful, but during the rest of the year (9–11 months) trees and shrubs provide the only green foliage.

The floristic composition of the herbaceous layer changes considerably from year to year. In a particular location, a given species may be dominant for one to several years and then almost disappear. This dynamism is very complex and

Fig. 8.6. Annual grassland of *Andropogon pseudapricus* and *Diheteropogon hagerupii* in the Soudano–Sahelian subzone of central Mali. The maximum standing-crop biomass of current year growth is about 300 g m^{-2}. The shrubs in the background are *Combretum ghazalense* and *C. glutinosum*. Soil is silt on iron pebble; annual precipitation 550 mm.

not fully understood. Apparently, complex interactions occur among: variability in amount and periodicity of rainfall, both in previous and present years; seed production; seed stocks in soils; seed consumption and storage by granivores; and seed hardness. Penning de Vries and Djiteye (1982) classified species according to hardness of seeds. When early rains are interspersed with dry periods, most "soft" and "semi-soft" seeds germinate, but the seedlings die. In this event species with "semi-hard" and "hard" seeds flourish. The results of a study in the Soudano–Sahelian subzone in Mali (Hiernaux et al., 1978) suggest that less than half of the herbaceous species found in a given site can be expected to occur there regularly; nevertheless, more variation occurs in yield of biomass from year to year than in floristic composition.

Several studies have revealed that the rate of recovery of grasses and forbs is rapid after a drought of several successive years (Bille, 1977; Boudet, 1977; Cornet, 1981; Cornet and Rambal, 1981; Penning de Vries and Djiteye, 1982; Wilson, 1983; Barral et al., 1983; Barry et al., 1983;

Grouzis, 1984, 1988). Good stands usually develop by the second or third year after the resumption of rain. However, in heavily trampled areas, the soil surface sometimes is left bare because it has been sealed, with or without an encrustation of Cyanophyceae (Barbey and Conte, 1976; Dulieu et al., 1977).

PRIMARY PRODUCTION

In multi-layered vegetation which includes shrubs and trees, the proportion of radiant energy used in photosynthesis is much greater than in herbaceous communities occurring in the same macroclimatic conditions. In a study in the Sahelian subzone in Senegal, the estimated net energy conversion factor varied from 0.3 to 1.5% at different times during the rainy season and was 0.08% on a year-long basis (Bourlière, 1978). Under typical Soudano–Sahelian conditions in Mali, total annual biomass production (both above ground and under ground) is about 660 g m^{-2} in open herbaceous vegetation and about 2400 g m^{-2}

in multi-layered bushland (Boudet and Leclercq, 1970; Hiernaux et al., 1978; Penning de Vries and Djiteye, 1982). Annual net energy storage is estimated to be 1040 and 4140 J cm^{-2}, respectively, assuming an average content of 18.8 kJ g^{-1} of dry matter in woody species and 15.7 kJ g^{-1} in herbaceous species (Grouzis, 1988). Net photosynthetically active radiation averages about 290 to 335 kJ cm^{-2} yr^{-1}, and the conversion rate is 1.3 to 5.3% of incident energy during the growing season and 0.33 to 1.32% on a year-long basis (Le Houérou, 1980b). The greater photosynthetic efficiency of multi-layered ligneous/herbaceous vegetation is explained by: (1) higher efficiency of moisture use associated with a lower rate of evaporation and lower temperature of the soil surface due to more shade; (2) increased water intake and storage of water due to more stable and permeable soil structure; and (3) quicker and more intensive turnover of nutrients (Ca, N, P, K, Mg, S) (Charreau and Vidal, 1965).

Growth is rapid during the peak of the rainy season in August and September. Bille (1977) reported a daily increment of 2 g DM m^{-2} in the herbaceous layer in the Sahelian subzone in Senegal, whereas Penning de Vries and Djiteye (1982) found this value to be 3.5 g in natural grassland under Soudano–Sahelian conditions in Mali.

Mean above-ground maximum standing crop (which occurs during September–October) in the steppe of the Saharo–Sahelian subzone ranges between 15 and 60 g DM m^{-2}, depending on soil condition and vegetative type. In herbaceous communities of the Sahelian subzone this value ranges from 50 to 250 g, and in the Soudano–Sahelian subzone from 80 to 300 g. The lower values occur in well-drained positions, whereas the higher ones are in somewhat depressed areas. In flooded meadows of *Echinochloa stagnina* (in the Sahelian subzone) maximum standing crop reaches 500 to 1500 g DM m^{-2}.

Net above-ground production is greater than the maximum standing crop of herbage, since considerable biomass fails to accumulate due to rapid decay during the rainy season and to consumption by insects (Bille, 1977; Le Houérou, 1980b; Penning de Vries and Djiteye, 1982; Grouzis, 1988). For example, the sum of increments in the herbaceous layer in the Soudano–Sahelian subregion was reported to be 25 to 36% greater than the maximum standing crop (Bille, 1977; Le Houérou, 1980b; Penning de Vries and Djiteye, 1982).

A large proportion of canopy biomass disappears during the dry season due to the activities of ants, termites, grasshoppers, locusts, rodents and birds, as well as weathering. Shoots are especially subject to fragmentation because of low moisture content (5% or less) during the latter part of the dry season (March–May), when the relative humidity of the lower atmosphere drops to 10% or less every afternoon. In this condition the leaves and stems are very fragile and subject to damage by animal activities — especially by trampling of livestock — and by wind. In estimating carrying capacities for livestock from estimates of maximum standing crop, one must take these losses into account.

Detailed measurements of water use by grassland vegetation have been made at intensive study sites in Senegal and Mali. In the drier location (Senegal), annual above-ground biomass increments of 3.5 to 5.0 kg DM ha^{-1} were found to be associated with each millimetre of water that was evaporated and transpired (Cornet, 1981; Cornet and Rambal, 1981). This is equivalent to 2564 and 2000 kg of water per kilogram of dry matter, respectively (or 0.39 to 0.50 g per kg of water). Of the rain-water that entered the soil, 72% was transpired and 28% evaporated. In the moister region (Mali) about 80% of infiltrated water is transpired and 20% evaporates (Penning de Vries and Djiteye, 1982; Breman and De Wit, 1983). Transpiration amounted to 260 to 300 kg of water per kilogram of dry matter added to above-ground biomass.

The relationship between maximum standing crop and amount of rainfall provides a comparison of rain use efficiency (R.U.E.) in herbage production among the three subzones. In the Saharo–Sahelian subzone R.U.E. averages 1.0 to 4.0 kg of dry matter per hectare per millimetre of rainfall, in the Sahelian subzone 1.7 to 8.0, and in the Soudano–Sahelian subzone 1.5 to 6.0 (Le Houérou and Hoste, 1977; Le Houérou, 1984).

The relationship between rainfall and herbage yield has been studied in relation to prediction of carrying capacity for livestock. As in other dry regions, plant production is as dependent on the distribution of rainfall as on the total amount received. The correlation between herbage yield and "infiltrated rains" in a particular year is excellent (Cornet, 1981; Cornet and Rambal, 1981). However, measurements of water intake are highly site-dependent; they cannot be extrapolated over large areas due to extreme heterogeneity in distribution of rainfall. The correlation between "useful rains" and herbage yield is also very good ($r^2 = 0.75$ to 0.95) (Cornet, 1981; Boudet, 1984). "Useful rains" are those that are equal to or greater than 0.35 to 0.50 of the potential evapo-transpiration measured by Penman's (1948) method on a weekly basis (Le Houérou and Popov, 1981; Boudet, 1984). A value of 0.35 corresponds to 1.5 mm day^{-1} in September and 2.8 mm day^{-1} in May, whereas a value of 0.50 is equivalent to 2.2 mm in September and 4.0 mm in May.

Deficiencies of nitrogen and phosphorus were found to limit herbage yield when rainfall exceeded 200 to 300 mm yr^{-1} (Penning de Vries and Djiteye, 1982). Therefore these factors probably affect growth in the Saharo–Sahelian subzone in 1 or 2 years out of 10, in the Sahelian subzone about 6 years out of 10, and in the Soudano–Sahelian subzone about 9 years out of 10. Yields were doubled or tripled, even in large-scale trials, by combined applications of these two nutrients (Hiernaux et al., 1978), and the yield of dry matter per kilogram of water transpired was considerably increased (Penning de Vries and Djiteye, 1982).

Seed production in the herbaceous layer is highly variable, and averages from 10 to 15% of the maximum standing crop of herbage. *Digitaria exilis* and *Panicum laetum* produce such an abundance of seeds that they are collected for human consumption from the rangeland by pastoralists, and they are sometimes seeded in small fields near villages (Wilson et al., 1983). Seed yields in natural stands are several hundred kilograms per hectare in good years. About 10% of the seeds are eaten by birds, rodents, ants and insects (Bourlière, 1978) and their availability accounts for the range of distribution of a pest bird *Quelea quelea* (millet eater) (Morel, 1968a, b; Gaston and Lemarque, 1976).

FAUNA

The only intensive multi-disciplinary ecosystem study in the Sahel was in the Fété Olé savanna in the northern Ferlo region of northern Senegal (Bourlière, 1978; Poulet, 1982). The study site is little modified by man, except for the fact that large animals are extinct. These originally included the following:

Addax nasomaculatus	addax antelope
Alcelaphus buselaphus	hartebeest
Gazella dorcas	dorcas gazelle
Gazella leptoceros	rhim gazelle
Gazella rufifrons	red-fronted gazelle
Giraffa camelopardalis	giraffe
Kobus defassa	defassa waterbuck
Loxodonta africana	elephant
Oryx dammah (= *O. algazel*)	scimitar-horned oryx
Panthera leo	lion
Panthera pardus	leopard
Struthio camelus	ostrich
Tragelaphus scriptus	bushbuck
Tragelaphus strepsiceros	greater kudu

These have been replaced by livestock populations of perhaps somewhat equivalent biomass. The indigenous ungulate biomass was probably no greater than 2000 kg live weight km^{-2} (Coe et al., 1976). This site is typical of the Sahelian subzone, with highly variable rainfall (long-term annual mean of 400 mm, but 300 mm during the period of study). There is no permanent source of water. Because of high variability and seasonality of precipitation and plant growth, range exploitation is of the migratory type. Studies of the fauna have not been sufficient, as yet, to permit an overall evaluation of secondary production. Populations and biomass of various consumers in this site were discussed by various specialists in another volume in this series (Bourlière, 1983), and figures for arthropods and birds are given by Monod (1986). Le Houérou and Grenot (1989) have also discussed the fauna of the Sahel.

Concentration of rainfall in a period of 1 to 3 months results in productivity constraints and low secondary productivity. The low level of available food towards the end of the long and severe dry season (high temperatures and deficiency of water and nutrients) greatly restricts the sedentary populations of consumers. Because the amount of food energy available during the rainy season is far beyond that which resident consumers can utilize fully, migratory and nomadic species are abundant during this period. Basically, the arid climate results in a longer period of conservation of plant tissues than in more humid regions.

Invertebrates in the canopy and upper layers of soil are mainly Arthropoda (Monod, 1986). No Lumbricidae or Mollusca occur (Gillon and Gillon, 1973, 1974a, b). The number of vertebrate species is few (Monod, 1986); mammals, especially rodents, have the greatest impact on plants (Poulet, 1982).

The stocking rate of wild ungulates in the Sahel is not well documented. An aerial survey throughout 250 000 km^2 in the Republic of Sudan suggested that the average biomass of large herbivores was only 1 kg live-weight km^{-2}, with concentrations in limited areas of from 8 to 11 kg km^{-2}. Eighteen species were represented (Wilson, 1979a, b). The biomass of wild ungulates has decreased greatly in the last three decades because of over-hunting and poaching. Bourlière (1962) suggested that stocking rates as high as 190 kg km^{-2} occurred in the Soudano–Sahelian subzone of the Ferlo region of Senegal prior to the 1960s. Gillet (1965) found the biomass of the four major ungulates existing in the Sahelian subzone of Chad in the late 1950s to be as high as 80 kg km^{-2}. *Addax nasomaculatus* and *Oryx dammah*, which were formerly common over large areas of the Saharo–Sahelian and Sahelian subzones (Monod, 1958; Valverde, 1968), are now nearly extinct. They have suffered as a result of unlawful hunting, even within game preserves, which has become more efficient because of the use of mechanical devices, including helicopters. Because of their adaptive and demographic strategies, Sahelian animal species are able to recover rapidly from losses during drought. The indigenous mammalian occupants of the Sahel are listed in Tables 8.3 and 8.4 (Le Houérou, 1989).

LAND USE AND MANAGEMENT

The impact of the 1970–1984 drought on land resources was aggravated by increases in populations of humans and livestock. Between 1950 and 1983 these populations increased in the seven Sahelian countries [which include 96% of the area of the Sahel (Table 8.1)] by about 121%, an exponential rate of about 2.4% per annum (FAO, 1950/83). By 1983 the average density of humans in the Sahelian countries, including very large uninhabited tracts of desert, was 7.7 km^{-2}. When the desert zones of these countries are excluded, the population density is of the order of 15 to 25 inhabitants km^{-2}. These increases in population have resulted in greatly increased pressure on land resources, with severe ecological consequences.

Rangelands

The predominant systems of livestock management practised by pastoralists is transhumant in character. These vary from long-range systems, in which the rangeland traversed annually reaches 1000 km or more, to medium-range systems of only 150 to 300 km. The moist season is spent in the northern region (Saharo–Sahelian and Sahelian subzones), where temporary ponds are available as watering places, and during the dry season the herds are kept in the south (Soudano–Sahelian subzone and Soudanian zone), where permanent water is available. These moves permit annual visits to saline areas to rectify their mineral supply, a practice that is considered by most pastoralists to be necessary to the health of their animals (Leprun, 1978; Bernus, 1981).

The grazing livestock populations of the seven Sahelian countries are equivalent to 51 million tropical livestock units (TLU)[1]. In terms of TLUs, the proportional representation of the various species is: cattle, 61%; sheep, 15%; goats, 12%; camels, 9%; and asses 3%. The average density in these countries is 6.53 TLU km^{-2}, but large expanses of desert exist which are not grazed. The average density over the 3 million km^2 of the Sahel is about 17 TLU km^{-2} or 5.9 ha per TLU. However, large numbers of animals are moved to the Soudanian zone during the dry season, so the stocking rate on a full-year basis is probably between 0.11 and 0.17 TLU ha^{-1} (about 30 to 40 kg live weight ha^{-1}). The distribution is very uneven, ranging from between 0.17 and 0.33 TLU ha^{-1} in some regions to 0.05 to 0.10 TLU ha^{-1} in areas where no permanent water source is available (Le Houérou, 1989).

Overgrazing of rangeland is widespread. The overall long-term carrying capacity for the Sahel

---

[1] Tropical Livestock Unit (TLU) is an abstract term defined as a mature zebu weighing 250 kg kept at maintenance level. The TLU while grazing is supposed to ingest 6.25 kg DM day^{-1} (Boudet and Rivière, 1968).

TABLE 8.3

Occurrence of mammals (excluding rodents) in the three ecoclimatic subzones[1] of the Sahel[2]

Species	Common name	Subzone		
		Saharo–Sahelian	Sahelian	Soudano–Sahelian
Acinonyx jubatus	cheetah	×	×	×
Addax nasomaculatus	addax antelope	+	–	–
Alcelaphus buselaphus	hartebeest	–	–	+
Alcelaphus caama	red hartebeest	–	×	×
Ammotragus lervia	barbary wild sheep	×	×	–
Canis adustus	side-striped jackal	–	×	×
Canis aureus	common jackal	×	×	×
Cercopithecus aethiops	vervet	–	×	×
Crocuta crocuta	spotted hyaena	–	×	×
Damaliscus korrigum	topi	–	+	×
Erythrocebus patas	red monkey	–	×	×
Felis caracal	African lynx	×	×	×
Felis libyca	African wild cat	×	×	×
Felis margarita	sand cat	×	×	–
Felis serval	serval	–	×	×
Galago senegalensis	lesser galago	–	×	×
Gazella dama	gazelle, mhorr	+	+	–
Gazella dorcas	dorcas gazelle	×	×	+
Gazella leptoceros	rhim gazelle	+	+	–
Gazella rufifrons	red-fronted gazelle	×	×	+
Genetta genetta	common genet	–	×	×
Giraffa camelopardalis	giraffe	–	×	×
Herpestes ichneumon	Egyptian mongoose	–	×	×
Hippotragus equinus	roan antelope	–	×	×
Hyaena hyaena	striped hyaena	×	×	×
Hippopotamus amphibius	hippopotamus	–	–	+
Ichneumia albicauda	white-tailed mongoose	–	+	×
Kobus defassa	defassa waterbuck	–	+	×
Kobus kob	kob waterbuck	–	–	+
Lepus capensis	Cape hare	×	×	×
Lepus crawshayi	Crawshay's hare	+	×	×
Loxodonta africana	elephant	–	–	×
Lycaon pictus	wild dog	+	×	×
Orycteropus afer	aardvark	–	×	×
Oryx dammah	scimitar-horned oryx	+	+	+
Ourebia ourebi	oribi	–	–	+
Panthera leo	lion	+	×	×
Panthera pardus	leopard	–	–	+
Papio anubis	anubis baboon	–	+	×
Poecilictis libyca	Libyan striped weasel	×	×	×
Phacochoerus aethiopicus	warthog	+	×	×
Procavia capensis	rock dassie	×	×	×
Redunca redunca	bohor reedbuck	–	×	×
Sylvicapra grimmia	Grimm's duiker	–	–	+
Syncerus caffer	African buffalo	–	–	+
Taurotragus derbyanus	giant eland	–	–	+
Tragelaphus scriptus	bushbuck	–	–	+
Tragelaphus strepsiceros	greater kudu	–	–	+
Vulpes pallida	sand fox, pale fox	×	×	+
Vulpes ruppelii	Ruppell's fox	×	×	–
Xerus erythropus	striped ground squirrel	–	×	×
Zorilla striatus	zorilla, striped polecat	+	×	×

[1] Ecoclimatic zones as defined by Le Houérou and Popov (1981).
[2] An interpretation of the indications given by Dekeyser (1955), Dorst and Dandelot (1970), Poulet (1972), Happold (1973), Delany and Happold (1979), and Haltenorth et al. (1985).

+ = rare or dubious occurrence; × = common; – = not present.

TABLE 8.4

Occurrence of rodents in the three ecoclimatic subzones[1] of the Sahel in Senegal (Poulet, 1982)

Species	Common name	Subzone		
		Saharo–Sahelian	Sahelian	Soudano–Sahelian
Arvicanthis niloticus	Nile rat	+	×	×
Desmodilliscus braueri	pouched gerbil	×	×	×
Euxerus erythropus	ground squirrel	–	×	×
Gerbillus gerbillus	gerbil	+	–	–
Gerbillus nanus	dwarf gerbil	+	–	–
Gerbillus pyramidum	Egyptian gerbil	+	–	–
Hystrix cristata	North African porcupine	–	×	×
Jaculus jaculus	jerboa	+	–	–
Mastomys erythroleucus	multimammate rat	–	×	×
Mastomys huberti	multimammate rat	–	×	×
Meriones libycus	Libyan jird	+	–	–
Mus haussa	haussa mouse	+	×	×
Pachyuromys duprasi	fat-tailed mouse	+	–	–
Psammomys obesus	sand rat	+	–	–
Taterillus arenarius	gerbil	×	×	–
Taterillus gracilis	gerbil	–	+	×
Taterillus pygargus	gerbil	+	×	×

[1] Ecoclimatic zones as defined by Le Houérou and Popov (1981).

+ = rare or dubious occurrence; × = common; – = not present.

region, as suggested from several experiments and stocking trials on state-owned ranches, as well as from many surveys, is of the order of 0.077 TLU ha^{-1} (10 ha per head of cattle) on a full-year basis. This rate of stocking is equivalent to a consumption rate of 200 kg DM ha^{-1} yr^{-1}. This is 20 to 25% of the maximum standing crop of the average Sahelian grassland or 15 to 20% of it when browse production is taken into account. The overall average level of stocking in the Sahel in the mid 1980s was probably about 30% above the carrying capacity, but locally it may reach two to three times the recommended density around permanent sources of water.

The average increase in stocking load in the Sahel from 1950 to 1983 was 137%, with the proportion represented by cattle increasing from 58 to 62%. Since the increase in livestock population was much greater than this in some countries (especially the Republic of Sudan) and lower in others (especially Chad), the vegetation has not been evenly subjected to heavier grazing. This rate of increase in populations of livestock is slightly greater than that of the human population (121%), so that the number of TLUs per person has increased from 0.9 to 1.0 during the last three decades. This difference, however, is probably not significant.

Range monitoring is receiving much attention (Justice, 1986; Le Houérou, 1988a, b). The use of satellite imagery to estimate biomass over large areas has given promising results (Tucker et al., 1985). This method is subject to two principal constraints: (1) the herbage must include at least 300 kg DM ha^{-1} of green-coloured biomass; and (2) the canopy of woody plants must occupy less than 10% of the surface of the terrain. Since these constraints do not apply to most of the Sahelian region, this approach may become a means of regional, national and international planning for more rational management of the grasslands (Le Houérou, 1988a, b).

Palatability is highly variable among species. Most grasses and many forbs are palatable in the early stages of their development. However, some grasses and forbs are seldom grazed. Most woody species are grazed during the dry season, species of Fabaceae and Capparidaceae being especially palatable. However, some woody species are unpalatable or poisonous.

Measured values of forage intake by livestock are as follows (in terms of kg DM ingested per day per 100 kg of live weight of animal): camels, 2.0 to 2.2; cattle, 2.5 to 2.6; sheep, 3.0 to 4.0; and goats, 5.0 to 6.0 (Dicko, 1980, 1985; Le Houérou, 1980c; Wilson, 1983). However, apparent intake (the difference between the amount of forage available and that refused) is usually from 2 to 3 times as great. This indicates that from one-half to two-thirds of the available forage is wasted in the grazing process, as shown in experiments conducted in Mali (Hiernaux et al., 1978; Dicko and Sangaré, 1984). On the other hand, the proportion of the maximum above-ground standing crop biomass that is consumable varies from 30 to 70%, depending on range type.

The proper use factor has not been determined accurately. However, there is a growing consensus that only 25 to 30% of the maximum standing crop of herbage should be considered as consumable forage when carrying capacity is calculated, rather than the level of 40 to 50% that was used until recently.

The nutritive quality of herbage is high during the rainy season, when crude protein content ranges between 8 and 14%, but it is not adequate for maintenance of livestock during the dry season (crude protein 1.8 to 3.5%).

Browse is of paramount importance to management of rangelands in the Sahel, not primarily as a source of energy but because it provides a source of protein, carotene and phosphorus for livestock during the dry season when the nutrient quality of herbage is very low (Le Houérou, 1980b, c). Yield studies of browse species have been made only recently, because their nutritive importance was not appreciated and because measurements of browse production are difficult to undertake. The available information has been synthesized by Le Houérou (1980b, c) and Von Maydell (1983). About 100 important browse species occur. The annual increment to woody biomass above ground averages about 55%, and the deciduous parts compose about 45% of the yield (Bille, 1978). Mean yield of browse in 1980 ranged from about 15 g m^{-2} in the Saharo–Sahelian subzone to 50 g m^{-2} in the Soudano–Sahelian subzone. About half of the annual yield is estimated to be available to livestock. Mean content of crude protein (nitrogen × 6.25) in browse averages 12.5%, mineral content 10.9%, and phosphorus 0.15%. Underground biomass is nearly equal to above-ground biomass, fibrous roots and large roots being of nearly equivalent weight.

Poor health conditions combine with inadequate nutrition to cause low reproductive rates, poor weight gains, and high mortality in livestock. Cattle do not reach their adult weight before the age of 5 to 6 years. Numerous serious diseases and internal parasitism prevail. Curbing of disease outbreaks by large-scale vaccination campaigns during the last three decades is largely responsible for increases in livestock populations. Contrary to the situation in East Africa, tick-borne diseases are not of great significance. Internal parasitism flourishes because water holes are continuously reinfected. Unlike East Africa, predation is not significant, but *Canis aureus* (jackal) and *Hyaena hyaena* kill small stock and, occasionally, young calves and sick animals.

Milk is the main production goal of Sahelian pastoralists. Wilson (1984) estimated that 75% of cattle, goats and camels in tropical Africa are maintained for milk production. The amount of milk produced is very low, averaging from 500 to 800 kg yr^{-1} in cows, 110 to 150 kg yr^{-1} in goats, 120 to 150 kg yr^{-1} in ewes, and 750 to 1500 kg yr^{-1} in dromedaries (Wilson, 1983; Wilson et al., 1983). The milk is shared equally between the herders and suckling young livestock, resulting in very low rates of growth and high mortality rates (30–40%) among the young animals.

Meat production is very low, with annual harvests of 10 to 15% in cattle, 20 to 30% in small stock, and 6 to 10% in camels (Wilson et al., 1983). Most meat is from culled animals and is of poor quality.

Rearing of livestock is undertaken for several other reasons as well. Pack animals are the principal means of transport for rural people. Draught animals are used for ploughing and in lifting water from wells. Hides, skins, wool and hair are important in trade, and are used for tent construction, clothing, carpets, bedding, saddlery, craftwork and manufacture of containers. Many hides and skins are of poor quality due to brand marks, poor curing practices, and damage from insects. The value of the skin is from 5 to 20% of that of the animal. Wool is of little importance, wool sheep yielding only about 700 g per head annually. An adult male hair sheep produces about 200 g annually. Dung is not yet used for fuel, but for fertilization of cropland. Also, it is occasionally mixed with straw and used as mortar in building. Farmers arrange with pastoralists for their stubble fields to be grazed, as a means of increasing their fertility. Unlike the situation in East Africa, blood is never consumed by Sahelian pastoralists. The social value of owning herds and the privileges derived therefrom have great significance, apart from the monetary value of the animals.

Woodlands

Wood production is important because it contributes 90 to 95% of the energy used by the rural human population (Le Houérou, 1979a; Keita, 1982). The annual increment of aerial woody biomass is about 40 to 60% of total increase in biomass of the canopy, the remainder being deciduous leaves, flowers, fruits and bark. The mean annual wood production has been estimated to be 5 g m^{-2} in the Saharo–Sahelian subzone, 20 g m^{-2} in the Sahelian subzone, and 50 g m^{-2} in the Soudano–Sahelian subzone (Bailly et al., 1982; Keita, 1982).

Croplands

The area seeded to rain-fed crops (mostly millet, *Pennisetum glaucum*) has expanded at an average exponential rate of 2 to 3% annually in the two southern subzones over the past half century (Gaston, 1975, 1981; Lamprey, 1975; Le Houérou, 1976a, 1979b; De Wispelaere and Toutain, 1976, 1981; De Wispelaere, 1980; Haywood, 1981; Barral et al., 1983; Barry et al., 1983; Peyre de Fabrègues and De Wispelaere, 1984; Le Houérou and Gillet, 1986). This has been achieved partly by reducing the length of the fallow period between successive periods of cropping and partly by expanding cropland to areas less desirable for crop production because of soil or climatic constraints. In the past, land was cropped for 2 or 3 years and then rested for a decade or more, before the next period of cropping, to permit restoration of fertility. Due to pressure for increased production of food, the fallow period has been shortened to 1–3 years or abandoned altogether. This has resulted in decreases in crop yields and has caused expansion of crop production to areas that are increasingly less suited to cropping. For example, since the 1950s the northern limit of crop production in Niger has moved northward from the 400-mm isohyet (15°N) to the 250-mm isohyet (16°20'N), a distance of 150 km — 100 km beyond the legal limit allowed under a 1961 ordinance (which was not enforced). This increased the rate of crop failure and has resulted in the destruction of large tracts of natural grassland (Bernus, 1981). Thus, competition between pastoralism and farming has become more and more acute.

IMPACT OF MAN AND LIVESTOCK

Archaeological evidence indicates that there has been a strong human impact on the environment in the Sahel for more than 2000 years (Mauny, 1956, 1961; Monod and Toupet, 1961; McIntosh and McIntosh, 1981a, b, 1983). Livestock husbandry has been practised for at least 7500 years. Holocene rock paintings and engravings in the Saharan mountains and in the Sahel indicate that a "bovidian" pastoral system existed in the desert and its southern fringe from 5000 to 1000 B.C. (Lhote, 1958; Mori, 1965; Hugo, 1974). All cattle depicted by this art are of the *Bos taurus* (*B. africanus*, *B. ibericus*) type. The zebu (*B. indicus*) type, which originated in the Indian subcontinent and now represents more than 95% of the Sahelian cattle population, was not introduced until about 400 A.D. (Epstein, 1971; Mason, 1984; Reed, 1984).

The human population averages about 5 km^{-2} in the Saharo–Sahelian subzone, from 5 to 10 km^{-2} in the Sahelian subzone, and from 10 to 15 km^{-2} in the Soudano–Sahelian subzone. These populations are much denser than they were at mid-century. Since then, the exponential rate of demographic growth among pastoralists has been 1.5 to 2.5% per annum, and 3 to 3.5% among settled farmers (Le Houérou, 1985). These population levels are far above the estimated potential carrying capacities for humans under low or intermediate inputs. The Saharo–Sahelian and Sahelian subzones are mainly occupied by nomadic and transhumant pastoralists. Some are long-range nomads; others are short- and long-range transhumants. The transhumants are the most common. They move from dry-season ranges in the Soudano–Sahelian subzone and northern Soudanian zone to rainy-season ranges in the Sahelian or Saharo–Sahelian subzones (Dupire, 1962; Gallais, 1967, 1975; Tubiana and Tubiana, 1977; Bernus, 1981; Bonte et al., 1981). The inhabitants of the Soudano–Sahelian subzone are predominantly settled farmers, with dry-season incursions of transhumant pastoralists. Both sedentary farming and semi-nomad farming systems are based on subsistence staple food crops in the clay depressions, which are alternately cropped and fallowed. After harvest in September, crop residues are grazed (and land is manured) by resident and nomadic herds. As the human population has grown, less land has been fallowed, resulting in declines in soil fertility and crop yields.

Increasingly heavy use of the grassland for grazing by livestock, combined with the effects of drought, has caused extremely severe degradation. Desertization has expanded over very extensive areas. In the Republic of Sudan, for example, Lamprey (1975) reported that the southern limit of the Sahara desert shifted 80 to 100 km southward between 1958 and 1975. The effect of overgrazing is to: (1) eliminate perennial grasses (species of *Andropogon*, *Aristida* and others); (2) replace mesic annual grasses of good forage value (species of *Brachiaria*, *Digitaria*, *Panicum*, *Pennisetum*, *Setaria*, *Sorghum* and *Urochloa*) with more xeromorphic and less palatable ones (species of *Aristida*, *Cenchrus*, *Chloris*, *Eragrostis* and *Stipagrostis*); and (3) permit invasion of less productive forbs of lower nutritive value (species of *Blepharis*, *Borreria*, *Cassia*, *Gisekia*, *Limeum*, *Mollugo*, *Tribulus* and *Zornia*) and of weedy shrubs (*Calotropis procera* and *Nicotiana glauca*) (Le Houérou and Gillet, 1986).

The soil surface becomes denuded surrounding watering places and other areas where livestock congregate (Fig. 8.7). For instance, the area of bare soil in central Mali increased from 4 to 26% between 1952 and 1975 (Le Houérou, 1979a, b; Haywood, 1981), and in Chad the herbaceous cover in the Sahel decreased by 32% between 1954 and 1974, while erosion increased by 28% and the southern limit of drifting sands moved 50 km southward (Gaston, 1975, 1981; Gaston and Dulieu, 1975). The surface of denuded sandy and silty soils often becomes covered by a crust of

Fig. 8.7. Depleted annual grassland in the Soudano–Sahelian subzone of central Mali. The large patches of bare ground are sealed on the surface with a biological crust composed of a species of *Scytonema* (Cyanophyceae). The woody layer (in poor condition) is of *Acacia seyal* and *Pterocarpus lucens*. The mean annual precipitation is about 550 mm.

algae (*Scytonema* spp.: Cyanophyceae) which renders it almost totally impervious to water (Barbey and Conte, 1976; Dulieu et al., 1977).

The effect of overuse by man and his livestock on the woody layers of shrub and woodland is even more catastrophic than on herbaceous communities. The cover of trees and shrubs has decreased by about 1% per year, in relative terms, since the 1950s over vast areas (Gaston, 1975, 1981; De Wispelaere, 1980; Le Houérou, 1980b; De Wispelaere and Toutain, 1981; Haywood, 1981; Barral et al., 1983; Peyre de Fabrègues, 1985). The regression of woody species is especially acute on hard ferruginous pans, surface-sealed silty soils, gravel plains, and pediments. In depressed topographic situations, however, ligneous vegetation sometimes increases in vigour and extent because of increased run-off from surrounding slopes (Haywood, 1981). On the other hand, in some locations trees and shrubs are killed by prolonged flooding. Another effect of overuse is the replacement of relatively mesic broad-leaved malacophyllous species by spiny microphyllous or sclerophyllous xerophytes.

Depletion of the woody vegetative layer is very serious from the point of view of range management, because this is the only source of protein, carotene, phosphorus and minerals for livestock during the dry season. Destruction of the woody elements of the vegetation renders rangeland unusable during the dry season, even if herbage is available, because satisfying nutritive deficiencies by feeding of supplemental concentrates to livestock is too costly to be economically viable (Le Houérou, 1979c).

IMPROVEMENT OF GRAZING LANDS

Improvement of depleted rangeland and better maintenance of all rangeland by improved management techniques are not possible under the prevailing socio-political constraints. These relate to the fact that land and water are communally owned, whereas the livestock are privately owned. This leads to a "looting strategy" (Le Houérou, 1976a) — also referred to by Hardin (1968) as "the tragedy of the commons" — whereby each individual's interest is to take immediate and full advantage of available resources regardless of the long-term consequences. Another difficulty is that low prices for animal products result from the monopolistic position of stock traders. But, even if these difficulties were removed, the high cost of fencing and of water development would interfere with implementation of controlled-grazing systems that are essential to recovery and maintenance of the rangeland. Fertilization also is not economically feasible. The only range management practice that is used widely is the maintenance of fire breaks. These are mandatory for any successful development scheme and are a charge of about 15% of the gross output of the range.

Improvement of depleted rangeland by reseeding to native or introduced forage species is not feasible with presently available technology, since at least 600 mm of annual rainfall is necessary for successful establishment of stands. Similarly, no long-term success has been achieved in establishing species of browse in areas with less than 400 mm of rainfall.

REFERENCES

Baasher, M.M., 1961. Range and livestock problems facing the settlement of nomads. *Proc. Sudan Philos. Soc.*, Khartoum.

Bagnouls, F. and Gaussen, H., 1953. Période sèche et végétation. *C. R. Acad. Sci.*, (Paris), 236: 1076–1077.

Bailly, C., Barbier, J., Clement, J., Goudet, J.P. and Hamel, O., 1982. *The Problems of Satisfying the Demand for Wood in the Dry Regions of Tropical Africa: Knowledge and Uncertainties*. Mimeo., CTFT, Nogent-sur-Marne, 24 pp.

Barbey, C. and Conte, A., 1976. Croûtes à Cyanophycées sur les dunes du Sahel Mauritanien. *Bull. IFAN, Ser. A*, 38: 732–736.

Barral, H., Benefice, E., Boudet, G., Denis, J.P., De Wispelaere, G., Diaïté, I., Diaw, O.L., Dieye, K., Doutre, M.P., Meyer, J.F., Noel, J., Pateut, G., Piot, J., Planchenault, D., Santoir, C., Valentin, C., Valenza, J. and Vassiliades, G., 1983. *Systèmes de production d'élevage au Sénégal dans la région du Ferlo*. GERDAT/ORSTOM, Paris, 172 pp.

Barry, J.P., Boudet, G., Bourgeot, A., Celles, J.C., Coulibaly, M., Leprun, J.C. and Manière, R., 1983. *Etudes des potentialités pastorales et de leur évolution en milieu sahélien du Mali*. GERDAT/ORSTOM, Paris, 116 pp.

Bernus, E., 1981. *Touaregs nigériens: unité culturelle et diversité régionale d'un peuple de pasteurs*. Mém. 94, ORSTOM, Paris, 508 pp.

Bille, J.C., 1977. Etude de la production primaire nette d'un écosystème sahélien. *Trav. Doc. ORSTOM*, 65, Paris, 82 pp.

Bille, J.C., 1978. Woody forage species in the Sahel: their biology and use. *Proc. 1st Int. Rangelands Congress*, Denver, Colo., pp. 392–395.

Bonte, P., Bourgeot, A., Digeard, J.P. and Lefebure, C., 1981. Human occupation: pastoral economies and societies. In: A. Sasson (Editor), *Tropical Grazing-Land Ecosystems*. Natural Resources Research 16, UNESCO, Paris, pp. 260–302.

Boudet, G., 1977. Désertification ou remontée biologique au Sahel. *Cah. ORSTOM, Ser. Biol.*, 12: 293–300.

Boudet, G., 1984. Recherche d'un équilibre entre production animale et ressources fourragères au Sahel. *Bull. Soc. Languedoc. Geogr.*, 18: 167–177.

Boudet, G. and Leclercq, P., 1970. *Etude agrostologique pour la création d'un ranch d'embouche dans la région de Niono (République du Mali)*. Etud. Agrostol. 19, IEMVT, Maisons-Alfort, 269 pp.

Boudet, G. and Rivière, R., 1968. Emploi pratique des analyses fourragères pour l'appréciation des pâturages tropicals. *Rev. Elev. Med. Vet. Pays Trop.*, 21: 227–266.

Bourlière, F., 1962. Les populations d'ongulés sauvages africains: caractéristiques écologiques et implications économiques. *Terre Vie*, 109: 150–160.

Bourlière, F., 1978. La savane sahélienne de Fété-Olé, Sénégal. In: M. Lamotte and F. Bourlière (Editors), *Problèmes d'écologie: structure et fonctionnement des écosystèmes terrestres*. Masson, Paris, pp. 187–229.

Bourlière, F. (Editor), 1983. *Tropical Savannas*. Ecosystems of the World, 13, Elsevier, Amsterdam, 730 pp.

Breman, H. and De Wit, C.T., 1983. Rangeland productivity and exploitation in the Sahel. *Science*, 221: 1341–1347.

Charreau, C. and Vidal, P., 1965. Influence d'*Acacia albida* Del. sur le sol, la nutrition minérale et les rendements des mils *Pennisetum* au Sénégal. *Agron. Trop.*, 20: 600–626.

Clanet, J.C. and Gillet, H., 1980. *Commiphora africana*: a browse tree of the Sahel. In: H.N. Le Houérou (Editor), *Browse in Africa*. ILCA, Addis Ababa, pp. 443–448.

Cochemé, J. and Franquin, P., 1967. *Etude agroclimatologique de l'Afrique sèche au Sud du Sahara en Afrique Occidentale*. Projet conjoint, OMN/UNESCO/FAO, FAO, Rome, 325 pp.

Coe, M.J., Cumming, D.H. and Phillipson, J., 1976. Biomass and production of large African herbivores in relation to rainfall and primary production. *Oecologia*, 22: 341–354.

Cornet, A., 1981. *Le bilan hydrique et son rôle dans la production de la strate herbacée de quelques phytocoenoses sahéliennes au Sénégal*. Thèse Dr.-Ingr., Université des Sciences et Technologique du Languedoc, Montpellier, 353 pp.

Cornet, A. and Rambal, S., 1981. Simulation de l'utilisation de l'eau et de la production végétale d'une phytocoenose sahélienne du Sénégal. *Oecol. Plant*, 4: 381–397.

Dancette, C., 1979. Agroclimatologie appliquée à l'économie de l'eau en zone Soudano–Sahélienne. *Agron. Trop.*, 34: 331–355.

Dancette, C., 1985. Contrariétés pédoclimatiques et adaptation de l'agriculture à la sècheresse en zone intertropicale. Actes du Colloque, *Résistance à la sècheresse en milieu tropical: quelques recherches à moyen terme*. CIRAD/GERDAT/ISRA, Montpellier–Dakar, pp. 27–41.

Davy, E.G., Mattei, F. and Solomon, S.F., 1976. *An evaluation of climate, water resources for development of agriculture in the Sudano–Sahelian zone of west Africa*. Special Report 9, WMO, Geneva, 289 pp.

Dekeyser, P.L., 1955. *Les mammifères de l'Afrique noire française*. IFAN, Dakar, 426 pp.

Delany, M.J. and Happold, D.C.D., 1979. *Ecology of African Mammals*. Longman, London.

Delwaulle, J.C., 1973a. Désertification au Sud de Sahara. *Bois For. Trop.*, 149: 51–68.

Delwaulle, J.C., 1973b. L'érosion au Niger. *Bois For. Trop.*, 150: 15–37.

De Wispelaere, G., 1980. Les photographies aériennes témoins de la dégradation du couvert ligneux dans un écosystème Sahélien Sénégalais. Influence de la proximité d'un forage. *Cah. ORSTOM*, 18: 155–166.

De Wispelaere, G. and Toutain, B., 1976. Estimation de l'évolution du couvert végétale en 20 ans, consécutivement à la sècheresse dans le Sahel voltaïque. *Rev. Photointerpret.*, 76: 3/2.

De Wispelaere, G. and Toutain, B., 1981. Etude diachronique de quelques géosystèmes sahéliens en Haute Volta septentrionale. *Rev. Photointerpret.*, 81: 1/1–1/5.

Dicko, M.S., 1980. Measuring the secondary production of pasture: an applied example in the study of an extensive production system in Mali. In: H.N. Le Houérou (Editor), *Browse in Africa*. ILCA, Addis Ababa, pp. 247–254.

Dicko, M.S., 1985. *Nutrition des bovins du système agropastoral du mil en zone sahélienne: comportement et ingestion volontaire*. Multigr., CIPEA/ILCA, Niamey.

Dicko, M.S. and Sangaré, M., 1984. Feeding behaviour of domestic ruminants in Sahelian zones. *Proc. 2nd Int. Rangelands Congress*, Adelaide, Australia, pp. 388–390.

Dorst, J. and Dandelot, P., 1970. *A Field Guide to the Larger Animals of Africa*. Delochaux and Nestlé, Neuchâtel, 286 pp.

Dulieu, D., Gaston, A. and Darley, J., 1977. La dégradation des pâturages de la région de N'Djamena (République du Tchad), en relation avec la présence de Cyanophyceae psammophiles. Etude préliminaire. *Rev. Elev. Med. Vet. Pays Trop.*, 30: 181–190.

Dupire, M., 1962. Peuls nomades — étude descriptive des Woodabe du Sahel Nigérien. *Trav. Mem. Inst. Ethnol.*, 64, Paris, 336 pp.

Epstein, H., 1971. *The Origin of the Domestic Animals of Africa*. Two Vols. Afr. Publ. Corp., New York, N.Y., 573 and 719 pp.

FAO, 1950/1983. *Production Yearbooks*. FAO, Rome.

FAO/UNESCO, 1973. Soil map of Africa. *Soil Map of the World* 1/5 000 000: sheets VI-1 and VI-2, FAO, Rome/UNESCO, Paris.

Fauck, R., 1973. Les sols rouges sur sables et sur grès de l'Afrique Occidentale. *Contribution à l'étude des sols des régions tropicales*, Mem. 61, ORSTOM, Paris, 258 pp.

Gallais, J., 1967. *Le delta intérieur du Niger, étude de géographie régionale*. Mém. 79, IFAN, Dakar, 621 pp.

Gallais, J., 1975. *Pasteurs et paysans du Gourma: la condition sahélienne*. Mem., CEGET/CNRS, Paris, 239 pp.

Gaston, A., 1975. *Étude des pâturages du Kanem après la sècheresse de 1973*. Multigr., IEMVT, Farcha, Tchad, 24 pp.

Gaston, A., 1981. *La végétation du Tchad, évolutions récentes sous les influences climatiques et humaines*. Thèse Doct. Sci.,

Université de Paris XII et IEMVT, Maisons-Alfort, 333 pp.
Gaston, A. and Dulieu, D., 1975. *Pâturages du Kanem. Effets de la sècheresse du 1973*. IEMVT, Maisons-Alfort, 175 pp.
Gaston, A. and Lemarque, G., 1976. Bilan de quatre années de travaux phytoécologiques en relation avec la lutte contre *Quelea quelea*. *Etud. Agrostol. en sous-traitance*, 25, IEMVT, Maisons-Alfort, 203 pp.
Gillet, H., 1965. L'Oryx algazelle et l'Addax au Tchad. *Terre Vie*, 3: 257–272.
Gillet, H., 1967. Essai d'évaluation de la biomasse végétale en zone sahélienne. *J. Agric. Trop. Bot. Appl.*, 14: 123–158.
Gillon, Y. and Gillon, D., 1973. Recherches écologiques sur une savane sahélienne du Ferlo septentrional, Sénégal: données quantitatives sur les arthropodes. *Terre Vie*, 27: 297–323.
Gillon, Y. and Gillon, D., 1974a. Recherches écologiques sur une savane sahélienne du Ferlo septentrional, Sénégal: données quantitatives sur les ténébrionides. *Terre Vie*, 28: 296–306.
Gillon, D. and Gillon, Y., 1974b. Comparaison du peuplement d'invertébrés de deux milieux herbacés Ouest-Africains: sahel et savane préforestière. *Terre Vie*, 28: 429–474.
Grouzis, M., 1984. *Pâturages sahéliens du Nord du Burkina-Faso*. ORSTOM, Ouagadougou, 35 pp.
Grouzis, M., 1988. *Structure, productivité et dynamique des systèmes écologiques sahéliens (Mare d'Oursi, Burkina Faso)*. Mém. de thèses, Université de Paris-Sud, Mém. et Doc. de l'ORSTOM, Paris, 333 pp.
Haltenorth, T., Diller, H. and Cusin, M., 1985. *Mammals of Africa and Madagascar*. Delochaux and Nestlé, Neuchâtel, 397 pp., 287 figs.
Happold, D.C.D., 1973. The distribution of large mammals in West Africa. *Mammalia*, 37: 90–93.
Hardin, G., 1968. The tragedy of the commons. *Science*, 162: 1243–1248.
Haywood, M., 1981. *Evolution de l'utilisation des terres et de la végétation dans la zone Soudano–Sahélienne du Projet CIPEA au Mali*. Multigr. Doc. de Trav. 3, Int. Livestock Centre for Africa, Addis Ababa, 187 pp.
Heusch, B., 1975. *La conservation des eaux et des sols dans la haute vallée de Keita (République du Niger)*. Multigr., SOGREAH, Grenoble, 31 pp.
Hiernaux, P., Cissé, M.I. and Diarra, L., 1978. *Rapport de la section d'écologie*. Multigr., Diff. Restr., ILCA/CIPEA, Bamako, 92 pp.
Hugo, H.J., 1974. *Le Sahara avant le désert*. Hespérides, Paris, 343 pp.
Hunting Technical Services, 1964. *Land and Water Use Survey in the Kordofan Province of the Republic of Sudan*. Hunting Technical Services, London; FAO, Rome, 349 pp.
Justice, C.O., 1986. Monitoring the grasslands of semi-arid Africa, using NOAA/AVHRR data. *Int. J. Remote Sensing*, 7: 1383–1622.
Keita, M.N., 1982. *Les disponibilités de bois de feu en région sahélienne de l'Afrique Occidentale: situation et perspectives*. Mimeo., Misc. 82/15, Dept. of Forestry, FAO, Rome, 79 pp.
Lamprey, H.F., 1975. *Report on the desert encroachment reconnaissance in northern Sudan*. Mimeo., UNEP/UNESCO, Nairobi, 14 pp.
Lawton, R.M., 1980. Browse in the Biombo woodland. In: H.N. Le Houérou (Editor), *Browse in Africa*. ILCA, Addis Ababa, pp. 25–34.
Le Houérou, H.N., 1976a. Nature and causes of desertization. In: M. Glantz (Editor), *Desertification In and Around Arid Lands*. Westview Press, Boulder, Colo., pp. 17–38.
Le Houérou, H.N., 1976b. *Nature et désertisation*. Cpte-Rend., Consultation sur la Forestrerie au Sahel. CILSS/UNESCO/FAO, FO:RAF/305/3, FAO, Rome, 21 pp.
Le Houérou, H.N., 1977. The grassland of Africa: classification, production, evolution and development outlook. *Proc. 13th Int. Grassland Congress*. Akademie Verlag, East Berlin, pp. 99–116.
Le Houérou, H.N., 1979a. Écologie et désertisation en Afrique. *Trav. Inst. Geogr. Reims*, 39–40: 5–26.
Le Houérou, H.N., 1979b. La désertisation des régions arides. *Recherche*, 99: 336–344.
Le Houérou, H.N., 1979c. Le rôle des arbres et arbustes dans les pâturages sahéliens. In: *Le rôle des arbres au Sahel*. Cpte-Rend., Colloque de Dakar, IDRC/CRDI, Ottawa, pp. 19–32.
Le Houérou, H.N. (Editor), 1980a. *Browse in Africa: The Current State of Knowledge*. ILCA, Addis Ababa, 491 pp. (also available in French).
Le Houérou, H.N., 1980b. The role of browse in the Sahelian and Sudanian zones. In: H.N. Le Houérou (Editor), *Browse in Africa*. ILCA, Addis Ababa, pp. 83–102.
Le Houérou, H.N., 1980c. Chemical composition and nutritive value of browse in West-Africa. In: H.N. Le Houérou (Editor), *Browse in Africa*. ILCA, Addis Ababa, pp. 261–290.
Le Houérou, H.N., 1981. The rangelands of the Sahel. *J. Range Manage.*, 33: 41–46.
Le Houérou, H.N., 1982/1984. An outline of the bioclimatology of Libya. *Bull. Soc. Bot. Fr.*, 131, Actual. Bot., 2–4: 157–178.
Le Houérou, H.N., 1984. Rain-use efficiency: a unifying concept in arid land ecology. *J. Arid Environ.*, 7: 213–247.
Le Houérou, H.N., 1985. The impact of climate on pastoralism. In: R.W. Gates, J.H. Ausubel and M. Berberian (Editors), *Climate Impact Assessments*. SCOPE 27, Wiley, New York, N.Y., pp. 155–185.
Le Houérou, H.N., 1986. *Surveillance continue des écosystèmes pâtures sahéliens au Sénégal*. Rapport technique de synthèse, FAO, Rome/UNEP, Nairobi, 50 pp.
Le Houérou, H.N., 1988a. *Inventory and Monitoring of Sahelian Pastoral Ecosystems*. FAO, Rome/UNEP, Nairobi, 144 pp., 69 figs., 19 tables.
Le Houérou, H.N., 1988b. *Introduction au Projet Écosystèmes Pastoraux Sahéliens*. FAO, Rome/UNEP, Nairobi, 146 pp., 69 figs., 19 tables.
Le Houérou, H.N., 1988c. Forage species diversity in Africa: an overview of the plant resources. In: F. Attere, H. Zedan, N.Q. Ng and P. Errino (Editors), *Crop Genetic Resources in Africa*, Vol. I. International Board for Plant Genetic Resources (Rome), UNEP (Nairobi), IITA (Ibadan) and CNR (Rome), pp. 99–117.
Le Houérou, H.N., 1989. *The Grazing Land Ecosystems of the*

African Sahel. Ecological Studies, 75, Springer-Verlag, Heidelberg, 282 pp.
Le Houérou, H.N. and Gillet, H., 1986. Conservation versus desertization in African arid lands. *Proc. 2nd Int. Conf. Conservation Biology*. University of Michigan, Ann Arbor, pp. 444–461.
Le Houérou, H.N. and Grenot, C.J., 1989. Wildlife. In: *The Grazing Land Ecosystems of the African Sahel*. Ecological Studies, 75, Springer-Verlag, Heidelberg, pp. 113–123.
Le Houérou, H.N. and Hoste, C., 1977. Rangeland production and annual rainfall relations in the Mediterranean basin and in the African Sahelian and Sudanian zones. *J. Range Manage.*, 30: 181–189.
Le Houérou, H.N. and Naegelé, A.F.G., 1972. *The Useful Shrubs of the Mediterranean Basin and the Arid Tropical Belt South of the Sahara*. AGPC: Misc. 24, FAO, Rome, 20 pp. (Also in: C.M. McKell, J.P. Blaisdell and J.R. Goodin (Editors), *Wildland Shrubs — Their Biology and Utilization*. Gen. Rept., INT-I, USDA, Forest Service, Intermountain Forest and Range Exp. Sta, Ogden, Utah, pp. 26–36.)
Le Houérou, H.N. and Norwine, J.R., 1987. Ecoclimatic investigations in southern Texas. *Proc. Int. Arid Lands Conference*. Office of Arid Lands Studies, University of Arizona, Tucson, pp. 417–443.
Le Houérou, H.N. and Popov, G.F., 1981. *An Ecoclimatic Classification of Inter-Tropical Africa*. Plant Production and Protection Paper 31, FAO, Rome, 40 pp.
Leprun, J.C., 1978. *Lutte contre l'aridité en milieu tropical, étude de l'évolution d'un système d'exploitation sahélien au Mali; Compte rendu de fin d'étude sur les sols et leur susceptibilité à l'érosion, les terres de cure salée, les formations de brousse tigrée dans le Gourma*. DGRST/ORSTOM, Paris, 51 pp.
Lhote, H., 1958. *Á la découverte des fresques du Tassili*. Arthaud, Paris, 268 pp.
Mainguet, M., 1972. *Le modelé des grès. Problèmes généraux*. Two Vols., Inst. Géogr. National, Paris, 657 pp.
Mainguet, M., 1976. Propositions pour une nouvelle classification des édifices sableux éoliens d'après les images Landsat I, Gemini, et NOAA 3. *Z. Geomorphol.*, N.F. (Berlin–Stuttgart), 20: 275–296.
Mainguet, M., 1984. A classification of dunes based on aeolian dynamics and the sand budget. In: F. Elbaz (Editor), *Deserts and Arid Lands*. Martinus Nijhoff, The Hague, pp. 31–38.
Mainguet, M., Canon, L. and Chemin, M.C., 1980. Le Sahara: géomorphologie et paléogéomorphologie éolienne. In: M.A.J. Williams and H. Faure (Editors), *The Sahara and the Nile*. A.A. Balkema, Rotterdam, pp. 17–35.
Mason, I.L., 1984. *Evolution of Domesticated Animals*. Longman, London, 452 pp.
Mauny, R., 1956. Préhistoire et zoologie: la grande faune éthiopienne du Nord-Ouest Africain du Paléolithique à nos jours. *Bull. IFAM*, 18(1): 246–249.
Mauny, R., 1961. *Tableau géographique de l'Ouest-Africain au Moyen-age, d'après les sources écrites, la tradition et l'archéologie*. Mém. 61, IFAN, Dakar, 587 pp.
McIntosh, R.J. and McIntosh, S.K., 1981a. The inland Niger delta before the Empire of Mali: evidence from Jenne-Jeno. *J. Afr. Hist.*, 22: 1–22.
McIntosh, S.K. and McIntosh, R.J., 1981b. West-African prehistory from 10,000 B.C. to 1,000 A.D. *Am. Sci.*, 69: 602–613.
McIntosh, S.K. and McIntosh, R.J., 1983. Current directions in West-African prehistory. *Ann. Rev. Anthropol.*, 12: 215–258.
Michel, P., 1973. *Les bassins des fleuves Sénégal et Gambie*. Three Vols. Mém. 63, ORSTOM, Paris, 752 pp.
Monod, T., 1958. *Majabat al Koubrâ. Contribution à l'étude de l'"Empty quarter" Ouest saharien*. Mém. 52, IFAN, Dakar, 407 pp.
Monod, T., 1986. The Sahel zone north of the equator. In: M. Evenari, I. Noy-Meir and D.W. Goodall (Editors), *Hot Deserts and Arid Shrublands*. Ecosystems of the World, 12B, Elsevier, Amsterdam, pp. 203–244.
Monod, T. and Toupet, C., 1961. Land-use in the Sahara–Sahel region. In: L. Dudley-Stamp (Editor), *A History of Land-Use in Arid Regions*. Arid Zone Research, Vol. 17, UNESCO, Paris, pp. 239–253.
Morel, G., 1968a. *Contribution à la synécologie des oiseaux du Sahel Sénégalais*. Mém. 29, ORSTOM, Paris, 180 pp.
Morel, G., 1968b. L'impact écologique de *Quelea quelea* sur les savanes sahéliennes: raison du pullulement de ce plocéide. *Terre Vie*, 22: 69–98.
Mori, F., 1965. *Tadrart–Acacus: arte rupestre e culture del Sahara prehistorico*. Einaudi, Torino, 257 pp.
Naegelé, A.F.G., 1971. Etude et amélioration de la zone pastorale du Nord-Sénégal. *Étude Pâturages et Cultures Fourragères*, 4, AGPC, FAO, Rome, 163 pp.
Penman, H.L., 1948. Natural evaporation from open water, bare soil and grass. *Proc. R. Soc. London*, A, 193: 120–146.
Penning de Vries, F.W.T. and Djiteye, A.M. (Editors), 1982. *La productivité des pâturages sahéliennes*. PUDOC, Wageningen, 525 pp.
Peyre de Fabrègues, B., 1985. Quel avenir pour l'élevage au Sahel? *Rev. Elev. Med. Vet. Pays Trop.*, 38: 500–508.
Peyre de Fabrègues, B. and De Wispelaere, G., 1984. Sahel: fin d' un monde pastoral? *Marchés Tropicaux*, 12 Oct. 1984, pp. 2488–2491.
Pias, J., 1970. *Les formations sédimentaires Tertiaires et Quaternaires de la cuvette tchadienne et les sols qui en dérivent*. Mém. 43, ORSTOM, Paris, 407 pp.
Piot, J. and Milogo, G., 1980. *Etude du ruissellement et à l'érosion dans l'Oudalan*. CIFT/ORSTOM, Ouagadougou, multigr., 33 pp.
Poulet, A.R., 1972. Recherches écologiques sur une savane sahélienne du Ferlo septentrional, Sénégal: les mammifères. *Terre Vie*, 22: 440–472.
Poulet, A.R., 1982. *Pullulation de rongeurs dans le Sahel. Mécanismes et déterminismes du cycle d'abundance de Taterillus pygargus et d'Arvicanthis niloticus (Rongeurs Gerbillides et Murides) dans le Sahel du Sénégal de 1975 à 1977*. ISBN 0638-4, ORSTOM, Paris, 368 pp.
Reed, C.A., 1984. The beginning of animal domestication. In: I.L. Mason (Editor), *Evolution of Domesticated Animals*. Lampman, London, pp. 1–6.
Riou, C., 1975. *La détermination pratique de l'évaporation. Application à l'Afrique Centrale*. Mém. 80, ORSTOM, Paris, 236 pp.

Rodier, J.A., 1964. *Régimes hydrologiques de l'Afrique Noire à l'Ouest du Congo*. Mém. 6, ORSTOM, Paris, 138 pp.

Rodier, J.A., 1975. *Évaluation de l'écoulement annuel dans le Sahel Tropical Africain*. Trav. Doc. 46, ORSTOM, Paris, 122 pp.

Roose, E.J., 1977. *Erosion et ruissellement en Afrique de l'Ouest. Vingt années de mesures en petites parcelles experimentales*. Trav. Doc. 78, ORSTOM, Paris, 108 pp.

Roose, E.J., 1981. *Dynamique actuelle des sols ferrallitiques et ferrugineux tropicaux d'Afrique Occidentale*. Trav. Doc. 130, ORSTOM, Paris, 569 pp.

Roose, E.J. and Piot, J., 1984. Runoff, erosion and soil fertility restoration on the Mossi Plateau (Central Upper-Volta). *Proc., Harare Symp. Challenges in African Hydrology and Water Resources*. Publ. 144, IAHS, pp. 485–498.

Shepherd, W.O., 1968. *Range and pasture management*. Report to the Government of Sudan, TA Rept. 2468, FAO, Rome, 49 pp.

Shepherd, W.O. and Baasher, M.M., 1966. Forage resources of the Sudan savanna: potentials and problems. *Proc., UNDP Meeting on Savanna Development*, Khartoum, FAO, Rome, pp. 136–145.

Tubiana, M.J. and Tubiana, J., 1977. *The Zaghawa from an Ecological Perspective*. A.A. Balkema, Rotterdam, 119 pp.

Tucker, C.J., Vanpraet, C., Sharman, M.J. and Van Ittersum, G., 1985. Satellite remote sensing of the total herbaceous biomass production in the Senegalese Sahel: 1980–1984. *Remote Sensing Environ.*, 17: 233–249.

Valverde, J.A., 1968. *Ecological Bases for Fauna Conservation in Western Sahara*. Paper 18, Int. Symp., IBP/CT, Hammamet, Tunisia, and London, mimeo., 14 pp.

Vanpraet, C. (Editor), 1985. *Méthodes d'inventaire et de surveillance continue des écosystèmes pastoraux sahéliens: application au développement*. Minist. Rech. Sci. Technol., Dakar, Sénégal, 439 pp.

Von Maydell, H.J., 1983. *Arbres et arbustes du Sahel, leurs caractéristiques et leurs utilisations*. Schriftenr. 147, GTZ, Eschborn, 531 pp.

Walker, B.H., 1980. A review of browse and its role in livestock production in southern Africa. In: H.N. Le Houérou (Editor), *Browse in Africa*. ILCA, Addis Ababa, pp. 7–24.

Walter, H. and Lieth, H., 1967. *Klimadiagram Weltatlas*. Fischer Verlag, Jena, 200 pp.

White, F., 1983. *The Vegetation of Africa*. Nat. Res. No. XX, UNESCO, Paris, 356 pp.

Wilson, R.T., 1979a. Recent resource surveys for rural development in southern Darfur, Sudan. *Geogr. J.*, 145: 452–460.

Wilson, R.T., 1979b. Wildlife in southern Darfur, Sudan: distribution and status at present and in recent past. *Mammalia*, 43: 323–338.

Wilson, R.T., 1983. *The Camel*. Lampman, London, 223 pp.

Wilson, R.T., 1984. Demography, vital statistics and productivity of domestic animals under traditional management in arid and semi-arid northern tropical Africa. In: H.N. Le Houérou (Editor), *Advances in Desert and Arid Land Technology and Development: Range and Livestock in Africa*. Harvard Press, New York.

Wilson, R.T., De Leeuw, P.N. and De Haan, C. (Editors), 1983. *Recherches sur les systèmes des zones arides du Mali: résultats préliminaires*. Rapport de recherches 5, CIPEA/ILCA, Addis Ababa, 189 pp.

Chapter 9

GRASSLANDS OF EAST AFRICA [1]

D.J. HERLOCKER, H.J. DIRSCHL and G. FRAME

INTRODUCTION

East Africa is environmentally diverse. Situated astride the Equator, its groups of mountain ranges and highlands, rift valleys, large lakes and high plateaus adjoin the Zaïre basin on the east. This varied topography in East Africa breaks up the east–west trending vegetation belts, which parallel the Equator in West and central Africa and reflect increasingly drier conditions northward and southward.

A combination of these topographic factors, together with regional patterns of air currents related to the monsoonal system of the inter-tropical convergence zone, has resulted in a number of distinct ecological regions.

Each region is characterized by a particular set of environmental conditions (including those related to topography, geology and soils), but the most important factor influencing ecosystem development is the regional climate and, more specifically, the amount and seasonal distribution of rainfall. Proximity to large bodies of water (Indian Ocean and Lake Victoria), elevation above sea level, and slope aspect exert a strong influence on precipitation patterns. The nature of the vegetation is influenced by geological factors through their influence on soil development, but this influence is usually on too local a scale to be readily apparent at the regional level.

Man has had a pronounced influence on the vegetation of East Africa through such practices as clearing, burning and grazing. Many plant communities that are currently widespread have resulted from vegetation changes directly or indirectly caused by man's actions.

For the purpose of this description, the major vegetation regions of East Africa were delineated by overlaying and comparing the regional vegetation maps of Edwards (1940), Pichi-Sermolli (1955), Heady (1960), Rattray (1960), Langdale-Brown et al. (1964), Trapnell et al. (1966, 1969, 1976), Trapnell and Langdale-Brown (1969), Resource Management and Research (RMR) (1979, 1981, 1985), and White (1983). From this analysis, eight regions emerged, within which grassland communities are major components. The detail available on the structure, composition and dynamic relationships of the various grassland communities differs widely from region to region. The vegetation of Ethiopia has been discussed only briefly because of the paucity of information (Reilly, 1978). The *miombo* woodland region of southern Tanzania is essentially a southern African biome and, therefore, has not been dealt with in this chapter.

Grasslands free from any significant growth of woody plants are extremely limited within East Africa. Where such grasslands do occur, they are the consequence of local edaphic or pyric conditions. Thus, most grasslands include a woody component and appear to have been derived from woody communities through burning and clearing (Michelmore, 1939; White, 1962; Langlands, 1967; Vesey-FitzGerald, 1970, 1973; Jackson, 1978).

Abundant, seasonally well-distributed rainfall is reflected in natural climax formations of evergreen forest. Semi-deciduous forest characterizes somewhat drier conditions. Where the forest cover

---

[1] Manuscript submitted in May, 1982; revised in March, 1988.

has been destroyed, tall grasses (*Hyparrhenia*[1] spp.) and giant (high) grasses (*Pennisetum purpureum*) are characteristic of secondary succession in warmer sites at lower elevations. Grasses of medium height (*Eleusine jaegeri* and *Pennisetum sphacelatum*) and a short grass (*P. clandestinum*) are typical of secondary succession under cooler highland conditions.

Warm regions, with somewhat less rainfall and a definite (often short) dry season, support climax vegetation ranging from evergreen bushland to broad-leaved *Combretum–Butyrospermum–Terminalia* woodland and woodland savanna. Tall *Hyparrhenia* grasses are associated with these woody formations.

Medium to high elevations with cooler average temperatures and with a rainfall regime that is drier and more variable, and which experience long dry seasons (about six months), support climax vegetation ranging from low, dry evergreen forest to deciduous *Acacia–Commiphora* bushland. Medium-height *Themeda triandra* and tall *Hyparrhenia filipendula* are the associated grass species.

Low, hot drylands with low, extremely variable rainfall, high evaporation and a very long dry season support vegetation ranging from deciduous *Acacia–Commiphora* bushland, with either a *Chrysopogon plumulosus* or *Leptothrium senegalense* grass layer, to dwarf shrubland and annual grassland in which dwarf and short *Aristida* species are the associated grasses.

The grass species and genera used to typify each region characterize the grasslands of that region when they are at their most widespread. This is when they are subjected to light to moderate grazing and are burned periodically. Removal of fire from the system causes increasing dominance by shrubs and trees. Only in the drier regions, where fire seldom occurs, do these grasses comprise an important part of the climax vegetation.

Contributions to other volumes in this series relate to the discussion in this chapter. Menaut (1983) has treated the vegetation of the African savannas as a whole in Volume 13 ("Tropical Savannas"), and Monod (1986) has included parts of East Africa in his treatment of the Sahelian zone in Volume 12 ("Hot Deserts and Arid Shrublands").

PENNISETUM MID-GRASS REGION

Elevation and climate

This region (Figs. 9.1 and 9.2), which comprises the highlands, ranges from 1650 to 3350 m in elevation and occurs as scattered units over most of East Africa. In Ethiopia, it occurs down to about 750 m elevation.

Various parts of the region experience mean annual precipitation ranging from 890 to 2540 mm. Considerable variability occurs also in seasonal distribution of rainfall. Thus, the northern and southern extremities of this region experience a single rainy season, from about November to May in the Southern Highlands of Tanzania and about July to September in the Ethiopian highlands. Along the Equator in Kenya (particularly in eastern Kenya), rainfall occurs during two seasons (roughly about March–May and again in November–December) (Griffiths, 1969; Kenworthy, 1977; Nieuwolt, 1977). The mountains in Uganda and in south-western Ethiopia have rain nearly year-around (Wolde-Mariam, 1970).

Mists are frequent. Frost occurs occasionally in the more southerly and northerly areas and above 2400 m elevation on the Equator (Edwards, 1940; Gilliland, 1952; Van Rensburg, 1952; Heady, 1960; Rattray, 1960; Langdale-Brown et al., 1964; Griffiths, 1969).

Geology and soils

Throughout most of Ethiopia, Kenya and north-central Tanzania, the highland areas are the result of Tertiary to Recent volcanic activity. The better-known mountains (Kenya, Elgon and Kilimanjaro) are, in fact, volcanic cones. Elsewhere, the highlands consist of uplifted blocks of sedimentary and metamorphic rocks, some of which — such as the gneiss complex of Uganda — are of great antiquity (Saggerson, 1969; Wolde-Mariam, 1970).

---

[1] Nomenclature of plants primarily follows the still incomplete "Flora of Tropical East Africa" (Hubbard et al., 1952–81), except for that of domesticated plants which primarily follows Lötschert and Beese (1981).

GRASSLANDS OF EAST AFRICA

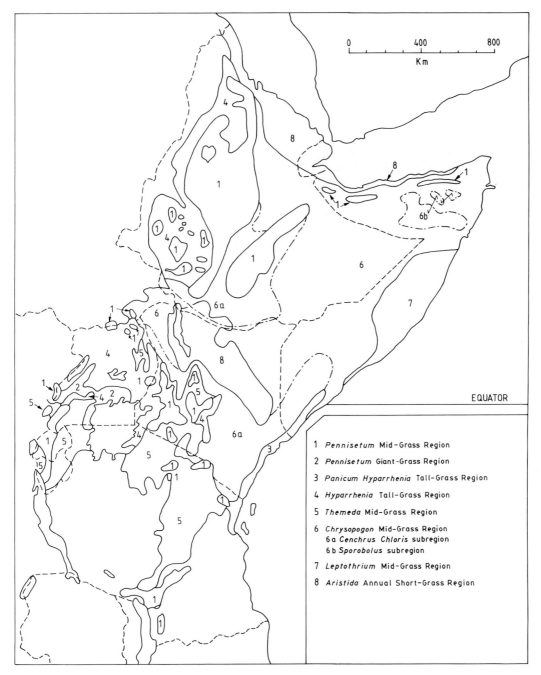

Fig. 9.1. Grassland regions of East Africa.

Altitudinal zonation of soil types is very evident in the highest mountains. The most common altitudinal sequence begins with dark brown loams occurring just below the alpine meadow zone, followed by dark-red friable clays on the mid-slopes and red friable clays on the lower slopes. All these soils are well drained and are ando-types at higher elevations and latosolic at lower elevations. The soil types typical of the middle and lower levels of the highest mountains are found over most of the remainder of the highland areas (Scott, 1969).

Fig. 9.2. Elephants passing through a stand of *Eleusine jaegeri*, an invader following heavy grazing, in the *Pennisetum* mid-grass region.

Vegetation

The grasslands of this region consist primarily of communities that are successional to evergreen montane forest and (in the drier sites) to evergreen bushland. Frequent mists may be responsible for the existence of the drier montane forests owing to fog drip and the reduction in evapotranspiration which they bring about. Cutting, fire, and grazing have created, and are now maintaining, these grassland communities (Edwards, 1940; Van Rensburg, 1952; Edwards, 1956).

Most of this process of forest destruction and subsequent formation of highland grasslands has occurred over the past two or three centuries, as human population densities and the need for agricultural land have increased (Van Rensburg, 1952). For instance, although the first agricultural tribes reached Mount Kenya about 1300 A.D., most forest clearing took place only during the period 1850 to 1900 (Were and Wilson, 1978). The beginnings of large-scale forest clearance in other parts of the East African highlands are of greater antiquity; as early as 1500 B.C. in northern Ethiopia (Phillipson, 1979) and around 950 A.D. in western Uganda (Lind and Morrison, 1974).

Because of the apparent ability of highland evergreen forest communities to occupy edaphically extreme sites, few natural edaphic grasslands have developed. Unlike at lower elevations, fire-tolerant trees and shrubs do not easily invade in this region following forest destruction (Michelmore, 1939). Instead, fire-subclimax grasslands develop which are dominated by grass species characteristic of lower elevations. This differentiates this zone floristically from the grasslands of the more elevated afro-alpine ericaceous belt, where mainly indigenous species are dominant (White, 1978).

Where soils are fertile and rainfall exceeds 1000 mm annually, forest destruction is commonly followed by short-grass communities dominated by *Pennisetum clandestinum* (Edwards, 1935, 1956). This is a densely matted stoloniferous and rhizomatous grass, less than 50 cm high (Bogdan, 1958a). Both rhizomes and stolons root freely from the nodes. Stems are usually short and leaves long, so that flowering can occur even when the sward is closely grazed. This grass establishes well both from seeds and from cuttings, is highly nutritious (crude protein is seldom under 12%), and responds well to fertilization (Bogdan, 1977b).

Cynodon dactylon and *Trifolium semipilosum* are commonly associated with *P. clandestinum*. However, these species are less tolerant of close grazing, so that heavily grazed swards are usually composed almost entirely of *P. clandestinum* (Edwards and Bogdan, 1951; Vesey-FitzGerald, 1974; Bogdan, 1977b). Ungrazed pastures are often invaded by *Digitaria abyssinica* (Edwards, 1940).

At elevations above 2000 m, grasslands dominated by *Pennisetum sphacelatum* and, more locally, by *Eleusine jaegeri* are extensive. These coarse-tufted bunch grasses, 120 to 180 cm high, are thought to invade grasslands as a result of intensive grazing (Fig. 9.2) or because soil fertility has declined owing to the extended period elapsed since the forest was cleared (Edwards, 1935, 1956; Bogdan, 1958a; Van Rensburg, 1960). These species are also found at lower elevations on sites too dry for *P. clandestinum* (Edwards, 1940; Van Rensburg, 1960). Being unpalatable to livestock, the tussocks of *E. jaegeri* and *P. sphacelatum* commonly form an intermittent layer above a close-cropped mat of species of *Andropogon*, *Cynodon*, *Digitaria* and *Exotheca* and of *Pennisetum clandestinum* (Edwards, 1956; Herlocker and Dirschl, 1972; O'Rourke et al., 1975a). However, grazing by *Syncerus caffer* (buffalo) sometimes causes *Setaria sphacelata* var. *aurea* to spread and smother tussocks of *Eleusine* in forest glades (Vesey-FitzGerald, 1974). There is some evidence that continued heavy grazing of *Pennisetum sphacelatum* stands leads to gradual dominance by *Eleusine jaegeri* which, once established, is extremely difficult to eradicate (Heady, 1960; Glover, 1961). Although two years of complete protection from grazing has, at least locally, led to replacement of *E. jaegeri* by other grasses (Makacha and Frame, 1979). In general, the *Pennisetum sphacelatum*–*Eleusine jaegeri* grassland can be considered to be a successional stage of relatively long duration within the successional dynamics of the highland evergreen forest. Thus, often, the only practical way of returning *P. clandestinum* to dominance is to plough, clear and fertilize the land. *Pennisetum clandestinum* will then regenerate naturally, but regeneration can be accelerated by planting runners (Edwards, 1935). Continual cultivation, however, results in dominance by *Digitaria abyssinica*, a rhizomatous perennial grass, which, although highly palatable, lacks vigour (Bogdan, 1958a; Van Rensburg, 1960).

On the other hand, lightly grazed sites with lower soil fertility, which experience periodic grass fires, are generally dominated by *Themeda triandra*. When the incidence of grass fires is significantly reduced through increased grazing, *Pennisetum sphacelatum* quickly invades and replaces *Themeda triandra* as the dominant (Edwards and Bogdan, 1951).

Loudetia kagerensis is a major component of highland grasslands on shallow lateritic soils in Uganda and Tanzania (Van Rensburg, 1952; Trapnell and Langdale-Brown, 1969), whereas *L. simplex* forms natural climax communities on infertile granitic and sandstone soils in the Southern Highlands of Tanzania, an area that experiences severe frost (Van Rensburg, 1960; Trapnell and Langdale-Brown, 1969). The *Loudetia* grasslands are of poor quality and are suitable for grazing only for a few months each year (Van Rensburg, 1960).

In the Ethiopian highlands, *Themeda quadrivalvis* and *Eleusine floccifolia* take the place of *T. triandra* and *E. jaegeri*, whereas *Pennisetum riparium* is abundant in waterlogged areas (Rattray, 1960). An extensive peripheral zone in the western highlands of Ethiopia, transitional between montane forest and *Hyparrhenia* grasslands of lower elevations, is characterized by such species as *Eleusine floccifolia*, *Eragrostis braunii*, *Panicum deustum*, *Pennisetum sphacelatum*, *Setaria sphacelata* and *Sporobolus agrostoides*.

A spur of the Ethiopian highlands, that extends northward into Eritrea, supported montane forest in ancient times when human settlement first occurred. The present vegetation is highly degraded and grows on a stony, eroded granitic substrate (Rattray, 1960; Beals, 1968). *Aristida* spp., *Bothriochloa insculpta*, *Digitaria abyssinica*, *Heteropogon contortus* and *Sporobolus* spp. are the principal grasses present (Rattray, 1960).

Fauna

The grassland mosaic within the highland evergreen forest region supports many of the same species of large mammals found in the grasslands of the adjacent regions, but especially the *Pennisetum* giant-grass, *Hyparrhenia* tall-grass, and

Themeda mid-grass regions) (Fig. 9.1). The most typical herbivores[1] are:

Cephalophus spp. (duikers)
Diceros bicornis (black rhinoceros)
Kobus spp. (waterbucks)
Loxodonta africana (elephant)
Potamochoerus porcus (bush pig)
Sylvicapra grimmia (duiker)
Syncerus caffer (buffalo)
Tragelaphus scriptus (bushbuck)

Prominent carnivores include:

Acinonyx jubatus (cheetah)
Canis spp. (jackals)
Crocuta crocuta (spotted hyaena)
Felis serval (serval)
Genetta spp. (genets)
Herpestes spp. (mongooses)
Ichneumia albicauda (mongoose)
Lycaon pictus (African wild dog)
Panthera leo (lion)
Panthera pardus (leopard)

Such migratory ungulates as:

Alcelaphus buselaphus (Bubal hartebeest)
Connochaetes taurinus (wildebeest)
Equus burchelli (Burchell's zebra)
Gazella granti (Grant's gazelle)
Gazella thomsoni (Thomson's gazelle)
Taurotragus oryx (eland)

sometimes visit the region in search of palatable forage during the dry season, but utilize other, less elevated, regions during the rains. The highlands of Ethiopia are home to such endemics as *Canis simensis* (Simien fox), *Capra walie* (walia ibex), *Tragelaphus buxtoni* (mountain nyala) and *Theropithecus gelada* (gelada baboon) (Blower, 1970; Morris and Malcolm, 1977). Distribution maps of the large mammals were presented by Dorst and Dandelot (1970), and a more detailed review of mammal anatomy, behaviour, ecology and distribution was prepared by Kingdon (1971, 1974, 1977, 1979).

Among smaller mammals that are notably conspicuous in the grasslands are the mole rats (*Tachyoryctes* spp.), which are responsible for extensive excavations (Jarvis, 1969; Yalden, 1975).

---

[1] Nomenclature of wildlife species follows that of Mackworth-Praed and Grant (1957), Dorst and Dandelot (1970) and Clements (1978).

The avifauna of this region is largely distinct from that of lower elevations (Stuart, 1981). In the grasslands, several of the larger characteristic bird species are *Aquila verreauxi* (Verreaux's eagle), *Buteo rufofuscus augur* (augur buzzard), *Gypaetus barbatus* (lammergeyer) and *Struthio camelus* (ostrich). Endemic birds of the Ethiopian highlands are *Bostrychia carunculata* (wattled ibis), *Columba albitorques* (white-collared pigeon) and *Corvus crassirostris* (thick-billed raven) (Blower, 1970). In Tanzania, the numerous bird species of the high-altitude grasslands were described by Moreau and Sclater (1937, 1938) and by Elliott and Fuggles-Couchman (1948).

Glossina spp. (tsetse flies) are absent from these high elevations. Temperature, moisture and soil conditions are important determinants of termite distribution (Pomeroy, 1977, 1978), this probably being the explanation of why termites are scarce or absent in this region.

Land use

The most productive uses of land in this region are (in decreasing order of importance) intensive agriculture, forestry and range.

The lower slopes of the wetter areas are the most productive and support maize (*Zea mays*), bananas (*Musa* spp.), sweet potatoes (*Ipomoea batatus*), beans (*Phaseolus vulgaris*) and finger millet (*Eleusine coracana*). The last is presently being replaced by more productive crops. Arabica coffee (*Coffea arabica*) is an important cash crop. The higher slopes support cabbages (*Brassica oleracea* var. *capitata*) and brussels sprouts (*B. oleracea* var. *gemmifera*). Tea (*Camellia sinensis*) and pyrethrum (*Chrysanthemum cinerariaefolium*) are cash crops. Wheat (*Triticum* spp.) is grown in the drier rain-shadow areas (Morgan, 1973).

Although the area of natural forest is now very small, this region is the centre for the East African forest industry, much of it based on extensive plantations of exotic tree species (*Cupressus* spp. and *Pinus* spp.), which are often planted in areas of poor soils (Parry, 1953; Pratt and Gwynne, 1977).

Cattle are kept within the region, often in large numbers. On the lower slopes, these are often tethered and fed supplementary feed, often by bringing grass to them from the plains below

(Morgan, 1973) or from forest glades above.

Dairy farming is important on the higher slopes; estimates of potential stocking rates range from 304 kg (live weight) ha^{-1} (Bogdan, 1977b) to 2400 kg ha^{-1} (Sombroek et al., 1982). There is a trend toward production of milk from small farms, and local zebu cattle — a form of *Bos indicus* — are being replaced with higher-grade exotics (Morgan, 1973; Pratt and Gwynne, 1977).

Pennisetum clandestinum is the pasture grass used on dairy farms (Edwards, 1940). Legume–grass leys are important in the drier areas where wheat, barley (*Hordeum* spp.), pyrethrum and sheep (*Ovis aries*) are raised. Clovers, such as *Trifolium repens*, and grasses such as *Setaria sphacelata* and *Chloris gayana*, are the best ley components. *Setaria sphacelata* and legumes such as species of *Cajanus* and *Leucaena* are planted in wetter areas for stall feeding (Van Rensburg, 1969).

Intensive management of such pastures is required to maintain high levels of primary and secondary productivity (Edwards, 1940). Forage yields reach 1500 g DM m^{-2} yr^{-1} (Bogdan, 1977b). Fertilization is necessary to maintain satisfactory yields. On the less fertile non-volcanic soils, heavy manuring, composting and phosphate fertilizers are required to maintain productivity (Van Rensburg, 1952, 1969). Legume production is maintained directly by fertilization, and grass production by increased cover of nitrogen-fixing legumes (Van Rensburg, 1969).

Overused sites dominated by *Eleusine jaegeri* and *Pennisetum sphacelatum* must be cultivated and planted with *P. clandestinum* to bring this species back to dominance or, alternatively, they must be protected from grazing and burned to bring back *Themeda triandra* (Edwards, 1942: Edwards and Bogdan, 1951). Removal of *Eleusine* stumps or treatment with herbicides is also effective (O'Rourke et al., 1975b).

PENNISETUM GIANT-GRASS REGION

Elevation and climate

This region occurs along the shores of Lake Victoria and in western Uganda between about 1000 and 2000 m elevation (Fig. 9.1). Rainfall averages 1140 to 1650 mm annually and there is only a short dry season (Heady, 1960; Langdale-Brown et al., 1964; Griffiths, 1969).

Geology and soils

The typically redissected topography of inselberg plains and flat-topped hills is composed of two ancient Precambrian crystalline formations. The oldest is the gneiss complex of Uganda made up primarily of metamorphic and granitic rocks. The intensively folded Buganda–Toro system consists of slates, schists, quartzites, basic volcanics and gneisses (Saggerson, 1969; Scott, 1969; Morgan, 1973).

Extensive areas are often covered by a single soil catena. In the Lake Victoria area one of the more important catenas is characterized by shallow, stony soils with rock outcrops or by brown to yellow-red, sandy clay loams with laterite horizons on hill tops, ridges and on lower slopes. These soils are often acidic and of low fertility. Soils of the upper slopes are well drained, dark, friable clays with laterite horizons (latosols) (Scott, 1969). At mid-slope, red to dark-red friable clays with laterite horizons are commonly found. In the valley bottoms, soils are of low fertility and are derived from either alluvium or Recent lacustrine deposits (Thomas, 1945, 1946; Scott, 1969). Peaty swamps are also extensive.

The soils of mid-slope and, particularly, upper-slope locations tend to be quite fertile. Unusually high concentrations of phosphorus occur frequently on hill tops. This phenomenon may have resulted from past volcanic activity in western Uganda (Lang Brown and Harrop, 1962). However, immediately north and north-east of Lake Victoria, phosphorus enrichment undoubtedly reflects the long-term human occupation of the mid- to upper-slope parts of the soil catena (Thomas, 1945, 1946). Agricultural settlement in both areas dates from approximately 1000 A.D. and pastoral use from about 1350 to 1600 A.D. (Were and Wilson, 1978).

Persistent dumping of household refuse and the long-term cultivation of deep-rooted banana plants, which pump nutrients to the soil surface, has changed the original infertile grey soils (derived from acidic gneiss rocks) to fertile, reddish soils through concentration of bases. Conversely, the soils of pastures of adjacent valley bottoms, which are covered by *Cymbopogon nardus* grassland, have probably become even less fertile over centuries of use. The practice (of the Buganda people) of grazing cattle on these pastures during the day and stabling them near households at night appears to result in a net transport of nutrients upward along the soil catena.

Vegetation

The overall pattern of vegetation in this region has resulted from the clearing of the original *Celtis–Piptadeniastrum* evergreen forests and *Celtis–Chrysophyllum* semi-deciduous forests, followed by repeated sequences of slash-and-burn agriculture. The present vegetation is a mosaic of forest remnants in valley bottoms, invading savanna trees (especially on much-cultivated or poorer soils), *Pennisetum purpureum* grasslands, and grassy communities in early stages of succession following abandonment of cultivation (Langdale-Brown et al., 1964). *Imperata cylindrica* var. *africana* is a common species of early successional stages; it gradually gives way to *Hyparrhenia rufa* and *Panicum maximum*. These species, in turn, are succeeded by *Pennisetum purpureum*, which is subsequently maintained by frequent grass fires (Langdale-Brown et al., 1964).

Pennisetum purpureum is a large perennial grass which forms clumps up to 4 m in height (Edwards and Bogdan, 1951). When growing under optimal conditions (fertile soils, warm temperatures and rainfall over 1000 mm annually), it is typically vigorous, competitive, palatable and highly productive. It often forms almost pure stands (Thomas, 1945, 1946; Bogdan, 1977b). *Pennisetum purpureum* tolerates a wide range of environments, but with somewhat reduced vigour and productivity under sub-optimal conditions. This species has been transplanted as a forage grass widely throughout the tropics and subtropics (Bogdan, 1977b).

Pennisetum purpureum retains its dominant role only under light to moderate grazing pressure. Intensive grazing leads rapidly to dominance by *Imperata cylindrica* (Thomas, 1940) or *Digitaria abyssinica*. Whereas *I. cylindrica* can be readily eliminated by burning in order to re-establish *P. purpureum*, it is very difficult to eradicate *D. abyssinica* once it has become established. *Digitaria abyssinica* lies dormant when over-shaded by *P. purpureum*, but becomes prominent when the overhead cover is burned off. However it is a good fodder grass, particularly when young; in cultivated fields it becomes a persistent weed, very difficult to eliminate (Thomas, 1946; Haarer, 1951; McIlroy, 1978). Complete protection from grazing soon results in colonization of stands of *Pennisetum purpureum* by savanna tree species (Thomas, 1940, 1946; Langdale-Brown et al., 1964).

Soil fertility both influences and is influenced by the growth of *Pennisetum purpureum*. On the more fertile soils, growth of this species tends to restore fertility levels (through improved soil structure) that were reduced by agricultural cropping (Nye, 1937; Thomas, 1940, 1945, 1946). The ease with which this species colonizes old cultivated sites (within two or three years after abandonment), its effect in restoring soil fertility, and its ability to thrive under annual burning are considered to be important factors in the ability of the area north of Lake Victoria to support dense human populations (Nye, 1937; Thomas, 1940, 1946).

Post-cultivation communities in the forest zone of western Uganda, between 1340 and 2165 m elevation, consist primarily of *Hyparrhenia cymbaria*, *H. diplandra*, *Pennisetum unisetum* and *Pteridium aquilinum*. On very poor forest soils *Cymbopogon nardus* is an abundant post-cultivation species (Langdale-Brown et al., 1964), although such soils are not often tilled (Thomas, 1940).

Fauna

The fauna of this small region of high rainfall consists most noticeably of (Wing and Buss, 1970):

Hippopotamus amphibius (hippopotamus)	*Phacochoerus aethiopicus* (warthog)
	Potamochoerus porcus
Kobus spp.	*Syncerus caffer*
Loxodonta africana	*Tragelaphus scriptus*

Connochaetes taurinus and *Equus burchelli* (zebra), which are typical of short-grass and mid-grass areas, are absent from this densely vegetated and densely settled region.

Termites (*Macrotermes* spp.) and ants (*Paltothyreus* spp.) are abundant. Termite mounds usually number 2 to 8 or more per hectare (Pomeroy, 1977).

Land use

Human population densities in most parts of this region are very high, particularly near Lake Victoria. The land is privately owned and most of the natural forest has been cleared to support intensive subsistence agriculture. Bananas and maize are the main food crops, whereas sugar (*Saccharum* spp.), coffee (*Coffea robusta*) and cotton (*Gossypium hirsutum*) are important as cash crops. Forestry is concerned primarily with the management of indigenous tree species. Livestock raising is also a productive activity, but both arable agriculture and forestry bring greater returns (Langdale-Brown et al., 1964; Pratt and Gwynne, 1977). Communally owned grasslands are severely overgrazed, and milk production is low (Van Rensburg, 1969). There is, however, a good potential to improve livestock production through the use of intensive range-management techniques, including bush-control measures, establishment of planted pastures, proper stocking rates, and preservation of seasonal fodder reserves (Van Rensburg, 1969; Pratt and Gwynne, 1977). *Pennisetum purpureum*, although very productive and palatable when young, soon grows too high for efficient grazing (Edwards and Bogdan, 1951; Pratt and Gwynne, 1977; Bogdan, 1977b), and shades out shorter plant species which are important as livestock fodder (Langdale-Brown et al., 1964). Therefore, periodic mowing of the sward (and preparation of ensilage for stall feeding of high-grade cattle) is becoming recognized as a valuable practice for maintaining long-term productivity. Herbage yields are very high, ranging from 488 g DM m^{-2} yr^{-1}, to at least 1600 g (Mugera and Ogwang, 1976; Bogdan, 1977b) and is even as high as 2000 g m^{-2} where nitrogenous fertilizers are applied (Stephens, 1967). Planting *P. purpureum* in mixture with such legumes as *Desmodium uncinatum* also results in significant increases in yield (Bogdan, 1977b).

In areas of bush vegetation, the abundance of tsetse flies (*Glossina* spp.) often limits grazing by cattle. In the early 1900s a severe outbreak of this pest caused the resident tribes to abandon the northern and north-eastern shoreline and near-shore islands of Lake Victoria. Subsequently, dense forest became re-established very quickly (Lind and Morrison, 1974). Even though fly control programs have been undertaken since 1964, the region remains understocked relative to its carrying capacity for livestock (Peberdy, 1969; Morgan, 1973). However, even with these limitations, the zone supports more cattle than do the semiarid pastoral lands (Morgan, 1973). Intensive management can bring about livestock carrying capacities ranging in live weight from 234 to 350 kg ha^{-1} (Bogdan, 1977a).

PANICUM–HYPARRHENIA TALL-GRASS REGION

Elevation and climate

This region extends in a narrow belt along the coast of the Indian Ocean in Kenya, north-eastern Tanzania and south-eastern Somalia (Figs. 9.1 and 9.3). Elevation ranges from sea level to 400 m.

Annual rainfall within this belt ranges from 500 to 1500 mm and occurs throughout most of the year. There is a short (2–3 months) dry season (Griffiths, 1969). Rainfall is higher here than in the hinterland and coastal areas to the north and south because the south-east monsoon reaches this part of the coast after a long fetch over the Indian Ocean. The coastal areas to the south receive less precipitation owing to the rain-shadow effect of Madagascar, whereas those to the north are affected by directional changes of the monsoonal air masses (from a north-west direction south of the Equator to a north-east direction north of the Equator) (Nieuwolt, 1977).

Geology and soils

A wide variety of soil parent materials exists within this region. The local geological formations occur in bands parallel to the coast and to the major physiographic features, so that the effects of geology and orographic rainfall reinforce each other.

A narrow coastal plain consists of soils derived from Pleistocene corals and coastal sand deposits. These soils are often deep, but have low moisture retentivity. A plateau to the west, which rises abruptly to an elevation of 140 m, consists of Jurassic shales, mudstones and limestones. Arid types of vegetation occur on soils derived from

Fig. 9.3. *Hyphaene coriacaea* (palm) forming the overstorey in *Panicum–Hyparrhenia* tall grassland of the coastal region of Kenya. (Photograph by P. Kuchar.)

shale, because of their low moisture permeability and infiltration rates. However, soils developed on limestone are deep, permeable and highly fertile. Further westward, a low range of hills rises abruptly to an elevation of about 400 m. Sandstone parent materials have given rise to reddish to yellow-reddish loamy sands on hill slopes and, on lower slopes and valleys, to brown, yellow, light-grey or white-mottled loamy sands with laterite horizons. Beyond this an extensive plateau, which composes most of eastern Kenya, rises very gradually westwards. Almost level, this plateau consists of sandstone on which are developed brown clay and brown to yellow-red sandy clay loam soils with laterite horizons (Edwards, 1956; Moomaw, 1960; Scott, 1969).

Vegetation

This region supports a mosaic of plant communities, most of which are successionally related to lowland forests (Heady, 1960; Moomaw, 1960; Rattray, 1960). The East African coast has been settled to some degree for at least 2000 years, but more intensively during the last 1000 years during which the development of the Swahili coastal city states took place (Were and Wilson, 1978). Most of the climax lowland forest has been cleared or burned during the present millenium.

In general, grassland is less important within the coastal region than elsewhere because cleared lands tend to revert rapidly to bushland consisting of species of *Combretum*, *Commiphora*, *Grewia* and *Lantana* (Pratt and Gwynne, 1977). However, some rather small areas of grassland, composed of tall and mid grasses, do occur within a mosaic of stands of relict forest and bushland (Heady, 1960; Moomaw, 1960; Rattray, 1960; Pratt and Gwynne, 1977).

Near the coast, a savanna grassland is expanding in area as forest types are altered or eliminated by fire, grazing and cultivation. It is varied in composition, containing species of *Heteropogon*, *Hyparrhenia*, *Imperata*, *Panicum*, *Pennisetum* and *Sporobolus* (Heady, 1960; Moomaw, 1960; Rattray, 1960). Where forest has been cleared recently, *Digitaria* spp. and *Panicum maximum* dominate (Edwards, 1956). Areas of deep, porous and infertile coastal sand, essentially ungrazed, commonly have a cover of *Bothriochloa bladhii*,

Echinochloa pyramidalis, Hyparrhenia spp., *Panicum infestum, P. maximum, Setaria sphacelata* and *Themeda triandra*. Moderate grazing results in the replacement of these species by *Heteropogon contortus*, whereas heavy grazing causes *Bothriochloa insculpta* and *Digitaria milanjiana* to become dominant. Common associated annual grasses include *Brachiaria leucantha, Chloris virgata, Eragrostis ciliaris* and *Perotis hildebrandtii* (Hendy, 1975).

Although Sombroek et al. (1982) estimated potential annual herbage yield to range from 300 g DM m^{-2} yr^{-1} in the driest part of this region to 2000 g m^{-2} yr^{-1}, the wettest natural pastures of these deep, infertile sandy soils have relatively low yields of about 430 g m^{-2} yr^{-1} under an average annual rainfall of 1200 mm. Although palatable when young, the grasses soon become coarse, unpalatable and of low nutrient content. Leguminous species compose only a small portion of the standing crop, are unpalatable and apparently add little nitrogen to the soil. Invasion by bush species is rapid, even under low stocking rates (Hendy, 1975).

Fauna

The distribution of animals within this region appears to be quite variable and has not been systematically studied. The fauna of the Shimba Hills area, near Kenya's southern coast, was reviewed by Risley (1966) and Glover (1969). Some of the ungulates present are:

Hippotragus niger (sable antelope)
Kobus spp.
Nesotragus moschatus (suni)
Redunca redunca (Bohor reedbuck)
Rhynchotragus kirkii (dik-dik)
Tragelaphus scriptus
Cephalophus spp. (duikers)

Because of their decline in the *Chrysopogon* mid-grass region (Fig. 9.1), a number of *Hippotragus equinus* (roan antelope) were captured there and released in the Shimba Hills as a conservation measure (Sekulic, 1978). *Damaliscus korrigum* (topi), *Phacochoerus aethiopicus* and *Potamochoerus porcus* also occur in this region (Dorst and Dandelot, 1970). The biomass of large mammals is low in this coastal region, as compared to the *Themeda* mid-grass region.

The birds of the Shimba Hills, as well as of other localities in this region, were described by Williams and Fennessy (1967).

Glossina spp. are common in bushland areas (Glover, 1969). The common termites (species of *Cubitermes, Macrotermes* and *Odontotermes*) build large conspicuous mounds (Glover, 1969).

Land use

Where rainfall and soils permit, as on sandy soils overlying Jurassic shales, the most common use of land is for production of tree crops [coconut (*Cocos nucifera*), mango (*Mangifera indica*), cashew (*Anacardium occidentale*) and *Citrus* spp.]. Cassava (*Manihot esculenta*), maize, and cotton are grown in small fields throughout the region. The cultivation of sisal (*Agave sisalana*), once the major commercial crop, has greatly diminished as a consequence of competition from synthetic fibres (Peberdy, 1969; Morgan, 1973; Pratt and Gwynne, 1977). Where rainfall and soils are less suitable for growing of these crops (as on the northern coast of Kenya and in southern Somalia), settlement is less dense, and the commoner forms of land use are cattle raising, hunting, and culture of millet (*Sorghum vulgare*) and maize using traditional slash-and-burn techniques with a long-term bush fallow (Hendy, 1985).

The main deterrents to successful livestock enterprises in this region have been: (1) shortage of water; (2) the occurrence of extensive bush thickets and the associated tsetse flies; and (3) low soil fertility (on the deep sandy soils of the narrow coastal strip) (Peberdy, 1969; Hendy, 1975; Pratt and Gwynne, 1977). However, livestock (mostly cattle) ranches are now being developed in the dry part of the coastal region. Estimated stocking rates range from 57 to 110 kg (live weight) ha^{-1}, depending on rainfall (Pratt et al., 1966; Peberdy, 1969). Although potential carrying capacity is estimated by Sombroek et al. (1982) to range from 75 to 1300 kg (live weight) ha^{-1}, these levels may not be achieved in areas of higher rainfall because of low soil fertility. Some of the large sisal estates on the northern Tanzanian coast are also being converted to managed rangeland (ranches) (Pratt and Gwynne, 1977). Due to the low productivity of natural grasslands on the coastal sands, the use of improved grass and legume species may become necessary (Hendy, 1975).

Fig. 9.4. *Hyparrhenia* tall-grass vegetation in an Uganda national park, with elephants (left) and waterbuck (right).

HYPARRHENIA TALL-GRASS REGION

Elevation and climate

This region, which ranges from 2000 to 3000 m elevation, includes central and northern Uganda, south-western Kenya and western Ethiopia (Figs. 9.1 and 9.4).

Annual rainfall ranges from about 1500 mm in the wetter areas (south-western Ethiopia and central Uganda) to approximately 500 mm in the driest parts (north-eastern Uganda and north-western Ethiopia). There is a single rainy season with a dry period of 2 to 4 months (December–February) in the wetter areas (possibly even shorter in Ethiopia), but which extends to 5 to 7 months in the drier areas. Mean annual evaporation ranges from 1500 to 2500 mm and most of the region has a precipitation/evapo-transpiration ratio of from 0.60 to 0.75. Therefore, the climate throughout most of the region is subhumid, whereas the driest parts are semiarid (Langdale-Brown et al., 1964; Wolde-Mariam, 1970; Morgan, 1973; Kenworthy, 1977; UNESCO, 1979).

Geology and soils

The part of this region that is situated within Uganda is underlain by rocks of the early Precambrian gneiss complex, except for local deposits of Quaternary sediments in the north-east and along the River Nile. The topography of this very old land surface typically is redissected, level to gently undulating with inselberg plains and occasional flat-topped hills (Scott, 1969; Morgan, 1973). Both old crystalline rocks (schists, granites) and Tertiary trappean lavas are common in Ethiopia (Wolde-Mariam, 1970).

In Uganda, many soils have been formed from pre-weathered material (Radwanski and Ollier, 1959). In the wetter south and central areas, red friable clays and yellow-red sandy clay loams (both latosols) occupy the well-drained, upper catenary levels, whereas red to dark-red friable clays and brown to yellow-red sandy clay loams (both with laterite horizons and seasonally slightly impeded drainage) occur on the mid- and lower slopes. Alluvium, lacustrine deposits, and peaty swamps are common in valley bottoms (Scott, 1969).

To the north-east, where the topography is gentler, these latosolic clays and clay loams extend over most of the catena, whereas dark-grey to black clays (grumosols) with impeded drainage occupy the depressions (Scott, 1969). Shallow, stony soils with rock outcrops are widespread in north-western Uganda. Although soils are deepest in the wetter southern and south-western areas (Scott, 1969), the more fertile soils are found in the drier north-east (Thomas, 1943).

Vegetation

This region is characterized by woodlands and wooded grasslands, the composition, status and successional relationships of which reflect various positions on a gradient of decreasing rainfall. For instance, as rainfall decreases, the height of the woody overstorey decreases from 15 m to 3 m and the grass height decreases from 2.6 m to 0.5 m (Langdale-Brown et al., 1964).

The wettest portion of the moisture gradient, which extends north-west across Uganda from the Kenya border at Lake Victoria, is occupied by woodlands dominated by species of *Combretum*, *Sapium* and *Terminalia*. These woodlands are successional to climax types ranging from forest through semi-deciduous thicket to woodland, and are maintained by burning, cutting and cultivation. *Hyparrhenia rufa* is the dominant grass species throughout, whereas *Imperata cylindrica* dominates an earlier stage in post-cultivation succession.

Generally, burning, cutting, and cultivation are all required to arrest succession of the vegetation back to climax (Fiennes, 1940; Langdale-Brown et al., 1964). However, in the drier areas, burning alone may be adequate (Lock, 1977). Repeated long-term burning is thought to create wooded grasslands dominated by *Hyparrhenia diplandra* and *H. filipendula* (Langdale-Brown et al., 1964).

Although the dominant species of *Hyparrhenia* are tall, the herbage does not form a dense ground cover. As a result, soil erosion is common during the rainy season. Following burning, an interval of approximately 5 weeks is required for sufficient foliage to develop to provide for adequate soil cover and grazing (Masefield, 1948). This herbage, and that of the associated *Imperata cylindrica*, is valuable for grazing only when young or when constantly grazed. The nutrient content of the *Hyparrhenia* species, in particular, varies widely (Bogdan, 1977b).

Optimal grazing keeps the height of *Hyparrhenia filipendula* and *H. rufa* low, and encourages the occurrence of other valuable forage grasses, such as *Cynodon dactylon*, *Panicum maximum* and *Setaria sphacelata*. Overgrazing at first increases the importance of *Cynodon dactylon* and the annual grasses *Digitaria velutina* and *Urochloa panicoides*, but prolonged grazing brings about dominance of *Sporobolus pyramidalis* and the occurrence of soil erosion (Fiennes, 1940).

Drier conditions occur north-east and south-west of this zone. These areas are occupied by *Butyrospermum* and *Combretum* woodlands, which are primarily fire-derived, and by wooded grasslands. *Butyrospermum* communities are very similar to the climax vegetation, but the drier *Combretum* communities are successional to mixed deciduous and evergreen thicket. *Hyparrhenia filipendula* and *Hyperthelia dissoluta* are the dominant grass species. *Cymbopogon nardus* and *Loudetia arundinacea* are also important (Langdale-Brown et al., 1964).

The driest part of the moisture gradient occurs in north-eastern Uganda (Karamoja District) and in the west along the River Nile. These areas are dominated by *Combretum* wooded grassland in which *Hyperthelia dissoluta* is the principal grass species. They are degraded forms of *Combretum* wooded grassland and woodland, created by burning and overgrazing (Langdale-Brown et al., 1964).

Chrysopogon plumulosus, *Setaria incrassata* and *Themeda triandra* are locally dominant. *Setaria* is restricted to heavy clay soils. Intensive grazing results in increased importance of *Cynodon plectostachyus* and *Pennisetum ramosum* (on clays) and *Chloris gayana*, *Cynodon plectostachyus* and *Sporobolus pyramidalis* on other soils (Thomas, 1943; Langdale-Brown et al., 1964). *Cynodon dactylon*, one of the most drought-resistant grasses in Uganda, revegetates areas where cattle have been driven out by tsetse flies (Thomas, 1940).

Continued overgrazing creates first *Acacia* bushland (*A. brevispica*, *A. mellifera* and *A. nilotica*) which retains an important *Hyperthelia dissoluta* component, and then *Acacia reficiens–Commiphora* thicket with shorter (often annual) grasses and much bare soil. Finally, as around settlements, only annual grasses and succulent thicket (*Euphorbia–Adenium*) occur (Thomas, 1940, 1943; Langdale-Brown et al., 1964). Because of the nature of the underlying rock, erosion ultimately leads to the formation of stone mantles, porous and receptive to rainfall, which are then colonized by species typical of *Combretum* wooded *Hyperthelia* grasslands (Thomas, 1940, 1943; Langdale-Brown et al., 1964). Such extensive areas of overgrazed land are a relatively new phenomenon in north-eastern Uganda, as they apparently did not exist 50 to 60 years ago (Wilson, 1960).

Woodlands and wooded grasslands dominated by *Combretum* and *Terminalia* tree species and by *Hyparrhenia filipendula*, *H. rufa* and *Hyperthelia dissoluta* also occur in central Kenya, as well as extensively from 450 to 1500 m elevation on broken topography throughout western Ethiopia. That part of the region in northern Ethiopia is extremely degraded as a consequence of a very long period of dense human settlement (Rattray, 1960; Beals, 1968; Bolton, 1969–1972).

Fauna

The grasslands of this region support many of the same large mammal herbivores that occur in the *Themeda* and *Chrysopogon* mid-grass regions (Fig. 9.1). Some of the commonest are (Ross, 1969; Bolton, 1973; Edroma, 1975):

Alcelaphus buselaphus	*Loxodonta africana*
Equus burchelli	*Syncerus caffer*
Giraffa camelopardalis (giraffe)	*Taurotragus oryx.*

Hippotragus equinus, which is rare in the *Themeda* mid-grass region, is more common here (Allsopp, 1979).

The biomass of large herbivores in parts of the region is quite high. For example, the area north-east of Lake Mobutu Sese Seko (Lake Albert) supports 133 kg ha^{-1} (live weight) (Coe et al., 1976). Within Uganda's national parks, *Loxodonta africana* was identified as being one of the main factors contributing to the transformation of woodland into grassland (Ross et al., 1976), but the recent decline in its population has reversed this trend (Eltringham and Malpas, 1980).

The common birds of the parks and reserves in this region were described by Williams and Fennessy (1967).

The abundance of *Glossina* spp. has kept portions of the region free from human settlement (Allsopp, 1979). *Macrotermes* spp. are abundant and important in speeding the decomposition of litter (Pomeroy, 1978). Their large mounds, 1 m or more in height, generally number 2 to 4 ha^{-1}, but in some localities there are more than 8 mounds ha^{-1} (Pomeroy, 1977).

Land use

In general, this region has a high agricultural potential with room for further expansion in crop and livestock production (Pratt et al., 1966; Morgan, 1973). Until recently, land ownership was predominantly communal. However, where population pressure is severe, communally granted land rights tend to become permanent, and consequently individual land ownership is now well established (Peberdy, 1969; Pratt and Gwynne, 1977).

Mixed farming of annual crops is commonest, with maize, finger millet, sorghum and pigeon pea (*Cajanus cajan*) the main food crops. Cotton is extensively grown as a cash crop. Cattle raising is also important, with planted/seeded pastures of potential importance on arable lands. The proportion of the land that is cultivated by ploughing is higher than anywhere else in East Africa, but shifting cultivation continues to be practised in the dry areas of northern Uganda where the human population is less dense (Morgan, 1973; Pratt and Gwynne, 1977). Some silviculture is practised in the wettest areas, and wildlife conservation is important throughout the region.

Cultivation and human population density coincide only poorly with rainfall (Morgan, 1973), hence the major controlling factors are probably those of communications, soils and distribution of the tsetse fly. Tsetse infestations are particularly common in western and northern Uganda (Morgan, 1973). Human population shifts and heavy grazing during the colonial period brought about an expansion of dense bush in parts of north-eastern Uganda, which was followed by an increase in tsetse. The affected areas have had to be abandoned as range for livestock, and this in turn has led to the overstocking of tsetse-free areas (Deshler, 1972).

Large areas of this region are now used for livestock raising and potential productivity is high, particularly in the moister areas. Herbage yield of *Hyparrhenia rufa* ranges from 300 g DM m^{-2} yr^{-1} (Bogdan, 1977b) to about 1200 g (Wendt, 1970), but can be as high as 1700 g when phosphorus fertilizer is applied (Wendt, 1970). Peak aboveground biomass (live plants) of the drier (500–900 mm of rain yr^{-1}) and more fertile soils of this region is estimated to be from 350 to 750 g m^{-2} (Deshmukh, 1984). Potential annual yield is estimated by Sombroek et al. (1982) to range from 1200 g m^{-2} yr^{-1} (drier areas) to 3000 g m^{-2} yr^{-1} (wetter areas). However, soil fertility is a major constraint to attaining these levels of production.

Hyparrhenia filipendula, *H. rufa* and *Hyperthelia dissoluta* are generally palatable only when young. Crude protein levels vary widely for both *Hyparrhenia rufa* (3–15%) and *H. filipendula* (7–17%). *Hyperthelia dissoluta* is less valuable as a forage than the other two species (Edwards and Bogdan, 1951; Dougall et al., 1964; Bogdan, 1977b).

The most efficient grazing practice for *H. rufa* is to graze frequently, but lightly, so as to maintain the sward at a height of at least 30 cm (Bogdan, 1977b). Resting and burning the range every 4 to 5 years controls woody regrowth and allows shorter, more palatable grasses to compete (Pratt and Gwynne, 1977).

Potential carrying capacities of from 600 to 2400 kg (live weight) ha^{-1} were estimated by Sombroek et al. (1982), the major constraint being low soil fertility. Actual year-long carrying capacity has been measured to range from 64 to 300 kg (live weight) ha^{-1} (Bogdan, 1977a). Short-term carrying capacities may be greater, as in more productive *Hyparrhenia rufa* grasslands where 454 kg ha^{-1} is common over an 8-month growing season. However, this can be increased even further (to 757 kg ha^{-1}) when *H. rufa* is mixed with *Stylosanthes* (Bogdan, 1977a). *Hyparrhenia rufa* is grown for hay (Edwards and Bogdan, 1951; Bogdan, 1977b) and planted pastures, and reseeding of areas which have been degraded by overgrazing is economically feasible throughout the region (Pratt and Gwynne, 1977).

The high-rainfall portions of this region have the highest cattle densities in all of East Africa, with up to 91 to 132 kg (live weight) ha^{-1} on good-quality pastures (Peberdy, 1969; Morgan, 1973). But even under these circumstances, cattle raising must compete with other economic activities. There is no doubt, however, that these high-potential areas could be developed to become a major source of breeding and fattening stock for the less well-stocked range areas of the region (Peberdy, 1969).

The tsetse fly remains as a serious constraint to full utilization of the potential for livestock production in this region. Infestations often limit the use of bushland as rangeland, because of the danger to both livestock and humans from the tsetse-transmitted flagellate protozoan *Trypanosoma* sp. which causes trypanosomiasis (sleeping sickness), particularly in the drier areas. Although carrying capacities could readily be doubled through proper range-management practices within these areas, the high costs of bush control now limit rangeland use. Hence, in Uganda, efforts to control tsetse and to develop ranches have centred on the more productive portions of the region (Peberdy, 1969).

THEMEDA MID-GRASS REGION

Elevation and climate

This region (Figs. 9.1, 9.5 and 9.6) includes the most productive grazing land in East Africa. Elevations range from 1170 to 2330 m. Average precipitation ranges from less than 500 to 1500 mm annually. Rainfall occurs primarily during a single season (Kenworthy, 1977), which locally, however, may include two peaks (Norton-Griffiths et al., 1975). There is a 2- to 7-month dry period. The driest part of the area is in the lee of the Southern Highlands and other highland ranges (Uluguru, Usambara and Pare mountains) along the region's eastern edge where annual rainfall is less than 500 mm and a 7-month dry season occurs. The wettest area, with a rainfall of 1000 to 1500 mm yr^{-1} and a dry season of only 2 months duration lies west of Lake Victoria in north-western Tanzania and south-western Uganda (Kenworthy, 1977).

Geology and soils

The geology of this region is quite varied. Late Precambrian and Cambrian crystalline rocks of

Fig. 9.5. *Themeda triandra* grassland with scattered low shrub (*Acacia drepanolobium*) on the floor of the Rift Valley in Kenya.

Fig. 9.6. Fires are an important element in the maintenance of *Themeda triandra* grassland.

the Mozambique Belt (intensively folded metamorphic gneisses and quartzites) occur in the central and eastern part, whereas Tertiary to Recent volcanic rocks predominate in Kenya and northern Tanzania. East of Lake Victoria bed-rock consists of Quaternary sediments and granites, migmatites and associated acid gneisses, whereas to the west of the Lake sediments overlying the Nyanza Shield and incompletely metamorphosed sediments with intrusive granites are extensive (Saggerson, 1969).

The overlying soils, which are as varied as the bed-rock upon which they have developed, range from well-drained red soils to heavy black clays in locations of impeded drainage (Scott, 1969). These soils not only reflect the underlying bed-rock, but also rainfall and position along the soil catena. Thus, soils on the upper part of the catena are generally well-drained clays and clay loams, but tend to be rocky and sandy under higher rainfall conditions, and loams or sandy and silty loams under lower rainfall. Soil texture becomes heavier as elevation on the catena decreases. Soils lowest on the catena are heavy black to grey clays, which are often seasonally inundated and are occasionally saline or alkaline (Scott, 1934; Milne, 1937, 1947; Burtt, 1942; Herlocker and Dirschl, 1972; Herlocker, 1975; Peterson, 1978).

The soils of this region differ from those of the *Brachystegia* woodland (miombo) region of southern Tanzania in that they tend to be more calcareous, compact and less freely drained (Scott, 1934; Milne, 1937, 1947; Burtt, 1942).

Vegetation

Several distinct plant communities cover large tracts within this region. Their position on the soil catena and the morphology of the catenas themselves are related to the rainfall gradient (Fig. 9.7). Generally, *Themeda triandra* grassland occurs on the well-drained soils of the upper part of the catena, and *Pennisetum mezianum* grassland occupies the poorly drained heavy clay soils at the base of the catena. The exceptions are at the two extremes of the rainfall gradient, where *Themeda triandra* grassland occurs on the mid or lower catena slopes. Where rainfall is highest, the upper catenary soils are dominated by *Hyparrhenia rufa* grassland; where rainfall is lowest, *Aristida ad-*

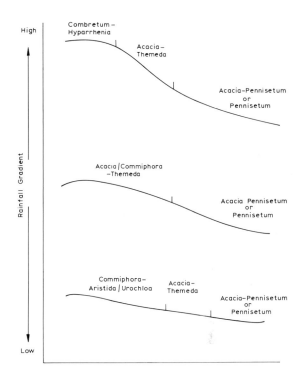

Fig. 9.7. Schematic distribution of plant communities in the *Themeda* mid-grass region, in relation to topographic position and rainfall gradient.

scensionis and *Urochloa mosambicensis* grassland dominates this catenary position (Peterson, 1978).

In locations of higher rainfall, *Hyparrhenia rufa* is replaced by *Loudetia kagerensis* on the less fertile soils and by *Themeda triandra* on the more fertile volcanic soils. *Panicum maximum* is abundant in the transition to the *Brachystegia* woodland (miombo) in southern Tanzania (Scott, 1934; Burtt, 1942; Glover and Trump, 1970; Schmidt, 1975a; Peterson, 1978). Together with *Hyparrhenia rufa*, these grass species form the herb layer for *Combretum* woodland (*C. molle* and *C. zeyheri*). The successional status of these woodlands is unclear. In general, they are thought to be persistent seral stages which are fire-derived from low, dry evergreen forest (Herlocker, 1975) or semi-evergreen bushland (Glover and Trump, 1970) or deciduous thicket (Phillips, 1930; Scott, 1934; Burtt, 1942). However, some authors have regarded *Combretum* woodland as a climax community (Burtt, 1942; Verdcourdt, 1962; Peterson, 1978).

Themeda triandra grassland is usually associated with woodlands dominated by *Acacia clavigera*, *A. gerrardii* and *A. nilotica* in the high-rainfall areas and woodland, bushland and thicket dominated by *Commiphora schimperi* and *Acacia* spp. elsewhere (Phillips, 1930; Scott, 1934; Burtt, 1942; Greenway and Vesey-FitzGerald, 1969; Schultz, 1969; Herlocker and Dirschl, 1972; Herlocker, 1975).

Associated grass species vary with the amount of rainfall and include: *Hyparrhenia* spp. (in wet areas); *Digitaria macroblephara* and *Panicum coloratum* (in areas of moderate rainfall); and *Digitaria milanjiana* (in dry areas) (Langdale-Brown et al., 1964; Qvortrup and Blankenship, 1975; Schmidt, 1975a; Peterson, 1978). *Cynodon dactylon* and *C. plectostachyus* occur on volcanic soils (the former on the wetter sites), and *Panicum maximum* is abundant in the early stages of degradation of *Commiphora* thicket (Edwards and Bogdan, 1951; Greenway and Vesey-FitzGerald, 1969; Glover and Trump, 1970; Herlocker and Dirschl, 1972).

In the driest part of the region, *Themeda triandra* occurs lower on the catena, together with *Cynodon dactylon* and *C. plectostachyus*, on well-drained sandy or loamy alluvial benches in *Acacia tortilis* woodland, or with *Pennisetum mezianum*, in association with *Acacia mellifera* and *A. seyal*, on somewhat lower sites with poorly drained heavy black clays (Phillips, 1930; Peterson, 1978). Here, where the rainfall is lowest, the upper slopes of the catena are dominated by *Aristida adscensionis* and *Urochloa mosambicensis* grasses, which are an annual component of *Commiphora schimperi* bushland or thicket. Within this community, *Cenchrus ciliaris* and *Cynodon plectostachyus* occur on outwash from termite mounds. The transition zone between this community and the lower *Pennisetum mezianum* grassland is often composed of hard-pan, formed by illuviation from above. This supports *Commiphora campestris* or *Lannea–Dalbergia–Commiphora* bushland with a low *Aristida* spp.–*Microchloa kunthii* grassland.

Regardless of rainfall, *Pennisetum mezianum* grassland dominates the lowest slopes of the catena, occurring on heavy black to grey clay soils that are seasonally inundated for varying periods and are subsequently poorly drained. *Sporobolus helvolus*, *S. ioclados* and *Themeda triandra* are important associated grasses. These grasslands are treeless or support *Acacia drepanolobium* (Fig. 9.5), *A. mellifera* or *A. seyal* shrubland. In the higher-rainfall areas, these are sometimes fire-derived communities successional to deciduous to semi-evergreen bushland (Phillips, 1930; Scott, 1934; Burtt, 1942; Greenway and Vesey-FitzGerald, 1969; Peterson, 1978).

Some extensive areas of natural grasslands having little or no woody vegetation occur within the *Themeda triandra* grassland region on volcanic ash, pumice soils (poorly drained because of either a high water table or shallow underlying calcareous hard-pan), alkaline, saline or calcareous soils.

Species of *Sporobolus* dominate in many short-grass communities in saline or alkaline sites. These sites may be either poorly or excessively drained because of high water tables or underlying shallow calcareous hard-pans (Anderson and Talbot, 1965; Greenway and Vesey-FitzGerald, 1969; Glover and Trump, 1970; Vesey-FitzGerald, 1973). *Sporobolus consimilis* and *S. spicatus*, associated with *Dactyloctenium aegyptium*, occupy alkaline sites, and *S. ioclados* (with *Chloris gayana*) more saline sites. *Cynodon dactylon* and *Digitaria macroblephara* are tolerant of salinity and alkalinity in dry soils, and also occur together on intermediate soils of better structure and drainage (Anderson and Talbot, 1965; Anderson and Herlocker, 1973). *Cynodon dactylon* often occupies disturbed sites (Verdcourdt, 1962; Greenway and Vesey-FitzGerald, 1969; Glover and Trump, 1970); it also occurs with *C. plectostachyus* on poorly drained soils with a high, fresh water table. Except for *S. consimilis*, these species are short- to mid-height stoloniferous perennials with shallow, spreading root systems, well adapted to high salt and low nutrient concentrations (Schmidt, 1975b).

Rhizomatous and stoloniferous perennials of medium height become more abundant under conditions of: greater depth of soil and of depth of rooting; increased rate of weathering; freer drainage; increased soil maturity; leaching of salts down through the soil profile; and of stronger soil structure. These species include *Andropogon greenwayi*, *Cynodon plectostachyus* and *Digitaria abyssinica*, as well as the ever-present *Cynodon dactylon* and *Digitaria macroblephara* (Anderson and Talbot, 1965; Anderson and Herlocker, 1973). *Andropogon greenwayi* and *Cymbopogon caesius* have particularly close association with higher nutrient levels. Foliage cover and vigour decrease rapidly with a drop in levels of calcium and nitrogen (Schmidt, 1975a).

Themeda triandra dominates the most mature and best-drained soils where nutrient factors are not limiting. *Bothriochloa insculpta* and *Digitaria macroblephara* are common associates (Anderson and Talbot, 1965; Glover and Trump, 1970; Anderson and Herlocker, 1973).

Frequent dry-season burning of the grasslands is necessary for the continued abundance of *Heteropogon contortus*, *Themeda triandra*, and the most palatable *Hyparrhenia* species (Edwards, 1940; Heady, 1966; Langlands, 1967). The two former species often increase under a regular (annual) burning regime (Edwards, 1942; Thomas and Pratt, 1967; Glover and Trump, 1970; Harrington and Ross, 1974). *Themeda* is judged to recover better from hot fires late in the dry season than from early, cool fires (Masefield, 1948).

In the absence of burning, *Themeda triandra* is replaced by coarse grasses with deep rhizomes, such as *Cymbopogon caesius* and *Digitaria abyssinica*. Eventually, woody vegetation becomes dominant (Langlands, 1967). *Themeda triandra* persists only on some poorly drained heavy clay soils, where it may be regarded as a possible "climax" species (Schmidt, 1975a, b).

Thus, *Themeda* grasslands are largely fire-maintained plant communities that are successional to various woody communities including *Combretum–Acacia* bushland and woodland, deciduous and evergreen thicket, and the drier types of montane forest (Phillips, 1930; Scott, 1934; Langdale-Brown et al., 1964; Glover and Trump, 1970; AGROTEC, 1974; Herlocker, 1975; Schmidt, 1975a; Blankenship et al., 1976; Pratt and Gwynne, 1977).

Themeda triandra communities are tolerant of light to moderate grazing, which may even increase primary productivity somewhat (McNaughton, 1976). However, if grazing becomes too intensive, grassland composition quickly changes (Edwards, 1940; Heady, 1966; Langlands, 1967; Pratt and Gwynne, 1977). In fact, compositional differences within *Themeda* grassland, in some cases, are due more to different intensities of grazing than to differing conditions of soil (Heady, 1966). The relative paucity of *Themeda triandra* in Masailand of north-central Tanzania reported by Lamprey (1963), Peterson (1978) and Kahurananga (1979) is probably due to heavy grazing of this area.

Heavy grazing increases the abundance of grass species which are either less palatable or better adapted to being grazed than is *Themeda triandra*. For instance, grasslands dominated by this species in southern Uganda have been invaded by trees of *Acacia gerrardii* and *A. hockii* and by *Cymbopogon nardus*, a tall, coarse, tufted grass of little grazing value (Ivens, 1958; Harker, 1959; Pratt and Gwynne, 1977).

Overgrazing also causes reduced vegetative cover, an increase in annual grasses, in annual and perennial forbs of low nutritive status and palatability, and in woody vegetation. Overall, increased grazing pressure results in a vegetative chronosequence similar to that resembling the vegetation pattern that parallels a soil gradient ranging from deep, mature, well-drained and well-structured soils to shallow, poorly drained structure (Table 9.1). *Themeda* recovers under protection from grazing, except sometimes on heavy black (vertisolic) clay soils where *Pennisetum mezianum*, the other principal grass in such areas, becomes dominant (Heady, 1960).

Fauna

This region of dry forest and bushland supports the largest concentrations of large wild ungulates in East Africa, and perhaps in the entire world (Lamprey, 1969). These are mostly the same as those in the adjacent highland *Pennisetum* midgrass region (Fig. 9.1). However, some mammals that are conspicuously absent from most of the latter region are common here, including:

Aepyceros melampus (impala)
Damaliscus korrigum
Felis caracal (caracal)
Giraffa camelopardalis
Helogale parvula (dwarf mongoose)
Hippopotamus amphibius
Lepus capensis (Cape hare)
Lepus crawshayi (Crawshay's hare)
Mellivora capensis (ratel)
Mungos mungo (banded mongoose)
Otocyon megalotis (bat-eared fox)
Phacochoerus aethiopicus

The animals in northern Tanzania's Serengeti ecosystem (east of Lake Victoria) have been stud-

TABLE 9.1

Some grass and forb species characteristic of areas within the *Themeda triandra* grassland region subjected to different levels of grazing pressure

Grazing pressure	Plant growth form, palatability and foliage cover	Species
Light to moderate	Mid to tall perennial bunch grasses. Palatability and cover generally high.	*Bothriochloa insculpta* *Hyparrhenia filipendula* *H. rufa* *Pennisetum mezianum* (heavy clay soils) *Pennisetum stramineum* *Themeda triandra*
Moderate to heavy	Medium height rhizomatous and stoloniferous perennial grasses. Palatability and cover are variable.	*Cynodon dactylon* *Heteropogon contortus* *Panicum turgidum* (Ethiopia) *Pennisetum mezianum* (heavy clay soils) *Pennisetum stramineum* *Rhynchelytrum repens* *Setaria verticillata*
Heavy (over-grazing)	Annual and perennial short grasses. Generally unpalatable. Cover is seasonally low.	*Aristida adscensionis* *Aristida keniensis* *Chloris pycnothrix* *Eragrostis tenuifolia* *Harpachne schimperi* *Microchloa kunthii* *Sporobolus festivus* *Tragus berteronianus*
Continued heavy (over-grazing)	Annual and perennial forbs. Palatability and cover are generally low.	*Indigofera* spp. *Ipomoea* spp. *Solanum* spp. *Tribulus terrestris* *Zaleya pentandra*

ied in considerable detail (Sinclair and Norton-Griffiths, 1979). The larger ungulates migrate annually onto the short-grass plains in the rainy season, and return to the *Acacia* woodlands in the dry season (Pennycuick, 1975). When these migratory herds are on the short-grass plains (from December through May), the ungulate biomass is as high as 181 kg (live weight) ha^{-1} (Sinclair, 1975). The entire Serengeti ecosystem has an average ungulate biomass (throughout the year) of 84 kg ha^{-1} (Coe et al., 1976). But these are figures of over a decade ago, and the biomass became even greater subsequently because of the increasing populations of ungulates (Sinclair and Norton-Griffiths, 1979).

Estimates of the biomass of large herbivores within this region range from 39 kg (live weight) ha^{-1} in central Tanzania to 199 kg ha^{-1} in Uganda (Coe et al., 1976; Kahurananga, 1981). In various small localities of Uganda over a decade, the large mammal biomass density ranged from 51 to 365 kg (live weight) ha^{-1} (Bourlière, 1965; Petrides and Swank, 1965; Field and Laws, 1970). For the year-round standing-crop biomass of large herbivores in the typical bushed grassland, the estimate is 295 kg (live weight) ha^{-1} (Field and Laws, 1970). Estimates in Nairobi National Park (Kenya) vary from 52 to 126 kg (live weight) ha^{-1} over a 6-year period (Foster and Coe, 1968). Large herbivores constitute 31% of total livestock–large herbivore biomass within the Kenya part of this region (Dirschl et al., 1978).

The *Themeda* mid-grass communities support at least three or four times the year-long ungulate biomass found in European and North American grasslands (Talbot and Talbot, 1963). Such high biomass densities of herbivores are possible because of high rainfall, fertile soils and complementarity in grazing (McNaughton, 1976). The constraint preventing still higher densities of herbivores, both large and small, is the temporal spatial availability of nutritious forage (Sinclair, 1975).

Among the many rodent species of this region, one of the most abundant is *Arvicanthis niloticus* (diurnal unstriped grass mouse), with biomass values as high as 309 kg (live weight) ha^{-1} (Moehlman, 1978). On the Serengeti plains in the dry season, *Lepus* spp. (hares) provide a biomass equivalent to 2.1% of the ungulate biomass (Frame and Wagner, 1981). These hares are important prey for three species of *Canis*, as well as for *Acinonyx jubatus*, *Felis caracal* and *Felis libyca* (African wild cat).

The avifauna of the parks and reserves within this region has been briefly described by Williams and Fennessy (1967). A more thorough account of the birds of the Serengeti was compiled by Schmidl (1982). The grasslands are an important habitat for such Palaearctic migrants as *Charadrius asiaticus* (Caspian plover), *Falco* spp. (kestrels), *Hirundo* spp. (swallows) and *Philomachus pugnax* (ruff) (Williams and Fennessy, 1963). Six species of vultures are common here (Kruuk, 1967). These are:

Gyps bengalensis (white-backed vulture)
Gyps ruppelli (Ruppell's vulture)
Necrosyrtes monachus (hooded vulture)
Neophron percnopterus (Egyptian vulture)
Torgos tracheliotus (Nubian vulture)
Trigonoćeps occipitalis (white-headed vulture)

Termites are especially conspicuous throughout the region because of the tall mounds constructed by *Macrotermes* spp. (Lamprey, 1963). There are also less-noticeable species of termites, such as *Trinervitermes bettonianus*, which is the main food of *Proteles cristatus* (aardwolf) (Kruuk and Sands, 1972).

Macrotermitinae are thought to be the dominant termite group in the region where rainfall is less than 1000 mm yr^{-1} (Deshmukh, 1987). Termites remove up to 40% and 15% of grass litter and standing-dead grass, respectively (Deshmukh, 1985). This compares with an estimated consumption by large mammals of about 10% of primary production (Deshmukh, 1986a). However, as Macrotermitinae prefer dead to live grass, they are unlikely to compete with livestock except when dead (litter and standing-dead) grass is in short supply, as during drought periods or on overgrazed rangelands. Deshmukh (1987) developed a model of a savanna ecosystem with a primary production typical of this region to show that termites are the most important secondary producers; production is four times that of herbivorous mammals and insects combined. Only 15% of primary productivity is not consumed. This slowly accumulates as litter until it is eventually burned by grass fires.

Land use

Wildlife is an important source of revenue in this region, and national parks and reserves have been established throughout the region. Historically, sport hunting has been an important activity, but has recently been outlawed (or severely curtailed) in conjunction with an attempt to bring increased poaching activities for skins and ivory under control. Commercial game cropping has had only limited success (Lamprey, 1969; Morgan, 1973), but still holds considerable promise as an economically viable land use (Hopcraft, 1986).

Historically, competition for forage between wildlife and the livestock of pastoral tribes was not severe. However, in recent decades, increasing use of the limited area of dry-season ranges by rapidly growing livestock populations, and loss (in some cases) of grazing areas to expanding cultivation, have considerably decreased wildlife numbers (Morgan, 1973). In 1960, in the Serengeti National Park (north-central Tanzania), where grazing by livestock is prohibited, the population of large ungulates was 19 km^{-2}, whereas in adjacent parts of Tanzania the population of these animals was only 0.82 km^{-2} (Lamprey, 1964, 1969; Morgan, 1973; Sinclair, 1979).

Livestock husbandry is the most important land use in the region. Cattle are emphasized and constitute 81% of the total livestock biomass (Dirschl et al., 1978). The potential for crop production on arable land is high only in some localities of north-western Tanzania and southern Uganda, but rangeland of high grazing potential is found throughout the region (Pratt et al., 1966; Pratt and Gwynne, 1977). For instance, the settled area south-east of Lake Victoria shares with adjacent areas of the *Hyparrhenia* tall-grass region in south-western Kenya the highest cattle population density in East Africa (Peberdy, 1969).

Deshmukh (1984) developed a highly significant predictive relationship between peak biomass of green herbage and rainfall (applicable to areas with less than 900 mm yr^{-1}) using published information from Walter (1971), Harris (1972), Braun (1973), Hillman and Hillman (1977), Seely (1978), Strugnell and Pigott (1978), Herlocker and Dolan (1980) and Owaga (1980). This relationship, which may apply more to fertile than to infertile soils, estimates peak biomass levels of from 350 to 750 g DM m^{-2} yr^{-1} in the drier parts of this region. Studies by Cassady (1973), AGROTEC (1974), Harris and Fowler (1975) and Karue (1975) showed similar peak live-plant biomass/rainfall relationships.

A study which accounted for plant mortality and removal of biomass by herbivores, as well as for change in above-ground green biomass, resulted in a net primary production of 1071 g m^{-2} yr^{-1} for a rainfall of 520 mm (Deshmukh, 1986b).

Primary productivity (above- and underground) and forage quality are closely associated with the degree of grazing by herbivores. Overgrazing by large wild herbivore populations has almost eliminated *Themeda triandra* from Queen Elizabeth National Park in western Uganda (Lock, 1972; Edroma, 1985), yet *Themeda* plants clipped infrequently and lightly produced more dry matter than unclipped plants (Edroma, 1985). Similarly, clipping studies by Braun (1973) reduced yield (peak biomass of green plants) of shorter grasses (*Kyllinga–Sporobolus–Andropogon*) less than the taller *Themeda triandra*. McNaughton (1979) reported short-term rates of above-ground primary productivity of from 20 to 40 g DM m^{-2} day^{-1} in similar grasslands grazed by wild ungulates. Under such conditions, grasslands of this region may be some of the most productive in the world.

Many authors consider the nutritional value of *Themeda triandra* to be relatively low as a forage species for livestock and wildlife (Dougall et al., 1964; Marshall and Bredon, 1967; Karue, 1975; Kreulen, 1975), particularly if it is not grazed regularly (AGROTEC, 1974; Long et al., 1969). However, Bogdan (1958a, 1977b) believed it to be a valuable forage plant, and that *Themeda* grasslands are capable of attaining a reasonable level of animal production. Field et al. (1973) have shown *Themeda* to be generally palatable to cattle, and studies by AGROTEC (1974) have suggested that grazing during, or burning prior to, the growing season produces high-quality herbage (Fig. 9.6).

In the absence of grazing, grasses mature and senesce even under continuing rainfall. Crude protein declines from as much as 20–30% in the early rainy season to under 7% or even 4% in the dry season 25 weeks later (Braun, 1973; McNaughton, 1979; Tessema, 1986). Minimum levels of crude protein needed to sustain normal livestock intake and digestibility (7–8%) may be reached in as little as from 8 to 12 weeks. However, this is compensated to some extent by the selectivity of livestock grazing; the content of crude protein is greater in livestock-selected forage than in clipped vegetation (Tessema, 1986). In addition, a short but intense period of grazing by large concentrations of *Connochaetes taurinus* (wildebeest) can create dense lawn-like mats of low, tillering grasses that persist in a productive state throughout the following dry season. These mats support and are maintained by herds of *Gazella thomsoni* (McNaughton, 1979). Under such conditions *Themeda triandra* grasslands are highly nutritious. Frequent clipping (every two weeks) is sufficient to maintain crude protein levels above 10% (Braun, 1973).

Where grass fires are prevented, as in Nairobi National Park, dry matter may accumulate over several years even when the grassland is supporting as much as 83 kg (live weight) of large herbivores ha^{-1} yr^{-1}. This situation may result in *Themeda triandra* being replaced by *Setaria incrassata* (Deshmukh, 1986b).

Estimates of carrying capacity range from 15 to 300 kg (live weight) ha^{-1} (AGROTEC, 1974; Karue, 1975; Qvortrup and Blankenship, 1975; Bogdan, 1977b; Pratt and Gwynne, 1977; O'Rourke, 1978; Sombroek et al., 1982). This wide range of values reflects habitat differences within the region, as well as different methods of calculating carrying capacity. Actual stocking rates in the region are in the order of about 37 and 17 kg (live weight) ha^{-1} for livestock and large wild ungulates, respectively (Dirschl et al., 1978). These concentrations compare with those of the major national parks, where stocking of large game animals ranges from 30 to 225 kg (live weight) ha^{-1} (Leuthold and Leuthold, 1976).

Semi-nomadic pastoralism has been the major land use of the *Themeda triandra* grassland region for a very long time. Pastoral tribes have occupied the entire region since before the 16th century (Were and Wilson, 1978); they have made use of the Rift Valley area at least since 1000 B.C. (Phillipson, 1979).

Except for a few commercial ranches located mostly in Kenya, the majority of the land has been under customary tribal law. Grazing in many areas is limited by the presence of tsetse-infested bush, whereas lands that are free of tsetse flies often carry excessive numbers of livestock (Peberdy, 1969; Brooke, 1972; Morgan, 1973). Because of the limited use made by livestock of tsetse-infested areas, several of these areas have been protected as wildlife reserves and national parks (Morgan, 1973).

Despite the generally low potential for arable agriculture, human population pressures have caused an expansion of subsistence cropping into the region, and brought about serious land degradation (Peberdy, 1969; Pratt and Gwynne, 1977). More recently this situation has been exacerbated by increased mechanized cropping. In north-central Tanzania, soil erosion was apparent as early as 1910, and has continued to worsen since then (Brooke, 1972).

Government-sponsored development efforts have subsequently concentrated on reform of land tenure and use. In Kenya, individual and group ranches are being allocated, and some land is being registered under private ownership. In Tanzania, emphasis is on settlement schemes and co-operatives, such as those organized under the framework of the Masai Rangeland Development Commission, which has powers to demarcate and regulate use of grazing areas. Several large government-owned ranches have also been set up in Tanzania, such as the one at Kongwa on the site of a previous unsuccessful groundnut (*Arachis hypogaea*) scheme (Peberdy, 1969; Pratt and Gwynne, 1977).

Large private ranches are now confined primarily to Kenya. Covering about 2% of the total range area, they average 10 000 ha in size and exhibit high standards of animal husbandry. Indigenous cattle have been developed into a registered breed (Boran). Ranch breeding and fattening is often supplemented by feeding of immature stock from northern Kenya. Merino sheep are being raised, in addition to cattle, so as to increase utilization of ranges (Peberdy, 1969; Pratt and Gwynne, 1977).

Some smaller ranches of about 1000 ha have been established in Uganda. Although cattle raising has been a traditional cultural pursuit (rather than an economic enterprise), cattle are now being supplied to butchers in neighbouring areas (Morgan, 1973).

CHRYSOPOGON MID-GRASS REGION

Elevation and climate

This region covers a large area which includes most of the Horn of East Africa (Somalia and eastern Ethiopia) and northern and eastern Kenya (Figs. 9.1, 9.8 and 9.9). Elevations range from sea level to about 2000 m, but most of the area is less than 400 m above sea level. The landscape is mainly gently rolling to level.

The climate is arid to semiarid (UNESCO, 1979). The monsoonal air masses are typically divergent, stable, warm and dry. The moist south-eastern monsoon, which is characteristic south of the Equator, here turns 90° to the north-east to become the dry south-western monsoon. Under the influence of a persistent strong high-pressure area over northern India it picks up speed and moves parallel to the Somalia coast. As there is little land above an elevation of 1000 m, orographic lifting to produce rainfall is almost absent. Furthermore, a secondary low-pressure area over the Ethiopian highlands and a funnelling of air masses westward between the Kenyan and Ethiopian highlands lead to increased divergence and dryness of air masses (Nieuwolt, 1977).

As a result, rainfall is low and highly variable. Two rainy seasons occur: April–May and October–November. The first period usually experiences the greatest rainfall. To the north of

Fig. 9.8. Large tussocks of *Chrysopogon plumulosus* in rangeland in good condition in eastern Kenya. The shrub is *Acacia reficiens*.

Fig. 9.9. *Chrysopogon plumulosus* grassland in eastern Kenya, with an open canopy of *Acacia reficiens* and *A. tortilis*.

5°N, there is a tendency towards a single rainy season from July to September (Wolde-Mariam, 1970; Nieuwolt, 1977). Mean annual rainfall is less than 500 mm, but drier parts of the region, especially the Horn, receive 200 mm or less (Hemming, 1966; Wolde-Mariam, 1970; Fullard, 1977; Somali National Range Agency (SNRA), 1977).

Geology and soils

Limestones, gypsum deposits and sandstones of old marine sediments are very extensive; schists, granites, gneisses and lavas also occur (Gillett, 1941; Gilliland, 1952; Wolde-Mariam, 1970). Southern Ethiopia is principally underlain by crystalline rocks, schists and granites (Wolde-Mariam, 1970). Tertiary volcanics, early Precambrian gneisses and Palaeozoic sediments are widespread in Kenya. Quaternary sediments form the featureless plains of east-central Kenya and southern Somalia (Morgan, 1973).

Over most of the region, erosion rates normally exceed soil formation, so most soils are immature or truncated. This situation is often exacerbated by accelerated soil erosion caused by overuse. Therefore, most soils clearly reflect their geological origin, and stone mantles and exposed bed-rock (the final stages of erosion) are found extensively within the driest part of the region (Hemming, 1973; Van Wijngaarden and Van Engelen, 1985).

Calcareous soils often exhibit abrupt microtopographical boundaries caused by the interbedding of thin layers of limestone and gypsum parent materials. These soils are generally overlain by quartz sand of variable, but usually shallow, depth and of variable erodibility. Gypsum soils have developed by weathering of the surface of anhydrite deposits. These soils commonly range from sandy loams to silty clays; they are frequently poorly drained and often saline (Hemming, 1966, 1973). Once exposed by erosion, gypsum soils have poor permeability and lack cohesiveness; therefore, they are easily degraded and blown away (Hemming, 1972b).

Soils developed on Plio–Pleistocene sediments are typically imperfectly (to poorly) drained, medium to deep grey-brown, moderately to strongly sodic clay loams, often with a sealed surface (Sombroek et al., 1982). They are low in organic matter, but high in exchange capacity (Scott, 1969). Such soils predominate in southeastern Kenya. Elsewhere in eastern Kenya soils are deep, well-drained, red sandy clay loams to clays (western fringe) or loose fine loamy sands to sandy clay loams (north-east) (Sombroek et al., 1982). These soils often have a calcareous crust (Gillett, 1941) and are very susceptible to erosion; even on gentle slopes, limited rainfall produces sheet erosion and surface sealing (AGROTEC, 1974).

There are few permanent streams, but old alluvial valleys without active stream channels and local areas (depressions) of impeded drainage and clay soils are common throughout the region (Hemming, 1966; Van Wijngaarden, 1985).

In general, the soils of this region are inherently fertile, but deficient in nitrogen. Phosphorus occasionally is unavailable to plant growth because of the preponderance of calcium. In depressions where water collects and evaporates, salinity is common. Gypsum soils exhibit a greater influence on composition and structure of the vegetation than do soils developed from any other parent material.

Vegetation

Chrysopogon-dominated grasslands occur in association with the following three major types of woody vegetation:

(1) *Commiphora–Acacia* bushland occurs primarily below an elevation of 1000 m but may also extend up to 1500 m. Rainfall ranges from 200 to 700 mm (Gilliland, 1952; Hemming, 1966; AGROTEC, 1974; Le Houérou and Haywood, undated). Most woody plants are shrubs of *Acacia* and *Commiphora* 1 to 6 m in height with a canopy cover up to 60%. This woody layer is species-rich, especially in terms of the genus *Commiphora*. The most abundant *Acacia* species are *A. edgeworthii*, *A. horrida* and *A. senegal*. Trees, up to 10 m high, often scattered but sometimes with a canopy cover up to 40%, emerge above the shrub layer. These are usually *Acacia tortilis* (Fig. 9.9), *Albizia anthelmintica*, *Delonix elata* and *Gyrocarpus hababensis* (Hemming, 1966, 1973; Van Wijngaarden, 1985). *Commiphora* spp. generally dominate eluvial soils. These vary from well- to imperfectly drained, neutral to acidic, and tend

to be calcareous reddish loamy sands. Species of *Acacia*, such as *A. mellifera*, *A. nubica* and *A. reficiens* (Figs. 9.8 and 9.9), dominate alluvial soils, which are often poorly drained and may be alkaline or saline. However, trees are usually not abundant on poorly drained soils (Gillett, 1941; Hemming, 1966; Van Wijngaarden, 1985).

(2) *Acacia bussei* open woodland occurs at elevations from 600 to 1600 m in Somalia and Ethiopia. Annual rainfall ranges from 140 to 270 mm yr^{-1} (Hemming, 1966, 1973; Le Houérou and Haywood, undated). The trees are 3 to 8 m high, branching above 3 m, and are widely spaced (at intervals as great as 40 m). Roots have been found to extend as far as 18 m from the base of a tree. Tree density is only about 3% of that of the *Commiphora–Acacia* bushland (type 1), but the herb layer is six times as dense (Gillett, 1941).

(3) *Acacia etbaica* open woodland occurs from 1200 to 1800 m elevation, where annual rainfall is between 300 and 600 mm (Gillett, 1941; Hemming, 1968; International Livestock Centre for Africa (ILCA), 1984; Le Houérou and Haywood, undated). Trees average about 3 m in height.

Cenchrus–Chloris subregion

In the higher-rainfall areas of *Commiphora–Acacia* bushland (380–700 mm yr^{-1}), *Cenchrus ciliaris* and *Chloris roxburghiana* codominate with *Chrysopogon plumulosus* (Edwards and Bogdan, 1951; Bogdan, 1958b; Rattray, 1960; FAO, 1972; AGROTEC, 1974). This is an extensive subregion including eastern Kenya, southern Somalia and the lower southern foothills of the Ethiopian highlands. Unlike subunits of other regions, this area is large enough to be mapped (Fig. 9.1). *Chloris roxburghiana* generally dominates well-drained, sandy clay loams that are often acid (Van Wijngaarden, 1985; Kemei, 1986). *Panicum maximum* and *Digitaria macroblephara* dominate in the wettest part of the rainfall gradient.

Chrysopogon plumulosus (with *Tetrapogon bidentatus*) dominates on well- to imperfectly drained neutral soils (Van Wijngaarden, 1985). *Echinochloa haploclada* dominates on seasonally flooded, often alkaline/saline clay soils together with *Schoenefeldia transiens* and *Sporobolus helvolus* (Bogdan, 1958b; Andrews et al., 1975; Van Wijngaarden, 1985). Such sites are frequently pure grassland (Heady, 1960) and provide valuable grazing (FAO, 1972). *Sporobolus helvolus* dominates in fire-subclimax stands following removal of riverine evergreen high forest along the Juba river in southern Somalia (Deshmukh, 1990).

Sporobolus subregion

In the driest part of the region (100–300 mm yr^{-1}) *Chrysopogon* shares dominance with *Sporobolus ruspolianus* and *S. variegatus* on gypsic soils. Soils of increasing salinity are dominated first by *Sporobolus ruspolianus* and *S. variegatus* and then by *Limonium axillare* (a dwarf shrub). *Urochondra setulosa* colonizes shallow sands (Hemming, 1966). The boundaries of this large area are fairly well known (Hemming, 1966; RMR, 1981). This allows it to be presented here as another subunit of the *Chrysopogon* mid-grass region (Fig. 9.1).

Throughout the remainder of the region *Chrysopogon plumulosus* is the principal dominant regardless of the associated woody vegetation or the underlying geology. It is associated here with a large number of other species among which are (Gillett, 1941; Gilliland, 1952; Rattray, 1960; Hemming, 1966; AGROTEC, 1974; ILCA, 1984):

Cenchrus ciliaris   *Leptothrium senegalense*
Dactyloctenium scindicum *Tetrapogon cenchriformis*
Heteropogon contortus   *Tetrapogon villosus*

Chrysopogon grasslands are very sensitive to grazing, which can quickly change them from largely perennial to largely annual in composition (Pratt et al., 1966; Pratt and Gwynne, 1977). Overgrazing first leads to increased cover of less palatable or more grazing-resistant perennial grasses, such as *Dactyloctenium scindicum*, *Sporobolus ruspolianus* and *S. variegatus*, and, where water accumulates, *Andropogon kelleri*. Continued overuse results in increased abundance of annual grasses, including: the genera *Aristida*, *Eragrostis* and *Tragus*; forbs and dwarf shrubs of the genera *Cassia*, *Hypoestes*, *Ipomoea*, *Indigofera*, *Solanum* and *Tribulus*; and unpalatable succulents of the genera *Aloe*, *Cissus* and *Sansevieria*. At this stage on saline gypsic soils, widely spaced halophytic succulent species of *Lagenantha* and *Suaeda* occur (Gilliland, 1952; Rattray, 1960). Ultimately, either unpalatable or grazing-resistant species dominate on bare loose or surface-capped soil (Gillett, 1941; Gilliland, 1952; Hemming, 1966, 1973).

On seasonally waterlogged black cracking clays, overgrazing results in a degradational sequence leading from *Sporobolus helvolus*, a valuable perennial forage grass, to *Sorghum purpureosericeum* to *Eragrostis* annual grassland and, ultimately, to bare soil and/or bushland or thicket (Pratt, 1962). Overuse of much of the region, particularly of the *Acacia bussei* and *A. etbaica* types, has caused both *Chrysopogon plumulosus* and the major tree species to become widely scattered and scarce, and eroded soils to be common (Gillett, 1941; Gilliland, 1952; Hemming, 1973; Cossins and Bille, 1984). The trees often occur as heavily lopped or hedged shrubs, and many of the more shallow-rooted ones, particularly *Acacia bussei*, are thought to have been killed by lack of infiltration of rain into capped, impermeable soils (Hemming, 1966, 1973).

Infiltration capacity is related more to herbaceous vegetative cover than to soil type (Van Wijngaarden and Van Engelen, 1985), probably through the funnelling effects of plants, the prevention of sealing, and/or the breaking up of the sealed surface (Glover et al., 1962). Surface-sealing occurs on all soil types where they have been overgrazed or otherwise denuded. This results in substantial surface run-off, which usually does not occur when herbaceous ground cover is high (Van Wijngaarden, 1985).

Overgrazing on gentle topography throughout the region has often created unique vegetation patterns ("steppe tigré") in the form of arcs and stripes (Gilliland, 1952; Boaler and Hodge, 1962, 1964; Hemming, 1965; AGROTEC, 1974). These are associated with sheet erosion and surface capping, in which fine soil particles are stirred into suspension by rain water and redeposited as an impermeable skin on the surface of the soil.

Generally, between 250 and 500 mm of annual rainfall, only some of the perennial grasses are expected to actually behave as perennials. The remainder last a season and then die before becoming fully established. Below 250 mm yr^{-1}, perennial grasses persist if they are in good sites and are only lightly grazed (Bogdan and Pratt, 1967). Depending on the degree of shrub cover and the competition with shrubs for available soil moisture, perennial and annual grass cover within the *Chrysopogon* mid-grass region may comprise up to 80% and 35%, respectively, of the total herbaceous cover during years of good rainfall (Van Wijngaarden, 1985). Although soil moisture only becomes available for annual grasses when not fully utilized by perennial grasses (Van Wijngaarden, 1985), Bogdan (1958b) reports that annual grasses may comprise up to 50% of the total herbaceous cover during years of poor rainfall.

According to Cassady (1973), the principal grass species, but especially those of the subregion 6a, are characterized by rapid growth and early maturity and senescence; they are able to withstand a high degree of defoliation while dormant. Cassady also stated that nutritious fodder is not carried from one rainy season to the next, yet Van Wijngaarden (1985) and Kamau (1986) considered that forage quantity during the dry season is not limiting. Unusually long rainfall periods do not appear to produce a significant increase in forage (Cassady, 1973; Van Wijngaarden, 1985). Fires occasionally occur within parts of this region, especially where shrub canopy cover is under 40% and grazing is not heavy (Van Wijngaarden, 1985; Hendy, 1985) on well-drained soils. The grassland is derived by burning of woody communities, from their destruction by elephants, or by decay through senility. The balance between woody and grassland types is very unstable. If not burned at least every five years, the grasslands revert to bushland or thicket (FAO, 1972). Therefore intensive grazing, which reduces or eliminates the incidence of fires, results in rapid bush encroachment and in a reduction of carrying capacity for livestock (FAO, 1972; AGROTEC, 1974; Hendy, 1985).

Continued annual burning can reduce infiltration rates, increase sediment production, reduce perennial grass cover and increase the amount of annual grasses, forbs and bare soil (Cheruiyot, 1984; Van Wijngaarden, 1985). However, prescribed burning, especially just prior to the rains (when growth of grasses is not adversely affected) has the least impact on infiltration and sediment production and enhances forage quality (Cheruiyot, 1984; Ali et al., 1986; Mbui and Stuth, 1986). Thus, fire is a valuable management tool within this region.

Fauna

The large mammals in this region of deciduous bushland and shrubland with mid grasses are many of the same species that occur in the regions of higher elevation. There is a shift, however, toward greater abundance of species better adapted to drier conditions, such as: *Hippotragus equinus*, *Litocranius walleri* (gerenuk), *Oryx beisa* (beisa oryx), *Rhynchotragus kirkii* (dik dik) and *Tragelaphus* spp. (kudu). *Damaliscus hunteri* (Hunter's antelope) is confined to an area near the Kenya–Somalia border (Bunderson, 1977). The following are still present in Somalia and Ethiopia, but only in small numbers (Blower, 1968; Vos, 1970; Simonetta, 1971; RMR, 1981; Wilson, 1981):

Alcelaphus buselaphus swaynei (Swayne's hartebeest)
Ammodorcas clarkei (Clarke's gazelle or dibatag)
Dorcatragus megalotis (beira)
Equus asinus somalicus (Somali wild ass)
Gazella pelzelni (Pelzelni's gazelle)
Gazella soemmeringi (Soemmering's gazelle)

In most of the region located within Kenya and north-eastern Tanzania, the biomass of large herbivores ranges from 11 kg to 65 kg (live weight) ha^{-1} (Coe et al., 1976). *Loxodonta africana* and *Syncerus caffer* account for as much as 90% of the total biomass in north-eastern Tanzania (Harris, 1972), and the former composes nearly 75% of the biomass in eastern Kenya (Leuthold and Leuthold, 1976). Movements of these two species in and out of particular localities results in large fluctuations in biomass of from 8 to 180 kg (live weight) ha^{-1} (Leuthold and Leuthold, 1976). Movements are often over large distances. Elephant home ranges are as large as 43 000 km^2 (Cobb, 1976). Large herbivores constitute 24% of the total large wild herbivore–livestock biomass within the Kenyan part of this region (Dirschl et al., 1978; Peden, 1987), but only 1 and 5%, respectively, in northern and southern Somalia (RMR, 1981, 1985).

However, intensive poaching since the late 1970's has almost eliminated rhinoceros and greatly reduced elephant herds. Poaching and human settlement have resulted in immigration of elephants into Tsavo National Park in south-eastern Kenya. Subsequent, less effective, use of the habitat by livestock has reduced total herbivore biomass in dry-season grazing areas outside of the Park (Van Wijngaarden, 1985).

Where they still exist in sufficient numbers, elephants are the driving force in the ecosystem. Their opening up of dense shrubland stands increases the availability of grass; subsequently, grazers increase whereas browsers then decrease. Increased grass fuels burning which further increases grassland –and grazers –at the expense of shrubland and browsers. Reduction of elephant numbers through poaching reverses the trend (Van Wijngaarden, 1985).

Among the smaller mammals, *Heliophobius argenteocinereus* (solitary mole rat) and *Heterocephalus glaber* (colonial mole rat) burrow extensively in the grasslands (Jarvis, 1969; Jarvis and Sale, 1971).

Struthio camelus is common (Vos, 1970). Other birds common in the parks and reserves of this region were listed by Williams and Fennessy (1967).

The conspicuous and economically important termites are species of *Hodotermes*, *Macrotermes* and *Odontotermes* (Glover, 1966, 1967b). Microtermitinae comprise 90% of the total termite biomass. Their relative importance increases (even as total consumption of plant biomass decreases) as rainfall decreases within the region (Buxton, 1981). A description of common ants in Somalia was provided by Glover (1967a).

Land use

Because of low and highly variable rainfall throughout this region, nomadic or semi-nomadic pastoralism, based primarily on the camel, has been the primary land use (Pratt et al., 1966; Pratt, 1969; Morgan, 1973; Naylor, 1977; Pratt and Gwynne, 1977; SNRA, 1977). Livestock husbandry is the most important land use and forms the principal economic resource for Somalia, most of which lies within the region (Le Houérou, 1972).

The livestock (camel, sheep, goats and cattle) spend the wet seasons widely dispersed over the extensive areas of otherwise waterless plains and bushland, but during the dry season concentrate around occasional sources of permanent water. Thus, dry-season grazing areas are quite limited compared to the vast areas available for use during the wet season. In the past this imbalance was a major check on the growth of livestock herds

(Pratt and Gwynne, 1977; Bille and Selassie, 1983; Cossins and Bille, 1984).

Livestock are privately owned, but land is usually not. There is also little restriction on livestock movements, other than the availability of water and forage, and the occurrence of the tsetse fly. During periods of severe drought, livestock may even be moved beyond traditional tribal areas (Lawrie, 1954; Peberdy, 1969; Morgan, 1973; Pratt and Gwynne, 1977). Because the main objective of traditional livestock management is the amassing of as large a herd as possible (for prestige and for protection against losses during the next drought), overgrazing is common particularly near permanent water (Peberdy, 1969; FAO, 1971a, 1972; Hemming, 1972b; Morgan, 1973; Naylor, 1977; Pratt and Gwynne, 1977; Bourn, 1978; Hendy, 1985).

Agriculture is generally only possible through local irrigation projects (Pratt et al., 1966; Naylor, 1977; SNRA, 1977). Reliable cultivation is considered to be possible only in areas with more than 700 mm yr^{-1} rainfall (Bille and Selassie, 1983), but in the wetter situations of the region sorghum, maize and cow-peas (*Vigna sinensis*) are cultivated under rain-fed conditions in conjunction with the raising of livestock (Lawrie, 1954; Cossins and Bille, 1984; RMR, 1985; Hendy, 1985). Approximately 5% of southern Somalia is in rain-fed cropland, most of which receives under 600 mm rainfall per annum (RMR, 1985).

Subsistence hunting has been important for about 8000 years (Thorbahn, 1984) and, until recent times, has probably been the dominant land use throughout much of the area in the dry season when stock are removed to better-watered areas (Van Wijngaarden, 1985). Poaching remains an active and lucrative land use within the region.

In recent years, however, the pattern of land use has started to change. Permanent settlements are arising around newly established watering points. An expansion of subsistence agriculture, private enclosure of land (RMR, 1981, 1985), increased use of cattle, and a decline in the use of camels are associated phenomena. In Kenya, cattle now constitute 55% and camels only 25% of the total livestock biomass (Dirschl et al., 1978). The cutting of trees for fuel and building materials, together with overgrazing by the water-dependent cattle, result in rapid and severe environmental degradation of the lands surrounding these watering points (Hemming, 1972b; Hendy, 1985). Increasing year-round use of traditionally dry-season grazing areas by primarily subsistence agriculturalists — who are either also local pastoralists or outsiders who have moved from the densely settled areas elsewhere — has exacerbated the situation (Le Houérou and Haywood, undated; FAO, 1972; Bille and Selassie, 1983; Cossins and Bille, 1984; Hendy, 1985). At the same time crop residues and irrigated fallows are important dry-season range resources (Hendy, 1985). Some pastoralists have built surface-water collection tanks to enable them to remain year-long in their traditional wet-season grazing areas. About 70% of the rangelands where such tanks have been built have subsequently been overgrazed (Cossins and Bille, 1984).

Other areas have remained under-utilized because they lack permanent water, support dense bushland, are infested by the tsetse fly, or are subject to political instability (Edwards, 1956; FAO, 1971a; Pratt and Gwynne, 1977). Bush encroachment caused by overgrazing, and the elimination of elephant populations by poaching, have reduced livestock stocking rates in yet other areas (Assefa et al., 1985; Van Wijngaarden, 1985). Hence, livestock populations large enough to require the support of the entire region are concentrated on portions of it only.

This ecologically unbalanced situation results in massive livestock losses and subsequent impoverishment of their owners during drought. This happened to about two-thirds of the pastoral population of Somalia following the 1973–75 drought (SNRA, 1977).

However, despite extensive overuse, most of the region's rangeland is still potentially productive. About 80% could be improved through proper grazing management (Naylor, 1977; SNRA, 1977). Similarly, livestock condition is surprisingly good, even where range condition is poor, because rapid desiccation of forage in the dry season produces excellent standing herbage. Livestock also have a relatively light load of internal parasites (Edwards, 1956; Le Houérou, 1972).

Recommended improvements include: (1) introduction of forage legumes into agricultural fallows to reduce dry-season grazing of adjacent pastures and increase the yield of crops in the next rotation; (2) establishment of additional water points (Cossins and Bille, 1984); (3) development of more formalized rotational grazing systems based on reliable access to dry-season grazing areas (Naylor, 1977; Bille and Selassie, 1983); (4) prescribed burning to control bush encroachment and maintain supplies of grass forage; and (5) control of stocking rates (Hendy, 1985). The latter is required if the area is to be burnt (Pratt et al., 1966; Van Wijngaarden, 1985). Burning will be especially needed if poaching is not controlled and elephant numbers remain low (Van Wijngaarden, 1985).

Stocking rates of livestock range from 20 to 60 kg (live weight) ha^{-1}, with southern Somalia sustaining the highest rates. Cattle are most abundant in Kenya (60% of the total livestock biomass), whereas goats and sheep are most abundant in southern Somalia (35% and 60% of total livestock biomass, respectively). The proportion of wild ungulates in the total large-herbivore biomass ranges throughout the region from 1 to 24% (Dirschl et al., 1978; Peden, 1987; RMR, 1981, 1985). Stocking rates of livestock on dry-season ranges may be 17 times those on more extensive wet-season ranges (FAO, 1971a, b; Pratt and Gwynne, 1977). Carrying capacity has been estimated at from 10 to 77 kg ha^{-1} (Pratt and Gwynne, 1977; Mwova, 1977; Sadera, 1986).

Standing crops of herbage from 10 to 870 g DM m^{-2} have been measured within the region by Le Houérou and Haywood (undated), Cassady (1973), SNRA (1977), Cossins and Bille (1984), ILCA (1984), Assefa et al. (1985) and Van Wijngaarden (1985). This range reflects differences in rainfall, soil type and range condition. Based on a rainfall range of 200 to 700 mm yr^{-1}, a range of peak biomass of from 150 to 600 g m^{-2} is predicted for this region (Deshmukh, 1984).

Throughout the region, development efforts by governments emphasize increased sedentarization of nomads, better distribution and control of range and water resources, and a switch from traditional pastoralism to economic production of beef for the national markets. Development of improved marketing and transport systems is an essential component of such plans. Indications are that under such modern systems of management of the range resources, there will be no need to increase livestock numbers (FAO, 1971a, b; Le Houérou, 1972; SNRA, 1977).

Because of the low carrying capacity of these grasslands, economic livestock enterprises are quite large in area, and their management must be extensive rather than intensive. Depending on the type of unit (commercial ranch, group ranch, or grazing block[1]) and the productive potential of the land, those units now in operation range from 65 to 5700 km^2 in area (Peberdy, 1969; FAO, 1971a; Pratt and Gwynne, 1977; Langat, 1986; Muriuki, 1986; Sadera, 1986).

LEPTOTHRIUM MID-GRASS REGION

Elevation and climate

This region comprises most of central Somalia, a large plain with few strong topographic features, which slopes gently upward and westward from sea level to about 400 m elevation at the Ethiopian border (Figs. 9.1, 9.10 and 9.11). The only two significant topographical features are the Shebelli river valley, which forms the southern boundary of the region, and the coastal ridge, which is 300 m high and 30 km wide, about 20 km inland from the coast.

The climate is aridic. Rainfall occurs primarily in April–August and October–November and ranges from less than 150 to 400 mm yr^{-1}. Average monthly temperatures during the dry seasons may exceed 30°C and rainfall amounts to 3 to 20% of evaporative demand. Depending on the season, the monsoonal wind system blows from the south-west or the north-east. Winds from the south-west are the strongest (monthly mean of 30 km hr^{-1}) and are the principal influence on direction of dune movements (Fig. 9.11). Wind speed increases and rainfall decreases northward. Thus, the north tends to be more arid than the south. Wind erosion is common (RMR, 1979; UNESCO, 1979; United Nations Sahelian Office (UNSO), 1984).

---

[1] In the sense used here, a **grazing block** is a large unit of arid and/or semiarid communal rangeland for which the government provides basic infrastructure (water, roads, etc.) and guidance/assistance/direction in range/livestock management activities for the pastoralist occupants. In theory, a grazing block is self-contained (like a ranch), but is communally owned (almost like a group ranch).

Fig. 9.10. Coastal-plain grassland of central Somalia in which the dominants (*Cenchrus ciliaris*, *Indigofera intricata* and *Leptothrium senegalense*) have been invaded by *Aristida kelleri*.

Fig. 9.11. Mobile sand dunes encroaching on grassland dominated by *Cenchrus ciliaris*, *Indigofera intricata* and *Leptothrium senegalense*.

Geology and soils

Geologically the region consists of limestones overlain (often shallowly) by sands. Anhydrite deposits occupy ancient drainages which stagnated when they were blocked from the sea by the uprising of the coastal ridge (RMR, 1979).

Stabilized sand dunes extend 100 km inland from the southern and central portions of the coastline. This includes a 20-km wide coastal plain of coarse white sands of recent marine origin and finer, reddish sands and silty sands of the adjacent coastal ridge (RMR, 1979). These soils are low in moisture-holding capacity, organic matter and fertility (Barker et al., 1989b; Herlocker et al., 1986a). Mobile sand dunes (Fig. 9.11) occur throughout this area, probably due largely to normal geomorphological processes, but in part aided and abetted by the activities of man (UNSO, 1984). Moving dunes significantly modify the floristic composition and structure of the existing vegetation (Barker et al., 1989a). There are also several extensive dune fields, 5- to 10-km wide and up to 90 km in length (RMR, 1979).

In the interior there are large areas of exposed limestone and reddish sands of varying depths overlying limestone. Whitish gypsum soils have formed at the surface of anhydrite deposits (Hemming, 1972a; RMR, 1979). These soils are often saline, powdery when dry. When wet, the area is impassable to motorized vehicles and of limited accessibility to camels.

The character of the vegetation of the region probably reflects the predominance of sands as compared to the generally finer and heavier soils of the adjacent, and climatically similar, *Chrysopogon* mid-grass region.

Vegetation

The coastal plain, which is about 17,000 km² in area and extends along most of the coastline of this region, is a natural grassland. Total shrub cover is only 2% — this despite the fact that grass fires are neither frequent nor large (Herlocker and Ahmed, 1986). The greater suitability of grasses (with their finer root systems) in this environment than shrubs (which usually have deeper, and coarser, root systems) may be related to the interaction of three factors: (1) the combination of highly porous infertile soils through which rainfall can quickly infiltrate and percolate into the underlying limestone; (2) long periods of high winds which increase evaporative stress; and (3) the ability of grasses to quickly utilize nutrients from rapidly decomposing litter.

Shrubs dominate elsewhere in the region, but especially in the wetter areas in the south and on the coastal ridge. Dwarf shrubland and grassland predominate in drier areas to the north. Shrub height ranges from 6 to 8 m, with canopy cover from 40 to 60%, in the south to under 1 m and 20%, respectively, in the north. The predominant genera are *Acacia* and *Commiphora*. *Acacia edgeworthii*, *A. horrida*, *A. nilotica*, *A. reficiens* and *A. senegal* are locally dominant. *Commiphora* species are multitudinous and some, such as *C. chiovendana*, dominate over extensive areas. Evergreen shrubs, such as *Balanites orbicularis* and *Boscia minimifolia* in the drier north and *Cordeauxia edulis* in the south, are important sources of dry-season forage for livestock and wildlife. *Indigofera ruspolii* is the most abundant dwarf shrub (Naylor and Jama, 1984; Herlocker and Ahmed, 1985; Kuchar et al., 1985).

The grassland component varies from large areas of natural grassland with little woody-plant admixture to sparse cover in dense shrubland. Perennial grasses dominate throughout the region although short-lived forms of the same species become more common in the driest areas. Forb cover peaks early in succession. Forbs, including species of *Blepharis*, *Endostemmon*, *Heliotropium*, *Psilotrichum* and *Tephrosia*, are abundant in shrublands but not in the coastal-plain grassland. Annual grasses are surprisingly uncommon in the region (Herlocker and Ahmed, 1986; Herlocker et al., 1987, 1988). Grass height is about 30 to 40 cm and cover ranges from 1.5 to 8% in shrublands and 6 to 50% in natural coastal-plain grassland. In the grassland both the herbaceous cover (Naylor and Jama, 1984; Herlocker and Ahmed, 1986; Barker et al., 1989b) and litter cover (which ranges from 0.6 to 20%) are positively related to range condition (Barker et al., 1989b), but litter cover also varies seasonally. Dry grasses and forbs fall to the ground within 6 months, and grass and forb litter is completely decomposed within 1 year. Therefore, litter does not substantially accumulate over time, even in an ungrazed situation (Thurow,

1989). This may partially explain the relative absence of bush fires in the wetter parts of the region.

The principal dominant grass species throughout the region, in order of abundance, are *Leptothrium senegalense*, *Cenchrus ciliaris*, *Aristida sieberiana* and *Sporobolus somalensis* (Hemming, 1972b; Naylor and Jama, 1984; Herlocker and Ahmed, 1986; Herlocker et al., 1987, 1988; Barker et al., 1990). All are highly palatable perennials, except for *Aristida sieberiana* (Herlocker and Kuchar, 1986). The enclaves of gypsic soils support a herbaceous vegetation similar to that found on gypsic soils in the *Chrysopogon plumulosus* midgrass region (*Chrysopogon plumulosus*, *Sporobolus ruspolianus*, *S. spicatus* and *Urochondra setulosa*) (Hemming, 1972b; Naylor and Jama, 1984; D.J. Herlocker, unpublished). Therefore, this community will not be discussed further here.

Cenchrus ciliaris and *Leptothrium senegalense* are components of mature (climax) stands of grassland throughout the region. On the deeper sands within 100 km of the coast, these highly palatable bunch grasses respond positively to grazing and increase for a time in relative cover, whereas the less abundant climax components decrease in relative cover (Naylor and Jama, 1984; Herlocker and Ahmed, 1986; Herlocker et al., 1986a, b, 1987, 1988; Barker et al., 1989b). These include in the coastal plain:

Afrotrichloris martini   *Cyperus* sp.
Aristida kelleri   *Enneapogon schimperanus*
Chrysopogon plumulosus   *Heteropogon contortus*

and on the coastal ridge: *Afrotrichloris hyaloptera* and *Aristida sieberiana*. Although not as abundant, *Cenchrus ciliaris* is larger than *Leptothrium senegalense*, produces more herbage per plant, and is highly tolerant of grazing.

Associated "increaser" species in the coastal-plain grassland are *Coelachyrum stoloniferum*, *Panicum pinnifolium* and the dwarf shrub *Indigofera intricata* (Herlocker and Ahmed, 1986; Barker et al., 1989b). *Brachiaria ovalis*, *Cynodon dactylon* and *Dactyloctenium scindicum* are common invaders of disturbed soils on the deeper sands around wells and villages within 100 km of the coast (Abdulle and Aden, 1986; Herlocker and Ahmed, 1986; Herlocker et al., 1986a, 1987, 1988; Barker et al., 1989b). *Cynodon dactylon* also dominates camp sites on the coastal plain where seasonal occupation over hundreds of years by nomads and their herds has significantly increased the organic-matter and nutrient levels of otherwise infertile, highly porous white sands (Barker et al., 1989b).

Cyperus chordorrhizus is the principal invader species in the coastal-plain grasslands (Raimondo and Warfa, 1980; Naylor and Jama, 1984; Barker et al., 1989b). This rhizomatous species colonizes the areas of recent and ongoing deposition of sand associated with mobile dunes, blowouts and wind-blown beach sands so common in this area. It is less abundant in areas from which sand is being blown, such as around wells and blowouts (Herlocker and Ahmed, 1986; Barker et al., 1989a, b).

On deep soils supporting shrubland (coastal ridge), *Aristida sieberiana* is a major invader of abandoned croplands which are abundant in this area. This perennial species of low palatability is also a vigorous competitor, and will persist for many years (60 or more) if lightly grazed. However, it is intolerant of grazing, and cover decreases as grazing levels increase (Herlocker et al., 1986a, 1987, 1988).

Little is known about the successional dynamics of the more aridic interior of the region. *Afrotrichloris hyaloptera*, *Cenchrus ciliaris* and *Leptothrium senegalense* probably comprise part of the climax herbaceous vegetation. However, because of the drier, more variable climate, the plants are less robust and their populations less stable and less tolerant of grazing. Thus, all three species are decreasers that soon disappear under moderate to heavy grazing. Most areas in fair range condition are dominated by a perennial species of *Aristida*. Even drier climate and heavier grazing result in dominance by the annual *Aristida adscensionis* and bare soil (Hemming, 1972b; Naylor and Jama, 1984).

Fauna

The most abundant wild ungulates are *Ammodorcas clarkei*, *Gazella spekei* (Speke's gazelle), *Phacochoerus aethiopicus* and *Struthio camelus* (RMR, 1979). *Gazella soemmeringi* (Soemmering's gazelle), *Oryx beisa* and *Tragelaphus imberbis* occur in shrubland areas.

Total wildlife biomass within the region averages 0.28 kg (live-weight) ha^{-1} during the dry season and 0.65 kg ha^{-1} during the wet season — about 1.6% of total large herbivore biomass. Biomass decreases south-westward as rainfall and shrub cover increase. The most productive habitat is coastal-plain grassland which supports a biomass of from 1.6 kg (dry season) to 3.2 kg (wet season) ha^{-1}, mostly *Gazella spekei* and *Struthio camelus* (RMR, 1979). There is no evidence of resource-partitioning between *Gazella spekei* and associated livestock species (cattle, goats, sheep and camels). This lack of competition between species reflects an abundance of forage caused by a natural grazing rotation driven by limited water supplies and seasonal outbreaks of biting flies (Thurow et al., 1987).

Within the region, wet-season wildlife biomass is 2 to 5 times that in the dry season — similar to seasonal changes in livestock biomass. This suggests some form of seasonal movement (RMR, 1979). Poaching has reduced wildlife numbers by a considerable amount.

Land use

The principal land use is nomadic or semi-nomadic pastoralism emphasizing the raising of camels and goats (50% and 24% of total biomass, respectively). This is a subsistence economy based on milk production. Cattle and sheep (16% and 8% of total biomass) are also raised, especially in the Shebelli river valley and coastal areas, and a lively cash economy has recently developed based on sale of live animals to the Arabian Gulf states. The average stocking rate increases from 23 kg ha^{-1} (live weight) in the north-east to 45 kg ha^{-1} in the south-west, although sheep densities decrease southward and goats are highest in the centre (RMR, 1979).

Somali pastoralists prefer to remain within traditional grazing areas based on a source of permanent water, although when necessary they are able to move over large distances in search of green forage. Depending on the rainfall and availability of permanent water, these areas vary from 500 to over 3000 km^2 (Herlocker et al., 1986a). The existence of a wet-season livestock biomass 2 to 3 times that in the dry season (RMR, 1979) indicates significant seasonal movements of stock within (and possibly also in and out of) the region.

About 20% of the region is over 20 km from permanent water. Tsetse fly — along the Shebelli river — and other biting flies (which proliferate seasonally in the thick bush of the coastal ridge) enforce seasonal rests of rangeland in these areas from use by livestock. However, tsetse-clearance programs along the Shebelli river are now opening up land for settlement.

Despite the low rainfall in the region, farming is surprisingly common. Only about 25% of the area shows no sign of ever having been cropped and about 10% supports relatively high densities of regular, albeit shifting, agriculture (RMR, 1979). Most farming is done by local people who were once full-time pastoralists. Cow-peas, sorghum and watermelons are grown. Land is cropped for from 5 to 8 years and rested for 20 to 30 years (Holt, 1985; Herlocker et al., in prep.).

Settlement is increasing and new villages are springing up even where permanent water is unavailable. An associated trend is the increasing area being fenced off for private grazing reserves, resulting in a *de facto* privatization of land, and the construction of large numbers of cement-lined tanks for collecting surface water in otherwise waterless areas.

Most of the region is in poor to fair range condition. Range condition improves with distance from permanent water. The rangelands in poorest condition are localized around villages, wells and areas where poor farming practices have occurred (Naylor and Jama, 1984; Herlocker and Ahmed, 1985; Kuchar et al., 1985; Barker et al., 1989b; D.J. Herlocker, unpublished).

Measurements of herbage biomass at the end of the growing season range from 117 to 310 g DM m^{-2} yr^{-1} (Herlocker et al., 1986b; Thurow and Hussein, 1989). These reflect not only the amount of rainfall, but also range condition and level of utilization. For example, the end-of-growing-season herbage biomass in moderately grazed grassland (stocked at 56 kg ha^{-1} for 4 days per month) exceeded that on plots protected from grazing (Thurow and Hussein, 1989).

Government-induced range and livestock management schemes are being initiated. These are based on traditional grazing areas and authorities. Improvements involve water development, rehabilitation of degraded areas around villages and simple forms of grazing management (such as the reservation of areas for grazing at particular seasons). These schemes must also consider the trend toward mixed farming and private enclosure of land.

ARISTIDA ANNUAL SHORT-GRASS REGION

Elevation and climate

Aristida short annual grasslands occupy two widely separated areas, the Red Sea coastal plain and northern Kenya (Figs. 9.1 and 9.12). Elevation ranges from sea level to about 1000 m, and the topography is generally gentle, except for the occurrence of occasional mountains and escarpments.

The climate is arid to hyperarid, with ratios of precipitation to evaporation ranging from 0.03 to 0.20. Winters are warm, the coldest month averaging 20 to 30°C; and summers hot, the warmest month averaging 30° (UNESCO, 1979). Rainfall is very low (50–250 mm yr^{-1}) and extremely variable in time and space (Gillett, 1941; Gilliland, 1952; Fullard, 1977; Edwards et al., 1979).

The climatic aridity of that part of this region, located in Ethiopia, Djibouti and Somalia, can be attributed to the rain-shadow position of the Red Sea coast, from which the moist south-western monsoon is blocked by the Ethiopian highlands (Hemming, 1961). The aridity of northern Kenya is due to its location in the gap between the Ethiopian and Kenyan highlands, where warm, dry air masses prevail (Nieuwolt, 1977; Edwards et al., 1979).

Geology and soils

Quaternary and Tertiary volcanics and Quaternary sediments prevail in northern Kenya, whereas much of the northern coastal plain of Ethiopia is covered with alluvium (Saggerson, 1969; Wolde-Mariam, 1970).

Throughout the region, soils are immature and closely reflect parent materials (Hemming and Trapnell, 1957; Hemming, 1961, 1972a; Makin, 1969; Herlocker, 1979). Sands overlay most of the region, except on lava where loams and clays have developed. In this extremely dry climate vegetation is more a function of water availability than of soil fertility (Makin, 1969). Sands have better

Fig. 9.12. *Aristida* annual short grassland at the peak of the rainy season in northern Kenya.

soil-moisture relations because they are more permeable and capable of storing the small amounts of rain that fall (Smith, 1949). The less permeable sands tend toward salinity. Stone mantles and exposed bed-rock are extensive throughout the region (Hemming and Trapnell, 1957).

Vegetation

Annual grasses (*Aristida adscensionis* and *A. mutabilis*) and forbs (*Heliotropium* spp. and *Tribulus* spp.) dominate both areas, particularly on lava and compacted, partly cemented sandy soils. However, on less compacted sands, some widely spaced perennial grasses are found as well. Of these, *Panicum turgidum* is particularly abundant on the deeper, looser sands of the Red Sea coastal plain, where it is associated with *Lasiurus hirsutus*. *Dactyloctenium bogdanii* and *Leptothrium senegalense* occur on sandy soils in north-central Kenya (Gillett, 1941; Gilliland, 1952; Pichi-Sermolli, 1955; Rattray, 1960; Hemming, 1961; Herlocker, 1979). Flood-plains and alluvial plains, which are extensive in northern Kenya, support such perennial grasses as *Echinochloa haploclada*, *Lintonia nutans* and *Sporobolus helvolus* (FAO, 1971a, b). On the whole, however, perennial grasses are insignificant in this region and are forced to remain in a drought-dormant condition for most of the year (Edwards, 1956).

Woody vegetation, consisting primarily of widely scattered small or stunted trees (often with multiple stems), occurs primarily along drainage lines on lava soils, but is more regularly distributed on sandy soils (Hemming, 1961; Herlocker, 1979). *Acacia reficiens* is a common tree species, often associated with the dwarf shrub *Indigofera spinosa*.

Fauna

Large mammals are able to subsist only in low densities in this arid region. The rare *Equus asinus somalicus*, *Gazella pelzelni* (Pelzeln's gazelle) and the more abundant *G. spekei* live in the Ethiopian and Somali portions of this region (Blower, 1968; RMR, 1981). Within Kenya, the two most common large herbivores of this region are *Equus grevyi* (Grevy's zebra) and *Oryx beisa* (Lewis, 1977).

Game biomass in the Kenyan portion of this region is only 0.7 kg (live weight) ha^{-1} or about 5% of the total large-herbivore biomass (Field, 1980). This compares with 3 kg ha^{-1} reported by Stewart and Zaphiro (1963). In the Ethiopian/Somalian part of the region wildlife comprise a very low proportion (about 1%) of the total herbivore biomass. Wild ungulates appear not to be competitive with livestock, but carnivores may be a greater problem because they rely mostly on livestock. Because of their low densities, wildlife have little economic value (Dirschl et al., 1978; RMR, 1981).

Bird species common to the arid area of East Africa are described by Williams and Fennessy (1963, 1967).

Of the total termite biomass, 99% is composed of fungus-growing Macrotermitinae, which consume over 80% of plant biomass (Bagine, 1982).

Land use

Land use in this region is tied closely to that of the *Chrysopogon* grasslands, which generally provide dry-season grazing for livestock from these *Aristida* grasslands. Because of the greater aridity, herbage yields and carrying capacity in this region are lower and more variable than those of the *Chrysopogon* grasslands. A range in herbage yield of from 30 to 94 g DM m^{-2} results from a range in rainfall from 50 to 250 mm yr^{-1} (Herlocker and Dolan, 1980; Lusigi et al., 1984). Carrying capacity, at best, is about 22 kg ha^{-1} live weight (Bogdan, 1977a). Stocking rates of 4 kg ha^{-1} and 15 kg ha^{-1} have been reported by Coe et al. (1976) and Field (1980), respectively. Approximately 20% of the region in Kenya is overstocked (Field, 1980). Cultivation of crops is not feasible in this arid region, where the only form of land use has been traditional camel-based nomadism. In fact, grazing use of these ranges is only possible during the rainy season, when some surface water can be found and herbage is green. During the dry season, the nomads move their animals to the adjacent, somewhat more productive, *Chrysopogon* grasslands. In Kenya, livestock composition on the basis of biomass has been estimated at 53% camels, 32% cattle and 14% sheep and goats by Dirschl et al. (1978), and 36% camels, 40% cattle and 18% sheep and goats by Field (1980). Donkeys comprise the remainder.

Range management enterprises, established under government range development schemes, are extensive and often include *Chrysopogon* grassland blocks as dry-season grazing areas (FAO, 1971a, b; Le Houérou, 1972).

In north-eastern Ethiopia the conversion of these lands to cropland by highland agriculturalists is calling into question the ability of the pastoral Afar tribe to continue to exist (Bille and Selassie, 1983).

ACKNOWLEDGEMENTS

The authors express their appreciation to Dr. Jan Gillett of the Kenya Herbarium, Nairobi, and to Dr. James O'Rourke, Department of Range Science, Utah State University, for reviewing this manuscript.

REFERENCES

Abdulle, A.M. and Aden, A.S., 1986. Characterization of initial stages of secondary succession on a cleared pasture near Afgoi, Somalia. *Somali J. Range Sci.*, 1(1): 17–18.

AGROTEC-CRG-/SESES, 1974. *Southern rangelands livestock development project*. Part II, Volume I, Report to the Ethiopian Imperial Government Livestock and Meat Board, Addis Ababa, 87 pp.

Ali, A.R., Smeins, F.E. and Trilica, M.J., 1986. Effects of burning on soil and plant water relations in a bushed grassland in Kenya. In: R.M. Hansen, M.W. Benson and R.D. Child (Editors), *Range Development and Research in Kenya*. Winrock International Institute for Agricultural Development, Morrilton, Ark., pp. 123–135.

Allsopp, R., 1979. Roan antelope population in the Lambwe Valley, Kenya. *J. Appl. Ecol.*, 16: 109–115.

Anderson, G.D. and Herlocker, D.J., 1973. Soil factors affecting the distribution of the vegetation types and their utilization by wild animals in Ngorongoro Crater, Tanzania. *J. Ecol.*, 61: 627–651.

Anderson, G.D. and Talbot, L.M., 1965. Soil factors affecting the distribution of the grassland types and their utilization by wild animals on the Serengeti Plains, Tanganyika. *J. Ecol.*, 53: 33–56.

Andrews, P., Grover, C.P. and Horne, J.F.M., 1975. Ecology of the lower Tana River flood plain (Kenya). *J. East Afr. Nat. Hist. Soc. Natl. Mus.*, 151, 31 pp.

Assefa, E., Bille, J.C. and Corra, M., 1985. *Ecological Map of South-western Sidamo*. International Livestock Centre for Africa, Addis Ababa, Ethiopia, 29 pp.

Bagine, R.K.N., 1982. *The Role of Termites in Litter Decomposition and Soil Translocation with Special Reference to Odontotermes in Arid Lands of Northern Kenya*. M.Sc. thesis, University of Nairobi, Kenya, 174 pp.

Barker, J.R., Herlocker, D.J. and Young, S.A., 1989a. Vegetal dynamics in response to sand dune encroachment within the coastal grasslands of central Somalia. *Afr. J. Ecol.*, 27: 277–282.

Barker, J.R., Herlocker, D.J. and Young, S.A., 1989b. Vegetal dynamics along a grazing gradient within the coastal grassland of central Somalia. *Afr. J. Ecol.*, 27: 283–289.

Barker, J.R., Thurow, T.L. and Herlocker, D.J., 1990. Ecology and vegetal associations of pastoralist campsites within the coastal grasslands of central Somalia. *Afr. J. Ecol.*, 28: 291–297.

Beals, E.W., 1968. Ethiopia. In: I. Hedburg and O. Hedburg (Editors), Conservation of Vegetation in Africa South of the Sahara. *Acta Phytogeogr. Suec.*, 54: 137–145.

Bille, J.C. and Selassie, A.H., 1983. The climatic risks to livestock production in the Ethiopian rangelands. Joint ILCA/RDP Ethiopian pastoral systems study. *Int. Livestock Centre for Africa, Addis Ababa, Res. Rep.*, 4, 33 pp.

Blankenship, L.H., Field, C.R. and Masheti, S., 1976. The vegetation of the Akira Ranch, Kenya. *East Afr. Agric. For. J.*, 42: 110–116.

Blower, J., 1968. The wildlife of Ethiopia. *Oryx*, 9: 276–285 plus plates 1–11.

Blower, J., 1970. The Simien — Ethiopia's new national park. *Oryx*, 10: 314–316 plus plates 14–16.

Boaler, S.B. and Hodge, C.A.H., 1962. Vegetation stripes in Somaliland. *J. Ecol.*, 50: 465–474.

Boaler, S.B. and Hodge, C.A.H., 1964. Observation on vegetation arcs in the northern region, Somali Republic. *J. Ecol.*, 52: 511–544.

Bogdan, A.V., 1958a. *A Revised List of Kenya Grasses*. Government Printer, Nairobi, 73 pp.

Bogdan, A.V., 1958b. Some edaphic vegetational types at Kiboko, Kenya. *J. Ecol.*, 46: 115–126.

Bogdan, A.V., 1977a. Grazing management. In: D.J. Pratt and M.D. Gwynne (Editors), *Rangeland Management and Ecology in East Africa*. Hodder and Stoughton, London, 310 pp.

Bogdan, A.V., 1977b. *Tropical Pasture and Fodder Plants (Grasses and Legumes)*. Longman, London, 475 pp.

Bogdan, A.V. and Pratt, D.J., 1967. *Reseeding Denuded Pastoral Land in Kenya*. Government Printer, Nairobi, 46 pp.

Bolton, M., 1969–1972. *Ecological Surveys within Ethiopia: Rift Valley, Gemu Gofa, southwest Ethiopia, Mago Valley, Cuchia Wereda, and western Danakil*. Reports to the Ethiopia Ministry of Agriculture, Forestry and Wildlife Conservation Development Organization, Addis Ababa, 31 pp.

Bolton, M., 1973. Hartebeest in Ethiopia. *Oryx*, 12: 99–108.

Bourlière, F., 1965. Densities and biomasses of some ungulate populations in eastern Congo and Rwanda, with notes on population structure and lion/ungulate ratios. *Zool. Afr.*, 1: 199–207.

Bourn, D., 1978. Cattle, rainfall and tsetse in Africa. *J. Arid Environ.*, 1: 49–61.

Braun, H.M.H., 1973. Primary production in the Serengeti; purpose, methods, and some results of research. *Annales de l'Université d'Abidjan*, Serie E: Ecologie, Tome VI, Fascicule 2. Comte rendu du Colloque de Lamto, pp. 171–188.

Brooke, C., 1972. Types of food shortages in Tanzania. In: R. Mansall Prothero (Editor), *People and Land in Africa South of the Sahara*. Oxford University Press, London, pp. 173–190.

Bunderson, W.T., 1977. Hunter's antelope. *Oryx*, 14: 174–175.

Burtt, B.D., 1942. Some East African vegetation communities. *J. Ecol.* (Burtt Memorial Supplement, edited by C.H.N. Jackson), 30: 65–146.

Buxton, R.D., 1981. Change in the composition and activities of termite communities in relation to changing rainfall. *Oecologia*, 51: 371–378.

Cassady, J.T., 1973. The effect of rainfall, soil moisture and harvesting intensity on grass production on two rangeland sites in Kenya. *East Afr. Agric. For. J.*, 39: 26–36.

Cheruiyot, S.K., 1984. *Infiltration Rates and Sediment Production of a Kenya Bushed Grassland as Influenced by Vegetation and Prescribed Burning*. M.Sc. Thesis, Texas A&M University, 88 pp.

Clements, J.F., 1978. *Birds of the World: A Check List*. The Two Continents Publishing Group Ltd., New York, 532 pp.

Cobb, S., 1976. *The Distribution and Abundance of the Large Herbivores of Tsavo National Park, Kenya*. Ph.D. Thesis, Oxford University (cited by Van Wijngaarden, 1985).

Coe, M.J., Cumming, D.H. and Phillipson, J., 1976. Biomass and production of large African herbivores in relation to rainfall and primary production. *Oecologia (Berlin)*, 22: 341–354.

Cossins, N.J. and Bille, J.C., 1984. *A Land-Use Study of the Jijiga Area, Ethiopia*. Joint ILCA/RDP Ethiopian Pastoral Systems Study Programme, International Livestock Centre for Africa, Addis Ababa, 15 pp.

Deshler, W., 1972. Livestock trypanosomiasis and human settlement in north-eastern Uganda. In: R.M. Prothero (Editor), *People and Land South of the Sahara*. Oxford University Press, London, pp. 50–57.

Deshmukh, I.K., 1984. A common relationship between precipitation and grassland peak biomass for east and southern Africa. *Afr. J. Ecol.*, 22: 181–186.

Deshmukh, I.K., 1985. Decomposition of grasses in Nairobi National Park, Kenya. *Oecologia*, 67, 147–149.

Deshmukh, I.K., 1986a. How do we manage? Errors and ignorance in savanna production ecology. *Int. Symp. African Wildlife*, Kampala.

Deshmukh, I.K., 1986b. Primary production of a grassland in Nairobi National Park, Kenya. *J. Appl. Ecol.*, 23: 115–123.

Deshmukh, I.K., 1987. How important are termites in the production ecology of African savannas? *Sociobiology*, 15: 155–168.

Deshmukh, I.K., 1990. *Terrestrial Ecology Baseline Studies*. Associates in Rural Development, Burlington, Vermont, 131 pp.

Dirschl, H.J., Mbugua, S.W. and Wetmore, S.P., 1978. *Preliminary results from an aerial census of livestock and wildlife of Kenya's rangelands*. Ministry of Tourism and Wildlife, Kenya Rangeland Ecological Monitoring Unit, Aerial Survey Technical Report Series 3, Nairobi, 17 pp.

Dorst, J. and Dandelot, P., 1970. *A Field Guide to the Larger Mammals of Africa*. Collins, London, 287 pp. plus 44 plates.

Dougall, H.W., Drysdale, V.M. and Glover, P.E., 1964. The chemical composition of Kenya browse and pasture herbage. *East Afr. Wildl. J.*, 2: 86–121.

Edroma, E.L., 1975. Wildlife count in a Uganda national park. *Oryx*, 13: 176–178.

Edroma, E.L., 1985. Effects of clipping on *Themeda triandra* Forsk and *Brachiana playnota* (K. Schum.) Robyns in Queen Elizabeth National Park, Uganda. *Afr. J. Ecol.*, 23: 45–51.

Edwards, D.C., 1935. The grasslands of Kenya, I. Areas of high moisture and low temperature. *Emp. J. Exp. Agric.*, 3: 153–159.

Edwards, D.C., 1940. A vegetation map of Kenya with particular reference to grassland types. *J. Ecol.*, 28: 377–385.

Edwards, D.C., 1942. Grass burning. *Emp. J. Exp. Agric.*, 10: 219–231.

Edwards, D.C., 1956. The ecological regions of Kenya; their classification in relation to agricultural development. *Emp. J. Exp. Agric.*, 24: 89–108.

Edwards, D.C. and Bogdan, A.V., 1951. *Important Grassland Plants of Kenya*. Pitman, Nairobi, 124 pp.

Edwards, K.A., Field, C.R. and Hogg, I.G.G., 1979. *A preliminary analysis of climatological data from the Marsabit District of northern Kenya*. UNESCO/IPAL Technical Report B-1, Nairobi, 44 pp.

Elliott, H.F.I. and Fuggles-Couchman, N.R., 1948. An ecological survey of the birds of the Crater Highlands and Rift Lakes, northern Tanganyika Territory. *Ibis*, 90: 394–425.

Eltringham, S.K. and Malpas, R.C., 1980. The decline in elephant numbers in Rwenzori and Kabalega Falls National Parks, Uganda. *Afr. J. Ecol.*, 18: 73–86.

FAO, 1971a. *Range Development in Isiolo District*. AGP:SF/Ken/11, working paper 9, Nairobi, 91 pp.

FAO, 1971b. *Range Development in Marsabit District*. AGP:SF/Ken/11, working paper 10, Nairobi, 134 pp.

FAO, 1972. *Range Development in the East Kitui Statelands*. AGP:SP/Ken 66/55, working paper 11, Nairobi, 26 pp.

Field, C.R., 1980. A summary of livestock studies within the Mt. Kulal study area. In: *Combatting Desertification and Rehabilitating Degraded Production Systems in Northern Kenya*. UNESCO Integrated Project on Arid Lands, Technical Report A-4, Nairobi, Kenya, pp. 89–223.

Field, C.R. and Laws, R.M., 1970. The distribution of the larger herbivores in the Queen Elizabeth National Park, Uganda. *J. Appl. Ecol.*, 7: 273–294.

Field, C.R., Harrington, G.N. and Pratchett, D., 1973. A comparison of the grazing preferences of buffalo (*Syncerus caffer*) and Ankole cattle (*Bos indicus*) on three different pastures. *East Afr. Wildl. J.*, 11: 19–29.

Fiennes, R.N.T.W., 1940. Grasses as weeds of pasture in northern Uganda. *East Afr. Agric. J.*, 5: 255–256.

Foster, J.B. and Coe, M.J., 1968. The biomass of game animals in Nairobi National Park, 1960–1966. *J. Zool.* (London), 155: 413–425.

Frame, G.W. and Wagner, F.H., 1981. Hares on the Serengeti Plains, Tanzania. In: K. Myers and C.D. MacInnes (Editors), *Proc. of the World Lagomorph Conf.*, University of Guelph, Guelph, Ont., 12–16 August 1979, pp. 790–802.

Fullard, H. (Editor), 1977. *Philip's Modern College Atlas for Africa*. George Philip and Son Ltd., London, 136 pp.

Gillett, J.B., 1941. The plant formations of western British Somaliland and the Harar Province of Abyssinica. *Kew Bull.*, 2: 37–318.

Gilliland, H.B., 1952. The vegetation of eastern British Somaliland. *J. Ecol.*, 40: 91–124.

Glover, P.E., 1961. *Report on the Ngorongoro Pasture Research Scheme*. Unpublished report to the Ngorongoro Conservation Unit, Tanzania, 19 pp.

Glover, P.E., 1966. Some notes on *Hodotermes erithreensis* (Sjostedt) *minor* Harris (Somali name: "Abor aro madu") in British Somaliland. *East Afr. Wildl. J.*, 4: 4–49.

Glover, P.E., 1967a. Notes on some ants in northern Somalia. *East Afr. Wildl. J.*, 5: 65–73.

Glover, P.E., 1967b. Further notes on some termites of northern Somalia. *East Afr. Wildl. J.*, 5: 121–132.

Glover, P.E., 1969. *Report on an Ecological Survey of the Proposed Shimba Hills National Reserve*. East African Wildlife Society, Nairobi, 148 pp. plus 4 appendices and 3 maps.

Glover, P.E. and Trump, E.C., 1970. *An Ecological Survey of the Narok District of Kenya Masailand, II. Vegetation*. Kenya National Parks, Nairobi, 175 pp.

Glover, P.E., Glover, J. and Gwynne, M.D., 1962. Light rainfall and plant survival in East Africa, II. Dry grassland vegetation. *J. Ecol.*, 50: 199–206.

Greenway, P.J. and Vesey-FitzGerald, D.F., 1969. The vegetation of Lake Manyara National Park. *J. Ecol.*, 57: 127–149.

Griffiths, P.J., 1969. Climate. In: W.T.W. Morgan (Editor), *East Africa: Its Peoples and Resources*. Oxford University Press, Nairobi, pp. 107–117.

Haarer, A.E., 1951. Grasses and grazing problems of East Africa. *World Crops*, 3: 8–10.

Harker, K.W., 1959. An *Acacia* weed of Uganda grassland. *Trop. Agric. (Trinidad)*, 36: 45–51.

Harrington, G.N. and Ross, I.C., 1974. The savanna ecology of Kidepo Valley National Park, I. The effects of burning and browsing on the vegetation. *East Afr. Wildl. J.*, 12: 93–105.

Harris, L.D., 1972. An ecological description of a semi-arid East African ecosystem. *Colo. State Univ. Range Sci. Dep., Sci. Ser.*, 11, 80 pp.

Harris, L.D. and Fowler, N.K., 1975. Ecosystem analysis and simulation of the Mkomazi Reserve, Tanzania. *East Afr. Wildl. J.*, 13: 325–346.

Heady, H.F., 1960. *Range Management in East Africa*. Government Printer, Nairobi, 125 pp.

Heady, H.F., 1966. Influence of grazing on the composition of *Themeda triandra* grassland, East Africa. *J. Ecol.*, 54: 705–727.

Hemming, C.F., 1961. The ecology of the coastal area of northern Eritrea. *J. Ecol.*, 49: 55–78.

Hemming, C.F., 1965. Vegetation arcs in Somaliland. *J. Ecol.*, 53: 57–67.

Hemming, C.F., 1966. The vegetation of the northern region of the Somali Democratic Republic. *Proc. Linnean Soc. London*, 177: 173–248.

Hemming, C.F., 1968. Somali Republic North. In: I. Hedburg and O. Hedburg (Editors), Conservation of Vegetation in Africa South of the Sahara. *Acta Phytogeogr. Suec.*, 54: 141–144.

Hemming, C.F., 1972a. The ecology of south Turkana: a reconnaissance classification. *Geogr. J.*, 138: 15–40.

Hemming, C.F., 1972b. *Ecological and Grazing Survey of the Mudugh Region, Somali Democratic Republic*. FAO working paper AGP: BF/SOM/70/512, Rome, 15 pp.

Hemming, C.F., 1973. *An Ecological Classification of the Vegetation of the Bosaso Region, Somali Democratic Republic*. FAO working paper AGP: SF/SOM/70/512, Rome, 65 pp.

Hemming, C.F. and Trapnell, C.G., 1957. A reconnaissance classification of the soils of the south Turkana desert. *J. Sci.*, 8: 167–183.

Hendy, C.R.C. (Editor), 1985. *Land Use in Tsetse Affected Areas of Southern Somalia*. Somali Democratic Republic National Tsetse and Trypanosomiasis Control Project. Land Resources Development Centre (LRDC), Overseas Development Administration, LRDC Report, 148 pp.

Hendy, K., 1975. Review of natural pastures and their management problems on the north coast of Tanzania. *East Afr. Agric. For. J.*, 41: 52–57.

Herlocker, D.J., 1975. *Woody vegetation of the Serengeti National Park*. Kleburg Studies in Natural Resources, Texas Agric. Res. Sta., RM 5/KS 1, College Station, 31 pp.

Herlocker, D.J., 1979. The vegetation of southwestern Marsabit District, Kenya. *UNESCO/IPAL Tech. Rep.*, D-1, 68 pp.

Herlocker, D.J. and Ahmed, A.M., 1985. Interim report on range ecology and management of Ceel Dhere District. *Somali National Range Agency/Central Rangelands Development Project, Tech. Rep.*, 8, 37 pp.

Herlocker, D.J. and Ahmed, A.M., 1986. The vegetation of the coastal plains, central Somalia. *Somali J. Range Sci.*, 1(2): 34–58.

Herlocker, D.J. and Dirschl, H.J., 1972. *Vegetation of the Ngorongoro Conservation Area, Tanzania*. Canadian Wildlife Service Rep. Ser. 19, Ottawa, 39 pp.

Herlocker, D.J. and Dolan, R., 1980. Primary productivity of the herb layer and its relation to rainfall. In: Proc. of a scientific seminar. Nairobi, Kenya. *UNESCO/IPAL Tech. Rep. Ser.*, A-3: 22–29.

Herlocker, D.J. and Kuchar, P., 1986. Palatability ratings of range plants in Ceel Dhere and Bulo Berte Districts. *Somali National Range Agency/Central Rangelands Development Project, Tech. Rep.*, 9, 18 pp.

Herlocker, D.J., Ahmed, A.M., Buh, B.B., Aden, M.O. and Ibrihim, A.M., 1986a. *Range Management Plans for the Range and Livestock Associations of Ceel Dhere District*. Ministry of Livestock, Forestry and Range, Central Rangelands Development Project, Misc. Tech. Rep., 125 pp.

Herlocker, D.J., Frye, D. and Khalif, H.M., 1986b. Result of two years (4 growing seasons) protection of coastal plain grassland. *Somali National Range Agency/Central Rangelands Development Project, Tech. Rep.*, 14, 9 pp.

Herlocker, D.J., Ahmed, A.M. and Thurow, T.L., 1987. Response of vegetation of the *Acacia reficiens/Dichrostachys* sp. shrubland range site to land use intensity, Ceel Dhere District, Somalia. *Somali J. Range Sci.*, 2(1): 10–24.

Herlocker, D.J., Ahmed, A.M. and Thurow, T.L., 1988. Response of vegetation of the *Acacia milotica/Dichrostachys* sp. shrubland range site to land use intensity, Ceel Dhere District, Somalia. *Somali J. Range Sci.*, 3(1): 1–11.

Herlocker, D.J., Thurow, T.L. and Buh, B.B., in prep. Post cultivation succession of a rainfed agricultural system in central Somalia.

Hillman, J.C. and Hillman, A.K.K., 1977. Mortality of wildlife in Nairobi National Park during the drought of 1973–1979. *East Afr. Wildl. J.*, 15: 1–18.

Holt, R.M., 1985. Agropastoralism and desertification in Ceel Dhere District: preliminary investigation and trial results. *Somali National Range Agency/Central Rangelands Development Project, Tech. Rep.*, 3, 130 pp.

Hopcraft, D., 1986. *Wildlife Land Use: A Realistic Alternative*. Wildlife Ranching and Research, P.O. Box Athi River, Kenya, 9 pp.

Hubbard, C.E., Milne-Redhead, E., Polhill, R.M. and Turrill, W.G. (Editors), 1952–1981. *Flora of Tropical East Africa*. Crown Agents for Overseas Governments and Administrations, London (incomplete; remainder in preparation).

International Livestock Centre for Africa (ILCA), 1984. A land use study of the Jijiga area, Ethiopia. *Joint ILCA/RDP Ethiopian Pastoral System Study Programme, Technical Annex 2. Environment and Land Use*. Addis Ababa, 112 pp.

Ivens, G.W., 1958. The effects of arboricides on East African trees and shrubs, I. *Acacia* species. *Trop. Agric. (Trinidad)*, 35: 257–271.

Jackson, G., 1978. Grasslands of Tropical Africa. In: R.J.M. McIlroy (Editor), *An Introduction to Tropical Grassland Husbandry*. Oxford University Press, London, pp. 123–133.

Jarvis, J.U.M., 1969. The breeding season and litter size of African mole-rats. *J. Reprod. Fertil., (Suppl.)*, 6: 237–248.

Jarvis, J.U.M. and Sale, J.B., 1971. Burrowing and burrow patterns of East African mole-rats *Tachyoryctes*, *Heliophobius*, and *Heterocephalus*. *J. Zool.* (London), 163: 451–479 plus plates 1–4.

Kahurananga, J., 1979. The vegetation of the Simanjiro Plains, northern Tanzania. *Afr. J. Ecol.*, 17: 65–83.

Kahurananga, J., 1981. Population estimates, densities and biomass of large herbivores in Simanjiro Plains, northern Tanzania. *Afr. J. Ecol.*, 19: 225–238.

Kamau, P.N., 1986. Effects of available forage on goat dietary selection and nutrition. In: R.M. Hansen, B.M. Woie and R.D. Child (Editors), *Range Development and Research in Kenya*. Winrock International Institute for Agricultural Development, Morrilton, Ark., pp. 103–106.

Karue, C.N., 1975. The nutritive value of herbage in semi-arid lands of East Africa, II. Seasonal influence on the nutritive value of *Themeda triandra*. *East Afr. Agric. For. J.*, 40: 372–387.

Kemei, I.K., 1986. Vegetation–environmental relationships at the National Range Research Station, Kiboko, Kenya. In: R.M. Hansen, B.M. Woie and R.D. Child (Editors), *Range Development and Research in Kenya*. Winrock International Institute for Agricultural Development, Morrilton, Ark., pp. 287–300.

Kenworthy, J., 1977. Climate. In: D.J. Pratt and M.D. Gwynne (Editors), *Rangeland Management and Ecology in East Africa*. Hodder and Stoughton, London, pp. 12–22.

Kingdon, J., 1971. *East African Mammals: An Atlas of Evolution in Africa, Vol. I*. Academic Press, New York, 446 pp.

Kingdon, J., 1974. *East African Mammals: An Atlas of Evolution in Africa, Vol. II. Part A (Insectivores and Bats) and Part B (Hares and Rodents)*. Academic Press, New York, 704 pp.

Kingdon, J., 1977. *East African Mammals: An Atlas of Evolution in Africa, Vol. III. Part A (Carnivores)*. Academic Press, New York, 476 pp.

Kingdon, J., 1979. *East African Mammals: An Atlas of Evolution in Africa, Vol. III. Part B (Large Mammals)*. Academic Press, New York, 436 pp.

Kreulen, D., 1975. Wildebeest habitat selection on the Serengeti Plains, Tanzania, in relation to calcium and lactation: a preliminary report. *East Afr. Wildl. J.*, 13: 297–304.

Kruuk, H., 1967. Competition for food between vultures in East Africa. *Ardea*, 55: 171–193.

Kruuk, H. and Sands, W.A., 1972. The aardwolf (*Proteles cristatus* Sparrman) 1783 as predator of termites. *East Afr. Wildl. J.*, 10: 211–227.

Kuchar, P., Omar, A.E. and Hassan, A.S., 1985. The rangelands and their condition in eastern Bulo Berte District. *Somali National Range Agency/Central Rangelands Development Project, Tech. Rep.*, 12, 59 pp.

Lamprey, H.F., 1963. Ecological separation of the large mammal species in the Tarangire Game Reserve, Tanganyika. *East Afr. Wildl. J.*, 1: 63–92.

Lamprey, H.F., 1964. Estimation of the large mammal densities: biomass and energy exchange in the Tarangire Game Reserve and the Masai Steppe in Tanganyika. *East Afr. Wildl. J.*, 2: 1–46.

Lamprey, H.F., 1969. Wildlife as a national resource. In: W.T.W. Morgan (Editor), *East Africa: Its People and Resources*. Oxford University Press, London, pp. 141–152.

Lang Brown, J.R. and Harrop, J.F., 1962. The ecology and soils of the Kibale grasslands, Uganda. *East Afr. Agric. For. J.*, 27: 264–272.

Langat, R.K., 1986. Commercial ranches in Kenya. In: R.M. Hansen, B.M. Woie and R.D. Child (Editors), *Range Development and Research in Kenya*. Winrock International Institute for Agricultural Development, Morrilton, Ark., pp. 39–46.

Langdale-Brown, I., Osmaston, H.A. and Wilson, J.G., 1964. *The Vegetation of Uganda and its Bearing on Land Use*. Government Printer, Entebbe, 159 pp.

Langlands, B.W., 1967. Burning in East Africa with particular reference to Uganda. *East Afr. Geogr. Rev.*, 5: 21–37.

Lawrie, J.J., 1954. Acacias and the Somali. *Emp. For. Rev.*, 33: 234–238.

Le Houérou, H.N., 1972. *Report on improvement of the rangelands and related problems in northern Somalia*. FAO working paper AGP: SF/SOMR (consultant report), Rome, 18 pp.

Le Houérou, H.N. and Haywood, M. (undated). *Ecological survey of the Jijiga area. Preliminary report to the Ethiopian Government on the arid and semi-arid programme*. International Livestock Centre for Africa, Addis Ababa, 70 pp.

Leuthold, W. and Leuthold, B., 1976. Density and biomass of ungulates in Tsavo East National Park. Kenya. *East Afr. Wildl. J.*, 14: 49–58.

Lewis, J.G., 1977. Report of a short-term consultancy on the grazing ecosystem in the Mount Kulal region, northern Kenya. *UNESCO/IPAL Tech. Rep.*, 5 E-3, Nairobi, 62 pp.

Lind, E.M. and Morrison, M.E.S., 1974. *East African Vegetation*. Longman, London, 257 pp.

Lock, J.M., 1972. The effects of hippopotamus grazing on grasslands. *J. Ecol.*, 60: 445–467.

Lock, J.M., 1977. Preliminary results from fire and elephant exclusion plots in Kabalega National Park, Uganda. *East Afr. Wildl. J.*, 15: 229–232.

Long, M.I.E., Thornton, D.D. and Marshall, B., 1969. Nutritive value of grasses in Ankole and the Queen Elizabeth National Park, Uganda, II. Crude protein, crude fibre and soil nitrogen. *Trop. Agric. (Trinidad)*, 46: 31–42.

Lötschert, W. and Beese, G., 1981. *Collins Guide to Tropical Plants*. Collins, London, 256 pp.

Lusigi, W.J., Nkurunziza, E., Awere-Gyeke, K. and Masheti, S., 1984. Range ecology. In: *Integrated Resource Assessment and Management Plan for Western Marsabit District, Northern Kenya*. UNESCO Integrated Project on Arid Lands, Tech. Rep., A-6, Nairobi, pp. 103–228.

McIlroy, R.J.M., 1978. *An Introduction to Tropical Grassland Husbandry*. Oxford University Press, London, 160 pp.

Mackworth-Praed, C.W. and Grant, C.H.B., 1957. *Birds of Eastern and Northeastern Africa*. Longmans Green and Co., London, Vol. 1, 806 pp.; Vol. 2, 1113 pp.

Makacha, S. and Frame, G.W., 1979. *A Summary of Conservation Areas in the Ngorongoro Conservation Area Authority, July, 1978 through June 1979*. Annual Report for the African Wildlife Workshop Foundation, Nairobi, 12 pp. (typescript).

Makin, J., 1969. Soil formation in the Turkana desert. *East Afr. Agric. For. J.*, 34: 493–496.

Marshall, B. and Bredon, R.M., 1967. The nutritive value of *Themeda triandra*. *East Afr. Agric. For. J.*, 32: 375–379.

Masefield, G.B., 1948. Grass burning: some Uganda experience. *East Afr. Agric. J.*, 13: 135–138.

Mbui, M.K. and Stuth, J.W., 1986. Effects of prescribed burning on cattle and goat diets. In: R.M. Hansen, B.M. Woie and R.D. Child (Editors), *Range Development and Research in Kenya*. Winrock International Institute for Agricultural Development, Morrilton, Ark., pp. 137–145.

McNaughton, S.J., 1976. Serengeti migratory wildebeest: Facilitation of energy flow by grazing. *Science*, 191: 92–94.

McNaughton, S.J., 1979. Grassland-herbivore dynamics. In: A.R.E. Sinclair and M. Norton-Griffiths (Editors), *Serengeti: Dynamics of an Ecosystem*. University of Chicago Press, Chicago, pp. 46–81.

Menaut, J.C., 1983. The Vegetation of African Savannas. In: F. Bourlière (Editor), *Tropical Savannas*. Ecosystems of the World, 13, Elsevier, Amsterdam, pp. 109–149.

Michelmore, A.P.C., 1939. Observations on tropical African grasslands. *J. Ecol.*, 27: 282–312.

Milne, G., 1937. Note on soil conditions and two East African vegetation types. *J. Ecol.*, 25: 254–255.

Milne, G., 1947. A soil reconnaissance journey through parts of Tanganyika Territory. *J. Ecol.*, 35: 192–265.

Moehlman, P.D., 1978. Jackals of the Serengeti. *Afr. Wildl. Leadership Foundation News*, 13: 2–6.

Monod, T., 1986. The Sahel zone north of the equator. In: M. Evenari, I. Noy-Meir and D.W. Goodall (Editors), *Hot Deserts and Arid Shrublands*. Ecosystems of the World, 12B, Elsevier, Amsterdam, pp. 203–243.

Moomaw, J.C., 1960. *A Study of the Plant Ecology of the Coast Region of Kenya Colony*. Government Printer, Nairobi, 52 pp.

Moreau, R.E. and Sclater, W.L., 1937. The avifauna of the mountains along the Rift Valley in north central Tanganyika Territory (Mbulu District), Part I. *Ibis*, Ser. 14, 1: 760–786 (plates 16–19).

Moreau, R.E. and Sclater, W.L., 1938. The avifauna of the mountains along the Rift Valley in north central Tanganyika Territory (Mbulu District), Part II. *Ibis*, Ser. 14, 2: 1–32.

Morgan, W.T.W., 1973. *East Africa*. Longman, London, 410 pp.

Morris, P.A. and Malcolm, J.R., 1977. The Simien fox in the Bale Mountains. *Oryx*, 14: 151–160.

Mugera, J.S. and Ogwang, B.H., 1976. Dry matter production and chemical composition of elephant grass hybrids. *East Afr. Agric. For. J.*, 42: 60–65.

Muriuki, R.M.M., 1986. Grazing block development in Kenya. In: R.M. Hansen, B.M. Woie and R.D. Child (Editors), *Range Development and Research in Kenya*. Winrock International Institute for Agricultural Development, Morrilton, Ark., pp. 29–38.

Mwova, M.B., 1977. *Guideline Information About Ranching Activities*. Range Management Division, Taita/Taveta District. Ministry of Agriculture, Kenya (cited by Van Wijngaarden, 1985).

Naylor, J.N., 1977. *Rangeland Conservation and Development: Somalia*. FAO draft technical report SF/SOM/72/003, Mogadiscio, 20 pp.

Naylor, J.N. and Jama, A.A., 1984. Ecological survey and initial management plans, Hobbio District. *Somali National Range Agency/Central Rangelands Development Project, Interim Rep.*, 60 pp.

Nieuwolt, S., 1977. *Tropical Climatology*. Wiley, London, 207 pp.

Norton-Griffiths, M., Herlocker, D.J. and Pennycuick, L., 1975. The patterns of rainfall in the Serengeti ecosystem. *East Afr. Wildl. J.*, 13: 347–373.

Nye, G.W., 1937. Preliminary notes on the use of elephant grass as a fallow crop in Buganda. *East Afr. Agric. J.*, 2: 186–190.

O'Rourke, J.T., 1978. Grazing rate and system trial over 5 years in a medium height grassland of northern Tanzania. *Proc., 1st Int. Rangeland Congr.*, pp. 563–566.

O'Rourke, J.T., Frame, G.W. and Terry, P.J., 1975a. Experimental results of *Eleusine jaegeri* Pilg. control in East African highlands. *East Afr. Agric. For. J.*, 41: 253–261.

O'Rourke, J.T., Frame, G.W. and Terry, P.J., 1975b. Progress on control of *Eleusine jaegeri* Pilg. in East Africa. *PANS*, 21: 67–72.

Owaga, M., 1980. Primary productivity and herbage utilization by herbivores in Kaputei Plains, Kenya. *Afr. J. Ecol.*, 18: 1–5.

Parry, M.S., 1953. Tree planting in Tanganyika, II. Species for the highlands. *East Afr. Agric. J.*, 19: 89–102.

Peberdy, J.R., 1969. Rangeland. In: W.T.W. Morgan (Editor), *East Africa: Its Peoples and Resources*. Oxford University Press, London, pp. 153–176.

Peden, D.G., 1987. Livestock and wildlife population distributions in relation to aridity and human populations in Kenya. *J. Range Manage.*, 40(1): 67–71.

Pennycuick, L., 1975. Movements of the migratory wildebeest population in the Serengeti area between 1960 and 1973. *East Afr. Wildl. J.*, 13: 65–87.

Peterson, D.D., 1978. *Seasonal Distributions and Interactions of Cattle and Wild Ungulates in Masailand, Tanzania*. M.Sc. Thesis, Virginia Polytechnic University, Blacksburg, Va., 166 pp.

Petrides, B.A. and Swank, W.G., 1965. Population densities and the range-carrying capacity for large mammals in Queen Elizabeth National Park, Uganda. *Zool. Afr.*, 1: 209–225.

Phillips, J., 1930. Some important vegetation communities in the Central Province of Tanganyika Territory. *J. Ecol.*, 18: 193–234.

Phillipson, D., 1979. The origins of prehistoric farming in East Africa. In: B.A. Ogot (Editor), *Ecology and History in East Africa*. Proc. of the 1975 Conf. of the Historical Society of Kenya, 7: 41–63.

Pichi-Sermolli, R.E.G., 1955. Tropical East Africa (Ethiopia, Somaliland, Kenya, Tanganyika). In: *Plant Ecology: Review of Research*, UNESCO, Paris, pp. 302–360.

Pomeroy, D.E., 1977. The distribution and abundance of large termite mounds in Uganda. *J. Appl. Ecol.*, 14: 465–475.

Pomeroy, D.E., 1978. The abundance of large termite mounds in Uganda in relation to their environment. *J. Appl. Ecol.*, 15: 51–63.

Pratt, D.J., 1962. The grazing areas of the lower Ewaso Ngiro Basin. In: *An investigation into the water resources of the Ewaso Ngiro Basin, Kenya*. Report of the Hydraulics Branch of the Ministry of Works, Nairobi, 4: 347–371.

Pratt, D.J., 1969. Management of arid grasslands in Kenya. *J. Br. Grassl. Soc.*, 24: 151–157.

Pratt, D.J. and Gwynne, M.D. (Editors), 1977. *Rangeland Management and Ecology in East Africa*. Hodder and Stoughton, London, 310 pp.

Pratt, D.J., Greenway, P.J. and Gwynne, M.D., 1966. A classification of East African rangeland with an appendix on terminology. *J. Appl. Ecol.*, 3: 369–382.

Qvortrup, S.A. and Blankenship, L.H., 1975. Vegetation of Kekopy, a Kenya cattle ranch. *East Afr. Agric. For. J.*, 40: 439–452.

Radwanski, S.A. and Ollier, C.D., 1959. A study of an East African catena. *J. Soil Sci.*, 10: 419–168.

Raimondo, F.M. and Warfa, A.M., 1980. Preliminary phytosociological research on synanthropical vegetation in southern Somalia. *Erbario Tropicale di Firenze, Publicazione*, 51. (*Not. Fitosoc.*, 15: 189–206).

Rattray, J.M., 1960. *The Grass Cover of Africa*. FAO Agric. Stud., 49, Rome, 168 pp.

Reilly, P.M., 1978. *Ethiopia*. Ministry of Overseas Development, Land Resources Bibliography 10, London, 280 pp.

Resource Management and Research (RMR), 1979. *Central Rangelands Survey*. National Range Agency. Somali Democratic Republic, Mogadishu, 4 volumes including maps (Vol. 1: Part 1, 127 pp.; Part 2, 10 maps; Part 3, 26 maps. Vol. 2: Part 2, 12 maps. Vol. 3: Part 2, 12 maps. Vol. 4, 90 pp.).

Resource Management and Research (RMR), 1981. *Northern Rangelands Survey*. National Range Agency. Somali Democratic Republic, Mogadishu, 4 volumes including maps.

Resource Management and Research (RMR), 1985. *Southern Rangelands Survey*. National Range Agency. Somali Democratic Republic, Mogadishu, 4 volumes including maps (Vol. 1, 180 pp., 52 pp., and 9 maps; Vol. 2, 93 pp. and 28 maps; Vol. 3, 130 pp. and 28 maps; Vol. 4., 241 pp.; Vol. 5, 630 pp.).

Risley, E., 1966. 'Multiple use' of the land. *Africana*, 2: 18–19 and 39.

Ross, I.C., 1969. Game distribution in the Kidepo Valley National Park. *East Afr. Wildl. J.*, 7: 171–174.

Ross, I.C., Field, C.R. and Harrington, G.N., 1976. The savanna ecology of Kidepo Valley National Park, Uganda, III. Animal populations and park management recommendations. *East Afr. Wildl. J.*, 14: 35–48.

Sadera, P.L.K. Ole, 1986. Group ranching in Kenya. In: R.M. Hansen, B.M. Woie and R.D. Child (Editors), *Range Development and Research in Kenya*. Winrock International Institute for Agricultural Development, Morrilton, Ark., pp. 17–18.

Saggerson, E.P., 1969. Geology. In: W.T.W. Morgan (Editor), *East Africa: Its Peoples and Resources*. Oxford University Press, London, pp. 68–94.

Schmidl, D., 1982. *The Birds of the Serengeti National Park, Tanzania: An Annotated Checklist*. British Ornithological Union, London.

Schmidt, W., 1975a. The vegetation of the northeastern Serengeti National Park, Tanzania. *Phytocoenologica*, 3: 133–145.

Schmidt, W., 1975b. Plant communities on permanent plots of the Serengeti Plains. *Vegetatio*, 30: 30–82.

Schultz, J., 1969. *Die Höhengliederung der Vegetation im nördlichen Tanzania*. Geographische Gesellschaft, München, 54: 153–168.

Scott, J.D., 1934. Ecology of certain plant communities of the Central Province, Tanganyika Territory. *J. Ecol.*, 22: 177–229.

Scott, R.M., 1969. Soils. In: W.T.W. Morgan (Editor), *East Africa: Its Peoples and Resources*. Oxford University Press, London, pp. 95–106.

Seely, M.K., 1978. Grassland productivity: the desert end of the curve. *S. Afr. J. Sci.*, 74: 295–297.

Sekulic, R., 1978. Roan translocation in Kenya. *Oryx*, 14: 213–217.

Simonetta, A.M., 1971. Somalia's wildlife. *Oryx*, 11: 25.

Sinclair, A.R.E., 1975. The resource limitation of trophic levels in tropical grassland ecosystems. *J. Anim. Ecol.*, 44: 497–520.

Sinclair, A.R.E., 1979. The eruption of the ruminants. In: A.R.E. Sinclair and M. Norton-Griffiths (Editors), *Serengeti: Dynamics of an Ecosystem*. University of Chicago Press, Chicago, Ill., pp. 82–103.

Sinclair, A.R.E. and Norton-Griffiths, M., 1979. *Serengeti: Dynamics of an Ecosystem*. University of Chicago Press, Ill., Chicago, 389 pp.

Smith, J., 1949. *Distribution of Tree Species in the Sudan in Relation to Rainfall and Soil Texture*. Sudan Ministry of Agric. Bull. 4, 64 pp.

Somali National Range Agency (SNRA), 1977. *Request by the Government of Somalia to the World Food Programme for the expansion of Project "Somalia 919"; reforestation and rangeland development*, Mogadiscio, 14 pp.

Sombroek, W.G., Braun, H.M.H. and van der Pouw, B.J.A., 1982. Exploratory soil map and agro-climatic zone map of Kenya, 1980, scale 1:1,000,000. *Kenya Soil Survey, Exploratory Soil Survey Report* E-1, 56 pp. plus maps.

Stephens, D., 1967. Effect of fertilizers on grazed and uncut elephant grass leys at Kawanda Research Station, Uganda. *East Afr. Agric. For. J.*, 32: 383–392.

Stewart, D.R.M. and Zaphiro, D.R.P., 1963. Biomass and density of wild herbivores in different East African habitats. *Mammalia*, 27: 483–496.

Strugnell, R.G. and Pigott, C.D., 1978. Biomass shoot production and grazing of two grasslands in the Rwenzori National Park, Uganda. *J. Ecol.*, 66: 73–96.

Stuart, S.N., 1981. A comparison of the avifaunas of seven East African forest islands. *Afr. J. Ecol.*, 19: 133–151.

Talbot, L.M. and Talbot, M.H., 1963. The high biomass of wild ungulates on East African savanna. *Trans., North Afr. Wildl. Nat. Resour. Conf.*, 28: 467–476.

Tessema, S., 1986. Chemical composition and in vitro dry matter digestibility of herbage. In: R.M. Hansen, B.M. Woie and R.D. Child (Editors), *Range Development and Research in Kenya*. Winrock International Institute for Agricultural Development, Morrilton, Ark., pp. 85–92.

Thomas, A.S., 1940. Grasses as indicator plants in Uganda, I. *East Afr. Agric. J.*, 6: 19–22.

Thomas, A.S., 1943. The vegetation of the Karamoja District, Uganda. *J. Ecol.*, 31: 149–177.

Thomas, A.S., 1945. The vegetation of some hillsides in Uganda; illustrations of human influences in tropical ecology, I. *J. Ecol.*, 33: 10–43.

Thomas, A.S., 1946. The vegetation of some hillsides in Uganda; illustrations of human influences in tropical ecology, II. *J. Ecol.*, 33: 153–172.

Thomas, B.D. and Pratt, D.J., 1967. Bush control studies in the drier areas of Kenya, IV. Effects of controlled burning on secondary thicket in upland Acacia woodland. *J. Appl. Ecol.*, 4: 325–335.

Thorbahn, P.F., 1984. Brier elephant and the brier patch. *Nat. Hist.*, 4: 71–78.

Thurow, T.L., 1989. Decomposition of grasses and forbs in coastal savanna of southern Somalia. *Afr. J. Ecol.*, 27: 201–206.

Thurow, T.L., in prep. Behavior and ecology of Speke's gazelle (*Gazella spekei*) on the coastal plain of central Somalia.

Thurow, T.L. and Hussein, A.J., 1989. Observations on vegetation responses to improved grazing systems in Somalia. *J. Range Manage.*, 42: 16–19.

Thurow, T.L., Herlocker, D.J., Shaabani, S.B. and Buh, B.B., 1987. Diets of large herbivores on the coastal plain of central Somalia. *Colloquio Produzione Foraggera del Somalia*. Universita Degli Studi di Firenze Italia, pp. 91–103.

Trapnell, C.G. and Langdale-Brown, I., 1969. Natural vegetation. In: W.T.W. Morgan (Editor), *East Africa: Its Peoples and Resources*. Oxford University Press, London, pp. 127–139.

Trapnell, C.G., Birch, W.R., Brunt, M.A. and Pratt, D.J., 1966. *Vegetation — land use survey of southwestern Kenya, Sheet 1*. Directorate of Overseas Surveys, London.

Trapnell, C.G., Brunt, M.A., Birch, W.R. and Trump, E.C., 1969. *Vegetation — land use survey of southwestern Kenya, Sheet 3*. Directorate of Overseas Surveys, London.

Trapnell, C.G., Birch, W.R., Brunt, M.A. and Lawton, R.M., 1976. *Vegetation — land use survey of southwestern Kenya, Sheet 2*. Directorate of Overseas Surveys, London.

UNESCO, 1979. *Map of the World Distribution of Arid Regions with Explanatory Note*. UNESCO, Paris, 54 pp.

United Nations Sahelian Office (UNSO), 1984. *Inventory of Sand Movement in Somalia*. UNSO/Des/Som/80/001, 100 pp.

Van Rensburg, H.J., 1952. Grass burning experiments on the Msima River stock farm, Southern Highlands, Tanganyika. *East Afr. Agric. For. J.*, 17: 119–129.

Van Rensburg, H.J., 1960. Ecological aspects of the major grassland types in Tanganyika. *Proc., 8th Int. Grassland Congr.*, pp. 367–370.

Van Rensburg, H.J., 1969. Management and utilization of pastures. *FAO Pasture and Crop Series*, 3, Rome, 118 pp.

Van Wijngaarden, W., 1985. Elephants — trees–grass-grazers: relationships between climate, soils, vegetation and large herbivores in a semi-arid savanna ecosystem (Tsavo, Kenya). *Int. Inst. Aerospace Surv. Earth Sci. Publ.*, 4, 159 pp.

Van Wijngaarden, W. and Van Engelen, V.W.P., 1985. Soils and vegetation of the Tsavo area. *Kenya Soil Surv., Reconnaissance Soil Surv. Rep.*, No. R 7.

Verdcourdt, B., 1962. The vegetation of the Nairobi Royal National Park. In: S. Heriz-Smith (Editor), *The Wild Flowers of the Nairobi Royal National Park*. East Africa Natural History Society, Nairobi, pp. 38–49.

Vesey-FitzGerald, D.F., 1970. The origin and distribution of valley grasslands in East Africa. *J. Ecol.*, 58: 51–75.

Vesey-FitzGerald, D.F., 1973. *East African Grasslands*. East African Publishing House, Nairobi, 95 pp.

Vesey-FitzGerald, D.F., 1974. Utilization of the grazing resources by buffaloes in the Arusha National Park, Tanzania. *East Afr. Wildl. J.*, 12: 107–134.

Vos, J.G., 1970. Somalia's wildlife. *Oryx*, 10: 304–305.

Walter, H., 1971. *Ecology of Tropical and Subtropical Vegetation*. Oliver and Boyd, Edinburgh, 539 pp.

Wendt, W.B., 1970. Response of pasture species in eastern Uganda to phosphorus, sulphur and potassium. *East Afr. Agric. For. J.*, 36: 211–219.

Were, G. and Wilson, D., 1978. *East Africa Through a Thousand Years*. Evans Brothers Ltd., London, 372 pp.

White, F., 1978. The afromantane region. In: M.J.A. Werger (Editor), *Bio-Geography and Ecology of Southern Africa*. Junk, The Hague, pp. 463–513.

White, F., 1983. The vegetation of Africa. *UNESCO Natural Resources Research Series*, XX, Paris, 356 pp. plus maps.

White, R.O., 1962. The myth of tropical grasslands. *Trop. Agric. Trinidad*, 39: 1–11.

Williams, J.G. and Fennessy, R., 1963. *A Field Guide to the Birds of East and Central Africa*. Collins, London, 288 pp. plus 40 plates.

Williams, J.G. and Fennessy, R., 1967. *A Field Guide to the National Parks of East Africa*. Collins, London, 352 pp. plus 32 plates.

Wilson, J.G., 1960. *The Vegetation of Karamoja District, Northern Province, Uganda*. Uganda Department of Agric., Memoirs of Research Division, Series 2, No. 5, Kampala, 159 pp.

Wilson, R.T., 1981. General Swayne's neglected namesake. *Anim. Kingdom*, 84: 36–41.

Wing, L.D. and Buss, I.O., 1970. Elephants and forests. *Wildlife Monograph No. 19*. The Wildlife Society, Washington, D.C., 92 pp.

Wolde-Mariam, M., 1970. *An Atlas of Ethiopia*. Poligrafico, Priv. Ltd. Co., Addis Ababa, 84 pp.

Yalden, D.W., 1975. Some observations on the giant mole-rat *Tachyoryctes macrocephalus* (Ruppell, 1842) (Mammalia, Rhizomyidae) of Ethiopia. *Monit. Zool. Ital.*, 15: 275–303.

Chapter 10

GRASSLANDS OF SOUTHERN AFRICA [1]

N.M. TAINTON and B.H. WALKER

INTRODUCTION

The area under consideration in this chapter is shown in Fig. 10.1. It encompasses the area covered by the maps of Acocks (1975) and of Wild and Grandvaux Barbosa (1967) (i.e., Botswana, Mozambique, Namibia, South Africa and the countries it surrounds, Zambia and Zimbabwe, as well as Angola, Malawi, and Tanzania). For much of the area, little information is available about the grasslands beyond descriptions (mostly qualitative) of floristic composition. The account of the productivity, dynamics and management of these grasslands, therefore, rests rather heavily on the results from only a few research sites, the majority of which are in South Africa and Zimbabwe.

CLASSIFICATION OF MAJOR COMMUNITIES

The grasslands of southern Africa fall into two major categories: (1) natural, pure (climax) grasslands; and (2) induced or wooded grasslands. By far the major portion of the region is wooded grassland, varying from grassland with sparse, scattered trees to virtually closed woodland (Fig. 10.1). The above-ground biomass of the woody plants in such savannas generally outweighs that of the herbaceous layer. However, it is the latter which is the more important in relation to secondary production and use of the area by man, since his concern is mostly with grazing rather than with browsing animals. The importance of savannas as rangeland justifies their inclusion in this account of the sub-continent's grasslands.

For ease of presentation the following discussion uses rather deterministic, Clementsian terminology. We acknowledge that the real situation is more complex, but for the objectives of this volume (aimed at a wide readership) the format adopted is more readily understandable.

Climax grasslands

The climax grasslands are those which, in the absence of interference by man, remain essentially treeless. They can be divided into four main types. Two of these are climatically determined, one is both climatically and edaphically determined, and the fourth is an edaphic climax. They are, in turn:

(1) hot, dry grasslands at low elevations (e.g. the Namib desert grasslands) (Fig. 10.2));
(2) the humid, but cold, mountain grasslands;
(3) the grasslands of the South African "high veld"[2]; and
(4) grasslands occurring at the lower end of the catena, which are seasonally inundated (known as *vleis* in southern Africa and *dambos* in central Africa) (Fig. 10.3) (Thompson and Hamilton, 1983), or having a seasonally high water table for other reasons. This type is scattered throughout the region and patches are often too limited in area to be mapped.

---

[1] Manuscript submitted in April, 1980; revised in January, 1990.

[2] "**Veld**" is a general term meaning "vegetation". "**High veld**" and "**low veld**" are local southern African terms which distinguish between the cooler, higher-rainfall regions above an altitude of approximately 1500 m and the hotter, more arid regions which occur at altitudes below about 750 m. The in-between regions are referred to as "**middle veld**".

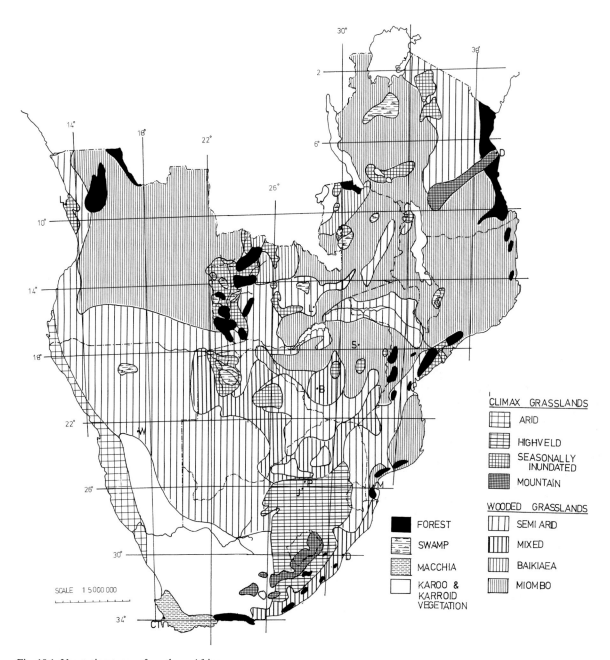

Fig. 10.1. Vegetation types of southern Africa.

Wooded or subclimax grasslands

These grasslands are climatically and edaphically suited to the development of woody vegetation, to a greater or lesser extent. Large areas, particularly in the drier regions, that have reached the full extent of their development of woody vegetation, still support a significant grass layer. Other areas, mainly in the higher-rainfall regions, are maintained at a lower stage of woody development (than the soils and climate would allow) by fire or other interference by man, or by the activities of large herbivores.

These wooded grasslands are very variable and

Fig. 10.2. Desert grassland, dominated by species of *Stipagrostis* (particularly *S. gonadostachys*) in gravel plains of the central Namib desert.

Fig. 10.3. Edaphic grasslands. A typical *vlei* in western Zimbabwe, with the lower, hydromorphic grassland, and an ecotone (of *Terminalia sericea*) merging into well-developed woodland. It is a typical habitat for *Hippotragus niger* (sable antelope), the species present.

cover a large range of different types of woody vegetation. For our purposes, however, we distinguish four main types in terms of their grass species composition and ecological characteristics. They are:

(1) semiarid savannas at low elevations (lower than about 750 m in elevation), usually on shallow, sandy loam to loamy soils, including the extensive areas of *Acacia* thorn savannas and the "low veld" of South Africa and Zimbabwe (Fig. 10.4);

Fig. 10.4. Semiarid savanna in south-western Zimbabwe. The density of woody vegetation has been slightly reduced by elephant activity. The dominant tree is *Colophospermum mopane*.

Fig. 10.5. Mesic savanna on sandy soil in the northern Transvaal (South Africa), dominated by *Burkea africana* (trees) and *Ochna pulchra* (shrubs), the dominant grass being *Eragrostis pallens*.

(2) mixed, more densely wooded, mesic savannas usually on sandy soils at medium elevations (700–1500 m) (Fig. 10.5);

(3) *Baikiaea plurijuga* associations on deep Kalahari sand deposits, in northern Botswana, western Zimbabwe, and in Zambia; and

(4) *miombo*[1] woodland associations, of which the grassy areas are all subclimax (Fig. 10.6).

---

[1] *Miombo* is a local, central African term describing those woodland regions dominated by trees in the genus *Brachystegia* or closely related genera.

GRASSLANDS OF SOUTHERN AFRICA

Fig. 10.6. The transition between wet savannas and the *miombo* woodlands of central Africa (near Salisbury, Zimbabwe). The dominant trees are *Brachystegia* spp. and *Julbernardia globiflora*.

CONTACTS WITH OTHER ECOSYSTEMS

The non-grassy biomes of the region are: desert scrub (usually a degraded form of xeric savanna); the sclerophyllous, Mediterranean-type vegetation of the Cape Province coastal region (known as the *fynbos* and in which the Poaceae are replaced by Restionaceae); the closed woodlands of the *karoo* (a xeric dwarf shrubland); and tropical and sub-tropical forest. These types of vegetation are discussed in other volumes of this book series (Boucher and Moll, 1981; Menaut, 1983; Werger, 1986).

Climax grasslands

Mountain grassland usually has a narrow ecotone with the forest, in protected hollows and on deeper soils — mostly at lower elevations; much of what is grassland today is in fact a biotic climax (or fire climax) induced and maintained largely by fire and to a lesser extent by grazing, in an environment otherwise suited to the development of forest. In the eastern highlands of Zimbabwe grassland grades into both sclerophyll scrub vegetation and a stunted form of *miombo* woodland. Small areas of mountain grassland occur at high elevations within the *fynbos*, and their ecotone is rather abrupt.

The grasslands of the South African high veld are bounded by the *karoo* (dwarf shrubland) on their south-western edge and by mixed savanna "bush veld" elsewhere, except where they grade into mountain grassland in the Drakensberg mountains.

The seasonally flooded or saturated grasslands have a localized distribution throughout the region. They are commonest in the savanna and woodland regions, and typically have a fringing, often unstable, ecotone of such tree species as *Terminalia sericea*, which in turn grades into the wooded vegetation of the mid-catena zone. Grasslands that owe their existence to a seasonally high water table occur over larger, often quite flat areas, where there is an impenetrable layer (usually of calcrete) close to the surface. The ecotone with the surrounding vegetation is abrupt and, within much of the *miombo* region, is marked by *Parinari curatellifolia* trees, which can withstand a relatively high water table.

The pure desert grasslands are restricted to the Namib desert in Namibia. They grade, very gradually, into a mixed scrub savanna along their eastern edge.

Wooded or subclimax grasslands

The semiarid savanna regions have a well-developed grass layer and constitute the most important ranching areas of the subcontinent. They occupy large parts of: Botswana; southern and central Mozambique; the northern Cape Province, northern Natal and northern and north-eastern Transvaal in South Africa; Namibia; and the south and south-west of Zimbabwe. On their dry fringes they grade into semi-desert scrub, or (in South Africa) into *karoo*, and at the wetter end they grade into the moist savannas. These moist savannas predominate over most of Angola, northern and central Mozambique, central Tanzania, Zambia, and the western and northern half of Zimbabwe. The *Baikiaea* associations occur mainly on the south-western fringe of the *miombo* region, on old Pleistocene sand ridges of Kalahari aeolian deposits, and they grade into the semiarid savannas.

DETERMINANTS OF THE MAIN VEGETATION TYPES

Climate

Apart from the few, small areas of mountain grassland in the Cape Province, where rainfall is largely confined to the winter period, the grass-dominated regions all experience summer rainfall and winter drought. The high veld and mountain grasslands in South Africa are also subjected to severe frosts in winter. Frost conditions extend into all of the wooded grasslands as well. They are not as severe as in the pure grasslands, and they decrease in frequency and severity with decreasing latitude and elevation. Nevertheless, they can be of considerable importance in determining the floristic and physiognomic structure of the vegetation in areas where only part of the landscape is affected by "frost pockets" (Rushworth, 1975). Within the savanna regions, the *vlei* (bottom of catena) grasslands are the areas subjected to most frost, as a result of cold-air drainage.

Representative climatic diagrams for each of the major grassland and savanna areas are presented in Fig. 10.7. They all show a period in the winter season when rainfall is negligible.

A feature that is not demonstrated by these average climates is the occurrence of intra-seasonal drought. It is commonest (not surprisingly) in the desert grasslands, is still of considerable significance in the savannas, and becomes less important in the high veld and *miombo* grasslands. The term "drought" refers here to periods when the soil moisture drops below the wilting point, and plant growth ceases.

Variation in rainfall between years increases with increasing aridity. Within Zimbabwe, for example, there is a gradient in rainfall from north-west to south-east, which corresponds with an increase in the coefficient of variation in the rainfall from ±20% in the north-west, with a mean annual rainfall of about 900 mm, to ±45% in the south-east, where the mean rainfall is about 500 mm (Anonymous, 1963). This high variation is a major problem in the successful management of the vegetation under ranching conditions, since the standing crop of herbage can vary by 500% from one year to the next (Kelly and Walker, 1976; Barnes, 1979).

Geology and soils

Generalization about the geology and soils of the grasslands and savannas is difficult. Soils range from deep aeolian sands, through rocky slopes with little or no top-soil, to very fine-textured, montmorillonitic ("self-ploughing" or "self-mulching") clays. The commonest, however, are the following:

(1) Aeolian deposits of Kalahari sand, covering extensive areas in Botswana, Namibia, Zambia and western Zimbabwe — *Baikiaea* woodlands occur where annual rainfall exceeds about 600 mm and deposits are more than 10 m deep, and a shrub and tree savanna (often dominated by *Terminalia sericea*) where conditions are less favourable.
(2) Granitic-derived sandy soils, supporting mesic savannas and *miombo* vegetation.
(3) Clay loams and clays derived from basalt and dolerite, occurring under all rainfall conditions at all elevations. At high elevations with rainfall in excess of 600 mm, and at latitudes higher than 20°S, *miombo* vegetation develops. At low elevations on finer-textured soils, trees and shrubs of *Colophospermum mopane* dominate over extensive areas. *Acacia* associations occur on the black clay soils of the northern Transvaal and Botswana.

General

The many different types of grassland and of savanna are a result of the interaction between climate and soil. In general, high rainfall leads to leaching and the development of either the moist, dystrophic savannas (Huntley, 1982) or "sour-veld"[1] grasslands, and low rainfall results in either

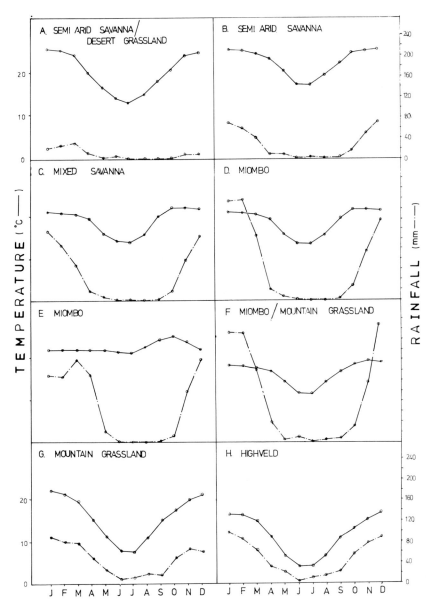

Fig. 10.7. Representative climatic diagrams for the major grassland and savanna vegetation types of southern Africa (after Wernstedt, 1972): (A) Keetmanshoop (Namibia); (B) Messina (S. Africa); (C) Bulawayo (Zimbabwe); (D) Harare (Zimbabwe); (E) Tabora (Tanzania); (F) Inyanga (Zimbabwe); (G) Mokhotlong (Lesotho); (H) Bloemfontein (S. Africa).

the xeric, eutrophic savannas or "sweet veld"[1]. The interaction of rainfall, temperature (particularly frost), fire and soil type determines the type of vegetation, but it is not always clear how or why some areas are pure grasslands and others not.

---

[1] These terms are explained below (pp. 273, 274).

An example is the high-veld grassland of South Africa. Acocks (1975) has stated that pure grass veld occurs on the upper plateaux and mountain tops at elevations between 1050 and 3050 m in regions which "are too dry and/or too frosty" for the development of forest, and that only on rocky hills are shrubs found. Roux (1969) states that the high veld is a grassland region because it has a moderately high rainfall, with summer rains and winter drought, and because it is subject to severe frost in winter. He states that fire, too, tends to hinder tree growth, except where rocky outcrops provide protection. An opposing view (Tinley, 1977) is that the high veld is grassland because of soil conditions — namely a podzol development with a gleyed horizon. This view would therefore give the high veld the same basic determinants as the seasonally flooded grasslands (anaerobic soil conditions which prevent tree growth). The full explanation is probably a mixture of both views.

THE GRASSLAND COMMUNITIES

As indicated earlier, the eight types of grassland recognized for this account have not been studied equally well. They have distinct flora and fauna, and the floristic composition of each will be dealt with separately. Other characteristics will be considered more generally.

Floristic composition

Climax grasslands

Desert grasslands. These are restricted to the gravel plains and inter-dune valleys of the Namib desert. These grasslands grade into complete desert on their western edge (no regular annual rainfall) and into scrub savanna on the east (annual rainfall exceeding 250 mm). They are short (15–20 cm high), sparse grasslands, dominated by very short perennial and annual species of *Aristida*, *Eragrostis* and *Stipagrostis* (*S. ciliata*, *S. gonadostachys* and *S. obtusa*). Included with them, and at about the same height, are several succulents (e.g. *Zygophyllum* spp.) and small forbs (*Ruschia* spp).

Mid-elevation grasslands. These occur in South Africa, both in the high veld and in Natal. Rainfall is fairly low (less than 800 mm annually), and summer temperatures are moderately high. The reason for exclusion of trees is still a debatable point, but (as described earlier) it is probably due to both climatic and edaphic conditions. This is a moderately tall (80 cm) tussock grassland dominated largely by the Andropogoneae (*Cymbopogon plurinodis*, *Diheteropogon filifolius*, *Heteropogon contortus* and *Themeda triandra*) and the Paniceae (*Brachiaria serrata*, *Digitaria eriantha* and *Setaria flabellata*), and with a variety of forbs (*Helichrysum* spp., *Senecio* and others) commonly associated with the grasses. Cool-season grass species (cf. *Bromus*, *Helictotrichon* and *Koeleria*) increase in abundance at higher elevations. Where heavy grazing pressure leads to a reduction in the density of the grasses of the climax community, they are replaced largely by:

Aristida congesta	*Eragrostis obtusa*
Chloris virgata	*Eragrostis plana*
Eragrostis capensis	*Sporobolus capensis*
Eragrostis chloromelas	*Sporobolus pyramidalis*
Eragrostis lehmanniana	*Urochloa panicoides*

Under such conditions these grassland communities are often invaded by trees and shrubs of the savanna, and particularly by species of *Acacia*, or by dwarf shrub species from the *karoo*, such as *Chrysocoma ciliata*, *Pentzia globosa* and (especially in the South African high veld) *Stoebe vulgaris*.

High-elevation grasslands. In these areas rainfall is relatively high (more than 750 mm annually) and winter temperatures are low. These are climatic-climax grasslands, characteristically short (less than 50 cm) and dense, with warm-season species such as *Diheteropogon filifolius* and *Trachypogon spicatus*, in addition to many of those of the climatic-climax grasslands of lower elevations (listed in the previous paragraph), but here the temperate element forms an important component of the community. Species of *Bromus*, *Festuca*, *Merxmuellera*, *Pentachistis* and others make up an increasing proportion of the flora as elevation increases. Degradation through mismanagement leads to replacement of the subtropical flora by temperate species, and elements of the temperate sclerophyllous *fynbos* (*macchia*) invade. Among these invaders, *Cliffortia* spp., *Erica* spp. and *Passerina montana* are often important (Acocks, 1975).

Seasonally inundated grasslands. These grasslands occur in areas climatically suited to the progression of succession to woodland or forest, but where seasonal waterlogging ensures retention of a grassland community. Whereas these grasslands are often treeless, they are also interspersed among forest patches in the more humid regions, where the forest patches are often confined to the cool southern slopes only. In areas of moderate rainfall (more than 700 mm), bush clumps composed of a mixture of tree and shrub species sometimes develop in areas where soil drainage has been improved (as around termite mounds) (Tinley, 1977). The majority of these grasslands occur as *vleis* or *dambos* at the lowest end of a catena. They may be waterlogged for only a few weeks, but this may be sufficient to prevent tree growth. The grassland that develops is usually tall (1 to 2 m) and dominated by such species as *Hyperthelia dissoluta*, *Loudetia simplex* and *Setaria sphacelata*. The fine, black montmorillonitic clay soils, typically occurring in this topographic situation in basalt areas, have a characteristic species composition dominated by *Ischaemum afrum*, *Sehima galpinii* and *Setaria woodii*.

Wooded grasslands

In the savanna regions a well-developed grass layer occurs together with the open shrub and tree layer as the climax vegetation. The ecosystem is water-limited, and the stable mixture of the two components has been ascribed to their different rooting depths and different abilities to take up soil water (Walter, 1971). The proportion of the vegetation contributed by grasses varies according to rainfall and to the frequency and intensity of fire.

In the moister woodland and forest areas, the existence of grassland in sites capable of supporting woody vegetation is due almost entirely to fire. These biotic or fire-climax grasslands are fairly tall (up to 2 m) and their species composition varies somewhat according to the type of woody vegetation.

Woodlands. Within the tropics, by far the most important and extensive of these wooded grasslands are those within the *miombo* woodlands. Much of this area is well suited to arable agriculture, and very little undisturbed woodland survives on potentially arable soils. In steeper or stony sites, mechanical clearing and fire have been used extensively to develop grassland for grazing by cattle. Under these conditions, succession is rapid, and frequent fires are necessary to maintain the grass layer. Uneaten grass accumulates rapidly in the absence of fire, and within a few years the sward becomes moribund. The resulting high fuel load is conducive to fire, and protection from fire becomes increasingly difficult. The likelihood of fire is reduced only when succession has advanced to the stage where the tree canopy has strongly suppressed grass growth. Fully mature *miombo* woodland is difficult to burn without some initial felling.

The characteristic species of these fire-climax grasslands include:

Andropogon gayanus	*Hyparrhenia cymbaria*
Brachiaria brizantha	*Hyparrhenia filipendula*
Brachiaria serrata	*Hyperthelia dissoluta*
Cymbopogon excavatus	*Hyperthelia newtonii*
Digitaria setivalva	*Themeda triandra*

Although it is a productive grassland, the species are generally rather coarse, and are high in protein content (and palatable) only for from 4 to 6 months of the year (hence the local South African term "sour veld").

Overgrazing leads to an increase in species that are unpalatable, either because of taste (*Bothriochloa insculpta*, *Cymbopogon* spp. and *Elionurus muticus*) or texture (*Eragrostis* spp. and *Sporobolus pyramidalis*), and to an overall decline in grass cover. Rainfall is too high for invasion of the karroid or savanna shrubs, and the result of misuse is usually severe soil erosion accompanied by increases in density and more rapid growth of the woody *miombo* species. The other significant grasslands derived from woodland are those on Kalahari sand deposits in the *Baikiaea plurijuga* region. The grass species are similar to those in areas of grassland derived from *miombo* woodlands on deep sands. This is a coarse, unpalatable grassland dominated by *Aristida meridionalis*, *A. stipitata* and *Eragrostis pallens*, together with *Digitaria eriantha* subsp. *pentzii*, *Panicum kalaharense*, *Schizachyrium sanguineum*, *Tricholaena monachne* and *Triraphis schinzii*. The large annual *Megaloprotachne albescens* is also prevalent.

The vegetation reverts rapidly to a dense shrub community in the absence of fire. Because of the very large under-ground woody biomass [as much as 9 times that of the canopy, according to Rushworth (1975)], frequent burning causes little loss of vigour to the woody plants.

Savanna. Fire is not an important feature in the arid savannas, as fuel loads are seldom sufficient to carry a fire. The soils are unleached and their high base saturation leads to very palatable grass (hence the local term "sweet veld"). The species composition is markedly different between the grass communities occurring beneath the canopies of trees and those that occur in the open. The sub-canopy grasses are more palatable and remain green longer into the dry season. The characteristic species are *Dactyloctenium* spp., *Panicum maximum* and *Urochloa mosambicensis*.

The characteristic species in the open areas on medium-textured soils are *Bothriochloa radicans*, *Brachiaria nigropedata* (very palatable), *Cenchrus ciliaris*, *Enneapogon cenchroides*, *E. scoparius*, *Eragrostis rigidior* and *Schmidtia pappophoroides*. On sandy soils *Aristida* spp., *Eragrostis lehmanniana*, *Pogonarthria squarrosa* and others become dominant.

Overgrazing leads to replacement of the perennial grasses by annual species of *Aristida* and other annual and short perennial grasses (such as *Eragrostis gummiflua* and *Oropetium capense*) and annual forbs (particularly *Tribulus terrestris*).

The mixed tree and shrub savanna, on sandy soils, is dominated in the woody layer by *Burkea africana*, *Combretum* spp., *Dombeya rotundifolia*, *Grewia* spp., *Ochna pulchra*, *Terminalia sericea* and many others. In the grass layer, the sub-canopy areas are dominated by *Digitaria eriantha* and *Panicum maximum*, whereas the dominant grasses in the open areas are *Digitaria eriantha*, *Eragrostis pallens*, *Heteropogon contortus*, *Hyparrhenia* spp., *Loudetia* spp., *Pogonarthria squarrosa* and *Schizachyrium sanguineum*. Where the woody component is largely removed by fire, the resulting fire-climax grassland is dominated by *Heteropogon contortus*, *Themeda triandra* and *Tristachya leucothrix*. *Hyparrhenia hirta* is usually prominent at lower elevations, and in southerly latitudes, particularly on northern slopes. A large number of species, and particularly those of the tribes Andropogoneae and Paniceae, are associated with these dominants, including in particular:

Alloteropsis semialata	*Elionurus muticus*
Andropogon schirensis	*Hyperthelia dissoluta*
Cymbopogon excavatus	*Loudetia simplex*
Diheteropogon amplectens	*Monocymbium ceresiiforme*
Diheteropogon filifolius	*Trachypogon spicatus*

With mismanagement, species of *Aristida* (e.g. *A. junciformis*), *Eragrostis* (e.g. *E. curvula* and *E. plana*) and *Sporobolus* (e.g. *S. capensis* and *S. pyramidalis*) increase in abundance, whereas in old cultivated sites the above-mentioned species of *Eragrostis* and *Sporobolus* generally dominate from about the third to the seventh year of fallow, to be replaced from about the eighth year onwards by *Hyparrhenia hirta*. The latter species forms an almost mono-specific community unless defoliation, by either fire or grazing, is infrequent. Where such *H. hirta* communities remain ungrazed or unburned, tall species of *Cymbopogon* (*C. validus*) and *Hyparrhenia* (*H. dregeana* and *H. rufa*) invade and are followed by shrubs and trees of the savanna as succession proceeds further.

Primary production[1]

In his review of primary production in southern Africa, Rutherford (1978) concluded that, in the grassland vegetation, annual herbage yield increases from less than 100 g m^{-2} in the drier western parts to about 100 to 200 g m^{-2} in the higher-rainfall but marginal grassland areas and high-veld vegetation and to considerably more (up to 400 g m^{-2}) in the mesic grasslands of Natal and the eastern Orange Free State. Because of these large differences in yield, it is necessary to consider the different vegetation types separately.

Climax grasslands

The desert grasslands exhibit extreme variability in annual yield, with accumulated maximum standing crop in the canopy ranging from zero in the dune and gravel plain areas to as much as 100 g m^{-2} (Walter, 1971). These grasslands are essentially ephemeral, and the standing crop of the annual species of *Aristida* and *Stipagrostis* mostly disappears within two years in the absence of rain as old leaves are dislodged and carried away.

In the permanent, perennial climax grasslands, grass yield is a function of soil water, temperature, and soil nutrient status. At low elevations, growth normally starts in September, when the temperature in the top 5 cm of soil rises to about 12°C, and continues through to April (Tainton, 1981b). The winter is too dry and too cold for growth, although the cool-season species and some of the forbs

---

[1] In this section the term "**yield**" is taken to mean the maximum above-ground biomass (standing crop) achieved during the growing season. "**Standing crop**" includes both live and dead material.

will grow at this time where moisture is available. Growth is usually most rapid in early summer and autumn. During mid-summer much of the rain falls in heavy downpours, so that little penetrates into the soil. Also, since evapo-transpiration rates are high, moisture stress severely limits plant growth in summer.

Herbage yields within these climax grassland communities are very much dependent on site characteristics and on previous management. Vorster (1975) quoted yields ranging from 59 to 195 g of dry matter m^{-2} for moderately arid climatic-climax grassland in a season of above-average rainfall (784 mm, where the 60-year mean is 523 mm). This large range in annual yield was occasioned by differences in prior grazing treatment, and indicates the degree to which such treatments can influence the vigour of the grass sward. Grunow et al. (1970) quoted yields as low as 47 g m^{-2} in one site, whereas Wiltshire (1978) recorded yield values of 900 g m^{-2} for mixed grass veld. He reported, however, that only a relatively small proportion of this material was living (green) at any particular time. A large proportion (220 to 420 g m^{-2}) occurred as standing dead and as much as 400 g as litter.

In the high-veld grassland at Frankenwald research station north of Johannesburg (rainfall about 750 mm yr^{-1}), Weinmann (1948) recorded annual herbage yields of 220 g m^{-2}. Under slightly higher rainfall (about 800 mm) but somewhat colder conditions, Rethman and Beukes (1973) measured yields between 200 and 230 g m^{-2}. In moist *vleis* (seasonally inundated grasslands) within this region, much higher yields than those cited above have been recorded by Wiltshire (1978). Here, over 2000 g m^{-2} of dry matter was recorded in March 1976, of which just over 450 g was green, 500 g was standing dead (in the canopy) and about 1100 g m^{-2} occurred as litter.

Although low rainfall is considered to be the factor that most limits production within this grassland region, yield responses have been noted to fertilizers, and particularly to nitrogen, even in moderately dry years (Grunow et al., 1970). It has been suggested that, as a general rule, 26.5 kg ha^{-1} of nitrogen may be applied as a standard dressing at the beginning of each growing season, and that additional applications of 26.5 kg N ha^{-1} should be applied for every 125 mm of rain which falls during summer. Grunow et al. (1970) estimated that an addition of about 80 kg N ha^{-1} in this way should more than double the herbage yield (from about 150 to 350 g m^{-2}) in dry years, and that 113 kg N ha^{-1} should raise the yield to 580 g m^{-2} in wet years. They noted, however, that it may also be necessary to apply phosphorus and potassium in order to make full use of the applied nitrogen.

There is also evidence that the response to nitrogen induces a change in species composition. Long-term experiments on the fertilization of climatic-climax grassland in the South African high veld, on sandy soils, showed a dramatic reversal in succession at high levels of nitrogen. The higher successional and climax grasses and associated forbs all disappeared, and were replaced by early seral species of *Cynodon* and *Eragrostis* (Roux 1969).

The productive capacity of the climatic-climax grasslands of colder regions at high elevations is severely limited by low temperature. The growing season is short and confined to the summer months. Yields are likely to be low (100 to 200 g m^{-2}).

Wooded and induced grasslands

Annual production within the biotic or fire-climax grasslands is extremely variable and depends not only on the rainfall in any season but also on the influence of past management treatments on the vigour and composition of the community. The annual yield produced by the fire-induced sward varies from 50 g m^{-2} on inherently infertile soils to about 250 g m^{-2} in wet years on inherently fertile soils. Where the grassland community has been degraded and is dominated by pioneer perennial species (*Aristida junciformis*, *Eragrostis curvula*, *E. plana*, *Hyparrhenia hirta* and *Sporobolus pyramidalis*), annual yields may exceed 500 g m^{-2} (Tainton et al., 1978b), but such material is usually relatively unacceptable to grazing animals.

Within any particular community the commencement of spring growth is controlled by both temperature and rainfall. In southern latitudes, temperatures are seldom sufficiently high to stimulate growth before September in pioneer communities and before October in fire-climax communities. Within the tropics, however, growth begins as soon as water is available. Frost often terminates growth in the autumn, before moisture is depleted, and some green material usually persists through winter. By spring, however, growth is very much dependent on moisture conditions, and significant growth does not occur before sufficient spring rain has fallen. Maximum growth rates in the northern fire-climax grasslands occur in November and December, and here as much as 80% of the season's yield is produced before mid-summer (Drewes, 1979). Further southward, however, growth continues into the autumn, so that production is spread more uniformly over the summer season.

Rainfall has a relatively strong influence on annual production, even within the areas of moderately high rainfall. For example, in the tall-grass veld of Natal, a high correlation was found between rainfall (in an area experiencing a long-term mean of 717 mm yr^{-1}) and annual yield, over an 18-year period.[1] These results also suggest that there was a residual effect of rainfall in one season on herbage yield in the following season.

The relationship between rainfall and growth is very marked in the arid savannas. The coefficient of variation in rainfall rises to around 40%, and the inter-seasonal variation in grass yield is as much as 500% (Barnes, 1979).

The production of grass in these savanna regions is strongly affected by the amount of woody vegetation, at least when tree density increases above 1000 trees ha^{-1} (Aucamp, 1979). Removal of the woody vegetation was reported by Barnes (1979) to lead to an increase of between 200 and 400% in grass growth, but this increase must be offset against changes that occur in the floristic composition of the grass layer. The grasses that occur under the canopies of trees are far more palatable than those occurring between the trees. Increase in grass growth when trees are removed is usually accompanied by a marked increase in the proportion of less palatable species, as shown by Grossman et al. (1980) who divided the grass species of a savanna (dominated by *Burkea africana*) into fodder and non-fodder species. Although total herbage yield of grasses in open areas was much greater than in the sub-canopy habitat, the yield of fodder grasses was approximately the same in the two sites.

The range in annual above-ground grass yield under various conditions of rainfall, soil type and woody vegetation is indicated in Table 10.1.

Little information is available concerning the under-ground biomass of the grasses, but from the work that has been done to date it would appear to be approximately the same as the maximum above-ground standing crop (excluding litter). Not surprisingly, it is greater in deep, sandy soils than in shallower, fine-textured soils. Rushworth (1975), working in a Kalahari sand region, found that the amounts of biomass under ground and above ground were almost exactly equal. In the South African Savanna Ecosystem Study at Nylsvley in northern Transvaal, also on sand, the standing crop of under-ground parts varied between sites and years. Satisfactory distinction between live and dead roots was not possible, nor was complete separation of the roots of grasses and woody species. Nevertheless, the estimated "root"/shoot ratio was about 1.3/1 (226 g m^{-2} of "roots" to a depth of 100 cm and 171 g m^{-2} of shoots) (Rutherford, 1978). The ratio may well be found to be lower than this when refinements are introduced in separating the woody and grass roots. In a much shallower and more loamy soil, Kelly and Walker (1976) found a much lower "root"/shoot ratio, of the order of 1/3 (71 g m^{-2} of "roots" to a depth of 100 cm and 208 g m^{-2} of shoots).

The maximum standing crop of litter in grassland ungrazed or lightly grazed by ungulates reaches a steady state approximately equal to the seasonal annual yield of herbage. The mass of accumulated litter is therefore largely a function of how fast it decomposes. The litter mass is greatest in the cooler, high-elevation grasslands and least in the hot, semiarid savannas. Kelly (1973) estimated an average turnover time of just over one year for above-ground standing crop in a semiarid savanna. Litter turnover was less than a year in sub-canopy habitats, and more than a year in the open. In the mixed savanna of the Nylsvley ecosystem study site, the turnover time of grass litter was estimated to be about 0.7 years, and of tree leaf litter to be from 3 to 4 years (Bezuidenhout,

---

[1] The values ranged from $r = 0.38$ ($P < 0.1$) and $r = 0.52$ ($P < 0.05$) in two sets of grassland plots mown once for hay in spring, to $r = 0.62$ ($P < 0.01$) and $r = 0.68$ ($P < 0.01$) in plots burned in late winter and spring and mown for hay in summer.

TABLE 10.1

Above-ground herbage biomass (maximum standing crop) in grasslands and savannas in southern Africa [1]

Vegetation type	Rainfall (mm)	Soil type	Biomass (g m^{-2}) ligneous	herbage
1 Semiarid, open grassland	218 (1971)	Loamy	Nil	37
	678 (1975)	Loamy	Nil	125
2 Semiarid savanna	414	Sandy	Cleared	208
Mesic savanna	579	Sandy	Cleared	193
Mesic savanna	566	Clay	Cleared	264
3 Semiarid savanna	551	Sandy loam	2137	208
4 Mixed (mesic) savanna	553	Granite sands	Cleared	156
	553	Red clay	Cleared	197
	553	Black clay	Cleared	226
5 Mesic savanna	630	Sand	Open	65–100
6 *Baikiaea* shrub	642	Kalahari sand	469	122
7 High veld	500	–	Nil	63–70
8 High veld	700	–	Nil	154

[1] References: 1. Le Roux (1979); 2. Barnes (1979); 3. Kelly (1973); 4. Mills, (1964); 5. Grunow and Bosch (1978); 6. Rushworth (1975); 7. Vorster and Mostert (1968); 8. Kruger and Smit (1973).

1978). Grunow and Bosch (1978) have presented the most complete picture of the pattern of grass growth and litter production in a savanna grassland (Fig. 10.8).

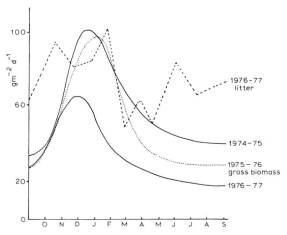

Fig. 10.8. Seasonal changes in grass biomass of canopy and litter in open grassland within *Burkea africana* savanna in the central Transvaal, South Africa (after Grunow and Bosch, 1978).

In addition to the effects of rainfall and (in the wooded grasslands) competition, annual grass production in these fire-climax grasslands is also very much influenced by soil fertility. This is especially the situation in woodland regions with relatively high rainfall and (therefore) with dystrophic soils. Application of fertilizer nitrogen, in particular, increases herbage yields by as much as 70% (Grunow et al., 1970), provided that sufficient phosphate is also applied. Where large amounts of nitrogen (200 kg ha^{-1}) are applied, annual yields are as great as 700 to 800 g m^{-2}. Fertilizer phosphorus by itself normally does not increase yields, except where nitrogen is also supplied (Booysen, 1981).

Where production is stimulated in this way, particularly on sandy soils, species changes occur (as in the climax grasslands), and the original species are replaced. A large number of annual weedy forbs are often associated with the induced subclimax perennial grass cover (Grunow et al., 1970; Booysen, 1981).

Mills (1964) reported on a long-term fertilizer experiment in fire-climax grasslands in Zimbabwe.

Where nitrogen was applied at more than about 120 kg ha^{-1} on sandy soils, the dominant climax species all disappeared or were greatly suppressed, and were replaced by a vigorous growth of *Eragrostis rigidior*. This change in floristic composition did not occur on clay soils.

Stability and resilience

Climax grasslands

The climatic-climax grasslands, in their climax condition, are inherently both unstable and non-resilient when grazed by livestock. Under mismanagement, they degenerate to pioneer communities dominated by perennial grasses (particularly species of *Aristida*, *Eragrostis* and *Sporobolus*) and annual grasses (including species of *Aristida* and *Chloris*), or they are invaded by woody species (e.g. of *Acacia*, *Chrysocoma*, *Elytropappus* and *Pentzia*). Once these areas have lost their erosion-resisting grass cover, they lose much of their top-soil through both wind and water erosion. Subsequent recovery is slow, for the soils then become too shallow to support the climax community. Also, once invading plants become well established, they offer severe competition to young seedlings of their own and other species, and may often permanently prevent the establishment of such seedlings. This competitive effect exerted by mature, well-established plants is further enhanced by their general unacceptability to animals. This ensures that they are seldom heavily grazed, whereas any plants of the climax sward that are present are normally severely overutilized by livestock. Therefore, the re-development of a climax grass community may be permanently suppressed unless the invaders are removed from the system or grazing is discontinued for a long time.

The edaphic-climax grasslands are considerably more stable and resilient than the climatic-climax grasslands for two reasons. Firstly, the invasion by woody plants is limited by the high water table, and, secondly, most of them are on fairly level ground at the lower end of a catena and are not easily eroded. If the grass cover is lost, however, these spongy areas are sometimes subject to gully erosion and subsequent drainage, leading to rapid loss of soil.

Fire-climax and wooded grasslands

The biotic or fire-climax grasslands are, in general, very much more stable than are the climatic-climax grasslands, particularly in their resistance to overutilization. However, if they are severely overused they degenerate, the climax community being replaced by pioneer grasses, mostly species of *Aristida*, *Eragrostis* and *Sporobolus*, grouped by Foran et al. (1978) in the category "increaser II". Their cover often remains dense, so that the rate of soil loss may not be materially affected, but such communities provide inferior grazing compared with the fire-climax community. Once these grasslands have been converted to the pioneer condition, they are extremely stable. Further degeneration is slow, but recovery to the climax condition is also extremely slow. Once again, the highly competitive pioneer species effectively prevent the reinvasion by climax species, even when the stocking rate is reduced or the area is protected from grazing.

A key factor in the stability of these grasslands is the relationship between grass cover and soil erosion. The impact of rain on bare soil leads to the development of a surface crust, which reduces infiltration to as little as one-tenth of that in a good sward (Kelly and Walker, 1976). Below some critical value of grass cover, therefore, these grasslands often degenerate rapidly. However, this effect is significant only on fine-textured soils, or soils prone to crusting. In sand soils the rate of infiltration is always high, and the system is able to recover. This distinction between sands and fine-textured soils extends to the reaction of the vegetation to heavy grazing and fire. Owing to their much larger under-ground components, and because water relations are always better in sands (in semiarid environments), the grass sward on sands can be subjected to quite severe grazing and burning with far less degradation than occurs in shallower, fine-textured soils.

Where these grasslands are under-utilized (i.e. grazing is light and infrequent and fire is excluded), the species of the fire climax are replaced by species of the post-fire climax such as *Alloteropsis semialata*, *Cymbopogon validus*, *Harpochloa falx*, *Hyparrhenia dregeana*, *H. rufa*, *Tristachya leucothrix*, grouped by Foran et al. (1978) under the rubric "increaser I". In such communities, the basal cover of the vegetation is often considerably lower than in the climax stand, but erosion is retarded through an accumulation of large quantities of canopy material and litter, which effectively dampens rain-drop impact and reduces the rate of water movement over the soil surface. Therefore, soils of such areas tend to remain stable, unless the protective cover is destroyed by, for example, untimely fires. In such situations, loss of soil is very marked, and this in turn affects the subsequent stability of the community. Controlled fires, in association with controlled grazing, may be effectively used to convert such disclimax communities to the climax condition, provided they have not developed through to more advanced communities of scrub and primary forest. In general, therefore, they are less stable than the pioneer communities that develop with overutilization. Even the scrub and primary forest communities are easily destroyed. Since by this time they may have lost the species of the fire-climax grassland community, re-establishment of the climax is often extremely slow. It is dependent on controlled use of the community to prevent excessive accumulation of canopy material, and on introduction of seed of the fire-climax grass species.

In savanna regions, where the woody vegetation is, for economic reasons, not actively cleared by man, the woody–grass vegetation complex is generally unstable, but resilient. Annual variation in the ratio of grass to woody production above ground, and in the woody vegetation cover, is sometimes quite large. However, the domain of attraction around the equilibrium values of the grass and woody vegetation is large, and the balance of woody to herbaceous components of the cover is relatively constant. Consequently, only when sustained overgrazing combines with drought, is the system likely to exceed the equilibrium boundaries, and degenerate into thicket and scrub. The situation in semiarid areas was examined by Walker et al. (1981; see also Knoop and Walker, 1985), and generally is in accord with the above observation. With increasing rainfall, however, the stable equilibrium combination of the grass and woody vegetation seems to be less resilient. Overgrazing leads easily to invasion by (or increase in) woody plants, and, once established, they are extremely difficult to eradicate. Their competitive position is then secure.

THE ANIMAL COMMUNITIES

Descriptions of the fauna of southern Africa occupy some 385 pages in Werger's volume (1978), and in the space allowed here we can deal only with those that have a major impact on the ecosystem. Therefore, this treatment is restricted to the large grazing herbivores, with some references and comparisons to other groups.

The high diversity of the large ungulate communities in the African plains is well known. The high biomass values suggested by Dasmann and Mossman (1960) are not easily verified, and at least in southern Africa such values are generally considered to have been overestimated (Mentis, 1977; Taylor and Walker, 1978; Walker, 1979). Perhaps the most characteristic feature of these populations is their marked spatial heterogeneity and movements. Failure to distinguish between overall (or crude) density and ecological density has led to the phenomenon of the "water-hole" effect, i.e. animals are counted in the vicinity of favoured areas and their numbers are then converted to a density estimate for that area, and invalidly extrapolated to the region as a whole.

The southern African rangelands appear to have been populated by a highly diverse community of large herbivores, comprised mainly of grazing, as opposed to browsing, animals. There is a continuum from individualistic, browsing animals that move very little, e.g. *Sylvicapra grimmia* (duiker) and *Tragelaphus strepsiceros* (kudu) on the one hand, to large, mobile herds of grazing animals on the other, e.g. *Connochaetes taurinus* (wildebeest) and *Syncerus caffer caffer* (buffalo) (Jarman, 1974). It is the latter group which is most important from the present point of view, since they comprise by far the largest proportion of the biomass, and they exert a much greater influence on the vegetation than the solitary species.

The non-ungulate fauna is equally diverse, albeit somewhat less spectacular. Rodent diversity is high, the avifauna is very rich, and insect populations include many thousands of species.

Composition and distribution of the large grazing herbivores

Climax grasslands

In the desert grasslands the important large

herbivores are *Antidorcas marsupialis* (springbok), *Equus burchelli* (zebra) and *Oryx gazella* (gemsbok) and, to a lesser extent, *Raphicerus campestris* (steenbok). All of these are pure grazers.

On the South African high veld the characteristic herbivore is *Damaliscus dorcas phillipsi* (blesbok). It was widespread and dominant in the past, and is still prevalent; it is actually increasing as a semi-domesticated farm animal. Its close relative, *D. d. dorcas* (bontebok) is restricted to the short-grass regions of south-western Cape Province. The only other antelope occurring to any extent on the high veld was *Ourebia ourebi* (oribi) and, along the margins, *Antidorcas marsupialis*.

In the mountain grasslands the main species were *Connochaetes gnou* (black wildebeest), *Equus zebra* (mountain zebra), *Ourebia ourebi*, *Pelea capreolus* (vaal ribbok), *Redunca fulvorufula* (mountain reedbuck) and *Taurotragus oryx* (eland). Bigalke (1978) claimed that *Pelea capreolus* is an ancient South African endemic. The relative abundance of these animals has changed greatly, and today *Taurotragus oryx* is completely dominant in the Drakensberg grasslands, whereas *Equus zebra* is confined to smaller areas in the Cape Province.

The edaphic seasonally waterlogged grasslands have a characteristic large-herbivore community. The extensive, flooded grasslands of northern Botswana and of Zambia are occupied by *Kobus leche* (lechwe) and *K. vardoni* (puku), with *K. ellipsiprymnus* (waterbuck) occurring largely along the margins. The short-grass areas on the banks of rivers and along the shores of large lakes or pans are intensively grazed by *Hippopotamus amphibius* (hippopotamus).

The characteristic species of the *vleis* (low-catena, hydromorphic grasslands) are *Redunca arundinum* (reedbuck) which typically remains within the tall grass communities in the *vleis*, and *Hippotragus niger* (sable antelope) (Fig. 10.3) and *Kobus ellipsiprymnus*, which tend to occupy the ecotone between woodland and grassland.

Wooded grasslands

The savannas contained a much larger variety, and a greater biomass, of large herbivores than did the pure grasslands. However, many of them were not obligate grazers. Those that graze significantly (the "intermediate feeders" of Hofmann and Stewart, 1972) are included here, but those that are essentially pure browsers have been omitted.

The most important species in the semiarid savannas, in terms of abundance and consumption of grass, were:

Aepyceros melampus (impala)
Alcelaphus buselaphus (red hartebeest)
Ceratotherium simum (white rhinoceros)
Connochaetes taurinus
Damaliscus lunatus (tsessebe)
Equus burchelli
Hippotragus equinus (roan antelope)
Loxodonta africana (elephant)
Taurotragus oryx

Most of these are pure grazers, the exceptions being *Aepyceros*, which is an opportunistic and highly successful mixed feeder; *Loxodonta*, which is a mixed feeder, grazing preferentially on grasses in spring and early summer; and *Taurotragus* (in the savanna), for which grass probably makes up less than half of the diet. *Damaliscus* and *Hippotragus* did not occur in the southernmost parts of the continent (south of the Orange river) (Bigalke, 1978). *Alcelaphus*, *Damaliscus* and *Hippotragus* are now much reduced in numbers, and in some areas are very rare or have disappeared. *Ceratotherium* exists only within game reserves.

In the moist savannas, *Equus burchelli* and *Loxodonta africana* were abundant, and the other important species were *Aepyceros melampus*, *Alcelaphus lichtensteini* (Lichtenstein's hartebeest) (a typically *miombo* species), *Ceratotherium simum*, *Hippotragus equinus*, *Hippotragus niger*, *Ourebia ourebi*, *Syncerus caffer caffer* and *Taurotragus oryx*. Now *Alcelaphus lichtensteini* is rare in southern Africa. Since most of the moist savannas are densely populated and farmed with domesticated livestock, the predominantly grazing wild herbivores are much reduced; the medium-sized *Aepyceros melampus* and the small *Raphicerus campestris* probably are the most abundant.

Standing crop and secondary production

Much of the earlier literature purported to show that standing crops and productivity of mixed communities of wild ungulates were greater than those of the present populations of domesticated livestock (Bigalke's review, 1978). However, as indicated above, recent detailed examinations of this subject (Mentis, 1977; Taylor and Walker, 1978; Walker, 1979) do not support this view. It is unlikely that, except in very arid regions where drinking water supplies are far apart, the indigenous game populations can produce as much meat as livestock on well-managed ranches. However, a carefully determined mixture of cattle and indigenous browsing animals probably will increase secondary production within the savannas.

As a final comment, we return to the point made at the beginning of this section. We have dealt with only the large, grazing herbivores. There are many other animal groups that play a significant role in these grasslands; as an example, Gandar (1979) showed that in the Nylsvley savanna, if cattle are excluded, grasshoppers consume approximately the same amount (4%) of grass herbage as do the large herbivores.

ACTIVITIES OF MICRO-ORGANISMS AND SOIL FAUNA

In the moist climax grasslands of southern Africa, turnover of organic matter proceeds much as in other temperate grasslands, and is apparently largely due to decomposition by Bacteria, Fungi and Actinomycetes. Decomposition rate is a function of temperature, moisture and substrate quality. With increasing aridity, the activities of micro-organisms and reducer organisms appear to assume prominence. Using graded mesh sizes of litter bags, Kelly (1973) showed that, in a semiarid savanna in south-eastern Zimbabwe, there was a marked reduction in the rate of litter disappearance in those bags which excluded organisms of the size of termites or larger.

The two most important reducer organisms in the savanna grasslands are dung-beetles and termites. Termites are active for much of the year, most species (e.g. *Microtermes* spp. and *Trinervitermes rhodesiensis*) being prominent in the wet season. *Hodotermes mossambicus* (harvester termite) occurs in fine-textured soils, and is most active during dry periods. The termites collectively reduce all dead (and some live) plant organic matter, as well as dry herbivore dung. Most of the dung, however, is reduced by the many species of dung beetles when still fresh. Endrody-Younga (1976) found that, in the beginning of the wet season, most dung was dispersed by dung rollers (large *Heliocopris* sp.), but by the end of the season most was being dug into the soil (by Coprinae and Aphodiinae).

The activities of *Hodotermes mossambicus* are greatly reduced by wet conditions in the nest and, conversely, they expand rapidly during drought. Removal of litter and organic matter around the nest enhances dry conditions by causing surface sealing of the soil, and therefore has a positive feedback on termite density (Barnes, 1979). In semiarid savannas, the activities of *H. mossambicus* constitute a major problem in range management, because during dry periods grass consumption by termites is sometimes greater than that by large herbivores. Grass is in any case less abundant at such times, and the effects of the termites are therefore heightened.

Decomposition of dead organic matter in savanna grasslands was described by Bezuidenhout (1978) for the *Burkea* savanna at Nylsvley. Soil counts showed Bacteria to range upwards from 162×10^{12} m^{-2}; Actinomycetes ranged from 0 to 1×10^{12} m^{-2}, and Fungi ranged from 2 to 4×10^9 m^{-2}. All three peaked in numbers during January and February and were at a minimum from August to November.

Bacterial counts in the litter of the two dominant grasses (*Digitaria eriantha* and *Eragrostis pallens*) were extremely low in late winter and early spring and rose sharply in December and January, to reach a peak of about 1×10^8 g^{-1} in July. Thereafter they declined rapidly for both species. Fungal counts followed a similar course, with maxima of about 1.5×10^7 g^{-1}.

Comparative counts of Bacteria and Fungi in the leaf litter of two woody species (*Burkea africana* and *Ochna pulchra*) showed bacterial and fungal numbers to be only about one-tenth and one-third, respectively, of those in the grass litter. This is in accord with the respective rates of litter decomposition. Grass leaves have lignin contents approximately equal to those of tree leaves, but have a higher initial C/N ratio, and decompose faster (Morris et al., 1982). Grass litter also breaks down faster than tree leaf litter, as described earlier.

THE NATURE OF HUMAN IMPACT

History of utilization

Southern Africa has been occupied by man for many thousands of years (more than 250 000), during which time the human population has increased steadily, and consequently the impact on the vegetation. Even in the early days there was some impact on the system, particularly through hunting activities, which probably influenced the nature of the indigenous game populations and their migratory habits. During this early stage, the impact of humans on the vegetation is likely to have been largely through the use of fire as a tool to assist in hunting operations. Fire was no doubt used to drive and concentrate game, to make them more visible (by burning off accumulated canopy growth), and to attract game to the new flush of growth developing on burned areas. As the human population increased, so did man's impact, particularly as the development of systems of shifting cultivation gained momentum. Since the Iron Age communities were hunters of migratory animals — rather than of the more solitary or non-migratory species (Maggs, 1976) — human communities are likely also to have been nomadic, following the migratory game from the essentially savanna regions in the winter to the grass veld in the summer. Cultivation was no doubt restricted to small areas of land at this time. Even immediately preceding permanent settlement of the colder humid grasslands, the nomadic populations appear to have confined their summer activity largely to areas of low-elevation grassland lying adjacent to the savannas. In Natal, for example, old village sites abound at elevations below 1400 m, but they are absent from regions of high elevation where temperatures are low, the soils more dystrophic, and the grassland less acceptable to stock for much of the year (i.e. sour veld).

As the human population increased further during the 17th and 18th centuries, man was forced to settle permanently in these regions of high elevation as well, and so the human impact on the vegetation was expanded quite considerably, and was reinforced by increasing numbers of domesticated livestock. Human influence has grown steadily in recent years, stimulated by the rapid development of the agricultural industry and of settled farming — an almost complete replacement of the natural herbivores by domestic stock (particularly by cattle and sheep), the cultivation of large areas for crops (particularly *Zea mays*) and for pastures, and control of the distribution of grazing animals at different seasons of the year. Hence, the normal migratory patterns of the animals which grazed these areas, have been completely disrupted. Provision of conserved feed during winter and during droughts has permitted the development of animal-production systems differing considerably from those which characterized these regions in pre-settlement days.

In the woodland regions of south-central Africa, rainfall is sufficient for arable dryland cropping, and very little potentially arable soil remains with its original vegetative cover. Induced (biotic) grasslands have been developed by fire and various mechanical means, such as ring-barking. In the semiarid savannas, tree removal is uneconomical and man's effects have been mainly associated with grazing by livestock

Modification of grassland by man

Human impact in the grasslands of southern Africa has been through large-scale replacement of the original community by sown pastures, by annual crops, and by degenerate pioneer communities in abandoned arable sites and on grazing land that has been either overgrazed or badly managed (e.g. selectively grazed). Indigenous forests, which are closely associated with the mesic grasslands (particularly on moist southern slopes), have apparently been greatly reduced in area in recent years, having been replaced in most sites by a tall grassland dominated by species of *Cymbopogon* and *Miscanthus*. Logging of many of the larger trees [particularly of *Podocarpus* spp. (yellow-woods) and *Ptaeroxylon obliquum* (sneezewood)] for building timber, and of saplings for fence posts, together with the increased grazing pressure (both in the forest margins and within the forests), has permitted the more ready entry of fire into these forest communities and an incipient destruction of their outer margins. Ryecroft (1944) estimated, for example, that the size of the Karkloof forests (in the mist-belt of Natal) declined to 25% of their original size (about 36 000 ha) over the period 1880 to 1944; Taylor (1963) reported that the Zwartkop forest (also in Natal), which in 1880 was estimated to cover 3240 ha and believed at that time to be a remnant of a much larger forest, had been reduced by 1963 to a mere 600 ha. This expansion of grassland into forest areas has, however, been more than compensated for by the destruction of grassland along its boundary with savanna, *macchia* and dwarf shrub (*karoo*) communities. Acocks (1975) estimated, for example, that over 5 million ha of climatic-climax grassland in the central region of South Africa had been converted to eroded *karoo*. There can be no doubt, also, that encroaching species of the savanna have converted large tracts (approximately 3.6 million ha in South Africa alone) of the more arid grassland into a false savanna (Danckwerts, 1979). In addition, increase in density of the bush component within the savanna areas themselves has been well documented, and large areas (approximately 700 000 ha) in the south and southwest, and at high elevation in interior regions, have been invaded by *macchia* (*fynbos*) (Danckwerts, 1979). The overall result of these invasions has, without doubt, been a large-scale reduction in the area of grassland.

Within the present grassland regions, only a relatively small proportion can be considered as being in a near-climax (either climatic or biotic climax) condition. The pioneer *Aristida junciformis* is an important constituent in no less than 29% of the total grassland area of South Africa (3.57 million ha), and over much of this area it is considered to be an invader. Even in those grasslands where it is normally a minor component, it has now increased greatly in abundance — to the extent that it is almost the sole species (Edwards et al., 1979).

CURRENT MANAGEMENT PRACTICES

Burning

Climax grasslands
Frequency of burning within these grasslands depends primarily on the palatability and nutritive value of the forage in a mature condition. Burning is seldom necessary in the arid regions, where the forage remains palatable and retains a relatively high nutritive value even when mature. However, as rainfall increases and as the soils become more highly leached, the feed value of mature forage declines. As a result, land managers are encouraged to burn all areas in which mature and moribund material has been allowed to accumulate. Unless this is done, subsequent new growth becomes mixed with old material, so that the quality of the material grazed is substantially reduced and selective grazing is encouraged (as animals attempt to select high-quality new growth).

In these more "sour" grasslands, the principles of using fire are very similar to those in the fire-climax grasslands (see below). In the more arid and generally "sweeter" veld (in which the herbage remains palatable and nutritive when mature), fire is not necessary in order to remove accumulated mature forage, since this may be removed by grazing — particularly if supplementary licks containing nitrogen are provided to the animals. However, more recent evidence suggests that early spring fires may be effectively used in these areas to suppress at least some of the major invading dwarf-shrub species (Trollope, 1978), so that its occasional use (even in these moderately arid areas) may be justified. However, post-fire management is likely to be crucial to the success of burning in these arid regions and spring-burnt veld should not normally be grazed until well into the following autumn.

Wooded and fire-climax grasslands
Fire-climax grasslands. Since these mixed and sour grasslands do not provide acceptable grazing to animals after the herbage has matured, fire is generally used widely to remove any accumulation of old growth that is carried over into the winter. Provided fire is applied with this objective in mind, it serves a useful purpose, since these grasslands degenerate (from a perspective of animal production) if they remain unutilized or under-utilized. Many of the most useful species in the fire-climax grasslands (notably *Heteropogon contortus* and *Themeda triandra*) tiller from basal nodes situated at or just above the soil surface. If material is allowed to accumulate on these plants, these tiller initials do not develop. Tillering is then confined to elevated nodes of the old flowering stems, and these tillers are short-lived. They are unable to develop their own root system, and are removed *in toto* by subsequent grazing or

burning treatments. When this happens the plant dies. However, some species in these grasslands react favourably to a build-up in canopy density. Such species as *Alloteropsis semialata*, *Harpochloa falx* and *Tristachya leucothrix* tiller from nodes below soil level, and are not capable of producing elevated tillers from stem nodes. These species appear to be insensitive to light conditions at the soil surface, and they are encouraged by lenient grazing and exclusion of fire. Generally, however, they produce herbage of lower quality than the species of the fire climax.

Assuming, then, that fire is necessary in these grassland areas when forage removal by grazing animals is incomplete, it is clear that the frequency of fire use should be dependent entirely on the extent to which mature forage accumulates. Where such material accumulates often, burning may need to be frequent; but where management is such that most of the forage is grazed before it matures each season, burning may even be unnecessary. Burning generally results in an appreciable reduction in yield in the season following the burn, and particularly in the yield of spring and early summer forage, so that the less frequently it is applied the higher will be the average carrying capacity of the veld (Tainton et al., 1977a, 1978a; Tainton, 1981a).

A further question that arises is the season in which burning may be most appropriately applied. It is now generally accepted that, in these areas of essentially summer rainfall, the veld should be burned at about the time of the first spring rains. Burning prior to these rains may lead to long periods of exposure of the bared soil, and on steep slopes in particular this may lead to soil erosion. Also, inherent in such a system is the temptation to graze the new flush of growth encouraged by the burn, so that early burning often leads to overgrazing in the early spring. Sometimes this practice may be deliberate in that the sole objective of the burn may be to induce a flush of out-of-season growth, which is then grazed intensively. This approach has led to the practice of burning in mid-summer or late autumn in an effort to produce a flush of growth for early-winter grazing. Such a practice cannot be sustained, because of the excessive grazing pressure which invariably follows, resulting in an extended period of soil exposure and a severe drain on the storage reserves of the grass plants, as they repeatedly produce new leaves that are then destroyed (by grazing, frost or winter drought). Such leaves import their energy requirements from the storage organs of the plant, and are removed or die without becoming exporters of energy substrates to the storage organs. This leads to a continual drain on the storage organs, and to a decline in plant vigour. However, neither should burning be delayed into the spring period. Once growth has started, burning results in severe damage to many of the species of the fire-climax sward. A large proportion of the active tillers are killed by such treatment (Tainton et al., 1977a), and regrowth must rely on the development of a new population of laterals. This process is generally slow, so the competitive ability of the plant is greatly reduced during the vital spring period, and this is likely to lead to the invasion of weeds. Therefore veld burning should be timed as far as possible to coincide with the spring rains, whereas grazing management should aim to limit the frequency at which unpalatable herbage accumulates so that the necessity of burning is reduced to a minimum.

Wooded grasslands. The main objective of burning within these grasslands is to control woody vegetation. Burning is necessary to maintain an open parkland in the wetter savannas. In the semi-arid savannas the use of fire is more difficult. If there is insufficient fuel, the fire is ineffective in controlling the developing scrub and serves only to weaken the grasses. Since most of the grass is palatable even when dry, the accumulation of fuel, to achieve a successful burn, represents a loss of grazing. It is therefore both impracticable and uneconomic to remove the woody vegetation completely and maintain an open grassland. The commonest aim is to reduce the ratio of woody vegetation to grass. Complete removal of woody vegetation in these semiarid areas may, in any case, be unwise (Walker, 1979).

Summer grazing

Summer grazing normally commences within 2 to 3 weeks of the first effective spring rains, and continues through to late autumn. Sweet veld is capable of supporting animal performance throughout the year, and summer grazing practices are such that sufficient material remains at the end of the autumn to supply the animals with winter feed. But in sour veld nutritive quality of herbage declines rapidly after mid-summer, and effective grazing is terminated in April or May. Where both sour veld and sweet veld are available, the latter is often reserved for winter grazing only; it may be grazed once in late spring to prevent an excessive accumulation of material which, when grazed in winter, will lack the quality of late-summer and autumn growth.

Throughout the grassland regions of southern Africa, various rotational grazing systems have been recommended, but not always practised. Continuous grazing in summer leads to excessive selective pressure on preferred areas or preferred species. Even though an area of grassland may appear uniform, there are invariably differences in acceptability to grazing animals brought about by differences in soil type, soil depth, aspect, steepness of slope and other factors. In addition, great differences in palatability exist among the species making up any community. Under all grazing systems, protection needs to be afforded to the most acceptable plants, and, at least in the areas of higher rainfall, this can be achieved only by some form of rotational grazing in which adequate periods of rest are provided. In the drier regions floristic composition is more strongly influenced by particular episodic events (unusually severe droughts or wet periods) which tend to override, or at least offset, the cumulative effects of selective grazing.

The appropriate system of grazing depends on the livestock system (i.e. species and mixture of age classes) and on the type of grassland. Generally, the length of the rotation cycle is reduced as rainfall increases. It should be sufficiently short to prevent animals from grazing newly formed regrowth. Also, as rainfall increases, the period of absence from a paddock is reduced because in more humid regions rainfall (and regrowth) is generally more reliable, and because the quality of the forage declines rapidly as it matures. In the humid regions (more than 1000 mm of rainfall yr^{-1}), rotations of 5 to 7 days grazing and 30 to 40 days resting are commonly recommended; in the more arid grasslands animals may remain in a paddock for 2 to 3 weeks, and are then removed for 3 to 4 months. Often additional periods of longer rest, of up to 12 to 18 months duration, are also provided at intervals of 3 to 4 years.

Considerable disagreement still exists as to the optimum rotation systems to use in the grassland regions of southern Africa, in spite of the extensive relevant research that has been undertaken (Booysen, 1966, 1969; Venter and Drewes, 1969; Roberts, 1970; Tainton et al., 1977b; Booysen and Tainton, 1978; Savory, 1978). As stated above, there is also evidence that in the drier regions rotational systems may not always be superior to continuous systems, at least in the short term (Barnes, 1979), and the short-term success or failure of a system may depend on the intra-seasonal (within the year) rainfall pattern. In particular, considerable controversy exists with reference to the degree to which the grass veld should be grazed in any paddock before the animals are moved to the next paddock in the rotation. Various systems have been developed to meet different objectives in different grass-veld types, the most prominent of which are: (1) the "high-utilization" or "non-selective" grazing systems, in which the grassland is well utilized before the animals are removed from the paddock; and (2) the "high performance" or "controlled selective" grazing systems, in which the grassland is only partially utilized before the area is withdrawn from grazing. The "short duration" grazing system emphasizes the need for short periods of grazing in the rotation cycle, and the "open camp" system emphasizes the need for flexibility.

In practice, continuous grazing is widely applied and this, together with often unrealistically high stocking rates, is mainly responsible for the widespread degeneration in the composition of the grasslands, particularly in areas of medium and high rainfall.

Winter grazing

Winter grazing is largely restricted to sweet-veld areas, and such areas are often reserved largely or entirely for winter use. Because there is negligible growth in these areas during winter, grazing may be continuous without detriment to the sward; but rotational methods of grazing have the advantage of rationing feed over the winter period. Where grazing is continuous, the animals utilize the choice forage in early winter, so that as late winter approaches they are forced onto a low quality diet of previously rejected forage.

In the sour veld, winter grazing is possible only if the animals are provided with nitrogen-containing supplements. These supplements, provided mainly in block form, increase the digestibility of the winter forage and permit its utilization. However, where grazing of such veld is necessary during winter, it should be rested at least through the late summer to save the herbage of acceptable species for use during winter. Species that are unacceptable to stock in summer are also invariably unacceptable during winter, and so provide poor grazing, even when supplemented with nitrogen-containing licks.

An alternative to nitrogen licks to supplement forage in sour veld in winter is the provision of browse, at least for those animal species that will supplement their winter intake with it. Legumes show promise in this respect because of their high protein content. Several forest precursors apparently provide useful browse for resident game species in the humid high-elevation grasslands. *Buddleja salvifolia* and *Leucosidea sericea* may be useful species in this respect, but other important species have doubtless not yet been identified, including some forbs. In the drier regions many shrub species are eaten in winter (Walker, 1980), and their most important role is to supply high-quality feed from August to November. This is the hot, pre-rainy season when the woody vegetation produces its new flush of leaves.

Hay production

Although by far the larger proportion of these grassland regions is used directly for grazing, parts of the more humid sour-veld regions in particular are mown regularly for hay. However, because hay yields are low (less than 3 t ha^{-1}), and because hay is difficult to cure in the wet summer period, harvesting seldom takes place before autumn. This ensures near-maximum yields from a single cut, but it results in poor-quality hay because the herbage is generally too mature to provide nutritious forage at this time. In spite of this, veld hay is often prized for winter feeding, particularly if it is from areas dominated by *Themeda triandra*.

Renovation of degenerate grassland

Methods of grassland renovation in southern Africa have for the most part had a strong ecological base. The methods used have essentially involved the design and application of management procedures aimed at promoting the redevelopment of the climax sward (described above under "Summer grazing", pp. 284–285). However, extended resting periods are also used as a renovation measure to encourage seed production of the climax species where they have survived, and to encourage the subsequent establishment of their seedlings (Tainton, 1972).

Only where the plant cover has degenerated to a degree endangering soil stability, or where sufficient numbers of climax species are no longer present, is artificial seeding used to re-establish the vegetative cover. This is preferably undertaken with the species of the climax community of the site; but few of these species seed freely, and many produce awned seed which is difficult to handle. Thatching (spreading of hay from climax communities containing mature seeds) has been used successfully, but this method is costly and is confined to small-scale operations. Where large areas must be revegetated, pioneer species are usually seeded and fertilizer is applied to facilitate establishment. Where active erosion has led to the development of gullies, construction of stone packs, concrete walls, contour banks and spillways is sometimes undertaken, but these are expensive and their effect may be no more than cosmetic. The major effort should be directed towards grassland management in the catchments rather than the gullies themselves, for a reduction in run-off from adjacent land often results in natural revegetation of gullies.

The use of direct methods of eradicating, or at least controlling, undesirable herbs which encroach into grassland areas has not received much attention in southern Africa. Such programmes are costly and are unlikely to be effective unless those management practices that caused these plants to invade or increase are corrected. Instead, the approach has been to alter management regimes in such a way as to encourage the development of a vigorous climax community, in which potential invaders cannot become established and those that have already entered are suppressed.

One notable exception to this general principle is the application of chemical methods of control — followed by fertilizer and seeding to *Stipa trichotoma*, an exotic grass species that has invaded areas of humid grassland.

The general attitude to the regeneration of arid grassland that has been invaded by the dwarf-shrub (*karoo*) species is essentially the same as that already described for invasion by pioneer grasses. However, where such invaders are tall shrubs and trees of the savanna, direct methods of control are usually used. These include slashing, ring-barking, chemicals, and fire. These treatments are followed by management programmes designed to encourage re-development of a dense grass cover. Justification for costly programmes (involving mechanical or chemical control) in such situations rests on the argument that these encroaching plants are foreign to the grassland community, and appropriate management should permanently prevent re-invasion once they have been removed and a vigorous grass cover has developed. Since a balance between grassland and savanna was apparently well established prior to permanent settlement, it is generally assumed that this balance can be maintained indefinitely by appropriate management. However, with the changed circumstances now existing within these regions, this assumption may be unreasonable. Consideration needs to be given to the alternative approach (to scrub and tree control) of accepting these species as part of the community and adjusting management practices accordingly. Hence, where there is a strong tendency for savanna species to invade, introduction of browsing animals (especially goats) may be advisable.

PRESENT AND FUTURE TRENDS

Climatic-climax grasslands

Within this region, both the area in pastures seeded to selected species and varieties of native and introduced grasses and the area under annual cropping (mostly maize and wheat) will probably increase where rainfall is sufficiently high and reliable to support high yields. However, the natural grasslands will undoubtedly continue to play a vital role in the economy, since limitations imposed by low rainfall or low temperatures in winter will probably make large-scale pasture production and annual cropping unrewarding. Therefore future development is likely to be concerned largely with improved veld management — division of grazing land into homogeneous units, better reticulation of stock water, better grazing management systems, and the use of optimum stocking rates. Large-scale reseeding programmes might greatly increase the productivity of much poor-quality grassland in this region, but such programmes are unlikely to be economical. Renovation must, it seems, depend largely on natural processes, encouraged as much as possible by appropriate grassland management.

Biotic or fire-climax grasslands

Except for the semiarid savannas, moisture is not normally the factor most limiting production in these regions. Rather, soil fertility sets the primary limit to herbage production, so that the addition of fertilizer — and particularly of nitrogen — will normally lead to a dramatic increase in herbage yield. However, such yield increases are somewhat restricted by the genetic constitution of the plants that inhabit these regions. The climax species, in particular, are only moderately responsive to improved fertility. As a result, they are gradually replaced by more responsive pioneer species (e.g. *Cynodon dactylon* and *Eragrostis curvula*) if fertilizers are applied for several years. Since these pioneer species (particularly *E. curvula*) are not particularly acceptable to animals, it is now generally agreed that use of fertilizers in these grasslands is appropriate only when combined with the introduction of selected pasture species (Booysen, 1981).

A considerable amount of research effort has been directed in recent years towards techniques of veld replacement and veld reinforcement. Such techniques involve, at the one extreme, the distribution of fertilizers and seed of selected pasture varieties over the area without any disturbance of the existing community. *Dactylis glomerata* has been successfully used for this purpose in the cool, humid parts of Natal. At the other extreme, the land is ploughed and a pasture is established. For this purpose grasses from both temperate (species of *Dactylis*, *Lolium* and *Festuca*) and tropical/

subtropical regions (*Chloris gayana*, *Cynodon* spp., *Digitaria eriantha*, *Eragrostis curvula* and *Pennisetum clandestinum*) and legumes (*Trifolium pratense* and *T. repens*) have proved successful. Between these two extremes are a large number of techniques, designed to achieve various degrees of soil disturbance, which destroy varying proportions of the existing community. These involve use of soil-disturbing equipment (e.g. sod-seeders), herbicides (to reduce the competition faced by seeded species), or "hoof cultivation" (which is achieved by concentrating animals on small areas of land, particularly during wet weather).

Forage production can be greatly increased and forage quality improved by the application of fertilizer, and by introducing responsive species which produce good-quality forage in these humid grassland areas. However, stability of these new communities is a problem. Temperate legumes (species of *Medicago* and *Trifolium*) do not persist well — largely, it seems, because they cannot be provided with sufficient phosphorus in the phosphorus-fixing soils that prevail. Tropical legumes (e.g. species of *Desmodium* and *Glycine*) do not require as much phosphorus, but they cannot tolerate the cold winters. Even when tropical/subtropical grass species are established in these areas, they are prone to invasion by pioneer weedy species of forbs and grasses. Such communities provide forage of poor quality. Regeneration of a productive pasture is expensive, whereas reversion back to the original sward occurs so slowly that it is unlikely to warrant consideration. Destruction of the original sward may therefore lead to a short-term increase in livestock production, but in the long term it results in reduced animal output. Destruction of natural grassland results in a degraded vegetative cover that is unlikely to regenerate to a climax sward.

In the semiarid savannas the inherently low production precludes improvements involving significant capital inputs. Improved veld management, with selective thinning of the woody vegetation and inclusion of browsing animals, is the current trend, and this will likely continue.

REFERENCES

Acocks, J.P.H., 1975. Veld types of South Africa, 2nd ed. *Mem. Bot. Surv. S. Afr.*, 40, Govt. Printer, Pretoria, 128 pp.

Anonymous, 1963. Rainfall Map No. 12A. In: *Atlas of the Federation of Rhodesia and Nyasaland*. Federal Govt. Printer, Salisbury (map).

Aucamp, A.J., 1979. *The Production Potential of the Valley Bushveld as Grazing for Angora Goats*. D.Sc. (Agric.) thesis, University of Pretoria, 234 pp.

Barnes, D.L., 1979. Cattle ranching in the semi-arid savannas of East and Southern Africa. In: B.H. Walker (Editor), *Management of Semi-Arid Ecosystems*. Elsevier, Amsterdam, pp. 9–54.

Bezuidenhout, J.J., 1978. *The Activity of Soil Micro-Organisms in a Savanna Ecosystem at Nylsvley*. M.Sc. thesis, University of Pretoria, S.A., 137 pp. (in Afrikaans).

Bigalke, R.C., 1978. Mammals. In: M.J.A. Werger (Editor), *Biogeography and Ecology of Southern Africa*. Junk, The Hague, pp. 983–1049.

Booysen, P. de V., 1966. A physiological approach to research in pasture utilisation. *Proc. Grassland Soc. Southern Afr.*, 1: 77–96.

Booysen, P. de V., 1969. An evaluation of the fundamentals of grazing systems. *Proc. Grassland Soc. Southern Afr.*, 4: 84–91.

Booysen, P. de V., 1981. Radical veld improvement. In: N.M. Tainton (Editor), *Veld and Pasture Management in South Africa*. Shuter and Shooter (in association with University of Natal Press), Pietermaritzburg, pp. 57–90.

Booysen, P. de V. and Tainton, N.M., 1978. Grassland management: principles and practice in South Africa. In: D.M. Hyder (Editor), *Proc. of the First International Rangeland Congress*, Denver, Colo., pp. 551–554.

Boucher, C. and Moll, E.J., 1981. South African Mediterranean shrublands. In: F. di Castri, D.W. Goodall and R.L. Specht (Editors), *Mediterranean-Type Shrublands*. Ecosystems of the World, 11, Elsevier, Amsterdam, pp. 233–248.

Danckwerts, J.E., 1979. *Recession of the Grassland Formation in South Africa*. Department of Pasture Science, University of Natal, Pietermaritzburg, 66 pp. (unpublished).

Dasmann, R.F. and Mossman, A.S., 1960. The economic value of Rhodesian game. *Rhod. Farmer*, 30: 17–20.

Drewes, R.H., 1979. *The Response of Veld to Different Winter Treatments*. M.Sc. Agric. thesis, Department of Pasture Science, University of Natal, Pietermaritzburg, 88 pp.

Edwards, P.J., Jones, R.I. and Tainton, N.M., 1979. *Aristida junciformis*: a weed of the veld. *Proc. 3rd Natl. Weeds Conf.*, S. Afr., pp. 25–32.

Endrody-Younga, S., 1976. *Dung-feeding Arthropods*. Report to the National Programme for Environmental Sciences, C.S.I.R, Pretoria, S. Afr., Council for Scientific and Industrial Research, 3 pp.

Foran, B.D., Tainton, N.M. and Booysen, P. de V., 1978. The development of a method for assessing veld condition in three grassveld types in Natal. *Proc. Grassland Soc. Southern Afr.*, 13: 27–33.

Gandar, N.V., 1979. Trophic ecology and plant-herbivore energetics. In: B.J. Huntley and B.H. Walker (Editors), *The Ecology of Tropical Savannas*, Ecol. Studies, Vol. 42, Springer-Verlag, Berlin, pp. 514–534.

Grossman, D., Grunow, J.O. and Theron, G.K., 1980. Biomass cycles, accumulation rates and nutritional characteristics of grass layer plants in canopied and uncanopied subhabitats of Burkea savanna. *Proc. Grassland Soc. Southern Afr.*, 15: 157–162.

Grunow, J.O. and Bosch, O.J.H., 1978. Above-ground annual dry matter dynamics of the grass layer in a tree savanna ecosystem. In: D.M. Hyder (Editor), *Proc. of the First International Rangeland Congress*, Denver, Colo., pp. 229–233.

Grunow, J.O., Pienaar, A.J. and Breytenbach, C., 1970. Long term nitrogen application to veld in South Africa. *Proc. Grassland Soc. Southern Afr.*, 5: 75–90.

Hofmann, R.R. and Stewart, D.R.M., 1972. Grazer or browser: a classification based on the stomach structure and feeding habits of East African ruminants. *Mammalia*, 36: 226–240.

Huntley, B.J., 1982. Southern African savannas. In: B.J. Huntley and B.H. Walker (Editors), *Ecology of Tropical Savannas*. Springer–Verlag, Berlin, pp. 101–119.

Jarman, P.J., 1974. The social organization of antelope in relation to their ecology. *Behaviour*, 48: 215–267.

Kelly, R.D., 1973. *A Comparative Study of Primary Productivity Under Different Kinds of Land-Use in South-Eastern Rhodesia*. Ph.D. thesis, University of Rhodesia, Salisbury, 250 pp.

Kelly, R.D. and Walker, B.H., 1976. The effects of different forms of land-use on the ecology of a semi-arid region in south-eastern Rhodesia. *J. Ecol.*, 64: 553–576.

Knoop, W.T. and Walker, B.H., 1985. Interactions of woody and herbaceous vegetation in a southern African savanna. *J. Ecol.*, 73: 235–253.

Kruger, J.A. and Smit, I.B.J., 1973. Reclamation of fallowed lands in the eastern Orange Free State by oversowing with *Digitaria smutsii*, *Eragrostis curvula* and *Themeda triandra*. *Agroplantae*, 5: 101–106 (in Afrikaans).

Le Roux, C.J.G., 1979. The grazing capacity of the plains in the Etosha National Park. *Proc. Grassland Soc. Southern Afr.*, 14: 89–94.

Maggs, T.M.O'C., 1976. Iron Age communities of the southern highveld. *Occas. Publ. Natal Mus.*, 2, 326 pp.

Menaut, J.C., 1983. The vegetation of African savannas. In: F. Bourlière (Editor), *Tropical Savannas*. Ecosystems of the World, 13, Elsevier, Amsterdam, pp. 109–149.

Mentis, M.T., 1977. Stocking rates and carrying capacities for ungulates on African rangelands. *S. Afr. J. Wildl. Res.*, 7: 89–98.

Mills, P.F.L., 1964. Effects of fertilizers on the botanical composition of veld grassland at Matopos. *Rhod. Agric. J.*, 61: 91–93.

Morris, J.W., Bezuidenhout, J.J. and Furniss, P.R., 1982. Litter decomposition. In: B.J. Huntley and B.H. Walker (Editors), *Ecology of Tropical Savannas*. Springer-Verlag, Berlin, pp. 535–554.

Rethman, N.F.G. and Beukes, B.H., 1973. Overseeding of *Eragrostis curvula* in north-eastern sandy Highveld. *Proc. Grassland Soc. Southern Afr.*, 8: 57–59.

Roberts, B.R., 1970. Why multicamp layouts? *Proc. Grassland Soc. Southern Afr.*, 5: 17–22.

Roux, E., 1969. *Grass. A story of Frankenwald*. Oxford University Press, London and New York, 212 pp.

Rushworth, J.E., 1975. *The Floristic, Physiognomic and Biomass Structure of Kalahari Sand Shrub Vegetation in Relation to Fire and Frost in Wankie National Park, Rhodesia*. M.Sc. thesis, University of Rhodesia, 163 pp.

Rutherford, M.C., 1978. Primary production ecology in Southern Africa. In: M.J.A. Werger (Editor), *Biogeography and Ecology of Southern Africa*. Junk, The Hague, pp. 623–659.

Ryecroft, H.B., 1944. The Karkloof forest, Natal. *J. S. Afr. For. Assoc.*, 11: 14–25.

Savory, A., 1978. A holistic approach to ranch management using short duration grazing. In: D.M. Hyder (Editor), *Proceedings of the First International Rangeland Congres*, Denver, Colo., pp. 555–557.

Tainton, N.M., 1972. An analysis of the objectives of resting grassveld. *Proc. Grassland Soc. Southern Afr.*, 6: 50–54.

Tainton, N.M., 1981a. Veld burning. In: N.M. Tainton (Editor), *Veld and Pasture Management in South Africa*. Shuter and Shooter (in association with University of Natal Press), Pietermaritzburg, pp. 363–382.

Tainton, N.M., 1981b. The ecology of the main grazing lands of South Africa. In: N.M. Tainton (Editor), *Veld and Pasture Management in South Africa*. Shuter and Shooter (in association with University of Natal Press), Pietermaritzburg, pp. 25–56.

Tainton, N.M., Groves, R.H. and Nash, R.C., 1977a. Time of burning and mowing veld: short term effects on production and tiller development. *Proc. Grassland Soc. Southern Afr.*, 12: 59–64.

Tainton, N.M., Booysen, P. de V. and Nash, R.C., 1977b. The grazing rotation: effects of different combinations of presence and absence. *Proc. Grassland Soc. Southern Afr.*, 12: 103–104.

Tainton, N.M., Booysen, P. de V., Bransby, D.I. and Nash, R.C., 1978a. Long term effects of burning and mowing on Tall Grassveld in Natal. *Proc. Grassland Soc. Southern Afr.*, 13: 41–44.

Tainton, N.M., Foran, B.D. and Booysen, P. de V., 1978b. The veld condition score: an evaluation in situations of known past management. *Proc. Grassland Soc. Southern Afr.*, 13: 35–40.

Taylor, H.C., 1963. A report on the Nxamalala forest. *For. S. Afr.*, 2: 29–35.

Taylor, R.D. and Walker, B.H., 1978. A comparison of vegetation use and condition in relation to herbivore biomass on a Rhodesian game and cattle ranch. *J. Appl. Ecol.*, 15: 565–581.

Thompson, K. and Hamilton, A.C., 1983. Peatlands and swamps of the African continent. In: A.J.P. Gore (Editor), *Mires: Swamp, Bog, Fen and Moor*. Ecosystems of the World, 4B, Elsevier, Amsterdam, pp. 331–373.

Tinley, K., 1977. *Framework of the Gorongosa Ecosystem*. D.Sc. thesis, University of Pretoria, 179 pp.

Trollope, W.S.W., 1978. Fire — a rangeland tool in Southern Africa. In: D.M. Hyder (Editor), *Proceedings of the First International Rangeland Congres*, Denver, Colo., pp. 245–247.

Venter, A.D. and Drewes, R.H., 1969. A flexible system of veld management. *Proc. Grassland Soc. Southern Afr.*, 4: 104–107.

Vorster, L.F., 1975. The influence of prolonged seasonal defoliation on veld yields. *Proc. Grassland Soc. Southern Afr.*, 10: 119–122.

Vorster, L.F. and Mostert, J.W.C., 1968. Trends in veld fertilization over one decade in the central Orange Free State. *Proc. Grassland Soc. Southern Afr.*, 3: 111–120 (in Afrikaans).

Walker, B.H., 1979. Game ranching in Africa. In: B.H. Walker (Editor), *Management of Semi-Arid Ecosystems*. Elsevier, Amsterdam, pp. 55–81.

Walker, B.H., 1980. A review of browse and its role in livestock production in Southern Africa. *Proc. Symp. on Browse in Africa*, International Livestock Centre for Africa, Addis Ababa, pp. 7–24.

Walker, B.H., Ludwig, D., Holling, C.S. and Peterman, R.M., 1981. Stability of semi-arid savanna grazing systems. *J. Ecol.*, 69: 473–498.

Walter, H., 1971. *Ecology of Tropical and Subtropical Vegetation*. Oliver and Boyd, Edinburgh, 539 pp. (translated by D. Meuller–Dombois.)

Weinmann, H., 1948. Effects of defoliation intensity and fertilizer treatment on Transvaal Highveld. *Emp. J. Exp. Agric.*, 11: 113–124.

Werger, M.J.A. (Editor), 1978. *Biogeography and Ecology of Southern Africa*. Junk, The Hague, 1439 pp.

Werger, M.J.A., 1986. The karoo and southern Kalahari. In: M. Evenari, I. Noy-Meir and D.W. Goodall (Editors), *Hot Deserts and Arid Shrublands*. Ecosystems of the World, 12B, Elsevier, Amsterdam, pp. 283–359.

Wernstedt, F.L., 1972. *World Climatic Data*. Climatic Data Press, Lemont, Pa.

Wild, H. and Grandvaux Barbosa, L., 1967. *Vegetation Map of the Flora Zambesiaca Area*. M.O. Collins (Pvt.) Ltd., Salisbury, Rhodesia, 71 pp.

Wiltshire, G.H., 1978. Nitrogen uptake on unfertilized pasture of the Willem Pretorius Game Reserve, Central Orange Free State. *Proc. Grassland Soc. Southern Afr.*, 13: 99–102.

Chapter 11

GRASSLANDS OF MADAGASCAR [1]

JEAN KOECHLIN

INTRODUCTION

Herbaceous vegetation occupies about 71% (42 million ha) of the area of Madagascar. It is almost entirely made up of secondary communities with grasses dominating. The floristic composition varies from region to region, depending on the climate and the nature of the soil (Koechlin et al., 1974). The importance of these grasslands to the economy is considerable, since livestock farming is the most important industry. The livestock population was estimated to be more than 10 million head of Bovidae in 1983, slightly more than the human population (FAO, 1984). The arid region of Madagascar has also been discussed in Volume 12B (Hot Deserts and Arid Shrublands) of this series by Rauh (1986), and the savannas have been discussed in Volume 13 (Tropical Savannas) by Menaut (1983).

ECOLOGICAL CONDITIONS

The major part of Madagascar is formed of a crystalline shield, composed of granites, migmatites and gneisses. Prominent mountain ranges occur, with peaks as high as 2876 m, some being volcanic (not presently active). The mountains are fringed by sedimentary deposits along the coast, narrower in the east than in the west. The landscape is markedly asymmetrical, rising steeply in the east into a plateau with an average altitude of 800 m, but sloping more gently downward towards the west coast.

The climate is under the influence of the trade winds and the monsoon (Rauh, 1986). The southeast trade wind, which blows almost all year, provokes a considerable amount of precipitation throughout the year along the east coast, the mean annual rainfall being 1600 mm. In contrast, the west coast receives rain only in summer (September to March), under the influence of the northwest monsoon. The amount decreases toward the south-west. The dry season in the west is accentuated by the foehn effect of the trade winds. It lasts for 6 to 7 months in the west, and for 8 to 9 months in the south-west with rainfall as low as 300 mm yr^{-1} along the coast. The distribution of precipitation is variable throughout the year and between years. The central region of the plateau is a transitional zone, being moister in the east (due to trade winds) and drier in the west, but everywhere with a well-marked 3-month dry season. The temperature sometimes falls below 0°C at the higher elevations.

The soils are ferralitic throughout most of the eastern and central regions. These are leached, acidic, low in nutrients and, except in the wooded areas, very much degraded by erosion, sometimes with iron duricrust. Tropical ferruginous soils usually occur to the west, under the influence of subhumid climate with a long dry season. These are more fertile and less eroded. The only soils that are suitable for arable agriculture are those developed on basalts or alluvium and certain types of hydromorphic soils (where water control is possible).

---

[1] Manuscript submitted in June, 1979; revised in 1988.

PHYTOGEOGRAPHICAL REGIONS

Two major phytogeographical regions can be distinguished: the moist eastern region, under the influence of the trade winds; and the drier western region. The climax vegetation over much of both regions was forest, but shrubs prevailed in the most arid areas and at higher elevations (Perrier de la Bathie, 1921).

The eastern region can be divided into four subregions (Fig. 11.1). The Eastern and Sambirano (in the north-western part of the island) Domains (occurring between 0 and 800 m in altitude) are the most humid, and the natural vegetation was a dense, humid evergreen forest. The Central Domain includes all parts of the plateau between 800 and 2000 m in elevation, where the average annual rainfall is about 1500 mm and a dry season occurs which increases in intensity westwards. In the eastern part of this Central Domain the climax vegetation was a dense humid forest differentiated by altitude, but this was replaced towards the west by a sclerophyllous forest. The High Mountain Domain, which is located above 2000 m, includes a successive series of vegetation types with increasing altitude: montane forest, thickets and then grasslands.

Within the western region, two subregions can be distinguished (Fig. 11.1). The Western Domain is located at elevations below 800 m in altitude. The climax vegetation was composed of dense, dry, deciduous forests that varied according to the texture of the soil (sand or clay) or the influence of limestone upon which some of the soil developed. The Southern Domain is located in the south-western part of the island. Rainfall is very irregular (from 250–500 mm yr^{-1}), and the original vegetation was a dense xerophilous thicket, well adapted to dry conditions (Rauh, 1986).

But in reality, in most areas, the natural forest has been replaced by secondary shrub vegetation or by grassland or savanna. Whereas forest still occupies a considerable portion of the Eastern Domain (excluding the coastal areas), it has almost totally disappeared in the Central Domain, especially in the western part. Similarly, grasslands now occupy most of the Western Domain, although some mountain slopes are forested, especially on the sandy and calcareous skeletal soils. In contrast, in the Southern Domain, thicket vegetation still prevails (Humbert and Cours Darne, 1965; Rauh, 1986).

Human activity is directly or indirectly responsible for replacement of the original vegetation by secondary types (Humbert, 1927). Forest has been destroyed (by cutting of trees and by fire) in order to provide cropland and natural rangeland. The latter is burned periodically to retard reinvasion of trees. Although the arrival of humans in Madagascar is recent (less than 2000 years ago), their impact on the natural vegetation has been intensified due to the nature of the soils and vegetation. Development of secondary forest is hindered by low fertility of the soils and by the absence of dynamic elements in the forest flora able to compete aggressively with the dominants of the grassland. Another factor is the high susceptibility to fire of dry forest of the west and of mountain forests.

Severe degradation of the soil is apparent nearly everywhere. This has accompanied degradation of the vegetation. Burning of grassland removes the organic layer, thereby exposing the soil to erosion. Steep slopes and heavy rain showers are important factors that account for severe gullying in many areas. The Southern Domain, which is too dry for arable agriculture, has been less affected by these processes. In addition, the xerophilous thicket vegetation, which includes many succulent species, resists fire remarkably well (Rauh, 1986).

A balance has been established in which fire is utilized in maintaining the herbaceous cover, but this balance is fragile and is subject to disruption, with further environmental degradation, by mismanagement.

HERBACEOUS VEGETATION

The original deforestation and subsequent management practices imposed to retard reinvasion of woody species have exposed the soil to

---

Fig. 11.1. Map of the vegetation of Madagascar: *1* = evergreen forest of the Eastern, Sambirano and Central Domains; *2* = gramineous formations of the Eastern Domain; *3* = gramineous formations of the Central Domain; *4* = high-mountain vegetation; *5* = deciduous forest of the Western Domain; *6* = gramineous formations of the Western Domain; *7* = thicket vegetation of the Southern Domain; *8* = gramineous vegetation of the Southern Domain.

GRASSLANDS OF MADAGASCAR

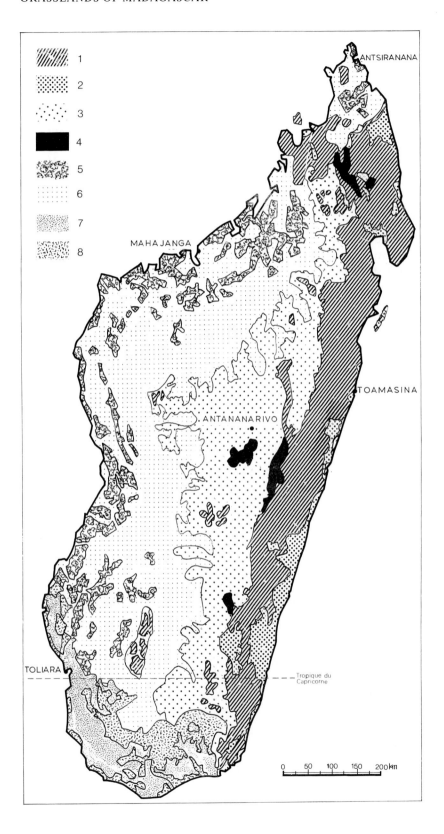

erosive forces. These have removed soil from upper slopes and deposited the material as alluvium on lower slopes and in the valleys. Consequently, a series of more or less extensive ecological niches exist, each characterized by a different type of vegetative cover.

In the deep, fertile soils of level areas various species of Andropogoneae (particularly *Heteropogon contortus*, *Hyparrhenia cymbaria* and *H. rufa*) are the main components of the herbaceous layer of the herbaceous savanna or somewhat shrubby savanna. On the degraded and continually eroded slopes, grasses of the genera *Aristida* and *Loudetia* dominate sparse herbaceous savannas. The colluvium and alluvium soils are densely populated by species of Andropogoneae and Paniceae.

In the Eastern and Central Domains, eroded soils occupy nearly every topographic position due to the combined effect of rain and slope. These soils support scattered, small bunches of grasses with narrow and tough leaves of little fodder value. The dominants are species of *Aristida*. The Andropogoneae type occupies only a very restricted area. In contrast, in the Western Domain the situation is reversed. The weaker erosive action, associated with less-pronounced topographic relief, permits the development of Andropogoneae savannas (Figs. 11.2 and 11.3). In the Southern Domain degradation of the thicket gives way to herbaceous vegetation similar in physiognomy and biological organization to that of the Sahelian steppes (see Chapter 8).

Eastern coastal region

Along the coast, deforested sandy dunes are occupied by a grassland of low stature dominated

Fig. 11.2. Tree savanna of *Stereospermum variabile* and *Heteropogon contortus* in the Western Domain. (Photo by Ph. Morat.)

Fig. 11.3. Savanna of *Medemia nobilis* and *Hyparrhenia* spp. in the Western Domain.

by *Digitaria didactyla*, *Panicum umbellatum* and *Stenotaphrum dimidiatum*, as undergrowth in the plantations of coconut (*Cocos nucifera*) and she-oak (*Casuarina equisetifolia*).

In the low hills which border the forested shelf of the plateau, the destruction of the natural forest, and then of the secondary forest, results in gramineous formations which are modified rapidly, due to fire and water erosion. The first community is a dense stand of *Andropogon eucomus*, *Hyparrhenia rufa*, *Imperata cylindrica* and *Panicum* spp. This is rapidly replaced by sparser grassland dominated by *Aristida similis*, which is the community that presently occupies most of the area (Fig. 11.4). This successional trend is reversible. If degradation of the soil is not too severe, protection against fire facilitates the return of *Hyparrhenia* savanna and the progressive establishment of forest elements.

Central plateau and mountains

In the Central Domain the successional relationships of the vegetation are the same as those described for the coastal grassland and also result in the development of very degraded vegetation types, which are regionally distinct.

A community of low-growing, scattered sclerophyllous grasses (*Aristida rufescens* and species of *Ctenium*, *Elionurus* and *Sporobolus*) predominates in the "Tanety", a vast group of hills (from 1200 to 1500 m in elevation on high plateaus) in which the soil is considerably degraded, very compact, almost impervious, and scoured by extensive erosion (Fig. 11.5). *Aristida rufescens* is particularly abundant. This species, which normally colonizes bare soils scoured by erosion, is the only one capable of surviving in the most degraded areas. Moreover, this degradation is so severe that

Fig. 11.4. Herbaceous savanna dominated by *Aristida similis* in the Eastern Domain, with patches of trees.

a return to the *Andropogon* savanna or the forest no longer seems possible.

Grass savanna dominated by *Loudetia simplex* subsp. *stipoides* occurs in the "Tampoketsa", a vast and nearly flat plateau, with poor ferralitic soils and cool climate. Because of the altitude, the temperature in winter may be as low as 0°C. This savanna occupies most sites, except for some forested patches and peaty depressions. Accompanying species vary according to soil conditions. *Elionurus tristis* and *Trachypogon spicatus* occur on the low and less-degraded slopes, and *Aristida rufescens* in the most eroded locations. The flora of this area is remarkably poor due to the selective effect of fire, only 34 species of spermatophytes being found.

The western slopes represent a transitional sector between the Central and the Western Domains. This is another region where the forest has nearly disappeared and is now occupied by vast savannas more or less devoid of woody species. A progressive transition of the vegetation of the Central Domain to that of the Western Domain occurs as the altitude decreases. *Loudetia* and *Aristida* are progressively replaced by *Heteropogon contortus* and *Hyparrhenia rufa*, whereas such truly western species as *Hyperthelia dissoluta* appear.

On some mountain ranges, the pristine forest has given way to grassland, particularly on the Ankaratra (2643 m) and Andringitra (2658 m) (Fig. 11.6). Above 2000 m, ferralitic soils are replaced by soils rich in humus (sometimes peaty), and the savanna gives way to true grasslands. The flora of these has holarctic affinities. The same genera are important as in African mountain ranges. These include *Anthoxanthum*, *Alchemilla*, *Brachypodium*, *Epilobium*, *Poa* and *Trifolium*. Many species are endemic to Madagascar, sometimes even to a single mountain range.

Fig. 11.5. "Tanety" vegetation dominated by *Aristida rufescens* in the Central Domain.

Western savannas

Whereas the herbaceous formations of the Eastern Domain are practically devoid of woody layers, those of the west are true savannas with shrubs or trees, and resemble (physiognomically and biologically) the African savannas. All forms of transition exist between the vegetation of the Western and Central Domains.

These western savannas occupy large areas (80% of the western region, more than 200 000 km^2) and the forest survives as isolated patches under very special ecological conditions. Many observations support the concept that the forest represents the climax vegetation. But the distribution of forests and savannas, under present climatic and soil conditions, represents an artificial balance maintained by human activities and by fire (usually as a management tool). The remarkably low species diversity and the homogeneity of the flora, even throughout diverse habitats, supports the concept that the savanna is secondary vegetation. Of the 300 plant species occurring in the savanna, only 85 are true savanna species, and among these only 18 are endemic to Madagascar. The others are of foreign origin, with a more or less extensive geographical distribution, or are plants of forest origin adapted to the savanna environment.

The grasses generally are not tall (less than 1 m in height), due either to the poor quality of the soil or to overgrazing. The majority of herbaceous species are perennials (hemicryptophytes). The annuals (which are very small) occupy only a restricted area. The frequent presence of a woody layer clearly distinguishes the western formations from those of the Central and of the Eastern Domains. The tree layer includes endemic palms (*Borassus madagascariensis*, *Hyphaene shatan* and *Medemia nobilis*),

Fig. 11.6. Herbaceous vegetation at 2200 m in the Andringitra mountains.

species of forest origin (*Stereospermum variabile* and *Tamarindus indica*), some of African origin (*Gymnosporia linearis* and *Poupartia caffra*), and some strictly endemic savanna species (*Acridocarpus excelsus* and *Dicoma* spp.) (Figs. 11.2 and 11.3).

Most species are distributed throughout the Western Domain and many overlap into the Central Domain. The latter include:

Aristida rufescens　　　*Hyparrhenia rufa*
Eriosema psoraloides　　*Stereospermum euphorioides*
Hyparrhenia cymbaria　　*Trachypogon spicatus*

Others (*Eragrostis chapelieri*, *Hyparrhenia rufa*, *Imperata cylindrica* and *Waltheria indica*) occur even in the herbaceous formations on the east coast. However, some differences are evident in the characteristic species of the northern and southern parts of the Western Domain (climates being moist and warm in the north and drier with cool winters in the south).

Northern species:
　Acacia farnesiana
　Themeda quadrivalvis
　Viguierella madagascariensis
　Ziziphus mauritiana

Southern species:
　Aristida congesta
　Enneapogon cenchroides
　Loudetia filifolia subsp. *humbertiana*
　Panicum pseudovoeltzkowii

The detailed phytosociological structure of the savanna is still little known, except in the southwest, where some phytosociological units have been distinguished (Morat, 1973), characterized by those species which find their optimum ecological conditions here.

The *Loudetia filifolia* subsp. *humbertiana* community occupies poor tropical ferruginous soils, between 500 and 900 m in elevation, which are sandy on the surface, but clayey and compact in the lower layers. Where a woody layer exists, it is always of low density with *Hyphaene shatan* in the driest parts and *Stereospermum euphorioides* when the destruction of the forest is more recent.

The *Loudetia simplex* community occupies thin ferralitic soils on ferruginous crusts or sandstones. This is an herbaceous savanna of low stature and sparse cover (45% bare soil).

The *Heteropogon contortus* community is the most important one. It occupies extensive areas under 600 m in altitude on various types of soils and with a mean annual rainfall above 600 mm. Various facies occur in which *H. contortus* is always present due to its wide ecological amplitude. These have developed due to edaphic differences or to varying degrees of human influence. For example, the *Medemia nobilis* (palm) savanna is indicative of the presence of hydromorphic conditions in deeper layers of the profile, whereas the presence of tree formations with *Stereospermum variabile* on ferruginous tropical soils indicates previously cultivated or overgrazed situations.

Southern region

Gradually, with decreased rainfall, the *Heteropogon* savannas of the west give way to formations similar to the characteristic steppes of African arid zones. This change is notable mostly through the increased proportion of annual species (*Aristida adscensionis* and *A. congesta*) or small perennials (*Cenchrus ciliaris* and *Panicum pseudovoeltzkowii*) which constitute very open, low-growing formations (35–40% of the soil exposed) in which fire spreads with difficulty (Fig. 11.7). The woody layer is usually poorly represented, either by true savanna species (*Hyphaene shatan* and *Poupartia caffra*) in the northern part or by xerophilous thicket species in the extreme south-west.

Azonal (edaphic) vegetation

Various types of marshy grasslands occur in the humid zone, due to the hydromorphic condition of the soil. These are dominated by tall species of Cyperaceae (*Cyperus latifolius* and *C. madagascariensis*) on peaty soils. Ponds and closed depressions are inhabited by various communities, depending on the degree of flooding. These include Poaceae (*Echinochloa* spp., *Leersia hexandra* and *Panicum* spp.), Cyperaceae and various halophytic or hydrophytic species of the genera *Aponogeton*, *Jussieua*, *Nymphaea*, *Pistia* and *Salvinia*. Dense herbaceous formations of *Hyparrhenia rufa*, *Hyperthelia dissoluta*, *Imperata cylindrica* or *Phragmites mauritianus* replace riparian forests on sandy banks of rivers.

The saline soils, especially along the west coast, are occupied by herbaceous communities composed of species of Chenopodiaceae (of the genera *Arthrocnemum*, *Atriplex* and *Salicornia*) and

Fig. 11.7. Steppe of annual grasses and *Didierea trollii* in the Southern Domain.

of Poaceae (*Paspalum virgatum*, *Sclerodactylon macrostachyum* and *Sporobolus virginicus*).

FAUNA

The fauna of Madagascar, like the flora, has a very high degree of endemicity. It has evolved in isolation from a very ancient foundation starting from well before the Cretaceous, when the island was not yet isolated. This explains the absence of various groups of more evolved animals. Moreover, this fauna lives essentially in the forest. That of the herbaceous formations is very poor and includes almost no original species, which adds further support to the view that these areas were formerly forested.

In the grasslands and savannas few rodents occur and ungulates, poisonous snakes, and many other groups are absent. Most species that are present are sylvan ones, adapted secondarily to life in the open landscape. Others have been introduced from abroad. Still others, particularly birds, are native to the arid south-west or to areas that are more or less open because of the peculiar conditions of the soil. Consequently, wild fauna has a limited role, especially in comparison to the very important impact of humans and their domesticated animals.

AGRICULTURE AND STOCK FARMING

Arable agriculture is both of the shifting type, by burning of forest vegetation, and rice (*Oryza sativa*) production on flooded soils. The area under crop at any one time represents, according to various authors, from 1.6 to 2.8% of the total geographical area of the island (Rouveyran and Chavanes, 1969). However, FAO data (FAO, 1984) suggest that 4 to 5% is arable — land that has been cultivated within the last 5 years. About half of this cropland is planted with rice, which is grown in flooded artificial terraces on the slopes and in the valleys. Away from the forest soils, where water is insufficient for the cultivation of rice, only the best soils are tilled. These are mainly cropped with groundnuts (*Arachis hypogaea*) or cassava (*Manihot esculenta*), especially in the west, and *Sorghum* spp. and *Zea mays* in the south.

Cattle comprise more than 95% of the grazing load by livestock on the vegetation. In the Western and Southern Domains, extensive stock farming is practised on the uncultivated grasslands. The system is essentially sedentary, since moving of herds is always limited. Only the *Heteropogon* savanna is utilizable. The two *Loudetia* communities are of no value for cattle raising because of low palatability and nutritive value. The grazing load in this savanna is probably between 0.4 and 0.5 animal per ha. This intensity of use is too great, especially considering the length of the dry season. Thus, the rangeland is overgrazed. However, little is known about biomass and productivity of these semi-natural grazing lands.

On the plateau of the Central Domain, livestock farming is less important and is linked to the needs of arable agriculture for draught animals, although some meat animals are intensively fed and milk is produced. Grazing is provided by untilled herbaceous communities of low nutritive value (mostly dominated by species of *Aristida*) and by rice fields after harvesting (Bosser, 1969).

Fire is regularly used in all grazing lands to rid them of old growth not consumed during the wet season, to promote the growth of grass during the dry season, and to retard invasion by woody species. The consequences of these practices are considerable. The soil is made susceptible to erosion and gullying, both by the trampling of the animals and by removal of the protective vegetative cover by fire. Fire is highly selective in removing from the community the most valuable forage species. Annuals are eliminated first, and then the most sensitive perennials, such as species of *Hyparrhenia* and *Heteropogon*, until only species of *Aristida* survive.

Overgrazing is similarly effective in modifying the flora by selection against those species most sought after by cattle. These are grazed too early in the growing season or too frequently, with the result that they become weakened and finally disappear. *Hyparrhenia rufa* is very sensitive to this process and disappears first. It is followed by *Heteropogon contortus*, leaving *Imperata cylindrica* and *Aristida rufescens* as dominants.

Itinerant agriculture causes "wearing out" (loss of fertility and breakdown of structure) of the soils. Production of groundnuts in the south-western savanna is particularly effective in this regard. After several years of rest from cultivation, the grass cover recovers to a degree, but sparse stands of relatively short grasses (such as *Mollugo nudicaulis* or *Sporobolus festivus*) prevail in place of the much more vigorous *Heteropogon contortus* savanna that preceded cultivation.

Overgrazing by livestock and "wearing down" of the soil by cultivation both favour encroachment of herbaceous communities by woody plants of the forest. These processes decrease the height and density of the herbaceous cover, so that it is less able to compete with woody invaders and provides insufficient fuel to permit a fire of sufficient intensity to kill woody plants. Important woody invaders include both native species (*Sarcobotrya strigosa* and *Stereospermum variabile* in the Western Domain, and *Helichrysum gymnocephalum* in the Central Domain) and introduced species (*Acacia farnesiana*, *Lantana camara* and *Ziziphus mauritiana* in the Western and Southern Domains).

The above observations demonstrate that the herbaceous communities used as grazing lands are subjected by mismanagement to successional trends in both directions — towards degraded grassland and back towards forest — both of which decrease the capacity of the range to support livestock (Granier, 1967). Overgrazing and regular burning induce a progressive degradation of the herbaceous layer until it reaches the *Aristida* stage, which is both of low forage value and ineffective in protecting the soil against erosion. On the other hand, overgrazing reduces the effectiveness of fire in controlling the invasion of woody species.

The natural grazing lands, particularly in the western region, are usually exploited without any management. They are exposed to uncontrolled use of fire and insufficient management inputs, such as provision for complementary forage and use of rotational grazing systems. Experiments have demonstrated huge possibilities for improvement. In extensive range operations, these include the application of rational techniques in relation to: the use of fire, controlled intensity of grazing, and rotational grazing. More intensive approaches that are applicable include the improvement of forage quality in grazing lands by the introduction of alternative species, and the establishment of arable grasslands for use as pastures or for harvesting fodder, hay or silage. A challenging objective remains to make these findings available for use by pastoralists.

REFERENCES

Bosser, J., 1969. *Graminées des pâturages et des cultures à Madagascar*. ORSTOM, Paris, 439 pp.

FAO, 1984. *1984 FAO Production Yearbook*. Vol. 37. Food and Agriculture Organization of the United Nations, 320 pp.

Granier, P., 1967. *Le rôle écologique de l'élevage dans la dynamique des savanes à Madagascar*. Institut d'Élevage et de Médecine Vétérinaire des Pays Tropicaux, Tananarive, 80 pp.

Humbert, H., 1927. La destruction d'une flore insulaire par le feu. Principaux aspects de la végétation à Madagascar. *Mém. Acad. Malgache*, Tananarive, 5, 78 pp.

Humbert, H. and Cours Darne, G., 1965. Carte internationale du tapis végétal (1/1 000), Madagascar. *Trav. Inst. Fr. Pondichery*, hors serie no. 6, 3 feuilles, notice 162 pp.

Koechlin, J., Guillaumet, J.L. and Morat, Ph., 1974. *Flore et végétation de Madagascar*. J. Cramer, Vaduz, 687 pp.

Menaut, J.C., 1983. The vegetation of African savannas. In: F. Bourlière (Editor), *Tropical Savannas*. Ecosystems of the World, 13, Elsevier, Amsterdam, pp. 109–149.

Morat, Ph., 1973. *Les savanes du Sud-Ouest de Madagascar*. ORSTOM, Paris, 235 pp.

Perrier de la Bathie, H., 1921. La végétation malgache. *Annales du Musée colonial de Marseille*, 3me ser., 9, 268 pp.

Rauh, W., 1986. The arid region of Madagascar. In: M. Evenari, I. Noy-Meir and D.W. Goodall (Editors), *Hot Deserts and Arid Shrublands*. Ecosystems of the World, 12B, Elsevier, Amsterdam, pp. 361–377.

Rouveyran, J.C. and Chavanes, B., 1969. *Approche descriptive et quantitative de l'agriculture malgache*. Annales Faculté de Droit et Sciences Economiques, Tananarive, 239 pp.

Chapter 12

OVERVIEW OF THE GRASSLANDS OF OCEANIA

A.N. GILLISON

INTRODUCTION

The grasslands of Oceania are distributed in fragmented units over a region extending some 105 degrees longitudinally [from Australia (120°E) to Pitcairn Island (135°W)] and 78 degrees latitudinally [from the Northern Mariana Islands (23°N) to the sub-Antarctic islands (55°S)]. Within this vast territory, the land area of approximately 8 509 400 km^2 supports a human population of about 25 060 000, the area per head ranging from nearly 0.3 ha in Tuvalu and Nauru to 14 ha in Papua New Guinea and 44 ha in Australia. Although the proportion of land which has been converted to grassland tends to vary directly with population pressure, direct relationships are confounded by variation in climate, soils, terrain, island size and patterns of land use. Some of the land areas treated here are outside Oceania *sensu stricto* and include "outliers" such as Hawaii, the Philippines and the sub-Antarctic islands. These have been included to provide a more appropriate regional biogeographic framework. A broad indication of the regional distribution of major grass tribes is given in Table 12.1.

Comparative studies of grasslands within Oceania are made complex by extreme variation in physical environments: although semiarid Australia contains extensive areas of "pure" grassland, the south-west Pacific island grasslands are mostly fragmented mosaics distributed along locally variable — often steep — physical environmental

TABLE 12.1

Composition of the grass flora (% contributed by each major tribe) in selected localities in and near Oceania, compared with the global average [1]

Locality	Agrosteae	Andropogoneae	Aveneae	Eragrosteae	Festuceae	Paniceae	Others
Hawaii (below *c*. 1300 m)	5.7	11.4	6.8	8.0	17.0	35.2	15.9
Luzon, Philippines	2.3	33.7	1.1	7.4	2.9	33.7	18.9
Papua New Guinea	5.5	29.6	4.6	8.3	7.4	22.2	22.2
Kimberley Dist., W. Australia	–	28.2	10.9	9.1	3.6	29.1	19.1
De Grey, Western Australia	1.1	15.7	10.1	22.5	–	27.0	23.6
South-western Queensland	4.4	10.6	8.0	21.3	1.8	18.6	35.3
Southern S. Australia	10.4	4.5	12.3	2.0	26.0	9.7	35.1
Tasmania	21.2	3.0	14.1	2.0	27.3	4.0	28.4
South-western New Zealand	25.3	–	29.3	–	30.7	–	14.7
Average	8.4	15.2	10.8	9.0	13.0	19.9	23.7
Normal distribution spectrum (global)	8.2	11.9	6.3	8.1	16.5	24.7	24.3

[1] The values used are from Hartley (1950), except for New Guinea, which are based on lists of grass genera extracted from Henty (1969). Hawaii and the Philippines are not parts of Oceania *sensu stricto*, but are of interest for biogeographic comparison.

gradients which may be variously associated with woody species. Similarly, in the sub-Antarctic islands, grasslands are components of ecosystems rather than ecosystems in their own right.

The following four chapters are concerned with the grasslands of Oceania: Australia (Chapter 13), New Zealand (Chapter 14), the sub-Antarctic islands (Chapter 15) and the south-west Pacific (Chapter 16.)

ORIGINS AND AFFINITIES

Uncertainty surrounds the origins of grassland in Oceania. Whatever the pre-human evolutionary determinants, the close interaction between man and grassland in both Australia (up to 40 000 years or more with fire) and Papua New Guinea (at least 9000 years with slash-and-burn agriculture) suggests it is logical to include humans as a "natural", albeit recent, evolutionary component for these areas (Gillison, 1972).

As with large areas of south-eastern Asia and the Philippines, in recent times many grasslands have been initiated by slash-and-burn agriculture and by fire-based hunting practices. In Australia, before the arrival of European man in the 1780s, aboriginal hunters manipulated the spatial extent and structural and floristic composition of many woodland savannas and grasslands by "patch"-burning in order to concentrate groups of grass-eating animals such as kangaroos and wallabies. Although the use of fire for hunting is also widespread in Melanesia and Polynesia, there is no parallel in Australia for aboriginal slash-and-burn subsistence agriculture.

This background, together with the uncertain origin of many so-called exotics, also creates problems in defining "naturalness". There is continuing debate about the origins of the New Zealand alpine and grassland biotas that have developed only recently and continue to evolve rapidly (see Chapter 14). The New Zealand grassland vegetation developed following a major orogeny in the Pliocene and Late Tertiary and was later subject to limiting climates under Pleistocene glaciation. The general lack of drought in New Zealand, and the relatively cool oceanic climate, distinguish it from the continental environment of Australia and give it a closer affinity with Tasmania and the sub-Antarctic islands. In the tropical Pacific, on the other hand, the closest affinities of New Zealand are with the upper montane and subalpine formations of New Guinea.

Partly because of a sparse fossil record, the origin of the gene pool from which Oceanic grasslands have been derived is unclear. Takhtajan (1969) and others have argued a case for likely centres of origin for the Angiosperms, one focal area being north-eastern Australia — a hypothesis which tends to be supported by the more recent findings of Audley-Charles (1987). A numerical analysis of world grass genera by Simon and Jacobs (1990) postulates that, with the exception of the pooids, the currently recognized groups of the Poaceae are Gondwanan in origin. The analysis by Simon and Jacobs (1990) has demonstrated close affinities between the tropical Pacific grasses and those of the cold temperate (mesotherm/microtherm) oceanic islands, whereas grasslands of microtherm/mesotherm Australia and New Zealand have closer ties with Hawaii and the Antarctic zone. The affinities of the sub-Antarctic island grasslands with South America (Gremmen, 1982; see also Chapter 15) are also reflected in the findings of Simon and Jacobs, where, at the generic level, there are evident links between tropical Asia, Australia and parts of the western Pacific.

Although providing a useful focus for regional geographic studies, phytogeographic analyses of grass genera alone are inappropriate for ecological interpretation. There is a need to take into account variability in environment as well as in plant functions and floristics in order to provide clearer insights into evolution and development and into responses to environmental change. Throughout Oceania there are "grasslands" which are closely associated with, or sometimes dominated by, cyperaceous or other graminoids. The pervasive distribution patterns of these graminoids and their close ecological ties with true grasses suggest that they should be included in studies involving grasslands.

Instead of concentrating solely on grass taxa (e.g. genera), some of the questions concerning grassland origins may be better focused using multivariate analytical techniques involving identifiable, response-based attributes of graminoids as a whole. Functional groups in graminoids can be characterized in part by specific forms of acid decarboxylation and associated enzymic pathways (Prendergast, 1989) and by other functional characteristics (Gillison, 1988). By examining the spatial and temporal distribution of these functional groups along defined physical environmental gradients, potentially useful models could be derived which would assist in interpreting graminoid origins. For Australia and the larger islands (New Zealand, New Guinea) and parts of the sub-Antarctic, various authors have argued that historical climatic fluctuations, and in particular oscillatory glaciation, have profoundly influenced the establishment and maintenance of grasslands. Periglacial activity in parts of the sub-Antarctic (e.g. Macquarie Island) also affects the establishment of grasses (Löffler et al., 1983; see also Chapter 15). These processes fuel speculation about grassland origins. Apart from geomorphic processes, grassland domains have a broad history of expansion and maintenance by man-made fire, except perhaps in the sub-Antarctic islands. In New Zealand it is argued that the origin of grasslands following the arrival of early Polynesian settlers over 1000 years ago may have been accompanied by climate change, and not be purely pyrogenic. In Papua New Guinea, the connection of lowland with upper montane and subalpine grasslands through the agency of man-made fire has facilitated the spread of lowland graminoids and associated introduced plant and animal species. Little is known about the likely origins of sub-Antarctic grassland. Although the earliest vegetation is recorded from the Tertiary and Quaternary, there is some evidence in the Îles de Kerguelen of loss in overall floristic richness since the Miocene. For the islands in the sub-Antarctic convergence it is likely that the climate has remained relatively constant over the past 10 000 years (see Chapter 15).

VEGETATION

Bioclimate classification

Within Oceania, present-day climates range from extremes of seasonal tropical lowland on the one hand and alpine on the other to cold non-seasonal regimes in the sub-Antarctic. Because there are broad climatic similarities between high-altitude/low-latitude and low-altitude/high-latitude domains, the use of terms such as "tropical" and "temperate" can mislead when comparing grassland climates between regions (Oliver, 1979).

The bioclimatic framework used in Chapter 16 (see also Gillison, 1983) to compare the grasslands of the south-west Pacific has been extended in the following discussion to apply to Oceania as a whole. This is based on broad categories of optimal growth responses of the whole plant to defined thermal domains. The use of a response-based approach of this kind makes possible more meaningful ecological comparisons between grassland regions with broadly similar climates than can be obtained with other, more arbitrary methods (see Nix, 1983, and Table 12.2)). Nix's approach has made possible a simplistic but relatively uniform comparison between grasslands in high and low latitudes with similar bioclimates. Although there are broad similarities in such contrasting

TABLE 12.2

The bioclimatic classes used by Nix (1983)

Bioclimatic class	Photosynthetic pathway	Optimum temp. (°C)	Lower temp. threshold (°C)	Upper temp. threshold (°C)
Megatherm	C_4 and CAM	30–32	10	46
Megatherm	C_3	26–28	10	36–38
Mesotherm	C_3	19–22	5	33
Microtherm	C_3	12–14	0	25
Hekistotherm	C_3	6–8	−10	25

climates of the sub-Antarctic convergence and tropical alpine areas, significant differences exist in rainfall seasonality, solar radiation intensity, day-length and wind. These differences are reflected in different plant functional characteristics expressed in the same (pooid) genera; the sub-Antarctic grasses are mesomorphic and the tropical alpine grasses xeromorphic. Other functional characteristics related to the photosynthetic pathway of grasses reveal a strong concentration of C_4 acid decarboxylation types (especially andropogonoid taxa) towards the megatherm seasonal bioclimate and a converse increase in the proportion of C_3 types along a decreasing thermal cline to complete C_3 (pooid) dominance in the sub-Antarctic. In applying this classification, a bioclimate domain is defined as "seasonal" if the coefficient of variation in annual rainfall during the year exceeds 60%.

Megatherm grasslands

In general terms the strongly seasonal megatherm grasslands include many CAM- and C_4-based vegetation types. For grasslands this domain is dominated by C_4 Andropogoneae (cf. Johnson and Tothill, 1985). In northern tropical Australia, perennial, chloridoid "spinifex" hummock grasses, especially *Plectrachne schinzii* and *Triodia basedowii*, occupy the drier seasonal interior north of 22°S, whereas in areas with more than 1200 mm mean annual rainfall there tends to be a divergence into "seasonal" and "perennial" andropogonoids, for example within the genus *Sorghum*. Some of the most extensive tussock grasslands within this Australian zone are those on cracking clays dominated by *Astrebla* spp. (Mitchell grass). These are relatively pure grasslands where the botanical composition is strongly influenced by rainfall seasonality. Like the spinifex-dominated grasslands of the more arid interior, they are unique within Oceania. It is not surprising that, in northern megatherm seasonal Australia, graminoids with Indo-malesian affinities occur towards the warmer, more humid extremes. These include in particular, members of the tribes Andropogoneae (*Arthraxon*, *Capillipedium*, *Chrysopogon*, *Dichanthium*, *Eulalia*, *Heteropogon*, *Imperata*, *Ischaemum*, *Saccharum*, *Sorghum* and *Themeda*) and Paniceae (*Alloteropsis*, *Brachiaria*, *Cenchrus*, *Digitaria*, *Melinis*, *Panicum*, *Pennisetum* and *Setaria*). In this region, semi-permanent to permanent swamp formations commonly contain indigenous oryzoid C_3 "wildrice" genera (*Leersia* and *Oryza*) and arundinoid *Phragmites*.

Two andropogonoid C_4 grasses *Imperata cylindrica* and *Themeda triandra* (formerly *T. australis*) typify much of the humid Australian and Papua New Guinean megatherm grasslands. For this reason their presence in southern mesotherm/microtherm Australia, where C_3 grasses predominate, is difficult to explain on functional grounds. In seasonally or permanently moist environments, members of the andropogonoid sub-tribe Saccharinae (*Miscanthus* and *Saccharum*) are conspicuous. It is puzzling that, although *Miscanthus* is widespread throughout the south-west Pacific, it is not indigenous to Australia despite the presence of an apparently suitable bioclimate. As with some other grass genera, the distribution of *Miscanthus* may be due largely to historical trading activities (cf. Simon, 1988). Maritime megatherm grasslands are the norm for many south-west Pacific islands, often with *Imperata cylindrica*, *Ischaemum* spp. and tall rottboellinoid species of *Rottboellia* and *Ophiuros*. Deliberate annual firing of these grasslands by man is common.

The seasonal megatherm woodland savannas of southern and south-western Papua New Guinea are characterized by *Eucalyptus* spp. and *Melaleuca* spp. This reflects a structural and floristic continuum with the savannas of Cape York Peninsula in Australia. These mixed grasslands are unique within Oceania. In New Caledonia where *Melaleuca* ("Niaouli") savannas occur without *Eucalyptus*, taxa (including *Melaleuca quinquenervia*) which are common to both Australia and New Caledonia exhibit much wider environmental amplitude. The structure and floristic composition of grasslands in the smaller islands of the south-west Pacific reflect the combined effects of oceanic climatic buffering, relatively lower floristic richness in the island domains, increased vulnerability to invasive exotics, and varying and often intense land use.

The fragmented nature of many megatherm grasslands makes difficult their comparison within the region. Apart from localized, repetitive patterns of mid-height to tall grasslands (typically of species of *Heteropogon*, *Miscanthus*, *Themeda* and members of the Rottboellinae), there are non-grass graminoid components which occur repeatedly throughout the region. Apart from typical herbaceous cyperaceous elements, these are woody, rosulate graminoids represented mainly by Pandanaceae (*Pandanus* spp. and *Sararanga* spp.) and palms, especially *Livistona*. Whereas Pandanaceae are distributed throughout Indo-malesia and the megatherm Pacific generally, they appear to be particularly well developed in south-west Pacific grasslands. The genus *Sararanga* is restricted to the Philippines, the Solomon Islands and the New Ireland Province of Papua New Guinea where it occurs in typically humid habitats. *Pandanus* ranges across a wide spectrum of physical environments from seasonal megatherm (Figs. 16.3, 16.10, 16.19 and 16.24) to cooler mesotherm situations.

The broad-leaf fan palm *Livistona* is represented in Papua New Guinea by *L. brassii* (Fig. 16.10), and another ten species (approximately) occur in Australian savannas. The distribution pattern of this palmoid grassland formation is repeated in similar environments around the megatherm seasonal tropics in various, presumably functional, analogues (e.g. *Borassus*, *Mauritia* and *Sabal*). As with *Pandanus*, the global distribution of this grassland type has yet to be studied in detail. Another feature of Australian and southwest Pacific grasslands is the frequent association of other woody rosulate growth forms and, in particular, their distribution along thermal gradients. In the lowland megatherm grasslands these are represented by the genus *Cycas* (*C. media* and *C. rumphii*) (Fig. 16.11) and in montane, mesotherm/microtherm grasslands by tree-ferns of the genus *Cyathea* (Figs. 16.14, 16.15 and 16.16). Apart from the *Cyathea* association which appears to be unique to New Guinea, the *Cycas* association is pantropical; yet its ecological connection with grassland does not appear to have been examined in a global context.

Mesotherm grasslands

Within the warmer part of the mesotherm range there is considerable spatial overlap between grasses with mesotherm and cooler megatherm growth optima. Whereas the latter tend to be dominated by C_4 andropogonoid and panicoid groups, in semiarid Australia the largest proportion of grasses is represented by the chloridoid xerophytic hummock (*Plectrachne* and *Triodia*) and tussock (*Astrebla*) grasses, which otherwise are megatherms. Within this broad region, soil type is a critical determinant in grassland distribution. This is especially evident in respect to species of *Astrebla* (see Chapter 13).

There is an evident climatic correlation with vegetation change throughout most of the area of mesotherm/megatherm overlap. The "semiarid crescent", which stretches from south-western Western Australia through to the rain shadow west of the Great Dividing Range to northern Australia, lies within this area. This constitutes an ecotone between the arid shrublands and grasslands too dry for arable agriculture and the higher-rainfall temperate and tropical woodlands in which crops can be grown. The drier boundary of the crescent in southern and eastern Australia approximates the limit for introduced pasture species (see Chapter 13).

Towards the cooler mesotherm region there is a gradual change from C_4 to C_3 grasses (especially Arundineae, Stipeae and Aveneae). In lowland semiarid southern Australia these may be variously associated with halophytic chenopod shrubs — the saltbushes (*Atriplex* spp.) and bluebushes (*Maireana* spp.) — often with scattered mallee-form *Eucalyptus* spp. (Chapter 13; Carnahan, 1990). These southern xerophytic grasslands are often invaded by Mediterranean annuals (e.g. *Hordeum leporinum* and *Medicago* spp.). In Australia, temperate (mesotherm) short-grass vegetation occurs between 27° and 42°S, with the proportion of Arundineae, Pappophoreae and Stipeae varying inversely with mean annual air temperature. In the montane and upper montane grasslands of Papua New Guinea, there is a parallel development in C_3 groups — especially Aveneae (*Anthoxanthum*, *Deyeuxia*, *Dichelachne*, *Hierochloë*), Arundineae (*Chionochloa*) and Isachneae (*Isachne* species. The "trunked" nature of some grasses, such as *Chionochloa archboldii*, is conspicuous in moist environments that are fired episodically. This trunked habit has some structural and possibly functional affinities with rosulate semi-ligneous plants [*Cyathea* spp. (Fig. 16.15) and *Pandanus* spp.] associated with upper montane and subalpine grasslands. Although the adaptive aspects of this form require further study, there are clear implications for adaptive convergence in a wide range of taxa — including grasses — in other tropical high mountains in the world (such as Africa and South America).

Other montane mesotherm grasslands in the south-west Pacific are relatively few. But where they exist — for example on Mount Tabwemasana (1800 m) in Espiritu Santo in Vanuatu — they are typically C_4 megatherm types with tall grasses (Rottboellinae) and andropogonoid *Imperata cylindrica* and *Themeda triandra* (see Fig. 16.8). On the island of Savaii in Western Samoa, *Imperata cylindrica* occurs on montane lava flows with a short herb and moss layer, as well as on a montane cinder cone and ash plain. Other associated species are *Oplismenus compositus* and *Paspalum orbiculare* (Whistler, 1978). The apparent elevational "depression" of vegetation in these smaller islands may be related to the Massenerhebung effect[1].

The New Zealand mesotherm grasslands resemble in floristic and structural profile the southernmost temperate short grasslands of Australia (cf. Chapter 13). Differences derive from the predominantly oceanic climate of New Zealand and the less buffered continental climate of Australia. Whereas in the North Island precipitation is relatively non-limiting, within the South Island of New Zealand there is a marked rainfall gradient, which decreases to the east in the form of a rain shadow from the Southern Alps. This exerts a major influence on the structure and composition of grasslands and associated vegetation types. Drought is not a feature of New Zealand vegetation, "desertic vegetation" being the only major plant formation absent from the two islands. Below the New Zealand alpine grasslands lie mesotherm tussock or bunch grasses dominated by species of Poeae (*Festuca* and *Poa*), which reflect the growth forms of tropical mesotherm and sub-Antarctic microtherm grasses.

Microtherm/hekistotherm grasslands

These grasslands are restricted to very small areas of the high mountains of southern Australia, and the New Zealand and New Guinean alpine and subalpine zones. The close bioclimatic affinities between the hekistotherm sub-Antarctic grasslands and the essentially microtherm grasslands of these alpine and subalpine areas suggest they should be treated as one entity. The broader-ranging microtherm grasses (mainly C_3 arundinoid and pooid groups) are common to the alpine, subalpine and high-montane zones of Australia, New Zealand and New Guinea. These form tussock formations of varying height (0.2–1.2 m), and their growth form and cover are influenced by soil and aspect (this is not always so in Papua New Guinea — see Smith, 1977). Although the ecological affinities of the high-altitude New Guinean grasslands are with high-latitude (30° to 55°S) southern Oceania, they contain woody elements — especially Ericaceae (*Rhododendron* spp.) — which have strong biogeographic links with the Himalayas.

Although there are closely overlapping thermal ranges, the only grass genera common to microtherm Australia, New Zealand and Papua New Guinea are the C_3 pooids, *Agrostis*, *Festuca* and *Poa*. Despite the apparent climate similarities between the sub-Antarctic islands and the alpine grasslands of the other areas, the rolled, xeromorphic leaves of the New Zealand pooids suggest closer affinities with the high-montane New Guinean tussock grasses rather than with the mesomorphic forms of the sub-Antarctic species, with their broad, flattened laminae. Limited studies of shoot/root ratios indicate that these are much higher in tropical alpine zones than elsewhere (cf. Chapter 15). Other features of the sub-Antarctic islands, that are unique in Oceania, are the large-leaved forbs associated with grassland. On Macquarie Island these are represented by *Pleurophyllum hookeri* and *Stilbocarpa polaris* and, on the Îles de Kerguelen, by *Pringlea antiscorbutica*. It is worth noting the convergence in form between the rhizomatous *Stilbocarpa polaris* (Macquarie Island cabbage) and *Rheum rhabarbarum* (domestic rhubarb) which is widespread in Iceland. Another large forb common in Arctic and sub-Arctic Fennoscandia is *Archangelica*. Al-

---

[1] "**Massenerhebung effect**" refers to situations on large islands where vegetation, occurring at high elevation inland, also occurs at increasingly lower elevation towards the coast (see Chapter 16, p. 437).

though large forbs are absent from New Guinea microtherm grassland habitats (except *Gunnera* spp. along forest edges in moist sites), they occur in alpine and subalpine areas of Africa and South America (e.g. *Lobelia* and *Espeletia*, respectively).

Despite some similarities in tropical alpine and sub-Antarctic climates, there are significant differences (see Chapter 15). Radiation loads are considerably less in the sub-Antarctic zone, and photoperiodic amplitude is much wider. In the sub-Antarctic islands, precipitation is non-limiting for plant growth and relative humidity is always high. There are also strong year-round winds arising from anti-cyclonic high-pressure systems which exert a dominant effect on the aspect of vegetation (Löffler et al., 1983). Although winds reach gale force in the high mountains of New Guinea, they are without the same intensity or persistence as those in the sub-Antarctic convergence. A notable feature of the sub-Antarctic climate is its relative stability throughout the year. Although Hnatiuk (Chapter 15) comments that tropical alpine areas of Papua New Guinea also show little climatic variation, the high-montane, south-eastern extremities of the main cordillera of Papua New Guinea (e.g. Mount Suckling at 3300 m elevation) do exhibit distinct seasonality (personal observation).

Grassland productivity

There are no comparative studies of grassland production within Oceania. Most productivity data are available in different forms, which complicates cross-tabulation. The data presented in Chapters 13 to 15 (summarized in Table 12.3) indicate regional bioclimatic trends in herbage production in defined grassland types. Despite the paucity of data, the mesotherm natural and introduced grassland species appear to out-perform their megatherm and microtherm counterparts.

Natural nutrient input varies with bioclimate and geographic location. Nutrient inputs appear to increase away from megatherm seasonality towards the highly buffered, oceanic, microtherm/hekistotherm environments. This may be due mainly to nutrient input in the latter from marine animals and sea birds and to increased atmospheric deposition of nutrients, especially nitrogen and phosphorus from ocean-borne nutrients. Significant atmospheric transfers of nutrients from the sea to the land — through aerosols and precipitation — occur in oceanic climates. Manuring by animals and sea birds also contributes significantly, especially in the sub-Antarctic zone. Some limited studies suggest that there also may be significant nitrogen input to the soil by blue-green algae. Nutrient pathways have been little studied, but, due to the paucity of vertebrate herbivores in some areas, energy may be channelled directly through the detritus food chain (Chapter 15). Nutrient transfers from vegetation to soil are reduced in the megatherm seasonal environments, where annual burning of grassland removes significant amounts of labile nutrients — especially nitrogen, potassium and sulphur — whereas phosphorus and other macronutrients tend to be lost in ash run-off. In New Zealand, earthworms play a significant role in soil-nutrient mobility (Chapter 14). The conspicuous presence of earthworm casts in seasonally flooded grasslands of the Sepik Plains in Papua New Guinea and on the north-east coast of Queensland also suggests that these invertebrates play a significant role in nutrient relationships, where the store of soil nutrients is otherwise impoverished.

Despite this apparent range of nutrient inputs, a limited set of productivity data of annual herbage production from grassland types across a range of bioclimates (Table 12.5) indicates that mesotherm grasslands are the most productive.

Effects of humans and other animals

In recent times man has exerted a powerful influence in the conversion of forest to grassland by fire and agriculture. In the south-west Pacific the rate of spread of anthropogenic grasslands is increasing with population pressure and forest removal.

In Oceania humans have been major agents in extending and maintaining grassland. Influences range from primitive slash-and-burn agriculture throughout the south-west Pacific, to purposive firing to drive game or else modify vegetation for subsistence (e.g. firing to increase productivity of the edible fern *Pteridium esculentum* by Maori). In some areas — such as highland Papua New Guinea — as population pressure has increased,

TABLE 12.3

Net annual herbage production (g m^{-2}) in various grassland types in selected bioclimates within Oceania [1]

Region	Bioclimate	Grassland type	Production
Australia	megatherm seasonal	tropical tall grass *Themeda triandra* *Themeda/Sorghum* *Heteropogon contortus*	112–170 120 250–750
	megatherm	xerophytic tussock (*Astrebla* spp.)	128–188
	megatherm/ mesotherm	xerophytic hummock grass *Triodia pungens*	94
	mesotherm	"Mediterranean" pasture (+ *Trifolium subterraneum*)	594
New Zealand	mesotherm	short tussock grass *Chionochloa crassiuscula* *C. pallens* *C. rigida* *C. macra* *C. oreophila*	949 718 539 346 180
		tall tussock grass *Chionochloa*	750–1042
Sub-Antarctic: [2] Marion Is.	microtherm	tussock grass *Poa cookii*	465
South Georgia	microtherm	*Poa flabellata*	6025

[1] Based on assumed total above-ground production in one growing season.
[2] Based on total above- and under-ground estimates in one growing season (data from Moore, Ch. 13; Mark, Ch. 14; Hnatiuk, Ch. 15).

slash-and-burn agriculture has changed to sedentary methods with accompanying replacement of tall grasses by relatively short grasses (e.g. *Imperata cylindrica* and *Themeda triandra*) — and often with increases in invasive weeds and overall loss in plant production.

In Australia, the extent of grassland has increased and undergone modification since the arrival of Europeans in 1788. Carnahan (1990) has estimated some of these changes (Table 12.4), of which the conversion to graminoid crops (cereals and sugar cane) and pasture with introduced species is the most significant. Introduced grazing animals in both New Zealand and Australia have had a major impact on grassland. Vast tracts of semiarid Australia have been modified under grazing pressure over the last century (see Chapter 13). In New Zealand, deer (*Cervus elaphus*) are three times as efficient in energy use as sheep (Chapter 14). The impact of feral deer species in New Zealand has been greatly reduced in recent years by culling for meat. In southern Papua New Guinea the population of introduced rusa deer (*Cervus timorensis*) has risen to a point where it is competing significantly with the indigenous agile wallaby (*Macropus agilis*) for a highly seasonal grassland food resource. In a Hawaiian coastal grassland, Mueller-Dombois (1981) reported results from a grazing exclosure where, after one decade he identified four dynamic categories of plant responses. One form ("persisters") could be divided into stable and oscillatory types. The latter can be compared with species showing similar behaviour in Mitchell grass (*Astrebla*) lands in

TABLE 12.4

Extent (10^3 km^2) of Australian grassland formations — past and present [1]

Structural form	Before 1780 [2]		At present	
	area	% [3]	area	%
NATURAL GRASSLANDS				
Hummock grassland	45	0.6	45	0.6
Closed grassland/sedgeland	14	0.2	25	0.3
Tussock grassland/sedgeland	359	4.7	326	4.2
Open tussock grassland	115	1.5	348	4.5
Sparse grassland	14	0.2	15	0.2
Subtotal	547	7.2	859	9.8
GRASS UNDERSTOREY IN WOODLANDS				
Hummock grasses	2045	26.9	2065	26.8
Tussock grasses	2385	31.0	2308	30.0
Subtotal	4430	57.9	4373	56.8
ARTIFICIAL GRASSLANDS				
Sown pasture	0	0.0	436	5.6

[1] From Carnahan (1990).
[2] That is, before the arrival of European settlers.
[3] % of total land area.

semiarid megatherm/mesotherm overlap areas of Australia.

The influence of native animals on vegetation in the sub-Antarctic islands is far more benign, as none is a herbivore and most (e.g. seals and penguins) contribute significantly to the nutrient resource. The introduced animals tell a different story. Mostly introduced as a source of food by early sealers and whalers, these are many and varied, including cattle, cats, horses, mice, mules, pigs, rabbits, rats, reindeer, sheep, and flightless rails (wekas). Whereas some populations have been reduced, others persist. As with mainland Australia, on Macquarie Island rabbits have created profound changes in grassland floristics and structure, and a concerted program of eradication using a myxoma virus has been under way for several years.

Vulnerability to exotic species

As a general rule, the relative impact on indigenous biota by invasive exotics increases with decreasing island size and distance from mainland sources. In this respect, at least for plants, the sub-Antarctic islands are no different to the islands of the tropical south-west Pacific. In the former, 32 species of invasive plants have become persistent or naturalized (Chapter 15). Of these, *Acaena magellanica* continues to spread with disturbance and grazing by introduced animals, especially rabbits. In the south-west Pacific (Chapter 16), apart from introduced weeds, which also increase under grazing pressure by cattle, continual firing and removal of woody vegetation by logging or gardening have paved the way for invading exotic grasses. The two most significant are *Pennisetum polystachyon* and *Panicum maximum*. The former continues to spread rapidly (Figs. 16.7 and 16.26) with increasing land degradation. Associated with these grasslands are two conspicuous woody weeds, *Leucaena leucocephala* (Fig. 16.7) and *Psidium guayava*, which, together with *Muntingia calabura*, rapidly occupy overgrazed and frequently fired lands.

Although many exotic species have invaded the "natural" grasslands of New Zealand, the richer and more diverse flora in Australia has rendered the grasslands less subject to the entrance of invaders. However, the massive invasion of prickly pear (*Opuntia* spp.) in the early 1900s is an exception to this pattern, and more recently other

TABLE 12.5

Summary of tribes in Australia with more than 1% of the total number of grass entities [1], showing the relative proportion of naturalized exotics (in descending order) [2]

Tribe	Native entities		Naturalized entities	
	No.	%	No.	%
Paniceae	187	68.8	85	31.2
Eragrostideae	154	88.0	21	12.0
Andropogoneae	130	88.4	17	11.6
Danthonieae	107	95.5	5	4.5
Agrostideae	76	77.6	22	22.4
Aristideae	64	100.0	0	0.0
Stipeae	61	92.4	5	7.6
Poeae	55	55.0	45	45.0
Chlorideae	30	75.0	10	25.0
Pappophoreae	20	100.0	0	0.0
Sporoboleae	16	76.2	5	23.8
Phalarideae	15	60.0	10	40.0
Aveneae	8	21.6	29	78.4
Micraireae	8	100.0	0	0.0
Triticeae	4	16.7	20	83.3
Bromeae	1	4.0	23	96.0

[1] Entities include species and infra-specific taxa.
[2] From Simon (1981).

invaders have appeared (Chapter 13). Although Australian grasslands have suffered invasion by native woody weeds, there are some conspicuous exotics. The grasslands of northern Australia are currently being invaded by the Madagascan rubber vine (*Cryptostegia grandiflora*), and in swampy areas in the north-west by *Mimosa pigra*, for which urgent control measures are being sought. Simon (1981) has provided an analysis which shows the relative proportion of the existing grass flora that is made up of naturalized exotics (Table 12.5).

CONCLUDING REMARKS

The future for grasslands in the region is not promising given the present and likely continuing levels of disturbance. To develop better management tools, a more comprehensive and spatially referenced regional data base is required for inventory and monitoring. This should include response-based models built from a basic knowledge of how grasslands respond to environmental change along local as well as regional gradients.

In this respect a minimum set of biophysical attributes using attributes of plant function as well as structure and floristics should be developed which can be directly related to temporal and spatial changes in the physical environment.

REFERENCES

Audley-Charles, M.G., 1987. Dispersal of Gondwanaland: relevance to evolution of the Angiosperms. In: T.C. Whitmore (Editor), *Biogeographical Evolution of the Malay Archipelago*. Clarendon Press, Oxford.

Carnahan, J.A., 1990. Vegetation. In: J.A. Carnahan (Compiler), *Atlas of Australian Resources*. Third Series, 6, Australian Land Surveying and Land Information Group, Dept. of Administrative Services, Canberra.

Gillison, A.N., 1972. The tractable grasslands of Papua New Guinea. In: J. Ward (Editor), *Change and Development in Rural Melanesia*. Fifth Waigani Seminar, University of Papua New Guinea and Australian National University, Port Moresby, pp. 161–172.

Gillison, A.N., 1983. Tropical savannas of Australia and the southwest Pacific. In: F. Bourlière (Editor), *Tropical Savannas*. Ecosystems of the World, 13, Elsevier, Amsterdam, pp. 183–243.

Gillison, A.N., 1988. A plant functional attribute proforma for dynamic vegetation studies and natural resource surveys. *CSIRO Aust. Division of Water Resources, Canberra, Tech. Memo.*, 88/3, 32 pp.

Gremmen, N.J.M., 1982. *The Vegetation of the Subantarctic Islands Marion and Prince Edward*. M.J.A. Werger (Editor), Geobotany, 3, Junk, The Hague, 149 pp.

Hartley, W., 1950. The global distribution of tribes of the Gramineae in relation to historical and environmental factors. *Aust. J. Agric. Res.*, 1: 355–373.

Henty, E.E., 1969. A manual of the grasses of New Guinea. *Papua New Guinea Dep. For. Bot. Bull.*, 1, 215 pp.

Johnson, R.W. and Tothill, J.C., 1985. Definition and broad geographic outline of savanna lands. In: J.C. Tothill and J.J. Mott (Editors), *Ecology and Management of the World's Savannas*. Australian Academy of Science, Canberra, in association with Commonwealth Agricultural Bureaux, Farnham Royal, Bucks., pp. 1–13.

Löffler, E., Sullivan, M.E. and Gillison, A.N., 1983. Periglacial landforms on Macquarie Island, subantarctic region. *Z. Geomorphol.*, N.F., 27: 223–236.

Mueller-Dombois, D., 1981. Vegetation dynamics in a coastal grassland of Hawaii. In: P. Poissonet, F. Romane, M.P. Austin, E. van der Maarel and W. Schmidt (Editors), *Vegetation Dynamics in Grasslands, Heathlands and Mediterranean Ligneous Formations*. Junk, The Hague. (Repr. from *Vegetatio*, 46: 131–140.)

Nix, H.A., 1983. Climate of tropical savannas. In: F. Bourlière (Editor), *Tropical Savannas*. Ecosystems of the World, 13, Elsevier, Amsterdam, pp. 37–62.

Oliver, J., 1979. A study of geographical imprecision: the tropics. *Aust. Geogr. Stud.*, 17: 3–17.

Prendergast, H.D.V., 1989. Geographical distribution of C_4 acid decarboxylation types and associated structural variants in native Australian grasses (Poaceae). *Aust. J. Bot.*, 37: 253–273.

Simon, B.K., 1981. An analysis of the Australian grass flora. *Austrobaileya*, 1: 356–371.

Simon, B.K., 1988. The biogeography of the Australian grasses. *Proc. Ecol. Soc. Aust.*, 15: 267–269.

Simon, B.K. and Jacobs, S.W.L., 1990. Gondwanan grasses in the Australian flora. *Austrobaileya*, 3: 239–260.

Smith, J.M.B., 1977. Vegetation and microclimate of east- and west-facing slopes in the grasslands of Mt. Wilhelm, Papua New Guinea. *J. Ecol.*, 65: 39–53.

Takhtajan, A., 1969. *Flowering Plants, Origin and Dispersal.* Oliver and Boyd, Edinburgh, 310 pp.

Whistler, W.A., 1978. Vegetation of the montane region of Savai'i, Western Samoa. *Pac. Sci.*, 32: 79–94.

Chapter 13

GRASSLANDS OF AUSTRALIA [1]

R.M. MOORE

INTRODUCTION

General descriptions of Australian vegetation have been given by Williams (1955), Moore and Perry (1970), Specht (1970) and Carnahan (1976). Australian grasslands have been described by Moore (1970a) and Groves and Williams (1981), and an account of the pastoral industries was given by Alexander and Williams (1973). Australian soils have been described on a continental basis by Stace et al. (1968) and by Hubble (1970); land forms and structures have been described by Mabbutt and Sullivan (1970); and herbivorous wild animals have been discussed by Main (1969) and Frith (1970).

Most of the herbaceous vegetation of Australia is used directly to some degree, or has been modified or replaced, for purposes of livestock production. Within 100 years of European settlement the sheep population alone had reached 100 million. The resultant increases in grazing pressures produced marked changes in the vegetation, particularly in southern and temperate regions (Moore, 1962, 1966, 1967a, b, 1970a, b). Grazing by domestic livestock, and in many areas the felling of forests, thinning of woodland trees, application of fertilizers and the introduction (accidentally or deliberately) of exotic species, have produced a wide variety of man-made grasslands.

In conformity with usage elsewhere (Moore, 1970a), all the resultant herbaceous communities used for livestock production are considered as grasslands in this chapter. Those composed essentially of native species are classed as "native grasslands" and are mapped in Fig. 13.1. Communities of exotic species, both sown and volunteer, are classed as "exotic grasslands or pastures". The areas within which introduced species are established or, as in the northern and tropical areas particularly, are thought to be of potential value in animal production, are shown in Fig. 13.2.

Introduced annual pasture species, notably *Medicago* spp. and *Trifolium* spp. in temperate Australia and *Stylosanthes humilis* in the tropics, are naturalized in modified native grasslands. The "temperate short-grass grasslands" (Fig. 13.1, no. 9), for example, have a wide variety of herbaceous communities, some almost wholly of native short-grass species, some of mixed native and exotic species, and others wholly of introduced species.

In addition to the areas where exotic species are important or potentially so, Fig. 13.2 shows the inland boundaries of some of the sown and volunteer species important in livestock production.

The area of Australia is 7.7 million km^2. Of this, some two-thirds is arid. Waterless sandy deserts, and rugged rocky outcrops difficult of access, cover large areas of Western Australia and the Northern Territory. With the exception of widely scattered but economically important mining operations, and a small but burgeoning tourist industry, livestock production is the only enterprise enabling large areas of the arid zone to be occupied and used for commercial purposes. Because stock-carrying capacities are low, holdings are large, and both human and livestock populations are relatively small. Only 14% of the Australian human population live in the countryside, but most of these are outside of the arid zone, in which Alice Springs is the only non-mining town with a

---

[1] Manuscript received in June, 1979; revised in 1988.

population above 10 000. Less than 20% of the country's livestock are carried on arid lands and, because of the absence of permanent rivers, there seems little prospect of any other industry occupying substantial areas.

Arid grasslands have very low productivity per unit area and generally low productivity per animal, but productivity per unit of labour and capital is higher than grasslands in more mesophytic areas (Perry, 1977). The economic history of the pastoral industry in the arid zone has been outlined by Barnard (1969).

The terms under which land has been leased for pastoral purposes and the sizes of leases in arid lands in the various Australian states have been reviewed by Heathcote (1969).

The Australian arid zone has been discussed by Williams and Calaby (1985), with emphasis on its desertic characteristics, in Volume 12A of this series.

The arid zone proper grades into the farming areas through a zone of shrub woodlands, low woodlands and eucalypt shrublands referred to by Moore (1973) as the "semi-arid crescent". The crescent is of ecological interest and significance to conservation because it may be considered as an ecotone between the arid shrublands and grasslands too dry for arable agriculture, and the higher-rainfall temperate and tropical woodlands where crops can be grown. The south-western part of the semiarid crescent in Western Australia and the eastern part in southeastern Australia are marginal for cereal growing. The drier boundary of the crescent in southern and eastern Australia is roughly the limit to which introduced pasture species are grown or have become naturalized. In northern Australia, the dry limit of most of the introduced pasture species presently available for commercial use appears generally to be at a higher mean annual rainfall than the high-rainfall boundary of the crescent (Moore, 1970a); possible exceptions are species of *Cenchrus*. *Cenchrus ciliaris* in particular, appears to have become naturalized in several small areas within the semiarid crescent, and during the unusually wet season of 1974–75 spread to floodplains and run-on areas in many parts of the arid zone proper.

Other features of the semiarid crescent are: it is the outer boundary of a large area of internal drainage (Mabbutt and Sullivan, 1970); it corresponds roughly to the floristic interzone between the eremean and higher-rainfall floristic zones of Burbidge (1960); and it is the zone beyond which, with increasing aridity, shrubs replace trees, and *Acacia* replaces *Eucalyptus* as the predominant genus of woody plants in shrublands other than the shrub steppe. It is also the approximate distribution limit of some native animals — the arid limit of *Macropus antilopinus* (antilopine kangaroo) of the north, and the high-rainfall limit of *Macropus rufus* (= *Megaleia rufa*) (red kangaroo) of the interior (Frith, 1970).

Outside the semiarid crescent in the higher-rainfall woodlands and forests, beef, mutton and wool are produced in grasslands of native grasses and forbs, on mixtures of native and naturalized introduced species, and on sown pastures.

The herbaceous understoreys of subtropical and temperate woodlands are the most productive of Australia's native grazing lands, and it is in these woodlands that conversion to croplands and pastures of introduced species has mostly occurred and is most likely to take place in the future. The principal croplands of southern Australia are within the drier fringe of the temperate woodland zone, and on most farms livestock production is an adjunct to cereal growing.

The mean precipitation in Australia is less than that of any other continent, and the amount that runs off is also less. For most of the arid zone the average annual run-off is calculated to be less than 12.5 mm yr^{-1} (Anonymous, 1967). About 40% of mainland Australia is within the tropics; the rainfall in the northern one-third of the continent occurs mainly in summer (December to February) and that in the southern one-third mostly in winter (June to August). In between, there is a transition zone receiving rain in both winter and summer.

The highest rainfall is on the tropical north-east coast of Queensland, where South Johnstone receives an average of more than 4450 mm annually. The lowest rainfall is in the vicinity of Lake Eyre in South Australia, where the mean annual rainfall is about 100. Rainfall over most of Australia is highly variable, and the variability increases inland with decreasing rainfall. Except in very small areas, rainfall does not exceed potential evaporation in any month of the year (Hounan, 1961).

Most of the continent receives at least 3000 hours of sunshine each year. A feature of importance for forage quality, and hence animal production, is that frosts occur with varying frequency over the whole of Australia excepting the coastal strip, the northern part of the Northern Territory and north Queensland. This is of particular importance in subtropical Queensland, where the standing dry herbage of warm-season grasses declines rapidly in nutritive value after the first frosts of winter.

Australian climates were described in detail in "Climate and Meteorology, Australia" (Anonymous, 1965), and a publication on Australian climatology of particular relevance to grasslands and pastures is that of Fitzpatrick and Nix (1970).

THE GRASSLANDS

Xerophytic hummock-grass communities (*Plectrachne–Triodia*)

Xerophytic hummock-grass (spinifex) communities (Fig. 13.1, no. 1) occupy large expanses of strongly weathered Quaternary sand-plains formed by wind-sorting of sandy residuals of a former climate (Mabbutt, 1962). They were described by Perry and Lazarides (1962), and their composition in central Australia was related to soils by Winkworth (1967). Hubble (1970) described hummock grassland soils as mostly red earths and yellow earthy sands, occurring where the mean annual rainfall is between 200 and 400 mm.

Xerophytic hummock-grass communities are characterized by perennial species of *Plectrachne* and *Triodia* (Fig. 13.3), the so-called "spinifex" grasses, which form hummocks (mounds) varying in diameter from less than 1 m in some species to 6 m in *Triodia longiceps* (Perry, 1970a). In some species, notably *Triodia basedowii*, the centre of the hummock dies after some years, leaving an annulus of living shoots about 0.5 m wide; the whole, including the bare centre, is often as large as 2.5 m in diameter.

The structure of hummock grasslands and the distribution of the various species of *Plectrachne* and *Triodia*, and of the associated inter-hummock species, varies regionally and topographically. Although classified as true grassland (Williams, 1955; Moore and Perry, 1970), trees and shrubs 1–2 m tall — or more rarely taller trees — are scattered sparsely throughout much of these extensive grasslands (Perry, 1970a).

The spinifex grasses are harsh and, except for inflorescences and seedlings, are mostly unpalatable to livestock. The morphology of the highly lignified leaves of *Triodia* species was described by Burbidge (1946).

The principal value of spinifex grasses is their perenniality and the stability they give to the sandy soils on which they commonly grow. In the most arid of the hummock grasslands the space between the hummocks is bare, except after rain when annual grasses and forbs become established. Without these inter-hummock species, spinifex grasslands are of little value for livestock (Perry, 1970a).

In the most northerly part of their distribution zone, the hummock grasslands intergrade with eucalypt low woodlands and shrublands, the ground vegetation of which is similar to that of the adjacent hummock grassland.

The so-called "hard spinifex", mapped by Perry (1960), covers some 58 000 km^2 in central Australia, where the annual rainfall is between 300 and 350 mm. The dominant species are *Plectrachne schinzii* and *Triodia basedowii*, the latter in the south and the former north of latitude 22°S (Winkworth, 1967).

Types of hummock grasslands in the west Kimberleys (Western Australia) and their condition classes were described by Payne et al. (1974). In the Kimberleys and Pilbara in the north-west of Western Australia, *Triodia pungens* is the principal species. These grasslands were initially more productive than most other spinifex grasslands because of more palatable species of *Chrysopogon* and *Eragrostis* growing between the spinifex hummocks. Overgrazing, particularly in the Pilbara, has now largely eliminated these more nutritious species.

In both the Pilbara and the Kimberleys, individual properties are large (80 000–400 000 ha) and mostly unfenced. It is common to burn spinifex communities when mustering sheep for shearing in autumn, a practice which, according to Nunn and Suijdendorp (1954), destroys seeds produced by species growing between tus-

Fig. 13.3. A xerophytic hummock-grass community dominated by *Triodia intermedia*, with an overstorey of dwarf *Eucalyptus brevifolia* (in the Kimberleys of Western Australia). [Photo from the Commonwealth Scientific and Industrial Research Organization (CSIRO).]

socks[1] of spinifex during the summer; as a consequence, the ground remains bare for some time. Delaying burning until just before summer rains allows more palatable species — *Chrysopogon latifolius*, *Eragrostis eriopoda* and *Eriachne obtusa* — to re-establish from seed; but, unless grazing is controlled during the regeneration period, the young stand is damaged and its productivity is again reduced. If grazing is deferred until the growing season following regeneration, a mixed stand of spinifex and grasses of higher nutritive value develops and can be maintained for about three years. During this period the grassland is able to support breeding ewes. When *Triodia pungens* again becomes dominant, the community can

be grazed by wethers for 5 to 6 years, after which it is usually burned again. In a later publication, Suijdendorp (1981) claimed that greater productivity can be achieved by grazing in the summer following burning, if the grazing is controlled and the grassland is protected during the establishment of grass seedlings. Maximum production, 94 g m^{-2}, in a *T. pungens* grassland was recorded in the third year after burning. Burning during winter can result in the establishment of *Acacia translucens*, an unpalatable shrub.

Soil type is a major determinant of vegetation in arid Australia; as a result there is considerable variation within the broad vegetation types shown in Fig. 13.1. Pockets of finer-textured soils occur among the coarse-textured soils of the hummock grasslands, and these carry outliers of other grassland types. In the north-west of Western Aus-

---

[1] The term "**tussock**" is used in Australia and New Zealand to describe the bunch-grass growth form.

tralia, the tropical tall grasses *Dichanthium* spp. and *Themeda australis*[1] once occupied these finer-textured soils, but as a result of overgrazing these areas are now dominated by the unpalatable *Triodia longiceps* (Wilcox, 1972).

In the parts of the Kimberleys where annual rainfall is between 400 and 500 mm, clay soils in summer-flooded areas support xerophytic tussock grasslands of *Astrebla* spp. Sandy rises within the same area carry "pindan", a low woodland of species of *Acacia*, with *Lysiphyllum* (= *Bauhinia*) and *Eucalyptus*. The herbaceous vegetation of "pindan" is mainly *Plectrachne pungens*; *Chrysopogon fallax*, a xerophytic mid-grass species, is a frequent associate where the sands have a higher clay content. On valley floors between the granitic hills of the Pilbara and the Hamersley Range, *Chrysopogon fallax* is found with *Eragrostis* spp. and *Eriachne* spp. *Plectrachne schinzii* and *Triodia pungens* occupy the sandy slopes and *T. lanigera* and *T. wiseana* the shallower soils of the uplands (Wilcox, 1972).

Stock-carrying capacities of hummock grasslands vary with grazing-land type; the *Eragrostis–Chrysopogon fallax* type will support a sheep on 2 ha and can be used for breeding animals. *Triodia pungens* supports one dry sheep on 8 to 10 ha, but *T. lanigera* and *T. wiseana* grassland is of value only for the ephemeral herbage species present for a short period after rain (Wilcox, 1972).

Two introduced species, *Aerva javanica* [= *A.tomentosa* (kapok bush, a shrub)] and *Cenchrus ciliaris* (buffel grass) have become established in small areas of hummock grassland near Port Hedland in the north-western part of Western Australia (see "Xerophytic perennial pastures", p. 347).

In South Australia, spinifex grasslands extend south of latitude 31°S, the commonest species being *Triodia basedowii*. The most abundant annuals (in decreasing order of importance) in the inter-hummock spaces after rains are the grasses, *Dactyloctenium radulans*, *Aristida contorta* (= *A. arenaria*) and *Eragrostis* spp., and the forbs, the so-called "herbage" species, *Calandrinia balonensis*, *Salsola kali* and *Tribulus* spp.

Large areas of arid hummock grassland, particularly where *Triodia basedowii* is dominant, are waterless deserts and of little or no value for grazing. In the Simpson Desert (a sand-ridge desert), the most constant of the perennial species are *Triodia basedowii* and *Zygochloa paradoxa*. The soils and vegetation of this region were described by Crocker (1946), Boyland (1970) and Wiedemann (1971).

Xerophytic tussock-grass communities (*Astrebla–Iseilema*)

Arid tussock grasslands (Fig. 13.1, no. 2), composed largely of *Astrebla* spp. (Mitchell grasses), occur on extensive plains of grey, brown and red, cracking clay soils from the north of Western Australia to southern Queensland. They are the most productive of the arid grazing lands, occupying only 10% of the arid zone, but in most seasons carrying half of the livestock (Perry, 1970a).

Productivity, as in hummock grasslands, depends largely on annual grasses and forbs which grow in the inter-tussock spaces following summer rains. The commonest of the annuals grazed preferentially by livestock are (in order of declining abundance[2]): *Iseilema membranaceum* (Figs. 13.4 and 13.5), *Iseilema vaginiflorum*, *Dactyloctenium radulans*, *Brachyachne convergens*, *Boerhavia repleta* (= *B. diffusa*) and *Sida* spp.

Much of the tussock grassland is true grassland (Figs. 13.4 to 13.6), but there are large areas with scattered trees and shrubs, particularly in the Victoria River district of Western Australia. The common woody species in these north-western areas are: *Terminalia arostrata*, *T. volucris*, *Acacia farnesiana*, *Atalaya hemiglauca*, *Ventilago viminalis*, *Carissa lanceolata* and *Lysiphyllum cunninghamii*.

Astrebla pectinata (Fig. 13.6) is a dominant in much of the northern xerophytic tussock grassland including the Barkly Tableland (Fig. 13.1). The communities were described by Perry (1960). Individual tussocks are 20 to 45 cm high, 15 to 30 cm in diameter, and about 60 cm apart. Where the mean annual rainfall exceeds 375 mm, the other perennial grasses associated with *A. pectinata* are *Aristida latifolia*, *Astrebla elymoides*, *A. squarrosa* and

---

[1] The latest revision suggests that this species is not distinguishable from *T. triandra*.

[2] Where species are not listed in alphabetical order, the order given indicates their relative abundance or importance.

Fig. 13.4. Grazed xerophytic tussock grassland dominated by *Astrebla lappacea* and *Iseilema membranaceum* (near Cunnamulla, Queensland). (Photo from CSIRO.)

Panicum decompositum. The companion perennials, where the annual rainfall is less than 375 mm, are principally *Aristida latifolia* and *Eragrostis xerophila*. In depressions where more soil moisture is stored than on the plains, *Astrebla elymoides* forms communities with the more mesophytic tropical grasses *Bothriochloa* spp., *Dichanthium* spp. and *Eulalia aurea* (= *E. fulva*).

On the fine-textured clay soils characteristic of the *Astrebla* tussock grassland, the new season's growth does not commence until 40 to 50 mm of rain has fallen. The cycle of productivity in the northern tussock grasslands was described by Perry (1960) as follows: at the onset of the wet season, the perennial tussocks produce new shoots, and annuals germinate and establish in the inter-tussock areas. During this early growth period, the grassland is productive, and, as the food value of the annuals is high, cattle gain weight. Towards the end of the wet season, after the more nutritious annuals have been eaten, cattle have only the perennial tussock grasses to graze. *Astrebla* spp. by this time have matured and are so low in nutrients that cattle lose weight until the beginning of the next wet season.

Dry-matter yields of grass herbage in a northern *Astrebla* grassland are from 125 to 188 g m^{-2} (Perry, 1960). This compares with the 195 g m^{-2} maximum biomass recorded by Christie (1981) in the middle of the wet season in a less arid tussock grassland in Queensland.

Northern *Astrebla* grasslands are stocked at one beast[1] to 18 to 20 ha, but Perry (1960) believed

---

[1] In Australia "**beast**" refers to a cattle unit including bulls, cows and bullocks, generally more than 4 to 5 years old. "Bullock" is a castrated male more than 3 years old, too old to be called a steer. In arid Australia cattle are virtually feral — usually males, females, neuters and calves are not separated.

Fig. 13.5. Ungrazed xerophytic tussock grassland dominated by *Astrebla lappacea* and *Iseilema membranaceum* (near Cunnamulla, Queensland). (Photo from CSIRO.)

that, with sufficient watering points to ensure that cattle are never more than 5 km from water, the number of cattle carried could be trebled.

North of latitude 18°S, where the rainfall exceeds 635 mm annually, the tussock grasslands intergrade with more mesophytic tropical tall grasses. The commonest species of *Astrebla* in these grasslands is *A. squarrosa*, a robust species of lower nutritional value than *A. pectinata*. The carrying capacity of *A. squarrosa* grassland was estimated by Christian and Stewart (1954) to be one beast to 26 to 30 ha.

There are about 40 million ha of xerophytic tussock grasslands in Australia, of which 75% are in Queensland. *Astrebla lappacea* (Figs. 13.4 and 13.5) is generally the dominant species and is often associated with *A. elymoides* and *A. pectinata*, the former in the wetter sites. In years of high rainfall, *Dichanthium sericeum* (Fig. 13.7) is prominent, and it was the change in dominance from *Astrebla* in dry years to *Dichanthium* in wet that led Blake (1938) to promulgate his concept of a "fluctuating climax".

The influence of rain and its season of incidence on the composition of *Astrebla* grassland has been emphasized by Orr (1981), who found that botanical composition was influenced more by trends in seasonal rainfall than by grazing pressures.

In Queensland south of latitude 20°S, xerophytic tussock grasslands are grazed mainly by sheep. Grazing studies by Roe and Allen (1945) at Cunnamulla in south-western Queensland (mean annual rainfall about 300 mm) indicated that rain in summer was the principal factor governing herbage production. Winter rains, as in other arid and semiarid grazing lands, produced a "herbage" of comparatively short-lived but nutritious nongramineous species in the inter-tussock spaces.

Fig. 13.6. Xerophytic tussock grassland dominated by *Astrebla pectinata* in the Northern Territory. (Photo from CSIRO.)

The common forbs at Cunnamulla, *Boerhavia repleta*, *Ipomoea reptans*, *Portulaca oleracea*, *Salsola kali* and *Sida corrugata*, are comparatively high in crude protein, calcium and phosphorus.

In their grassland management study at Cunnamulla, Roe and Allen compared two rotational systems — winter grazing/summer rest and summer grazing/winter rest — with continuous grazing. Under a summer-grazing regime response of the grassland to summer rains was poor and the proportion of *Astrebla* species declined. There were no significant differences in live weight of sheep or in yield of herbage between plots grazed continuously and those grazed rotationally at comparable rates. Differences between stocking at a sheep to 3 ha and a sheep to 2 ha were slight, but at the heaviest rate (a sheep to 1 ha), the proportion of *Astrebla* in the maximum seasonal above-ground biomass was reduced. Continuous grazing at a sheep to 2 ha maintained animal productivity without grassland deterioration.

Everist (in Everist and Webb, 1975) cited work by himself and by others suggesting that damage to *Astrebla* grasslands may be avoided if they are not grazed heavily within 6 weeks of the commencement of new growth. In the same paper, he advanced a hypothesis that the *Astrebla* grassland ecosystem may have been created by aboriginal fires, followed by intermittent but intensive grazing by marsupials, and then maintained subsequently by European man and his domestic animals. This is an interesting hypothesis, but the fine texture of the soil and the consequent shallow depth of wetting in a low-rainfall/high-evaporation environment would seem adequate to account for the absence of trees (Moore, 1970a). This is supported by the fact that trees and shrubs are usually present in more favourable "run-on" areas within *Astrebla* grassland (where livestock concentrations

Fig. 13.7. True grassland dominated by *Dichanthium sericeum* and *Eulalia aurea* (near Emerald, Queensland). (Photo by W.H. Burrows.)

are likely to be highest) and in patches of soils of coarser texture.

Whatever their origin, *Astrebla* grasslands are now remarkably stable and, of all Australian plant communities, seem least altered by European man, despite heavy grazing during recurrent droughts. Species of *Astrebla* are not particularly resistant to defoliation because intensive grazing removes the lower nodes, from which axillary tillers normally develop (Jozwik et al., 1970).

In Scanlon's (1980, 1983) studies of the effects of burning *Astrebla* grassland, one of the interesting findings was that burning in the dry season increased the proportion of seed-bearing tillers in the following growing season. Regeneration of *Astrebla* grasslands by seedlings following death of tussocks is an important factor in their stability (Williams and Mackey, 1983).

The stability of tussock grasslands would seem to result largely from the "protection" afforded by the relatively low palatability of *Astrebla* spp. and from the resistance of the fine-textured clay soils to wind erosion. Damage occurs mainly around watering points and places where cattle concentrate. Where grazing is particularly heavy, *Aristida latifolia* increases at the expense of *Astrebla pectinata*.

The largest area of degraded xerophytic tussock grassland is in the Ord River catchment of the east Kimberleys in Western Australia. Here *Astrebla–Chrysopogon* grasslands have deteriorated as a result of 60 years of overgrazing. An Ord River Regeneration Project has been established, and attempts are being made to restore productivity by sowing perennial pastures of *Aerva javanica* and *Cenchrus ciliaris* (Fitzgerald, 1968). Small areas of *Cenchrus ciliaris* have been established also in parts of the tussock-grass grazing lands of Queensland (Fig. 13.2), and Orr and Holmes (1984) considered that there is some possibility of introducing exotic legumes where climates are more favourable.

Within *Astrebla* grasslands, particularly on the Barkly Tablelands but to a lesser extent also in other areas including Alice Springs, there are seasonally flooded lands in which *Chenopodium auricomum* (northern bluebush) is dominant. This community is mostly a treeless shrubland, but in places there is an overstorey of *Eucalyptus microtheca*. *Chenopodium* occurs in almost pure stands, particularly where the annual rainfall is less than 400 mm, but where rainfall is higher it intergrades with grasslands of *Astrebla elymoides* and *A. squarrosa*. *Chenopodium auricomum* is a palatable, nutritious and drought-resistant shrub 50 to 120 cm high, with a foliage cover from 50 to 100 cm in diameter. Bushes are commonly 125 to 250 cm apart. This species withstands heavy grazing, and Christian and Stewart (1954) estimated the carrying capacity of this community to be one cattle beast to 2.5 to 4 ha.

Acacia shrub–short-grass (*Eragrostis–Aristida–Eriachne–Enneapogon*) community

The *Acacia* shrublands (Fig. 13.1, no. 3) occupy about one-third of Australia's arid lands (Moore and Perry, 1970; Perry, 1970a). In most of these shrublands, the principal shrub is *Acacia aneura* (mulga[1]), a tall shrub or low tree, usually 3 to 7 m high but in places as low as 2 m and in others as tall as 10 m (Figs. 13.8 and 13.9).

The structure and distribution of communities of *Acacia aneura* in different parts of Australia have been described by several authors (Melville,

---

[1] "**Mulga**" refers to either this species or the community in which it dominates.

Fig. 13.8. An *Acacia*-shrub community. The dominant grasses are *Monachather paradoxa*, *Thyridolepis mitchelliana* and *Aristida* spp.; the woody components are *Acacia aneura* (mulga) and *Eremophila gilesii*. (Photo by W.H. Burrows, near Charleville, Queensland.)

Fig. 13.9. An *Acacia*-shrub–short-grass community. The dominant grasses are *Eragrostis* spp. and *Enneapogon polyphyllus*; the shrub is *Acacia aneura* (mulga) (in southern Queensland). (Photo from CSIRO.)

1947; Beadle, 1948; Everist, 1949, 1972; Jessup, 1951; Wilcox, 1960, 1963; Perry, 1960, 1970a, b; Perry and Lazarides, 1962; Burrows and Beale, 1969; Preece, 1971a, b; Boyland, 1973). The shrub layer varies; it is commonly mono-specific, but *A. cambagei* in Queensland, *A. papyrocarpa* (= *A. sowdenii*) in South Australia, *A. kempeana* in central Australia, *A. georginae* in the Northern Territory, and *A. pruinocarpa* in Western Australia are often associated with *A. aneura* in extensive areas of the *Acacia* shrubland. Shrub densities range from 1 to 8000 bushes ha^{-1}.

The shrub communities vary widely in floristic composition. Perry and Lazarides (1962) recorded 18 distinct ground-storey communities under *Acacia aneura*; the commonest of these was found under 12 other upper-storey communities.

Jessup (1951) recorded seven species of *Acacia*, three of *Cassia*, eight of *Eremophila* and nine chenopods in an *Acacia aneura* community in South Australia. The soil of this community was described as an arid red earth, but was later designated by Litchfield (1962) as shallow texture-contrast[1] with a siliceous hard-pan. Total soluble salts and pH of the surface soil were 0.2% and 8.8, respectively.

In the driest of the *Acacia* shrublands, north of latitude 24°S in central Australia, *A. aneura* commonly grows in groves receiving run-off from

---

[1] "**Texture-contrast**" is well-established in Australian soil terminology. These soils (alfisols) are characterized by marked differences in texture between the A and B horizons. The change in texture is usually clear and often abrupt. The A horizons vary from 2 to 50 cm in depth and range in texture from sandy loam to clay loam; the B horizons vary from clay loam to heavy clay. Red and yellow podzolics, solodics, solodized-solonetz, red-brown earths and non-calcic brown soils are all texture-contrast soils.

sparsely vegetated intergrove areas. Near Alice Springs (Northern Territory), Perry (1970b) found that 96% of the *Acacia* shrubs were within groves, and during one study period 95% of the grass–forb dry matter was produced in these groves. In effect, the inter-grove areas functioned as water-collecting surfaces for the groves.

Acacia aneura is frequently associated with smaller shrubs in communities having different values for grazing. Common species in these lower shrub layers in Western Australia include: *Eremophila forrestii* (= *E. leucophylla*), *E. fraseri*, *Acacia tetragonophylla*, *A. linophylla*, *A. sclerosperma* and *A. craspedocarpa*.

The principal grasses in Western Australian *Acacia* shrub–short-grass communities include: *Eragrostis eriopoda*, *Monachather paradoxa* (= *Danthonia bipartita*), *Eragrostis lanipes*, *E. dielsii*, *Eriachne helmsii* and *Aristida contorta*. The common forbs are species of *Calotis* and *Helipterum*, with *Calandrinia* spp. and *Ptilotus exaltatus* abundant after winter rains (Perry, 1960). On banks of sand overlying hard-pans, known as "Wanderrie banks", *Thyridolepis mitchelliana* is associated with *Monachather paradoxa* (Fig. 13.8), *Eragrostis xerophila* and *Eriachne helmsii* in the herbaceous communities under *Acacia aneura*. Overgrazing of this community results in dominance of the annual grasses *Aristida contorta* and *Eriachne aristidea* (D. Wilcox, pers. commun., 1976).

In some *Acacia aneura* communities in Western Australia, notably those in groves with an understorey of *Eremophila gilesii* (Fig. 13.8), the principal grass is *Eragrostis eriopoda* or *Eriachne helmsii*, but scattered hummocks of *Plectrachne melvillei* and *Triodia basedowii* sometimes occur. Where *Eremophila compacta* is the common undershrub, the latter two spinifex grasses predominate.

Stocking rates in communities of *Acacia aneura* in Western Australia vary widely, depending on the composition of the grass layer. Most properties run a sheep to 16 to 20 ha, but in others stocking is often as low as a sheep to 40 ha. Even at these low stocking rates *Acacia* shrub–short-grass communities have suffered from overuse. Land-use surveys in the Gascoyne River catchment (Western Australia) (Fig. 13.1) revealed that some 9400 km^2 were badly eroded and a further 33 000 km^2 were degraded and showed some erosion (Wilcox and McKinnon, 1972).

In the Northern Territory, *Acacia* shrublands are found on red earths, texture-contrast, and calcareous desert soils. Common shrubs, in addition to *A. aneura*, are *A. calciola*, *A. estrophiolata*, *A. kempeana* and *A. georginae*. These shrubs are palatable and, unlike *A. georginae*, which is poisonous under certain conditions, provide useful "top feed" for cattle (Perry, 1960).

The ground vegetation of *Acacia* shrublands in the Northern Territory consists of several short-lived grasses and forbs, the commonest of which (Perry, 1962) are *Aristida contorta*, *Enneapogon* spp., *Helipterum charsleyae*, *H. floribundum*, *Ptilotus helipteroides* and *Stenopetalum* spp. (Perry, 1962).

The herbage yield of *Acacia* shrub–short-grass communities varies widely with species, location, rainfall and shrub densities. In the *Acacia* shrub–short-grass communities of southern Queensland, carrying capacity is rated at a sheep to 3 ha or more. Studies at Charleville by Beale (1972) and Burrows (1972) indicated that net production of herbage in *Acacia aneura*–short-grass communities was of the order of 0.25 g m^{-2} day^{-1}. On the basis of 50% utilization by grazing animals, Beale (1972) estimated that the amount of herbage available for animal consumption was approximately 14 g m^{-2} yr^{-1}. In some parts of Queensland, shrubs in these communities are very dense and grass herbage production is reduced. At Charleville, Beale (1972) found that reducing the density of *A. aneura* from 160 to 15 shrubs ha^{-1} increased grass density by 33% at one site and by 20% at another.

In other studies in Queensland on land completely cleared of *Acacia aneura*, Ebersohn (1970) obtained a yield of herbage of 132 g m^{-2} in autumn and Christie (1978) a maximum mid-season yield of herbage biomass of 122 g m^{-2}.

Everist (1949) suggested that a density of 175 shrubs ha^{-1} in areas receiving 500 mm of rain annually would allow good production of herbage and, at the same time provide a satisfactory reserve of shrubs for feeding sheep in times of drought. In Queensland, *Acacia aneura* and other shrubs are lopped or, more commonly, pushed over with a bulldozer blade to give sheep access to foliage in droughts. The value of *A. aneura* for this purpose was discussed by Everist et al. (1958).

Cenchrus ciliaris has been established in parts

of an extensive area of *Acacia*–short-grass communities in Queensland, principally where *Acacia cambagei* (gidgee) is the predominant shrub (Fig. 13.2).

In parts of both South Australia and Western Australia, *Acacia* shrublands frequently have understoreys of species of *Atriplex* (saltbushes[1]) and other species of the shrub steppe. In the north-eastern corner of South Australia, *Atriplex rhagodioides* and *Chenopodium auricomum* are associated with *Acacia cambagei* in a community combining some of the characteristics of *Acacia* shrubland, shrub woodland and shrub steppe; the ground vegetation has affinities with *Atriplex*–xerophytic mid-grass communities.

Saltbush–xerophytic mid-grass communities (*Atriplex–Maireana–Chloris–Stipa*)

Shrub steppes, and the understoreys of some eucalypt shrublands and semiarid shrub and low woodlands, are classified as saltbush–xerophytic mid-grass communities (Fig. 13.1, no. 4) (Moore, 1970a). They are characterized by chenopodiaceous shrubs, mostly less than 1 m high, belonging to the genera *Atriplex*, *Maireana* (formerly included in *Kochia*), *Sclerolaena* (formerly *Bassia*), *Rhagodia* and *Enchylaena* (Figs. 13.10 and 13.11), and occupy extensive areas south of the Tropic of Capricorn where mean annual rainfall is between 125 and 400 mm, of which 30 to 50% falls in winter.

Densities of shrubs range from 15 to 500 ha^{-1} (Perry, 1970a). The space between the shrubs is usually bare in dry weather, but is covered with annual grasses and forbs after rains. The common grass species of the south-eastern saltbush–xerophytic mid-grass communities are *Danthonia caespitosa*, *Stipa variabilis* and *Chloris truncata*. The most prevalent forbs are species of *Calandrinia*, *Ptilotus* and *Sclerolaena*. In the more northerly areas the grass species are mainly those of the northern xerophytic–mid-grass communities (Fig. 13.1, no. 5): *Enneapogon polyphyllus* (Fig. 13.9), *E. avenaceus*, *Eragrostis dielsii*, *Aristida* spp., *Monachather paradoxa* (Fig. 13.8) and *Dactyloctenium radulans*.

---

[1] Most species of *Atriplex* are referred to as "**saltbushes**" in Australia.

Ecological studies of the effects of grazing on shrub steppes have been reported by Osborn et al. (1932), Beadle (1948), Carrodus and Specht (1965), Carrodus et al. (1965), Leigh and Noble (1969), Perry, 1970a), Barker and Lange (1970), Everist (1972) and Leigh (1972).

Atriplex vesicaria (Figs. 13.10 to 13.12) is one of the most widespread and important species of the shrub steppe. Aspects of its biology and that of some other species of *Atriplex* were discussed by various authors, notably Beadle (1952), Carrodus (1962), Carrodus and Specht (1965), Leigh and Wilson (1970) and Leigh (1972).

Between 1945 and 1950 in north-western South Australia, Jessup (1951) assessed the condition of an *Atriplex vesicaria–Maireana sedifolia* (= *Kochia sedifolia*) (Fig. 13.10) community within the dog fence erected, in South Australia, and joined to similar fences in New South Wales and Queensland, to keep dingoes (*Canis familiaris dingo*) out of sheep-grazing lands. The assessment was based on densities of the perennial species of *Atriplex* and *Maireana*. A re-survey by Lay (1972) indicated that in most of the area shrubs had died even when stocked conservatively at less than one sheep to 10 ha. However, there were some areas in which there had been no appreciable loss of shrubs even though grazing had been continuous and just as heavy as where shrubs had died. Death of shrubs was invariably greatest around watering points, where trampling and grazing pressures were most intensive. It was concluded that the extent of shrub loss depended on the number of animals using a watering point. In one station where there were 300 to 350 sheep per watering point, there were few shrub losses over 22 years. In others, where watering points were fewer and densities of sheep per watering point were higher, *Atriplex* spp. and *Maireana* spp. had not regenerated.

Lay (1972) confirmed Jessup's (1951) observation that *Atriplex vesicaria* was more sensitive to grazing than either *Maireana astrotricha* or *M. sedifolia*. Jessup had noted that the population of *Maireana* spp. on grazed areas increased at the expense of *Atriplex* spp. He observed also, that *Atriplex* spp. regenerated from seed more prolifically than *Maireana* spp.; consequently, densities of *Atriplex* spp. increased relative to *Maireana* spp. in rested areas.

Fig. 13.10. A saltbush–xerophytic mid-grass community dominated by *Atriplex vesicaria* and *Maireana sedifolia* (in the Northern Territory). (Photo from CSIRO.)

Atriplex communities on texture-contrast and coarse-textured calcareous soils seem particularly sensitive to overuse (Condon, 1978). There is evidence to suggest that losses of the shrub components of saltbush–xerophytic mid-grass communities on coarse-textured soils occurred soon after pastoral occupation, notably during droughts when grazing pressures on shrubs are highest. Reduction of shrub cover on coarse-textured soils leads to loss of surface soil (sand drifts) and in many instances to exposure of the clay layer (scalds) (Fig. 13.12). Degradation of grasslands in this way in the low-rainfall parts of South Australia was described by Ratcliffe (1936).

At Koonamore (South Australia) re-establishment of *Atriplex vesicaria* was slow, even in the absence of livestock, where surface soil had been lost (Hall et al., 1964). Stannard and Condon (1958) found that revegetation of eroded scalds in western New South Wales took place only after long wet periods, and even then was slow.

In much of the saltbush–xerophytic mid-grass communities of Western Australia, there have been losses of the more palatable species of *Atriplex* and *Maireana* (bluebushes[1]), and increases of *M. pyramidata* and *Frankenia* spp., which are unpalatable. The annual *Atriplex lindleyi* subsp. *inflata* is another indicator of overuse.

In places, saltbush–xerophytic mid-grass communities form the lowest stratum in shrub woodlands. For example, in parts of the south of Western Australia and New South Wales chenopodiaceous species form a low shrub layer beneath woody species of the genera *Eremophila*, *Acacia*, *Heterodendrum* or *Myoporum*. These species in

---

[1] Most species of *Maireana* are referred to as "**bluebushes**" in Australia.

Fig. 13.11. A saltbush–xerophytic mid-grass community. The dominant shrub is *Atriplex vesicaria* and the tree in the background is *Eucalyptus salmonophloia* (in Western Australia). (Photo from CSIRO.)

turn may be under an upper stratum of species of *Eucalyptus* or *Casuarina*.

In the Eastern Goldfields of Western Australia, saltbushes and bluebushes, mainly *Atriplex vesicaria* and *Maireana sedifolia*, form an understorey in *Eucalyptus salmonophloia–E. transcontinentalis* communities (Fig. 13.10) in which *Acacia* spp., *Casuarina cristata* and *Eremophila* spp. are in a shrub layer of medium height. This community has affinities in structure and composition with the shrub woodlands of eastern Australia (Moore et al., 1970). It is interspersed with shrublands of *Eucalyptus oleosa* (mallee[1]), which also have understoreys of species of *Atriplex* and *Maireana*. The mean annual rainfall here is about 250 mm, and the carrying capacity of the *Atriplex*–xerophytic mid-grass type is about a sheep to 12 ha.

In the more northerly saline alluvial plains of the Wiluna region of Western Australia, *Maireana pyramidata* (= *Kochia pyramidata*) is the understorey in an *Acacia aneura* community, which farther south usually has a well-developed grass layer and relatively few chenopods. In the northern community, the principal grasses are *Aristida contorta* and *Eragrostis dielsii*; the forbs are commonly *Ptilotus exaltatus*, *P. alopecuroideus*, *Sclerolaena* spp. and semi-succulents, such as *Calandrinia* spp. *Maireana pyramidata* is associated also with the shrub *Grevillia maculata* in a community in which *Atriplex rhagodioides* and *A. vesicaria* are also common. The main grasses are *Eragrostis eriopoda*, *Monachather paradoxa* and the annual *Aristida contorta*.

Atriplex nummularia, the tallest of the chenopod shrubs, is found mainly in south-eastern and higher-rainfall parts of the shrub steppe. The bushes commonly reach a height of 185 cm and have foliage diameters of 125 to 217 cm. In the Riverina of New South Wales (Fig. 13.1),

---

[1] Several species of short (3- to 6-m high) multi-stemmed shrub-like species of *Eucalyptus* are referred to as "**mallee**". The term also applies to the community in which these species dominate.

Fig. 13.12. A saltbush–xerophytic mid-grass community in which "scalds" are developing. The dominant shrubs are *Atriplex vesicaria* and *Maireana aphylla* (Riverina region, New South Wales). (Photo from CSIRO.)

this species once formed an extensive shrub community in association with *Acacia pendula*, but has now largely disappeared as a result of grazing. *A. pendula* was seemingly the dominant on red-brown earths, and *Atriplex nummularia* on grey and brown clays (Moore, 1953). Associated small shrubs were commonly *Maireana aphylla* (Fig. 13.12), *Rhagodia spinescens* and *Enchylaena tomentosa*. The herbaceous layer is a *Stipa variabilis–Danthonia caespitosa* grassland. The proportions of *Stipa* and *Danthonia* vary with soil texture; *Stipa* predominates on the coarser-textured and *Danthonia* on the finer-textured soils. *Chloris truncata* is common on both soil types where grazing pressures have been heavy.

In the south-eastern shrub steppe, erosion commonly exposes the clay of the subsoil following loss of protection by shrubs; the indicators are *Sclerolaena paradoxa*, *S. divaricata*, *S. ventricosa* and *S. obliquecuspis* (Moore, 1959, 1962). A high presence of *Nitraria billardieri* is also an indicator of overuse of south-eastern saltbush grazing lands (Beadle, 1948; Noble and Whalley, 1978a, b).

Leigh and Mulham (1964, 1966a, b, 1967) studied the dietary preferences of sheep in an *Atriplex*–xerophytic mid-grass type in south-eastern Australia. The presence of the chenopod shrubs *Atriplex vesicaria* and *Maireana aphylla* in a *Stipa–Danthonia* grassland appeared to confer little direct benefit to animal production. Examination of the extrusa from oesophageal fistulae revealed that sheep preferred grass and other species to *Atriplex vesicaria* and *Maireana aphylla* (Leigh and Mulham, 1966a). A further study by Leigh and Mulham (1966b) of a *Maireana aphylla–Stipa variabilis–Danthonia caespitosa* community confirmed the preference of sheep for grasses and species other than those of the family Chenopodiaceae.

Leigh et al. (1968) removed the shrubs mechanically from half an experimental area of a *Maireana aphylla*–grassland community with an initial shrub density of 875 ha^{-1}. Removal had no effect on body weight or wool growth of sheep grazing at either 1.8 or 0.9 ha^{-1}; at the higher stocking rate *M. aphylla* was eaten in the control plots, but apparently made no difference to animal production.

In a comparison of wool production from an *Atriplex vesicaria* shrub community with that from a *Stipa–Danthonia* grassland without shrubs, Wilson and Leigh (1970) found that at a sheep density of 1.2 ha^{-1} more wool was produced in the first year from the shrub community, but in the second year production was half that of the grassland because of the failure of *Atriplex* spp. to recover fully from the initial grazing. In another study, Wilson et al. (1969) found that *A. vesicaria* was killed by complete defoliation; it was eliminated in 3 years by sheep grazing at a rate of 2 to 5 ha^{-1} and reduced to 2% of its initial population when grazed by sheep at 1.2 ha^{-1}. Effects of defoliation on the longevity of *A. vesicaria* are summarized by Leigh and Mulham (1971).

Despite the preference of livestock for grass, shrubs are valuable components of saltbush–xerophytic mid-grass communities. In the more arid and more extensive shrub steppe, grass is normally less abundant than in the Riverina outlier in southern New South Wales where the above studies were performed. *Atriplex* spp. are eaten to a greater extent in arid communities, and increase the nutritive levels of the forage intake, particularly when grasses are dry and of poor quality. Of even greater importance are the increased stability which shrubs contribute to soil and vegetation, and the reserve of feed which they provide for livestock in dry periods. There are many areas in the southern arid zone where continuing animal production would seem to depend largely on saltbushes and bluebushes.

Northern xerophytic mid-grass (*Aristida–Chloris–Bothriochloa*) communities

These grasslands (Fig. 13.1, no. 5) are the herbaceous communities of the extensive semiarid low woodlands and shrub woodlands which together form the northern arc of the semiarid crescent between the agricultural lands and the arid interior (Figs. 13.13 and 13.14).

In the Northern Territory and the northern part of Western Australia where the winters are virtually rainless, semiarid woodlands occur mainly between the 500- and 750-mm annual isohyets, but are found also in some areas receiving only 350 mm. The soils are red and yellow earths which have gradational or uniform texture profiles. Shrub woodlands and low woodlands have been described, in the Northern Territory by Perry and Christian (1954) and Perry (1960), and in the north of Western Australia by Speck et al. (1964).

West of longitude 138°E in northern Australia there are low woodlands of *Eucalyptus argillacea*, *E. brevifolia* (Fig. 13.3), *E. dichromophloia* and *E. pruinosa*. The trees are often only 6 m tall, and the sparse herbaceous understorey is characterized by a high constancy of species of *Aristida*, which are high in fibre and low in nutritive value. Carrying capacities of grasslands in which species of *Aristida* predominate are as low as 1 cattle beast to 65 ha (Perry, 1960).

The dominant perennial grass in *Eucalyptus dichromophloia* low woodlands of the Northern Territory is *Aristida pruinosa*. Between the perennial tussocks of this species there are annuals, such as *Sporobolus australasicus* and short-lived species of *Aristida*. The growing period of these grasslands is from 15 to 20 weeks, and their stock-carrying capacity is about one beast to 22 ha. The carrying capacity of the understorey grasslands of *Eucalyptus argillacea* and *E. microneura* low woodlands of northern Queensland is of the same order (Perry, 1964).

Towards their northern limits in the Northern Territory and Queensland, xerophytic mid-grass communities of the low woodlands merge with tropical tall-grass of the higher-rainfall area. In their drier margins they intergrade with xerophytic hummock-grass communities of *Triodia mitchellii* or *T. pungens*. The maximum carrying capacity of these ecotonal grasslands, in the wet season (3 months), is only a beast to 50 ha (Perry, 1964) and their main value for livestock is their rapid growth response to light falls of rain.

Within the northern xerophytic mid-grass communities, there are tracts of rugged landscape inaccessible to grazing animals. There are also flood-plains (the so-called "frontage woodlands") with much higher carrying capacities than the surrounding country.

Fig. 13.13. *Sorghum* spp. and annual *Schizachyrium fragile* as an understorey in woodland (a northern xerophytic mid-grass community, Cape York, Queensland).

South of the Gulf of Carpentaria, rainfall in the xerophytic mid-grass communities is 500 to 750 mm annually, and the growing period of *Aristida* spp. is about 20 weeks in summer. The carrying capacity is one beast to 22 ha (Perry, 1964).

Farther south in Queensland near latitude 24°S, the woodlands are of *Eucalyptus melanophloia* (Fig. 13.14), *E. populnea* and *E. similis*, and the principal grasses are *Bothriochloa decipiens*, *Eriachne* spp. and *Aristida* spp. Sheep as well as cattle are run, and the average carrying capacities are a wether to 2.5–6.5 ha or a beast to 16–28 ha. To the west, *Triodia mitchellii* increases in density, and here the carrying capacity declines to one beast to 250 ha (Story, 1967).

In southern Queensland and northern New South Wales, sheep for wool form the main grazing enterprise in the shrub woodlands and, although cattle numbers increased with the decline in wool prices in 1973–74, they declined subsequently when wool again became more profitable. The principal tree species is *Eucalyptus populnea*, with *Callitris columellaris* commoner on the sandier rises. Common grasses include:

Bothriochloa decipiens	*Thyridolepis mitchelliana*
Chloris acicularis [1]	*Monachather paradoxa*
Enneapogon polyphyllus	*Paspalidium constrictum*

Mean annual rainfall is of the order of 350 mm; carrying capacities are a sheep to 1.5–3.5 ha in land cleared of trees, and 5–6 ha in densely timbered country.

The vegetation and soils of the *Eucalyptus populnea* semi-desert woodlands of the Bollon district

---

[1] Throughout this chapter, *Chloris acicularis* is referred to in the broad sense (*sensu lato*). *Enteropogon acicularis* and *Chloris ramosa* now have been split off from *C. acicularis*.

GRASSLANDS OF AUSTRALIA

Fig. 13.14. A northern xerophytic mid-grass community in *Eucalyptus melanophloia* woodland (in central Queensland). The dominant grasses are species of *Aristida* and *Bothriochloa*.

in southern Queensland were described by Holland and Moore (1962).

High densities of trees and shrubs, particularly of *Eremophila mitchellii* and of *Eucalyptus populnea*, create problems in animal husbandry and are among the principal factors limiting livestock production (Moore, 1972). Tree densities are commonly 140 ha^{-1} and are often as high as 400 following re-establishment by seed after earlier ring-barking[1].

Methods of thinning *Eucalyptus populnea* shrub woodlands, without disturbance of the soil or the herbaceous layer, were described by Robertson (1966) and Robertson and Moore (1972).

The unpalatable *Eremophila mitchellii* is the most widespread shrub species; its control by chemicals was demonstrated by Robertson (1965).

A strong positive relationship between percentage thinning of trees and shrubs and herbage yield of grasses was shown by Walker et al. (1972). Although there is an immediate herbage response to thinning, there is also — on the coarser-textured soils particularly — a gradual increase in shrubs, particularly of the genera *Eremophila*, *Acacia*, *Cassia*, *Dodonaea* and *Myoporum*. Moore and Walker (1972) showed experimentally that regeneration of shrubs could be controlled by surface-sowing of an exotic species (*Cenchrus ciliaris*), grazing and burning. *Cenchrus*, when established, provided fuel for dry-season burning for control of the regenerating shrubs. Another interesting feature of the experiments was that shrubs commonly regarded as unpalatable — and some known to be poisonous when mature — were eaten readily and apparently without ill effects

---

[1] "**Ring barking**" or girdling refers to cutting a ring of bark (including the living phloem) 10 to 15 cm wide completely around the trunk of a tree to prevent downward translocation of nutrients from leaves to roots. This causes the death of the trunk and canopy, but many eucalypts resprout later from lignotubers.

Fig. 13.15. Stems of *Acacia aneura* (mulga) that have been pushed over for use in feeding sheep during drought (in an *Acacia*-shrub–short-grass community in Western Australia).

when resprouting or in the seedling stages of growth.

Some of the edible woody plants of the shrub woodlands, notably *Brachychiton populneus*, *Canthium oleifolium*, *Casuarina cristata*, *Heterodendrum oleifolium* and *Ventilago viminalis*, which are mostly inaccessible to animals when mature, are lopped or pushed over with bulldozers to provide feed for sheep and cattle during droughts (Fig. 13.15). Trees and shrubs are generally fairly high in crude protein (10–15%), but are low in digestibility and phosphorus content (Everist and Young, 1967).

Pastures of introduced species, mainly *Cenchrus ciliaris*, have been sown, usually with minimal soil preparation, in the eastern and higher-rainfall sections of the semiarid woodlands and shrublands of Queensland and some parts of northern New South Wales (Fig. 13.2).

Southern xerophytic mid-grass (*Stipa–Chloris–Aristida*) communities

The *Eucalyptus populnea* woodlands, which in Queensland have a well-developed shrub layer, separate into two types south of the New South Wales–Queensland border. To the east the community becomes less shrubby and resembles the adjacent temperate woodland of *E. microcarpa* (=*E. woollsiana*). In the drier western part the shrub layer predominates, and *E. populnea* is confined largely to drainage lines and "run-on" areas receiving additional water. The herbaceous understorey of the eastern *E. populnea* woodland might justifiably be classed as temperate short-grass grassland, but because of the high constancy of *Aristida jerichoensis* and of chenopod species it was classified as a southern xerophytic mid-grass community (Fig. 13.1, no. 6). (Moore, 1970a).

The autecology of some of the common herbaceous species of the eastern portion of the southern xerophytic mid-grass communities was studied by Biddiscombe et al. (1954). In earlier work in a woodland at Trangie (New South Wales, mean annual rainfall 445 mm) Biddiscombe (1953) found that *Stipa falcata*[1] increased in density at a stocking rate of 1 sheep to 0.8 ha, but as the intensity of grazing increased it was replaced, first by *Chloris acicularis* and *Stipa setacea*, and then — at 1 sheep to 0.2 ha — by *Chloris truncata*. On adjacent finer-textured soils of a former *Acacia pendula* shrub woodland, *Stipa aristiglumis* was replaced at the higher stocking rate, first by *Sporobolus caroli*, then by *Stipa setacea*, and finally — as on the coarse-textured soils of the *Eucalyptus populnea* woodlands — by *Chloris truncata*. At the *Chloris truncata* stage and even earlier, the two grassland types are commonly invaded by volunteer Mediterranean annuals (*Hordeum leporinum*[2] and *Medicago* spp.) (Fig. 13.1, no. 6b).

Sheep and wool production in a *Stipa falcata*[1]–*Chloris acicularis*–*Digitaria coenicola* disclimax grassland of a former *Eucalyptus populnea* woodland were measured by Biddiscombe et al. (1956) at Trangie (New South Wales). In the experiment three rates of stocking with 2-year old merino wethers, 1 animal to 0.4, 0.6 and 0.8 ha, were combined with three grazing-management systems (continuous, autumn-deferred and spring-deferred). During the five years of the study, rainfall was above average and there was only one period when sheep at the highest stocking rate lost more weight and produced less wool than those at the lowest rate. Sheep live weights and wool production were no higher in deferred than in continuously grazed plots at the same stocking rates.

In the arid west of New South Wales the xerophytic mid-grass communities of *Eucalyptus populnea* woodland intergrade with *Acacia* shrub–short-grass communities in two types of landscape known popularly as "soft red" and "hard red". The former is characterized by alluvial sandy brown acid soils, and *Acacia aneura* with *A. excelsa* or *Heterodendrum oleifolium* and some *Eucalyptus populnea*. The "hard red" country is hilly and undulating; the common shrubs are: *Acacia homalophylla*, *Eremophila mitchellii*, *Cassia artemisioides*, *Eremophila longifolia*, *Dodonaea* spp. and *Capparis mitchellii*. The "hard red" soils are acid red-brown loams and clay loams with compact subsurface horizons. The grasses are principally *Eragrostis dielsii*, *Thyridolepis mitchelliana*, *Aristida contorta* and *A. jerichoensis*, with varying proportions of *Sclerolaena convexula* and *S. uniflora*. Carrying capacities of "hard red" country are between 0.17 and 0.25 sheep per hectare. "Soft red" country responds quickly to light falls of rain; the characteristic grass is *Eragrostis eriopoda*, which is commonly associated with *Aristida contorta*, *A. jerichoensis* and *Enneapogon* spp. The carrying capacity varies between 0.2 and 0.33 sheep per hectare (Moore et al., 1970). In overgrazed areas *Eragrostis eriopoda* is replaced by *Aristida contorta* and the forbs *Blennodia lasiocarpa*, *Calotis* spp. and *Helipterum* spp.

The grasslands of the more arid of the eucalypt shrublands ("mallee") are classed also as xerophytic mid-grass communities (Fig. 13.16, but vary widely in composition. They include sparsely grassed lands of little or no grazing value, elements of xerophytic hummock grasslands, and, in Western Australia, include many species of the saltbush–xerophytic mid-grass communities (Fig. 13.1, nos. 1 and 4). The latter include *Atriplex stipitata*, *A. vesicaria*, *Maireana sedifolia* and species of *Sclerolaena*, and the grasslands are ecotonal with saltbush–xerophytic mid-grass vegetation. Carrying capacities of "mallee"-type xerophytic mid-grass communities range from 0.12 to 0.29 dry sheep per hectare. On coarse sandy soils the *Atriplex* and *Maireana* species give way to *Triodia irritans* (Fig. 13.16), a species of very low value for grazing. Where *T. irritans* dominates, as much as 15 ha may be required to support one dry sheep. Mallee grazing lands have been described by Choate (1989). Sizes of land holdings (sheep "properties" or "stations") in arid mallee country average about 26 000 ha, and the principal output is wool.

In some parts of the less arid mallee country, grazing values are much enhanced by the presence of naturalized annual species of *Medicago* (the so-called "burr medics") and *Hordeum leporinum* (Fig. 13.2).

---

[1] In a later revision this was determined as *S. variabilis*.

[2] *Hordeum leporinum* (barley grass) = *Critesion murinum* subsp. *leporinum*.

Fig. 13.16. A southern xerophytic mid-grass community of *Triodia irritans* and *Sclerolaena* spp. under *Eucalyptus salubris* (mallee) (in Western Australia). (Photo from CSIRO.)

Frontage woodlands of *Eucalyptus camaldulensis* and *E. largiflorens* occur along the main watercourses of the xerophytic mid-grass grasslands of eastern Australia. These are more productive than the adjacent grasslands and carry about a sheep to 2.5 ha (Moore et al., 1970).

Tropical tall-grass (*Themeda–Sorghum–Heteropogon* and *Bothriochloa–Dichanthium–Eulalia*) communities

The grassy understoreys of tropical woodlands (modified to different degrees by man and his grazing animals), and associated patches of treeless grasslands on fine-textured soils, constitute the tropical tall-grass communities (Fig. 13.1, no. 7) which extend in an arc across northern Australia from Western Australia (Fig. 13.17).

Tropical and subtropical grasslands have been discussed by Mott and Tothill (1984) and Tothill and Mott (1985). The composition of these grassland communities varies with soil type, rainfall, and length and seasonality of the wet season. Mott et al. (1985) distinguished three types of tropical tall-grass communities (subhumid savannas) — monsoon tall-grass (the monsoon grasslands of Carnahan, 1976), tropical tall-grass, and subtropical tall-grass communities.

In the Northern Territory and northern Western Australia, tropical grasslands extend from the coastal lowlands to the semiarid crescent.

Monsoon grasslands, so called because of the short (22 weeks) and distinctly seasonal growing season, are of two main types depending on the soil.

The principal species on texture-contrast soils is *Themeda australis*, which is commonly associated with the perennial *Sorghum plumosum*. This grassland was described by Perry (1960). It occurs in semi-deciduous woodlands of *Eucalyptus latifolia* and *E. tectifica* on red earths and yellow podzolic soils. Annual rainfall ranges from 625 to

Fig. 13.17. A tropical tall-grass community in partly cleared eucalypt woodland (near Rockhampton, Queensland). The dominant grass is *Heteropogon contortus*.

1250 mm. Species commonly associated with the two dominants include *Chrysopogon fallax*, *Heteropogon contortus*, *Sehima nervosum* and *Aristida pruinosa*. The inter-tussock areas often have a sparse cover of *Brachyachne convergens* and forbs.

Mott et al. (1985) stated that the above-ground biomass in monsoon grasslands, where *Themeda* predominates, is between 112 and 270 g m^{-2}. Perry (1960) had cited an annual herbage yield of the same order, 120 g m^{-2}, from tropical grassland in the Northern Territory. Perennial tropical grasses grow rapidly following rains in November and December and, except for a short period at the beginning of growth, the nutritive value of the grasses is low. In the dry winter season at Katherine (Northern Territory), cattle lose about 20% of their peak weight and their net annual live-weight gain is less than 50 kg (Norman and Arndt, 1959). Carrying capacity is less than 1 beast to 25 ha, and steers are usually 5 to 7 years old when slaughtered (Norman, 1965).

On soils of coarser texture (sands and skeletal soils) the principal species are *Sorghum intrans* and *S. stipoideum*. These tall annuals form the understorey of *Eucalyptus tetrodonta–E. miniata* woodlands in areas receiving annual rainfall in excess of 750 mm; they grow from 1.5 to 3 m in height and are usually associated with perennials, notably *Themeda australis* and *Heteropogon contortus*.

Where annuals predominate, nutritive value is low and carrying capacity is less than a beast to 65 ha. It is common practice to burn grasslands of annual *Sorghum* spp. at the end of the wet season, but Stocker and Sturtz (1966) showed that burning soon after the beginning of the wet season killed *Sorghum* seedlings, and the resultant ash provided a seed-bed in which the introduced annual legume

Stylosanthes humilis could be established by aerial sowing. This species is adapted to most tropical woodlands accessible to cattle north of latitude 24°S, except at their dry margins (see "Tropical annual pastures", pp. 346–347).

On volcanic soils *Themeda australis* is associated with species of *Bothriochloa* spp., *Dichanthium* spp. *Chrysopogon fallax* and *Sehima nervosum*. These grasslands have a higher nutritive value than the tall-grass communities on coarse-textured soils; but, as in much of the northern woodlands, extensive areas are not useable for livestock production because of rugged inaccessible topography and lack of water.

At the dry margins of the tall-grass region in the Northern Territory, *Plectrachne pungens* becomes common as aridity increases, and the tropical tall-grass community grades into hummock-grass vegetation (Fig. 13.1, no. 1).

In Queensland, the most northerly of the monsoon tall-grass communities have a very short growing season, and there is a change in dominance from *Themeda* and *Heteropogon* to mid-grass species of *Aristida, Eriachne* and the annual *Schizachyrium fragile* (Fig. 13.13). These grasslands, which have lower carrying capacities for cattle than the *Themeda–Heteropogon* type, are found in much of the woodland of Cape York Peninsula (Fig. 13.1).

South of the Gulf of Carpentaria on grey cracking clays, there are treeless grasslands of tropical tall grasses dominated by *Dichanthium fecundum, D. tenuiculum* and *Eulalia aurea* (Fig. 13.1). Annual species of *Iseilema* and *Sorghum* are common in the wet season. The carrying capacity of this grassland is a beast to 15 ha (Mott et al., 1981).

On coarser-textured soils in the same region, there are woodland communities with understoreys of tropical tall-grass species, sometimes mixed with xerophytic mid grasses and the less arid of the hummock-grass species. Among the most widespread of these northern woodlands are those of *Eucalyptus crebra* and *E. drepanophylla*. The herbaceous vegetation is composed of perennial tussock grasses 90 to 120 cm high. The principal grasses are *Bothriochloa ewartiana, Heteropogon contortus* and *Themeda australis*, with *Aristida* spp. becoming more common towards drier margins on coarse-textured soils and in ecotones with semiarid woodlands of *Eucalyptus brownii* and *E. microneura*. Carrying capacity is a beast to 13–14 ha (Perry, 1964).

Farther west, in areas with a mean annual rainfall of 500 to 700 mm and on fine-textured soils, woodlands of *Eucalyptus argillacea* have understoreys of *Aristida* spp., *Chrysopogon fallax, Heteropogon contortus, Sehima nervosum* and *Themeda australis*. These grasslands, too, are ecotonal between tropical tall-grass and xerophytic mid-grass communities; their carrying capacity is about a beast to 22 ha.

In more southerly parts of Queensland to about latitude 26°S, the indicator of tropical tall-grass grassland is *Heteropogon contortus* (Fig. 13.17).

This is the subtropical tall-grass community of Mott et al. (1985). In this so-called "spear-grass region", *Themeda australis* was formerly more common than at present but, as in temperate Australia, it has been replaced, though less completely and dramatically, by other species as a result of grazing.

As in summer-rainfall areas generally, the pattern of herbage production in grassland dominated by *Heteropogon contortus* is of rapid growth in summer, followed by maturation and decline in nutritive value in autumn, and a further drastic fall in value in winter, when cattle lose weight (Shaw and Norman, 1970).

In central Queensland, Shaw and Bisset (1955) obtained herbage yields of *H. contortus* ranging from 250 to 750 g m^{-2}. Crude protein seldom exceeded 7%, and fell to 2 to 3% in winter months. Net live-weight gains of cattle varied, but were of the order of 54 to 136 kg yr^{-1} (Shaw and Norman, 1970). Cattle from this region of central Queensland are not usually marketed until they are 3 to 4 years old. Farther north, the net annual live-weight gain of cattle in similar *H. contortus* grassland was of the order of 90 kg (Shelton, 1956). For the most part, livestock production on tropical tall-grass is in uncleared eucalypt woodlands.

Studies by Gillard (1979) and Walker et al. (1986) in subtropical woodlands showed that trees depress grass production within and beyond their projected canopy areas and that thinning of tree stands increased grass productivity. In a recent survey of the effects of reducing tree densities in several different woodlands, Scanlon and Burrows (1990) found that the lower the potential of a site for herbage production (for example, deficiency

Fig. 13.18. *Danthonia pallida* and *Dichelachne* spp. in eucalypt woodland (a temperate tall-grass community, New South Wales). (Photo by R. M. Moore.)

of soil moisture), the greater the depression of grass yields by trees and the greater the response to tree thinning. The inference is that increases in grass production from reduction in tree densities are more likely in the subtropics where dry periods within the summer growing season are more common than in the wet tropics where such occurrences are rare.

These studies revealed that the biomass of andropogonoid species increased while that of panicoid species decreased with reductions in the basal cover of trees. It is of interest that the species commonly sown in exotic pastures in tropical and subtropical woodlands are mostly panicoid — *Brachiaria*, *Panicum*, *Paspalum*, *Pennisetum* — whereas the native climax species are andropogonoid — *Dichanthium*, *Eulalia*, *Heteropogon*, *Sorghum*, *Themeda*.

Tropical tall-grass communities are normally fenced to segregate the cattle herd into sexes and age groups, but seldom for management of grasslands, which are uniformly stocked for most of the year. Cattle are run with little management other than dipping for control of *Boophilus microplus* (cattle tick). Steers are generally marketed when 4- to 5-years old. Except for relatively small areas sown to *Stylosanthes humilis*, and the provision of phosphate and urea licks, annual burning is the only form of grassland husbandry. In the most northerly areas — where the wet season is short — burning at the beginning of the dry season is often done to stimulate the perennial grasses to sprout again and so make use of any residual soil moisture. Farther south, burning is usually delayed until the end of the dry season; the purpose is to remove standing herbage and to allow cattle ready access to the new season's growth.

The use of fire as a management tool in tropical grasslands was discussed by Tothill (1971).

Within the *Acacia harpophylla* ("brigalow"[1]) lands of central Queensland, there are patches of true grassland on basaltic black earths. The principal species are the so-called "blue grasses", *Bothriochloa* spp. and *Dichanthium sericeum* (Fig. 13.7). These grasslands support a beast to 4 ha, but increasingly are being ploughed for winter and summer crops. When *Acacia harpophylla* forest is cleared, the land is of little value for grazing and is planted to various crops, including peanuts (groundnuts). The brigalow community was de-

---

[1] "**Brigalow**" refers to either this species or the community in which it dominates.

scribed by Coaldrake (1970) and methods of controlling regeneration of *A. harpophylla* have been discussed by Johnson (1964).

Temperate tall-grass (*Themeda–Poa–Dichelachne–Danthonia*) communities

Temperate rain forests, wet sclerophyll forests and heaths, like their tropical counterparts, are of little or no value for grazing except where the canopy is open enough to allow grasses, such as *Dichelachne* spp., *Poa labillardieri* and *Themeda australis*, to form a discontinuous ground layer (Figs. 13.1 (no. 8) and 13.18). The same species extend into the more open dry sclerophyll forests which occur on hilltops and slopes throughout the temperate woodland zone in southern Australia. Here, grass densities, though still very low, are higher than in wet sclerophyll forests, and one of the commoner species is the coarse tussock grass *Danthonia pallida* (Fig. 13.18). At best, mature grasses in forests and heaths provide only rough grazing; carrying capacities range from 0.6 to 1.8 sheep per hectare, depending on species composition and density. Grazing after burning, to stimulate new growth, reduces grass cover (Leigh and Holgate, 1979).

Disturbance of forests, by logging for fence posts, railway sleepers and firewood, has allowed ingress of introduced species, notably *Aira caryophyllea*, *Bromus sterilis*, *Hypochaeris radicata*, *Trifolium arvense* and *T. angustifolium*. These species are also of low value for grazing, and the basis of livestock production in forest and heath grazing lands is sown pastures of temperate perennials in cleared rain forests and wet sclerophyll forests, and of Mediterranean perennial and annual species of grasses and legumes in dry sclerophyll forests. These are discussed later in this chapter (p. 340).

Temperate short-grass (*Danthonia–Stipa–Enneapogon*) communities

Temperate woodlands extend from latitude 27°S in southern Queensland to the lower south-east of South Australia. There is a narrow strip north and south of Adelaide and a large area in the south-west of Western Australia. In Tasmania the woodlands are in the north-east and in the midlands as far south as latitude 42°S.

The rainfall regime in the woodland zone varies from pronounced winter dominance in the south-west of Western Australia to summer domi-

Fig. 13.19. Grazed temperate short-grass grassland composed of *Stipa falcata*, *Danthonia carphoides*, *D. auriculata* and *Enneapogon nigricans* (in the southern tablelands of New South Wales). (Photo by R.M. Moore.)

Fig. 13.20. A temperate short-grass disclimax community dominated by *Danthonia caespitosa* and *Stipa variabilis* (near Deniliquin, New South Wales). (Photo by R. Burdon.)

nance in northern New South Wales and southern Queensland, where, however, there is still a fairly high proportion of rain in winter (36% in south-eastern Queensland). The annual rainfall ranges from 400 to 760 mm in eastern Australia and from 500 to 690 mm in the south-west of Western Australia.

From southern Queensland southwards, temperate woodlands and grasslands (Figs. 13.19 and 13.20) occur on both the eastern and drier (western) sides of the 500-mm isohyet. They are found on five great soil groups: podzols; solods and solodized solonetz; red-brown earths; black earths; and grey and brown cracking clays. Podzolic, solodic and solodized solonetzic soils are the common soils of temperate woodlands in Tasmania and on the eastern and wetter side of the 500-mm isohyet, from central New South Wales to Victoria.

On the western and drier side of the 500-mm isohyet, the soils are mainly red-brown earths and grey and brown cracking clays. These finer-textured soils are leached to a lesser extent than the podzolics and solodics and contain calcium carbonate at least in their lower horizons (Moore, 1970b).

Temperate short-grass communities (Fig. 13.1, no. 9) are mostly the understoreys of temperate woodlands modified by clearing and grazing by livestock and rabbits (*Oryctolagus cuniculus*). Small areas of treeless grassland occur on both major soil types throughout the woodland zone. Originally both woodland understoreys and true grassland were composed mainly of taller warm-season grasses: *Themeda australis*, *Stipa bigeniculata* and *Poa labillardieri* on texture-contrast and other coarse-textured soils in the more mesophytic areas; and *Stipa aristiglumis* and *Themeda ave-*

nacea in the drier areas on fine-textured soils with calcium in their profiles (Moore, 1970a, b).

As a result of grazing pressures since European settlement, the tall warm-season species on texture-contrast soils have been replaced by the short cool-season species *Danthonia auriculata, D. carphoides, Stipa falcata*, and *Chloris truncata* (Fig. 13.19).

East of the 500-mm annual isohyet, the southeastern temperate grasslands of short, cool-season perennial species of *Danthonia* and *Stipa* support 2 to 4 sheep ha^{-1} year-long. These grasslands, which receive rain in both winter and summer, are relatively stable at these stocking rates. The grasslands, however, have been greatly modified by the regular use of superphosphate, accompanied by annual legumes, notably *Trifolium subterraneum*, either sown or as volunteers (Fig. 13.2). The degree to which the short-grass community has been changed varies from farm to farm, and within farms from paddock to paddock. As a result, there is a mosaic of grassland communities in which the proportions of introduced species relative to native species vary from 0 to 100%. In some farms perennial grass–annual legume pastures have been established by ploughing small areas and sowing the perennial grasses *Lolium perenne* or *Phalaris aquatica* (= *P. tuberosa*) (Fig. 13.2).

The woodland understoreys and true grasslands on more fine-textured soils on the drier side of the 500-mm annual rainfall line have affinities with the xerophytic mid-grass communities of the former *Acacia pendula* shrub woodlands. *Stipa aristiglumis* has persisted in some areas of finer-textured clay soils — notably the Liverpool Range and Macquarie region of New South Wales. On red-brown earths, both of these tall climax species are replaced under intensive grazing by shorter grasses, *Danthonia caespitosa, Stipa variabilis, Chloris acicularis, C. truncata, Sporobolus caroli* and *Stipa setacea* (Fig. 13.20). Volunteer species on previously cropped land are *Hordeum leporinum*, annual species of *Medicago*, and *Erodium cygnorum*.

In contrast to south-eastern Australia, the perennial grasses of the temperate woodlands of the south-west of Western Australia, where the rainfall is almost wholly during winter, disappeared rapidly following the introduction of domestic animals. Animal production in the south-west of Western Australia, in both forest and woodland areas, is now dependent entirely on sown or volunteer Mediterranean annuals, the principal species being *Trifolium subterraneum* and the volunteer annuals *Arctotheca calendula, Erodium botrys, Bromus rigidus* and *Vulpia myuros* (Rossiter and Ozanne, 1970). The carrying capacity of these introduced annual pastures is 5 to 7 sheep per hectare, depending on rainfall. As in the drier temperate woodland on less leached soils in eastern Australia, the main species in comparable woodlands in Western Australia are volunteer annuals, *Hordeum leporinum* and *Medicago* spp., the most widespread of which are *M. polymorpha* and *M. minima*.

It is because of the general occurrence of short, disclimax species and the scarcity of climax species that temperate woodlands and grassland communities are classified as temperate short-grass. Details of the changes resulting from grazing by livestock and rabbits and from application of phosphate fertilizer have been discussed by Moore (1959, 1962, 1966, 1967a, b, 1970a, b).

The true grasslands of the temperate areas are too small to be mapped at the scale of Fig. 13.1. The largest of them are west of Melbourne in the Western Districts of Victoria, west of Canberra in the Monaro Highlands, south and west of Tamworth on the Liverpool Plains in New South Wales and west of Lawes on the Darling Downs in Queensland, where there was originally a mixture of tropical and temperate species.

Subalpine sod tussock-grass (*Poa–Themeda–Danthonia*) community

Australian alpine and subalpine lands are restricted to small areas of New South Wales, Victoria and Tasmania. Their lower altitudinal limits are 1370 to 1525 m in New South Wales and Victoria and 915 m in Tasmania. Annual rainfall ranges from 750 mm to 2550 mm. The soils are acid.

The sod tussock-grass grasslands (Fig. 13.1, no. 10) are the understoreys of subalpine woodlands of *Eucalyptus pauciflora* subsp. *niphophila* (= *E. niphophila*) *E. stellulata* on the mainland, and of *E. coccifera–E. gunnii* in Tasmania. They are also the treeless communities of the high plains in the three States.

The woodland understoreys and the treeless grassland communities are similar floristically, the dominant grasses being species of *Poa*[1], with varying densities of *Danthonia nudiflora* and *Themeda australis*. In heavily disturbed sites, *Lolium perenne*, *Poa pratensis*, *Trifolium repens* and such introduced weeds as *Hypochaeris radicata*, *Rumex angiocarpa* (=*R. acetosella*) and *Taraxacum officinale* have become naturalized.

Historically, subalpine grasslands have been used for summer grazing by sheep and cattle. It was common practice to burn them, usually at the beginning of summer, to stimulate species of *Poa* to produce early green shoots. Today, more emphasis is being placed on water yields and conservation of the high country, and grazing is subject to restrictions, which vary from State to State. Grazing values of subalpine grasslands and damage resulting from overuse were discussed by Costin (1970).

Aquatic and saline (*Eleocharis–Fimbristylis–Sporobolus*) communities

A wide range of saline, brackish and freshwater communities occurs, mostly in very small areas (Fig. 13.1, no. 11). Where these are used, they are usually grazed as part of an adjoining grassland.

Among the largest is the seasonally inundated "grassland" of the coastal plains near Darwin (Northern Territory), and extending discontinuously eastwards to the Gulf of Carpentaria. These plains are the natural habitats of the introduced *Bubalus bubalis* (water buffalo). The freshwater communities are composed largely of species of *Eleocharis* and *Fimbristylis*, with varying proportions of *Leersia hexandra*, *Oryza australiensis* and *Pseudoraphis spinescens*.

In the numerous but individually small coastal saline communities, the common species are *Sporobolus virginicus*, *Suaeda australis*, *Arthrocnemum* spp., and *Xerochloa imberbis* or *X. barbata*. Other species often present include *Diplachne fusca* and *Leptochloa neesii* (=*L. brownii*).

LANDS WITHOUT NATIVE GRASSLANDS

Some forests, notably the rain forests, the denser of the wet sclerophyll forests, the brigalow (*Acacia harpophylla*), parts of the eucalypt shrublands (mallee), and the heaths and sedgelands, have few edible native species (Fig. 13.1, no. 12). They are of little or no value for livestock unless replaced by communities of sown pasture species or, as in the case of the open mallee communities, they are colonized by introduced forage species.

Some rain forests have been felled and, depending on location, sown to tropical or temperate pasture species. The brigalow has been cleared for crops and for sowing to xerophytic perennial pastures (Coaldrake, 1970). Large areas of heathlands in Queensland have been sown to tropical perennial pastures (Bryan, 1970) or, in temperate areas, to Mediterranean perennial-grass, annual-legume or Mediterranean annual pastures (Newman, 1970). Parts of the mallee have been cleared and sown to, or simply colonized by, Mediterranean annual pasture species (Barrow and Pearson, 1970; Rossiter and Ozanne, 1970).

As mentioned in discussing temperate short-grass communities (p. 344) (Fig. 13.1, no. 9) a large area of temperate forest and woodland in the south-west of Western Australia, where the rainfall predominantly falls in winter, is now virtually without native grasslands and is incapable of livestock production in the absence of introduced pasture species.

EXOTIC GRASSLANDS (PASTURES)

A major factor in increasing both crop and livestock production in Australia has been introduced legumes and, in many cases, concomitant use of fertilizers, particularly phosphorus. The accidental introduction and subsequent spread, both naturally and by sowing, of the now widely distributed annual species of *Medicago*, *Stylosanthes* and *Trifolium* has benefited animal production by increasing the amount and nutritive quality of the forage available during the growing season and the following dry season, both in temperate and tropical Australia.

Fig. 13.2 shows exotic pasture regions where species have been sown, or have become naturalized following deliberate or accidental introduction and, in many instances, land clearing and/or soil amendment. The proportions of these areas in which introduced species are established vary

---

[1] The snow grass *Poa* species are: *P. costiniana*, *P. fawcettiae* and *P. hiemata*.

but, excepting for the temperate grasslands, are generally small. Exotic grasslands range in floristic composition from sown pastures, in which all species are introduced, to mixtures of varying proportions of native and introduced species. Sowing pastures may involve complete destruction of the original plant community (as in croplands), partial modification by varying intensities of surface tillage (with or without removal of a woody plant overstorey), or surface broadcasting of seed and fertilizer without tillage of any kind.

Introduced species of value for animal production have become naturalized in native grasslands without the addition of fertilizers, as in the tropical tall-grass region and the temperate short-grass region (Fig. 13.1, nos. 7 and 9), or they have established themselves following the use of superphosphate, as in the higher-rainfall areas of the temperate short-grass region.

Woody legumes have been less successful. *Leucaena leucocephala* is promising in parts of central Queensland, but another introduction, *Acacia nilotica*, has become a serious weed problem in parts of the tussock grass community in eastern Queensland.

Despite the low nitrogen status of most Australian soils, little nitrogen fertilizer is applied. Symbiotic fixation with legumes is the main source of nitrogen, both for pastures and most field crops. Williams and Andrew (1970) estimated that pasture legumes were contributing 1 500 000 t of nitrogen to Australian soils annually. Much research has been devoted to the correct nutrition of legumes, and trace-element deficiencies, particularly of molybdenum, copper and zinc, have been found in several different soil types (Williams and Andrew, 1970; Rossiter and Ozanne, 1970).

Tropical perennial pastures

Tropical and subtropical rain forests, wet sclerophyll forests and heaths extend from about latitude 15°S in north Queensland to latitude 32°S in northern New South Wales. These communities are of little value for grazing unless cleared and sown to pastures of introduced species (Fig. 13.2, no. 1).

In the wet tropics north of latitude 21°S, where annual rainfall is as much as 3560 mm, high yielding introduced species of potential value for pastures include the following (Davies and Hutton, 1970):

GRASSES	LEGUMES
Brachiaria mutica	*Centrosema pubescens*
Panicum maximum	*Calopogonium mucunoides*
Digitaria decumbens	*Leucaena leucocephala*
Melinis minutiflora	*Desmodium heterophyllum*
Pennisetum purpureum	*Pueraria phaseoloides*

Carrying capacities of a cattle beast to 0.4–0.8 ha and live-weight gains of 730 kg ha^{-1} yr^{-1} have been achieved in perennial pastures in the wet tropics. However, costs of establishment and maintenance are high. Rain-forest soils, initially high in fertility, are rapidly leached, and costs of maintaining pasture production are generally too high for livestock industries.

In the high-rainfall areas (900–1500 mm annually) of the subtropics, the common pastures in cleared forest lands are of *Paspalum dilatatum* and *Trifolium repens*, or *Pennisetum clandestinum* and *Trifolium repens*. These pastures initially carry a dairy cow to 0.4 ha, but degenerate rapidly to a sward dominated by *Axonopus affinis*, *Lantana camara* and other tropical weedy species, which is capable of supporting only a cow to 1.25 ha (Bryan, 1970).

There are strong arguments on conservation, ecological and economic grounds for the reestablishment of forests in much of the former rain-forest lands cleared for dairying (Moore, 1976).

Tropical annual pastures

Pastures composed solely of annuals are rare in the tropics but *Stylosanthes humilis* (Townsville stylo), an introduced annual legume, is naturalized in many parts of the tropical tall-grass region. It is found commonly in grazed *Heteropogon contortus* grassland, from the Northern Territory to about latitude 25°S in Queensland, where the annual rainfall exceeds 635 mm (Fig. 13.2). The value of *Stylosanthes humilis* in raising protein levels of herbage and improving the nutrition of cattle in the dry season was shown by Norman and Arndt (1959). The potential of legumes, and specifically of *S. humilis*, for increasing beef production in tropical grasslands was demonstrated also by Shaw and Norman (1970) and Shaw and t'Mannetje (1970). Steers stocked at 0.8 to 1.2

ha per animal on annual pasture sown to this species reached slaughter weight in less than 3 years. In comparison, steers at a stocking rate of 1 to 3.6 ha per head on a native grassland of *Heteropogon contortus* took 4 to 5 years to reach marketable weight. Responses in animal production to be expected on leguminous pastures have been discussed by Gillard and Winter (1984) and by Tothill et al. (1985).

The continued use of *Stylosanthes humilis* for the improvement of tropical pastures for animal production received a setback through its susceptibility to anthracnose (*Colletotrichum gloeosporioides* (Irwin et al., 1984; Mott, 1986).

Fortunately there are other legumes of potential but, as yet, unproven commercial value in tropical grasslands. The most promising include the perennials *Stylosanthes hamata* cv Verano, which is tolerant of anthracnose, and *S. scabra* cv Seca (Fig. 13.21) which is resistant. *Macroptilium atropurpureum* is another legume of potential value (Tothill and Jones, 1977), particularly in central and southern Queensland where it appears to be better adapted than *Stylosanthes scabra* (Burrows, 1990).

Ecological and economic problems of establishing legumes in tall-grass communities have been pointed out by Mott (1986).

Xerophytic perennial pastures

Xerophytic perennial pasture species are found both in pure stands and in communities of native species. *Cenchrus ciliaris*, like so many of Australia's widely distributed and most successful pasture species, first came to notice as an accidental introduction. Subsequently, several cultivars have been introduced; these were described by Davies and Hutton (1970). This species is naturalized in many small areas of the northern arid and semiarid areas and is being sown to a small, but increasing, extent in the *Acacia* shrub–short-grass, xerophytic mid-grass and tropical tall-grass regions (Fig. 13.1, nos. 3 and 5–7), and to a lesser extent in parts of the hummock-grass and tussock-grass areas (Fig. 13.1, nos. 1 and 2).

Fig. 13.21. *Stylosanthes scabra* in *Heteropogon contortus* grassland (in central Queensland). (Photo by W.H. Burrows.)

Cenchrus ciliaris is adapted to clay soils with at least 25 ppm of total phosphorus (Christie, 1975) and has spread naturally in northern areas receiving 300 mm or more rain annually. In years of above-average rainfall, when forage is plentiful in shrub woodlands, *C. ciliaris* has potential for controlling unwanted woody plants by providing fuel for fires. A related species, *C. setigerus*, is naturalized in hummock-grass grassland near Port Hedland (Western Australia) (Fig. 13.2).

Cenchrus ciliaris (buffel grass) has been sown extensively following clearing of *Acacia harpophylla* (brigalow) and *A. cambagei* (gidgee). Cattle grazing *Cenchrus ciliaris* pasture initially gain about 0.5 kg (live weight) per head per day, but in the absence of an associate legume the daily gain declines to 0.4 kg (Burrows, 1990).

Other xerophytic perennial pasture species sown to a limited extent in more mesophytic areas are *Chloris gayana*, *Medicago sativa*, *Panicum maximum* var. *trichoglume* and *Sorghum almum*. Coaldrake and Smith (1967), in a year-long study of a pasture of *Panicum maximum* var. *trichoglume* and *Sorghum almum*, found that yearling steers gained 178 kg per head when stocked at 1 beast to 0.9 ha. Rainfall in the central brigalow region during the period of the experiment was 533 mm.

Temperate perennial pastures

Pastures of *Trifolium repens* (white clover) with *Lolium perenne* (perennial rye-grass) or *Pennisetum clandestinum* or *Paspalum dilatatum* are sown in cleared areas of temperate rain forests and wet sclerophyll forests and in temperate heaths from Tasmania northwards through Victoria to northern New South Wales. Average annual rainfall is above 900 mm, and is well distributed throughout the year.

Superphosphate is generally applied at the rate of 28 kg P ha^{-1} at the time of sowing, and regular annual applications are made subsequently of from 15 to 18.5 kg P ha^{-1}. The other nutritional amendments needed in various forest and heathland soils were discussed by Paton and Hosking (1970).

Dairying, fat-lamb production and cattle fattening are the main enterprises. The most productive dairy farms carry just under 2 cows ha^{-1}; butterfat production in southern New South Wales, Victoria and Tasmania ranges from 100 to 225 kg ha^{-1} yr^{-1}. In sheep properties, between 7 and 12 fat-lamb ewes are carried ha^{-1}. Beef production is from 110 to 450 kg (live weight) ha^{-1} yr^{-1}, and most cattle are sold when 18 to 24 months old and at 200 to 275 kg dressed weight (365 to 560 kg live weight) (Paton and Hosking, 1970).

Other pasture species sometimes sown are *Dactylis glomerata* and *Trifolium pratense*. In northern New South Wales and southern Queensland, where the proportion of the annual rainfall received in the summer is higher, *Paspalum dilatatum* is the common grass species in temperate perennial pastures.

Temperate perennial-grass–annual-legume pastures

The annual legume *Trifolium subterraneum* replaces the perennial *T. repens* as an associate of *Lolium perenne* where the annual rainfall is below about 650 mm — that is, in the lower-rainfall area of the temperate tall-grass region (Fig. 13.1, no. 8) and the higher-rainfall area of the temperate short-grass region (Fig. 13.1, no. 9) in Victoria and Tasmania (Fig. 13.2, no. 5a).

The establishment of perennial-grass–annual-legume pastures in dry sclerophyll forests is dependent on correction of soil-nutrient deficiencies of molybdenum, phosphorus and sulphur; in the dry heathlands, copper and zinc are required as well (Newman, 1970).

Acid soils of both forests and heaths require the addition of lime and the inoculation of seed with *Rhizobium* for successful nodulation and establishment of legumes (Newman, 1970).

Growth commences when autumn rains germinate the legume seeds. Herbage yield in winter seldom exceeds 68 g m^{-2}; peak growth occurs in the spring when 70 to 75% of the total annual herbage is produced in 10 to 12 weeks. Sheep farming for wool and fat-lamb production is the main enterprise. Because animals are carried year-long on pasture, the depression of plant growth in summer and autumn limit year-long carrying capacity to between 6 and 12 sheep ha^{-1}.

Phalaris aquatica (= *P. tuberosa*) displaces *Lolium perenne* in perennial-grass–annual-legume pastures where the rainfall is less than 610 mm annually in Victoria, Tasmania and southern New

South Wales (Fig. 13.2, no. 5b). Stocking rates in *Phalaris–Trifolium subterraneum* pastures average 8 sheep ha^{-1}, but there is evidence that, except on dry shallow soils, the pastures will support 12 sheep ha^{-1} without supplementary feeding with hay or other conserved forage. For comparison, the year-long carrying capacity of the native short-grass community is 2.5 to 4 sheep ha^{-1}.

Mediterranean annual pastures

As has been pointed out, throughout much of the higher-rainfall parts of the temperate short-grass region, where the soils are solodized solonetz and solodic, and acid in reaction, the character of the grassland has been changed by the incorporation of *Trifolium subterraneum* with minimal tillage and the addition of phosphorus and, on some soils, molybdenum and other trace elements. The effect of regular annual applications of phosphorus and continued grazing has been to increase soil nitrogen levels, decrease the densities of native perennial grasses, and increase numbers of volunteer introduced annuals (Moore and Biddiscombe, 1964). As a consequence, much of the short-grass area is in process of conversion to Mediterranean annual pastures (Fig. 13.2 and Table 13.1).

Species of high frequency in annual volunteer pasture communities in south-eastern Australia, in addition to *Trifolium subterraneum*, are *Bromus rigidus*, *Erodium cicutarium*, *Trifolium glomeratum*, *Vulpia bromoides* and, where fertility is high as a result of several years of clover growth and annual applications of phosphorus, *Carduus pycnocephalus* and other nitrophilous species.

A wide range of *Trifolium subterraneum* cultivars is sown, ranging in maturity times from early through mid-season to late, depending on the length of the rainfall season (Donald, 1970). Most frequently the legume is sown alone, but sometimes it is combined with the annual grass *Lolium rigidum*.

In eastern Australia, Mediterranean pastures based on *Trifolium subterraneum* have supported between 5 and 10 sheep ha^{-1} without loss of productivity per head.

In the south-west of Western Australia the commonest species other than *Trifolium subterraneum* are *Arctotheca calendula*, *Bromus rigidus*, *Bromus hordeaceus* (= *Serrafalcus mollis*), *Erodium botrys* and *Vulpia myuros*. The carrying capacity of this type of pasture at Kojonup (Western Australia), where the rainfall averages 560 mm yr^{-1}, was 12 wether sheep ha^{-1}, yielding 54 kg ha^{-1} of greasy wool (Rossiter and Ozanne, 1970). Average annual herbage yield of a Mediterranean annual pasture dominated by *Trifolium subterraneum* at Kojonup (6.5 months growing season) was 594 g m^{-2}.

In drier temperate woodlands (less than 500 mm annual rainfall) in both eastern and western Australia, *Hordeum leporinum*, *Medicago laciniata*, *M. minima*, *M. polymorpha* and *Vulpia bromoides* are naturalized on soils which are neutral or alkaline and which contain calcium carbonate in their profiles. These, with the native annual *Erodium cygnorum*, are the principal species of the Mediterranean annual pastures in temperate cereal-growing areas (Fig. 13.2, no. 6b). In recently cropped land these species, sometimes with a few other introduced annuals, are usually the only species present. In uncropped land the same species are present with the remnants of the native short-grass community.

Carrying capacities of *Hordeum–Medicago* pastures vary with rainfall and farming practices, particularly the application of phosphorus, but stocking rates of 1 to 2 sheep ha^{-1} are common in areas of 400 to 450 mm annual rainfall.

On the drier side of the 250-mm annual rainfall line, *Hordeum leporinum* and annual *Medicago* spp. extend into xerophytic mid-grass communities of the eucalypt shrublands and shrub woodlands. The *Medicago* species, in particular, are commonly found in saltbush–xerophytic-mid-grass and *Acacia* shrub–short-grass communities (Fig. 13.1, nos. 4 and 3), with native species of *Sclerolaena*.

The commonly occurring species of *Medicago* (the burr medics) have recurved spines on their seed pods which adhere strongly to wool and markedly reduce its commercial value. In mixed cereal-farming/livestock-producing areas, species of *Medicago* with straight non-adhesive spines have been sown to an increasing extent. Among these are *M. littoralis*, *M. rugosa*, *M. scutellata* and *M. truncatula*. *Lolium rigidum* is often sown with annual species of *Medicago* but, except on self-mulching soils, seldom re-establishes until the land is cultivated again for crops, when it is often a troublesome weed.

The ecology of Mediterranean annual pastures was discussed by Rossiter (1966), and the distribution of introduced temperate pasture species in relation to Australian climate and soils was examined by Donald (1970).

GRASSLAND COMMUNITY INTER-RELATIONSHIPS

The genera, and in some instances the species, characterizing the principal grasslands in the various geographical regions of Australia are shown in Fig. 13.1. Some genera have wide distributions, notably *Plectrachne* and *Triodia* in arid sand-plains, *Astrebla* on fine-textured cracking clays in regions of summer rainfall, and *Themeda* on a wide range of soils from the tropics to temperate and subalpine areas. Both *Danthonia* and *Stipa* are found in extensive areas of temperate Australia, as are the chenopods *Atriplex*, *Maireana* and *Sclerolaena*. Species of *Eragrostis* occur throughout arid and semiarid areas, and *Chloris*, *Enneapogon*, *Aristida* and *Chrysopogon* are widespread in both semiarid and mesophytic regions.

Species change, or their dominance varies, with the texture and depth of soil horizons and with topographic site, as well as with the amount of rainfall and its seasonality. This is particularly evident in arid areas. *Triodia pungens* is widespread in the arid hummock grasslands of northern Australia, particularly in the north-west where it is commonly associated with *Plectrachne pungens*. In central Australia, *Plectrachne schinzii* is the principal hummock-grass species to the north and *Triodia basedowii* is the main species to the south. These two species, which do not usually occur together, contribute 96 to 98% of canopy biomass on an area estimated to be 150 000 km² (Winkworth, 1967).

In the margins of their ranges, and particularly in arid areas where there are gradual changes in depth and texture of soils, grasslands become more varied in floristic composition. *Plectrachne pungens*, for example, is associated with northern xerophytic mid-grasses *Aristida hygrometrica*, *A. inaeguiglumis* and *Chrysopogon fallax* and with the *Acacia*-shrub–short-grass species *Enneapogon polyphyllus* on skeletal granitic soils in the northwest of Western Australia. In the same area the

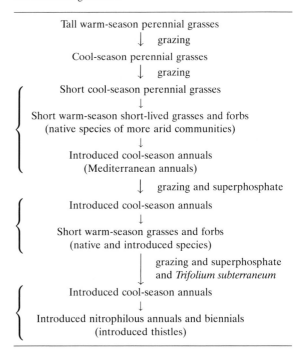

TABLE 13.1

The sequence of changes in floristic composition of a temperate woodland grassland [1]

Tall warm-season perennial grasses
↓ grazing
Cool-season perennial grasses
↓ grazing
{ Short cool-season perennial grasses
↓
Short warm-season short-lived grasses and forbs
(native species of more arid communities)
↓
Introduced cool-season annuals
(Mediterranean annuals) }

↓ grazing and superphosphate
{ Introduced cool-season annuals
↓
Short warm-season grasses and forbs
(native and introduced species) }

↓ grazing and superphosphate
and *Trifolium subterraneum*
{ Introduced cool-season annuals
↓
Introduced nitrophilous annuals and biennials
(introduced thistles) }

[1] After Moore and Biddiscombe (1964).

species found with *Plectrachne pungens* on shallow soils of basalt outcrops are *Sehima nervosum* and the annual tropical tall grass *Sorghum stipoideum*. Where soils are deeper, the tropical perennial tall grasses *Dichanthium fecundum* and *Sorghum plumosum* are usually also present.

Some species apparently have wide tolerance ranges and are found in several types of grassland and under various overstorey species. Among these in northern Australia are several species of *Aristida*, *Chloris acicularis*, *Chrysopogon fallax*, *Monachather paradoxa* and *Thyridolepis mitchelliana*. The short grasses *Chloris truncata*, *Enneapogon avenaceus*, *E. polyphyllus*, *Eragrostis setifolia* and *E. xerophila* also have broad environmental tolerances and are found in a wide range of grasslands.

Where the grassland dominants are unpalatable, as in xerophytic hummock-grass and xerophytic tussock-grass communities (Fig. 13.1, nos. 1 and 2), heavy and continuous grazing decreases species diversity by eliminating the more palatable inter-hummock and inter-tussock species. In those grasslands in which the original dominants do not survive grazing, as in the temperate grasslands of southern Australia [now designated temperate short-grass (Fig. 13.1, no. 9)], diversity appears to increase initially because of an ingress of introduced species and native species from adjacent drier communities. Ultimately, under heavy and continuous grazing pressures, diversity tends to decrease (Moore, 1962, 1966).

Communities of native forbs become established in most arid and semiarid grasslands after winter rains. Among the common species with wide distributions are:

Boerhavia repleta	Helipterum charsleyae
Blennodia lasiocarpa	Neptunia monosperma
Calandrinia spp.	Portulaca oleracea
Calotis hispidula	Salsola kali
Cleome viscosa	Senecio gregorii
Euphorbia drummondii	Sida spp.
Helichrysum spp.	

Forbs increase the quality rather than the quantity of forage available to livestock.

Despite the predominantly winter incidence of rain in southern Australia, temperate grasslands on low-nutrient soils in regions receiving more than 400 mm of rain annually were originally dominated by tall summer-growing species, notably *Themeda australis*, a species with a C_4 photosynthetic pathway widespread also in tropical Australia. *Themeda* grasslands evolved under light and intermittent grazing by native marsupials, and periodic burning. Under this regime they were remarkably stable; the evidence is that undisturbed *Themeda* communities were not invaded by introduced species. The equilibrium of a system growing partially out of phase with the rainfall regime is apparently easily upset, as *Themeda* disappeared rapidly from temperate grasslands when sheep and rabbits were introduced. The most rapid and complete disappearance of *Themeda*, and indeed all other native perennial grasses, occurred in the south-west of Western Australia where rainfall is almost wholly during winter. Loss of perennials under grazing has been paralleled in other Mediterranean climatic environments, such as California where grasslands are now composed almost wholly of introduced annuals (see Chapter 16 of Volume 8A).

The change from tall, warm-season perennial grasses *Themeda australis*, *Stipa bigeniculata* and *Poa labillardieri* and to dwarf, cool-season perennials *Danthonia auriculata* and *D. carphoides*, as a result of grazing by sheep and rabbits, was accompanied by an ingress of short, warm-season species from other and drier grassland communities — *Enneapogon nigricans*, *Chloris truncata* and *Panicum effusum* — and of cool-season annuals of Mediterranean origin — *Trifolium glomeratum*, *Vulpia bromoides* and *Erodium cicutarium*. Where phosphorus was applied regularly to encourage the growth of Mediterranean annual legumes, and particularly where the more productive *Trifolium subterraneum* was deliberately introduced, the grassland has become composed almost entirely of introduced cool-season annuals and, following rains in summer, of short-lived perennials and annuals, both introduced and native (Table 13.1) (Moore, 1967a, b).

In areas where rainfall is mainly during winter but there is some summer rain, the trend under grazing in temperate grasslands is towards xerophytism in the summer community and, with the accretion of nitrogen from legumes, to therophytism and nitrophily in the winter community (Moore, 1966). The visual evidence of the latter is the invasion by annual and biennial thistles belonging to the genera *Carduus*, *Onopordum* and *Silybum*, the ingestion of which by livestock occasionally results in nitrate poisoning. The ecological consequences of increasing nitrification from introduced legumes and increasing soil phosphorus and acidification from continued use of phosphoric fertilizer are, as yet, incompletely understood.

In temperate Australia, soil nitrogen is mineralized during the summer and, in land devoid of vegetation, reaches peak levels at the end of summer. Moore (1967b) found that nitrate-nitrogen in the top 10 cm of soils under *Themeda australis* grassland did not exceed 5 ppm at the end of summer, whereas levels of more than 36 ppm were recorded under the short, cool-season perennials *Danthonia auriculata* and *D. carphoides*. As a consequence, when the winter growing season commenced, there was a much larger pool of labile nitrogen in the subclimax *Danthonia* community than in the climax *Themeda* site; introduced cool-season annuals invaded the former, but not the undisturbed *Themeda* community. By growing when mineralization processes are actually or potentially active, and utilizing or otherwise limiting the accumulation of labile nitrogen in the soil surface, *Themeda*, seemingly, gives stability to the grassland community.

Levels of available nitrogen in the soil under grassland rise with disturbance and are accompanied by changes in the floristic composition of the community, if the disturbance factor is of sufficient intensity and duration to lead to the disappearance of *Themeda*.

Themeda has persisted better under cattle grazing in tropical and subtropical northern Australia than under sheep grazing in winter-rainfall areas of the temperate zone. Nevertheless, *Themeda* was found to be sensitive to high grazing pressures in the early months of the wet season at Katherine in the Northern Territory and has disappeared from much of the eastern subtropical grasslands now dominated by *Heteropogon contortus* (Mott and Tothill, 1984).

Grassland succession, in the absence of sedentary grazing, may be conceptualized as a process of vegetation change towards a stable assemblage of species that maintains a relatively high level of organic matter but a low level of labile nitrogen in the soil. It is, perhaps, significant that *Themeda* and many other so-called "climax" grasses produce a relatively large biomass low in protein, and their above-ground litter has a high C/N ratio. It is noteworthy, too, that in the absence of sedentary grazing *Themeda* grassland in temperate areas is given renewed vigour by periodic wildfires, which produce a short-term flush of mineralization of soil nitrogen.

TRUE GRASSLANDS

A classification of Australian grasslands was given by Moore (1970a). The most extensive of the "true" grasslands, as distinct from the herbaceous understoreys of shrublands and woodlands, are the xerophytic hummock-grass grasslands (spinifex) of *Plectrachne* and *Triodia* (Fig. 13.1, no. 1). As has been pointed out, there are widely spaced trees and shrubs over much of the hummock grasslands. But large areas also are treeless, and in a structural and physiognomic sense, spinifex communities are closer to grassland than to any other formation.

Treeless grasslands are found on fine-textured cracking clay soils in areas of low rainfall and are subject to regular and long seasonal drought. The most extensive of these grasslands are the xerophytic tussock grasslands in which the dominant species belong to the genus *Astrebla*. These grasslands are mainly between longitudes 124° and 146°E and latitudes 16° and 28°S.

The boundary between the hummock grasslands of the sand-plain soils and the tussock-grass grasslands on fine-textured cracking clay soils (Fig. 13.1, no. 2) is often abrupt, but toward their limits the two communities have some subordinate species in common.

The *Astrebla* and other treeless grasslands form a continuum of genera and related species from north to south in eastern Australia on both cracking clays and texture-contrast soils.

South of the Gulf of Carpentaria between longitudes 140°E and 145°E, the *Astrebla* tussock-grass grasslands intergrade in places with more mesophytic subhumid tropical grasslands, the Bluegrass–Browntop Downs, of *Dichanthium fecundum*, *Eulalia aurea* and *Bothriochloa* spp. Farther south, in central Queensland near Emerald, related treeless grasslands (Fig. 13.7) — the Bluegrass Downs of *Dichanthium sericeum* and *Bothriochloa erianthoides* on similar fine-textured soils — have species in common (notably *Dichanthium sericeum*) with the drier Mitchell Grass Downs to the west of the 500-mm annual rainfall line.

At latitude 28°S and longitude 150°E, where there is a winter component to the predominantly summer rainfall regime, there are remnants of a former grassland, now intensively cropped, thought to have been composed originally of both low-latitude species — *Dichanthium sericeum*, *Bothriochloa decipiens* and *Eulalia aurea* (Fig. 13.7) — and high-latitude species — *Stipa aristiglumis* and *Themeda avenacea*. The latter two species seem to have been the original dominants on small areas of clay soils west of the 500-mm rainfall line through the Liverpool Plains of New South Wales to the Wimmera region of Victoria. These areas from 30°S to 38°S are now largely grasslands of *Danthonia caespitosa* and *Stipa variabilis* (Fig. 13.20).

In higher-rainfall areas in New South Wales and

Victoria, small areas of treeless grassland occur in elevated tablelands on texture-contrast soils. Here the principal species were originally *Themeda australis*, associated with *Stipa bigeniculata* in dry sites and *Poa labillardieri* in wet sites; the predominant species now are the short grasses *Danthonia auriculata* and *D. carphoides* with varying amounts of *Stipa falcata*.

The absence of trees on fine-textured clays would appear to be a result of shallow penetration of water in environments with relatively low rainfall and high evaporation, where tree seedlings would be subjected to intense competition from surface-rooted grasses and other herbaceous species. In areas of higher rainfall, poor aeration following temporary waterlogging after winter rains may also be a factor in the absence of trees on fine-textured clay soils, such as those of the basalt plains west of Melbourne (Victoria).

In subalpine areas in New South Wales, Victoria and Tasmania there are patches of subalpine sod tussock-grass grasslands (Fig. 13.1, no. 10) below the tree line. The principal components are species of *Poa*[1], *Danthonia nudiflora* and *Themeda australis*.

On coarser-textured soils of elevated valleys and plains, very low temperatures following nocturnal inversions have been advanced to account for the absence of trees on valley floors in the subalpine areas — the treeless subalpine and tussock-grass communities (Costin, 1970; Moore and Williams, 1976).

MANAGEMENT OF GRASSLANDS

Fire has been a factor in Australian vegetation from prehistoric times. Wildfires are still frequent and, in addition, fire is used in the management of grasslands — particularly those of the subtropics and tropics (Shaw, 1957; Tothill, 1971) and semi-arid and arid areas (Noble, 1984). Walker (1981) mapped fuel-dynamic (fire-liable) regions based on climate and the fuel characteristics of vegetation. A general account of the effects of fire on plant and animal life in Australia has been given by Gill et al. (1981).

---

[1] See footnote on p. 345.

Cattle and sheep are the principal commercial livestock on Australian grasslands, but native marsupials and introduced feral animals increase grazing pressures locally and regionally and add to problems of management and the prevention of overuse of the grassland resource.

Australia has large populations of native herbivorous animals. The most important to grasslands are the marsupials of the family Macropodidae, the kangaroos and wallabies. *Macropus rufus* (red kangaroo) is found throughout semiarid and arid grazing lands. *Macropus antilopinus* (antilopine kangaroo) is much more restricted in its range, being confined to the tropical grasslands of north and north-eastern Australia. The two grey kangaroo species, *M. giganteus* and *M. fuliginosus*, inhabit forests and woodlands — the former in both tropical and temperate areas of eastern Australia, and the latter in temperate woodlands and shrublands of southern Australia. Both species overlap to some degree with *M. rufus* at the drier limits of their ranges. *M. robustus* (hill kangaroo, euro) lives in stony, hilly habitats throughout most of Australia, except the most northerly tropics and the most southerly woodlands and forests.

According to Frith (1970) kangaroos and euros have increased in numbers and their ranges have changed locally as a result of thinning of woodlands by European man and the modification of grasslands by domestic livestock. In times of drought there is competition for food between native herbivores and domestic livestock and, on occasion, this has prompted efforts to control kangaroo numbers locally. In general, native herbivores compete less with livestock than introduced animals, such as *Oryctolagus cuniculus* (European rabbit) in southern Australia and *Bubalus bubalis* (Timor water buffalo) in the coastal and subcoastal plains of the Northern Territory.

Other important competitors with sheep and cattle are feral animals. These include *Capra hircus* (goat) in semiarid and arid shrublands, *Camelus dromedarius* (camel) in arid shrublands and grasslands, *Equus caballus* (wild horse) in the Northern Territory, and *E. asinus* (wild donkey) in north-western Western Australia and the Northern Territory.

Behavioural patterns of animal grazing were discussed by Foran (1984) and Wilson and Harrington (1984). Kinds of livestock and even breeds influence management strategies. Brahman cattle, for example, are better adapted to tropical environments than British breeds but exert more prolonged grazing pressures on grasslands.

Northern tropical grasslands are grazed mainly by cattle, and southern temperate grasslands are grazed by both sheep and cattle. On account of their closer grazing habits, it is argued by some that sheep are better suited than cattle to southern grasslands. One of the reasons advanced is that the amount of green material available at a particular time is often as low as 10 g m^{-2} (Leigh et al., 1968; Wilson et al., 1969) and that the herbage is then too short for cattle. Sheep graze more closely and selectively than cattle, and are able to extract a diet of higher quality than the average nutritive value of the forage present. This is good for the nutrition of sheep but, depending on the grassland type, may be damaging to the grassland. Among the advantages of sheep in arid areas is their capacity to continue producing a saleable product, wool, in times of drought. Wool, too, is usually sufficiently valuable to meet the cost of transportation over long distances and can be stored without deterioration.

In shrub woodlands and shrublands, sheep are useful, in fact essential, for controlling the regeneration of unwanted trees and shrubs, but unless their grazing is controlled they can also prevent regeneration of useful *Atriplex* spp. and *Maireana* spp., on which the continuing productivity and stability of saltbush–short-grass communities depend.

There are two broad objectives to the management of Australian grasslands. In man-made grasslands (pastures) priority is given to the production of wool, meat and milk; but in native grasslands — particularly those in such sensitive environments as semiarid and subalpine areas — management regimes are designed, or should be designed, to conserve the resource, the soil and vegetation.

Australian studies on systems of management of pastures have been reviewed by Willoughby (1970) and Myers (1972). Rotational grazing has not shown to advantage in either pasture or livestock production when compared with continuous grazing at similar stocking rates per unit area (Moore et al., 1946; Rossiter, 1952). Six-monthly rotation of grazing on arid tussock grassland (Roe and Allen, 1945), 4-paddock rotation on temperate short-grass (Roe et al., 1959), and 6-weekly deferment on xerophytic mid-grass (Biddiscombe et al., 1956) were no more advantageous to animal production or to continued productivity of the grassland resource than continuous grazing.

Animal numbers per unit area are the main determinant of meat, milk and wool production, and control of grazing pressure is the key factor in maintenance of grassland productivity. Effective management, in essence, is the way in which these two factors are reconciled in a sedentary pastoral system.

Management of arid and semiarid grasslands has been discussed in broad terms by Moore (1960), Perry (1968, 1972, 1977), Williams (1968), Box and Perry (1971), and Wilson (1977).

In arid lands, recurring droughts make difficult the avoidance of periodic overuse and damage to the resource. The greatest damage occurs through stock concentrations around watering points, an area designated by Lange (1969) as the "piosphere". Even in seasons of good rainfall and at moderate average stocking over the whole area, grazing pressures around the widely dispersed watering points common in arid-land holdings are excessive and damage to vegetation is high. Provision of more strategically placed water points would reduce the area of heavy disturbance and ensure more effective use of the grassland by making possible a lower and more uniform level of utilization over the whole area. In the Alice Springs area, Stewart and Perry (1962) estimated that, of 100 000 km^2 of useable grazing land, only 44 000 km^2 were within 5 km of water and therefore able to be used effectively.

The principal aim of management in arid communities must be the stability of the grassland, and stability depends on the perennial vegetation components, both edible and non-edible. Loss of perennials through overstocking usually results in loss of surface soil and development of "bare scalds". Charley and Cowling (1968) showed that most of the nutrients are in the top 10 cm of arid soils, and are lost when erosion occurs. Top-soil is the most important resource in arid lands, and management should be directed to its preservation. Only while the top-soil remains is regeneration of the original grassland community possible.

The problem is to match animal numbers (domestic and feral) to sustainable use and conservation in environments where droughts are frequent and often prolonged. Trends in grassland condition and ways of forecasting carrying capacity have been discussed by Newman and Condon (1969) and Friedel (1981a, b). Scattini and Orr (1987) suggested that stock numbers should be based on the amount of forage available at the end of each growing season.

Under-utilization, while not without problems of stock management and danger from uncontrolled fires in good seasons, would seem to be the key to the maintenance of the grassland resource in arid areas. Properties are large in area, but many could be even larger to enable sections to be unstocked for periods sufficient to allow useful species to regenerate and, in some instances, to provide fuel for the control of unwanted shrubs by fire. Increasing the size of arid-land leases, without increasing stock numbers, would permit a degree of "nomadism" within properties and a more controlled use of grassland types differing in floristic composition and productivity. Condon (1978) has indicated the significance of the conditions of land tenure on the management and conservation of arid lands.

One of the potentially most vulnerable areas in Australia is the semiarid crescent of shrub woodlands and shrublands forming a barrier between the arid communities of the interior on the one hand, and the agricultural lands and temperate and tropical grasslands of the higher-rainfall areas (that is, those above 500 mm of annual rainfall in the north and above 250 mm in the south) on the other. This area, which is gradually being diminished by land-clearing on its higher-rainfall margins for crop production in good seasons, serves as a protective barrier against encroaching aridity. The extent and direction of its further use should depend on ecological studies of the implications of such use for the communities of the crescent and adjoining grasslands and croplands.

REFERENCES

Alexander, G. and Williams, O.B., 1973. *The Pastoral Industries of Australia*. Sydney University Press, Sydney, 567 pp.

Anonymous, 1965. *Climate and Meteorology of Australia*. Bureau of Meteorology, Australia Bulletin No. 1, Government Printer, Melbourne, 62 pp.

Anonymous, 1967. Surface water resources. *Atlas of Australian Resources*. 2nd Ser., Dept. of Nat. Development, Canberra.

Barker, S. and Lange, R.T., 1970. Population ecology of *Atriplex* under sheep stocking. In: R. Jones (Editor), *The Biology of Atriplex*. CSIRO Division of Plant Industry, Canberra, pp. 105–120.

Barnard, A., 1969. Aspects of the economic history of the arid land pastoral industry. In: R.O. Slatyer and R.A. Perry (Editors), *Arid Lands of Australia*. Australian National University Press, Canberra, pp. 209–228.

Barrow, P.M. and Pearson, F.B., 1970. The mallee and mallee heaths. In: R.M. Moore (Editor), *Australian Grasslands*. Australian National University Press, Canberra, pp. 219–227.

Beadle, N.C.W., 1948. *The Vegetation and Pastures of Western New South Wales with Special Reference to Soil Erosion*. Government Printer, Sydney, 281 pp.

Beadle, N.C.W., 1952. Studies in halophytes, I. The germination of the seeds and establishment of the seedlings of five species of *Atriplex* in Australia. *Ecology*, 33: 49–62.

Beale, I.F., 1972. *The Effect of Thinning on the Productivity of Two Mulga (Acacia aneura F. Muell.) Communities in Southwestern Queensland*. M. Agr. Sc. Thesis, University of Queensland, Brisbane.

Biddiscombe, E.F., 1953. A survey of the natural pastures of the Trangie District, New South Wales, with particular reference to the grazing factor. *Aust. J. Agric. Res.*, 4: 1–28.

Biddiscombe, E.F., Cuthbertson, E.G. and Hutchings, R.J., 1954. Autecology of some natural pasture species of Trangie, New South Wales. *Aust. J. Bot.*, 2: 69–98.

Biddiscombe, E.F., Hutchings, R.J., Edgar, G. and Cuthbertson, E.G., 1956. Grazing management of natural pastures at Trangie, New South Wales. *Aust. J. Agric. Res.*, 7: 233–247.

Blake, S.T., 1938. The plant communities of western Queensland and their relationships, with special reference to the grazing industry. *Proc. R. Soc. Queensl.*, 49: 156–204.

Box, T.W. and Perry, R.A., 1971. Rangeland management in Australia. *J. Range Manage.*, 24: 167–171.

Boyland, D.E., 1970. Ecological and floristic studies in the Simpson Desert National Park, south-western Queensland. *Proc. R. Soc. Queensl.*, 82: 1–16.

Boyland, D.E., 1973. Vegetation of the mulga lands with special reference to south-western Queensland. *Trop. Grassl.*, 7: 35–42.

Bryan, W.W., 1970. Tropical and sub-tropical forests and heaths. In: R.M. Moore (Editor), *Australian Grasslands*. Australian National University Press, Canberra, pp. 101–111.

Burbidge, N.T., 1946. Morphology and anatomy of the Western Australian species of *Triodia* R. Br. *Trans. R. Soc. S. Aust.*, 70: 221–234.

Burbidge, N.T., 1960. The phytogeography of the Australian region. *Aust. J. Bot.*, 8: 75–212.

Burrows, W.H., 1972. *Studies in the Ecology and Control of Green Turkey Bush (Eremophila gilesii F. Muell.) in South-west Queensland*. M. Agr. Sci. Thesis, University of Queensland, Brisbane.

Burrows, W.H., 1990. Prospects for increased production in the north-east Australian beef industry through pasture improvement and management. *J. Aust. Inst. Agric. Sci.*, 3: 21–24.

Burrows, W.H. and Beale, I.F., 1969. Structure and association in the mulga (*Acacia aneura*) lands of south-western Queensland. *Aust. J. Bot.*, 17: 539–552.

Carnahan, J.A., 1976. Natural vegetation. *Atlas of Australian Resources.* 2nd Ser., Dept. of Nat. Resources, Canberra.

Carrodus, B.B., 1962. *Some Aspects of the Ecology of Arid South Australia: The relative distribution of Atriplex vesicaria Heward ex Benth. and Kochia sedifolia F.v.M.* M.Sc. Thesis, University of Adelaide, Adelaide.

Carrodus, B.B. and Specht, R.L., 1965. Factors affecting the relative distribution of *Atriplex vesicaria* and *Kochia sedifolia* (Chenopodiaceae) in the arid zone of South Australia. *Aust. J. Bot.*, 13: 419–433.

Carrodus, B.B., Specht, R.L. and Jackman, M.E., 1965. The vegetation of Koonamore Station, South Australia. *Trans. R. Soc. S. Aust.*, 89: 42–57.

Charley, J.L. and Cowling, S.W., 1968. Changes in soil nutrient status resulting from overgrazing and their consequences in plant communities of semi-arid areas. *Proc. Ecol. Soc. Aust.*, 3: 28–38.

Choate, J.H., 1989. Pastoralism. In: J.C. Noble and R.A. Bradstock (Editors), *Mediterranean Landscapes in Australia. Mallee Ecosystems and their Management.* CSIRO, Australia, pp. 307–317.

Christian, C.S. and Stewart, G.A., 1954. The land-use groups of the Barkly region. *CSIRO Aust., Land Res. Ser.*, 3: 150–180.

Christie, E.K., 1975. A study of phosphorus nutrition and water supply on the early growth and survival of buffel grass grown on a sandy red earth from south-east Queensland. *Aust. J. Exp. Agric. Anim. Husb.*, 15: 239–249.

Christie, E.K., 1978. Ecosystem processes in semi-arid grasslands, I. Primary production and water use of two communities possessing different photosynthetic pathways. *Aust. J. Agric. Res.*, 29: 775–787.

Christie, E.K., 1981. Biomass and nutrient dynamics in a C_4 semi-arid Australian grassland community. *J. Appl. Ecol.*, 18: 907–918.

Coaldrake, J.E., 1970. The brigalow. In: R.M. Moore (Editor), *Australian Grasslands.* Australian National University Press, Canberra, pp. 123–140.

Coaldrake, J.E. and Smith, C.A., 1967. Estimates of animal production from pastures on brigalow land in the Fitzroy Basin, Queensland. *J. Aust. Inst. Agric. Sci.*, 33: 52–54.

Condon, R.W., 1978. Land tenure and desertification in Australia's arid lands with particular reference to New South Wales. *Search*, 9: 261–264.

Costin, A.B., 1970. Sub-alpine and alpine communities. In: R.M. Moore (Editor), *Australian Grasslands.* Australian National University Press, Canberra, pp. 191–198.

Crocker, R.L., 1946. The soils and vegetation of the Simpson Desert and its borders. *Trans. R. Soc. S. Aust.*, 70: 235–258.

Davies, J.G. and Hutton, E.M., 1970. Tropical and subtropical pasture species. In: R.M. Moore (Editor), *Australian Grasslands.* Australian National University Press, Canberra, pp. 273–302.

Donald, C.M., 1970. Temperate pasture species. In: R.M. Moore (Editor), *Australian Grasslands.* Australian National University Press, Canberra, pp. 303–320.

Ebersohn, J.P., 1970. Herbage production from native pastures and sown pastures in south-west Queensland. *Trop. Grassl.*, 4: 37–41.

Everist, S.L., 1949. Mulga (*Acacia aneura* F. Muell.). *Queensl. J. Agric. Sci.*, 6: 87–139.

Everist, S.L., 1972. Australia. In: C.M. McKell, J.P.Blaisdell and J.R. Goodin, *Wildland Shrubs: Their Biology and Utilization.* USDA Forest Serv. Gen. Tech. Rep. INT-1, pp. 16–25.

Everist, S.L. and Webb, L.J., 1975. Two communities of urgent concern in Queensland; Mitchell grass and tropical closed forests. In: F. Fenner (Editor), *A National System of Ecological Reserves in Australia.* Australian Academy of Science, Canberra, pp. 39–52.

Everist, S.L. and Young, R.B., 1967. Fodder trees in sheep lands. *Queensl. Agric. J.*, 93: 54–546.

Everist, S.L., Harvey, J.M. and Bell, A.T., 1958. Feeding sheep on mulga. *Queensl. Agric. J.*, 84: 352–361.

Fitzgerald, K., 1968. The Ord River regeneration project, 2. Dealing with the problem. *J. Agric. West. Aust.*, 9: 90–95.

Fitzpatrick, E.A. and Nix, H.A., 1970. The climatic factor in Australian grassland ecology. In: R.M. Moore (Editor), *Australian Grasslands.* Australian National University Press, Canberra, pp. 3–26.

Foran, B.D., 1984. Central arid woodlands. In: G.N. Harrington, A.D. Wilson and M.D. Young (Editors), *Management of Australia's Rangelands.* Commonwealth Scientific and Industrial Research Organization, Australia, pp. 299–315.

Friedel, M.H., 1981a. Studies of central Australian semi-desert rangelands, I. Range condition and biomass dynamics of the herbage layer and litter. *Aust. J. Bot.*, 29: 219–231.

Friedel, M.H., 1981b. Studies of central Australian semi-desert rangelands, II. Range condition and the nutrient dynamics of the herbage layer, litter and soil. *Aust. J. Bot.*, 29: 233–245.

Frith, H.J., 1970. The herbivorous wild animals. In: R.M. Moore (Editor), *Australian Grasslands.* Australian National University Press, Canberra, pp. 74–83.

Gill, A.M., Groves, R.H. and Noble, I.R., 1981. *Fire and the Australian Biota.* Australian Academy of Science, Canberra, 582 pp.

Gillard, P., 1979. Improvement of native pasture with Townsville stylo in the dry tropics of sub-coastal northern Queensland. *Aust. J. Exp. Agric. Anim. Husb.*, 19: 325–336.

Gillard, P. and Winter, W.H., 1984. Animal production for *Stylosanthes* based pastures in Australia. In: H.M. Stace and L.A. Edye (Editors), *The Biology and Agronomy of Stylosanthes.* Academic Press, London, pp. 408–430.

Groves, R.H. and Williams, O.B., 1981. Natural grasslands. In: R.H. Groves (Editor), *Australian Vegetation.* Cambridge University Press, Cambridge, pp. 293–316.

Hall, E.A., Specht, R.L. and Eardley, C.M., 1964. Regeneration of the vegetation on Koonamore Vegetation Reserve 1926–1962. *Aust. J. Bot.*, 12: 205–264.

Heathcote, R.L., 1969. Land tenure systems: past and present. In: R.O. Slatyer and R.A. Perry (Editors), *Arid Lands of Australia*. Australian National University Press, Canberra, pp. 185–208.

Holland, A.A. and Moore, C.W.E., 1962. The vegetation and soils of the Bollon district in south-western Queensland. *CSIRO Aust., Div. Pl. Ind. Tech. Pap.*, 17, 31 pp.

Hounan, C.E., 1961. Evaporation in Australia. *Bull. Bur. Meteorol. Aust.*, 44, 88 pp.

Hubble, G.D., 1970. Soils. In: R.M. Moore (Editor), *Australian Grasslands*. Australian National University Press, Canberra, pp. 44–58.

Irwin, J.A.G., Cameron, D.F. and Lenne, J.M., 1984. Responses of *Stylosanthes* to anthracnose. In: H.M. Stace and L.A. Edye (Editors), *Biology and Agronomy of Stylosanthes*. Academic Press, London, pp. 295–309.

Jessup, R.W., 1951. The soils, geology and vegetation of north-western South Australia. *Trans. R. Soc. S. Aust.*, 74: 189–273.

Johnson, R.W., 1964. *Ecology and Control of Brigalow in Queensland*. Queensland Department of Primary Industries, Government Printer, Queensland, 92 pp.

Jozwik, F.X., Nicholls, A.O. and Perry, R.A., 1970. Studies on the Mitchell grasses (*Astrebla* F. Muell.). In: M.J. Norman (Editor), *Proceedings of the 11th International Grassland Congress*. University of Queensland Press, Brisbane, pp. 48–51.

Lange, R.T., 1969. The piosphere: sheep track and dung patterns. *J. Range Manage.*, 22: 396–400.

Lay, B.G., 1972. *Ecological Studies of Arid Rangelands in South Australia*. M.Sc. Thesis, University of Adelaide, Adelaide.

Leigh, J.H., 1972. Saltbush and other chenopod browse shrubs. In: N. Hall (Editor), *The Use of Trees and Shrubs in the Dry Country of Australia*. Department of National Development, Canberra, pp. 284–288.

Leigh, J.H. and Holgate, M.D., 1979. The responses of the understorey of forests and woodlands of the Southern Tablelands to grazing and burning. *Aust. J. Ecol.*, 4: 25–45.

Leigh, J.H. and Mulham, W.E., 1964. Dietary preferences of sheep in two semi-arid pastoral ecosystems. *Proc. Aust. Anim. Prod.*, 5: 251–255.

Leigh, J.H. and Mulham, W.E., 1966a. Selection of diet by sheep grazing semi-arid pastures of the Riverine Plain, 1. A bladder saltbush (*Atriplex vesicaria*)–cotton bush (*Kochia aphylla*) community. *Aust. J. Exp. Agric. Anim. Husb.*, 6: 460–467.

Leigh, J.H. and Mulham, W.E., 1966b. Selection of diet by sheep grazing semi-arid pastures of the Riverine Plain, 2. A cotton bush (*Kochia aphylla*)–grassland (*Stipa variabilis–Danthonia caespitosa*) community. *Aust. J. Exp. Agric. Anim. Husb.*, 6: 468–474.

Leigh, J.H. and Mulham, W.E., 1967. Selection of diet by sheep grazing semi-arid pastures on the Riverine Plain, 3. A bladder saltbush (*Atriplex vesicaria*)–pigface (*Disphyma australe*) community. *Aust. J. Exp. Agric. Anim. Husb.*, 7: 421–425.

Leigh, J.H. and Mulham, W.E., 1971. The effect of defoliation on the persistence of *Atriplex vesicaria*. *Aust. J. Agric. Res.*, 22: 239–244.

Leigh, J.H. and Noble, J.C., 1969. Vegetation resources. In: R.O. Slatyer and R.A. Perry (Editors), *Arid Lands of Australia*. Australian National University Press, Canberra, pp. 73–92.

Leigh, J.H. and Wilson, A.D., 1970. Utilization of *Atriplex* species by sheep. In: R. Jones (Editor), *The Biology of Atriplex*. CSIRO Division of Plant Industry, Canberra, pp. 97–104.

Leigh, J.H., Wilson, A.D. and Mulham, W.E., 1968. A study of merino sheep grazing a cotton bush (*Kochia aphylla*)–grassland (*Stipa variabilis–Danthonia caespitosa*) community on the Riverine Plain. *Aust. J. Bot.*, 19: 947–961.

Litchfield, W.H., 1962. Soils of the Alice Springs area. In: Lands of the Alice Springs Area. *CSIRO Aust. Land Res. Ser.*, 6: 185–207.

Mabbutt, J.A., 1962. Geomorphology of the Alice Springs area. In: Lands of the Alice Springs Area. *CSIRO Aust. Land Res. Ser.*, 6: 163–184.

Mabbutt, J.A. and Sullivan, M.E., 1970. Landforms and structure. In: R.M. Moore (Editor), *Australian Grasslands*. Australian National University Press, Canberra, pp. 27–43.

Main, A.R., 1969. Native animal resources. In: R.O. Slatyer and R.A. Perry Editors), *Arid Lands of Australia*. Australian National University Press, Canberra, pp. 93–104.

Melville, G.F., 1947. An investigation of the drought pastures of the Murchison district of Western Australia. *J. Dept. Agric. West. Aust.*, 24: 1–29.

Moore, C.W.E., 1953. The vegetation of the south-eastern Riverina, New South Wales, 1. The climax communities. *Aust. J. Bot.*, 1: 485–547.

Moore, R.M., 1959. Ecological observations on plant communities grazed by sheep in Australia. In: A. Keast R.L. Crocker and C.S. Christian (Editors), *Biogeography and Ecology in Australia*. Series Monographiae Biologicae, Vol. 8, Junk, The Hague, pp. 500–513.

Moore, R.M., 1960. The management of native vegetation in arid and semi-arid regions. In: Plant–Water Relationships in Arid and Semi-Arid Regions. *UNESCO Arid Zone Res.*, 15: 173–190.

Moore, R.M., 1962. Effects of sheep grazing on Australian vegetation. In: A. Barnard (Editor), *The Simple Fleece: Studies in the Australian Wool Industry*. Melbourne University Press, Melbourne, pp. 170–183.

Moore, R.M., 1966. Man as a factor in the dynamics of plant communities. *Proc. Ecol. Soc. Aust.*, 1: 106–110.

Moore, R.M., 1967a. Interaction between the grazing animal and its environment. *Proc. 9th Int. Congr. Anim. Prod.*, pp. 188–195.

Moore, R.M., 1967b. The naturalisation of alien plants in Australia. *IUCN Publ., New Ser.*, 9: 82–97.

Moore, R.M., 1970a. Australian grasslands. In: R.M. Moore (Editor), *Australian Grasslands*. Australian National University Press, Canberra, pp. 85–100.

Moore, R.M., 1970b. South-eastern temperate woodlands and grasslands. In: R.M. Moore (Editor), *Australian Grasslands*. Australian National University Press, Canberra, pp. 169–190.

Moore, R.M., 1972. Trees and shrubs in Australian sheep grazing lands. In: J.H. Leigh and J.C. Noble (Editors), *Plants for Sheep in Australia*. Angus and Robertson, Sydney, pp. 56–64.

Moore, R.M., 1973. Australia's arid shrublands. In: D.N. Hyder (Editor), *Arid Shrublands*. Proc. 3rd Workshop United States/Australia Rangelands Panel, Tucson, Ariz., pp. 6–11.

Moore, R.M., 1976. The Border Ranges — a land-use conflict. In: R. Munroe and N.C. Stevens (Editors), *The Border Ranges*. Publ. R. Soc. Queensl., pp. 79–81.

Moore, R.M. and Biddiscombe, E.F., 1964. The effects of grazing on grasslands. In: C. Barnard (Editor), *Grasses and Grasslands*. Macmillan, London, pp. 221–235.

Moore, R.M. and Perry, R.A., 1970. Vegetation. In: R.M. Moore (Editor), *Australian Grasslands*. Australian National University Press, Canberra, pp. 59–73.

Moore, R.M. and Walker, J., 1972. *Eucalyptus populnea* shrub woodlands; control of regenerating trees and shrubs. *Aust. J. Exp. Agric. Anim. Husb.*, 12: 437–440.

Moore, R.M. and Williams, J.D., 1976. A study of a sub-alpine woodland-grassland boundary. *Aust. J. Ecol.*, 1: 145–153.

Moore, R.M., Barrie, N. and Kipps, E.H., 1946. Grazing management: continuous and rotational grazing by Marino sheep, I. A study of the production of a sown pasture in the Australian Capital Territory under three systems of grazing management. *Counc. Sci. Ind. Res. Aust. Bull.*, 201: 7–82.

Moore, R.M., Condon, R.W. and Leigh, J.H., 1970. Semi-arid woodlands. In: R.M. Moore (Editor), *Australian Grasslands*. Australian National University Press, Canberra, pp. 228–245.

Mott, J.J., 1986. Planned invasions of Australian tropical savannas. In: R.H. Groves and J.J. Burdon (Editors), *Ecology of Biological Invasions. An Australian Perspective*. Australian Academy of Science, Canberra, pp. 89–96.

Mott, J.J. and Tothill, J.C., 1984. Tropical and sub-tropical woodlands. In: G.N. Harrington, A.D. Wilson and M.D. Young (Editors), *Management of Australia's Rangelands*. Commonwealth Scientific and Industrial Research Organization, Australia, pp. 255–269.

Mott, J.J., Tothill, J.C. and Weston, E.J., 1981. Animal production from the native woodlands and grasslands of Northern Australia. *J. Aust. Inst. Agric. Sci.*, 47: 132–141.

Mott, J.J., Williams, J., Andrew, M.H. and Gillison, A.N., 1985. Australian savanna ecosystems. In: J.C. Tothill and J.J. Mott (Editors), *Ecology and Management of the World's Savannas*. Australian Academy of Science, Canberra, in co-operation with Commonwealth Agricultural Bureaux, Farnham Royal, Bucks., pp. 56–82.

Myers, L.F., 1972. Effects of grazing and grazing systems. In: J.H. Leigh and J.C. Noble (Editors), *Plants for Sheep in Australia*. Angus and Robertson, Sydney, pp. 183–192.

Newman, J.C. and Condon, R.W., 1969. Land use and present condition. In: R.O. Slatyer and R.A. Perry Editors), *Arid Lands of Australia*. Australian National University Press, Canberra, pp. 105–132.

Newman, R.J., 1970. Dry temperate forests and heaths. In: R.M. Moore (Editor), *Australian Grasslands*. Australian National University Press, Canberra, pp. 159–168.

Noble, J.C., 1984. Mallee. In: G.N. Harrington, A.D. Wilson and M.D. Young (Editors), *Management of Australia's Rangelands*. Commonwealth Scientific and Industrial Research Organization, Australia, pp. 223–240.

Noble, J.C. and Whalley, R.D.B., 1978a. The biology and autecology of *Nitraria* L. in Australia, I. Its distribution, morphology and potential utilization. *Aust. J. Ecol.*, 3: 141–164.

Noble, J.C. and Whalley, R.D.B., 1978b. The biology and autecology of *Nitraria* L. in Australia, II. Seed germination, seeding establishment and response to salinity. *Aust. J. Ecol.*, 3: 165–177.

Norman, M.J.T., 1965. Seasonal performances of beef cattle on native pasture at Katherine, N.T. *Aust. J. Exp. Agric. Anim. Husb.*, 5: 227–231.

Norman, M.J.T. and Arndt, W., 1959. Performance of beef cattle on native and sown pastures at Katherine, N.T. *CSIRO Aust. Land Res. Req. Surv. Tech. Pap.*, 4, 12 pp.

Nunn, W.M. and Suijdendorp, H., 1954. Station management — the value of "deferred grazing". *J. Dept. Agric. W. Aust.*, 3: 385–387.

Orr, D.M., 1981. Changes in the quantitative floristics in some *Astrebla* spp. (Mitchell grass) communities in south-western Queensland in relation to trends in seasonal rainfall. *Aust. J. Bot.*, 29: 533–545.

Orr, D.M. and Holmes, W.E., 1984. Mitchell grasslands. In: G.N. Harrington, A.D. Wilson and M.D. Young (Editors), *Management of Australia's Rangelands*. Commonwealth Scientific and Industrial Research Organization, Australia, pp. 241–254.

Osborn, T.G.B., Wood, J.G. and Paltridge, T.B., 1932. On the growth and reaction to grazing of the perennial saltbush, *Atriplex vesicarium*. An ecological study of the biotic factor. *Proc. Linn. Soc. N.S.W.*, 57: 377–402.

Paton, D.F. and Hosking, W.J., 1970. Wet temperate forests and heaths. In: R.M. Moore (Editor), *Australian Grasslands*. Australian National University Press, Canberra, pp. 141–158.

Payne, A.L., Kubicki, A. and Wilcox, D.G., 1974. *Range condition guides for the West Kimberley area, W.A.* Western Australian Department of Agriculture.

Perry, R.A., 1960. Pasture lands of the Northern Territory, Australia. *CSIRO Aust. Land Res. Ser.*, 5: 1–55.

Perry, R.A., 1962. Natural pastures of the Alice Springs area. In: Lands of the Alice Springs area. *CSIRO Aust. Land Res. Ser.*, 6: 240–258.

Perry, R.A., 1964. Land use in the Leichhardt-Gilbert area. *CSIRO Aust. Land Res. Ser.*, 11: 192–213.

Perry, R.A., 1968. Australia's arid rangelands. *Ann. Arid Zone*, 7: 243–249.

Perry, R.A., 1970a. Arid shrublands and grasslands. In: R.M. Moore (Editor), *Australian Grasslands*. Australian National University Press, Canberra, pp. 246–259.

Perry, R.A., 1970b. The effects on grass and browse production of various treatments on a mulga community in central Australia. *Proc. 11th Int. Grassl. Congr.*, Australia, pp. 63–66.

Perry, R.A., 1972. Native pastures used by sheep in South

Australia and the Northern Territory. In: J.H. Leigh and J.C. Noble (Editors), *Plants for Sheep in Australia*. Angus and Robertson, Sydney, pp. 25–37.

Perry, R.A., 1977. Rangeland management for livestock production in semi-arid and arid Australia. In: Impact of Herbivores on Arid and Semi-Arid Rangelands. *Proceedings of the Second United States/Australia Rangeland Panel*. (Adelaide), Australian Rangeland Society, Perth, pp. 311–316.

Perry, R.A. and Christian, C.S., 1954. Vegetation of the Barkly region. *CSIRO Aust. Land Res. Ser.*, 3: 78–112.

Perry, R.A. and Lazarides, M., 1962. Vegetation of the Alice Springs area. In: Lands of the Alice Springs area, Northern Territory. *CSIRO Aust. Land Res. Ser.*, 6: 208–236.

Preece, P.B., 1971a. Contributions to the biology of mulga, I. Flowering. *Aust. J. Bot.*, 19: 21–38.

Preece, P.B., 1971b. Contributions to the biology of mulga, II. Germination. *Aust. J. Bot.*, 10: 39–49.

Ratcliffe, F.N., 1936. Soil drift in the arid pastoral areas of South Australia. *Counc. Sci. Ind. Res. Aust. Pamp.*, 64.

Robertson, J.A., 1965. The chemical control of *Eremophila mitchellii*. *Aust. J. Exp. Agric. Anim. Husb.*, 5: 299–304.

Robertson, J.A., 1966. The effect of basal injections of 2,4,5-T on mature trees of *Eucalyptus populnea*. *Aust. J. Exp. Agric. Anim. Husb.*, 6: 344–349.

Robertson, J.A. and Moore, R.M., 1972. Thinning *Eucalyptus populnea* woodlands by injecting trees with chemicals. *Trop. Grassl.*, 6: 141–150.

Roe, R. and Allen, G.H., 1945. Studies on the Mitchell grass association in south-western Queensland. *Counc. Sci. Ind. Res. Aust. Bull.*, 185.

Roe, R., Southcott, W.H. and Turner, H.N., 1959. Grazing management of native pastures in the New England region of New South Wales, I. Pasture and sheep production with special reference to systems of grazing and natural parasites. *Aust. J. Agric. Res.*, 10: 530–554.

Rossiter, R.C., 1952. The effect of grazing on a perennial veldt grass–subterranean clover pasture. *Aust. J. Agric. Res.*, 3: 148–159.

Rossiter, R.C., 1966. Ecology of the Mediterranean annual-type pasture. *Adv. Agron.*, 18: 1–56.

Rossiter, R.C. and Ozanne, P.G., 1970. South-western temperate forests, woodlands, and heaths. In: R.M. Moore (Editor), *Australian Grasslands*. Australian National University Press, Canberra, pp. 199–218.

Scanlon, J.C., 1980. Effects of spring wildfires on *Astrebla* (Mitchell grass) grasslands in north-west Queensland under varying levels of growing season rainfall. *Aust. Rangelands J.*, 2: 162–168.

Scanlon, J.C., 1983. Changes in tiller and tussock characteristics of *Astrebla lappacea* (curly Mitchell grass) after burning. *Aust. Rangelands J.*, 5: 13–19.

Scanlon, J.C. and Burrows, W.H., 1990. Woody overstorey impact on herbaceous understorey in *Eucalyptus* spp. communities in central Queensland. *Aust. J. Ecol.*, 15: 191–197.

Scattini, W.J. and Orr, D.M., 1987. The long view is the best view in natural pastures. *Queensl. Agric. J.*, 13: 58–60.

Shaw, N.H., 1957. Bunch spear grass dominance in burnt pastures in south-eastern Queensland. *Aust. J. Agric. Res.*, 8: 332–368.

Shaw, N.H. and Bisset, W.J., 1955. Characteristics of a bunch spear grass [*Heteropogon contortus* (L) Beauv.] pasture grazed by cattle in sub-tropical Queensland. *Aust. J. Agric. Res.*, 6: 539–552.

Shaw, N.H. and Norman, M.J.T., 1970. Tropical and sub-tropical woodlands and grasslands. In: R.M. Moore (Editor), *Australian Grasslands*. Australian National University Press, Canberra, pp. 112–122.

Shaw, N.H. and t'Mannetje, L., 1970. Studies on a speargrass pasture in central coastal Queensland — the effect of fertilizer, stocking rate and oversowing with *Stylosanthes humilis* on beef production and botanical composition. *Trop. Grassl.*, 4: 43–56.

Shelton, J.N., 1956. The performance of beef cattle in North Queensland. *Proc. Aust. Soc. Anim. Prod.*, 1: 130–137.

Specht, R.L., 1970. Vegetation. In: G.W. Leeper (Editor), *The Australian Environment*, 4th ed. CSIRO and Melbourne University Press, pp. 44–67.

Speck, N.H., Fitzgerald, K. and Perry, R.A., 1964. Pastoral lands of the west Kimberley area. *CSIRO Aust. Land Res. Ser.*, 9: 175–191.

Stace, H.C.T., Hubble, G.D., Brewer, R., Northcote, K.H., Sleeman, J.R., Mulcahy, M.J. and Hallsworth, E.G., 1968. *A Handbook of Australian Soils*. Rellim Technical Publications, Glenside, S.A., 435 pp.

Stannard, M.E. and Condon, R.W., 1958. Further fodder trees and shrubs of western N.S.W. *J. Soil Conserv. Serv. N.S.W.*, 14: 73–83.

Stewart, G.A. and Perry, R.A., 1962. Introduction and survey description of the Alice Springs area. *CSIRO Aust. Land Res. Ser.*, 6: 9–19.

Stocker, G.C. and Sturtz, J.D., 1966. The use of fire to establish Townsville lucerne in the Northern Territory. *Aust. J. Exp. Agric. Anim. Husb.*, 6: 277–279.

Story, R., 1967. Vegetation of the Isaac-Comet Region. *CSIRO Aust. Land Res. Ser.*, 19: 108–168.

Suijdendorp, H., 1981. Responses of the hummock grasslands of northwestern Australia to fire. In: A.M. Gill, R.H. Groves and I.R. Noble (Editors), *Fire and the Australian Biota*. Australian Academy of Science, Canberra, pp. 417–424.

Tothill, J.C., 1971. A review of fire in the management of native pastures with particular reference to north-eastern Australia. *Trop. Grassl.*, 5: 1–10.

Tothill, J.C. and Jones, R.M., 1977. Stability in sown and oversown Siratro pastures. *Trop. Grassl.*, 11: 55–65.

Tothill, J.C. and Mott, J.J., 1985. Australian savannas and their stability under grazing. In: M.G. Ridpath and L.K. Corbett (Editors), Ecology of the Wet-Dry Tropics. *Proc. Ecol. Soc. Aust.*, 13: 317–322.

Tothill, J.C., Nix, H.A., Stanton, J.P. and Russell, M.J., 1985. Land use and production potential of Australian savanna lands. In: J.C. Tothill and J.J. Mott (Editors), *Ecology and Management of the World's Savannas*. Australian Academy of Science, Canberra, pp. 125–141.

Walker, J., 1981. Fuel dynamics in Australian vegetation. In: A.M. Gill, R.H. Groves and I.R. Noble (Editors), *Fire and the Australian Biota*. Australian Academy of Science,

Canberra, pp. 101–128.

Walker, J., Moore, R.M. and Robertson, J.A., 1972. Herbage response to tree and shrub thinning in *Eucalyptus populnea* shrub woodlands. *Aust. J. Agric. Res.*, 23: 405–410.

Walker, J., Robertson, J.A., Penridge, L.K. and Sharpe, P.J.H., 1986. Herbage response to tree thinning in a *Eucalyptus crebra* woodland. *Aust. J. Ecol.*, 11: 135–140.

Wiedemann, A.M., 1971. Vegetation studies in the Simpson Desert, N.T. *Aust. J. Bot.*, 19: 99–124.

Wilcox, D.G., 1960. Studies in the mulga zone, 2. Some aspects of the value of the mulga scrub. *J. Agric. W. Aust.*, 1: 581–586.

Wilcox, D.G., 1963. The pastoral industry of the Wiluna–Meekatharra area. *CSIRO Aust. Land Res. Ser.*, 7: 195–212.

Wilcox, D.G., 1972. The use of native pastures by sheep in Western Australia. In: J.H. Leigh and J.C. Noble (Editors), *Plants for Sheep in Australia*. Angus and Robertson, Sydney, pp. 45–54.

Wilcox, D.G. and McKinnon, E.A., 1972. *A Report on the Condition of the Gascoyne Catchment*. Department of Lands and Surveys, Perth, W.A., 335 pp.

Williams, C.H. and Andrew, C.S., 1970. Mineral nutrition of pastures. In: R.M. Moore (Editor), *Australian Grasslands*. Australian National University Press, Canberra, pp. 321–338.

Williams, O.B., 1968. Pasture management in the pastoral zone — a review. *Wool Tech. Sheep Breed.*, July, pp. 45–48.

Williams, O.B. and Calaby, J.H., 1985. The hot deserts of Australia. In: M. Evenari, I. Noy-Meir and D.W. Goodall (Editors), *Ecosystems of the World, 12A. Hot Deserts and Arid Shrublands*. Elsevier, Amsterdam, 365 pp.

Williams, O.B. and Mackey, B., 1983. Easy-care, no-hassle conservation — Mitchell grass (*Astrebla*). In: J. Messer and G. Mosley (Editors), *What Future for Australia's Arid Lands?* Australian Conservation Foundation, Melbourne, pp. 141–145.

Williams, R.J., 1955. Vegetation regions. In: *Atlas of Australia*. Commonwealth Dept. of National Development, Canberra.

Willoughby, W.M., 1970. Grassland management. In: R.M. Moore (Editor), *Australian Grasslands*. Australian National University Press, Canberra, pp. 392–397.

Wilson, A.D., 1977. Grazing management in the arid areas of Australia. *Proceedings of the Second United States/Australia Rangeland Panel*, (Adelaide). Australian Rangeland Society, Perth, pp. 83–92.

Wilson, A.D. and Harrington, G.N., 1984. Grazing ecology and animal production. In: G.N. Harrington, A.D. Wilson and M.D. Young (Editors), *Management of Australia's Rangelands*. Commonwealth Scientific and Industrial Research Organization, Australia, pp. 63–77.

Wilson, A.D. and Leigh, J.H., 1970. Comparisons of the productivity of sheep grazing natural pastures of the Riverine Plain. *Aust. J. Exp. Agric. Anim. Husb.*, 10: 549–554.

Wilson, A.D., Leigh, J.H. and Mulham, W.E., 1969. A study of merino sheep grazing a bladder saltbush (*Atriplex vesicaria*)–cotton bush (*Kochia aphylla*) community in the Riverine Plain. *Aust. J. Agric. Res.*, 20: 1123–1136.

Winkworth, R.E., 1967. The composition of several arid spinifex grasslands of central Australia in relation to rainfall, soil water relations, and nutrients. *Aust. J. Bot.*, 15: 107–130.

Chapter 14

INDIGENOUS GRASSLANDS OF NEW ZEALAND[1]

A.F. MARK

INTRODUCTION

Mainland New Zealand consists of two large islands, North (114 740 km^2) and South (151 120 km^2) as well as several smaller ones, the archipelago extending some 1500 km between 34° and 47°S in the south-west Pacific Ocean. The New Zealand landscape is essentially one of extreme complexity and diversity, with about half the land area classified as steep, one-fifth as moderately hilly and less than one-third flat to gently rolling. North Island is generally hilly and locally mountainous, whereas South Island has much sharper relief and overall higher elevation, giving it a generally mountainous topography dominated by the Southern Alps running parallel and close to the western coast-line through the central portion of the Island and fanning out both northwards and southwards (Fig. 14.1). In the lee of the Alps, outlier ranges give characteristic basin and range topography with the basins filled with outwash gravels, often distinctly terraced, and grading eastward into low hills or, in the central waist of the island, into the extensive alluvial Canterbury Plains. The permanent snow-line varies from about 2400 m in central North Island to about 2000 m in the far south; there is an extensive nival zone in South Island concentrated around the highest peak, Mt. Cook (3764 m) (Fig. 14.2).

Apart from a generally oceanic climate, the main features are: (1) a pronounced west-to-east moisture gradient, particularly in South Island where the continuous mountain chain of the Southern Alps intercepts the prevailing westerly winds; and (2) a north-to-south temperature gradient amounting to 6°C in mean annual air temperature (15°C in the far north to 9°C in the south). Annual precipitation varies from 330 mm in the basins of central South Island to >10 000 mm on the windward western slopes of the Southern Alps (McSaveney, 1978).

Desert is the only major plant formation absent from New Zealand. The general vegetation pattern of forest, shrubland, grassland and alpine communities (Fig. 14.3) was described by Cockayne (1928) and mapped in broad detail by Holloway (1959a). Forests are all evergreen temperate, either mixed coniferous-hardwood or pure hardwood. Apart from *Agathis australis*[2] ("kauri", Araucariaceae) in the far north the conifers belong chiefly to the Podocarpaceae, whereas the most important hardwood genus is *Nothofagus* of the Fagaceae. Four species of this genus (typically with small leaves) dominate extensively, sometimes with conifers co-dominant, but often alone. Pure *Nothofagus* forest is typically associated with the more severe climates of inland subcontinental regions and the subalpine zone, but there are some anomalies, mostly explicable in terms of incomplete readjustment

---

[1] Manuscript submitted in May, 1979; revised in November, 1987 and June, 1990.

[2] Nomenclature follows Cheeseman (1925) for grasses, Allan (1961) for gymnosperms and dicotyledons, Moore and Edgar (1970) for monocotyledons (except grasses) and Brownsey and Smith-Dodsworth (1989) for pteridophytes, apart from recent updates (Connor and Edgar, 1987).

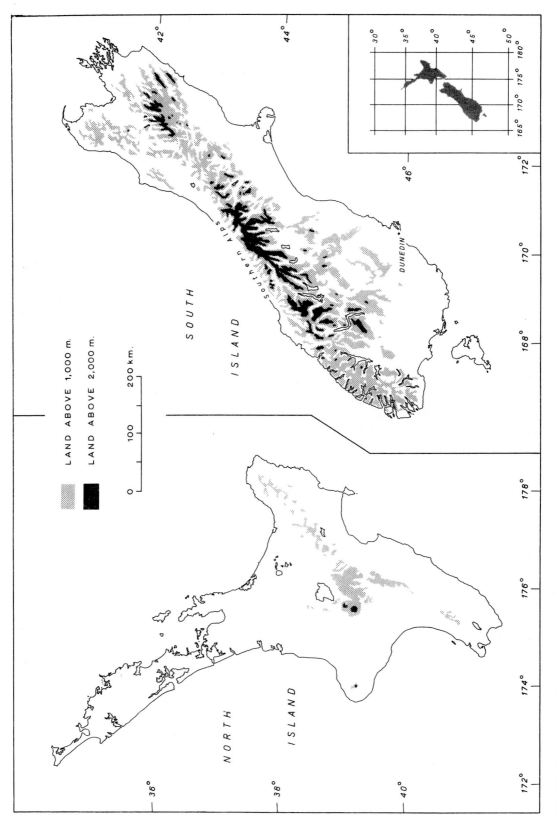

Fig. 14.1. Physical map of New Zealand (reproduced from Mark, 1969).

Fig. 14.2. View of the south-western slopes of Mount Cook (3764 m) from *Chionochloa flavescens* (broad-leaved snow tussock) grassland at about 1000 m elevation in the Hooker Valley.

following restrictions in plant distribution during the most recent of the Pleistocene glaciations (Wardle, 1963, 1980). Other explanations have also been offered (McGlone, 1985; Haase, 1990). Shrublands range from quasi-climax lowland tall scrub in semiarid regions and some wetlands to seral scrub under forest climates (Burrows et al., 1979), most typically dominated by *Kunzea* and *Leptospermum* (Myrtaceae) (Burrell, 1965). A mixed subalpine scrub, with Araliaceae, Asteraceae, Epacridaceae, Podocarpaceae and Rubiaceae conspicuous, usually substitutes for forest where tree line is depressed, often because *Nothofagus* is absent (Burrows et al., 1979).

Apart from alpine grasslands above the limits of forest or tall scrub (Fig. 14.4), the grasslands that confronted the first European settlers in New Zealand some 150 years ago were extensive in the rain-shadow regions east of the Southern Alps in South Island (Figs. 14.3 and 14.5), whereas in North Island, where precipitation is everywhere sufficient for forest, grassland was confined to the limited alpine regions and the pumice areas of the central volcanic region (Fig. 14.3), where both volcanism and Polynesian fires played a role (Rogers, 1988). Throughout these grasslands, with few exceptions, the dominants have a bunched or tussock growth form that elsewhere characterizes tropical high mountains and sub-Antarctic regions (Hnatiuk, 1978). In the drier, lower-altitude areas below 700 to 1000 m in South Island, the characteristic dominants were tussock species of *Elymus*, *Festuca* and *Poa* about 50-cm tall with rolled leaves (Fig. 14.5), often with some savanna-like thorny or microphyllous shrubs (Fig. 14.6).

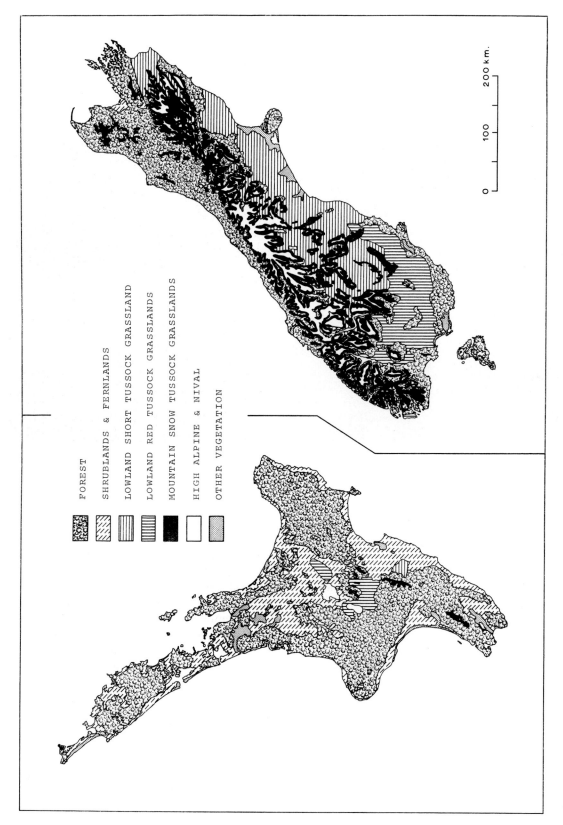

Fig. 14.3. Map of pre-European vegetation of New Zealand (modified from Mark, 1969).

Fig. 14.4. A typically abrupt tree line of *Nothofagus menziesii* (silver beech) at about 1200 m on the South Island's Southern Alps above Haast Pass, South Westland. Tall *Chionochloa* grassland with scattered shrubs occupies the low-alpine zone of this north-west facing slope (foreground).

The upland grasslands were dominated by one or more species of snow tussock, the common name of the largely indigenous genus *Chionochloa* (formerly *Danthonia*). Here low temperature (rather than insufficient moisture) precludes forest. On many of the drier South Island foothills and outlier ranges east of the Southern Alps, however, the grassland is relatively recent and largely of cultural origin (Molloy et al., 1963; Molloy, 1977). In some of the drier intermontane basins such as central Otago, however, destruction of forest by fire and establishment of grassland preceded human settlement by several centuries, implicating natural fires (McGlone, 1989). Forest cover increases with approach to the Southern Alps from the east, so that along and west of the main chain and on the higher North Island mountains the tall grasslands occur above a usually abrupt tree line (Fig. 14.4) at 950 to 1450 m (depending largely on latitude), and dominate a c. 300-m wide vertical zone as a low-alpine vegetation type (Wardle, 1964). *Chionochloa* grasslands of similar life form, but floristically different, occur at similar altitudes on the drier mountains to the east of the Southern Alps where, apart from some areas of persisting forest, they merge into the lowland short-tussock grassland (Fig. 14.7).

THE GRASSLAND ENVIRONMENT

Biogeography and geology

The beginning of New Zealand's geological and biogeographical history can be traced back to

Fig. 14.5. Montane short-tussock grassland dominated by *Festuca novae-zelandiae* (fescue or hard tussock) covering glacial outwash gravels at about 750 m in the Ahuriri Valley, east of the Southern Alps, central South Island.

Gondwanaland[1] (Fleming, 1975; Enting and Molloy, 1982; Stevens et al., 1988), and many ancestors of the present indigenous forest components reflect this origin (Fleming, 1977). Increasing isolation from the upper Cretaceous onwards was an important biogeographic factor and, although grasses first appeared during the Eocene (Fleming, 1975), a temperate to subtropical forest cover persisted throughout the Tertiary. Grassland vegetation probably developed only as a combined result of the Kaikoura orogeny during Pliocene times and late Tertiary climatic deterioration which culminated in the Pleistocene ice ages.

New Zealand owes much of its present geography to these events. High elevation and tectonic activity along the main alpine chain exposed schist to erosion and sculpturing by several major glaciations along, and west of, the main Southern Alps, and locally elsewhere. Extensive outwash gravel plains and terraces, together with substantial deposits of loess, developed east of the Alps.

The impressive volcanic mountains of central North Island (Ruapehu, 2796 m; Ngauruhoe, 2290 m; and Tongariro, 1986 m), together with the isolated cone of Mt. Egmont (2518 m) on its west coast (Fig. 14.1) are of more recent origin than the others, reaching their present height only during or since the last Pleistocene glaciation (Stevens, 1985).

At the peak of the last (Otiran) glaciation, the snow-line was lower than today by some 1000 m

---

[1] This refers to the ancient supercontinent of the Southern Hemisphere which consisted of Antarctica, Australia, Africa, India, New Guinea, New Zealand and South America. These separated at various times from 180 to 60 million years ago, drifting apart on crustal plates through sea-floor spreading (Stevens et al., 1988).

INDIGENOUS GRASSLANDS OF NEW ZEALAND

Fig. 14.6. Montane short-tussock grassland dominated by *Festuca novae-zelandiae* and *Poa cita*, with occasional microphyllous thorny shrubs of *Discaria toumatou* (Rhamnaceae), particularly in gully on left (at *c*. 600 m on the lower slopes of the schistose Old Man Range in south-central South Island). Sheep graze in the foreground, and small snow banks persist along the lee of the summit ridge at about 1500 m.

(Willett, 1950; Gage, 1965), but vegetation zones were lower by some 800 m to the east of the heavily glaciated main divide mountains (Wardle, 1963). The prevailing westerly winds caused the snow-line to descend westward. Rain-shadows east of the main ranges substantially limited the extent of glaciation, although periglacial conditions obviously were extensive (Soons, 1962). Grassland and shrubland were widespread in subalpine and lowland areas of South Island and over parts of North Island during early Pleistocene times (Fleming, 1975). Although the extent of deforestation during glaciation continues to be debated (Wardle, 1963; Fleming, 1975; Wardle and McKellar, 1978; McGlone, 1985; Stevens et al., 1988), it is generally agreed to have been very extensive in South Island. Certainly palynological studies from several areas (the more recent ones supplemented with ^{14}C dates) have revealed successions through grassland or grassland–shrubland to taller woody vegetation in the post-glacial (Aranuian stage) sequences from the wetter, currently forested regions of New Zealand (Moar, 1971; Fleming, 1975), beginning about 14 000 years ago (Suggate, 1965).

The alpine zone, now dominated by *Chionochloa* spp. in its lower reaches, was probably established during the Pliocene, when rising mountains first exceeded the limits of tree growth. The driest eastern grasslands at low elevations in the South Island interior could have originated simultaneously with those above tree line, as the rain-shadow effects developed in conjunction with mountain building. However, the linkage of the two grassland types in many areas appears to be of recent origin, coinciding with forest destruction by

Fig. 14.7. Subalpine mixed *Festuca novae-zelandiae–Chionochloa rigida* (narrow-leaved snow tussock) grassland (transitional between montane short-tussock and tall *Chionochloa* grasslands) at about 900 m on Old Man Range in south-central South Island.

the fires of early Polynesian settlers less than 1000 years ago (Molloy et al., 1963).

Although the New Zealand alpine and grassland biotas have developed in relatively recent times and are continuing to evolve rapidly, their possible origins continue to be debated (Wardle, 1963, 1968; Raven, 1973; Fleming, 1975; Wardle, 1978).

New Zealand's mountain landscapes, then, have resulted from tectonic activity associated with their location along the boundary of the Pacific and Indian–Australian crustal plates (Stevens et al., 1988), as well as from geologically recent volcanic activity. These, together with the widespread effects of glaciation, have caused land forms in and close to the mountains to show high relief, but also to be geologically unstable. Apart from the volcanoes, the mountains that are being uplifted (at the rate of 5 to 10 mm yr^{-1}) consist of siliceous rocks — greywackes, argillites and schists — neither markedly acidic nor basic. However, differences in degree of metamorphism are important today in differentiating their erodibility. The slightly indurated sedimentary rock (greywacke) is particularly prone to frost-shattering and accelerated erosion (Fisher, 1962). This type extends down the North Island mountain chain and continues in South Island along and east of the Southern Alps (Fig. 14.8) to about 45°S, where it merges with chemically similar, but strongly metamorphosed and much more stable, schist (Fig. 14.6). The latter also extends northwards as a narrow strip on the western flank of the Southern Alps (Fig. 14.4). The mountains of north-western and south-western South Island generally consist of much harder and less erodible plutonic rocks. Substantial ultramafic outcrops, supporting characteristic stunted and open vegetation (Brooks, 1987, pp. 375–399; Lee, 1992), occur in both northern and southern South Island. These

Fig. 14.8. Low-alpine *Chionochloa macra* (slim snow tussock) grassland at about 1250 m on the Craigieburn Range, central South Island. Note the eroding, over-steepened slopes[1] of greywacke in the background, the abundant flowering of *Chionochloa* and the large flowering heads of *Aciphylla scott-thomsonii* (Apiaceae) ("speargrass"). (February, 1974.)

---

[1] This refers to slopes steepened beyond the angle of repose for rock debris (i.e. >32° slope).

have been spread some 420 km apart by movement along the Alpine Fault which coincides with the boundary of the tectonic plate and the main mountain chain. These areas support limited areas of grassland (Lee, 1992).

Climate

New Zealand's climate is predominantly oceanic. Its location in the mid-latitude belt of westerly winds, with frequent alternation of depressions and anticyclones throughout the year, means that the general climate is greatly influenced by this predominant wind flow. The country's shape, orientation and generally sharp relief result in much of the precipitation being orographic with striking west-to-east gradients paralleled by trends in sunshine, temperature extremes, humidity and evaporation rates. These contrasts are much less pronounced in North Island, which everywhere receives sufficient and adequately distributed moisture to support some form of woody vegetation. In South Island, however, a highly favourable forest environment in the western and southern lowlands becomes marginal in the lee of the Southern Alps, particularly in the intermontane basins and eastern plains where annual precipitation may be less than 800 mm and moisture demand is high (Coulter, 1973).

In the wetter regions, the alpine grasslands are delimited below by the tree line (Fig. 14.4), which is characterized (as in most regions) by a mean summer isotherm of 10°C and is probably controlled by some feature of tree growth or foliage hardening (Wardle, 1971, 1985a). Tree-line elevation declines with latitude from about 1450 m in central North Island to about 900 m in the far south (Wardle, 1973). At the upper limit of continuous grassland (Figs. 14.9 and 14.10) [sometimes referred to as the "grass-line" (Burrows, 1967), separating the low-alpine and high-alpine zones (Mark and Adams, 1973)], summer temperatures again appear to be important (Mark, 1965a; Mark and Bliss, 1970; Wilson, 1976; Meurk, 1978, 1984; McCracken, 1980). In the absence of any official climatological stations in these areas, however, only short-term data are available (Coulter, 1973). Summers are generally cool but reasonably sunny. Average global radiation varies from 475–575 langleys per day in summer to 100–200 in winter, the geographical variation reflecting both latitude and topography (Coulter, 1973); in upland regions daily values reach 700–875 langleys in mid-summer. Although the growing season extends for at least 5 months in the low-alpine zone and exposed sites in the high-alpine zone, freeze–thaw cycles are frequent throughout the year so that the frost-free period is usually less than 2 weeks (Mark and Bliss, 1970; Bliss and Mark, 1974; Williams, 1977).

Drought is not a feature of the New Zealand alpine grasslands. This has been confirmed by comparisons of measurements of soil moisture with precipitation data and estimates of potential evapo-transpiration. Indeed, substantial surpluses of water occur during most months, even on the rain-shadow mountains of South Island (Archer and Collett, 1971; Mark and Rowley, 1976; Meurk, 1978; Holdsworth and Mark, 1990). Here, interception gains from fog by the long, fine leaves of the dominant grasses may make significant contributions to the water yield (Holdsworth and Mark, 1990). At lower altitudes, however, summer droughts are an important feature of the natural and induced grasslands. Frequency, duration and intensity of droughts can be related to grassland types and the associated soils (Coulter, 1973; Molloy, 1988).

Mean annual air temperatures throughout the natural grasslands range between 0 and 11°C, with annual extremes between −20 and 38°C (Mark, 1965a; Meurk, 1984). Williams (1977) pointed out that many New Zealand tussock grasslands fall within the broad climatic characteristics of the coniferous forest biome, and outside those of arctic–alpine tundras and grasslands as defined by Hammond (1972). Hot summers are usually associated with drought when air temperatures frequently reach 30°C and occasionally 40°C. Winters are not cold by continental standards (McCracken, 1980; Wardle, 1985b), air temperatures rarely falling below −15°C, although at higher altitudes soil may freeze to a depth of 50 cm (Bliss and Mark, 1974). In the alpine zone, snow cover is important in affecting soil freeze–thaw cycles, particularly when it is redistributed by persistent wind (Mark and Bliss, 1970; Burrows, 1977b). Frequent freeze–thaw cycles cause frost lifting which ejects seedlings from moist uninsulated soil (Gradwell, 1960).

Fig. 14.9. Low-alpine *Chionochloa macra* grassland at about 1800 m, close to its upper limits (the grass-line) on The Remarkables in south-central South Island. *Chionochloa oreophila* (snow patch grass), here at its eastern limit, dominates the low depressions around the "tarns" and is still emerging from some of the receding snow banks. Most of the slopes in the background, including the peak on the left (Double Cone, 2324 m), are in the high-alpine zone occupied by fell-field and snow-bank vegetation. (January 20, 1976.)

Soils

Most New Zealand soils, being of Quaternary origin, are relatively young and therefore weakly developed, particularly at higher altitudes. Nevertheless most are zonal in that they reveal the combined effects of vegetation, climate and topography on the predominantly siliceous rocks or the drift material derived from them (Leamy and Fieldes, 1975; Molloy, 1988). Alluvium, colluvium, loess and solifluction debris are mineralogically similar. In the mountain grasslands such zonal soils are much less widespread, since most soils on steep slopes (lithosols or skeletal soils) are sufficiently unstable to be considered azonal. Traditionally, the zonal soils in New Zealand have been named according to the colour of the subsoil, although technical names to characterize the "basal form of the soil body" have also been offered (Taylor and Pohlen, 1962).

Sitiform (brown-grey) soils [approximately equivalent to aridisols (USDA, S.C.S., 1975)] are associated with lowland, semiarid grasslands in the floors of the intermontane basins of central South Island where annual precipitation ranges from 330 to 550 mm. These relatively fertile soils have characteristically brownish-grey sandy-loam top soils overlying compact brown silt-loam subsoils with clay and calcium carbonate accumulations. Their natural productivity is limited largely by moisture deficiency.

Subhumid, seasonally dry lowland grasslands (annual rainfall 500–1000 mm) are associated with palliform (yellow-grey) soils [equivalent to the fragic great groups of inceptisols (USDA, S.C.S., 1975)]. Top soils are uniformly silt loams, often

Fig. 14.10. Low-alpine *Chionochloa crassiuscula* (curled snow tussock) grassland at about 1350 m on the schistose mountains of the Southern Alps above Haast Pass. The upper limit of the low-alpine zone, the grass-line, shows beneath the main crest of the Alps in the background at about 1800 m.

of loess, with fragile aggregates. These are prone to wind and sheet erosion and, where the loess is deep, also to tunnel gullying. Most subsoils are distinctly iron-stained and mottled, with a weakly structured B horizon over a massive fragipan.

Under humid conditions (1000–2000 mm annual precipitation) fulviform (yellow-brown) soils or dystrochrepts (USDA, S.C.S., 1975) have formed. These are associated with both indigenous forest and various types of upland tussock grassland, particularly those which have replaced forest within the last millennium. Although well drained, moderately to strongly leached and of low fertility, these soils show little clay or iron illuviation. Their subsoils are yellow-brown to brown in colour.

As precipitation increases up to 2500 mm annually, in areas of higher elevation and lower temperatures, the natural *Chionochloa* grasslands of inland South Island are associated with eldefulvic (high-country podzolized yellow-brown) soils and podic soils (podzols) which are both very strongly leached and slowly draining. On gentle slopes these soils have dark-grey, firm, silty-loam top soils over a pale grey, mottled, silty loam, which is separated from the yellowish-brown silty-clay subsoil by a thin (6 mm), but distinctive, iron pan.

Chionochloa grasslands under superhumid conditions (annual precipitation >2000 mm) of western South Island and North Island mountains are associated with high-country gley and organic soils which form complex mosaics. The soil profiles are variable, with evidence of the gley process functioning beneath a blanket of peat. Variants of these main soil groups on steep slopes are typically both shallower and stonier with less differentiated profiles than their more stable and fully developed counterparts (Leamy and Fieldes, 1975).

Application of the American taxonomic system

of soil classification (USDA, S.C.S., 1975) to New Zealand soils reveals that most are grouped into the inceptisol order because of their generally youthful feature of weak horizon development. Soils in regions of high elevation and/or high rainfall, including those associated with the mountain *Chionochloa* grasslands, show the effects of strong leaching, including subsoil horizons of accumulation typical of the spodosol order. Soils of the low-elevation semiarid grassland climates of the South Island interior are classed in the aridisol order; those on flood plains with minimum horizon development fall into the entisol order; whereas those on organic-rich parent materials belong to the histosol order. The central North Island soils formed on recent (less than 40 000 years old) volcanic ash have distinctive properties associated with the domination of amorphous materials in the solum. They are classified in the inceptisol order and distinguished at suborder level as andepts, although a possible elevation of this distinction to order level, to recognize these soils as andisols, is currently being considered (M.L. Leamy, pers. commun., 1985).

VEGETATION

Vegetation pattern at time of settlement by Europeans

The pattern of grassland vegetation at the time of European settlement (*c.* 1840) is reasonably well-known (Fig. 14.3), even in the absence of detailed descriptions from that time. There is some variation of opinion as to the degree of natu-

Fig. 14.11. Forest dimples (small mounds and associated hollows) in strongly modified subalpine *Chionochloa rigida* grassland at about 700 m on the western slope of Maungatua Range in south-eastern South Island. The dimples were created by wind-throwing of *Nothofagus menziesii* trees which probably occupied the site prior to Polynesian settlement.

ralness and stability that was exhibited by the various grasslands at that time. The widespread presence of surface and buried logs and charcoal, and of forest podzols and forest dimples (Fig. 14.11), within the South Island rain-shadow grasslands up to an elevation of about 1000 m (Molloy et al., 1963), clearly indicate a previous forest cover. Radio-carbon dates of the logs and charcoal imply widespread deforestation about the 12th century, coinciding with the assumed arrival in South Island of the earliest settlers from Polynesia (Cumberland, 1962; Stevens et al., 1988). However, evidence is also claimed for a concurrent change in climate (Holloway, 1954) which may have aided replacement of forest by grassland. According to diary comments by some of the early European explorers (James Cook in particular in 1769), these areas were being burned up to the time of European settlement. More recent evidence for periodic natural fires preceding settlement by Polynesians has been accumulating (Molloy, 1977; McGlone, 1989). However, the fact that the influence of burning was not accompanied by that of mammalian grazing is significant. Grazing by ungulates is an important additional factor introduced by the European settlers (Holloway, 1959b; Howard, 1959, 1965), to which species of *Chionochloa* in particular were ill-adapted (Mark, 1965d; Williams and Meurk, 1977; Mills et al., 1989).

Tussock grassland of three broad kinds covered about one-third of the country (Fig. 14.3).

Lowland short-tussock grassland, dominated chiefly by *Elymus rectisetus* (formerly *Agropyron scabrum*), *Festuca novae-zelandiae* and *Poa cita* (formerly *P. laevis*) about 50-cm tall, covered the arid and semiarid regions of eastern and central South Island at elevations below about 900 m (Fig. 14.5). The relative importance of these species remains unclear (O'Connor, 1986; Mark et al., 1989). Thorny shrubs of *Discaria toumatou* up to 5 m tall probably imparted a savanna-like character in places, particularly on nutrient-rich outwash gravel fans and near water courses (Fig. 14.6). A few small saline depressions in the driest parts of the intermontane basins supported a sparse cover of halophytes.

Lowland tall-tussock grassland, typically dominated by a distinctive reddish-coloured species, *Chionochloa rubra* (red tussock) up to 1.5 m tall,

occurred mostly at elevations below the regional tree line on poorly drained, often peaty soils of south-eastern and southern South Island, as well as locally above tree line in similar sites. *Chionochloa rubra* was also extensive on the volcanic uplands of North Island.

Upland tall-tussock grassland, dominated by one or more of the several alpine species of *Chionochloa*, from 0.1 to 1.5 m tall, occupied the low-alpine zone and, in the absence of forest or scrub, descended locally to sea level — but more generally merged with one or other of the previous two grassland types. There are significant regional differences in distribution of the dominant *Chionochloa* species (Zotov, 1963, 1970), related in part to the present environment but with anomalies that appear to reflect historical factors, particularly Pleistocene glaciation (Burrows, 1965). Four

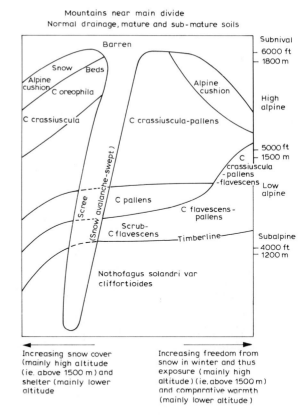

Fig. 14.12. Distribution of vegetation in the alpine zone near Arthur's Pass, central Southern Alps, in relation to altitude and snow cover. Four species of *Chionochloa* (snow tussock) are involved. (Reproduced with permission from Burrows, 1977c.)

Fig. 14.13. Low-alpine *Chionochloa* grasslands of various types at about 1200 m on the rugged Murchison Mountains of central Fiordland in south-western South Island. Relatively stable well-drained slopes above tree line (left distance) and on the bluffs on the right are dominated by *Chionochloa flavescens* and shrubs. The stabilized debris fan (middle right) is dominated by *C. pallens* (midribbed snow tussock), whereas the poorly drained flats and depressions (left mid-ground) are covered with *C. crassiuscula* (curled snow tussock).

species dominate extensively under high-rainfall conditions along the Southern Alps, where their distribution can be related to elevation, topography and edaphic factors (as affected by soil development) (Fig. 14.12). The typical altitudinal pattern in stable sites with mature soils is for the largest alpine species, *C. flavescens* or *C. rigida* (up to 1.2 m tall), to dominate the zone within 200 m of tree line, often with codominant shrubs and tall forbs (Figs. 14.2 and 14.4). With increased altitude or impeded drainage, or on cold slopes, dominance shifts to the smaller *C. crassiuscula* (*c*. 40 cm tall) which extends to the upper limits of the low-alpine zone, except for snow hollows where the smallest species, *C. oreophila* (*c*. 10 cm tall), replaces it in company with typical high-alpine species. Young or immature, well-drained soils, such as on stabilized debris fans throughout the low-alpine zone, are usually dominated by another alpine species, *C. pallens* (Williams, 1975; Williams et al., 1976) (Figs. 14.13 and 14.14), which may reach 1 m in height. Of these four species, *C. crassiuscula* and *C. oreophila* do not reach the northern South Island mountains. *Chionochloa flavescens* extends only to the mountains of southern North Island, but is replaced in the south-west by *C. rigida* extending from the east (Connor and Purdie, 1981). This leaves *C. pallens* to dominate the alpine grasslands of the remaining non-volcanic North Island mountains (Zotov, 1963). A small carpet-forming species, *C. australis*, is localized, but important, in the mountains of north-western South Island. Three other species (*C. acicularis*, *C. ovata* and *C. teretifolia*) are endemic to moorland-type alpine grasslands of the Fiordland mountains in its south-western corner (Zotov, 1963), where one of them also dominates a coastal moorland on an apparently

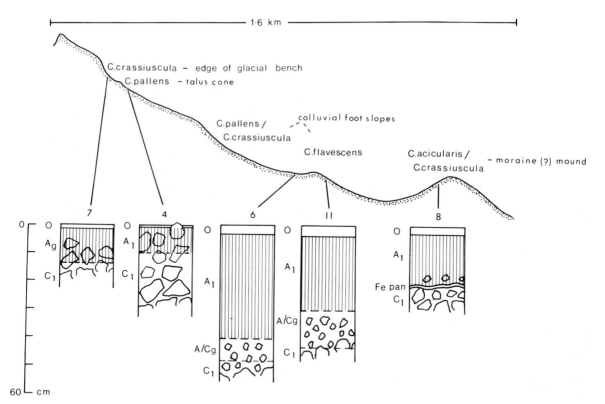

Fig. 14.14. Distribution of species of *Chionochloa* in relation to landscape components in selected sites along a diagrammatic transect of the head of Chester Burn, Murchison Mountains, Fiordland in south-western South Island. Depth of profile humification (vertical hatching) and stones or rocks are indicated. Altitudes range from 996 m (site *11*) to 1264 m (sites *7* and *4*). (Reproduced with permission from Williams et al., 1976.)

unglaciated peneplain (Fig. 14.15) (Wardle et al., 1973). *Chionochloa pungens* is confined to the low-alpine moorlands of Stewart Island in the far south (Wells and Mark, 1966). Two other species, *C. defracta* and *C. lanea*, were described from ultramafic outcrops of northern South Island and from Stewart Island, respectively (Connor and Edgar, 1987). Similar restrictions occur in other biota and are assumed to reflect differential Pleistocene survival (Wardle, 1963; Burrows, 1965).

Small areas of a fourth type, maritime grassland, dominated by tussock species of *Poa*, characterize the off-shore islands of southern New Zealand (Fineran, 1966; Johnson, 1975, 1976a) and represent a northern attenuation of the sub-Antarctic maritime grasslands (Cockayne, 1928) (see Chapter 15).

On the drier rain-shadow mountains of South Island, the western *Chionochloa* species merge with one or both of the two dominant eastern species, *C. macra* and *C. rigida* (Fig. 14.16) (Connor, 1965; Connor and MacRae, 1969; Burrows, 1977c; Meurk, 1982), although chemo-taxonomic studies indicate that *C. rigida* extends to the western mountains in the southern South Island (Connor and Purdie, 1981). This species is restricted geographically to the southern South Island, where, in the rain-shadow region, it dominates in the low-alpine and subalpine zones in the south-east (Fig. 14.17) and even reaches sea level locally (Greer, 1979). It is replaced above about 1400 m by *C. macra*, which extends up to about 2000 m in the high-alpine zone (Meurk, 1978) (Fig. 14.9) and, beyond the northern limits of *C. rigida*, also to relatively low altitudes (Connor and MacRae, 1969).

Early qualitative descriptions of these communities (such as those of Cockayne, 1928), while valuable accounts of floristics and general physiognomy, were limited because of subsequent tax-

Fig. 14.15. Coastal moorland-type grassland of regionally endemic *Chionochloa acicularis* (needle-leaved snow tussock), with scattered shrubs of *Leptospermum scoparium*, on an apparently unglaciated peneplain about 80 m above sea level at West Cape, Fiordland, in south-western South Island. View eastwards across the peneplain to the more typical glaciated and forested country of Fiordland.

onomic revisions (particularly of *Chionochloa*). This ecologically important genus was distinguished only recently (Zotov, 1963), but much unresolved variation remains within a number of species (Mark, 1965c; Burrows, 1967; Zotov, 1970; Connor and Purdie, 1976; Williams et al., 1977a; Connor, 1985; Connor and Edgar, 1987)[1]. The early accounts of Cockayne and others stressed the important role of large forbs in the high-rainfall regions prior to the establishment of several introduced mammals, *Cervus elaphus* (red deer) in particular.

After more than a century of European occupation very few areas of indigenous grassland remain unmodified. Introductions of both domestic and game animals into ecosystems that had developed in the absence of mammalian grazing, together with the use of fire to encourage more palatable regrowth for the use of livestock, initiated major changes both in structure and composition of the grasslands. More recent descriptions from the wet western mountains, which were rarely if ever grazed by domestic stock (Wraight, 1960; Mark and Burrell, 1966; Wilson, 1976; Burrows, 1977a, b; Mark, 1977), mostly have reported floras depleted of their more palatable species but with the plant cover reasonably complete, despite locally heavy grazing by wild game animals. Seriously deteriorating trends in vegetative condition and catchment stability on both North and South Island mountain ranges have prompted intensive scientific evaluations (Holloway et al., 1963), ani-

---

[1] Now largely resolved by Connor's (1991) recent revision (which post-dated the last revision of this paper).

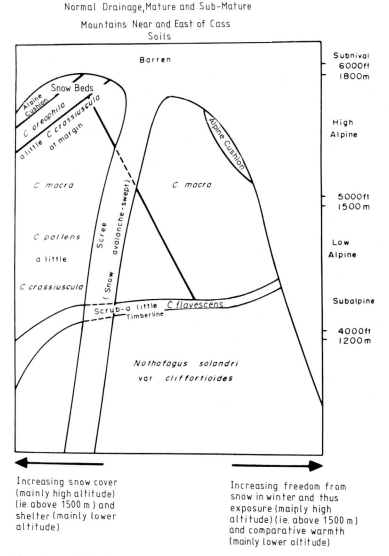

Fig. 14.16. Distribution of vegetation in the alpine zone of the Craigieburn Range in the rain-shadow region of central South Island. Five species of *Chionochloa* are involved. (Reproduced with permission from Burrows, 1977c.)

mal research and eradication (or at least control), plus revegetation programs mostly using exotic plant species. The Protection Forestry division of the New Zealand Forest Service was particularly involved in these activities (Riney, 1956; Morris, 1965; Orwin, 1978; Holloway, 1982).

The initial pattern of occupation in the traditional pastoral "run" country (rangeland) in the *Chionochloa* grasslands of the South Island rain-shadow mountains has been outlined by O'Connor and Kerr (1978), whereas evolution of the pastoral communities and their management practices in two high-country regions have been described by McCaskill (1969) and O'Connor (1978). European settlement has resulted in up to 150 years of periodic burning combined with extensive grazing, initially by sheep, with few official restrictions. This led to serious deterioration in the condition of these grasslands. In the more vulnerable regions this was accompanied by serious erosion, with which soil and water conservation practices, at considerable cost to the nation, are now at-

INDIGENOUS GRASSLANDS OF NEW ZEALAND

Fig. 14.17. Low-alpine *Chionochloa rigida* (narrow-leaved snow tussock) grassland at 1220 m on the schistose Old Man Range of south-central South Island. *Chionochloa* still provides much of the cover, despite at least a century of sheep grazing and periodic burning. (April, 1967.)

tempting to deal (McCaskill, 1973; Moore, 1976; Holloway, 1982). More effective land tenure, and official control of burning and stocking, have been achieved within the last 40 years, but not before substantial modification had occurred to the original cover of the tussock grassland (Figs. 14.18 to 14.20). The situation was described for the northern regions of South Island by Wraight (1963), for the central regions by Connor and co-workers (Connor, 1964, 1965; Connor and MacRae, 1969) and by Barker (1953), and in several unpublished government reports, and for the more southerly areas by Daly (1967) and Mark (1965d). Connor (1964, 1965) in particular has attempted to trace the derivation of present grassland communities on the basis of their floristics, using the Braun-Blanquet phytosociological approach (Fig. 14.21).

Rabbits (*Oryctolagus cuniculus*) introduced and liberated soon after settlement, increased steadily to reach plague proportions over much of the run country by the late 1940s, before being brought under reasonable control with an government extermination policy during the 1950s (Howard, 1958; Holloway, 1982). Meanwhile rabbits contributed substantially to the serious deterioration of lowland tussock grassland in the driest areas and its replacement by a semi-desert type of vegetation (Zotov, 1938) dominated by an indigenous cushion composite, *Raoulia australis* (Fig. 14.22), which typically pioneers river-bed successions. This control of rabbit numbers in most areas allowed tussock grassland to be re-established (Fig. 14.23), but a recent resurgence of rabbits in many of these areas (Fraser, 1988) has resulted in serious depletion of the grassland and a concurrent "explosion" of the exotic rosette weeds *Hieracium* spp. (Scott, 1984; Makepeace et al., 1985). Elsewhere the lowland short-tussock grassland has been substantially reduced by cultivation of the gentle slopes since European occu-

Fig. 14.18. Short *Poa colensoi* (blue tussock) grassland induced by periodic burning and severe grazing by sheep from a low-alpine *Chionochloa macra* grassland at about 1370 m on the Old Man Range in south-central South Island. Relict tussocks of *Chionochloa* testify to their former importance.

pation, whereas the steeper areas have also been modified by aerial applications of fertilizers and seed of exotic grasses and legumes, by periodic burning, and by almost continuous grazing by livestock.

Seasonal aspects

Since all of the dominant tussock species and most of the accompanying forbs are evergreen, it is only in mid-summer, when the many forbs that characterize the *Chionochloa* grasslands of high-rainfall regions are in flower, that there is any obvious seasonal aspect to the grasslands (apart from winter snow cover). The few geophytes in the grasslands [various ground orchids (*Caladenia* and *Prasophyllum*) and *Bulbinella* (Liliaceae) in partic-

ular] are not usually in sufficient numbers to impart a vernal aspect to a vegetation that responds with relative slowness to the increasing temperatures, but typically changeable weather, which characterize the New Zealand spring. Irregular flowering of the *Chionochloa* species (Fig. 14.24) (and some other genera, such as *Aciphylla*, *Celmisia* and *Phormium*) results in even the dominant species inconsistently revealing the passage of summer. The irregular flowering is synchronized within *Chionochloa* (but not necessarily between genera), being dependent on a cumulative effect of high temperatures during the long-day period of the preceding summer. This is followed by autumn initiation and over-wintering of partially formed inflorescences (Mark, 1965b, 1968; Connor, 1966; Scott, 1978). Seasonal changes in

Fig. 14.19. An induced cover of *Festuca matthewsii* (alpine fescue tussock), *Poa colensoi* (blue tussock) and *Aciphylla aurea* (Apiaceae) ("speargrass" or "spaniard") of low grazing value, which has replaced low-alpine *Chionochloa rigida* grassland at 1160 m on a north-west facing slope in the Dunstan Mountains of south-central South Island.

canopy colour of the tussock grasslands are relatively slight [albedo (i.e. reflectance) values for *Chionochloa* grassland range around 0.24], despite the increased leaf die-back in autumn (Mark, 1965a; Rowley, 1970; Williams, 1977).

Relatively long growing seasons (of 5 to 8 months) have been indicated by a few detailed phenological studies of tussock-grassland species and associated flora (Scott, 1960, 1978; Bliss and Mark, 1974) and confirmed by measurements of vegetative growth (Fig. 14.25) (Mark, 1965a, c; Rowley, 1970; Bliss and Mark, 1974; Williams, 1977; Scott, 1978). Measurements of a few grassland species showed a slight to pronounced peak in vegetative growth rates in mid-summer (December–January), followed by a steady autumnal decline (Mark, 1965a, c; Rowley, 1970; Williams, 1977; Scott, 1978). Variations in growth rates are most closely related to soil and air temperatures (Mark, 1965a; Scott, 1978; Greer, 1979).

Plant–soil relationships

The broad distribution patterns of the main grassland types, reflecting equally broad patterns of both soil and climate, are well established in the literature. However, some intensive studies have investigated climo-sequences and topo-sequences of tussock-grassland soils, embracing both low-rainfall and high-rainfall regions. From the drier high-country yellow-brown earth (dry-hygrous eldefulvic) region of the South Island interior, Ives and Cutler (1972) described a topo-sequence in which increasing soil development is related to both increasing altitude (700 to 1600 m) and decreasing slope gradient. The chemistry of a range

Fig. 14.20. Low-alpine *Chionochloa rigida* (narrow-leaved snow tussock) grassland burned one year earlier (at about 1070 m on St. Bathans Range in south-central South Island). To the right of the fence only a small area was burnt and recovery is being seriously impaired by concentrated grazing on this area. There is negligible inter-tussock cover on either side of the fence. (October, 1958.)

of South Island tall-tussock soils was described by Williams et al. (1978d), but they reported few significant differences in relation to either altitude (from montane to high-alpine) or precipitation (from <1000 to >2500 mm annual precipitation).

A very intensive study by several soil specialists demonstrated the importance of climate in development of tussock-grassland soils along a wide climo-sequence across the southern South Island. In particular, forms of phosphorus, together with extractable iron and aluminium, were used as criteria for ranking the soils in a development sequence (brown-grey earths < yellow-grey earths < high-country and alpine yellow-brown earths < upland yellow-brown earths < podzolized upland and high-country yellow-brown earths) consistent with a zonal classification (Molloy and Blakemore,

1974). Several related studies dealing with invertebrates, microbial and enzyme activities, carbohydrates, and litter decomposition in the nine soils

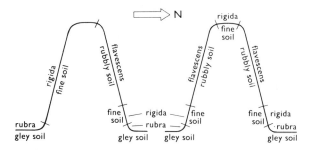

Fig. 14.21. Diagrammatic zonation of *Chionochloa* spp. from valley bottoms and terraces to the mountain tops in the middle Rakaia Valley of central South Island. (Reproduced with permission from Connor, 1965; for "*C. rigida*" read "*C. macra*".)

Fig. 14.22. Montane induced semi-desert dominated by *Raoulia australis* ("scabweed", Asteraceae) on a dry slope at about 670 m on the Old Man Range in south-central South Island. The original cover of short-tussock grassland dominated by *Festuca novae-zelandiae* persists in the slightly moister depressions whereas a gravel "pavement" covers eroded soil between the scabweed cushions. Photographed near the site shown in Fig. 14.23, but in 1959, before rabbits (*Oryctolagus cuniculus*) had been more adequately controlled.

along this sequence are discussed later (pp. 396–398). Other studies have shown similar patterns along an altitudinal gradient (Molloy, 1988) or more local plant–soil correlations that are important in accounting for mosaics or small-scale vegetation zonations. Connor (1965) related the distribution of three tall-tussock species and their associated floras to slope, aspect and soil development in the rain-shadow of central South Island (Fig. 14.21). A predominantly western species, *Chionochloa flavescens*, favours sunny slopes with coarse-textured stony soils, and occurs on shady slopes only where coarse gravels are exposed or covered with shallow, coarse-textured soils. *Chionochloa macra* (previously included in *C. rigida*), a predominantly eastern species, occupies fine-textured soils on shaded slopes or flattish ridges, whereas *C. rubra*, the lowland red tussock species, prefers "flats, terraces, moraines and basins below the mountain slopes" on permanently or seasonally wet sites with gleyed soils.

Subsequent studies in the high-rainfall alpine regions of southern North Island and south-western South Island have confirmed, for alpine *Chionochloa* species and their associated flora, the importance of certain edaphic factors, particularly drainage, organic carbon, carbon/nitrogen ratios, and available nutrients (especially phosphorus), and have revealed the importance of local chronosequences related largely to topographic variation (Figs. 14.13, 14.14 and 14.26) (Williams, 1975; Williams et al., 1976). Levels of macro-elements in the dominant tussocks vary between regions, within plants (green blades, sheaths, stems and

Fig. 14.23. Induced semi-desert of *Raoulia australis* ("scabweed") vegetation being reinvaded, particularly by grasses [including indigenous tussocks of *Festuca novae-zelandiae* and annual adventives (*Aira* spp. and *Vulpia dertonensis*)]. Photographed near the site shown in Fig. 14.22, but in 1965, several years after rabbits (*Oryctolagus cuniculus*) had been more adequately controlled.

roots), between species, and even within species locally on different sites. However, the variation generally reflects the levels of nutrients available within the habitats preferred by the species in question (Williams et al., 1978a–e).

A greenhouse study with seedlings of four alpine *Chionochloa* species grown reciprocally on their respective soils, confirmed that soils strongly influence growth. Moreover, the various species proved to be adequate assessors of the soil differences between their natural habitats (Molloy and Connor, 1970).

A study of macro-element pools and fluxes in indigenous tussock grasslands involved stands of subalpine *Chionochloa rigida* at 884 m and low-alpine *C. macra* at 1257 m in the rain-shadow mountains of central South Island (Williams et al., 1977b). Above-ground pools of elements were invariably larger in the subalpine stand (range between stands was 143–64.4 kg ha^{-1} for nitrogen, 24.4–9.5 kg ha^{-1} for phosphorus, and 4.0–1.5 kg ha^{-1} for sodium), with relatively large proportions in the substantial dead component (*C. rigida*, from 76% for calcium to 35% for potassium; *C. macra*, from 77% for calcium to 23% for potassium). Under-ground pools were generally similar to those above ground (nitrogen, 154 and 100 kg ha^{-1}; phosphorus, 19.0 and 15.8 kg ha^{-1}; sodium, 10.4 and 7.3 kg ha^{-1} — for the *C. rigida* and *C. macra* stands, respectively). The order of total element concentration also varied somewhat between the two species, being:

C. rigida   K > N > Ca > P > Mg > S > Na
C. macra   K > N > P > Ca > S > Mg > Na

Estimates of annual element uptake for each species were derived from seasonal differences

Fig. 14.24. Heavy flowering of *Chionochloa rigida* (narrow-leaved snow tussock), the dominant in low-alpine grassland, at about 1200 m on the mid-slopes of Mt. Cardrona in south-central South Island. Flowering is in response to an abnormally warm summer in the previous growing season (photographed in February, 1979).

between shoot weights plus elements lost in litter fall. Among the major elements, values (in kg ha^{-1}) were:

	K	N	Ca	P
C. rigida	35.0	19.1	9.1	7.1
C. macra	11.7	9.5	6.5	3.7

The schematic model of element pools in live and dead compartments of these grasslands in a non-flowering state are given in Fig. 14.27, together with transfer pathways derived from this study. The relatively large pools above ground, particularly of dead plant material — both attached and unattached — are apparent, as is the general similarity in pool sizes of live blades, leaf sheaths, stems and roots. About one-fourth of each live pool comprises production of the current growing season.

Stability

Much of the New Zealand plant–soil mantle, including the tussock grasslands, shows obvious features of current instability, which can usually be related to one of two major causes — vast areas of naturally over-steepened terrain associated with their recent origin and current tectonic activity; or some form of direct or indirect human interference. A range of palynological and geomorphic information (Molloy, 1962; Suggate, 1965; Moar, 1971; O'Connor, 1984; McSaveney and Whitehouse, 1989) indicates that the current instability is not a recent phenomenon in many areas of tussock grassland.

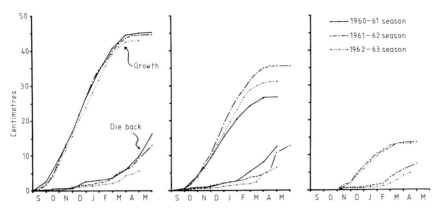

Fig. 14.25. Mean cumulative leaf elongation and die-back in the youngest exposed leaves (from three tillers in each of 10 tussocks of *Chionochloa*) for three consecutive seasons at three altitudes on Old Man Range in south-central South Island. Values for the 910 m (left) and 1220 m (centre) sites are for *C. rigida*, those for the high-altitude (1590 m) site (right) are for *C. macra*. (Reproduced from Mark, 1969.)

Fig. 14.26. Patterns of low-alpine *Chionochloa* grassland at an elevation of 1370 m on the Tararua Range in southern North Island. *Chionochloa flavescens* dominates on mature soils in foreground and at the right, whereas *C. pallens* along with the silvery forb *Astelia nervosa* (Liliaceae) and dark shrubs of *Olearia colensoi* (Asteraceae) (leatherwood) cover the better-drained younger soils on the left. This is the area studied by Williams (1975).

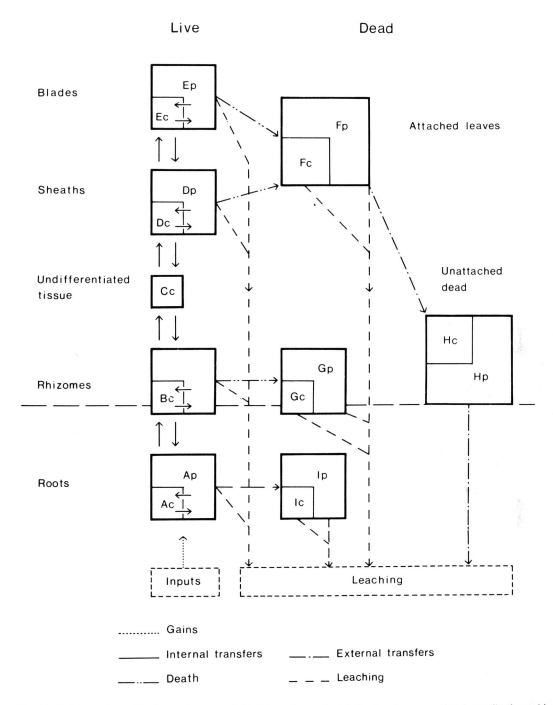

Fig. 14.27. Schematic model of part of a perennial tall-tussock grassland indicating the proportional contribution to biomass at the time of maximum standing crop of each of the major live and dead compartments (A to I) and identifying transfer pathways of materials. Subscript "c" indicates materials in the current new live tissue or litter, and "p" indicates those present in compartments from previous growing seasons and periods of litter formation. (Reproduced with permission from Williams, Nes and O'Connor, 1977.)

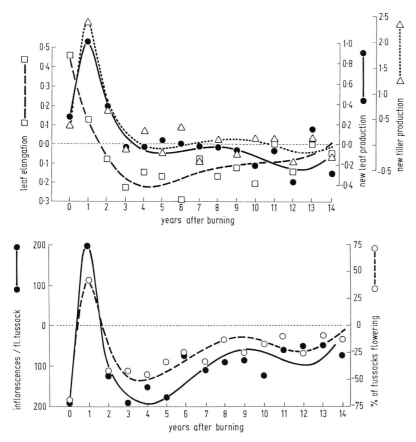

Fig. 14.28. Long-term effects of a single spring-time fire on vegetative (above) and reproductive (below) behaviour of narrow-leaved snow tussock (*Chionochloa rigida*) at 1220 m on the Old Man Range in south-central South Island. The values are relative to those of unburned plants at the same site. Vegetative growth is expressed as total seasonal leaf elongation of the youngest leaf per tiller, new leaf production per tiller, and new tiller production per tiller. Reproductive response is expressed as number of inflorescences per flowering tussock and percentage of tussocks flowering. Excluding the season of burning (year 0), the curves are computer-drawn fourth-degree polynomials. (Redrawn with permission from Payton and Mark, 1979.)

Several studies have been made of soil development and associated primary successions (Calder, 1961; Molloy, 1964; Archer, 1973; Archer et al., 1973; Sommerville et al., 1982), but secondary successions have received only cursory attention (Calder and Wardle, 1969; Wardle, 1972; Roxburgh et al., 1988) despite their significance. Secondary successions related to human interference are widespread because of grazing and burning, particularly in the eastern South Island grasslands. Successions within the *Chionochloa* grasslands are likely to be relatively slow because of the slow growth rates and extreme longevity of the individual tussocks and several associated species (Lough et al., 1988), features that prompted the appropriate comment some 30 years ago that tussock grassland "has many of the characteristics of a forest and few of those of a short-rotation pasture. Like a forest, it is the product of a long slow development and like a forest it is much easier to destroy than to rebuild" (Moore, 1955).

The hypothesis of a climatic change in the 13th century was invoked by a special committee of government scientists (including J.T. Holloway, L.B. Moore and J.D. Raeside) to explain the apparently inherent instability and "relict status" of extensive areas of tall-tussock grassland on the South Island rain-shadow mountains (Tussock Grassland Research Committee, 1954). This hypothesis was developed largely on the basis of apparent forest anomalies in South Island (Holloway, 1954), together with the occurrence of outlier areas of "relic" tussock-grassland soils and vegetation beyond the upper limits of continuous plant cover on some of the interior South Island mountains (Raeside, 1948). Features that were stressed to justify such relict status were the "notable rarity of seedling snow tussocks" and their "failure to survive". However, subsequent detailed autecological studies (within part of the region they considered) showed this reasoning to be untenable (Mark, 1965b). On the contrary, seeding, germination and seedling establishment, although intermittent, were found to be adequate for such potentially long-lived species. Rather, the deteriorating trends in the condition of the grasslands can now be ascribed to an intolerance of grazing, which periodic burning had aggravated by increasing the palatability of herbage to livestock (Mark, 1965d) through reallocation of nutrients from root to leaf tissue (Williams and Meurk, 1977; Payton et al., 1986).

Both vegetative growth and flowering of *Chionochloa macra* and *C. rigida* are stimulated within the first 2 years following fire (Mark, 1965b, d; Rowley, 1970; Payton and Brasch, 1978). An extended study revealed that, following the initial stimulus, vegetative and reproductive performance in *C. rigida* was impaired for as much as 14 years (Fig. 14.28) (Payton and Mark, 1979; Payton et al., 1986).

Standing crop

Values for above-ground biomass (including dead material) are available for six community types. Short turf-forming carpet grass (*Chionochloa australis*) from an altitude of 1370 m in northern South Island gave a value of 5.9 kg m^{-2} (Wraight, 1965). Single values for four monospecific stands of *Chionochloa* (including dead attached material but not dead detached leaf bases) on the central Southern Alps at altitudes of 1220 to 1800 m were (in kg m^{-2}):

C. pallens	(40% cover)	2.20
C. flavescens	(30% cover)	1.80
C. crassiuscula	(50% cover)	1.58
C. oreophila	(45% cover)	0.26

For a stand of *C. macra* farther east (40% cover) the value was 1.83 kg m^{-2} (Burrows, 1977c). This latter value is at the lower end of the range obtained by Williams (1977) and Meurk (1978) for total above-ground phytomass (1.60–6.62 kg m^{-2}) of denser *C. macra* alpine grasslands in central and south-central South Island, respectively, at altitudes of 1260 to 2000 m, whereas their three values for the taller *C. rigida* grassland from the same regions but at lower elevation (880–1000 m) ranged from 3.8 to 8.7 kg m^{-2}. A distinctive feature of these phytomass values is the large proportion contributed by the dead component: 55 to 79% in three stands of *C. rigida* and 56 to 77% in five stands of *C. macra*. Therefore, the values for live above-ground phytomass (i.e. green leaf plus stem) are within the range of 360 to 1850 g m^{-2} (Williams, 1977; Meurk, 1978).

Only the latter two authors have presented estimates of root (under-ground) phytomass (Williams, 1977; Meurk, 1978). These are generally in the order of 1.5 to 3.1 kg m^{-2} for dead plus live roots. Accordingly, root/shoot ratios, based on live plus dead compartments, range from 0.32 to 0.82 for *C. rigida* and from 0.32 to 1.39 for *C. macra*. These values are relatively very low compared to grasslands in other countries, where ratios are often around 2 or more. Low root/shoot ratios also appear to characterize high-alpine vegetation in New Zealand (Bliss and Mark, 1974) and may reflect the comparatively long growing seasons and mild winters experienced here.

Meurk (1978) pointed out that, although the above-ground standing crops of New Zealand mountain *Chionochloa* grasslands are substantially greater than those of most natural grasslands elsewhere, the relatively low root/shoot ratios result in the total phytomass in these New Zealand communities being within the range of those published for closed herb–shrub grasslands in other regions, in particular climax low arctic-alpine shrub tundra, prairies and temperate disclimax grasslands.

Productivity

Whereas phytomass determinations in New Zealand tussock grasslands offer no particular problems and several have now been made, measurements of productivity and turnover encounter difficulty. Unfavourable response to clipping (Mark, 1965d) in the large-statured but slow-growing, evergreen *Chionochloa* grasslands has precluded use of the traditional harvest method and has necessitated considerable innovation.

The tiller has been used as the unit of measurement in studies of biomass production in these tussock grasslands, because the large size of individual tussocks and their evergreen habit makes the use of whole tussocks impracticable. Mature tussocks of most *Chionochloa* species usually contain several hundred tillers. Wraight (1965) and Burrows (1977c) adopted simplified approaches in their studies of pure stands of *Chionochloa*. For *C. australis*, Wraight estimated annual net production of leaves from three parameters: (1) the number of new leaves produced annually per tiller (mean = 2.05); (2) values for mean annual stem elongation as indicated by nodal leaf scars (4.06 to 11.93 mm depending on "turf type"); and (3) weights of oven-dried green leaves separated from representative turfs (15 × 15 cm). His estimate was 268 g m^{-2}. Since weight determinations were made only once, in mid-autumn, with no measurement or allowances for any losses, the net production value derived must be assumed to be only approximate.

Burrows (1977c) made a crude estimate of annual net above-ground production of *Chionochloa* in three stands on the basis of three parameters: mean number of new leaves produced annually per tiller; mean leaf dry weight; and number of mature tillers per unit area. Although there were some errors in the published values, the following estimates of production during a 6.5-week midsummer period were confirmed (pers. commun., 1979):

C. crassiuscula	949 g m^{-2}
C. pallens	718 g m^{-2}
C. oreophila	180 g m^{-2}

These values included most of a year's production, but were considered to be somewhat low because of the "cool stormy summer".

A more sophisticated method was adopted by Williams (1977) in two relatively undisturbed pure *Chionochloa* stands on a rain-shadow mountain of central South Island. One was of a *C. rigida* stand (91 cm tall) with a mean tussock density of 1.48 plants m^{-2} and mean basal area per plant of 13.6 dm^2 at an elevation of 884 m; the other was of a shorter (70 cm) *C. macra* stand with a greater density (2.83 tussocks m^{-2}) but smaller basal area (7.1 dm^2) at 1257 m. Net production values were determined from 5 marked tillers in each of 10 randomly selected tussocks per site, adjusted to a unit area on the basis of measured tiller density. Values for periodic measurements of leaf elongation and die-back, numbers of new leaves appearing on marked tillers, and numbers of new tillers, were converted to net production on the basis of dry-weight values for several components [dead leaf tips, green blades, sheaths, live rhizomes, dead rhizomes, and roots (Fig. 14.29)]. These were derived from destructive sub-sampling of 10 tillers

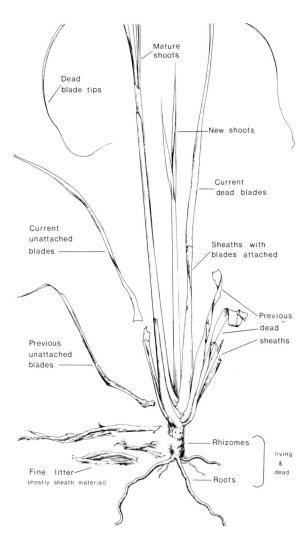

Fig. 14.29. Semi-diagrammatic sketch illustrating the structure of *Chionochloa rigida* tillers and the terms used to describe the components. (Reproduced with permission from Williams, 1977.)

from individual, randomly selected tussocks during late summer. Additional periodic destructive sub-sampling of individual tussocks was used to follow the seasonal pattern of tiller composition and weight-to-length ratios of the various components. Green-leaf standing crops of tussocks in both stands increased significantly between the early-spring (October) and autumn (March) sampling dates, but the peak values, based on number of leaves produced, were assumed to have occurred during late summer (February). Annual shoot net production (leaves only) was calculated to be 518 g m^{-2} for *C. rigida* and 330 g m^{-2} for *C. macra*, based on a 7.5-month growing season. Total annual above-ground net productivity for the tussocks in these stands was estimated to be 539 g m^{-2} and 346 g m^{-2}, respectively. Total net production (including under-ground portions) was estimated to be 1083 g m^{-2} and 726 g m^{-2}, respectively (Williams, 1977). Daily rates were in the order of 3.1 g and 2.0 g m^{-2} (leaves only), respectively. For both species, leaf sheaths continued to increase in weight as their blades died back from the tips, a process which increased steadily through summer to peak during autumn. Williams presented a schematic model to indicate biomass distribution in the major compartments of both stands. That for *C. rigida* is presented in Fig. 14.30. The pattern for *C. macra* is similar, but with lower values.

A somewhat more critical study of net primary productivity, involving corrections of simulated harvest data for "invisible" losses, was made by Meurk (1978, 1982). He examined seven stands of tall *Chionochloa* grassland involving the same two dominant species studied by Williams, together with shorter herbfield, cushion and turf grassland stands, at 920 to 1998 m on the rain-shadow mountains in south-central South Island. In pure stands of tall *Chionochloa*, lowest production values were at the lowest elevation and highest values at intermediate elevations (Table 14.1). Growing season efficiency of fixation of usable solar energy (0.6 to 1.2%; mean = 1.1%) tended to increase with altitude. This reflected an increase in average above-ground daily net production of the *Chionochloa* grasslands with altitude from about 4 to 7 g m^{-2}, with peak values of 5.7 to 10.6 g m^{-2} day^{-1} in midsummer (late December). Meurk considered these to be relatively high values which suggested

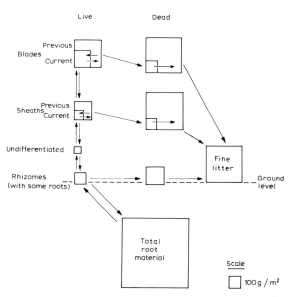

Fig. 14.30. Summary of the distribution of biomass (g m^{-2}, including ash) in a stand of *Chionochloa rigida* at 884 m in the upper Ashburton catchment, central South Island. Above-ground represents *Chionochloa* components only, whereas "total root material" includes all roots in the stand, both live and dead. Arrows indicate inferred directions of carbon flows. The "undifferentiated" shoot tissue and "current" sub-compartments indicate the estimated size of the annual dry matter (including ash) fluxes. Note scale at bottom right. (Reproduced with permission from Williams, 1977.)

adaptation to the "short, climatically erratic summers". Comparable values for the shorter vegetation types were generally only one-half to two-thirds those of the *Chionochloa* stands. The seasonal net production of above-ground biomass per degree-day similarly attained greatest values at the highest altitudes, both in *Chionochloa* (0.4 to 1.61 g d-d^{-1}) and in the shorter vegetation (0.35 to 0.78 g d-d^{-1}). The degree-day values (0°C base) for the snow-free season ranged from 2064 at 1000 m elevation to 522 at 1998 m (Meurk, 1978, 1982).

Meurk attempted to predict net productivity for each of his study sites, given a "normal" climatic regime, by relating the significantly different shoot-production values from three growing seasons of investigation to the degree-day and rainfall/evapo-transpiration characteristics of those same growing seasons. Extrapolations were then made, based on long-term normals for these environmental parameters, to derive estimates for: "normal" total net primary production; net en-

TABLE 14.1

Estimates of annual primary production and energy flow in five tall *Chionochloa* stands at various elevations [1]

	Elevation above sea level (m)		
	920	intermediate	1998
Net biomass production (g m^{-2})			
Above-ground	550	760–840	628
Under-ground	406	370–500	422
Total	956	1211–1263	1050
Net energy fixed (kJ m^{-2})	17610	22870–23974	19757
Length of growing season (days)[2]	267	225	148
Number of degree-days (0°C base)	2063		522

[1] After Meurk (1978). Mature tussock densities ranged from 0.7 to 1.67 m^{-2} for *C. rigida* and from 2.7 to 5.3 m^{-2} for *C. macra*, with about 370 to 1400 mature primary tillers m^{-2}. Inter-tussock species accounted for 20 to 80% of the ground cover, but their productivity was only estimated indirectly. Total above- and under-ground phytomass ranged from about 5.1 to 8.7 kg m^{-2} (92000–167000 kJ m^{-2}) in *Chionochloa* to 1.9 to 3.9 kg m^{-2} (34000–69000 kJ m^{-2}) in the single-layered shorter vegetation (short grassland, herbfield, cushion).
[2] Based on periodic measurements of leaf elongation in *Chionochloa*.

ergy fixation; ecological efficiency (on the basis of snow-free season, and photosynthetically active radiation); and net primary production per degree-day (snow-free season). Meurk suggested that "normalized" total annual net production values for tall *Chionochloa* grassland in the low- and high-alpine zones ranged from 750 to 1040 g m^{-2} or from 13600 to 19100 kJ m^{-2}, with similar values for a low-alpine turf grassland of *Poa colensoi* that had been induced (by burning) from a tall *Chionochloa* grassland (1000 g m^{-2} and 17500 kJ m^{-2}). A multiple regression between net energy fixation (*NEF*: kJ m^{-2} yr^{-1}), seasonal accumulated degree-days (*D*), and annual estimated evapo-transpiration/mean annual precipitation (*E*) indicated a curvilinear relationship between net annual production in tussock grassland and the correlates of altitude in the region. The regression:

$$NEF = 2.674 + 1.221 D - 0.00053 D^2 + 1699.7 E - 1467.5 E^2$$

(with $P = 0.03$) accounted for 99% of the variation. Thus, according to Meurk, "the dominant effect of increasing accumulated temperature and the correlated evapo-transpiration ratio is for *NEF* to increase. However, above a certain point (1.15) the *E* ratio becomes critical and thereafter increasing water stress (and the concomitant temperature) has a depressing effect on growth." He admitted, however, that the case for this latter effect rests on somewhat questionable data for the lowest site and requires verification. Meurk suggested that his estimates of phytomass and productivity were higher than those of Williams (1977) (for similarly mature stands of the same species) because of differences in methodology and his higher values for tiller density (370–1400 primary tillers m^{-2} compared with Williams' values of 250–450). He claimed that, despite few supporting data as yet, annual net production values approaching 1000 g m^{-2} yr^{-1} can no longer be considered atypical for *Chionochloa* grasslands in the subalpine or low-alpine zones.

Meurk (1978, 1982) observed that total phytomass and productive capacity of the New Zealand mountain *Chionochloa* grasslands approximate those of the herbaceous layers of tropical savannas, semi-aquatic rush–sedge stands, sub-Antarctic tussock grasslands, and lush herbfields in other parts of the world. As alpine vegetation, however, the *Chionochloa* grasslands are physiognomically atypical of many tundras. This feature, together with low root/shoot ratios and high retention of photosynthates in the canopy (about 63% in *Chionochloa* grassland and 56% in shorter communities) probably reflects the oceanic influence, even in the New Zealand interior. Meurk's study area was in New Zealand's most continental region, yet had a continentality index of only 10 to 15 on Conrad's (1946) 1-to-100 scale from extremely oceanic to extremely continental (in which 30–40 is considered typically continental) (Meurk, 1978).

The results of studies of biomass production confirm those relating to eco-physiology (Mark, 1975; Greer, 1984), seasonal growth patterns, and environments. Both approaches support the view that the ecological success of the genus *Chionochloa* relates to: the evergreen habit and long life of the individual plants; their slow but steady growth over a relatively long growing season, coupled with low energy and nutrient demands; and tolerance of low temperatures. Combined, these features obviously achieve an effective adaptation to the cool, moist, windy, generally oceanic environments that characterize most New Zealand mountains.

Turnover

Turnover times for snow-tussock leaf blades on the ground were estimated from litter-bag studies to be about 10 yr for the *C. rigida* stand, and 20 yr for *C. macra* at the higher altitude (Williams et al., 1977b). Mean turnover times corrected for "invisible" degradation losses for all vegetation types studied by Meurk (1978, 1982) were similar to those of Williams et al. (1977b), being 10.9 ± 2.4 yr, with a maximum of 21.8 yr at the highest site.

FAUNA

Vertebrates

The thirteen species of extinct flightless moas in several genera of Dinornithiformes (Anderson, 1989), although widespread, probably played a minor role in tussock grassland, since they appear to have been browsers rather than grazers (Greenwood and Atkinson, 1977) and more important in forest than grassland (Simmons, 1967; Burrows et al., 1981). Indeed, their decline appears to have coincided with the widespread destruction of forest and expansion of grassland associated with early Polynesian settlement (Fleming, 1975; King, 1984; Anderson, 1989). Both of the remaining avian grazers, *Notornis mantelli* (flightless rail or "takahe") and *Strigops habroptilus* (flightless parrot or "kakapo"), are restricted to remote areas in southern New Zealand and both are seriously endangered. Each utilizes *Chionochloa* species in a distinctive manner (Johnson, 1976b; Mills and Mark, 1977) and has suffered in competition with introduced *Cervus elaphus* (red deer) and through predation by mustelids, particularly *Mustela erminea* (stoat) (King, 1984). *Notornis mantelli* is now restricted to a 650-km^2 area of Fiordland National Park. Mills (1976) described the population dynamics of banded individuals in a limited area and revealed a continuing decline. Mills and Mark (1977) accounted for this decline, in part, by the ability of the co-existing *Cervus elaphus*, like *Notornis*, to select, both inter- and intra-specifically, those plants with relatively high nutrient and carbohydrate levels among the five alpine *Chionochloa* species present (Figs. 14.31 to 14.33). Moreover, it has been revealed that the mode of feeding of *Notornis* (pulling out tussock tillers from the base) caused no significant decline in tillering rates in its preferred *Chionochloa pallens*, even with up to 50% of the tillers removed; however, severe clipping to simulate deer grazing invariably caused a decline in tillering (Mills et al., 1989). With only 10% of the tillers removed (about the normal level of feeding by *Notornis*) tillering rates were enhanced (W.G. Lee, pers. commun., 1988). This could be a case of co-evolution between *N. mantelli* and *Chionochloa pallens*.

The sketchy information on secondary production in the tussock grasslands is probably due to the lack of an indigenous grazing mammalian fauna and the minor role played by the endemic avifauna (Clarke, 1970). The latter may be associated with irregular flowering and an unreliable supply of seed in the genus *Chionochloa* (Mark, 1969), which in itself may represent a case of predator satiation by allowing a high percentage of seed to escape predation in flowering years, as has been suggested for masting in certain tree species (Silverton, 1980; Norton and Kelly, 1988).

The mountain *Chionochloa* grasslands are among the poorer grazing ranges of the world. Extensive grazing of *Chionochloa* grasslands on

Fig. 14.31. A dense stand of *Chionochloa pallens* at about 1100 m in the Murchison Mountains of Fiordland in south-western South Island showing droppings of the endangered *Notornis mantelli* (flightless takahe) and signs of their feeding (the small pile of cut tillers of Chionochloa at left in the inter-tussock area). (Ice axe handle is 70 cm long.)

the South Island rain-shadow mountains by domestic animals, chiefly sheep, probably results in an annual consumption of only 10 g of herbage m^{-2} (based on a stocking rate of about 0.5 ewe equivalents ha^{-1} for approximately 4 months, and on a mean intake of 600 kg $sheep^{-1}$ yr^{-1}) (Meurk, 1978). Additional consumption by grasshoppers is as much as 5 g m^{-2} yr^{-1} (White, 1978).

Information on the composition of the diet of sheep foraging the tussock grasslands was based initially on direct observations (Cockayne, 1919), followed by qualitative assessments from faecal analyses, with identification of plant cuticles (Croker, 1959). Errors involved with differential digestibility between species in the diet may be reduced by using rumen samples, as in a recent study of the diet of *Cervus elaphus* in the Fiordland mountains of south-western South Island (Lavers, 1978), where these animals are being intensively hunted because of their competition with *Notornis mantelli*.

No field studies have been made of secondary productivity of any wild animals introduced into New Zealand. However, recent measurements of secondary production of *Cervus elaphus* confined on high-producing exotic pasture indicate a significantly greater efficiency than that of sheep [about 9.5 kg vs. 30 kg of pasture dry matter per kg of carcass gain for deer and lambs, respectively (Drew, 1979)]. The much smaller fat content of venison (10 to 12%) than in most commercial lamb and beef carcasses (25 to 35%) means that its energy value is correspondingly less (6.3 kJ g^{-1} of fresh venison leg; 14.6 kJ g^{-1} of rump steak; 11.3 kJ g^{-1} of lamb leg); but a greater energy-use efficiency for deer is clearly indicated (K.R. Drew, pers. commun., 1979).

Dispersal rates in the wild have been assessed for nine species of introduced ungulates (Caughley, 1963). Among these, *Rupicapra rupicapra* (chamois) shows a rate (8.6 km yr^{-1}) which is substantially faster than for *Hemitragus jemlahi-*

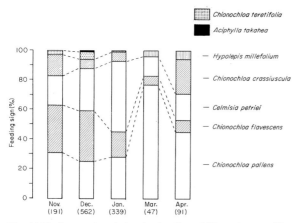

Fig. 14.32. Seasonal variation in the diet of *Notornis mantelli*, as determined from numbers of cut tillers along line transects in low-alpine *Chionochloa* grassland on the Murchison Mountains of Fiordland in south-western South Island. The number of tillers counted are given in brackets along the bottom. (Reproduced with permission from Mills and Mark, (1977.)

cus (Himalayan thar) (1.8 km yr^{-1}) and several species of deer (0.6 to 1.6 km yr^{-1}), including *Cervus elaphus* which has now spread virtually throughout the mountain country. Caughley assessed the roles of population density and/or environment in determining these rates. Subsequently, field studies of population dynamics of some of the introduced ungulates in New Zealand mountain regions have included the eruption of *Hemitragus jemlahicus* (Caughley, 1970) (studied by sampling populations at different distances from the point of liberation some 50 years previously) and the eruption and decline of a herd of *Cervus elaphus* in northern South Island (Clarke, 1976). Here the decline was attributed to the combined effects of depleted food supplies and periodic harsh winters.

A study of a population of *Lepus europaeus* (European hare) over an area of 1250 ha containing forest, shrubland and alpine tussock grassland (at

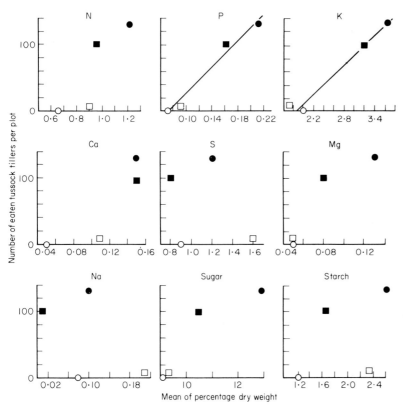

Fig. 14.33. Preference for four species of *Chionochloa* by *Notornis mantelli*, in relation to chemical composition. Usage is based on the mean number of tillers removed in 10 x 5 m plots (10 replicates) of each species. The value of each nutrient is a mean of the percentage dry weight for plants in that plot. Solid circle = *C. pallens*; solid square = *C. flavescens*; hollow square = *C. crassiuscula*; hollow circle = *C. teretifolia*. (Reproduced with permission from Mills and Mark, 1977.)

790–2250 m in northern South Island) by Flux (1967) revealed some eight animals limited to about 120 ha of tussock grassland and scrub on a sunny slope, where they utilized chiefly tussocks of *Chionochloa* and shrubs in winter and the small tussock grass *Poa colensoi* in summer. Abundant faecal pellets created the false impression of a higher hare density because of the large number produced (mean of 388 pellets day^{-1} based on captive animals) and the long half-life of pellets (exceeding 3 yr at 1520 m). Flux considered that the impact of hares was minor in relation to that of the 40 to 60 *Cervus elaphus* and *Rupicapra rupicapra* in the study area.

Invertebrates

Overall, the invertebrate fauna of New Zealand grasslands (both indigenous tussock grasslands and improved pastures) is "fairly meagre" in diversity, though not necessarily in numbers (Harrison and White, 1969).

The widespread distribution of nematodes in tussock grasslands is revealed by their occurrence among eight soils of a climo-sequence from 80 to 1550 m (Yeates, 1974). Although 70 species were recorded, only one occurred at all sites and 22 were each found at only one site. Abundance and diversity were related to altitude, precipitation and chemical factors (pH, organic C, soil N) of the soil. Wood (1973) described the feeding relationships of the common soil nematodes from subalpine tussock-grassland soil of central South Island. Almost all of the 23 species of stylet-bearing nematodes fed on a wide range of plant and animal foods and showed variable levels of food specialization. Only the 16 species of non-stylet-bearing nematodes fed on bacteria.

The ecological significance of nematodes in the tussock grasslands, however, appears to be minor compared with the beneficial effects of earthworms (Lee, 1958) or the destructive effects of several indigenous insects, particularly larval stages of *Costelytra zealandica* (Coleoptera) and of species of *Wiseana* (porina caterpillars), both important root-feeders, particularly in semiarid tussock grasslands and improved pastures derived from them (Dick, 1940; Kelsey, 1957, 1968). Now that the use of dichloro-diphenyl-trichloro-ethane (DDT) for control of such agricultural pests in New Zealand has been banned, and that of the less persistent chemicals discouraged because of both cost and effects on non-target species, recent emphasis has been given to ecological life-history studies and biological control, particularly with viruses. Sub-economic levels of control of some important root-eating pests have been achieved (Moore et al., 1974; Kalmakoff and Moore, 1975; Crawford and Kalmakoff, 1977).

White (1975c) established that larvae of three insect species (one a lepidopteran caterpillar, and the other two Diptera, one an unidentified Cecidomyiid gall midge) severely reduce seed production in several *Chionochloa* species. He concluded that "consistently high levels of damage over sequences of flowering years might affect *Chionochloa* regeneration."

Lee (1958) investigated earthworm populations as part of an inter-disciplinary study of organisms in soils of lowland-montane tussock grassland. The presence of higher populations beneath tussocks was ascribed to the fact that the soil beneath them is protected from frost.

Some detailed studies of population dynamics and energetics have been made of the indigenous alpine grasshoppers (Acrididae) in the *Chionochloa* grasslands of central South Island. Batcheler (1967) established that grasshopper populations may reach 4.9 m^{-2}, with a live biomass of 2.56 g m^{-2}, in alpine *Chionochloa* communities that had been seriously depleted by *Cervus elaphus* and *Rupicapra rupicapra*. Numbers were considerably less (0.11 m^{-1}) in dense stands of *Chionochloa* spp. Using a provisional calculation of their energy requirements based on other sources, primary production values of Wraight (1965) and energy values derived elsewhere, Batcheler estimated that food requirements for grasshoppers amounted to about 30% of the total annual leaf production in the most depleted sites (6% plant cover) and a "negligible proportion" of it on the least disturbed sites. He assumed a production/ingestion efficiency of 13% and a biomass increment rate of 33% per month over 6 months. The results of this study stress the possible commensal dependence of native grasshoppers on introduced ungulates (or natural disturbances) in the tussock grasslands, and the probability that grasshopper herbivory will contribute to further deterioration of the most precarious stands of *Chionochloa*.

Batcheler suggested that climatic change in recent decades has favoured grasshoppers by an extended snow-free season that permits reproduction of two generations of adults, as well as development of two generations of juveniles to about the fourth instar.

More recent and extended studies of energetics and dynamics of alpine grasshopper populations in the South Island *Chionochloa* grasslands have been made by White (1975a, b, 1978). Up to 25-fold variations in grazing pressure among 10 sites were demonstrated, reflecting combined effects of population density, consumption levels, vegetation composition, extent of plant cover, and orientation of slope (White, 1975a, b). Batcheler's early evidence for serious cumulative vegetation damage through grasshopper grazing was partially confirmed with White's conclusion that, although grasshoppers in tussock grassland remove only low volumes of herbage, the cumulative vegetation damage may be serious in some grasslands. This applies even in high-density sites, where production of grasshopper biomass may be up to 14.7 kg ha^{-1} yr^{-1} (compared to Batcheler's (1967) estimates of up to 360 kg) and consumption of only 3% of leaf production (compared to the 30% suggested by Batcheler). White considered Batcheler's assumed biomass increment rates to be excessive, particularly for the long-lived adult instar, which contributes most of the total grasshopper biomass.

More recent and precise information on energetics and consumption rates of a community of three species of alpine grasshoppers (in a central South Island *Chionochloa* grassland at 1400 to 1550 m) revealed "some very specific adaptations between grasshopper energetics and feeding behaviour" (White, 1978). Consumption was usually 1 to 2% of net primary above-ground production; on only one site did this figure exceed 6%, reaching 18% at 1920 m. Estimates of the proportion of phytomass consumed are dependent on the accuracy of assumptions concerning phytomass production, estimates of which may have been substantially too low in this study. White found that grazing, on a preferred species that provided up to 11% of total ground cover, removed as much as 59% of leaf biomass and caused death of plants. The energy budget data obtained by White places New Zealand alpine grasshoppers among the least productive of the Acrididae for which information is available.

MICRO-ORGANISMS

An early inter-disciplinary study of some microbiological activities and their seasonal fluctuations was made of three lowland-montane (750–900 m) tussock-grassland soils from contrasting regions in North and South Islands (Thornton, 1958a). Groups examined included fungi (Thornton, 1958b; di Menna, 1958), bacteria (Stout, 1958a; Ross, 1958a) and their nitrifying activity (Ross, 1958b), protozoa (Stout, 1958b), streptomycetes (Vernon, 1958), and algae (Flint, 1958). A more recent study along a climo-sequence of seven zonal tussock-grassland soils, involving 10 sites across southern South Island (Molloy and Blakemore, 1974), considered enzyme activities and nitrogen mineralization (Ross and McNeilly, 1975; Ross, 1975a, b, c; Speir, 1976, 1977a, b), structural and water-soluble carbohydrates (Molloy et al., 1977; Ross and Molloy, 1977), respiratory activities (Ross et al., 1975), litter decay (Molloy et al., 1978a, b; Speir and Ross, 1978), and decomposition of hemicellulose, cellulose, lignin and water-insoluble protein in kaolinite pellets buried in the soil (Molloy et al., 1978c; Ross et al., 1978). Between-site differences in many properties were often significant despite considerable within-site variation. Correlations were established with such factors as moisture availability, temperature, and physical and chemical factors of the ten soils. Few trends were regular, and single environmental factors rarely explained most of the variation. Activity, however, generally decreased in the driest and the coldest soils.

Weight losses of *Chionochloa rigida* litter in mesh bags amounted to 15 to 20% at most sites after the first year, but was much less in the second year (Molloy et al., 1978a, b, c). Seven of the sites showed only minor differences in weight losses and chemical changes, but decomposition was relatively rapid under *Chionochloa rubra* in the azonal soil of a super-humid lowland environment that generally supports rain forest. Some 55 to 60% of the variance in a series of multiple regressions related loss in weight of litter to temperature, moisture and soil chemical factors.

Of the nutrient elements in the litter, calcium, magnesium, potassium and sodium were lost most rapidly, particularly from the sites at higher elevation which have an environment conducive to leaching. Whereas loss of phosphorus was more variable, it tended to be greatest in these sites. Levels of sulphur decreased only slightly.

Lignin extracted from *Chionochloa* leaves and incorporated into kaolinite pellets showed negligible losses over the 2-year study period, whereas hemicellulose and cellulose decomposed rapidly in most soils. Only 10 to 20% of the substrate remained at most sites after 1 year. However, there was no significant correlation between loss rates of cellulose or hemicellulose from the pellets and from leaf litter exposed on the soil surface, although for cellulose, loss rates of buried standard (International Biological Programme) cellulose and that in the buried pellets were significantly correlated. Cellulose squares (25 cm^2 × 1 mm) in nylon mesh bags were used to compare decomposition on the soil surface (top of the O_2 (organic-rich) layer), a few centimetres away from the base of tussocks, with that in top soil at a depth of 5 cm. Weight losses at all sites were generally greater in the top soil, where most of the cellulose had disappeared after 2 years. Losses were least in the driest site and greatest at the super-humid lowland site. However, little of the variance could be accounted for by chemical or biochemical factors in the soil, or by climate. Rather, micro-environment was invoked as the major controlling factor. Weight losses of the cellulose squares tended to be positively (but not significantly) correlated with the weight loss of exposed *Chionochloa* litter and of its cellulose component, and with that of the pelleted preparation of cellulose buried in the top soil. The conclusion reached was that "decomposition rates of I.B.P. cellulose did not provide a reliable guide to the rate of plant litter breakdown or to the other soil biochemical activities at these sites" (Ross et al., 1978).

Individual studies of tussock-grassland soils have involved mycorrhizal associations, non-symbiotic nitrogen-fixing organisms and nematodes.

Crush (1973) (using combined glasshouse and field experiments along an altitudinal sequence of tussock grasslands in south-central South Island) established that, whereas endo-mycorrhizas are common in the grassland species, it is only at higher altitudes that levels of available phosphorus in the soil are likely to be sufficiently low — less than 8 ppm by the Truog (1930) test — for the dominant grasses to be mycotrophic. However, only two of the five tussock grasses studied (*Chionochloa rigida* and *Poa colensoi*) appeared to exploit the endo-mycorrhizas (*Endogone* spp. and *Glomus tenue*) that are generally ubiquitous in the tussock-grassland soils. *Chionochloa macra* appeared to be better adapted than *C. rigida* in obtaining phosphorus from poor soils without a mycorrhizal association, perhaps because of its longer root hairs (Crush, 1973). Resting spores of endo-mycorrhizal fungi (Endogonaceae), involving several species, are widespread in most New Zealand soils, but generally are more abundant under indigenous forest and scrub than in tussock grasslands (Hayman, 1978).

An intensive study of non-symbiotic nitrogen-fixing soil micro-organisms (Line and Loutit, 1973) confirmed an earlier assessment (Ross, 1958b) that such organisms make only minor contributions to nitrogen reserves in the soil. Three sites of an altitudinal sequence (500–1000 m) were examined, ranging from lowland *Festuca* tussock to low-alpine *Chionochloa* grasslands in south-central South Island, and single samples were examined from 21 widely spread sites in tussock grassland. The anaerobic *Clostridium butyricum* was confirmed as the most significant bacterium, but five facultatively anaerobic nitrogen-fixing species were also isolated in the intensive study (Line and Loutit, 1971), in numbers comparable with those reported in soils from other temperate regions. Blue-green algae (species of *Anabaena*, *Nostoc* and *Tolypothrix*) also contribute to the annual non-symbiotic nitrogen fixation, estimated, from acetylene-reduction assay of unamended soil samples, to be equivalent of up to 3 kg N ha^{-1} (Line and Loutit, 1973).

PRESENT AND FUTURE MANAGEMENT PRACTICES

It is perhaps surprising that there are few net production values available for the substantial areas of modified pasture that have been derived from the tussock grasslands. Probably typical are those for a hill pasture dominated by *Agrostis capillaris* (brown top) developed from lowland short-tussock grassland on yellow-brown earth at 290 to 440 m in central South Island. Here, annual herbage yields of 350 to 737 g m^{-2} were obtained using the trim technique, monthly, over three years (Radcliffe et al., 1977). Higher production was achieved on south-facing slopes than on north slopes (in association with reduced evapotranspiration and soil-moisture stress). By applying an energy value of 18.41 kJ g^{-1} of dry matter for gross energy of common pasture fodders (7.53 kJ of metabolizable energy) (Jagusch, 1973), an equivalent net energy fixation for herbage of 6440 to 13 660 kJ m^{-2} yr^{-1} was obtained. This is within the general range of above-ground net primary production of *Chionochloa* grassland (Meurk, 1978), though the comparison may be dubious because of different methodologies. The yield was double these values (1457 g m^{-2} yr^{-1}) in *Lolium perenne–Trifolium repens* pastures on high-quality arable soils in southern New Zealand (Monteath et al., 1977). Up to 1700 g m^{-2} (range 200 to 1700 g m^{-2}) annual yield was reported from short-tussock grassland that had been over-sown to *Trifolium* and top-dressed with fertilizer in a subhumid region of central South Island, the variation being significantly correlated with spring rainfall (O'Connor et al., 1968).

Because of the relatively favourable economic return from soil-nutrient amendment and pasture improvement of the lowland tussock grasslands, the areas that remain are under increasing pressure for conversion to exotic pasture. On slopes too steep for cultivation, this may be achieved with aerial over-sowing and top-dressing, a technique pioneered in New Zealand in the late 1940s (McCaskill, 1973). The high costs involved with full development, even on relatively level terrain, are rather high. In consequence, the Government until recently has played a major role by acquiring large, extensively managed holdings of tussock grassland at relatively low altitudes for development. Generally this has been for sub-division into intensively farmed economic units to be offered freehold for private settlement. In a few cases it has been for exotic forestry, whereas in the drier interior regions irrigation has been promoted by Government policies.

The tussock grasslands at higher altitude in the eastern and central South Island mountains probably will continue to be leased from the Crown (i.e. Government) as some 400 pastoral units (or "runs") to be extensively grazed, with certain restrictions, on a seasonal basis. The snow-prone high country will continue to be utilized during the summer and the lower, safer areas throughout the winter, with or without supplementary feed (O'Connor, 1977). Carrying capacities here generally range from 2 to 5 ha per stock unit.[1]

More efficient management of these areas is being encouraged, incorporating improved conservation practices. These include the use of soil and water conservation plans, available for individual farming units. The plans are organized, in collaboration with the run-holder, by professional soil conservators employed by regional authorities. Land-inventory surveys form the basis on which the land-use capability of individual runs is assessed, using a modification of the system described in the Soil Conservation Society Handbook and the Land Capability Classification of the United States Department of Agriculture (Ministry of Works, 1971). Improvements encouraged by Government subsidies have included increased sub-divisional fencing, fertilizer application, over-sowing with improved pasture species, irrigation, and (in the case of severely eroded land) retirement from grazing and perhaps surrender of the lease to the Crown. The amount of the subsidy has varied with time and also according to the degree of benefit to the run-holder. Over the 30 years of operating this scheme, the 2.29 million ha of Crown-owned tussock grasslands and mountain lands being used for pastoral purposes in South Island have absorbed some $NZ9.6 million in soil conservation and water management works, with about $NZ5.0 million of it funded by Government (representing 0.064% and 0.033%, respectively, of 1979 New Zealand gross national product). To date, some 39% of the pastoral area has been covered by management plans; livestock have been removed from some 300 000 ha (with agreement of land occupiers), and a further 400 000 ha has graz-

---

[1] The annual requirement of a 55 kg sheep which rears one lamb to weaning, generally is considered to be 550 kg of dry matter.

Fig. 14.34. A. Relatively undisturbed *Chionochloa rigida* grassland at about 850 m on rolling schistose topography of the Lammerlaw Range in south-eastern South Island. Periodic burning and light grazing, chiefly by sheep, has maintained this grassland in good condition. A peat bog with small drifts of snow crosses the mid-ground.

ing controlled under the terms of management plans.

Because of the deterioration in vegetation that had accompanied the traditional combination of sheep grazing and burning, use of cattle, alone or in conjunction with sheep, was promoted throughout the *Chionochloa* grasslands about two decades ago, both for diversification of stock as well as to obviate the need for periodic burning. However, in some cases heavy grazing by cattle (causing virtual destruction of *Chionochloa* spp. and serious impairment of wetlands) has been more damaging than a century of fire and sheep grazing (Fig. 14.34).

The significance of vegetation type and cover to water yield from the mountain tussock grasslands is important in relation to management, but unfortunately has received little attention. Pilot studies (using small non-weighing lysimeters at seven upland sites in south-eastern South Island) indicate that significantly greater yields of water are obtained from near-natural stands of pure *Chionochloa* grassland than from a range of modified types (Holdsworth and Mark, 1990). This is because of relatively low transpiration rates of *Chionochloa*, combined with interception gains from frequent fog and light rain in this region by the tall, fine foliage of *Chionochloa* spp. (Mark and Rowley, 1976; Holdsworth and Mark, 1990). These results have been generally confirmed by a catchment-based study in the region (Pearce et al., 1984) but conflict with results of a hydrological study of a depleted tussock grassland catchment in central South Island, which "cast serious doubt on the validity of water yield as an object of management for this and comparable catchments" (Hayward, 1976). Nevertheless, the substantial and increasing value of water from many New Zealand mountain catchments (for use in agriculture, town supply, or hydro-electric gen-

Fig. 14.34 (continued). B. Recent severe grazing by cattle on a similar adjoining site has devastated the *Chionochloa* cover, perhaps permanently.

eration) together with the implications of catchment management, justifies the establishment of well-monitored catchment studies in representative areas (O'Laughlin et al., 1984; Pearce et al., 1984).

Despite a clear obligation to protect the native plants, animals and natural communities of the New Zealand high country for their intrinsic worth, for our own enjoyment, and for the world at large (Billings, 1978), and in terms of relevant legislation (1977 Reserves Act), this had not been manifest in formal reservations. Rather, New Zealand's system of protected natural areas, which in 1984 amounted to more than 4.5 million ha (about 17% of the land area), had concentrated on protection-type vegetation of mountainous terrain, particularly for its ten national parks and its forest reserves (Mark, 1985). Alpine-tussock grasslands of South Island high-rainfall regions and of North Island volcanic regions were also well represented within this system, but the extensive and diverse indigenous grasslands of the South Island rain-shadow region were virtually unrepresented (Scott, 1979; O'Connor, 1982; Mark, 1985). This situation reflected the emphasis in earlier legislation on scenic landscapes together with the greater public support for forest conservation, and was highlighted in reserve surveys conducted by government botanists (Mark, 1985). Recent changes in legislation (Reserves Act, Forests Act, National Parks Act) have resulted in a concerted effort to redress this imbalance. These have emphasized the importance of representative ecological systems and scientific values, and reflect the adoption of a broader perspective by conservation groups, as well as some well-publicised disputes over reservation proposals for tussock grasslands (Mark, 1985). In particular, the Protected Natural Areas Programme (PNAP) (Kelly and Park, 1986; Dickinson and Mark, 1988), initiated by the National Parks and Reserves Authority in 1982, has concentrated its effort in the South Island on the

Fig. 14.35. Paired photographs of low-alpine *Chionochloa pallens* grassland at 1250 m in the Humboldt Mountains of Mount Aspiring National Park in south-western South Island. A. In 1970, the unpalatable large daisy, *Celmisia armstrongii* (Asteraceae) and the smaller trailing *C. walkeri* dominated the *Chionochloa*, which had been severely depleted by heavy grazing by introduced *Cervus elaphus* (red deer).

non-forest ecosystems of the pastoral leasehold land in the rain-shadow region. A biogeographic framework of 268 Ecological Districts (McEwen, 1987) provides the basis for PNAP surveys aimed at identifying those areas which best represent the full range of natural ecosystems that still persist (Kelly and Park, 1986). Conceived as a 10-year programme and as "the most important conservation initiative of the 1980's" by the Chairman of the National Parks and Reserves Authority, D. Thom, it has achieved only limited success during its first five years and is in urgent need of promotion and action (Dickinson and Mark, 1988). Meanwhile, indigenous forest conservation has made significant progress, with the establishment (in 1987, during the centennial year of National

Fig. 14.35 (continued). B. In 1977, a drastic reduction in deer numbers, beginning in the early 1970s, has permitted recovery of *Chionochloa*, which has regained dominance.

Parks) of two additional national parks (Paparoa and Whanganui) (108 000 ha) — both predominantly forest-covered. Several surveys have been completed in the tussock-grassland rain-shadow region (Brumley et al., 1986; Dickinson, 1988, 1989) and some better protection achieved. An ambitious proposal for a conservation park has been made for an area of 1460 km^2 of upland tussock grasslands, following completion of three surveys (Mark, 1990).

Introduced mammals, particularly *Cervus elaphus*, *Lepus europaeus*, *Mustela erminea* and *Trichosurus vulpecula* (Australian brush-tailed possum), have become established throughout most national parks and many reserves, whereas several others — *Capra hircus* (goat), *Hemitragus jemlahicus* (thar) and *Rupicapra rupicapra* are more local in distribution, but no less destructive. Unfortunately, eradication has proved to be impossible, and even adequate control has been frustrated in many areas by the combined difficulties of access, terrain and climate. During the last two decades a lucrative game-meat export market, particularly of venison, has allowed the use of extremely effective hunting techniques with helicopters that essentially eliminated deer from the alpine-tussock grasslands. However, the future welfare of the alpine *Chionochloa* grasslands will continue to be threatened as long as deer find refuge in adjoining forests. This is particularly so with the recent innovation of deer farming. Within 10 years the market demand for wild animals has been largely satisfied and has virtually rendered uneconomic the hunting of deer in the wild. The resilience of the flora and vegetation of the *Chionochloa* alpine grasslands in these high-rainfall regions became apparent with the decline of deer over the first two decades (Fig. 14.35) (Mark, 1989), allowing these grasslands to return to a condition probably approaching their natural state.

ACKNOWLEDGEMENTS

I wish to thank Colin Meurk for his helpful comments on the manuscript and also to acknowledge the substantial financial support and encouragement offered over the last three decades by the Miss E.L. Hellaby Indigenous Grasslands Research Trust for my own research into the New Zealand tussock grasslands and for studies by several others which have been referred to in this review.

REFERENCES

Allan, H.H., 1961. *Flora of New Zealand, Vol. I. Indigenous Tracheophyta: Psilopsida, Lycopsida, Filicopsida, Gymnospermae, Dicotyledones.* Government Printer, Wellington, 1085 pp.

Anderson, A.J., 1989. *Prodigious Birds — Moas and Moa Hunting in prehistoric New Zealand.* Cambridge University Press, Cambridge, 238 pp.

Archer, A.C., 1973. Plant succession in relation to a sequence of hydromorphic soils formed on glacio-fluvial sediments in the alpine zone of the Ben Ohau Range, New Zealand. *N.Z. J. Bot.*, 11: 331–348.

Archer, A.C. and Collett, G.I., 1971. Climatopes of the subalpine and alpine zones of the north-east Ben Ohau Range, New Zealand. *Proc. 6th Geogr. Conf.* New Zealand Geographical Society, Christchurch, pp. 216–236.

Archer, A.C., Simpson, M.J.A. and MacMillan, B.H., 1973. Soils and vegetation of the lateral moraine at Malte Brun, Mount Cook region, New Zealand. *N.Z. J. Bot.*, 11: 23–48.

Barker, A.P., 1953. An ecological study of tussock grassland; Hunter Hills, South Canterbury. *N.Z. Dept. Sci. Ind. Res., Bull.*, 107, 58 pp.

Batcheler, C.L., 1967. Preliminary observations of alpine grasshoppers in a habitat modified by deer and chamois. *Proc. N.Z. Ecol. Soc.*, 14: 15–26.

Billings, W.D., 1978. The rational use of high mountain resources in the preservation of biota and the maintenance of natural life systems. In: *The Use of High Mountains of the World.* Department of Lands and Survey, Wellington, pp. 209–223.

Bliss, L.C. and Mark, A.F., 1974. High-alpine environments and primary production on the Rock and Pillar Range, Central Otago, New Zealand. *N.Z. J. Bot.*, 12: 445–483.

Brooks, R.R., 1987. *Serpentine and its Vegetation: A Multidisciplinary Approach.* Dioscorides Press, Portland, Ore., 454 pp.

Brownsey, P.J. and Smith-Dodsworth, J.C., 1989. *New Zealand Ferns and Allied Plants.* David Bateman, Wellington, 168 pp.

Brumley, C.F., Stirling, M.W. and Manning, M.S., 1986. *Old Man Ecological District.* Survey Report for the Protected Natural Areas Programme. Lands and Survey Department, Wellington, 174 pp.

Burrell, J., 1965. Ecology of *Leptospermum* in Otago. *N.Z. J. Bot.*, 3: 3–16.

Burrows, C.J., 1965. Some discontinuous distributions of plants within New Zealand and their ecological significance. *Tuatara*, 13: 9–29.

Burrows, C.J., 1967. Progress in the study of South Island alpine vegetation. *Proc. N.Z. Ecol. Soc.*, 14: 8–13.

Burrows, C.J., 1977a. Distribution and composition of plant communities containing *Chionochloa oreophila*. *N.Z. J. Bot.*, 15: 549–564.

Burrows, C.J., 1977b. Alpine grasslands and snow in the Arthur's Pass and Lewis Pass regions, South Island, New Zealand. *N.Z. J. Bot.*, 15: 665–686.

Burrows, C.J., 1977c. Vegetation of the Cass District and its ecology: Grassland vegetation. In: C.J. Burrows (Editor), *Cass: History and Science in the Cass District, Canterbury, New Zealand.* University of Canterbury, Christchurch, pp. 185–213.

Burrows, C.J., McQueen, D.R., Esler, A.E. and Wardle, P., 1979. New Zealand heathlands. In: R.L. Specht (Editor), *Heathlands and Related Shrublands.* Ecosystems of the World 9A, Elsevier, Amsterdam, pp. 339–364.

Burrows, C.J., McCulloch, B. and Trotter, M.M., 1981. The diet of moas based on gizzard contents samples from Pyramid Valley, North Canterbury and Scaife's Lagoon, Lake Wanaka, Otago. *Rec. Canterb. Mus.*, 9(6): 309–336.

Calder, D.M., 1961. Plant ecology of subalpine shingle river beds in Canterbury. *J. Ecol.*, 49: 581–594.

Calder, J.W. and Wardle, P., 1969. Succession in subalpine vegetation at Arthur's Pass, New Zealand. *Proc. N.Z. Ecol. Soc.*, 16: 36–47.

Caughley, G., 1963. Dispersal rates of several ungulates introduced into New Zealand. *Nature*, 200: 280–281.

Caughley, G., 1970. Eruption of ungulate populations, with emphasis on Himalayan Thar in New Zealand. *Ecology*, 51: 53–72.

Cheeseman, T.F., 1925. *Manual of the New Zealand Flora*. Government Printer, Wellington, 1163 pp.

Clarke, C.M.H., 1970. Observations on population, movement and food of the kea (*Nestor notabilis*). *Notornis*, 17: 105–114.

Clarke, C.M.H., 1976. Eruption, deterioration and decline of the Nelson red deer herd. *N.Z. J. For. Sci.*, 5: 235–249.

Cockayne, L., 1919. An economic investigation of the montane tussock grassland of New Zealand, II. Relative palatability for sheep of the various pasture plants. *N.Z. J. Agric.*, 27: 321–331.

Cockayne, L., 1928. *The Vegetation of New Zealand*. Die Vegetation der Erde, Bd. XIV, 2nd ed. Engelmann, Leipzig, 456 pp.

Connor, H.E., 1964. Tussock grassland communities in the Mackenzie Country, South Canterbury, New Zealand. *N.Z. J. Bot.*, 2: 325–351.

Connor, H.E., 1965. Tussock grasslands in the middle Rakaia Valley, Canterbury, New Zealand. *N.Z. J. Bot.*, 3: 261–276.

Connor, H.E., 1966. Breeding systems in New Zealand grasses, 7. Periodic flowering of snow tussock, *Chionochloa rigida*. *N.Z. J. Bot.*, 4: 392–397.

Connor, H.E., 1985. Biosystematics of higher plants in New Zealand 1965–1984. *N.Z. J. Bot.*, 23: 613–644.

Connor, H.E., 1991. *Chionochloa* Zotov (Gramineae) in New Zealand. *N.Z. J. Bot.*, 29: 219–283.

Connor, H.E. and Edgar, E., 1987. Name changes in the indigenous New Zealand flora 1960–1980 and Nomina Nova IV. 1983–1986. *N.Z. J. Bot.*, 25: 115–170.

Connor, H.E. and MacRae, A.H., 1969. Montane and subalpine tussock grasslands in Canterbury. In: G.A. Knox (Editor), *The Natural History of Canterbury*. Reed, Wellington, pp. 167–204.

Connor, H.E. and Purdie, A.W., 1976. Triterpene methyl ether differentiation in *Chionochloa* (Gramineae). *N.Z. J. Bot.*, 14: 315–326.

Connor, H.E. and Purdie, A.W., 1981. Triterpene methyl ethers in *Chionochloa* (Gramineae): distribution in western South Island, New Zealand. *N.Z. J. Bot.*, 19: 161–170.

Conrad, V., 1946. Usual formulas of continentality and their limits of validity. *Trans. Am. Geophys. Union*, 27: 663.

Coulter, J.D., 1973. Ecological aspects of the climate. In: G.R. Williams (Editor), *The Natural History of New Zealand*. Reed, Wellington, pp. 28–60.

Crawford, A.M. and Kalmakoff, J., 1977. A host–virus interaction in a pasture habitat: *Wiseana* spp. (Lepidoptera: Hepialidae) and its baculoviruses. *J. Invertebr. Pathol.*, 29: 81–87.

Croker, B.H., 1959. A method of estimating the botanical composition of the diet of sheep. *N.Z. J. Agric. Res.*, 2: 72–85.

Crush, J.R., 1973. Significance of endomycorrhizas in tussock grassland in Otago, New Zealand. *N.Z. J. Bot.*, 11: 645–660.

Cumberland, K.B., 1962. Climatic change or cultural interference? — New Zealand in Moahunter times. In: M. McCaskill (Editor), *Land and Livelihood*. New Zealand Geographical Society, Christchurch, pp. 88–142.

Daly, G.T., 1967. Ordination of grassland and related communities in Otago. *Proc. N.Z. Ecol. Soc.*, 14: 63–70.

Di Menna, M.E., 1958. Biological studies of some tussock grassland soils, III. Yeasts. *N.Z. J. Agric. Res.*, 1: 939–942.

Dick, R.D., 1940. Observations on insect-life in relation to tussock-grassland deterioration. *N.Z. J. Sci. Technol.*, A22: 19–29.

Dickinson, K.J.M., 1988. *Umbrella Ecological District*. Survey Report for the Protected Natural Areas Programme. Department of Conservation, Wellington, 179 pp.

Dickinson, K.J.M., 1989. *Nokomai Ecological District*. Survey Report for the Protected Natural Areas Programme. Department of Conservation, Wellington, 139 pp.

Dickinson, K.J.M. and Mark, A.F., 1988. The New Zealand protected natural areas programme — a progress report. *Search*, 19: 203–208.

Drew, K.R., 1979. Physiological aspects of meat production from deer. *49th Conf. Aust. N.Z. Assoc. Adv. Sci. Abstr.*, 1: 410.

Enting, B. and Molloy, L.F., 1982. *The Ancient Islands: New Zealand's Natural Environments*. Port Nicholson Press, Wellington, 153 pp.

Fineran, B.A., 1966. The vegetation and flora of Bird Island, Foveaux Strait. *N.Z. J. Bot.*, 4: 133–184.

Fisher, F.J.F., 1962. Observations on the vegetation of screes in Canterbury, New Zealand. *J. Ecol.*, 40: 156–167 (erratum following p. 247).

Fleming, C.A., 1975. The geological history of New Zealand and its biota. In: G. Kuschel (Editor), *Biogeography and Ecology in New Zealand*. Junk, The Hague, pp. 1–86.

Fleming, C.A., 1977. The history of life in New Zealand forests. *N.Z. J. For.*, 22: 249–262.

Flint, E.A., 1958. Biological studies of some tussock grassland soils, IX. Algae: preliminary observations. *N.Z. J. Agric. Res.*, 1: 991–997.

Flux, J.E.C., 1967. Hare numbers and diet in an alpine basin in New Zealand. *Proc. N.Z. Ecol. Soc.*, 14: 27–33.

Fraser, K.W., 1988. Reproductive biology of rabbits, *Oryctolagus cuniculus* (L.), in Central Otago, New Zealand. *N.Z. J. Ecol.*, 11: 79–88.

Gage, M., 1965. Some characteristics of Pleistocene cold climates in New Zealand. *Trans. R. Soc. N.Z.*, 3: 11–21.

Gradwell, M.W., 1960. Soil frost action in snow-tussock grassland. *N.Z. J. Sci.*, 3: 580–590.

Greenwood, R.M. and Atkinson, I.A.E., 1977. Evolution of the divaricating plants in New Zealand in relation to moa browsing. *Proc. N.Z. Ecol. Soc.*, 24: 21–33.

Greer, D.H., 1979. Effects of long-term preconditioning on growth and flowering of some snow tussock (*Chionochloa* spp.) populations in Otago, New Zealand. *Aust. J. Bot.*, 27: 617–630.

Greer, D.H., 1984. Seasonal changes in photosynthetic activity of snow tussocks along an altitudinal gradient in Otago, New Zealand. *Oecologia*, 63: 271–274.

Haase, P., 1990. Environmental and floristic gradients in Westland, New Zealand and the discontinuous distribution of *Nothofagus*. *N.Z. J. Bot.*, 28: 25–40.

Hammond, A.L., 1972. Ecosystem analysis: biome approach to environmental research. *Science*, 175: 46–48.

Harrison, R.A. and White, E.G., 1969. Grassland invertebrates. In: G.A. Knox (Editor), *The Natural History of Canterbury*. Reed, Wellington, pp. 379–390.

Hayman, D.S., 1978. Mycorrhizal populations in sown pastures and native vegetation in Otago, New Zealand. *N.Z. J. Agric. Res.*, 21: 271–276.

Hayward, J.A., 1976. The hydrology of a mountain catchment and its implications for management. Proceedings, Soil and Plant Water Symposium. *N.Z. Dept. Sci. Ind. Res., Inf. Ser.*, 126: 126–136.

Hnatiuk, R.J., 1978. The growth of tussock grasses on an equatorial high mountain and on two sub-Antarctic islands. In: C. Troll and W. Lauer (Editors), Geoecological Relations Between the Southern Temperate Zone and the Tropical Mountains. *Erdwiss. Forsch.*, XI: 159–190.

Holdsworth, D.K. and Mark, A.F., 1990. Water and nutrient input:output budgets — effects of plant cover at seven sites in upland snow tussock grasslands of eastern and central Otago, New Zealand. *J. R. Soc. N.Z.*, 20: 1–24.

Holloway, J.T., 1954. Forests and climates in the South Island of New Zealand. *Trans. R. Soc. N.Z.*, 82: 329–410.

Holloway, J.T., 1959a. Pre-European vegetation of New Zealand. In: A.H. McLintock (Editor), *A Descriptive Atlas of New Zealand*. Government Printer, Wellington, 109 pp.

Holloway, J.T., 1959b. Noxious animal problems of the South Island alpine watersheds. *N.Z. Sci. Rev.*, 17: 21–28.

Holloway, J.T., 1982. *The Mountain Lands of New Zealand*. Tussock Grasslands and Mountain Lands Institute, Lincoln College, Christchurch, 105 pp.

Holloway, J.T., Wendelken, W.J., Morris, J.Y., Wraight, M.J., Wardle, P. and Franklin, D.A., 1963. Report on the condition of the forests, subalpine-scrublands and alpine grasslands of the Tararua Range. *N.Z. For. Serv. For. Res. Inst., Tech. Pap.*, 41, 48 pp.

Howard, W.E., 1958. The rabbit problem in New Zealand. *N.Z. Dept. Sci. Ind. Res., Inf. Ser.*, 16, 47 pp.

Howard, W.E., 1959. Introduced browsing mammals and habitat stability in New Zealand. *J. Wildl. Manage.*, 28: 421–429.

Howard, W.E., 1965. Control of introduced mammals in New Zealand. *N.Z. Dept. Sci. Ind. Res., Inf. Ser.*, 45, 96 pp.

Ives, D.W. and Cutler, E.J.B., 1972. A toposequence of steepland soils in the drier high-country yellow-brown earth (dry-hygrous eldefulvic) region, Canterbury, New Zealand. *N.Z. J. Sci.*, 15: 385–407.

Jagusch, K.T., 1973. Livestock production from pasture. In: R.H.M. Langer (Editor), *Pastures and Pasture Plants*. Reed, Wellington, pp. 229–242.

Johnson, P.N., 1975. Vegetation and flora of Solander Islands, southern New Zealand. *N.Z. J. Bot.*, 13: 189–213.

Johnson, P.N., 1976a. Vegetation and flora of Womens Island, Foveaux Strait, New Zealand. *N.Z. J. Bot.*, 14: 327–331.

Johnson, P.N., 1976b. Vegetation associated with kakapo (*Strigops habroptilus* Gray) in Sinbad Gully, Fiordland, New Zealand. *N.Z. J. Bot.*, 14: 151–159.

Kalmakoff, J. and Moore, S.G., 1975. The ecology of nucleopolyhedrosis virus in Porina (*Wiseana* spp.) (Lepidoptera: Hepialidae). *N.Z. Entomol.*, 6: 73–76.

Kelsey, J.M., 1957. Insects attacking tussocks. *N.Z. J. Sci. Technol.*, A38: 638–643.

Kelsey, J.M., 1968. Insects and tussock grassland development. *Tussock Grassl. Mountain Lands Inst. Rev.*, 15: 51–55.

Kelly, G.C. and Park, G.N. (Editors), 1986. *The New Zealand Protected Natural Areas Programme: A Scientific Focus*. Department of Scientific and Industrial Research, Wellington, 68 pp.

King, C., 1984. *Immigrant Killers — Introduced Predators and the Conservation of Birds in New Zealand*. Oxford University Press, Auckland, 224 pp.

Lavers, R.B., 1978. The diet of red deer (*Cervus elaphus*) in the Murchison Mountains: a preliminary report. In: *Seminar on the Takahe and Its Habitat*. Fiordland National Park Board, Invercargill, pp. 187–199.

Leamy, M.L. and Fieldes, M., 1975. Soil. In: I. Wards (Editor), *New Zealand Atlas*. Government Printer, Wellington, pp. 122–143.

Lee, K.E., 1958. Biological studies of some tussock grassland soils, X. Earthworms. *N.Z. J. Agric. Res.*, 1: 998–1002.

Lee, W.G., 1992. New Zealand ultramafics. In: B.A. Roberts and J. Proctor (Editors), *The Ecology of Areas With Serpentinized Rocks, A World View*. Junk, The Hague, pp. 375–418.

Line, M.A. and Loutit, M.W., 1971. Non-symbiotic nitrogen-fixing organisms from some New Zealand tussock grassland soils. *J. Gen. Microbiol.*, 66: 309–318.

Line, M.A. and Loutit, M.W., 1973. Studies on non-symbiotic nitrogen fixation in New Zealand tussock-grassland soils. *N.Z. J. Agric. Res.*, 16: 87–94.

Lough, T.J., Wilson, J.B., Mark, A.F. and Evans, A.C., 1988. Succession in a New Zealand alpine cushion community: a Markovian model. *Vegetatio*, 71: 129–138.

Makepeace, W., Dobson, A.T. and Scott, D., 1985. Interference phenomena due to mouse-ear and king devil hawkweed. *N.Z. J. Bot.*, 23: 79–90.

Mark, A.F., 1965a. The environment and growth rate of narrow-leaved snow tussock, *Chionochloa rigida* in Otago. *N.Z. J. Bot.*, 3: 73–103.

Mark, A.F., 1965b. Flowering, seeding, and seedling establishment of narrow-leaved snow tussock, *Chionochloa rigida*. *N.Z. J. Bot.*, 3: 180–193.

Mark, A.F., 1965c. Ecotypic differentiation in Otago populations of narrow-leaved snow tussock, *Chionochloa rigida*. *N.Z. J. Bot.*, 3: 277–299.

Mark, A.F., 1965d. Effects of management practices on narrow-leaved snow tussock, *Chionochloa rigida*. *N.Z. J. Bot.*, 3: 300–319.

Mark, A.F., 1968. Factors controlling irregular flowering in four alpine species of *Chionochloa*. *Proc. N.Z. Ecol. Soc.*, 15: 55–60.

Mark, A.F., 1969. Ecology of snow tussocks in the mountain grasslands of New Zealand. *Vegetatio*, 18: 289–306.

Mark, A.F., 1975. Photosynthesis and dark respiration in three alpine snow tussocks (*Chionochloa* spp.) under controlled environments. *N.Z. J. Bot.*, 13: 93–122.

Mark, A.F., 1977. Vegetation of Mount Aspiring National Park. *N.Z. Natl. Park Auth. Sci. Ser.*, 2, 79 pp.

Mark, A.F., 1985. The botanical component of conservation in New Zealand. *N.Z. J. Bot.*, 23: 789–810.

Mark, A.F., 1989. Responses of indigenous vegetation to contrasting trends in utilization by red deer in two southwestern New Zealand national parks. *N.Z. J. Ecol.*, 12 (Suppl.): 103–114.

Mark, A.F., 1990. Ecological and conservation values: the case for a conservation park. In: G.W. Kearsley and B.B. Fitzharris (Editors), *Southern Landscapes: Essays in Honour of Bill Brockie and Ray Hargreaves*. University of Otago, Dunedin, pp. 233–273.

Mark, A.F. and Adams, N.M., 1973. *New Zealand Alpine Plants*. Reed, Wellington, 262 pp.

Mark, A.F. and Bliss, L.C., 1970. The high-alpine vegetation of Central Otago, New Zealand. *N.Z. J. Bot.*, 8: 381–451.

Mark, A.F. and Burrell, J., 1966. Vegetation studies on the Humboldt Mountains, Fiordland, I. The alpine tussock grasslands. *Proc. N.Z. Ecol. Soc.*, 13: 12–18.

Mark, A.F. and Rowley, J., 1976. Water yield of low-alpine snow tussock grassland in Central Otago. *J. Hydrol. (N.Z.)*, 15: 59–79.

Mark, A.F., Dickinson, K.J.M., Patrick, B.H., Barratt, B.I.P., Loh, G., McSweeney, G.D., Meurk, C.D., Timmins, S.M., Simpson, N.C. and Wilson, J.B., 1989. An ecological survey of the central part of the Eyre Ecological District, northern Southland, New Zealand. *J. R. Soc. N.Z.*, 19: 349–384.

McCaskill, L.W., 1969. *Molesworth*. Reed, Wellington, 292 pp.

McCaskill, L.W., 1973. *Hold This Land: A History of Soil Conservation in New Zealand*. Reed, Wellington, 274 pp.

McCracken, I.J., 1980. Mountain climate in the Craigieburn Range, New Zealand. In: U. Benecke and M.R. Davis (Editors), *Mountain Environments and Subalpine Tree Growth*. New Zealand Forest Service Tech. Pap. 70, 288 pp.

McEwen, W.M. (Editor), 1987. *Ecological Regions and Districts of New Zealand*, 3rd ed. Publication No. 5, Ecological Resources Centre, New Zealand Department of Scientific and Industrial Research, Wellington, Part 1, 35 pp.; Part 2, 61 pp.; Part 3, 105 pp.; Part 4, 125 pp.

McGlone, M.S., 1985. Plant biogeography and the late Cenozoic history of New Zealand. *N.Z. J. Bot.*, 23: 723–749.

McGlone, M.S., 1989. The Polynesian settlement of New Zealand in relation to environmental and biotic changes. *N.Z. J. Ecol.*, 12 (Suppl.): 115–129.

McSaveney, M.J., 1978. The magnitude of erosion across the Southern Alps. *Proc., Conf. on Erosion Assessment and Control in New Zealand*. New Zealand Association for Soil Conservation, pp. 7–24.

McSaveney, M.J. and Whitehouse, I.E., 1989. Anthropic erosion of mountain land in Canterbury. *N.Z. J. Ecol.*, 12 (Suppl.): 151–163.

Meurk, C.D., 1978. Alpine phytomass and primary productivity in Central Otago, New Zealand. *N.Z. J. Ecol.*, 1: 27–50.

Meurk, C.D., 1982. *Alpine Phytoecology of the Rainshadow Mountains of Otago and Southland, New Zealand*. Ph.D. thesis, University of Otago, Dunedin, 774 pp.

Meurk, C.D., 1984. Bioclimatic zones for the Antipodes — and beyond? *N.Z. J. Ecol.*, 7: 175–182.

Mills, J.A., 1976. Population studies on takahe, *Notornis mantelli* in Fiordland, New Zealand. *Int. Counc. Bird Pres. Bull.*, XII: 140–147.

Mills, J.A. and Mark, A.F., 1977. Food preferences of takahe in Fiordland National Park, New Zealand, and the effect of competition from introduced red deer. *J. Anim. Ecol.*, 46: 939–958.

Mills, J.A., Lee, W.G. and Lavers, R.B., 1989. Experimental investigations of the effects of takahe and deer grazing on *Chionochloa pallens* grassland Fiordland, New Zealand. *J. Appl. Ecol.*, 26: 397–417.

Ministry of Works, 1971. *Land Use Capability Survey Handbook: A New Zealand Handbook for the Classification of Land*, 2nd ed. Ministry of Works, Wellington, 139 pp.

Moar, N.T., 1971. Contributions to the Quaternary history of the New Zealand flora, 6. Aranuian pollen diagrams from Canterbury, Nelson and north Westland, South Island. *N.Z. J. Bot.*, 9: 80–145.

Molloy, B.P.J., 1962. Recent changes in mountain vegetation and effects on profile stability. *Proc. N.Z. Soc. Soil Sci.*, 5: 22–26.

Molloy, B.P.J., 1964. Soil genesis and plant succession in the subalpine and alpine zones of Torlesse Range, Canterbury, New Zealand, 2. Distribution, characteristics, and genesis of soils. *N.Z. J. Bot.*, 2: 143–176.

Molloy, B.P.J., 1977. The fire history. In: C.J. Burrows (Editor), *Cass: History and Science in the Cass District, Canterbury, New Zealand*. Department of Botany, University of Canterbury, Christchurch, pp. 157–172.

Molloy, B.P.J. and Connor, H.E., 1970. Seedling growth in four species of *Chionochloa* (Gramineae). *N.Z. J. Bot.*, 8: 132–152.

Molloy, B.P.J., Burrows, C.J., Cox, J.E., Johnston, J.A. and Wardle, P., 1963. Distribution of subfossil forest remains, eastern South Island, New Zealand. *N.Z. J. Bot.*, 1: 68–77.

Molloy, L.F., 1988. *The Living Mantle: Soils in the New Zealand Landscape*. Mallinson Rendel and New Zealand Society of Soil Science, Wellington, 239 pp.

Molloy, L.F. and Blakemore, L.C., 1974. Studies on a climosequence of soils in tussock grasslands, 1. Introduction, sites, and soils. *N.Z. J. Sci.*, 17: 233–255.

Molloy, L.F., Bridger, B.A. and Cairns, A., 1977. Studies on a climosequence of soils in tussock grassland, 13. Structural carbohydrates in tussock leaves, roots and litter and in the soil light and heavy fractions. *N.Z. J. Sci.*, 20: 443–451.

Molloy, L.F., Bridger, B.A. and Cairns, A., 1978a. Studies on a climosequence of soils in tussock grassland, 15. Litter decomposition: weight losses and changes in contents of total N and organic constituents. *N.Z. J. Sci.*, 21: 265–276.

Molloy, L.F., Bridger, B.A. and Cairns, A., 1978b. Studies on a climosequence of soils in tussock grassland, 16. Litter decomposition: changes in the contents of some nutrient elements. *N.Z. J. Sci.*, 21: 277–283.

Molloy, L.F., Bridger, B.A. and Cairns, A., 1978c. Studies on a climosequence of soils in tussock grassland, 19. Decomposition of hemicellulose, cellulose and lignin in kaolinite pellets buried in the soils. *N.Z. J. Sci.*, 21: 451–458.

Monteath, M.A., Johnstone, P.D. and Boswell, C.C., 1977. Effects of animals on pasture production, 1. Pasture productivity from beef cattle and sheep farmlets. *N.Z. J. Agric. Res.*, 20: 23–30.

Moore, L.B., 1955. The plants of tussock grassland. *Proc. N.Z. Ecol. Soc.*, 3: 7–8.

Moore, L.B., 1976. The changing vegetation of Molesworth Station, New Zealand, 1944 to 1971. *N.Z. Dept. Sci. Ind. Res., Bull.*, 217, 118 pp.

Moore, L.B. and Edgar, E., 1970. *Flora of New Zealand, Vol. II. Indigenous Tracheophyta: Monocotyledones except Gramineae*. Government Printer, Wellington, 354 pp.

Moore, S.G., Kalmakoff, J. and Miles, J.A.R., 1974. An iridescent virus and a rickettsia from the grass grub *Costelytra zealandica* Coleoptera: Scarabaeidae). *N.Z. J. Zool.*, 1: 205–210.

Morris, J.Y., 1965. Climate investigations in the Craigieburn Range, New Zealand. *N.Z. J. Sci.*, 8: 556–582.

Norton, D.A. and Kelly, D., 1988. Mast seeding over 33 years by *Dacrydium cupressinum* Lamb. (rimu) (Podocarpaceae) in New Zealand: the importance of economies of scale. *Funct. Ecol.*, 2: 399–408.

O'Connor, K.F., 1977. Understanding hill-land ecology in New Zealand as a basis for management. *Proc., Int. Symp. on Hill Lands*. University of West Virginia, Morgantown.

O'Connor, K.F., 1978. Evolution of a high country pastoral community. In: D.C. Pitt (Editor), Society and Environment — The Crisis in the Mountains. *University of Auckland Working Papers in Comparative Sociology*, No. 8, pp. 182–224.

O'Connor, K.F., 1982. The implications of past exploitation and current developments to the conservation of South Island tussock grasslands. *N.Z. J. Ecol.*, 5: 97–187.

O'Connor, K.F., 1984. Stability and instability of ecological systems in New Zealand mountains. *Mountain Res. Dev.*, 4: 15–29.

O'Connor, K.F., 1986. The influence of science on the use of tussock grasslands. *Tussock Grassl. Mountain Lands Inst. Rev.*, 43: 15–78.

O'Connor, K.F. and Kerr, I.G.C., 1978. The history and present pattern of pastoral range production in New Zealand. *Proc., 1st Int. Rangelands Conf.*, Denver. pp. 104–107.

O'Connor, K.F., Vartha, E.W., Belcher, R.A. and Coulter, J.D., 1968. Seasonal and annual variation in pasture production in Canterbury and North Otago. *Proc. N.Z. Grassl. Assoc.*, 30: 50–63.

O'Laughlin, C.L., Rowe, L.K. and Pearce, A.J., 1984. Hydrology of mid-altitude tussock grasslands, upper Waipori catchment, Otago, I. Erosion, sediment yields and water quality. *J. Hydrology* (N.Z.), 23: 45–59.

Orwin, J. (Editor), 1978. Revegetation in the rehabilitation of mountain lands. *Forestry Research Inst. Symp.* No. 16. N.Z. Forest Service, Wellington, 244 pp.

Payton, I.J. and Brasch, D.J., 1978. Growth and non-structural carbohydrate reserves in *Chionochloa rigida* and *C. macra*, and their short-term response to fire. *N.Z. J. Bot.*, 16: 435–460.

Payton, I.J. and Mark, A.F., 1979. Long-term effects of burning on growth, flowering, and carbohydrate reserve in narrow-leaved snow tussock (*Chionochloa rigida*). *N.Z. J. Bot.*, 17: 43–54.

Payton, I.J., Lee, W.G., Dolby, R. and Mark, A.F., 1986. Nutrient concentrations in narrow-leaved snow tussock (*Chionochloa rigida*) after spring burning. *N.Z. J. Bot.*, 24: 529–537.

Pearce, A.J., Rowe, L.K. and O'Laughlin, C.L., 1984. Hydrology of mid-altitude tussock grasslands, upper Waipori catchment, Otago, II. Water balance, flow duration and storm runoff. *J. Hydrol.* (N.Z.), 23: 60–72.

Radcliffe, J.E., Young, S.R. and Clarke, D.G., 1977. Effects of sunny and shady aspects on pasture yield, digestibility, and sheep performance in Canterbury. *Proc. N.Z. Grassl. Assoc.*, 38: 66–77.

Raeside, J.D., 1948. Some post-Pleistocene climatic changes in Canterbury and their effect on soil formation. *Trans. R. Soc. N.Z.*, 77: 153–171.

Raven, P.H., 1973. Evolution of subalpine and alpine plant groups in New Zealand. *N.Z. J. Bot.*, 11: 177–200.

Riney, T., 1956. A zooecological approach to the study of ecosystems that include tussock grassland and browsing and grazing animals. *N.Z. J. Sci. Technol.*, B37: 455–472.

Rogers, G.M., 1988. *Landscape History of Moawhango District*. Ph.D. thesis, Victoria University, Wellington, 313 pp.

Ross, D.J., 1958a. Biological studies of some tussock grassland soils, V. Non-symbiotic nitrogen-fixing bacteria. *N.Z. J. Agric. Res.*, 1: 958–967.

Ross, D.J., 1958b. Biological studies of some tussock grassland soils, VI. Nitrifying activities. *N.Z. J. Agric. Res.*, 1: 968–973.

Ross, D.J., 1975a. Studies on a climosequence of soils in tussock grasslands, 5. Invertase and amylase activities of topsoils and their relationship with other properties. *N.Z. J. Sci.*, 18: 511–518.

Ross, D.J., 1975b. Studies on a climosequence of soils in tussock grasslands, 6. Invertase and amylase activities of tussock plant material and of soils. *N.Z. J. Sci.*, 18: 519–526.

Ross, D.J., 1975c. Studies on a climosequence of soils in tussock grasslands, 7. Distribution of invertase and amylase activities in soil fractions. *N.Z. J. Sci.*, 18: 527–534.

Ross, D.J. and McNeilly, B.A., 1975. Studies of a climosequence of soils in tussock grassland, 3. Nitrogen mineralisation and protease activity. *N.Z. J. Sci.*, 18: 361–375.

Ross, D.J. and Molloy, L.F., 1977. Studies of a climosequence of soils in tussock grasslands, 14. Water-soluble carbohydrates in tussock plant materials and in a light fraction from soil. *N.Z. J. Sci.*, 20: 453–459.

Ross, D.J., McNeilly, B.A. and Molloy, L.F., 1975. Studies of a climosequence of soils in tussock grasslands, 4. Respiratory activities and their relationships with temperature, moisture, and soil properties. *N.Z. J. Sci.*, 18: 377–389.

Ross, D.J., Molloy, L.F., Bridger, B.A. and Cairns, A., 1978. Studies on a climosequence of soils in tussock grasslands, 20. Decomposition of cellulose on the soil surface and in the topsoil. *N.Z. J. Sci.*, 21: 459–465.

Rowley, J., 1970. Effects of burning and clipping on the temperature, growth, and flowering of narrow-leaved snow tussock. *N.Z. J. Bot.*, 8: 264–282.

Roxburgh, S.H., Wilson, J.B. and Mark, A.F., 1988. Succession after disturbance of a New Zealand high-alpine cushionfield. *Arct. Alpine Res.*, 20: 230–236.
Scott, D., 1960. Seasonal behaviour of some montane plant species. *N.Z. J. Sci.*, 3: 694–699.
Scott, D., 1978. Plant ecology above timberline on Mt Ruapehu, North Island, New Zealand, II. Climate and monthly growth of five species. *N.Z. J. Bot.*, 15: 295–310.
Scott, D., 1979. Use and conservation of New Zealand native grasslands in 2079. *N.Z. J. Ecol.*, 2: 71–75.
Scott, D., 1984. Hawkweeds in run country. *Tussock Grassl. Mountain Lands Inst. Rev.*, 42: 33–48.
Silverton, J.W., 1980. The evolutionary ecology of mast seeding in trees. *Biol. J. Linn. Soc.*, 14: 235–250.
Simmons, D.R., 1967. Little Papanui and Otago prehistory. *Rec. Otago Mus. Anthropol.*, 4, 63 pp.
Sommerville, P., Mark, A.F. and Wilson, J.B., 1982. Plant succession on moraines of the upper Dart Valley, southern South Island, New Zealand. *N.Z. J. Bot.*, 20: 227–244.
Soons, J.M., 1962. A survey of periglacial features in New Zealand. In: M. McCaskill (Editor), *Land and Livelihood*. New Zealand Geographical Society, Christchurch, pp. 74–87.
Speir, T.W., 1976. Studies on a climosequence of soils in tussock grasslands, 8. Urease, phosphatase, and sulphatase activities of tussock plant material and of soil. *N.Z. J. Sci.*, 19: 383–387.
Speir, T.W., 1977a. Studies on a climosequence of soils in tussock grasslands, 10. Distribution of urease, phosphatase, and sulphatase activities in soil fractions. *N.Z. J. Sci.*, 20: 151–157.
Speir, T.W., 1977b. Studies on a climosequence of soils in tussock grasslands, 11. Urease, phosphatase, and sulphatase activities of topsoils and their relationships with other properties including plant available sulphur. *N.Z. J. Sci.*, 20: 159–166.
Speir, T.W. and Ross, D.J., 1978. Studies on a climosequence of soils in tussock grasslands, 18. Litter decomposition: urease, phosphatase, and sulphatase activities. *N.Z. J. Sci.*, 21: 297–306.
Stevens, G., 1985. *Lands in Collision: Discovering New Zealand's Past Geography*. Science Information Publishing Centre, Department of Scientific and Industrial Research, Wellington, 129 pp.
Stevens, G., McGlone, M.S. and McCulloch, B., 1988. *Prehistoric New Zealand*. Heinemann Reed, Auckland, 128 pp.
Stout, J.D., 1958a. Biological studies of some tussock grassland soils, IV. Bacteria. *N.Z. J. Agric. Res.*, 1: 943–957.
Stout, J.D., 1958b. Biological studies of some tussock grassland soils, VII. Protozoa. *N.Z. J. Agric. Res.*, 1: 974–984.
Suggate, R.P., 1965. Late Pleistocene geology from the northern part of the South Island, New Zealand. *N.Z. Geol. Surv., Bull.* (n.s.) 77, 190 pp.
Taylor, N.H. and Pohlen, I.J., 1962. Soil survey method — a New Zealand handbook for the field study of soils. *N.Z. Dept. Sci. Ind. Res., Soil Bur., Bull.*, 25, 242 pp.
Thornton, R.H., 1958a. Biological studies of some tussock grassland soils, I. Introduction, soils and vegetation. *N.Z. J. Agric. Res.*, 1: 913–921.
Thornton, R.H., 1958b. Biological studies of some tussock grassland soils, II. Fungi. *N.Z. J. Agric. Res.*, 1: 922–938.
Truog, E., 1930. Determination of readily available phosphorus in soils. *J. Am. Soc. Agron.*, 22: 874–882.
Tussock Grassland Research Committee, 1954. The high-altitude snow tussock grassland in South Island, New Zealand. *N.Z. J. Sci. Technol.*, A36: 335–364.
U.S.D.A., S.C.S., Soil Survey Staff, 1975. Soil Taxonomy: A Basic System of Soil Classification for Making and Interpreting Soil Surveys. *Agric. Handb.*, 436. U.S. Govt. Printing Office, Washington, D.C., 754 pp.
Vernon, T.R., 1958. Biological studies of some tussock grassland soils, VIII. Streptomycetes. *N.Z. J. Agric. Res.*, 1: 985–990.
Wardle, P., 1963. Evolution and distribution of the New Zealand flora, as affected by Quaternary climates. *N.Z. J. Bot.*, 1: 3–17.
Wardle, P., 1964. Facets of the distribution of forest vegetation in New Zealand. *N.Z. J. Bot.*, 2: 352–366.
Wardle, P., 1968. Evidence for an indigenous pre-Quaternary element in the mountain flora of New Zealand. *N.Z. J. Bot.*, 6: 120–125.
Wardle, P., 1971. An explanation for alpine timberline. *N.Z. J. Bot.*, 9: 371–402.
Wardle, P., 1972. Plant succession on greywacke gravel and scree in the subalpine belt of Canterbury, New Zealand. *N.Z. J. Bot.*, 10: 387–398.
Wardle, P., 1973. New Zealand timberlines. *Arct. Alpine Res.*, 5: 127–135.
Wardle, P., 1978. Origin of the New Zealand mountain flora, with special reference to trans-Tasman relationships. *N.Z. J. Bot.*, 16: 535–550.
Wardle, P., 1980. Ecology and distribution of silver beech (*Nothofagus menziesii*) in the Paringa district, South Westland, New Zealand. *N.Z. J. Ecol.*, 3: 23–36.
Wardle, P., 1985a. New Zealand timberlines, 3. A synthesis. *N.Z. J. Bot.*, 23: 263–271.
Wardle, P., 1985b. Environmental influences on the vegetation of New Zealand. *N.Z. J. Bot.*, 23: 773–788.
Wardle, P. and McKellar, M.H., 1978. *Nothofagus menziesii* leaves dated at 7490 yr B.P. in till-like sediments at Milford Sound, New Zealand. *N.Z. J. Bot.*, 16: 153–157.
Wardle, P., Mark, A.F. and Baylis, G.T.S., 1973. Vegetation and landscape of the West Cape District, Fiordland, New Zealand. *N.Z. J. Bot.*, 11: 599–626.
Wells, J.A. and Mark, A.F., 1966. The altitudinal sequence of climax vegetation on Mt Anglem, Stewart Island, 1. The principal strata. *N.Z. J. Bot.*, 4: 267–282.
White, E.G., 1975a. A survey and assessment of grasshoppers as herbivores in the South Island alpine tussock grassland of New Zealand. *N.Z. J. Agric. Res.*, 18: 73–85.
White, E.G., 1975b. Identifying population units that comply with capture–recapture assumptions in an open community of alpine grasshoppers. *Res. Popul. Ecol.*, 16: 153–187.
White, E.G., 1975c. An investigation and survey of insect damage affecting *Chionochloa* seed production in some alpine tussock grasslands. *N.Z. J. Agric. Res.*, 18: 163–178.
White, E.G., 1978. Energetics and consumption rates of alpine grasshoppers (Orthoptera: Acrididae) in New

Zealand. *Oecologia*, 33: 17–44.

Willett, R.W., 1950. The New Zealand Pleistocene snow line, climatic conditions, and suggested biological effects. *N.Z. J. Sci. Technol.*, B32: 18–48.

Williams, P.A., 1975. Studies of the tall-tussock (*Chionochloa*) vegetation/soil systems of the southern Tararua Range, New Zealand, 2. The vegetation/soil relationships. *N.Z. J. Bot.*, 13: 269–303.

Williams, P.A., 1977. Growth, biomass, and net productivity of tall-tussock (*Chionochloa*) grasslands, Canterbury, New Zealand. *N.Z. J. Bot.*, 15: 399–442.

Williams, P.A. and Meurk, C.D., 1977. The nutrient value of burnt tall-tussock. *Tussock Grassl. Mountain Lands Inst. Rev.*, 34: 63–66.

Williams, P.A., Grigg, J.L., Nes, P. and O'Connor, K.F., 1976. Vegetation/soil relationships and distribution of selected macro-elements within the shoots of tall-tussocks on the Murchison Mountains, Fiordland, New Zealand. *N.Z. J. Bot.*, 14: 29–53.

Williams, P.A., Mugambi, S. and O'Connor, K.F., 1977a. Shoot properties of nine species of *Chionochloa* Zotov (Gramineae). *N.Z. J. Bot.*, 15: 761–765.

Williams, P.A., Nes, P. and O'Connor, K.F., 1977b. Macro-element pools and fluxes in tall-tussock (*Chionochloa*) grasslands, Canterbury, New Zealand. *N.Z. J. Bot.*, 15: 443–476.

Williams, P.A., Grigg, J.L., Nes, P. and O'Connor, K.F., 1978a. Macro-element composition of *Chionochloa pallens* and *C. flavescens* shoots, and soil properties in the North Island, New Zealand. *N.Z. J. Bot.*, 16: 235–246.

Williams, P.A., Grigg, J.L., Nes, P. and O'Connor, K.F., 1978b. Macro-element levels in red tussock (*Chionochloa rubra* Zotov) shoots, and properties of associated volcanic soils, North Island, New Zealand. *N.Z. J. Bot.*, 16: 247–254.

Williams, P.A., Grigg, J.L., Nes, P. and O'Connor, K.F., 1978c. Macro-elements within shoots of tall-tussocks (*Chionochloa*), and soil properties on Mt Kaiparoro, Wairarapa, New Zealand. *N.Z. J. Bot.*, 16: 255–260.

Williams, P.A., Grigg, J.L., Mugambi, S. and O'Connor, K.F., 1978d. Soil properties beneath tall-tussocks (*Chionochloa*) in South Island. *N.Z. J. Sci.*, 21: 149–156.

Williams, P.A., Mugambi, S., Nes, P. and O'Connor, K.F., 1978e. Macro-element composition of tall-tussocks (*Chionochloa*) in the South Island, New Zealand, and their relationship with soil chemical properties. *N.Z. J. Bot.*, 16: 479–498.

Wilson, H.D., 1976. Vegetation of Mount Cook National Park. *N.Z. Natl. Parks Auth. Sci. Ser.*, 1, 138 pp.

Wood, F.H., 1973. Nematode feeding relationships: feeding relationships of soil-dwelling nematodes. *Soil Biol. Biochem.*, 5: 593–601.

Wraight, M.J., 1960. The alpine grasslands of the Hokitika River catchment, Westland. *N.Z. J. Sci.*, 3: 306–332.

Wraight, M.J., 1963. The alpine and upper montane grasslands of the Wairau River catchment, Marlborough. *N.Z. J. Bot.*, 1: 351–376.

Wraight, M.J., 1965. Growth rates and potential for spread of alpine carpet grass, *Chionochloa (Danthonia) australis*. *N.Z. J. Bot.*, 3: 171–179.

Yeates, G.W., 1974. Studies on a climosequence of soils in tussock grasslands, 2. Nematodes. *N.Z. J. Zool.*, 1: 171–177.

Zotov, V.D., 1938. Survey of the tussock grasslands of the South Island of New Zealand. *N.Z. J. Sci. Technol.*, A20: 212–244.

Zotov, V.D., 1963. Synopsis of the grass subfamily Arundinoideae in New Zealand. *N.Z. J. Bot.*, 1: 78–136.

Zotov, V.D., 1970. *Chionochloa macra* (Gramineae): a new species. *N.Z. J. Bot.*, 8: 91–93.

Chapter 15

GRASSLANDS OF THE SUB-ANTARCTIC ISLANDS [1]

R.J. HNATIUK

INTRODUCTION

The grasslands of the sub-Antarctic islands are very different from the natural grasslands in most other parts of the world. For many decades, interest in sub-Antarctic islands was almost exclusively the domain of whalers, sealers and biogeographers. More recently, these oil and fur hunters have been replaced largely by meteorologists, geologists, taxonomists and ecologists. These islands have unique characteristics to entice both the scientist and romanticist. The high rate of primary production of some of the grasslands is particularly interesting. Knowledge of the sub-Antarctic grasslands has increased greatly in the past 15 to 20 years. The stabilization of taxonomic nomenclature has permitted inter-island comparisons to be more firmly based.

The sub-Antarctic islands can be narrowly defined as those that occur near the Antarctic Convergence (Wace, 1960), that region of the ocean where cold, northward-moving, surface, Antarctic water gradually sinks below the warmer, southward-moving, subtropical water. The position of the Convergence is remarkably stable seasonally and from year to year (Ostapoff, 1965), but shifts slightly (up to 70 km) from time to time in relation to changes in wind and ice, whereas its long-term stability is related to changes in the ocean floor (Ostapoff, 1965; van Zinderen Bakker, 1970). Wace (1960) delimited the sub-Antarctic islands as those that have a closed herbfield vegetation, but which lack *Sphagnum* and cushion bogs. His classification is used here with the only amendment being the inclusion of Marion and Prince Edward Islands as sub-Antarctic rather than cool-temperate; this is in accord with Greene and Greene (1963), Wace (1965) and Smith (1976a). There are other definitions of the sub-Antarctic and its components (Cockayne, 1909; Holdgate, 1970; Meurk, 1977) that result in slightly different classifications. The islands taken as a whole are not homogeneous in their vegetation. Campbell Island (in the New Zealand sector) has extensive "sub-Antarctic" vegetation, but also significant areas of dense, tall scrub and low forest. It represents a transition between the sub-Antarctic and cool-temperate regions. Heard Island and South Georgia also have sub-Antarctic vegetation but support permanent glaciers and experience much stronger seasonality of climate than do islands such as Macquarie Island. They represent transitions between sub-Antarctic and Antarctic regions. For the purpose of the present review, Wace's modified definition is used to circumscribe the sub-Antarctic islands, although reference to the adjacent cool-temperate islands is made where possible.

The distribution of the sub-Antarctic islands is circum-south-polar (Fig. 15.1 and Table 15.1). They are isolated from other land, and the typical climate is highly oceanic (Table 15.2), although it increases in seasonality towards both the north and south. These islands are generally small, ranging from 6000 km^2 (Grande Terre, Îles de Kerguelen), to about 120 km^2 (Macquarie Island) (Table 15.1). They are mostly volcanic in origin and date from the Tertiary or Quaternary, although South Georgia and Kerguelen are of a continental, pre-Tertiary origin (van Zinderen Bakker, 1970). There is doubt about the time of emergence of

---

[1] Manuscript submitted in August, 1979; minor updating during 1987.

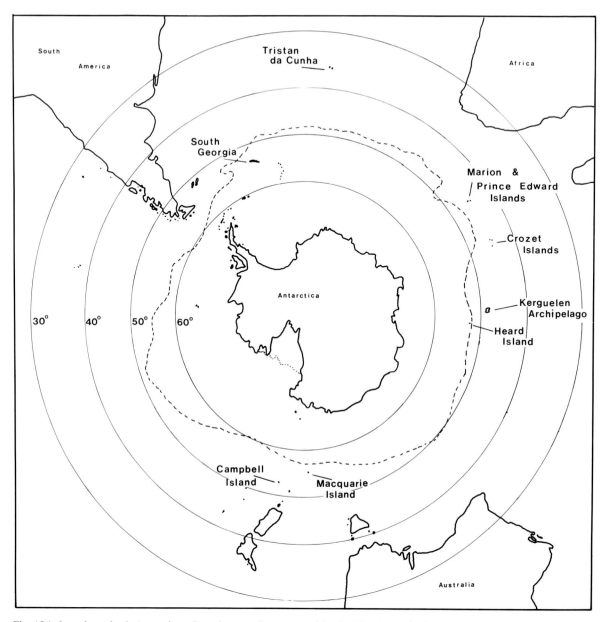

Fig. 15.1. Location of sub-Antarctic and southern cool-temperate islands. The Antarctic Convergence zone is indicated by the broken line. Its location is adapted from Ostapoff (1965) in a position south of the Kerguelen archipelago (as determined by IGY observations).

Macquarie Island which is thought to be late Tertiary (Quilty et al., 1973) or late Pleistocene (Colhoun and Goede, 1973). The great Quaternary glaciations affected the sub-Antarctic islands, several of which are still heavily glaciated (van Zinderen Bakker, 1970), although there is still doubt regarding the magnitude or time of occurrence of glaciers on Macquarie Island (Jenkin, 1975). The floristic composition (apart from recent alien introductions) appears to have been fairly constant since the end of the last glacial period (van Zinderen Bakker, 1970; Barrow, 1978), although the Miocene flora of the Îles de Kerguelen is reputed to have been rich in species that do not occur

TABLE 15.1

Sub-Antarctic islands: their location and other geographic data

	Crozet Islands	Heard Island	Îles de Kerguelen	Macquarie Island	Marion Island	South Georgia
Latitude	46°27′S	53°06′S	48°50′S	54°36′S	46°53′S	54°15′S
Longitude	52°00′E	73°30′E	68°70′E	158°55′E	37°45′E	36°45′E
Area (km^2)	505	c. 700	7000	c. 120	290	3756
Major rock types	Volcanic	Volcanic	Basaltic lavas	Crustal and volcanic	Volcanic lavas	Sedimentary and lavas
Presence of glaciers	No	Yes	Yes	No	Yes	Yes
Maximum elevation (m)	2000	2745	1960	433	1230	2934
Administered by	France	Australia	France	Australia	S. Africa	U.K.

TABLE 15.2

Selected climatic data for sub-Antarctic islands

	Heard Island[1]	Îles de Kerguelen[2]	Macquarie Island[3]	Marion Island[4]	South Georgia[5]
Temperature (°C):					
Mean annual	1.1	4.3	4.7	5.2	1.8
Range of monthly means	0.2–3.3	1.9–7.3	3.1–6.7	1.0–10.6	−1.5 to 5.3
Mean yearly maximum			6.2	8.1	5.2
Mean yearly minimum			2.8	2.9	−1.1
Mean relative humidity (%)		78	89	83	75
Mean wind velocity (m s^{-1})			8.6	8.9	4.3
Precipitation (mm):					
Mean annual		c. 1120[1]	926	2576	1405
Range of monthly means		75–150	67–99	165–249	77–149
Extreme yearly range				2203–2913	800–2237
Sunshine (hr):					
Total annual	525	1349	814	1272	1236
Daily mean	1.4	3.7	2.2	3.5	3.4
Range in monthly mean (d^{-1})	0.8–2.2	1.2–5.7	0.7–3.8	1.8–5.2	0.2–5.8
Mean day length				12.1	12.1
Number of days of snowfall:					
Total annual			106		221
Range in monthly mean					4–31

References: [1] Chastain (1958); [2] Fabricius (1957); [3] Jenkin (1975); [4] Schulze (1971); [5] Lewis Smith and Walton (1975).

there today (Young, 1971). There have been local, postglacial fluctuations in the relative position of various plant communities since the last major warming began (van Zinderen Bakker, 1969, 1973). Glacial refugia for plants during the Pleistocene, on at least some of the islands, have been suggested on the basis of palynological and geomorphic evidence for Kerguelen (van Zinderen Bakker, 1970), South Georgia (Barrow, 1978) and Marion Island (van Zinderen Bakker, 1973). The climate of the sub-Antarctic islands appears to have been similar to that of the present for the past 10 000 years or so (Schalke and van Zinderen Bakker, 1971).

Fig. 15.2. *Poa foliosa* high sub-Antarctic tussock grassland on Macquarie Island, near sea level (foreground) and extending up the slope (background). (Photo by C. Russell, 1974, courtesy of the Antarctic Division, Australia.)

The topography of the sub-Antarctic islands is generally rugged (Fig. 15.2), although extensive areas of outwash plains, raised beaches, and wave-cut platforms (clothed in peat and mire) smooth the surface of the landscape. Most of the islands rise steeply for several hundred metres out of the sea, their upper reaches either extending above the climatic vegetation line or passing through or into a tundra-like feldmark dominated by short grasses and forbs (e.g. *Azorella selago*). The coastal zone is usually occupied by maritime tussock grassland (*Poa* spp.), which lies adjacent to the narrow, sandy, pebble, or rocky beaches (Fig. 15.3). Some of these beaches are major resting places for *Artocephalus* spp. (fur seals), *Mirounga leonina* (elephant seals), *Phocarctos hookeri* (sea lions) and for species of *Aptenodytes*, *Eudyptes* and *Pygoscelis* (penguins). Trampling, as much as manuring, by these animals has profoundly destructive effects locally (Gillham, 1961). Abundant leaching and frequent rainfall appear to limit the build-up of salt.

The grassland of any individual sub-Antarctic island scarcely constitutes an ecosystem. Each island may best be viewed as an entire ecosystem, with the grasslands being but one component. As such, the island ecosystems contact only one other major ecosystem, the ocean.

CLIMATE

The sub-Antarctic region occurs in the Polar Maritime climate (EM) of Shear (1964). In the Southern Hemisphere it includes all of the islands classified above as sub-Antarctic, plus Campbell Island. It is a unique climate that is best characterized by Varney's (1926) description of Kerguelen:

Fig. 15.3. *Poa flabellata* tussock grassland near sea level and on low slopes on South Georgia. Wildlife includes *Artocephalus* sp. (fur seal), *Mirounga leonina* (elephant seal) and *Pygoscelis papua* (gentoo penguin) on the beach and among the tussocks. (Photo courtesy of R.I. Lewis Smith, British Antarctic Survey.)

"It is a climate so cold that snow may fall on any day of the year, though warm enough for the mean temperature of no month to be below freezing; so cold that frosts occur on 140 days annually, but so warm that the ground never freezes deeper than five cm.; cold enough to have the lowest temperature below freezing on 248 days annually, yet warm enough to have more rainy days than snowy." Although not applicable to other sub-Antarctic islands (e.g. South Georgia) this description gives the essence of much of the sub-Antarctic.

The climate of the sub-Antarctic is dominated by two major factors. Firstly, there is a continual procession of large, anti-cyclonic, high-pressure systems, frequently with cold fronts embedded in them. These systems move around the world in a west to east direction. They are uninterrupted by major land masses except for the southern part of South America and the Antarctic Peninsula. The result of this circular procession is a fairly regular alternation, at 4- to 8-day intervals, of periods of squalls, light rain or snow, followed by periods of heavy cloud and strong wind. Intermittent showers are seldom heavy on Macquarie Island (Delisle, 1965) or on Kerguelen (Chastain, 1958), but are often heavy on South Georgia (Lewis Smith and Walton, 1975). Orographic cloud usually forms about the mountains. The belt of cyclones/anticyclones moves from north to south on a seasonal basis, increasing the strength and frequency of storms in winter. Solar radiation at ground level is nearly always low due to the extremely high incidence of cloud. Daylight hours vary considerably with season (as expected at 45° to 55°S) from about 7 to 9 hrs in winter (June) to 15 to 17 hrs in summer (December).

The second major factor affecting the climate is directly related to the Antarctic and sub-Antarctic Convergence Zones. Over a belt some 10 degrees of latitude wide, sea-surface temperatures range in the south between 1 to 2°C in winter and 4 to 5°C in summer, and in the north between 10 to 14°C in winter and 14 to 18°C in summer (Deacon, 1964). The influx of warm, moist air into a cool environment results in the production of so much cloud that Fabricius (1957) described it as one of the cloudiest areas on earth. The result is that solar radiation received by the islands is greatly reduced — about 11% of annual potential sunshine falls on Heard Island, 18% on Macquarie Island, and 28% on South Georgia. There is also considerable seasonal variation between islands. For example, Macquarie Island receives only 27% as much sunshine at mid-winter and 66% as much at mid-summer as does Marion Island (Fabricius, 1957).

The thermal climate of some sub-Antarctic islands is relatively stable because of the buffering of the large mass and expanse of ocean (Ostapoff, 1965) and because of the generally overcast and windy conditions. Air temperatures have been described as being the most uniform on earth, even considering the lowland tropics (Troll, 1958). The thermoisopleths in Fig. 15.4 illustrate the thermal stability of the region.

Microclimates naturally vary according to how much direct sunlight reaches the surface. The same factors (noted above) which produce large-scale thermal stability also result in reduced thermal microclimatic variability. Whereas only very small variations in temperature between leaves and the air (less than 3°C) appear to be the norm on Macquarie Island (Jenkin, 1972), Clarke and Greene (1970) reported a greater range in bryophyte communities on South Georgia. Moseley (1875), having measured a 2.8°C higher temperature within an *Azorella selago* 'tuft' (as compared with the air) on Marion Island, commented on the potential benefits of the heat-storing capacity of the mounded growth form. Schulze (1971) reported that the mean minimum temperature just above ground level (on Marion Island) was about 1.7°C lower than that measured in a standard meteorological screen. Many detailed microclimatic studies were reported by Lewis Smith (1986). Although small, these 1 to 3°C differences may be significant for plant metabolism at the prevailing low temperatures. The relationship between increased performance of species in grasslands and increasing shelter was documented (for South Georgia) by Callaghan and Lewis (1971a, b) and by Lewis Smith and Walton (1975). Tussock density may be related to site aspect on Macquarie, where lesser densities appear to occur in sites more directly inclined towards the north (sunward aspect) (Hnatiuk, 1978). The role of wind in creating micro-habitat variation has been recognized by several workers (Huntley, 1971; Jenkin, 1972; Hnatiuk, 1975), but requires further study.

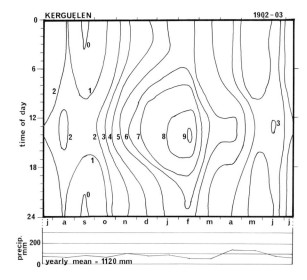
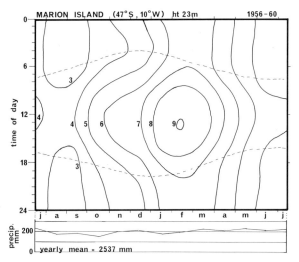
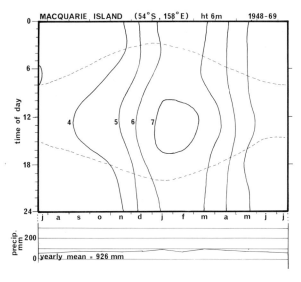

Precipitation is nowhere a major limiting factor to plant growth in the sub-Antarctic. Characteristically rain or snow falls on 70% to over 80% of the days of the year. Only rarely does snow lie on the ground at sea level on Macquarie Island (Jenkin, 1975) or on Marion Island (Schulze, 1971), but it can form a continuous cover on South Georgia in July and August (Lewis Smith and Walton, 1975). Short-term droughts occur very occasionally, but the peaty nature of soils on many of the islands acts to dampen their effects.

Relative humidity is high at nearly all times (Fabricius, 1957; Jenkin, 1975; Lewis Smith and Walton, 1975). For example, mean morning and mid-day values at Macquarie were reported as 92% and 91%, respectively (range 90 to 94% and 88 to 93%, respectively) (Anonymous, 1958). A somewhat wider range probably occurs on other islands where direct sunlight at ground level is received more frequently and for longer periods than on Macquarie Island.

VEGETATION

Major world classification systems do not specifically cater for the sub-Antarctic grasslands, but down to the level of Formation Sub-class and possibly Formation Group the UNESCO classification serves fairly well (Anonymous, 1973). Two major kinds of grassland are present on sub-Antarctic islands and may be classified as "Medium Tall Grassland" (*V.B.5b*), and "Short Grassland" (*V.C.6*) [or possibly Graminoid Tundra (*V.C.8*)]. The grassland vegetation of the sub-Antarctic islands shares certain similarities with northern polar vegetation (Wielgolaski, 1972; Bliss, 1979). Bliss's (1979) classification puts these southern grasslands into the High Sub-Antarctic–Maritime tussock grassland (tall tussocks, associated forbs and some shrubs in the north) and Grass Forb (for low tussocks). Smaller tufted grasses are classed as Antarctic Biome, Low Antarctic tundra–Grass forb.

Fig. 15.4. Thermoisopleths for three sub-Antarctic stations. Temperatures are in °C. Times of sunrise and sunset on Macquarie and Marion Islands are marked by broken lines. Precipitation (annual mean and distribution by month) is shown below each thermoisopleth (figures: N. Wace, pers. commun., 1978).

Ecologically, Bliss equated the tussock grasslands of the sub-Antarctic to shrubland in the Arctic. The greatest merits of Bliss's system are that it is simple and that it concentrates on the qualities of the vegetation. It removes the confusion that has existed in the definition of "sub-Antarctic islands" (see p. 411); islands may thus be sub-Antarctic because they have sub-Antarctic vegetation, but they may also contain vegetation that is other than sub-Antarctic.

The area that is covered by grassland on each of the islands is generally not known, but it appears to be a small fraction of the total area. On Marion Island, mire grassland occupies about one-third of the area of the island below 300 m altitude.

Grasslands occur along the sea coast on sand, gravel or rock, all of which are often overlain by peat, on steep slopes of up to 40° or more from the horizontal, or on flat to gently sloping peat bogs and mires. They extend up to about 200 to 300 m above sea level, the precise limit varying with exposure of the site and the proximity of the island to the Antarctic Convergence.

The sub-Antarctic short grassland communities are dominated by *Agrostis magellanica* and *Festuca contracta*. The *Festuca* grassland appears to reach its best development on South Georgia (Fig. 15.5) where, as a climax community, it forms locally extensive stands of closed grassland in moderately sheltered, fairly dry, well-drained slopes. In more exposed situations there is a large cryptogamic component (*Chorisodontium aciphyl-*

Fig. 15.5. *Festuca contracta* low sub-Antarctic tundra grassland on a gently sloping hillside, King Edward Cove, South Georgia. (Photo courtesy of R.I. Lewis Smith, British Antarctic Survey.)

lum, Cladonia spp. and *Polytrichum* spp.) (Lewis Smith and Walton, 1975). *Agrostis magellanica* is a more widespread sub-Antarctic species, occurring on Prince Edward Island, Marion Island, the Îles Crozet, the Îles de Kerguelen, and Macquarie Island (Greene and Walton, 1975). It appears to reach dominance in open grasslands on Marion Island that are very moist and described as soligenous mires (Huntley, 1972; Smith, 1976a). The wet conditions favour an abundance of sedges (such as *Uncinia dikei*) and rushes (such as *Juncus scheuchzerioides*). A dense carpet of bryophytes forms the understorey, with *Drepanocladus uninatus*, *Plagiochila crozetensis* and *Rhacomitrium lanuginosum* being most abundant.

Large tussock-forming grasses are the most characteristic feature of the southern cool-temperate and sub-Antarctic medium-tall grasslands (Figs. 15.2, 15.3 and 15.6). In the sub-Antarctic, three species of *Poa* dominate, namely *Poa cookii, P. flabellata* and *P. foliosa*, whereas in the cool-temperate zone *Chionochloa antarctica* and *Poa litorosa* occur on Campbell and Auckland Islands, *P. flabellata* on the Falkland Islands, Gough Island, Diego Ramirez, and Tierra del Fuego (N.M. Wace, pers. commun., 1979), and *Spartina arundinacea* on Tristan da Cunha, Gough Island, New Amsterdam, and St. Paul (Wace, 1960). The geographic distribution of these taxa illustrates the degree of floristic discontinuity between islands, but the habitats dominated by tussock grasses in each region are similar. The coastal tussock grassland is the most widespread and extensive grassland community. It was described as halophytic (Wace, 1965). Parts of it are frequently invaded seasonally by *Artocephalus* spp., *Mirounga leonina* and *Phocarctos hookeri* when they come ashore to breed and moult. It extends several hundred metres up the slopes that often rise steeply from narrow beaches. Although having good drainage, the slope habitats are often moist due to seepage of water from higher ground, aided by the high frequency of showers and the high water retention of tussock peat. On Marion Island, tussock communities occurring inland are associated with ground-nesting birds and consequent manuring (Smith, 1976b); but on Macquarie Island and Campbell Island, the inland occurrences of tussock grassland are not always associated with bird roosts.

Other types of sub-Antarctic grassland are usually of very limited area, but may form pure or nearly pure swards. For example, on South Georgia, *Phleum alpinum* occasionally dominates small (up to 50 m^2) stands typically on sheltered moist slopes and raised beach areas; rarely, *Alopecurus magellanicus* forms dense stands of up to 500 m^2 in extent (R.I. Lewis Smith, pers. commun., 1979). *Deschampsia antarctica* forms locally extensive, almost pure stands on moist, raised beach areas where peat accumulates to a depth of 2 m, and may also form a sward amongst *Poa flabellata* tussocks (R.I. Lewis Smith, pers. commun., 1979). *Deschampsia antarctica* forms an important phytogeographic link between the sub-Antarctic and maritime Antarctic botanic zones. Similar small localized grassland types prevail on other sub-Antarctic islands [e.g. the *Juncus scheuchzerioides–Agrostis magellanica–Deschampsia penicillata* association on Macquarie Island (Taylor, 1955)], but they have not yet been studied in detail.

The structure of sub-Antarctic grasslands is simple. In many instances the communities form a single stratum composed of a closed or nearly closed canopy of leaves of tussock grasses. Where the canopy is open, a few small forbs occur either between the tussocks or on their bases. Some tussock communities, as on Macquarie Island, are co-dominated by very large-leaved forbs (such as *Stilbocarpa polaris*, and to a lesser extent *Pleurophyllum hookeri*). However, most of the giant forbs and rosette plants, as well as smaller forbs, also dominate communities in which tussocks are either a minor component or non-existent (Cockayne, 1909; Taylor, 1955; Chastain, 1958; Wace, 1960; Jenkin, 1975; Jenkin and Ashton, 1979). In grasslands or wet sites (e.g. in which *Agrostis magellanica* dominates), a dense understorey of bryophytes is often present (Huntley, 1972; Smith, 1976a). The bryofloras of these communities (about 60 spp. in mire grassland on Marion Island: V.R. Smith, pers. commun., 1979), and their apparently narrow habitat preferences have only begun to be examined (Ashton, 1965; Seppelt, 1977).

The grasslands are part of a mosaic pattern of vegetation on the islands. Differences in soils, drainage, protection from wind, snow-lie, or exposure to sunlight favour the dominance of certain species over others at any particular site. Lewis

Fig. 15.6. *Poa cookii* high sub-Antarctic tussock grassland near the coast on Marion Island. (Photo courtesy of V. Smith, University of the Orange Free State.)

Smith and Walton (1975) have indicated the importance of soil-moisture status in determining composition of the vegetation on South Georgia, whereas Meurk (1980) noted the importance of depth to the water table as a major factor controlling the vegetation mosaic on Campbell Island.

Grazing effects by introduced herbivorous vertebrates are now drastically altering the previous balance that had evolved between the small numbers of plant species and the environment. No large native herbivores (mammals, birds or reptiles) were present and invertebrate herbivores have little impact (Jenkin, 1975; Smith, 1977b).

A characteristic structural feature of many of the tussock grasslands is the formation of a pedestal beneath the canopy of each tussock. These pedestals, reaching to over 1.5 m high, are composed of the living and dead leaf bases, stems and roots of the grass (Hnatiuk, 1975). They raise the canopy to heights of 3.5 m for *Spartina arundinacea* (Wace, 1960) and 2.5 m for *Poa flabellata* (Moore, 1968). Hooker (1847) likened the *P. flabellata* tussocks on the Falkland Islands to "groves of small low palm trees". The tussocks are commonly less than 1.5 m in height. The roots of the grasses, where tussocks are well developed, are most abundant in the upper part of the pedestal (Jenkin, 1975). Some roots extend into the inter-tussock soil, but no detailed studies of root distribution appear to have been made to date. The tallest tussocks occur along the coastal zone where *Mirounga leonina* forms wallows between pedestals. There is a progressive diminution of pedestal size with distance inland and with elevation above sea level, which is only partly a function of the occurrence of wallows in the low coastal habitats. Where wind exposure becomes very high, as near hill crests, tussocks become progressively smaller until a sward is formed (Hnatiuk, 1975).

Very few studies have been completed on the community dynamics of the grasslands. Under intense grazing by rabbits (*Oryctolagus cuniculus*), grassland is replaced by communities of mosses and small forbs on Macquarie Island (Costin and Moore, 1960), and by *Acaena adscendens* (now

A. magellanica) on Grande Terre, Îles de Kerguelen (Chastain, 1958; Greene and Walton, 1975). Detailed studies by Ashton (1965) on the regeneration pattern of *Poa foliosa* tussocks on Macquarie Island have suggested a cyclical development process through pioneer, building, mature, and degenerate phases. A gap-stage was identified, wherein small forbs, grasses and bryophytes or the large-leaved *Stilbocarpa polaris* occur. The turnover time of each cycle is not known, but individual dominant plants live for several decades.

Lack of descriptions of the vegetation of several islands impedes comparative analyses. Such studies are basic to an understanding of the region and should be given priority over detailed studies of individual non-dominant species.

AUTECOLOGY

Detailed autecological studies of a few species of grasses from South Georgia have shown the nature of some of the adaptations of the native plants to cool, humid environments (Callaghan and Lewis, 1971a, b). Studies of *Phleum alpinum* have shown that native soils from the fellfield are severely limiting in nutrients, a contrast to the situation with *Poa foliosa* on Macquarie Island (Jenkin, 1975). Most indicators of growth performance (relative growth rate, unit leaf rate, and specific leaf area) decreased as the microhabitat became increasingly severe, but leaf-area ratio (average leaf area per unit dry weight over a measured time interval) showed no simple relationship to site conditions (Callaghan and Lewis, 1971b). Vegetative reproduction was similar over a range of site conditions, but sexual reproduction declined markedly with increasing severity of habitat (Callaghan and Lewis, 1971a). Similar results were found in studies of *Festuca contracta* on South Georgia (Tallowin, 1977a, b; Tallowin and Lewis Smith, 1977). Comprehensive studies on the performance of other dominant species are needed.

SEASONAL VARIATION IN GROWTH

Seasonal changes in the vegetation of the sub-Antarctic grasslands are small in comparison with those of northern, continental grasslands. The degree to which seasonal changes are evident varies from island to island, but is largely dependent upon nearness to the Antarctic Convergence, and therefore the relative severity of the winter temperatures. Thus, South Georgia may be snow-covered for four months or more, whereas Macquarie and Marion Islands are only briefly and intermittently covered by light falls of snow.

Several adaptations to this uncommon environment are evident among the various species. *Stilbocarpa polaris*, a giant forb in tussock grasslands on Macquarie Island and Campbell Island, shows a pronounced dying back of the large, summer leaves and over-winters with relatively small leaves. Flowering for many species occurs over several months, but predominantly in summer, with seed dispersal in late summer through to mid-winter (Taylor, 1955; Chastain, 1958). At least three species (*Deschampsia antarctica*, *Festuca contracta* and *Poa flabellata*) pre-form inflorescences in the autumn on South Georgia (Walton, 1982), leading to flowering early in the following spring. Leaves on *Poa foliosa* shoots on Macquarie Island are reported to show little seasonal variation in production rate, individual leaves living for 8 to 10 months (Jenkin, 1975). The abundant understorey of bryophytes that characterizes the mire grasslands of Marion Island may exhibit a flush of growth, both before and after the main summer growth period of the vascular plants (V. Smith, pers. commun., 1979).

STANDING CROP AND PRIMARY PRODUCTIVITY

The International Biological Programme stimulated or assisted comparative studies of plant biomass and growth on South Georgia and Macquarie Island. These studies continue and have been expanded to include Marion Island, as well as cool-temperate Campbell Island. The studies on Macquarie Island were reported by Jenkin (1975) and Hnatiuk (1978), those on Marion Island by Smith (1976a, 1978b), and those on Campbell Island by Hnatiuk (1978) and Meurk (1980). However, the most detailed studies have been on South Georgia (Callaghan and Lewis, 1971a; Lewis Smith and Stephenson, 1975; Walton et al.,

TABLE 15.3

Comparative values of maximum standing crop (g m^{-2}) in grasslands at sub-Antarctic stations and on a tropical high mountain

Location and type of vegetation	Above-ground plant parts[1]									Bryophytes, lichens and algae	Litter	Under-ground parts			Total	Leaf area index (LAI)	
	Monocots[2]		Forbs		Woody spp.		Subtotal									summer	winter
	live	dead	live	dead	live	dead	live	dead				live	dead	total			
Macquarie Island[3]																	
Tussock grassland, elevation 45 m	637	2560	262	22	13	10	912	2592	6	101	1690	3110	4800	8411	7.6	4.5
elevation 235 m	321	936	1	0	6	4	328	940	5						1.7	
South Georgia[4]																	
Festuca contracta	208	1598	0	0	217	0	425	1598	512	140			1642	4317		
Poa flabellata tussock grassland, enriched	7500	5000	25	5	0	0	7525	5005	5	250			5000	17785		
Marion Island[5,6]																	
Poa cookii–Azorella: crest							804	3654	Tr				2001	6459		
Poa cookii tussock grassland: 1976							778	1649	230				3988	6645		
1978							868	1341	382				2483	5074		
Agrostis magellanica mire							117	303	219				2024	2663		
Campbell Island[7]																	
Chionochloa antarctica tussock grassland	634	1742	34	25	215	67	883	1834	0		1116	1206	2322	5039	5.5	
Mount Wilhelm, New Guinea[7]																	
Deschampsia klossii tussock grassland	378	2978	19	0	0	0	397	2978	56		85	336	421	3852		1.8[8]

[1] There is little seasonal difference in the standing crop of seasonal material (Jenkin, 1975). Living material included non-green structures — stems, leaf bases and young leaves.
[2] Monocotyledons were almost all grasses.
References: [3] Jenkin (1975), Jenkin and Ashton (1970); [4] Lewis Smith and Walton (1975); [5] Smith (1976a); [6] Smith (1978b); [7] Hnatiuk (1978).
[8] "Dry" season.

1975; Tallowin, 1977a; Callaghan, 1978).

A wide range of values for standing crop and net primary production in grasslands has been reported (Tables 15.3 and 15.4). Care must be used in making direct comparisons between these studies because of the differing methods used by each worker. Tussock grasslands are clearly very much more productive than short-grass grasslands. The highest standing crop and productivity (above- plus under-ground) were recorded for *Poa flabellata* tussock grassland, growing in an area enriched by nutrients from *Mirounga leonina* wallows (Lewis Smith and Walton, 1975). Comparable data from unenriched sites are very much less (R.I. Lewis Smith, pers. commun., 1980).

Two *Poa cookii* tussock grasslands on Marion Island were reported as having very similar total (living plus dead, above-ground plus under-ground parts) standing crops (Table 15.3) (Smith, 1976a), but only about one-third that of the enriched *P. flabellata*. One of the *P. cookii* sites receives nutriment from burrow-nesting birds (Smith, 1976b), but has only about 3% greater total standing crop (above- plus under-ground in all species) than the unenriched site. However, there is a great difference in community structure and distribution of dry matter above- and under-ground between the two sites. There is also variation in biomass of grassland vegetation between sites on Marion Island depending upon the nature of the lava substrate (Smith, 1977a). For *Agrostis magellanica* mires, on the older, preglacial, grey lavas, living biomass (both above- and under-ground) was only 73% of that on black, post-glacial lavas; for *Poa cookii* tussock grassland, the comparable value was 80%. For total standing crop including dead material these figures were respectively 85% and 81%. The

TABLE 15.4

Comparative values of annual net production of biomass (g m^{-2}) in grasslands at sub-Antarctic stations and on a tropical high mountain[1]

Location and type of vegetation	Above-ground plant parts					Under-ground parts	Total
	Monocots[2]	Forbs	Woody plants	Subtotal	Bryophytes, lichens and algae		
Macquarie Island[3]							
Poa foliosa tussock grassland (elevation 45 m)	1440	395	55	1890	21	3670	5581
South Georgia[4]							
Festuca contracta	160	0	180	340	152	c. 350	840
Poa flabellata tussock grassland, enriched	5000	20	0	5020	5	1000	6025
Marion Island[5]							
Poa cookii tussock grassland	352			352	113		465
Campbell Island[6]							
Chionochloa antarctica tussock grassland	1505			1505			
Mount Wilhelm, New Guinea[6]							
Deschampsia klossii tussock grassland	259			259			

[1] For Macquarie Island, South Georgia and Marion Island, production was estimated from the differences between harvests at different times of the year. For Campbell Island and Mount Wilhelm, production was estimated from the relationship of measured leaf-extension rates to weight classes for different sizes of leaves. Since no account is taken of losses between harvests in the first method or of increments in the second method, these values are only estimates of minimum production. The biomass in the "pedestal" of the tussocks was included as under-ground biomass.
[2] The monocotyledons were mostly grasses.
References: [3] Jenkin (1975); [4] Lewis Smith and Walton (1975); [5] Smith (1978a); [6] Hnatiuk (1978).

relative importance of age, nutrient status, and topography offered by the two lava substrates to the rate of plant growth has not yet been evaluated. The *Poa foliosa* tussock grassland on Macquarie Island, unenriched by manuring, had just under half the standing crop (living plus dead, above- and under-ground) of the enriched *P. flabellata* site, but over one-fourth more than the enriched *P. cookii* site.

The net annual addition of plant biomass, both above ground and under ground was about the same for *P. flabellata* as for *P. foliosa* (Table 15.4), although the latter had about twice as long a growing season. The growing season for Macquarie Island was based upon the period during which above-ground living matter increased [as measured by successive harvests (Jenkin, 1975)]. It is possible that the season would be found to be even longer if the growth period alone were to be measured rather than the period during which there was recorded a positive balance of living biomass between successive harvests. For South Georgia, the growing season was taken as the period during which leaf growth occurred (Lewis Smith and Walton, 1975). Meurk (1980) has reported a virtually year-long growing season for tussock grasses on Campbell Island (as indicated by growth of marked leaves). Growth of leaves appeared to stop only during irregular, short periods of particularly low temperatures during the short days of winter (C.D. Meurk, pers. commun., 1979).

Independent reports of net productivity from measurements of leaf growth have corroborated the high values of Jenkin (1975) on Macquarie Island (Hnatiuk, 1978). Mid-season production rates of 6.62 to 10.88 g m^{-2} d^{-1} compare favourably with Jenkin's mean annual rate of 8.99 g m^{-2} d^{-1} for *Poa foliosa* tussock grassland. The similarity of the values seems to imply that either the seasonal range of values is small or that any quiescent period is short. Emphasis is placed on the fact that these data do not come from atypical sites, such as those enriched by seal and sea-bird manure. Although Lewis Smith (1986) expressed scepticism with respect to these estimates of production in tussock grasslands, he provided no basis for this. The conditions under which high productivity occurs clearly need closer study.

Considerable variation exists in levels of annual total (shoot plus root) production reported for tussock grasslands on different islands (Table 15.4), ranging from 465 g m^{-2} for the small *Poa cookii* tussocks on Marion Island to 6025 g m^{-2} for the large *P. flabellata* tussocks on South Georgia. The latter values are from dense stands of tussocks of unusual productivity (R.I. Lewis Smith, pers. commun., 1980). The dynamics of seasonal changes in productivity in one community on an island with pronounced seasonal periodicity are perhaps best illustrated by Lewis Smith and Stephenson (1975) for *Festuca contracta* on South Georgia. There, a good correlation was found between the amount of plant matter produced and both the number of hours of sunshine received and the mean daily temperature. Attempts to carry out comparable studies with the much more robust tussock grasses have been frustrated by problems of sampling large biomass using limited labour resources.

Annual variations in living biomass within a single community have been indicated by some workers, though not thoroughly examined as yet. Jenkin and Ashton (1970) attribute variation in growth (both above- and below-ground) between consecutive summers as possibly due to slight variations in the temperature and light regimes between years, slightly warmer conditions with more light resulting in increased growth. A similar sensitivity to slight increases in temperature was reported for Campbell Island (Meurk, 1980). The variation in production of *Poa cookii* tussock grassland on Marion Island (Smith, 1978b) was attributed to possible sampling errors from small sample size (plots 50 cm by 50 cm, $n = 4$) in the first year. Heterogeneity in species distribution within communities is also noted as a factor affecting results on Marion Island (Smith, 1978b) and South Georgia (Walton et al., 1975). The differing growth forms and production of the dominant species, combined with their wide ecological tolerances and frequently mosaic distribution patterns, make it difficult to compare either production or productivity of different communities.

More detailed studies of ecosystems as a whole are required for more understanding of such factors as the prodigious growth of certain species. A beginning has been made on Marion Island, where Smith (1977b) has qualitatively defined the

NUTRIENT FLOW

Soil nutrients

The nutrient status of the soils of sub-Antarctic grasslands has not been investigated widely or in detail (Jenkin, 1975; Smith, 1976a), apart from largely unpublished data from South Georgia (R.I. Lewis Smith, pers. commun., 1979). It appears to be strongly influenced by salts coming from the sea via spray and precipitation, by inputs from manuring by sea birds and seals, and by the nature of the chemical balances in the highly organic materials comprising the many soils. Jenkin (1975) noted the importance of precipitation (on Macquarie Island) as a source of nutrients in tussock grassland. His work also indicated that nitrogen is redistributed by wind and water from penguin rookeries to nearby grassland vegetation. The direct effects of animal and bird manuring on grassland soils have been investigated on Marion Island, where significant amounts of nitrogen and phosphorus in the soil and plants have been shown to come from this source (Smith, 1976b, 1979). Nitrogen, rather than phosphorus, is considered to be the principal deficiency in the unenriched soil. Much phosphorus in the soil appears to be unavailable to plants because it is bound into iron and aluminium complexes (Smith, 1979). Microbial activity in the soil may be a major source of phosphorus to plants (Smith, 1979), but further study is required.

Plant nutrients

The nutrient status of plants of sub-Antarctic grasslands appears variable between islands, on the basis of the few preliminary results available (Macquarie Island: Jenkin, 1975; Marion Island: Smith, 1976a; South Georgia: Lewis Smith and Walton, 1975; Walton et al., 1975; Walton and Lewis Smith, 1980). However, the results of these studies are not completely comparable as different methods of analysis and presentation have been used. Furthermore, direct comparisons of nutrient levels between different plant taxa, each adapted to its own environment, can be misleading — for example, what is a deficiency of phosphorus for one species can be adequate for another.

Experimental work at South Georgia, in which nutrients were applied in pot trials to native species of *Festuca* grassland, showed that improved growth and "performance" resulted from additions of nitrogen, phosphorus and potassium (Lewis Smith and Walton, 1975). On Marion Island, Smith (1976b, 1978a) reported that the luxuriant, dark green vegetation dominated by *Poa cookii* in areas manured by animals contrasted with the drab yellow green of the surroundings. By studying the nutrient status of these *Poa* communities, and documenting the enriched nitrogen and phosphorus status of the soil and plants growing there, he showed that the source of these nutrients was the manure from the large number of burrow-nesting "prions" and small "petrels" (Family Procellariidae) or surface-nesting birds which lived in the community. I have seen a similarly dramatic, local greening of *Poa litorosa* tussocks adjacent to nests of *Diomedea epomophora* (royal albatross) on Campbell Island. On Macquarie Island, however, with the exception of the uncommon endemic *Poa hamiltonii*, which occurs only in areas affected by bird manuring (Taylor, 1955), no such conspicuous greening occurs in the vicinity of bird nests, rookeries, or seal wallows. The tussock grassland there is essentially 'green' everywhere, so that nutrients from nesting sites do not increase the greenness of the grass. Very substantial amounts of nutrients are reported to be deposited in natural precipitation on Macquarie Island, and nutrient deficiencies are thought not to be very significant there (Jenkin, 1975). It is difficult to understand why on Macquarie Island there are sufficient nutrients from precipitation for luxuriant tussock growth, whereas on the other islands under similar conditions verdure is much less.

The predominant source of atmospheric nutrients is the surrounding sea, but why the sea in some areas may supply more or less nutrients than others has yet to be documented. Answers might be sought in the relative abundance of marine life, including that which comes ashore, and in upwellings of nutrient-rich water in relation to the location of each of the sub-Antarctic islands. As a source of nutrients, the relative importance of the differing types of rock that compose each island also needs study, as does the efficiency of nutrient utilization by each plant species.

Blue-green algae may provide part of the nitrogen requirements of higher plants in some grasslands. Croome (1973) reported nitrogen fixation on Marion Island by *Calothrix* sp., *Nostoc commune* and *Tolypothrix* sp., but especially (on circumstantial evidence) by *Stigonema ocellatum*. He measured a maximum rate of 3.33 μg N mg^{-1} algal N for a 3-hr period. Fogg and Stewart (1968) showed a fixation rate for *Nostoc commune* on Signy Island of 0.48 μg N mg^{-1} algal N per day. The lower rate is possibly due to lower temperatures on Signy Island. The presence of these algae, either directly in grasslands and in up-stream water passing to grassland, suggests that they may be a nitrogen source for some grasslands. Further studies, however, are needed.

DECOMPOSITION AND SOIL FAUNA

Turnover rates and decomposition of various components of the grasslands have been studied less intensively and less extensively than have growth and standing crop. The leaves of the short grass *Festuca contracta* live for about one year or less. A proportion remains green over the winter but dies early in the following spring. The leaves of the tussock grass *Poa flabellata* grow to a height of over 1 m and die back seasonally, although the leaf bases remain green for 1 to 3 years (Lewis Smith and Walton, 1975). Leaves of *Poa foliosa* on Macquarie Island are reported to live for 10 months (Jenkin and Ashton, 1970).

The rate of decomposition varies greatly from taxon to taxon. For dicotyledons, Jenkin (1975) reported a 90% loss of senescent leaves of *Stilbocarpa polaris* in 1 year, whereas Lewis Smith and Walton (1975) reported a 90% dry-weight loss of *Acaena magellanica* leaves in 14 weeks, with very little litter accumulation.

Decomposition of grass leaves is variable, depending upon species and site. Dead leaves of *Festuca contracta* in the canopy (they normally remain attached to the plant) decreased in dry weight by 20% to 25% in one year, whereas green leaves placed in litter bags on the ground lost 56% dry weight (Lewis Smith and Stephenson, 1975). Leaves of *Poa foliosa*, however, lost about 63% during the first year when in contact with the ground as litter and 57% when attached to the plant; after 2 years, nearly 85% of the leaves on the ground had decomposed (Jenkin, 1975). The relatively deep accumulations of peat under some grasslands [e.g. 0.5–2 m on Macquarie Island (Taylor, 1955) and on South Georgia (Lewis Smith, 1981)] and the high rates of organic decomposition already noted, imply either a very long time during which accumulation was slow, or a faster rate of accumulation in the past than at present.

The numbers of soil-inhabiting microfauna and microflora appear to be quite high in the few studies that have been conducted. Lewis Smith and Walton (1975) described seasonal variations in bacterial numbers (on South Georgia) which appear to relate to microclimatological factors influenced by canopy closure during the growing season. In sheltered *Festuca* grassland in early summer, litter contained 0.08×10^6 bacteria per gram dry weight, and soil contained 0.01×10^6 per gram. By late summer the numbers had risen to 2.09×10^6 g^{-1} in litter and 0.03×10^6 g^{-1} in soil. Bacterial counts for *Poa flabellata* peat were higher, although litter counts were lower than in the *Festuca* site. *Acanthrodilus* sp. (a lumbricid earthworm) was relatively common in tussock peat on South Georgia (Lewis Smith and Walton, 1975). Sims (1971) reported on some 15 species of Oligochaeta from the soils of sub-Antarctic islands.

Jenkin (1975) summarized the available microbiological characteristics of some Macquarie Island soils. Bacterial numbers in tussock root masses there (Bunt and Rovira, 1955) appear high in comparison to those from South Georgia, but because the techniques used for sampling were different (both in the field and in the laboratory), the results may not be directly comparable. For the Macquarie tussocks, bacterial numbers ranged from 7.1 to 17.9×10^6 g^{-1}) Earthworms were common (Jenkin, 1975), as were fungi (Bunt, 1965). Nematodes were reasonably abundant, appearing to feed mostly on bacteria and algae (Bunt, 1954).

Smith (1977b) believed that, with the paucity of vertebrate herbivores, a major channelling of energy through the detritus food chain is a significant part of the sub-Antarctic island food web.

IMPACT OF NATIVE ANIMALS ON VEGETATION

Utilization of the sub-Antarctic grasslands by vertebrates was minimal prior to the introduction of such animals by man, both deliberate and accidental. The native animals take little or nothing from the grassland, but leave behind nutrients that are often in toxic concentrations locally, but in general are beneficial to plants by supplying nitrogen and phosphorus. Tussocks are used as nesting sites by several species of birds. These include: albatross and mollymawks (*Diomedea cauta, D. chrysostoma, D. epomophora, D. exulans, D. melanophrys*); giant petrels (*Macronectes giganteus, M. halli*); light-mantled sooty albatross (*Phoebetria palpebrata*); and gentoo penguin (*Pygoscelis papua*). The soil beneath tussocks was used in some places by ground-nesting prions and petrels (Huntley, 1971).

Nesting sometimes results in the death of tussocks through removal or crushing of leaves, or stimulates growth through fertilization by excreta. *Eudyptes chrysocome* (rockhopper penguin) utilizes the inter-tussock areas. The very large colonies of *Aptenodytes patagonicus* (king penguin) and *Eudyptes schlegeli* (royal penguin) spread into tussock grassland, destroying parts of it and fertilizing other parts (Fig. 15.7).

The presence of *Mirounga leonina* on beaches around the sub-Antarctic and southern cool-temperate zone also has a dual effect on tussock grasslands. Tussocks are flattened, uprooted or killed by the lounging and wallowing of the animals during breeding and moulting, but the grassland community as a whole receives large inputs of nutrients. The disturbance is mostly confined to the coastal zone, where it is evident as a mosaic of successional stages of wallows and infilling (Gillham, 1961). Quantitative studies of the grassland–marine nutrient system have yet to be undertaken.

MAN'S INTERFERENCE

The sub-Antarctic islands appear to have been free of interference and disturbance by man until the 17th century, when the massive, global, marine-based conquests by Europeans began. Although a chronicle of all early visits is impossible to construct, the islands offered a landfall first for shipwrecked sailors, then for sealers and whalers. Duration of visits was from a few days to several years. Holdgate and Wace (1961) and Detwyler (1971) have summarized the major attempts by people to occupy the islands. They stated that abortive attempts were made to colonize the Auckland Islands (Eden, 1955), Campbell Island (Eden, 1955) and Amsterdam Island (Aubert de la Rue, 1953; Reppe, 1957), whereas some livestock was kept on the Îles de Kerguelen (Aubert de la Rue, 1930; Jeannel, 1941; Reppe, 1957) and Marion Island (Marsh, 1948; Grange, 1954), Gough Island (Holdgate, 1958), and Inaccessible Island (Hagen, 1952). Since 1945 their value as sites for weather stations has led to the establishment of semi-permanent bases on Amsterdam Island (Aubert de la Rue, 1953; Reppe, 1957; Aubert, undated), Campbell Island (Eden, 1955), Gough (Holdgate, 1958), the Îles de Kerguelen (Reppe, 1957; Aubert de la Rue, 1953), Macquarie Island (Law and Burstall, 1956), and Marion Island (Marsh, 1948; Grange, 1954). Heard Island was occupied from 1947 until 1954 (Law and Burstall, 1956) and visited only intermittently since, and Auckland Island between 1941 and 1945 (Eden, 1955), whereas stations on Bouvetøya and the Îles Crozet (Reppe, 1957) were projected.

Dreux (1964) reported on the creation of a base on Ile de la Possession in the Crozet archipelago during the southern summer of 1961–62. Sheep and cattle were successfully introduced to Campbell Island (Wilson and Orwin, 1964; Taylor et al., 1970) and to South Georgia and Macquarie Island (Cumpston, 1968), although they were removed subsequently from the latter. Reindeer (*Rangifer* sp.) were introduced to the Îles de Kerguelen (Holdgate and Wace, 1961) and to South Georgia (Walton and Lewis Smith, 1973). Rabbits (*Oryctolagus cuniculus*) were deliberately and successfully established on several of the southern islands to provide a source of food for shipwrecked persons (Chastain, 1958; Cour, 1958; Holdgate and Wace, 1961; Dreux, 1964). A flightless New Zealand bird, *Gallirallus australis* (the weka) was also successfully introduced to Macquarie Island for the same purpose (Cumpston, 1968). Less deliberate introductions of rats (*Rattus* spp.), mice (*Mus musculus*) and cats (*Felis catus*) occurred

Fig. 15.7. *Poa flabellata* high sub-Antarctic tussock grassland on a coastal slope with *Aptenodytes patagonicus* (king penguins) in the foreground, Undine Harbour, South Georgia. (Photo courtesy of R.I. Lewis Smith, British Antarctic Survey.)

as a result of human occupation. The latter prey upon the indigenous bird species and are thought to be responsible for the extinction of several species, some endemic (Holdgate and Wace, 1961).

Numerous species of plants have been introduced, of which some have become aggressive invaders of the native vegetation. Walton (1975) listed 32 species as persistent and naturalized in the sub-Antarctic and adjacent New Zealand shelf islands. He noted that, whereas the native floras of the sub-Antarctic were not about to be over-run by alien plants, as has happened on sub-tropical islands such as St. Helena (Holdgate and Wace, 1961), they were not free of dangers posed by grazing animals through their effects on floristic composition.

Grazing by vertebrates introduced by man has profoundly affected the vegetation of some sub-Antarctic islands, including the grasslands. Whereas large tussock grasses are very productive, they are largely intolerant of grazing. Nearly all of the main island of the Kerguelen archipelago (Grande Terre) has been radically altered by rabbit grazing (Chastain, 1958), and the maintenance of sheep, mules, pigs, horses, cattle, and reindeer (Holdgate and Wace, 1961) has contributed to the grazing pressure.

The once extensive communities of *Poa cookii* and *Pringlea antiscorbutica* (the latter is a sub-

Antarctic generic endemic) have been replaced with a dense carpet of *Acaena magellanica* or have become barren, eroding areas (Chastain, 1958). The particular susceptibility of *Poa cookii* to grazing by sheep has been reported by Huntley (1971). Grazing by rabbits on Macquarie Island is steadily causing a degeneration of tussock grassland and herbfields (Costin and Moore, 1960; Jenkin, 1975). The initial, selective grazing removes the large succulent forbs and grasses. In the absence of effective colonization of denuded areas, erosion of peat begins. Erosion progressively spreads over wide areas under the influence of wind, water and frost. Excessive grazing by sheep has caused similar deterioration on Campbell Island (Wilson and Orwin, 1964).

The "new" element of grazing in the sub-Antarctic and nearby islands has not only resulted in erosion. Radical changes in species composition and structure of the vegetation have also occurred. *Poa* and *Pringlea* have been replaced by *Acaena* on Kerguelen (as noted earlier). On Macquarie Island, *Poa foliosa* tussocks are temporarily replaced by a secondary turf of mosses and minor forbs, a condition that sometimes results in erosion (Costin and Moore, 1960; Gillham, 1961). On South Georgia, grazing and trampling by reindeer is having a deleterious effect on *Festuca contracta* grassland (Lindsay, 1973), and in severely affected areas native vegetation is being replaced by nearly pure swards of *Poa annua* (Walton, 1975). *Poa flabellata* tussock grasslands do not provide preferred grazing except in winter, for reasons which are as yet unclear. However, severe degradation has been reported from grazing and trampling by reindeer (Knightly and Lewis Smith, 1976). The lichens suffer most severely from grazing and trampling.

Fire

The role of fire in sub-Antarctic grasslands is less than in continental grasslands in temperate and tropical regions. However, fire has been used in the management of sheep ranges on Campbell Island (Wilson and Orwin, 1964; Meurk, 1977). Localized fires have evidently occurred (started both by accident and on purpose), but fire does not generally play a significant role in the vegetation of the sub-Antarctic region.

Management

Management of the sub-Antarctic islands varies from island to island, according to the government which controls them. South Georgia (Britain) is managed much in accordance with the terms of the 1959 Antarctic Treaty, although it is not specifically covered by it (Lewis Smith and Walton, 1975). Macquarie Island is administered by the Tasmanian National Parks and Wildlife Service, whose aims should ensure that uncontrolled exploitation does not recur (Jenkin, 1975). Campbell Island is a reserve for the preservation of flora and fauna (Wilson and Orwin, 1964). Marion and Prince Edward Islands, which are under the control of the South African Government (Prince Edward Island Act No. 43 of 1948), have been declared nature reserves, so as to completely protect the plant and animal life (van Zinderen Bakker, 1971).

Management of these islands for the protection of flora and fauna is a complex matter. To date, the most important management practice has been the prohibition on the killing of seals and birds. With little else as yet of economic importance, this action has allowed the recovery of the stocks of animals, hitherto greatly depleted. Tourism, if uncontrolled, can pose a serious problem due to excessive trampling of vegetation and disturbance of breeding animals.

Much more difficult management decisions have yet to be made, however, that relate to the control or extermination of alien species. Where introduced species of animals have been naturalized for relatively long periods, new balances between them and the native biota may have been established. Programs that aim to control them individually may disclose hidden problems. For example, the removal of sheep from Campbell Island has been recommended in order to protect the native flora and the breeding ground of *Diomedea epomophora* (Taylor et al., 1970). However, conflicting evidence suggests that changes in vegetation brought about by sheep grazing may have greatly benefitted *D. epomophora* to the extent that removal of the sheep could result in a decrease in the numbers of this albatross. A major experiment to clear half the island of sheep, and leave the remainder, was undertaken in 1970–71, involving the construction of a trans-island fence.

Early results (Meurk, 1980) point to a good recovery of native plants at the expense of aliens, although recovery is patchy and appears related both to the proximity of seed from refugial sites and to the survival of under-ground plant parts.

On Macquarie Island, a trial program using the flea *Spilopsyllus cuniculi* as a vector for myxomatosis (*Myxoma leporipoxvirus*) to control rabbits is in operation (Sobey et al., 1973). The role of rabbits in the Macquarie ecosystem needs careful evaluation. They are the direct cause of massive degradation of vegetation and of soil instability (Costin and Moore, 1960), but are eaten by introduced cats (Jones, 1977). The effect on the cat population in relation to the well-being of ground-nesting birds needs consideration before a massive program for the extermination of rabbits is begun.

New and previously unimaginable uses may be considered for some sub-Antarctic islands (Sanders, 1987). The Îles de Kerguelen may be considered as the potential location for tests of nuclear weapons by France, and the South African Government is considering the location of an airfield on Marion Island to improve access. The latter is suspected also of being related to nuclear weapons research (Sanders, 1987). Airfields profoundly affect the immediate areas where they are established. By improving access they increase the pressure of human activity (trampling and waste disposal) on native ecosystems. Nuclear testing not only affects the immediate area, but contamination of the sea and air are always potential threats that can extend widely in the area affected by the westerly air streams and sea currents.

It would be valuable to continue the study of the little-disturbed ecosystems of some of the islands as well as the devastating effects of introduced animals and the resulting new balances that result. Southern islands and their grasslands may be small in area but they provide an interesting and sometimes unique facet of the biology of our planet.

RELATIONSHIP TO TROPICAL HIGH MOUNTAIN GRASSLANDS

The sub-Antarctic islands have been linked with the vegetation of high tropical mountains (Troll, 1960; Troll and Lauer, 1978). This initially surprising comparison results from the strong similarity in physiognomy — large tussock grasses and giant-leaved forbs (Hedburg, 1968; Wade and McVean, 1969; Hope, 1976). The comparison is strengthened by Shear's (1964) delineation of the Polar Marine climate in which he included that of the tropical high mountains. Van Zinderen Bakker (1971) noted that the similarities between the regions are superficial and of no use to ecologists. However, detailed comparative studies of the two widely separated regions are extremely few. Two such studies (Hnatiuk et al., 1976; Hnatiuk, 1978) examined the climate and growth of tussock grasslands in the high mountains of New Guinea and on Campbell and Macquarie Islands. Both of these areas experience greatly reduced thermal seasonality (i.e. are seasonally isothermal), but only the sub-Antarctic also has diurnally unchanging temperatures. Whereas air temperatures are cool (less than 10°C) in both regions, they differ in that the southern islands are very much windier, have seasonally large variations in day length, relatively low levels of solar radiation, and fairly constant and high relative humidity as compared to the tropical high mountains.

The total standing crop (i.e. both living and dead shoots and under-ground parts) in the most productive *Deschampsia klossii* tussock site on Mt. Wilhelm (New Guinea) was greater than that of the *Agrostis* mire on Marion Island, but less than that in the sheltered *Festuca* site on South Georgia; however, the standing crop in the former location was much greater than that of any of the tussock sites in the sub-Antarctic (Table 15.4). Net primary above-ground production on the tropical mountain was somewhat less than at the *Festuca* site. The distribution of living biomass (above- and under-ground) was dramatically different between the two regions (Hnatiuk, 1978; Smith, 1985). In equatorial high-mountain grassland, there were 21.2 m^2 of foliage per kilogram of live under-ground biomass, whereas the figure for Campbell Island was 4.6 m^2 and for Macquarie Island 4.5 m^2 (Table 15.3 and Hnatiuk, 1978). This is probably related to relatively high transpiration rates in the windier conditions of the sub-Antarctic site. The broad, flat leaves of the dominant sub-Antarctic tussock grasses are perhaps an adaptation to the low light regime and perpetually moist soils. Most of the tropical mountain tussock grasses have narrow or tightly rolled leaves, perhaps a response to the high radiation loads; but exceptions occur in both regions. There is clearly room for more research into the understanding of the differences between these ecosystems.

ACKNOWLEDGEMENTS

I am indebted to the following persons for their assistance and comments: Dr. W.L. Francis, Dr. J.W. Green, Dr. J.F. Jenkin, Mr. C.F.H. Jenkins, Dr. R.I. Lewis Smith, Dr. C. Meurk, Dr. V.R. Smith and Dr. N.M. Wace. I also wish to thank the following persons for their help in providing information about sub-Antarctic islands under the control of their respective governments: the officers of the French and South African embassies in Australia; the Manager, Marine and Earth Science Programmes, Council for Scientific and Industrial Research, Pretoria; Professor D.F. Toerien, Institute for Environmental Science, Bloemfontein; and Monsieur l'Administrateur Supérieur, Terres Australes et Antarctiques Françaises, Paris.

REFERENCES

Anonymous, 1958. *Tables of Temperature, Relative Humidity and Precipitation for the World, Part VI. Australasia and the South Pacific Ocean Including the corresponding Sectors of Antarctica.* Air Ministry, Her Majesty's Stationery Office, London, 54 pp.

Anonymous, 1973. UNESCO international classification and mapping of vegetation. *Ecol. Conserv.*, 6: 15–37.

Ashton, D.H., 1965. Regeneration pattern of *Poa foliosa* Hook. f. on Macquarie Island. *R. Soc. Victoria Proc.*, 79: 215–234.

Aubert, M., undated. L'éstablissement de la Nouvelle Amsterdam. *Terres Aust. Antarct. Fr.*, 6: 5–20.

Aubert de la Rue, E., 1930. *Terres françaises inconnues: l'archipel de Kerguelen et les possessions françaises Australes.* Société Parisienne d'édition, Paris, 189 pp.

Aubert de la Rue, E., 1953. *Les Terres Australes.* Presses Universitaires de France, Paris, 126 pp.

Barrow, C.J., 1978. Postglacial pollen diagrams from South Georgia (sub-Antarctic) and West Falkland Island (South Atlantic). *J. Biogeogr.*, 5: 251–274.

Bliss, L.C., 1979. Vascular plant vegetation of the Southern Circumpolar Region in relation to Antarctic alpine and arctic vegetation. *Can. J. Bot.*, 57: 2167–2178.

Bunt, J.S., 1954. The soil-inhabiting nematodes of Macquarie Island. *Aust. J. Zool.*, 2: 264–274.

Bunt, J.S., 1965. Observations on the fungi of Macquarie Island. *Aust. Natl. Antarct. Res. Exped. Rep.*, Ser. B (II), Publ. 78: 1–22.

Bunt, J.S. and Rovira, A.D., 1955. Microbiological studies of some sub-Antarctic soils. *J. Soil Sci.*, 6(1): 119–128.

Callaghan, T.V., 1978. Studies on the factors affecting the primary production of bi-polar *Phleum alpinum*. In: L.C. Bliss and F.E. Wielgolaski (Editors), *Primary Production and Production Processes: Tundra Biome.* I.B.P. Tundra Biome Steering Committee, Stockholm, pp. 153–167.

Callaghan, T.V. and Lewis, M.C., 1971a. The growth of *Phleum alpinum* L. in contrasting habitats at a sub-Antarctic station. *New Phytol.*, 70: 1143–1154.

Callaghan, T.V. and Lewis, M.C., 1971b. Adaptation in the reproductive performance of *Phleum alpinum* L. at a sub-Antarctic station. *Br. Antarct. Surv. Bull.*, 26: 59–75.

Chastain, A., 1958. La flore et la végétation des Îles de Kerguelen. *Mem. Mus. Natl. Hist. Nat. N. S., Ser. B, Bot.*, XI(1): 1–136.

Clarke, G.C.S. and Greene, S.W., 1970. Reproductive performance of two species of *Pohlia* at widely separated stations. *Trans. Br. Bryol. Soc.*, 6: 114–128.

Cockayne, L., 1909. The ecological botany of the sub-Antarctic Islands of New Zealand. In: C. Chilton (Editor), *The Sub-Antarctic Islands of New Zealand.* Philosophical Institute of Canterbury, Government Printer, Wellington, pp. 182–235.

Colhoun, E.A. and Goede, A., 1973. Fossil penguin bones, ^{14}C dates and raised marine terraces of Macquarie Island. Some comments. *Search*, 4: 499–501.

Costin, A.B. and Moore, D.M., 1960. The effects of rabbit grazing on the grasslands of Macquarie Island. *J. Ecol.*, 48: 729–732.

Cour, P., 1958. A propos de la flore de l'Archipel de Kerguelen. *Terres Aust. Antarct. Fr.*, 4–5: 10–32.

Croome, R.L., 1973. Nitrogen fixation in the algal mats on Marion Island. *S. Afr. J. Antarct. Res.*, 3: 64–67.

Cumpston, J.S., 1968. *Macquarie Island.* Antarctic Division, Department of External Affairs, Australia, 380 pp.

Deacon, G.E.R., 1964. The Southern Ocean. In: R.J. Priestley, R.J. Adie and G. de Q. Rubin (Editors), *Antarctic Research.* Butterworths, London, pp. 292–307.

Delisle, J.F., 1965. The climate of the Auckland Islands, Campbell Island and Macquarie Island. *N.Z. Ecol. Soc. Proc.*, 12: 37–44.

Detwyler, T.R., 1971. Chapter 14. *Man's Impact on Environment.* McGraw-Hill, New York, pp. 476–492.

Dreux, Ph., 1964. Observations sur la flore et la végétation de l'Ile aux Cochons (Archipel Crozet). *Soc. Bot. Fr. Bull.*, 8: 382–387.

Eden, A.W., 1955. *Islands of Despair.* Andrew Melrose, Stratford Place, London, 212 pp.

Fabricius, A.F., 1957. Climate of the sub-Antarctic Islands. In: M.P. van Rooy (Editor), *Meteorology of the Antarctic.* Government Printer, Pretoria, pp. 111–135.

Fogg, G.E. and Stewart, W.D.P., 1968. *In situ* determinations of biological nitrogen fixation in Antarctica. *Br. Antarct. Surv. Bull.*, 15: 39–46.

Gillham, M.E., 1961. Modification of sub-Antarctic flora on Macquarie Island by sea birds and sea elephants. *Proc. R. Soc. Victoria*, 74: 1–12, 2 pl.

Grange, J.J. la, 1954. The South African Station on Marion Island, 1948–55. *Polar Rec.*, 7(48): 155–158.

Greene, S.W. and Greene, D.M., 1963. Check list of the sub-Antarctic and Antarctic vascular flora. *Polar Rec.*, 11(73): 411–418.

Greene, S.W. and Walton, D.W.H., 1975. An annotated check list of the sub-Antarctic and Antarctic vascular flora. *Polar Rec.*, 17(110): 473–484.

Hagen, Y., 1952. Birds of Tristan da Cunha. *Res. Norw. Sci. Exped. Tristan da Cunha 1937–1938*, 20, 248 pp.

Hedburg, O., 1968. Taxonomic and ecological studies on the Afroalpine flora of Mt. Kenya. *Hochgebirgsforschung (High Mountain Research)*, H. 1: 171–194.

Hnatiuk, R.J., 1975. *Aspects of the Growth and Climate Tussock Grasslands in Montane New Guinea and sub-Antarctic Islands*. Ph.D. thesis, Australian National University, Canberra, 161 pp.

Hnatiuk, R.J., 1978. The growth of tussock grasses on an equatorial high mountain and on two sub-Antarctic islands. In: C. Troll and W. Lauer (Editors), Geoecological Relations Between the Southern Temperate Zone and the Tropical Mountains. *Erdwiss. Forsch.*, 11: 159–190.

Hnatiuk, R.J., Smith, J.M.B. and McVean, D.N., 1976. Mt. Wilhelm Studies, 2. The Climate of Mt. Wilhelm. *Research School of Pacific Studies, Department of Biogeography and Geomorphology Publications* BG/4, The Australian National University, Canberra, 76 pp.

Holdgate, M.W., 1958. *Mountains in the Sea*. Macmillan, London, 222 pp.

Holdgate, M.W., 1970. Introduction to part XII. Vegetation. In: M.W. Holdgate (Editor), *Antarctic Ecology*. Vol. 2, Academic Press, London, pp. 729–732.

Holdgate, M.W. and Wace, N.M., 1961. The influence of man on the floras and faunas of southern islands. *Polar Rec.*, 10: 475–493.

Hooker, J.D., 1847. *The Botany of the Antarctic Voyage of H.M. Discovery Ships Erebus and Terror in the Years 1839–1843, Vol. 1. Flora Antarctica*. Reeve Brothers, London, 574 pp.

Hope, G.S., 1976. Vegetation. In: G.S. Hope, J.A. Petersen, U. Radok and I. Allison (Editors), *The Equatorial Glaciers of New Guinea*. Balkema, Rotterdam, pp. 113–172.

Huntley, B.J., 1971. Vegetation. In: E.M. van Zinderen Bakker Sr., J.M. Winterbottom and R.A. Dyer (Editors), *Marion and Prince Edward Islands*. Report of the South African Biological and Geological Expedition 1965–1966. Balkema, Cape Town, pp. 98–160.

Huntley, B.J., 1972. Aerial standing crop of Marion Island plant communities. *J. S. Afr. Bot.*, 38(2): 115–119.

Jeannel, R.G., 1941. *Au seuil de l'Antarctique: croisière du 'Bougainville' aux îles des manchots et des éléphants de la mer*. Editions du Museum, Paris, 236 pp.

Jenkin, J.F., 1972. *Studies on Plant Growth in a Sub-Antarctic Environment*. Ph.D. Thesis, Department of Botany, University of Melbourne, 297 pp.

Jenkin, J.F., 1975. Macquarie Island, sub-Antarctic. In: T. Rosswall and O.W. Heal (Editors), Structure and Function of Tundra Ecosystems. *Ecol. Bull.* (Stockholm), 20: 375–397.

Jenkin, J.F. and Ashton, D.H., 1970. Productivity studies on Macquarie Island Vegetation. In: M.W. Holdgate, W. Mackay and J. Mackay (Editors), *Antarctic Ecology*. Academic Press, London, pp. 851–863.

Jenkin, J.F. and Ashton, D.H., 1979. Pattern in *Pleurophyllum* herbfields on Macquarie Island (sub-Antarctic). *Aust. J. Ecol.*, 4: 67–74.

Jones, E., 1977. Ecology of the feral cat, *Felis catus* (L.), (Carnivora: Felidae) on Macquarie Island. *Aust. Wildl. Res.*, 4: 249–262.

Knightly, S.P.J. and Lewis Smith, R.I., 1976. The influence of reindeer on the vegetation of South Georgia, 1. Long term effects of unrestricted grazing and the establishment of exclosure experiments in various plant communities. *Br. Antarct. Surv. Bull.*, 44: 57–76.

Law, P.G. and Burstall, T., 1956. Heard Island. *Aust. Natl. Antarct. Res. Exped. Interim Rep.*, 7, 53 pp.

Lewis Smith, R.I., 1981. Types of peat and peat-forming vegetation on South Georgia. *Br. Antarct. Surv. Bull.*, 53: 119–140.

Lewis Smith, R.I., 1986. Terrestrial plant biology of the sub-Antarctic and Antarctic. In: R.M. Laws (Editor), *Antarctic Ecology*. Vol. 1, Academic Press, London, pp. 61–162.

Lewis Smith, R.I. and Stephenson, C., 1975. Preliminary growth studies on *Festuca contracta* T. Kirk., and *Deschampsia antarctica* Desv. on South Georgia. *Br. Antarct. Surv. Bull.*, Nos. 41 and 42: 59–75.

Lewis Smith, R.I. and Walton, D.W.H., 1975. South Georgia, Sub- Antarctic. In: T. Rosswall and O.W. Heal (Editors), Structure and Function of Tundra Ecosystems. *Ecol. Bull.* (Stockholm), 20: 399–423.

Lindsay, D.C., 1973. Effects of reindeer on plant communities in the Royal Bay area of South Georgia. *Br. Antarct. Surv. Bull.*, 35: 101–109.

Marsh, J., 1948. *No Pathway Here*. Howard R. Timmins, Cape Town (for Hodder and Stoughton, London), 200 pp.

Meurk, C.D., 1977. Alien plants in Campbell Island's changing vegetation. *Mauri Ora*, 5: 93–118.

Meurk, C.D., 1980. Plant ecology of Campbell Island. In: L. Kenworthy (Editor), *Preliminary Reports of the Campbell Island Expedition 1975–1976*. Department of Lands and Surveys, Wellington, pp. 90–96.

Moore, D.M., 1968. The vascular flora of the Falkland Islands. *Br. Antarct. Surv. Sci. Rep.*, 60, 202 pp.

Moseley, H.N., 1875. On the botany of Marion Island, Kerguelen's land, and Yong Island of the Heard Group. *Linn. Soc. Proc. Bot.*, 14: 387–389.

Ostapoff, F., 1965. Antarctic oceanography. In: P. van Oye and J. van Mieghem (Editors), *Biogeography and Ecology in Antarctica*. Junk, The Hague, pp. 97–126.

Quilty, P.G., Rubenach, M. and Wilcoxon, J.A., 1973. Miocene ooze from Macquarie Island. *Search*, 4: 163–164.

Reppe, X.N., 1957. *L'aurore sur l'Antarctique*. Nouvelles Editions Latines, Paris, 221 pp.

Sanders, N.K., 1987. *Senate Weekly Hansard*. Australian Government Publishing Service, Canberra, No. 1: 141–142.

Schalke, H.J.W.G. and van Zinderen Bakker, E.M., Sr., 1971. History of the vegetation. In: E.M. van Zinderen Bakker, Sr., J.M. Winterbottom and R.A. Dyer (Editors), *Marion and Prince Edward Islands*. Report of the South African Biological and Geological Expedition 1965–1966. Balkema, Cape Town, pp. 89–97.

Schulze, B.R., 1971. The climate of Marion Island. In: E.M. van Zinderen Bakker, Sr., J.M. Winterbottom and R.A. Dyer (Editors), *Marion and Prince Edward Islands*. Report of the South African Biological and Geological Expedition 1965–1966. Balkema, Cape Town, pp. 16–31.

Seppelt, R.D., 1977. Studies on the bryoflora of Macquarie Island. Introduction and check list of species. *Bryologist*, 80: 167–170.

Shear, J.A., 1964. The polar marine climate. *Ann. Assoc. Am. Geogr.*, 54(99): 310–317.

Sims, R.W., 1971. Oligochaeta. In: E.M. van Zinderen Bakker, Sr., J.M. Winterbottom and R.A. Dyer (Editors), *Marion and Prince Edward Islands*. Report of the South African Biological and Geological Expedition, 1965–1966. Balkema, Cape Town, pp. 391–393.

Smith, J.M.B., 1985. Aboveground, belowground phytomass ratios in Venezuelan Paramo vegetation and their significance. *Arct. Alp. Res.*, 17(2): 189–198.

Smith, V.R., 1976a. Standing crop and nutrient status on Marion Island (sub-Antarctic) vegetation. *J. S. Afr. Bot.*, 42(2): 231–263.

Smith, V.R., 1976b. The effect of burrowing species of Procellariidae on the nutrient status of inland tussock grasslands on Marion Island. *J. S. Afr. Bot.*, 42(2): 265–272.

Smith, V.R., 1977a. Vegetation standing crop of the grey lava flows and of the eastern coastal plain on Marion Island. *J. S. Afr. Bot.*, 43(2): 105–114.

Smith, V.R., 1977b. A qualitative description of energy flow and nutrient cycling in the Marion Island terrestrial ecosystem. *Polar Rec.*, 18(115): 361–370.

Smith, V.R., 1978a. Animal–plant–soil nutrient relationship on Marion Island (Sub-Antarctic). *Oecologia* (Berlin), 32: 239–253.

Smith, V.R., 1978b. Standing crop and production estimates of selected Marion Island plant communities. *S. Afr. J. Antarct. Res.*, 8: 103–108.

Smith, V.R., 1979. The influence of seabird manuring on the phosphorus status of Marion Island (Sub-Antarctic) soils. *Oecologia* (Berlin), 41: 123–126.

Sobey, W.R., Adams, K.M., Johnston, G.C., Gould, L.R., Simpson, K.N.G. and Keith, K., 1973. Macquarie Island: the introduction of the European rabbit flea *Spilopsyllus cuniculi* (Dale) as a possible vector for myxomatosis. *J. Hyg.*, 71: 299–309.

Tallowin, J.R.B., 1977a. Vegetative proliferation in *Festuca contracta* T. Kirk on South Georgia. *Br. Antarct. Surv. Bull.*, 45: 13–17.

Tallowin, J.R.B., 1977b. Studies in the reproductive biology of *Festuca contracta* T. Kirk on South Georgia, II. The reproductive performance. *Br. Antarct. Surv. Bull.*, 45: 117–129.

Tallowin, J.R.B. and Lewis Smith, R.I., 1977. Studies in the reproductive biology of *Festuca contracta* T. Kirk on South Georgia, I. The reproductive cycles. *Br. Antarct. Surv. Bull.*, 45: 63–76.

Taylor, B.W., 1955. The flora, vegetation and soils of Macquarie Island. *Aust. Natl. Antarct. Res. Exped. Rep.*, Ser. B, Vol. 11, 192 pp.

Taylor, R.H., Bell, B.D. and Wilson, P.R., 1970. Royal albatross, feral sheep and cattle on Campbell Island. *N.Z. J. Sci.*, 13: 78–88.

Troll, C., 1958. Climate seasons and climatic classification. *Orient. Geogr.*, 2: 141–165.

Troll, C., 1960. The relationship between the climates, ecology, and plant geography of the southern cold temperate zone and of the tropical mountains. *Proc. R. Soc., Ser. B*, 152: 529–532.

Troll, C. and Lauer, W. (Editors), 1978. Geoecological relations between the southern temperate zone and the tropical mountains. *Erdwiss. Forsch.*, 11, 563 pp.

Van Zinderen Bakker, E.M., Sr., 1969. Quaternary pollen analytical studies in the southern hemisphere with special reference to the sub-Antarctic. *Palaeoecol. Afr.*, 5: 175–212.

Van Zinderen Bakker, E.M., Sr., 1970. Quaternary climates and Antarctic biogeography. *Antarct. Ecol.*, 1: 31–40.

Van Zinderen Bakker, E.M., Sr., 1971. Introduction. In: E.M. van Zinderen Bakker, Sr., J.M. Winterbottom and R.A. Dyer (Editors), *Marion and Prince Edward Islands*. Report of the South African Biological and Geological Expedition 1965–1966. Balkema, Cape Town, pp. 1–15.

Van Zinderen Bakker, E.M., Sr., 1973. The glaciation(s) of Marion Island (sub-Antarctic). In: E.M. van Zinderen Bakker, Sr. (Editor), *Palaeoecology of Africa and of the Surrounding Islands and Antarctica*. Vol. 8, Balkema, Cape Town, pp. 161–178.

Varney, B.M., 1926. Climate and weather at Kerguelen Island. *Monthly Weather Review*, LIV (10), U.S. Dept. of Agriculture, Weather Bureau.

Wace, N.M., 1960. The botany of the southern oceanic islands. *Proc. R. Soc. London, Ser. B*, 152: 475–490.

Wace, N.M., 1965. Vascular plants. In: P. van Oye and J. van Mieghem (Editors), *Biogeography and Ecology in Antarctica*. Junk, The Hague, pp. 201–266.

Wade, L.K. and McVean, D.N., 1969. Mt. Wilhelm Studies, I. The Alpine and sub-Alpine Vegetation. *Research School Pacific Studies, Department of Biogeography and Geomorphology Publication*, BG/1, Australian National University, Canberra, 225 pp.

Walton, D.W.H., 1975. European weeds and other alien species in the sub-Antarctic. *Weed Res.*, 15: 271–282.

Walton, D.W.H., 1982. Floral phenology in the South Georgian vascular flora. *Br. Antarct. Surv. Bull.*, 55: 11–25.

Walton, D.W.H. and Lewis Smith, R.I., 1973. Status of the alien vascular flora of South Georgia. *Br. Antarct. Surv. Bull.*, 36: 79–97.

Walton, D.W.H. and Lewis Smith, R.I., 1980. The chemical composition of South Georgian vegetation. *Br. Antarct. Surv. Bull.*, 49: 117–135.

Walton, D.W.H., Greene, D.M. and Callaghan, T.V., 1975. An assessment of primary production in a sub-Antarctic grassland on South Georgia. *Br. Antarct. Surv. Bull.*, 41 and 42: 151–160.

Wielgolaski, F.E., 1972. Vegetation types and plant biomass in tundra. *Arct. Alp. Res.*, 4(4): 291–305.

Wilson, P.R. and Orwin, D.F.G., 1964. The sheep population of Campbell Island. *N.Z. J. Sci.*, 7: 460–490

Young, S.B., 1971. Vascular flora of the Kerguelen Islands. *Antarct. J. U.S.*, 6: 110–111.

Chapter 16

GRASSLANDS OF THE SOUTH-WEST PACIFIC

A.N. GILLISON

INTRODUCTION

Scope

This chapter deals with grassland areas within the major islands of Melanesia, and to a much lesser extent, Micronesia and Polynesia (Fig. 16.1). Within Melanesia, most attention is focused on the eastern half of the island of New Guinea, Bougainville, New Britain and New Ireland, the Solomon Islands, New Caledonia and Vanuatu (formerly New Hebrides) and Fiji. Some minor Micronesian grasslands are described on the islands of the eastern and western Carolines; and the Polynesian islands of Western Samoa and Hawaii are briefly considered.

When compared with Australia and the North and South Islands of New Zealand, the study area contains few pure grasslands of any major extent, except in New Guinea. Furthermore, the often steep and irregular gradients of the physical environment and the fragmentary nature of Pacific island grasslands make difficult any simple comparison with pure grassland regions elsewhere in Oceania. Some are considered in a previous treatment of tropical savannas of Australia and the south-west Pacific (Gillison, 1983) which covered woody as well as graminoid vegetation.

Unlike their counterparts in New Zealand and Australia, grasslands in the south-west Pacific have not been studied in detail. For this reason the present account is more descriptive than quantitative. The purpose of this chapter is therefore to review the floristic, structural and functional aspects of some typical Pacific island grasslands and their uses. It also aims to highlight some of the dramatic processes currently under way in converting relatively productive land, under intact or secondary vegetation, to unproductive areas dominated by exotic graminoids. To address management-related issues requires a change in emphasis from the use of broad descriptive classifications to the application of response-based, biophysical variables which can be used directly in resource inventory and computerized geographic information systems.

A bioclimatic framework for Pacific grasslands

As with savannas in general, there are almost as many methods of classifying grassland as there are grassland ecologists. A seminal treatment by Johnson and Tothill (1985) used a range of climate–soil factors and broad floristic categories to define savannas better. But although their definition is useful as a broad geographic utility, it remains severely limited as a management tool for interpreting plant response to environmental change.

In recognizing limitations associated with "exclusive definition of savanna lands", Johnson and Tothill (1985) reviewed other combinations of plant–environment parameters. They included functional parameters based on C_3 and C_4 photosynthetic pathways, coupled with a bioclimate classification derived by Nix (1983) for tropical savannas. Although the parameters of the latter were the same as those used earlier by the author (see fig. 9.4 and table 9.2 in Gillison, 1983), in a provisional framework for the savannas of Australia and the south-west Pacific, that framework was regarded as inappropriate for grassland studies by Johnson and Tothill, since it included trees.

The question whether or not trees should

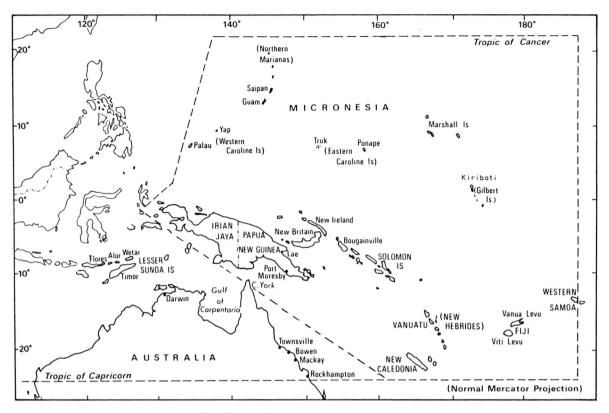

Fig. 16.1. The area of the south-west Pacific defined for this chapter.

be recognized as an integral part of a woody–graminoid continuum for the purpose of exclusively defining savannas, highlights a general but largely irrelevant problem. The focus has shifted from a preoccupation with a need for a better descriptive classification of grasslands — or savannas — to one now concerned with identifying a minimum, response-based set of biophysical variables. Such variables should be appropriate for rapid inventory where they can be used in a rapid and uniform way to gather information at a level which is relevant to management as well as research. The information recorded must be applicable to dynamic studies and provide clearer insights into the behaviour of organisms. Unlike the broad environmental parameters commonly used in attempting to define the elusive concept of savanna, the data must be relevant to generating useful predictive models at a management scale for specific taxa whether they fall inside or outside a savanna boundary as viewed through human eyes.

The bioclimatic framework outlined by Nix (1983) for tropical savannas provides a platform where fine-scale climate data can be aggregated with a gain rather than a loss of information. When used with other biophysical attributes related to plant functional response (cf. Gillison, 1988) using a spatially referenced database, the data have the potential to provide useful information at a range of specified spatio-temporal scales. The arrival of computerized geographic information systems has obviated the need for simplistic global descriptive approaches in dynamic studies. With such a system, "raw" site data need no longer be subsumed in a simplified descriptive matrix. Yet, provided appropriate descriptive attributes are included in the initial data set, the data may be used where necessary to generate descriptive classifications for geographic purposes. For these reasons the application of geographic information systems in plant ecology is gaining momentum.

As elsewhere in this series (Gillison, 1983), the classification in this chapter uses as its primary framework, broad thermal optima for plant growth, modified by rainfall seasonality. These have been found useful in interpreting vegetation response to environmental change (cf. Nix and Gillison, 1985). They are based on the classification by Nix (1974, 1982, 1983) and for present purposes employ three thermal optima as follows: "megatherm" (28 to 35°C), "mesotherm" (19°C) and "microtherm" (12°C). To these is added a "seasonality" factor suggested by H.A. Nix (pers. commun., 1982) if the coefficient of variation of soil moisture availability over the year is likely to exceed 60%.

Although seasonality patterns are a major contributing factor to the distribution of Pacific island vegetation, this potentially useful information tends to be suppressed in the climate classifications of Köppen (1918) and Thornthwaite (1948) (see also argument in Nix, 1983). The life zone system as employed by Holdridge et al. (1971) also fails to predict certain climatic forest limits in New Guinea (Smith, 1975). Existing biogeographic classifications (e.g. Dasmann, 1973; Udvardy, 1975) have little functional basis for describing vegetation in terms of the physical environment and are disregarded here for that reason.

Although a gradient-based approach is relevant to any vegetation study, the wide ranging distribution of grasslands along highly variable gradients of the physical environment works against a simple classification based on structure and/or floristics. At the regional level, most variability in grassland floristics, structure and function can be accounted for by bioclimate — and this is the reason for its application here. It is asserted that within this primary set of environmental determinants, other factors such as terrain and edaphic variability exert a secondary influence which, in turn, may be modified to varying degrees by human activity.

The adoption of a bioclimatic/plant-functional framework makes possible a more meaningful comparison between similar high-order environmental analogues (such as high latitude/lowland and low latitude/upland) than terms such as "tropical" and "temperate" (cf. Oliver, 1979). The argument that environmentally based classifications of vegetation are open to circular reasoning is less applicable where plant-functional response to shifts in bioclimate can be demonstrated, and where spatial extrapolation of vegetation based on bioclimate can be adequately tested. The development of mechanistic models which demonstrate boundary conditions for certain plant functional and morphological attributes (cf. Givnish, 1986) clearly illustrates the direction for improved definitions of vegetation domains.

Although bioclimatic models have been found useful in Australia, in island areas there can be interpretative problems where coasts form a significant proportion of the landscape. For example, despite the influence of extreme seasonality in some sub-coastal areas (such as the Hisiu coast near Port Moresby in Papua New Guinea), climates near the shoreline are strongly buffered by a marine environment. Another confounding factor is the so-called "*Massenerhebung*" effect (Richards, 1966; Whitmore, 1975) where, on large islands such as New Guinea, vegetation which occurs at high elevations inland, may occur at increasingly lower elevations towards the coast. As Grubb (1971) has pointed out, this phenomenon may also be further confounded by the distribution of soil types.

The data used to generate the provisional bioclimatic boundaries by Gillison (see fig. 9.4 in Gillison, 1983) were adequate for the Australian continent but were, and still are, limited for New Guinea. For the remainder of the south-west Pacific most climate data are restricted to populated coastal zones. For many unpopulated areas, especially at high altitudes, data are unavailable. Where these can be generalized (cf. Brookfield and Hart, 1966; Fitzpatrick et al., 1966; McAlpine et al., 1982) they have been used to approximate thermal regimes and rainfall seasonality patterns.

In anticipation of this volume, the earlier descriptive treatment by Gillison (1983) in this series specifically excluded "grasslands" *sensu stricto* from savannas which were defined as containing more than 2% woody plant cover. For descriptive purposes, grasslands are therefore defined here as *graminoid-dominated formations with less than 2% woody plant cover*. They include dryland as well as permanently or periodically inundated habitats. The descriptive component of this chapter is mainly a review of existing literature set in a broad bioclimatic context.

Floristic, structural and functional attributes

Floristic and structural criteria have long been the mainstay of savanna classifications (Bews, 1929; Beard, 1953; Cole, 1963; Eiten, 1982). The south-west Pacific is no exception (Brass, 1938, 1956, 1964; Van Royen, 1963a, b; Whitmore, 1969; Heyligers, 1965, 1967, 1972a, b; Robbins, 1960, 1964, 1970; Robbins and Pullen, 1965; Schmid, 1975; Smith, 1975, 1977; Paijmans, 1976; Hope, 1976a, b, 1980; Bellamy, 1986). Most, if not all structural classifications use more or less arbitrary parameters; in some cases height categories have been defined, in others not. Despite the term "grassland" most graminoid-dominated assemblages, even in the high mountains of New Guinea, have forbs and woody plants present even as micro-assemblages (Gillison, 1969; Kalkman and Vink, 1970). Even in tall lowland swamp associations (*Phragmites karka* and *Saccharum spontaneum*) forbs and ligneous species co-occur to varying degrees.

The most widespread grassland associations, apparently with single floristic dominants, such as *Imperata cylindrica* and *Themeda triandra* (formerly *T. australis*) — the "kunai" grasslands of New Guinea — are often very variable locally. In the south-west Pacific one cannot assume that grass genera always dominate "grassland" formations. Many mid- and high-altitude grasslands in New Guinea are dominated by sedges and sedge-like plants (e.g. *Gahnia javanica* and *Machaerina rubiginosa*) which may form dense tussocks (Hope, 1980). Lowland swamp assemblages are very often dominated by tall sedges such as *Gahnia* spp. and *Thoracostachyum sumatranum*. Species

Fig. 16.2. Lowland mosaic of fluctuating swamp conditions surrounding a meander pattern on a typical megatherm seasonal, coastal plain in Papua New Guinea. Riverine edges with *Saccharum robustum* grading inland to tall mixed andropogonoid (C_4) grassland dominated by *Saccharum spontaneum* and *Imperata cylindrica*. [Photo: P. Heyligers, Commonwealth Scientific and Industrial Resaerch Organization (CSIRO).]

Fig. 16.3. Open lowland, megatherm seasonal (C_4) grassland with *Pandanus* sp. in the Morehead river area of megatherm seasonal southern Papua New Guinea. (Photo: K. Paijmans, CSIRO.)

of the families Plagiogyraceae, Sparganiaceae and Typhaceae occur in moist habitats. Both in New Guinea and in the surrounding islands, bamboo graminoids occupy a structural category of their own. If the "graminoid" leaf were to be extended among other Monocotyledoneae, such as pandanoid and palmoid life forms, "grassland" areas would become dramatically extended. Both forms are commonly associated with formations otherwise accepted as "open grassland" (Figs. 16.2 and 16.3).

Much ecological significance has been attached to the geographic and environmental distribution of tribes within the Poaceae both in Australasia and elsewhere (Hartley, 1950, 1958a, b; Clifford and Simon, 1981; Simon and Jacobs, 1990). In this respect the work of Johnson and Tothill (1985) is relevant. These authors selected a series of useful environmental gradients based on soil texture, soil water relations and annual rainfall as a framework for better interpretation of the environmental distribution of grass genera and tribes in eastern tropical and subtropical Australia (Fig. 16.4). They showed clear trends which are paralleled in much of the south-west Pacific with similar physical environments.

For dynamic studies, functional aspects of the broad floristic groups are important. At present these are best reflected in the nature of the photosynthetic C_3 and C_4 pathways. For Australia this has been carefully examined by Hattersley and Watson (1976), Hattersley et al. (1982), and Hattersley (1983, 1984). A recent comprehensive treatment of the geographical distribution of native Australian C_4 grasses has been presented by Prendergast (1989). As Johnson and Tothill (1985) pointed out, there are differences in the reaction sequence for C_4 acid decarboxylation which parallel the floristic pattern in Fig. 16.4. Within a soil moisture gradient, tropical lowland species

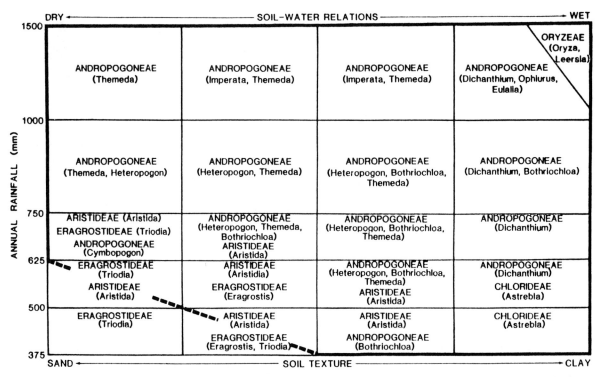

Fig. 16.4. Distribution of grass genera and tribes within savanna systems of eastern tropical and subtropical Australia (from Johnson and Tothill, 1985).

with *Kranz* parenchyma sheaths form aspartate on drier soils, whereas under increasing soil moisture there is an increase in species with *Kranz* metastome sheaths which form malate. Beyond a certain moisture threshold — for instance in permanent to semi-permanent swamps — the proportion of C_4 plants (oryzoid and panicoid grasses) to those with C_3 metabolism increases dramatically.

According to Vogel et al. (1986), for C_4 grasses in the Sinai, Negev and Judean deserts, malic enzyme species are more successful under conditions where moisture stress is not a dominating factor, and aspartate-forming plants are more frequent under xeric conditions. Prendergast (1989) found clear bioclimatic relationships for the distribution of certain C_4 types in northern Australia, but pointed out that the C_4 type is not necessarily always an adaptation to a particular bioclimatic regime; other factors such as microclimate and edaphic variability are important ecological considerations. Despite these caveats, there seems little doubt that the entire Andropogoneae which dominate tropical lowland grasslands are C_4 types.

Using the bioclimate classification of Nix (1982), Prendergast (1989) found temperature to be an important determinant in the distribution of certain C_4 types. This elemental factor was essentially ignored by Johnson and Tothill in their environmental model (Fig. 16.4), which is surprising given that with both elevational and latitudinal shifts there are important confounding effects brought about by changing thermal regimes. Such shifts appear to outweigh soil moisture effects beyond certain thermal thresholds within Australia and the south-west Pacific. The influence of elevational gradients on thermal regime and the associated distribution of C_3/C_4 grasses has been examined in Hawaii by Rundel (1980), who noted that nearly two-thirds of both the native and exotic species in Hawaii were C_4 types (Table 16.1). He found that C_4 taxa dominated species composition and cover of grasses up to 1000 m, whereas above 1400 m C_3 taxa predominated. At 1400 m (the point of floristic balance between C_3 and C_4) the monthly mean minimum temperature was about 9°C with a mean maximum temperature for the

TABLE 16.1

The distribution of the C_4 and C_3 syndrome among native Hawaiian grasses [1]

Genus	Number of species	C_4 species	C_3 species
Agrostis	2	–	2
Calamagrostis	2	–	2
Cenchrus	3	3	–
Deschampsia	2	–	2
Dissochondrus	1	1	–
Eragrostis	12	12	–
Festuca	1	–	1
Garnotia	1	1	–
Heteropogon	1	1	–
Isachne	2	–	2
Ischaemum	1	1	–
Lepturus	1	1	–
Panicum	30	19	11
Poa	3	–	3
Sporobolus	1	1	–
Trisetum	2	–	2
	65	40	25

[1] From Rundel (1980).

warmest month of about 21°C. This approximates the mid- to upper montane thermal regime for New Guinea, where there is a rapidly increasing shift in the proportion of C_3 to C_4 species. The general elevational trend for C_3/C_4 relationships is reflected across a range of habitats in the Hawaii Volcanoes National Park (Table 16.2).

For Hawaii, Rundel (1980) also found that, despite a shaded habitat, C_4 grasses were the most important in wet forests at intermediate elevations. This trend does not agree with the findings of Ehleringer (1978) in North America, which suggest that the quantum yield (initial slope of the photosynthetic light-response curve) for C_4 types places them at a disadvantage with C_3 types.

The "natural" origins of Pacific grasslands

In a work which is aimed primarily at natural grasslands, it is difficult in some cases to determine which species are "natural" and which are introduced. Although there is no doubt concerning the origins of some grass species (for instance, *Melinis minutiflora* and especially recent pasture introductions), the situation is less clear regarding others — such as *Imperata cylindrica* and *Pennisetum polystachyon*. Hancock and Henderson (1988) claim it is unlikely that both of the latter were part of the original climax vegetation in the Solomon Islands, where they "... are almost certainly anthropomorphic [anthropogenic] in origin". (present author's parentheses) From its rapid spread in both the Solomon Islands and Micronesia, *Pennisetum polystachyon* shows all the trends of an invasive exotic.

The case with *Imperata cylindrica* is less clear. Although *I. cylindrica* is probably indigenous to New Guinea, it is regarded in Australia generally as an introduced weed. This seems unfounded, given that its relatively stable ecological status in northern Australia parallels that in New Guinea. There, as in Australia, it occupies similar successional sequences and habitats, and in some cooler inland hill areas is being rapidly displaced by the African *Melinis minutiflora* (the Eungella and

TABLE 16.2

Elevational distribution and structural characteristics of major plant communities in Hawaii Volcanoes National Park [1]

	No. of stands	Elevation (m)	Mean coverage (%)				
			Trees	Shrubs	Forbs	C_4 grasses	C_3 grasses
Coastal scrub	13	15– 520	4	16	42	37	+
Grassland scrub	12	625–1340	16	34	44	37	+
Rainforest	17	690–1660	43	52	57	14	2
Mixed forest	13	1190–1875	36	21	82	19	33
Subalpine scrub	10	2030–3050	19	68	9	–	5

[1] From Rundel (1980).

Atherton tablelands in Australia and the Bulolo–Wau uplands in Papua New Guinea). If *Imperata cylindrica* is exotic, it would appear to have integrated rapidly with other "Indo-malesian" elements of the Austro–New Guinean flora. For these reasons I use the term "exotic" with some reservation.

From an ecological viewpoint, it seems illogical to exclude known exotics (such as *Panicum maximum*) from a discussion of natural grassland areas where it may be more or less stable or even actively invading. There are critical aspects of grassland dynamics which are central to management. Nonetheless, I have attempted to discriminate between the essential ecological nature of indigenous grasslands and that of recognized exotics.

The origins of the Pacific grassland flora have attracted a deal of debate. For the south-west Pacific, the most notable phytogeographic studies of present-day distributions have been made by Van Balgooy (1971) and Good (1960, 1963) [see also Gillison (1975a, b) for Vanuatu]. These have focused on woody plants rather than graminoids as bases for comparison between island plant assemblages. For the Pacific generally, most authors agree that the rate of spread of grasslands, over at least the past 1000 years, is due to interference by humans. The likely nature and extent of the original Tertiary climate or even that of relatively recent periods on the floristic and structural matrix of grassland is unknown. Changes in tree-line boundaries at high altitudes may not be useful indicators, given the paucity of fossil grass pollen and its usually short residence time in the pollen profile.

For Papua New Guinea, most studies of historical sequences have concentrated in the highlands (Smith, 1975, 1977, 1980; Löffler, 1979; Walker and Flenley, 1979). For the mountain grasslands of New Guinea, Smith (1982) has argued that the whole process of plant migration is too complex or cryptic and variable to allow interpretation of origins with any reasonable certainty. Walker and Flenley (1979) demonstrated the existence of significant changes in the Papua New Guinea highland vegetation over the past 2700 years, and even then within a relatively few canopy generations.

In New Guinea, Löffler (1979) asserted a close correlation between glacial movement and grassland development. He concluded that grassland ("paramo") is not climatically controlled because of the low elevational extent of glaciation. He argued also for a "quasi-natural" origin arising from two main factors, one being glacial erosion and the other the effect of man — who provided an invasive path for lowland graminoids. He also concluded that some natural grasslands exist in Mount Digimi in the Kubor range.

Inferential argument is widely used in determining origins. Johnson and Tothill (1985) commented that for India and South-East Asia, the widespread presence of "typical" combinations of trees of the Fabaceae and Combretaceae with andropogonoid grasses suggests that savanna vegetation predated the advent of man. Although it may be reasonable to interpret various grassland features from associated trees (Gillison, 1983), tree assemblages of the Fabaceae and Combretaceae are by no means restricted to India and South-East Asia. They occur in many tropical savannas and marginal forest landscapes where the grass component may vary from dominant to absent. In the south-west Pacific such a gradient is evident in megatherm seasonal Papua New Guinea in closed semi-deciduous forest (Fig. 16.5), with prominent *Combretum goldieanum* and *Terminalia* spp. (Combretaceae) and species of *Erythrina*, *Maniltoa* and *Pithecellobium* (Fabaceae), where graminoids are, for the most part, absent (Heyligers, 1965, 1972a; Gillison, 1970a).

In Papua New Guinea, studies of forest–grassland boundaries (Walker, 1966; Taylor, 1968; Gillison, 1970a, b) have highlighted the dynamism inherent in the ebb and flow of plant assemblages in the face of man-made fire. Despite these studies, no firm conclusion can be reached concerning the natural origin of the grassland component. Manner (1976) concluded that protection from fire can result in succession by woody vegetation with associated soil fertility. According to Manner, this provides strong evidence that tropical grasslands may be anthropogenic, an argument which tends to be supported by Havel (1960) and Robbins (1960). Eden (1974) studied grasslands in the Port Moresby region and, while he was uncertain about their origin, concluded that, apart from climatic fluctuations, man-induced grasslands appear to be dominant at present.

Fig. 16.5. Frequent man-made firing causes this typical retreat of semi-deciduous forest to gully refugia with the expansion of open grassland of mainly *Imperata cylindrica* and *Themeda triandra*. Under continued firing this megatherm seasonal grassland may become dominated by eucalypts. (Photo: P. Heyligers, CSIRO.)

PAPUA NEW GUINEA

Although the term "New Guinea" refers to the island as a whole, most attention is given here to the political entity of Papua New Guinea which forms the eastern half. There are few bioclimates within the south-west Pacific domain that are not found in the island of New Guinea. Grasslands range from high alpine complexes of ridge and bog to widespread montane valley systems that are either a mosaic of sedentary agriculture settled over many thousands of years, to more frequent slash-and-burn domains with patchier grassland formations.

Lowland grassland formations cover thousands of square kilometres of swamp mosaics as in the Fly and Sepik river systems, or else occur in increasingly widening areas within seasonal semi-deciduous forests along the south-central coast. Eucalypt savannas are most widespread south of the main cordillera, but extend north of the Owen Stanley range in the Northern Province. Near Port Moresby on the plains and foothills, there are interdigitating assemblages of eucalypt savannas and open grassland which are rapidly changing in floristics and structure with increased land-use intensity, in particular the use of fire (Fig. 16.6).

Eucalypt savannas are entirely absent in the Markham and Ramu valley systems and along the north coast and Sepik river flood-plains. In general, grasslands can be readily related to bioclimatic patterns, with variations caused by local edaphic changes and land use.

Although provisional bioclimatic provinces are used in the southern region, lack of data prevents a more detailed bioclimatic application for the

Fig. 16.6. Interdigitating megatherm seasonal eucalypt woodland savanna, open grassland or "kunai" and semi-deciduous forest in south-eastern Papua New Guinea.

TABLE 16.3

Vegetation zones[1] and approximate bioclimatic equivalents for Papua New Guinea

Vegetation zone	Elevation (m)	Approximate bioclimate (thermal optima)
Lowland	0– 700	Megatherm (35°–28°C)
Montane	700–3000	Megatherm/mesotherm (28°–19°C)
Subalpine	3000–4000	Mesotherm/microtherm (19°–12°C)
Alpine	>4000	Microtherm (below 12°C)

[1] After Johns (1982).

remainder of the Papua New Guinea. For descriptive purposes, and to relate literature records with the present review, vegetation zonations approximating those of Johns (1982) are used in conjunction with tentative major bioclimatic classes (Table 16.3). Paijmans (1975, 1976) has provided an excellent descriptive basis for the vegetation of Papua New Guinea; this is used here within a bioclimatic context.

Lowland megatherm seasonal grasslands

Although all of the lowlands are megathermal, the most seasonal are to the south-west of

Port Moresby — especially in the Hisiu coastal area and the Trans-Fly. The Fly–Digul region, in the extreme south-west, occupies the most extensive domain covering vast stretches of seasonally flooded savanna with mixed woody and graminoid complexes. These range from "wild rice" (*Leersia hexandra* and *Oryza sativa*) floating grass to *Pseudoraphis spinescens* fringing seasonally dry rivers in the Morehead–Bensbach area. They are surrounded by various woodlands dominated by species of *Acacia*, *Eucalyptus* and *Melaleuca*, which reflect the proximity of the Australian mainland vegetation across the Torres Strait with somewhat similar climate patterns (Nix and Kalma, 1972).

Sedgelands are locally common and are dominated by *Schoenus* sp. associated with the grasses *Eriachne* spp., *Germainia capitata* and *Ischaemum barbatum*. These occupy dense clay soils which are seasonally inundated (Paijmans, 1969, 1971, 1976). This south-west region is characterized by large termite mounds and other "Australian-like" assemblages represented by surrounding woody tree genera *Acacia*, *Banksia*, *Grevillea* and *Tristania* (=*Lophostemon*).

River meander patterns contribute to the complexity of grassland assemblages (Fig. 16.2), with *Saccharum robustum* along the river margins, grading into tall grassland of *Coelorhachis rottboellioides* and *Ophiuros tongcalingii*. Drier zones, usually fired, are commonly composed of *Imperata cylindrica*. Along the drier inland plains in the Wassi Kussa and Morehead river areas are shorter grasslands variously dominated by *Acacia* spp. and *Pandanus* spp. (Fig. 16.3). To varying degrees the *Pandanus*–grassland association

Fig. 16.7. Secondary succession of indigenous (*Alstonia muelleriana*, *Timonius timon*), and exotic (*Broussonetia papyrifera* and *Leucaena leucocephala*) tree species, in grassland dominated by *Pennisetum polystachyon*, coastal Guadalcanal Plain, Solomon Islands.

Fig. 16.8. Mesotherm (1800 m) mid-height to tall secondary grassland of (C_4) species of *Imperata cylindrica*, *Ischaemum* spp., *Miscanthus floridulus* and *Themeda triandra* on old garden site near Mount Tabwemasana, Espiritu Santo, Vanuatu.

is widespread throughout Indo-malesia and the south-west Pacific, and is typified by the "talasiga" or "sun-burnt" lands in Fiji and similar vegetation elsewhere (Figs. 16.7, 16.8 and 16.9).

Transitional swamp to dryland vegetation commonly contains a mixture of emergent palms (*Livistona brassii*) and *Melaleuca* spp. with tall sedges and grasses, often with *Phragmites karka* (Fig. 16.10). The formation has a wider pantropic distribution than the pandanoid type and is represented throughout the megatherm seasonal tropics as various functional analogues but with different taxa (e.g. with the palms *Borassus* and *Sabal*), often on heavy clay soils which dry out seasonally. In Papua New Guinea this transition zone grades into drier foot-slopes dominated by semi-deciduous forest, which is readily converted to grassland following man-made fire (Fig. 16.5).

Some of the most extensive grasslands in Papua New Guinea occur in the Markham and Ramu rift valley systems north of the main cordillera. The Markham valley represents a strong seasonality gradient away from the coast at Lae. At the western forest margin in the Nadzab area, forest gives way to tall grassland with woodland elements of *Albizia procera* and *Antidesma ghaesembilla*. Farther west the plains were once dominated by extensive grassland with *Cycas media* (Fig. 16.11), which has been removed to make way for cattle as the leaves are toxic to ungulates.

Some unusual plant growth forms occur in this area, notably the musaceous *Ensete calospermum* — a large banana-like plant which stores water, presumably as an adaptation to fire. The large aroid cryptophyte *Amorphophallus campanulatus* occurs throughout the valley and exhibits a large, single, quasi-compound leaf which alternates seasonally with a huge single flower which produces a carrion odour attracting pollinating vectors. The

Fig. 16.9. Secondary forest with fire-induced, tall grasses on fired hill-slopes near Nadi, Viti Levu, Fiji.

under-ground stem is rich in starch and is used as a famine crop by the Markham people.

As with other lowland grasslands in the island, swampy depressions may contain deciduous trees of *Nauclea coadunata* — sometimes surrounded by *Albizia procera*. Annual hunting fires reduce large areas of grassland to ash, but these recover quickly due to the finely tuned fire-adaptive mechanisms of the plant cover. In grasses these are characterized either by rhizomes (*Imperata cylindrica*) or else by annual seeding (*Themeda triandra*).

In the lower reaches of the Sepik river on the northern side of the island, grasslands are dominated by tall-grasses — *Saccharum spontaneum* and *Imperata cylindrica*. In the middle reaches these give way to vast plains of *Ischaemum barbatum* and *Themeda triandra* (Paijmans, 1976). The origin of these grasslands has been widely debated because it is difficult to discriminate between the influence of climate, edaphic features, and land use (Reiner and Robbins, 1960; Haantjens et al., 1965, 1967). Whatever the apparent influences, it represents a highly dynamic ecosystem with a large measure of edaphic control.

Patterns of flooding clearly influence the extent of subsequent firing. Despite the claim by Haantjens et al. (1965) that fires never enter or drive back rainforest, this may not always be true — being influenced more by patterns of land use. The interpretation of complex dynamic patterns of this kind can be assisted by gradient-based studies using fine-scale, spatially referenced, biophysical variables within the framework of a geographical information system as described earlier.

Active volcanic areas of Papua New Guinea support extensive tall grassland on ash deposits. The dominant species is commonly *Miscanthus floridulus*, often associated with *Saccharum spontaneum*. These are conspicuous on the ash and

Fig. 16.10. Transitional swamp mosaic of tall megatherm seasonal grassland dominated by palms (*Livistona brassii*) and *Pandanus* spp., with *Phragmites karka, Saccharum spontaneum* and tall sedges. (Photo: P. Heyligers, CSIRO.)

gravel fans below Mount Lamington towards the eastern end of the main cordillera, where the pioneering tree species *Gymnostoma papuanum* (formerly *Casuarina papuana*) exists as monospecific stands (see also Taylor, 1964a, b and Paijmans, 1967 for discussion on *Gymnostoma papuanum* succession). Tall grassland (mainly *Miscanthus floridulus*) is also found on the slopes of active volcanoes on the islands of New Britain and on Bougainville (Van Royen, 1963c; Heyligers, 1967). On the youngest lava flows on Pago volcano in New Britain, in the absence of fire and grazing, Paijmans (1973) found mixed vegetation with grassland dominated by *Eulalia leptostachys, Imperata cylindrica*, and *Thysanolaena maxima*. On Mount Bagana, west of Kieta, tall *Miscanthus floridulus* grassland combined with the fern *Dicranopteris linearis* covers much of the lee slopes under a volcanic plume laden with sulphur dioxide. Fallout from the plume clearly influences vegetation pattern. *Miscanthus floridulus* exists even on the upper slopes at about 900 m, albeit as a stunted form.

Montane, megatherm–mesotherm grasslands

Low montane grasslands are most extensive in distinctly seasonal zones such as the foothills and lower parts of the south-eastern Owen Stanley range and in the upper Ramu and Markham valleys and the Bulolo–Watut divide. Open grasslands tend to be dominated either by *Themeda triandra* or *Imperata cylindrica* (the two most common "Kunai" species). These tend to be distributed according to soil patterns, *Themeda* occupying the drier, stony soils and the rhizomatous *Imperata* the deeper and more nutrient-rich sites. Continual removal of *Imperata* will usually lead

Fig. 16.11. Scattered *Cycas media* with *Imperata cylindrica* mid-height grassland in the megatherm seasonal Markham Valley of Papua New Guinea. Increased firing and grazing has drastically modified these widespread grassy plains, which occur in both the Markham and adjoining Ramu valley systems.

to replacement by other grasses before *Themeda*. In lower foothill areas, on dry shallow soils, black spear grass (*Heteropogon contortus*) is predominant.

Even "open" grasslands contain woody elements, such as the trees *Antidesma ghaesembilla*, *Cathormion umbellatum*, *Glochidion* spp., *Nauclea coadunata* and *Timonius timon*. These are often precursors to eucalypt invasion in megatherm areas and in the warmer part of mesotherm environments such as the Sogeri plateau and Mount Brown (cf. Lane-Poole, 1925a, b) south of the Owen Stanley range. As such, they tend to resemble dryland analogues of the so-called "biological nomads" of rainforest succession, so named by Van Steenis (1958) as "plants whose life cycle and/or other ecological characters do not fit into the closed cover of the undisturbed climax and which can maintain themselves only under disturbed conditions either by nature or by man on the 'margin' of the pertaining climax". Mixed, tall grasses are common, including:

Arundinella setosa
Capillipedium parviflorum
Cymbopogon procerus
Eulalia leptostachys
Ischaemum spp.
Pseudopogonatherum irritans

Taller grasses include:

Apluda mutica
Coelorhachis rottboellioides
Ophiuros tongcalingii
Pennisetum macrostachyum
Saccharum spontaneum

Associated herb layers vary with seasonality. Seasonal cryptophytes are common, notably *Amorphophallus campanulatus*, *Curculigo orchioides*, *Curcuma longa* and *Pygmaeopremna sessilifolia*. Cryptophytic lianes (species of *Cissus*, *Dioscorea* and *Merremia*) are frequent towards the forest edge (Gillison, 1970a).

Some seasonal areas of the upper Ramu and Markham valleys have been subject to increased firing in recent years. This has promoted expansion of tall to mid-height grasses, with increasing dominance of *Albizia procera* which appears to be invading from the lowlands.

Montane, mesotherm grassland

Between about 800 m and 2000 m, wide expanses of grassland occur in mostly populated areas — such as the eastern and western highlands, the Bulolo–Watut divide (Figs. 16.12 and 16.13), the upper Markham, Irimu, and Ramu valleys and parts of the Finschhafen/Saruwaket (Huon) peninsula. In seasonal parts of the Bulolo–Watut area, forest cover and composition are highly variable and may be locally dominated by Araucariaceae (*Agathis robusta* subsp. *nesophila*, *Araucaria cunninghamii* and *A. hunsteinii*). The areas where these trees now dominate were, in many cases, previously grassland (according to local informants in the Manki area near Bulolo) (Gillison, 1970a). The more accessible *Araucaria* forests have been heavily logged and grasslands are now rapidly invading. Whitmore (1975) has commented on the conversion of analogous vegetation in Indo-malesia.

Montane, mesotherm grasslands are variously dominated by *Arthraxon* spp., *Eulalia leptostachys*, *Imperata cylindrica* and *Themeda triandra* — and often with sedgeland (*Rhynchospora rubra*) on

Fig. 16.12. Low montane (900 m) man-induced, mesotherm (C_4) grasslands of the Bulolo–Watut areas of Papua New Guinea dominated by *Imperata cylindrica* and *Themeda triandra*. Most of the area was formerly forest.

Fig. 16.13. Montane (2000 m) highly dynamic *Miscanthus/Imperata* mesotherm (mixed C_4, C_3) grassland mosaic in the eastern highlands of Papua New Guinea, with sedentary agriculture and cultivated *Casuarina oligodon* groves.

shallow soils on hill-slopes, and the tall grass *Miscanthus floridulus* on deeper soils, especially near forest margins. Moister sites support *Phragmites karka* and *Saccharum spontaneum*. Common woody emergents are *Decaspermum paniculatum*, *Omalanthus* spp., *Macaranga* spp., *Schuurmansia henningsii* and *Wendlandia paniculata*. More seasonal locations are occupied by *Guioa* spp., *Macaranga* spp., *Rhus taitensis* and *Timonius timon*. The majority of grass species are C_4 types but increasing proportions of C_3 species appear at about 2000 m.

Montane, mesotherm–microtherm grasslands

Over about 2500 m, the tree-line begins to fragment and the proportion of open grassland increases with altitude, andropogonoid C_4 grasses being increasingly replaced by C_3 types. Although there is higher relative humidity, occasional spells of dry, fine weather with intense sunshine place considerable stress on a wide range of plants, including grasses. The morphology and apparent function of the grasses differ from those in the lowlands, with smaller, narrower and often tightly rolled leaves. The proportion of trunk-forming grasses increases with altitude and prevailing moisture. Although these have close affinities with temperate, southern grasslands, shoot/

root ratios appear to be much greater (Hnatiuk, 1975; Smith, 1980 — also see Chapter 15). Tussock, trunk-forming grasses such as *Chionochloa archboldii* may represent an adaptation to both fire and low temperatures. As with some Andean and African high mountain plants of similar habit, the formation of a trunk may confer insulative advantages.

Short and tussock grasslands, both rich in forbs, abound within this domain. Legumes appear to be absent (Gillison, 1970b; Paijmans, 1976), although why this should be so is unclear. The dominant grasses are *Chionochloa archboldii*, *Deschampsia klossii*, *Deyeuxia brassii* and *Hierochloë redolens*. *Agrostis reinwardtii*, *Anthoxanthum angustum* and *Arundinella furva* dominate locally. Such short, ephemeral grasses as *Isachne globosa* and *I. myosotis* occupy transient fire patterns (Gillison, 1969). Species of *Danthonia* and *Dichelachne* are locally common together with sedges (*Machaerina rubiginosa* and *Rhynchospora rugosa*).

The floristically rich herb layer contains many hemi-cryptophytes and/or forbs (e.g. *Plantago aundensis* and *Xyris papuana*), whereas others are semi-reptant (*Hydrocotyle* spp. and *Potentilla* spp.). Suppressed woody species (such as *Xanthomyrtus* spp.) occur in irregularly fired open grasslands, and ericaceous heath occupies a patchwork pattern in the grasslands (Fig. 16.14), with species of *Dimorphanthera*, *Rhododendron*, *Styphelia* and *Vaccinium*. These woody plants are most common in thickets and typically occupy forest edges, where they layer to maintain their position within a fluctuating edge (Gillison, 1970b; Smith, 1980). Paijmans and Löffler (1974) and Paijmans (1976) described in detail the occurrence of vegetation dominated by tree-ferns (*Cyathea* spp.) in these upland areas (Fig. 16.14)

Fig. 16.14. Subalpine (3200 m) patchwork of fired microtherm (C_3) *Danthonia* tussock grassland with ericaceous heath and surrounding tree-fern (*Cyathea*) savanna. (Papua New Guinea; photo by J. Saunders, CSIRO.)

Fig. 16.15. Subalpine (3400 m) microtherm (C_3) grassland with grasses *Danthonia*, *Deschampsia*, *Deyeuxia* and *Hierochloë* with sedges and tree-ferns (*Cyathea* sp.). (Papua New Guinea; photo by R. Hoogland, CSIRO.)

(see also Gillison, 1983). This unique formation is clearly fire-maintained — if not fire-derived — and is commonly set within a high montane context of open grassland and ericaceous heath (Fig. 16.15).

Subalpine, microtherm grassland

From 3700 m to above 4000 m, the grassland structure changes to mostly short grassland with cushion forms as well as other short grasses and some tussocks (Fig. 16.16). With the environment becoming more limiting for plant growth, species are fewer and (not surprisingly) mainly C_3 types. Short, pooid grasses are represented by *Festuca papuana* and *Poa callosa*, with short tussock *Deschampsia klossii*. The cushion form of *Monostachya oreoboloides* and the similarly formed sedge *Oreobolus pumilio* indicate the climatic limitations and the close affinities with grasslands in similar climates in high latitudes.

The prevailing moist habitat is indicated by local bog formations with *Astelia papuana* and *Carpha alpina*. Tree-fern savanna gives way to the bizarre finger fern (*Papuapteris linearis*) common near the summit of Mount Wilhelm at about 4300 m. On upper slopes and in fired areas, the rhizomatous fern *Gleichenia vulcanica* is abundant as it is at lower elevations on poorer, frequently fired sites. Stunted woody species of *Coprosma*, *Eurya*, *Styphelia* and *Xanthostemon* are locally common in protected areas. Exposed sites towards summits above about 4000 m are limiting for graminoid forms and grade into more open stony ground with bryophytes. This is especially the case where frost is frequent and where there may be ephemeral snow.

Fig. 16.16. Subalpine to alpine (3800–4200 m) microtherm (C_3) *Chionochloa archboldii* tussock grassland near Mount Bangeta, Saruwaket plateau of Papua New Guinea, surrounded by fire-refugic thicket dominated by ericaceous species (including *Rhododendron* spp.) and by tree-ferns (*Cyathea* sp.).

SOLOMON ISLANDS

The Solomon Islands lie immediately to the east of Bougainville; they are a widely scattered chain of islands covering about 900 km. The bioclimate is primarily megatherm with few if any grasslands extending into the mountainous interior of the main islands. The geology is complex, dating from the Oligocene with active volcanoes in some areas, which have appeared in Pleistocene and Recent periods; there are also elevated terraces of marine limestone (Thompson and Hackman, 1969). Tectonic disturbances and episodic but relatively frequent damage from hurricanes continually create openings for grassland expansion, particularly if accompanied by fire. In comparison with Vanuatu, the Solomon Islands are richer in plant species (Henderson and Hancock, 1989) and therefore may be less vulnerable to invasive exotics.

Grasslands are well developed on Guadalcanal, and probably originated from man-made fire and gardening activities (Whitmore, 1969; Hancock and Henderson, 1988). The best developed are on the north coast on Pleistocene sediments in the shadow of the Kavo range (Fig. 16.17). To the east are mid-height to tall grasslands on the Guadalcanal plain. Here Whitmore (1969) described extensive tracts of almost pure *Themeda triandra* with small areas of *Imperata cylindrica*, and *Phragmites karka* and *Saccharum spontaneum* in wet areas. At the time he noted the occurrence of *Pennisetum polystachyon* in disturbed locations. Since then the species has vigorously invaded much of the grasslands on Guadalcanal (Figs. 16.7, 16.18 and 16.19).

Fig. 16.17. Man-induced, megatherm seasonal (C_4) mid-height to tall grasslands on sub-coastal zone north of Honiara, Guadalcanal, Solomon Islands. Dominated by exotic *Pennisetum polystachyon* with *Imperata cylindrica* and *Themeda triandra*.

In the absence of fire, vegetation succession is relatively rapid, with tree species *Alstonia muelleriana*, *Colona scabra Commersonia bartramia* and *Hibiscus tiliaceus* (Fig. 16.7). The exotic trees *Broussonetia papyrifera* and *Leucaena leucocephala* are rapidly spreading. There are few other grassland types in the Solomons other than occasional associations of *Imperata cylindrica* and *Themeda triandra*, with *Pandanus* spp. on coastal fired areas (Fig. 16.19).

Recent and extensive logging on several islands, including New Georgia and Malaita, will pave the way for further grassland invasion. Short grassland is already expanding in logged-over and heavily gardened areas of west Malaita. On Guadalcanal in the urban fringe of the capital Honiara, forest has been entirely replaced by rapidly degrading grassland dominated by *Pennisetum polystachyon* and introduced tree species. Garden productivity is already declining and in many cases has reached the Cassava or "Manioc" stage — cassava (*Manihot* spp.) being the last type of crop that can be grown without addition of fertiliser (Fig. 16.20).

NEW CALEDONIA

The megatherm savannas of New Caledonia occupy a wide range of substrates from shoreline limestone terraces to ultrabasic mountainous massifs. For the most part they are associated with the "niaouli" paperbark tree *Melaleuca quinquenervia* as a woodland savanna mosaic (Gillison, 1983). This vegetation has been variously described by several authors (Guillaumin, 1952; Virot, 1956; Jaffré, 1974; Jaffré and Latham, 1974; Holloway, 1979). On the rain-shadow slopes of the western massifs lie mixed *Melaleuca* and *Imperata* savannas. On the lower slopes of Mount Khogis are patches of sedgeland with *Baeckea virgata* maquis.

As with some areas on Vanuatu, coastal *Acacia spirorbis* forest is patchily associated with grasses (*Imperata cylindrica* and *Miscanthus floridulus*), being best developed near Anse Longue (Holloway, 1979). Holloway (1979) compared the "niaouli" formation with the "talasiga" of Fiji, but considered that this vegetation was not comparable to the *Melaleuca*-dominated grasslands

Fig. 16.18. Roadside aspect of coastal *Pennisetum polystachyon* on limestone with littoral vine *Ipomoea pes-caprae*, near Honiara, Guadalcanal, Solomon Islands.

of New Guinea. The New Caledonian *M. quinquenervia* is highly variable in form, and in the Thie valley can exist as a semi-prostrate woody shrub. One factor adding to the uniqueness of the "niaouli" is that, unlike Australia, there are no competing indigenous *Eucalyptus* spp. In Australia, along some relatively similar physical environmental gradients, grasslands associated with *Eucalyptus* and *Melaleuca* exhibit quite different floristic and structural complexes.

VANUATU

Vanuatu lies to the east and north-east of New Caledonia and west of Fiji. There are 11 main islands in a latitudinal range from 13° to 21°S. Although primarily maritime megatherm, there are upland areas containing grassland which approach mesotherm status (Mount Tabwemasana on Espiritu Santo). The geology is locally complex, ranging from early Miocene limestones to Pleistocene volcanics (cf. Lee, 1975; Mallick, 1975). Vanuatu lies within an active zone of the Pacific arc, and local tectonism has a direct influence on vegetation succession, especially on the west coast of Malekula. There, recent upraised marine terraces contain a littoral flora now dominated by the exotic tree *Leucaena leucocephala* with patches of *Imperata cylindrica* and scattered *Cycas rumphii*.

The flora is intrinsically poor (<400 angiosperm taxa), which has rendered it extremely susceptible to invasive exotics such as *Leucaena leucocephala*. Tectonism, combined with frequent hurricanes, further assists weedy invaders. Although there are no strongly seasonal zones, most of the larger islands (Aneityum, Erromango, Espiritu Santo, Malekula and Tanna) exhibit rainshadow effects on the lee side. Schmid (1975) described *Miscanthus floridulus* grasslands on Aneityum, mainly in the north and west. On the eastern side some scattered mid-height grassland occurs with *Imperata cylindrica*, especially surrounding old garden areas and some recent forest plantings.

On Erromango, widespread *Leucaena leucocephala* was almost totally defoliated by a het-

Fig. 16.19. Megatherm seasonal (C_4) fired coastal grassland mosaic dominated by *Pandanus spiralis*, with *Imperata cylindrica* (southern Guadalcanal, Solomon Islands).

eropsyllid insect (*Heteropsylla cubana*) in 1986, but this did little to affect the *Leucaena* population which is variously associated with *Imperata cylindrica* and *Miscanthus floridulus* (Fig. 16.21). As in New Caledonia, *Acacia spirorbis* forest–grassland mosaics are widespread in some parts of Vanuatu. On Erromango in particular, *Acacia* is being removed for cattle grazing. This has the immediate consequence of reducing available "natural" grass cover and allowing the ingress of exotic malvaceous and verbenaceous weeds, especially *Sida rhombifolia* and *Stachytarpheta urticaefolia* (Fig. 16.22).

Although most grasslands have developed in the rain-shadow zones, some appear in certain upland localities — for instance, on Mount Tabwemasana on the island of Espiritu Santo, and on remote Futuna island to the south-east of Erromango (Figs. 16.8 and 16.23). These quasi-mesotherm grasslands have been generated mainly as a result of gardening practices. On Mount Tabwemasana they are usually patches of tall grassland, mainly with *Imperata cylindrica*, *Ischaemum muticum* and *Miscanthus floridulus*. They tend to be fired annually and are expanding along spurs in several adjacent mountains due to the ingress of fire from lowland grasslands.

FIJI

There are over 500 islands in Fiji, of which the two largest are Viti Levu and Vanua Levu. The climate is mainly megatherm in areas where grasslands occur, often with a seasonal aspect in a climate domain with between 1600 mm and 2300 mm annual rainfall. The vegetation has been variously described by several authors, most notably by Twyford and Wright (1965), Parham (1972) and Smith (1979). Most agree that the former vegetation was closed forest, which has been converted to grassland by man. The exception may have been small areas of swamp or open water (Twyford and Wright, 1965).

Fig. 16.20. Rapidly degrading, fired and gardened, urban perimeter west of Honiara, Solomon Islands. Almost all the plant cover, including grasses, is exotic. Gardens have now reached the "Manioc" (*Manihot esculenta*) stage of low productivity, which will progressively decline.

Fiji presents a familiar pattern in the southwest Pacific where drier, leeward forest is being replaced by grassland and disclimax fernland. Twyford and Wright (1965) reported that, of the nineteen grasses recorded for Fiji by Seeman in 1918, only five were indigenous. About 140 species have been introduced since, the most notable being *Pennisetum polystachyon* or "Mission" grass (presumably imported by early missionaries), "wire" grass (*Sporobolus indicus*) and "Guinea" grass (*Panicum maximum*). Together with the exotic woody *Psidium guayava*, *Sporobolus indicus* dominates hills on Viti Levu and large areas of Macuata and Bua Provinces of Vanua Levu. Parham (1972) recorded *S. virginicus* and *Thuarea involuta* in littoral zones. According to Kirkpatrick and Hassall (1981), grasses on the Sigatoka sand dunes are introduced species which have been encouraged by firing and cutting. There, conspicuous species are *Panicum maximum* and *Pennisetum polystachyon*, and on open grassland *Brachiaria reptans* and *Panicum maximum*.

The best-known grasslands in Fiji are the "talasiga" or "sun-burnt" lands (Fig. 16.24). Here, *Miscanthus floridulus* has been replaced with rhizomatous *Dicranopteris linearis* ("fire" ferns) and *Pteridium esculentum* (bracken), *Sporobolus indicus* and *Pennisetum polystachyon*.

The dominant tree in many areas is *Casuarina equisetifolia* associated with *Pandanus odoratissimus* (Parham, 1972), and together with *Pennisetum polystachyon* typify "talasiga" on Viti Levu. Other grasses are *Dichanthium caricosum*, *Heteropogon contortus*, *Panicum maximum*, *Paspalum orbiculare* and *Sporobolus elongatus*, which (according to Parham, 1972) have become naturalized over large areas. If burning ceases for several years, successional species appear, including:

Acacia richii
Alphitonia zizyphyoides
Decaspermum fruticosum
Dodonaea viscosa
Hibbertia lucens

Leucopogon cymbulae
Morinda citrifolia
Mussaenda raiateensis
Syzygium richii

Fig. 16.21. Irregularly fired landscape on Erromango Island, Vanuatu, dominated by the cultivar escape, shade-tree *Leucaena leucocephala* with tall to mid-height megatherm seasonal (C_4) grassland (*Miscanthus floridulus/Imperata cylindrica*). On limestone soils.

The assemblage has affinities with similar successional patterns in eastern Papua New Guinea and north-eastern Queensland.

MICRONESIA

Most of the Micronesian islands lie to the north of New Guinea and are scattered over about 7.8 million km² of ocean with a total land area of about 1800 km². They are usually classified as "low" (coralline) or "high" (volcanic) islands which tend to have correspondingly different floras. Apart from some upland areas of Ponape which approach a megatherm–mesotherm bioclimate, most of the islands have a highly buffered maritime megatherm climate. Most are isothermal, with an average daily maximum of 27°C and a maximum annual deviation of 2°C.

No grasslands have been reported on "low" coral islands. On "high" islands they exist mostly as disclimax mosaics with floristically variable patches of low scrub (Safford, 1905; Bridge and Golditch, 1948; Glassman, 1952; Fosberg, 1960). The principal species are as follows (Fosberg, 1960):

Grasses:	*Dimeria* spp., *Heteropogon contortus*, *Miscanthus floridulus*
Sedges:	*Fimbristylus* spp., *Rhynchospora* spp., *Scleria* spp.
Ferns:	*Blechnum orientale*, *Cheilanthes tenuifolia*, *Dicranopteris linearis*, *Lycopodium cernuum*, *Lygodium scandens*
Shrubs, etc.:	*Eurya* spp., *Geniostoma* spp., *Glossogyne tenuifolia*, *Melastoma malabathrica*, *Morinda citrifolia*, *Myrtella benningseniana*, *Nepenthes mirabilis*, *Pandanus* spp.

The presence of *Casuarina equisetifolia* and *Pandanus* spp. under such conditions is analogous to

Fig. 16.22. Clearing of *Acacia spirorbis* secondary forest for gardens and cattle grazing, Erromango Island, Vanuatu. This will be transformed rapidly into a grassland mosaic dominated by *Imperata cylindrica* and exotic forbs.

the "talasiga" lands of Fiji. These, together with *Dicranopteris linearis*, form a conspicuous association (locally known as "tedh") on Yap Island (Fig. 16.25), which is supported on shallow acid soils (as indicated by *Eurya*, *Melastoma* and *Nepenthes* spp.).

Gillison (1983) recorded the invasion of *Pennisetum polystachyon* in grassland dominated by *Imperata cylindrica* on Moen Island in the Truk group (Fig. 16.26). It was associated with the grasses *Ischaemum muticum* and *Saccharum spontaneum*. There is speculation about the origin of these grasslands, and whether they are entirely derived from forest. Rather like Twyford and Wright (1965), who speculated that some grassland species in Fiji may have existed naturally around swamps and watercourses, Fosberg (1960) has argued that some savanna species may have existed as rare plants in similarly isolated niches in Micronesia, especially on peaks, crests and landslides.

"Savannization" (i.e. conversion of closed forest to grassland-dominated vegetation) is increasing in the main islands (Fig. 16.27). At Airai on Babelthuap Island (Belau), grassland with *Pandanus* spp. is expanding annually as a result of fire, which makes difficult the establishment and maintenance of forest plantations and subsistence gardens. The reduced productivity brought about by the expansion of fire disclimaxes is likely to continue in the absence of more effective land management.

Fig. 16.23. Coastal megatherm–mesotherm uplands of Futuna Island, Vanuatu, with tall grassland of *Imperata cylindrica* and *Miscanthus floridulus* following fire, surrounded by secondary *Acacia spirorbis* forest.

THE ROLE OF FIRE AND FROST IN PACIFIC GRASSLANDS

Fire

There seems to be general consensus that if fire is not concerned directly with the genesis of grasslands, it is at least responsible for their maintenance and subsequent spread. On high mountains, several authors have observed damage to forest by fire (Lane-Poole, 1925a, b; Archbold and Rand, 1938; Brass, 1964; Wade and McVean, 1969; Kalkman and Vink, 1970; Paijmans and Löffler, 1974). Various authors have considered the probable effect of natural fires on the vegetation. In Kalimantan, Seavoy (1975) suggested that grasslands may have been initiated by lightning; in New Guinea, Shaw (1968) documented lightning strike in coconut (*Cocos nucifera*) and cacao (*Theobroma cacao*) plantations in New Britain; in savannas in north-western Australia, in excess of 1000 ground strikes per month by lightning have been recorded within range of the horizon (R.J. Thistlethwaite, pers. commun., 1980). In the south-west Pacific, the relatively high frequency of hurricanes also increases the probability of fire following the drying out of wind-damaged vegetation. In a similar way, there are many subsistence gardening practices (Barrau, 1958, 1961) which may generate a fire-prone vegetation nucleus from which grasslands may radiate under a persistent fire regime.

For some Central American neotropical savannas, Kellman (1984) suggested that vegetation was inherently more fire-prone on infertile than on fertile soils, because of the slowness with which closed tree and shrub canopies are established on such sites. For the neotropics, Kellman suggested

Fig. 16.24. "Talasiga" (sun-burnt land) on Viti Levu, Fiji. Megatherm-seasonal (C_4) grassland mosaic with *Casuarina equisetifolia* and *Pandanus* sp.

that vulnerability to fire might also be enhanced by the lower mineral nutrient content of plant tissues on infertile soils. In Papua New Guinea, Gillison (1969) observed that fire patterns were related to apparent variability in the soil water table (see also discussion by Paijmans and Löffler, 1974). Fluctuations in the grassland–forest boundary due to fire and other environmental determinants have been examined by various authors (Walker, 1966; Taylor, 1968; Wade and McVean, 1969; Gillison, 1970a, b; Eden, 1974; Hope, 1976a, 1980; Smith, 1980). Certain plant assemblages along the forest boundary may act as a buffer to fire (Wade and McVean, 1969; Gillison, 1970b; Coode and Stevens, 1972).

It is arguable that the high level of volatile and essential oils in much of the forest-edge vegetation (species of the families Ericaceae, Lauraceae, Myrtaceae, Pittosporaceae, Rutaceae) confers resistance to physiological stress through either frost or drought. Often these "edges" will resist wild fire, but beyond a certain temperature threshold they reach a flash-point when fire may then penetrate the forest proper. With the aid of Kalkman and Vink (1970), Gillison (1970b) recorded detailed evidence for high vegetative mobility among groups of montane edge-dwelling species where the dynamics were clearly controlled by episodic fire.

There is some evidence for cyclic succession between closed forest and grassland in seasonal areas of the south-west Pacific. In the seasonal Bulolo–Manki area of eastern low-montane Papua New Guinea, Gillison (1970a) found evidence that *Araucaria*-dominated patches of forest may have passed through several cycles of open grassland and forest cover. Based on detailed site investigations and oral history from local inhabitants,

Fig. 16.25. Short fern (*Dicranopteris linearis*) and (C_4) megatherm ("talasiga-like") grassland with *Imperata cylindrica* and *Ischaemum* spp. on Yap Island, Micronesia.

Gillison (1970a) developed a hypothetical model which indicates how a cyclo-pyric succession of this kind might develop under varying fire frequencies (Fig. 16.28).

Evidence for similar fire-driven catastrophic sequences in seasonal areas (usually <1200 mm annual rainfall) are not unusual in the Pacific, and in some respects resemble conifer–broad-leaf cycles in higher latitudes. In parts of Australia, under the influence of fire, *Callitris* spp. (Cupressaceae) and *Araucaria* spp. (Araucariaceae) may replace *Eucalyptus* spp. and semi-deciduous vine thicket assemblages, respectively, sometimes within one generation of canopy species — and often with intermediate grassland stages (Gillison, 1983, fig. 9.8). Oscillations of this kind, with relatively high frequency, may have been further influenced by Quaternary climatic variation over a longer time base (Walker and Flenley, 1979).

Frost

Among the south-west Pacific islands considered in this study, New Guinea is the only land mass high enough to be exposed to snow and frost. The potential role of frost as an agent for modifying forest and/or generating natural grasslands in this area has attracted some attention. Brown and Powell (1974) documented extensive damage from both frost and drought to highland vegetation in Papua New Guinea in 1972. Frost damage was most evident in swamp and beech (*Nothofagus*) forest. Subsequent droughting following frost raises the probability of fire.

Ponding of cold air also assists in maintaining upland grasslands such as those in the extensive Neon basin near Mount Albert Edward (Paijmans and Löffler, 1974). At Lake Myola near Efogi in the Owen Stanley ranges north of Port Moresby,

Fig. 16.26. Megatherm (C_4) man-induced grassland on Moen Island, Truk group, Micronesia. *Imperata cylindrica*, *Ischaemum* spp. and *Saccharum* spp., with invading *Pennisetum polystachyon*.

at an elevation of about 2200 m, much of the in-filled lake area is graminoid swamp with surrounding dryland grasses — and thus subject to both irregular firing and cold-air ponding. Because of the relatively depressed temperature, some species (such as *Potentilla* spp. and *Ranunculus* spp.) normally found at much higher elevations are well established here. Farther to the east, grasslands exist at a similar elevation on the southern slopes of Mount Dayman, which lies towards the eastern end of the main cordillera. Under conditions where there is no apparent frost or ponding of cold air, the Mount Dayman grasslands include species that are common in lowland vegetation. Because of this variable floristic pattern at similar elevations, caution must be applied when assessing likely vegetation features from derived thermal regimes in mid-altitude Papua New Guinea.

LAND USE

Grasslands in the south-west Pacific generally have been considered by some authorities (e.g. Whyte, 1968) to be relatively useless or "intractable". Throughout the tropics and certainly in the south-west Pacific, progression to short grassland is usually accompanied with loss in soil fertility (Gillison, 1970b, 1983; Manner, 1976; Bleeker, 1983) and, presumably, productivity. Despite these negative associations, grasslands form an important element in society in most of the major islands, especially Papua New Guinea. Their uses have been documented by Gillison (1972) and Henty (1969, 1982), who recorded a wide variety of uses, especially in lowland grasslands. Larger grasses (*Saccharum spontaneum* and *Miscanthus floridulus*) are used as roofing thatch and as arrow shafts. In the Menyamya area of Papua

Fig. 16.27. Short fern (*Dicranopteris linearis*, *Pteridium* sp. and megatherm (C_4) grassland (mainly *Ischaemum* spp.) on Ponape island, Micronesia.

New Guinea, ash from *Saccharum spontaneum* and sometimes *Phragmites karka* is a source of much-needed salt. Some grasses produce edible inflorescences (*Saccharum edule*) or leafy shoots (*Setaria palmifolia*), which are locally important dietary components.

The grassland habitat has been in Papua New Guinea long enough for an identifiable grassland fauna to emerge. In the Trans-Fly region of the Western Province, grasslands support extensive animal populations which are hunted using fire-driving techniques (Bulmer, 1968). The most important are the agile wallaby or "magani" (*Macropus agilis*) and, more recently, the exotic rusa or Timor deer (*Cervus timorensis*) which has spread throughout most of the lowland Papuan savannas, to which it is presently restricted. The grasslands in the Markham and Ramu valleys north of the main cordillera also support significant amounts of game birds, bandicoots or "mumut" (Peramelidae: *Echymipera kalubu* and *Isoodon macrourus*), and large pythons or "moron" (species of *Liasis* and *Morelia*) which are both hunted annually by fire. In these large rift valley systems and in parts of the eastern and western highlands, human society has evolved around the grassland biome to the extent that reversion to forest would create a major societal impact.

The montane and subalpine grasslands also produce game, especially rats (*Uromys*) and bandicoots. Although the abundance of mammals appears to be much less than at lower elevations, they form important reservoirs of protein for the indigenous people who hunt them on a regular basis, mostly by fire.

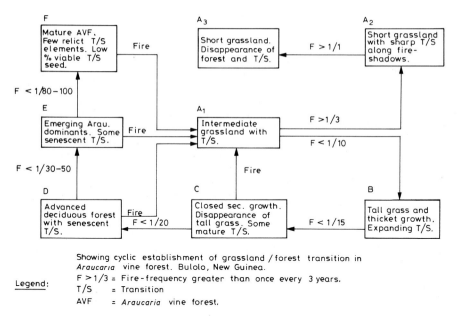

Fig. 16.28. Schematic diagram showing cyclic establishment of grassland/forest transition in *Araucaria* vine forest. Bulolo, Papua New Guinea. (From Gillison, 1970a.)

GRASSLANDS AS A PATHWAY FOR INVASIVE EXOTICS

There can be little doubt that, once established, grasslands provide a direct avenue for the spread of certain exotic weeds. The rate of spread of invasive species depends on the physical nature and floristic richness of the target area. Small island domains with few species are most vulnerable, especially if the grasslands lie at the seasonal end of the climate continuum (Fosberg, 1963).

Adamson and Fox (1982) predicted that weed infestation of the New Guinea vegetation will rise in direct proportion to disturbance of the forests. Although this may be so, it should be remembered that humans have been engaged in slash-and-burn agriculture in the Pacific for many thousands of years. In this context, man may be regarded as a "natural" component in ecosystems which have evolved relatively predictable cyclic successional phases (Gillison, 1970a).

Under circumstances where such "stability" is upset, and where soil fertility is reduced, weed invasion is facilitated. Grazing by cattle and other animals (cf. Mueller-Dombois, 1981, for Hawaii) and intensive logging practices contribute significantly to this process. Typical invaders are the tree species *Leucaena leucocephala* and *Muntingia calabura*. In Vanuatu and on the island of Guam, *Leucaena* has formed vast, dense thickets which are virtually irremovable by conventional means. The grass *Pennisetum polystachyon* is perhaps the most urgent threat to grasslands in the south-west Pacific where it continues unabated in its replacement of natural graminoids. Less imminent as a threat, but nevertheless significant, are exotic species of bamboo. These are gradually becoming established in several island areas, especially in southern New Caledonia and parts of megatherm seasonal Papua New Guinea and Fiji.

Although research is required to understand better the dynamics of natural grasslands, the trends are clear: unless more informed management practices are developed and implemented, under the increasing land pressures brought about by burgeoning human populations, the future scenario for the grassland biome is one of continuing degradation.

REFERENCES

Adamson, D.A. and Fox, M.D., 1982. Change in Australasian vegetation since European settlement. In: J.M.B. Smith (Editor), *A History of Australasian Vegetation*. McGraw-Hill, Sydney, pp. 109–146.

Archbold, R. and Rand, A.L., 1938. Results of the Archbold Expeditions, 7. Summary of the 1933–34 Papuan Expedition. *Bull. Am. Mus. Nat. Hist.*, 68: 527–529.

Barrau, J., 1958. Subsistence Agriculture in Melanesia. *B.P. Bishop Mus. Bull.*, 219, 111 pp.

Barrau, J., 1961. Subsistence Agriculture in Polynesia and Micronesia. *B.P. Bishop Mus. Bull.*, 223, 94 pp.

Beard, J.S., 1953. The savanna vegetation of Tropical America. *Ecol. Monogr.*, 23: 149–215.

Bellamy, J.A. (Editor and compiler), 1986. Papua New Guinea inventory of natural resources, population distribution and land use handbook. *CSIRO Aust. Div. Water Land Res. Nat. Resour. Ser.*, 149 pp.

Bews, J.W., 1929. *The World's Grasses: Their Differentiation, Distribution, Economics and Ecology*. Longmans, Green and Co., London, 408 pp.

Bleeker, P., 1983. *Soils of Papua New Guinea*. CSIRO Australia in association with Australian National Univ. Press, Canberra, 352 pp.

Brass, L.J., 1938. Results of the Archbold Expeditions, II. Notes on the vegetation of the Fly and Wassi Kussa Rivers, British New Guinea. *J. Arnold Arbor., Harv. Univ.*, 19: 174–193.

Brass, L.J., 1956. Results of the Archbold Expeditions, 75. Summary of the fourth Archbold Expedition to New Guinea (1953). *Bull. Am. Mus. Nat. Hist.*, 111: 83–152

Brass, L.J., 1964. Results of the Archbold Expedition to New Guinea (1959). *Bull. Am. Mus. Nat. Hist.*, 127: 147–216.

Bridge, J. and Golditch, S.S., 1948. *Preliminary Report on the Bauxite Deposits of Babelthuap Island. Palau Group*. Prepared by United States Geological Survey, Military Geology Section, under direction of Office of the Engineer. General Headquarters Far East Command, 46 pp.

Brookfield, H.C. and Hart, D., 1966. Rainfall in the Tropical Southwest Pacific. *Aust. Nat. Univ. Dept. Geogr. Publ.* G/3, Canberra, A.C.T.

Brown, M. and Powell, J.M., 1974. Frost and drought in the highlands of New Guinea. *J. Trop. Geogr.*, 38: 1–6.

Bulmer, R., 1968. The strategies of hunting in Papua New Guinea. *Oceania*, 38: 302–318.

Clifford, H.T. and Simon, B.K., 1981. The biogeography of Australian grasses. In: A. Keast (Editor), Ecological Biogeography of Australia. *Monogr. Biol.*, 41: 537–554.

Cole, M.M., 1963. Vegetation and geomorphology in Northern Rhodesia: an aspect of the distribution of the savanna of central Africa. *Geogr. J.*, 129: 290–310.

Coode, M.J.E. and Stevens, P.F., 1972. Notes on the flora of two Papuan mountains. *Proc. Papua New Guinea Sci. Soc.*, 1971, 23: 18–25.

Dasmann, R.F., 1973. A system for defining and classifying natural regions for the purposes of conservation. *I.U.C.N. Occas. Pap.*, 7, 48 pp.

Eden, M.J., 1974. The origin and status of savanna and grassland in southern Papua. *Trans. Inst. Br. Geogr., Publ.*, 63: 97–110.

Ehleringer, J.R., 1978. Implications for quantum yield differences on the distributions of C_3 and C_4 grasses. *Oecologia*, 31: 255–267.

Eiten, G., 1982. Brazilian savannas. In: B.J. Huntley and B.H. Walker (Editors), *Ecology of Tropical Savannas*. Springer-Verlag, Berlin, pp. 25–47.

Fitzpatrick, E.A., Hart, D. and Brookfield, H.C., 1966. Rainfall seasonality in the tropical southwest Pacific. *Erdkunde*, 20: 181–194.

Fosberg, F.R., 1960. The vegetation of Micronesia, I. General description, the vegetation of the Mariana Islands, and a detailed account of the flora of Guam. *Bull. Am. Mus. Nat. Hist.*, 119: 1–70.

Fosberg, F.R., 1963. Disturbance in island ecosystems. In: J.L. Gressitt (Editor), *Pacific Basin Biogeography*. 10th Pac. Sci. Congr., Univ. Hawaii (1961). Bishop Museum Press, Honolulu, pp. 557–561.

Gillison, A.N., 1969. Plant succession in an irregularly fired grassland area — Doma Peaks Region, Papua. *J. Ecol.*, 57: 415–428.

Gillison, A.N., 1970a. *Dynamics of Biotically Induced Grassland/Forest Transitions in Papua New Guinea*. M.Sc. thesis, Australian National University, Department of Forestry, Canberra, A.C.T., 145 pp.

Gillison, A.N., 1970b. Structure and floristics of a montane grassland/forest transition, Doma Peaks Region, Papua. *Blumea*, 18: 71–86.

Gillison, A.N., 1972. The tractable grasslands of Papua New Guinea. In: J. Ward (Editor), *Change and Development in Rural Melanesia*. 5th Waigani Seminar, University of Papua New Guinea and Australian National University, Port Moresby, pp. 161–172.

Gillison, A.N., 1975a. *Vegetation Types of the New Hebrides with Particular Reference to the Northern Islands*. (Unpubl. mimeo), 28 pp.

Gillison, A.N., 1975b. Phytogeography of the northern islands of the New Hebrides. *Philos. Trans. R. Soc. London*, B 272: 385–390.

Gillison, A.N., 1983. Tropical savannas of Australia and the southwest Pacific. In: F. Bourlière (Editor), *Tropical Savannas*. Ecosystems of the World 13, Elsevier, Amsterdam, pp. 183–243.

Gillison, A.N., 1988. A plant functional attribute proforma for dynamic vegetation studies and natural resource surveys. *CSIRO Aust. Div. Water Resour., Tech. Mem.* 88/3, Canberra, 32 pp.

Givnish, T.J., 1986. On the use of optimality arguments. In: T.J. Givnish (Editor), *On the Economy of Plant Form and Function*. Cambridge University Press, pp. 1–9.

Glassman, S., 1952. The flora of Ponape. *B.P. Bishop Mus. Bull.*, 209, 152 pp. (Repr. Kraus, New York, 5971).

Good, R., 1960. On the geographical relationships of the Angiosperm flora of New Guinea. *Bull. Br. Mus. Nat. Hist. Bot.*, 2: 205–226.

Good, R., 1963. On the biological and biophysical relationships between New Guinea and Australia. In: J.L. Gressitt (Editor), *Pacific Basin Biogeography*. 10th Pac. Sci. Congr., B.P. Bishop Museum Press, Honolulu, Hawaii, pp. 301–308.

Grubb, P.J., 1971. Interpretation of the 'Massenerhebung' effect on tropical mountains. *Nature*, 229: 44–45.

Guillaumin, A., 1952. Les caractères del la végétation néo-calédonienne. *C.R. Somm. Séances Soc. Biogéogr.*, 29: 82–86.

Haantjens, H.A., Mabbutt, J.A. and Pullen, R., 1965. Environmental influences in anthropogenic grasslands in the Sepik plains, New Guinea. *Pac. Viewpoint*, 6: 215–219.

Haantjens, H.A., Reynders, J.J., Mouthaan, W.L.P.J. and Van Baren, F.A., 1967. Major Soil Groups of New Guinea and Their Distribution. *Department of Agricultural Research, Communication 55*, Koninklijk Instituut voor de Tropen, Amsterdam, 87 pp.

Hancock, I.R. and Henderson, C.P., 1988. Flora of the Solomon Islands. *Ministry of Agriculture and Lands, Dodo Creek Research Station, Res. Bull.*, 7, Honiara, Solomon Islands.

Hartley, W., 1950. The global distribution of tribes of the Gramineae in relation to historical and environmental factors. *Aust. J. Agric. Res.*, 1: 355–373.

Hartley, W., 1958a. Studies on the origin, evolution and distribution of the Gramineae, I. The tribe Andropogoneae. *Aust. J. Bot.*, 6: 115–128.

Hartley, W., 1958b. Studies on the evolution and distribution of the Gramineae, II. The tribe Paniceae. *Aust. J. Bot.*, 6: 343–357.

Hattersley, P.W., 1983. The distribution of C_3 and C_4 grasses in Australia in relation to climate. *Oecologia*, 57: 113–128.

Hattersley, P.W., 1984. Characterization of C_4 type leaf anatomy in grasses (Poaceae). Mesophyll: bundle sheath area ratios. *Ann. Bot.*, 53: 163–179.

Hattersley, P.W. and Watson, L., 1976. C_4 grasses: an anatomical criterion for distinguishing between NADP-malic enzyme species and PCK or NAD-malic enzyme species. *Aust. J. Bot.*, 2: 297–308.

Hattersley, P.W., Watson, L. and Johnson, C.R., 1982. Remarkable leaf anatomical variations in *Neurachne* and its allies (Poaceae) in relation to C_3 and C_4 photosynthesis. *Bot. J. Linn. Soc.*, 84: 265–272.

Havel, J.J., 1960. Factors influencing the establishment of ligneous vegetation in mid-mountain pyro and anthropogenic grasslands. In: *Symp. on Impact of Man on Humid Tropics Vegetation*. Administration of T.P.N.G. and UNESCO Science Coop., Office for Southeast Asia, Goroka, pp. 119–122.

Henderson, C.P. and Hancock, I.R., 1989. *A Guide to the Useful Plants of the Solomon Islands*. Research Department, Ministry of Agriculture and Lands, Honiara, Solomon Islands.

Henty, E.E., 1969. A manual of the grasses of New Guinea. *Papua New Guinea Dept. For. Bot. Bull.*, 1, 215 pp.

Henty, E.E., 1982. Grassland and grassland succession in New Guinea. In: J.L. Gressitt (Editor), Biogeography and Ecology of New Guinea. *Monogr. Biol.*, 42: 459–473.

Heyligers, P.C., 1965. Vegetation and ecology of the Port Moresby–Kairuku Area. In: Lands of the Port Moresby–Kairuku Area, Papua New Guinea, *CSIRO Aust. Land Res. Ser.*, 14: 146–173.

Heyligers, P.C., 1967. Vegetation and ecology of Bougainville and Buka Islands. In: Lands of the Bougainville and Buka Islands, *CSIRO Aust. Land Res. Ser.*, 20: 121–145.

Heyligers, P.C., 1972a. Analysis of the plant geography of the semi-deciduous scrub and forest and eucalypt savanna near Port Moresby. *Pac. Sci.*, 26: 229–241.

Heyligers, P.C., 1972b. Vegetation and ecology of the Aitape–Ambunti Area. In: Lands of the Aitape–Ambunti Area, Papua New Guinea, *CSIRO Aust. Land Res. Ser.*, 30: 73–99.

Hnatiuk, R.J., 1975. *Aspects of the Growth and Climate of Tussock Grasslands in Montane New Guinea and Sub-Antarctic Islands*. Thesis, Australian National University, Canberra (quoted by G.S. Hope, 1980).

Holdridge, L.R., Grenke, W.C., Hatheway, W.H., Liang, T. and Tost, J.A., Jr., 1971. *Forest Environments in Tropical Life Zones: a pilot study*. Pergamon press, Oxford, 747 pp.

Holloway, J., 1979. A Survey of the Lepidoptera. In: *Biogeography and Ecology of New Caledonia*. Junk, The Hague, 588 pp.

Hope, G.S., 1976a. The vegetation history of Mt Wilhelm, Papua New Guinea. *J. Ecol.*, 64: 627–661.

Hope, G.S., 1976b. Vegetation. In: *The Equatorial Glaciers of New Guinea*. Balkema, Rotterdam, pp. 113–172.

Hope, G.S., 1980. New Guinea mountain vegetation communities. In: P. van Royen (Editor), *The Alpine Flora of New Guinea*. J. Cramer, Vaduz, 1: 153–222.

Jaffré, T., 1974. La végétation et la flore d'un massif des roches ultrabasiques de Nouvelle Caledonie: le koniambo. *Candollea*, 29: 427–456.

Jaffré, T. and Latham, M., 1974. Contribution à l'étude des relations sol–végétation sur un massif des roches ultra-basiques de la côte ouest de la Nouvelle Caledonie: le boulinda. *Adansonia*, Ser. 2, 14: 311–336.

Johns, R.J., 1982. Plant zonation. In: J.L. Gressitt (Editor), Biogeography and Ecology of New Guinea. *Monogr. Biol.*, 42: 309–330.

Johnson, R.W. and Tothill, J.C., 1985. Definition and broad geographic outline of savanna lands. In: J.C. Tothill and J.J. Mott (Editors), *Ecology and Management of the World's Savannas*. Australian Academy of Sciences, Canberra in association with C.A.B., Farnham Royal, Bucks., pp. 1–13.

Kalkman, C. and Vink, W., 1970. Botanical exploration in the Doma Peaks Region, New Guinea. *Blumea*, 18: 87–135.

Kellman, M., 1984. Synergistic relationships between fire and low soil fertility in neoptropical savannas: a hypothesis. *Biotropical*, 16: 158–160.

Kirkpatrick, J.B. and Hassall, D.C., 1981. Vegetation of the Sigatoka sand dunes, Fiji. *N.Z. J. Bot.*, 19: 285–297.

Köppen, W., 1918. Klassifikation der Klimate, nach Temperatur, Niederschlag, und Jahreslauf. *Petermanns Geogr. Mitt.*, 64: 193–203 and 243–248.

Lane-Poole, C.E., 1925a. *The Forest Resources of the Territories of Papua and New Guinea*. Government Printer, Melbourne, Vic., 209 pp.

Lane-Poole, C.E., 1925b. *Questionnaire by the British Empire Forestry Conference, 1923 regarding Papuan Forests*. (Unpubl. mimeo).

Lee, K.E., 1975. A discussion of the results of the 1971 Royal Society–Percy Sladen expedition to the New Hebrides: introductory remarks. *Philos. Trans. R. Soc. London*, B 272: 269–276.

Löffler, E., 1979. Origin and distribution of paramo grasslands in east New Guinea. *Erdkunde, Archive Wiss. Geogr.*, 33: 226–36 (in German).

Mallick, D.I.J., 1975. Development of the New Hebrides archipelago. *Philos. Trans. R. Soc. Lond.*, B 272: 277–285.

Manner, H.I., 1976. *The Effects of Shifting Cultivation and Fire on Vegetation and Soils in the Montane Tropics of New Guinea*. Ph.D. thesis, University of Hawaii. (Quoted by P. Bleeker, 1983).

McAlpine, J.R., Keig, G. and Falls, R., 1982. *Climate of Papua New Guinea*. CSIRO Australia and Australian National University Press, Canberra, 200 pp.

Mueller-Dombois, D., 1981. Vegetation dynamics in a coastal grassland of Hawaii. In: P. Poissonet, F. Romane, M.P. Austin, E. van der Maarel and W. Schmidt (Editors), *Vegetation Dynamics in Grasslands, Heathlands and Mediterranean Ligneous Formations*. Junk, The Hague. (Repr. from *Vegetatio*, 46: 131–140.)

Nix, H.A., 1974. Environmental control of breeding, post-breeding dispersal and migration of birds in the Australian region. In: *Proc. 16th Int. Ornithological Congress*. Australian Academy of Science, Canberra A.C.T., pp. 272–305.

Nix, H.A., 1982. Environmental determinants of biogeography and evolution in Terra Australis. In: W.R. Barker and P.J.M. Greenslade (Editors), *Evolution of the Flora and Fauna of Arid Australia*. Peacock publications in assoc. with Australian Systematic Botany Society and ANZAAS, South Australian Division Inc., pp. 47–66.

Nix, H.A., 1983. Climate of tropical savannas. In: F Bourlière (Editor), *Tropical Savannas, Ecosystems of the World 13*. Elsevier, Amsterdam, pp. 37–62.

Nix, H.A. and Gillison. A.N., 1985. Towards an operational framework for habitat and wildlife management. In: J. Kikkawa (Editor), *Wildlife Management in the Forests and Forestry — Controlled Lands in the Tropics and the Southern Hemisphere*. I.U.F.R.O., S1.08, Wildlife and its Habitats, pp. 39–55

Nix, H.A. and Kalma, J.D., 1972. Climate as a dominant factor in the biogeography of northern Australia and New Guinea. In: D. Walker (Editor), Bridge and Barrier — The Natural and Cultural History of Torres Strait. *Aust. National Univ. Dep. Biogeogr. Geomorphol. Publ.*, BG/3: 61–91.

Oliver, J., 1979. A study of geographical imprecision: the tropics. *Aust. Geogr. Stud.*, 17: 3–17.

Paijmans, K., 1967. Vegetation of the Safia–Pongani area. In: Lands of the Safia–Pongani Area, Papua New Guinea. *CSIRO Aust. Land Res. Ser.*, 17: 142–167.

Paijmans, K., 1969. Vegetation of the Kerema–Vailala area. In: Lands of the Kerema–Vailala Area, Papua New Guinea. *CSIRO Aust. Land Res. Ser.*, 23: 95–116.

Paijmans, K., 1971. Vegetation, forest resources and ecology of the Morehead–Kiunga area. In: Lands of the Morehead–Kiunga Area, Papua New Guinea. *CSIRO Aust. Land Res. Ser.*, 29: 88–113.

Paijmans, K., 1973. Plant succession on Pago and Witori volcanoes, New Britain. *Pac. Sci.*, 27: 26–268.

Paijmans, K., 1975. Explanatory Notes to the Vegetation Map of Papua New Guinea. *CSIRO Aust. Land Res. Ser.*, 35: 25 pp.

Paijmans, K., 1976. Vegetation. In: K. Paijmans (Editor), *New Guinea Vegetation*. CSIRO and Aust. National Univ. Press, Canberra, A.C.T, pp. 23–105.

Paijmans, K. and Löffler, E., 1974. High altitude forest and grasslands of Mt Albert Edward, New Guinea. *J. Trop. Geogr.*, 34: 58–64.

Parham, J., 1972. *Plants of the Fiji Islands*. Government Printer, Suva, rev. ed., 462 pp.

Prendergast, H.D.V., 1989. Geographical distribution of C4 acid decarboxylation types and associated structural variants in native Australian grasses (Poaceae). *Aust. J. Bot.*, 37: 253–273.

Reiner, E.J. and Robbins, R.G., 1960. The middle Sepik plains, New Guinea. A physiographic study. *Geogr. Rev.*, 54: 20–44.

Richards, P.W., 1966. *The Tropical Rain Forest* (3rd reprint). Cambridge University Press, 450 pp.

Robbins, R.G., 1960. The anthropogenic grasslands of Papua New Guinea. In: *Symposium on Impact of Man on Humid Tropics Vegetation*. Administration of T.P.N.G. and UNESCO Science Coop., Office for Southeast Asia, Goroka, pp. 313–329.

Robbins, R.G., 1964. The montane habit in the tropics. In: The Ecology of Man in the Tropical Environment. Proc. and Papers of the Ninth Techn. Meeting of I.U.C.N. *I.U.C.N. Publ.*, N.S., 4: 163–171.

Robbins, R.G., 1970. Vegetation of the Goroka–Mt Hagen Area. In: Lands of the Goroka–Mt Hagen Area, Papua New Guinea. *CSIRO Aust. Land Res. Ser.*, 27: 104–118.

Robbins, R.G. and Pullen, R., 1965. Vegetation of the Wabag–Tari area. In: Lands of the Wabag-Tari Area, Papua New Guinea. *CSIRO Land Res. Ser.*, 15: 100–115.

Rundel, P.W., 1980. The ecological distribution of C4 and C3 grasses in the Hawaiian islands. *Oecologia*, 45: 354–359.

Safford, W.F., 1905. The useful plants of Guam. *Conf. U.S. Natl. Herbariums*, 9: 1–416.

Schmid, M., 1975. La flore et la végétation de la partie meridionale de l'Archipel de Nouvelles Hébrides. *Philos. Trans. R. Soc. Lond.*, B 272: 329–342.

Seavoy, R.E., 1975. The origin of tropical grasslands in Kalimantan, Indonesia. *J. Trop. Geogr.*, 40: 48–52.

Shaw, D.F., 1968. Lighting strike of *Cacao* and *Leucaena* in New Britain. *Papua New Guinea Agric. J.*, 3: 75–84.

Simon, B.K. and Jacobs, S.W.L., 1990. Gondwanan grasses in the Australian flora. *Austrobaileya*, 3: 239–260.

Smith, A.C., 1979. *Flora Vitiensis Nova*, Vol. I. Pacific Tropical Botanical Garden, Lawai, Kauai, Hawaii, 495 pp.

Smith, J.M.B., 1975. Mountain grasslands of New Guinea. *J. Biogeogr.*, 2: 27–44

Smith, J.M.B., 1977. Vegetation and microclimate of east- and west-facing slopes in the grasslands of Mt Wilhelm, Papua New Guinea. *J. Ecol.*, 65: 39–53.

Smith, J.M.B., 1980. Ecology of the high mountains of New

Guinea. In: P. van Royen (Editor), *The Alpine Flora of New Guinea*, 1: 111–132.

Smith, J.M.B., 1982. An introduction to the history of Australasian vegetation. In: J.M.B. Smith (Editor), *A History of Australasian Vegetation*. McGraw-Hill, Sydney, pp. 1–31.

Taylor, B.W., 1964a. Vegetation in the Wanigela–Cape Vogel area: In: Lands of the Wanigela–Cape Vogel Area, Papua New Guinea. *CSIRO Aust. Land Res. Ser.*, 12: 69–83.

Taylor, B.W., 1964b. Vegetation of the Buna–Kokoda area. In: Lands of the Buna–Kokoda Area, Papua New Guinea. *CSIRO Aust. Land Res. Ser.*, 10: 89–98.

Taylor, B.W., 1968. Changes in the forest, savanna boundary in the per-humid lowland of New Guinea. In: T.L. Hills and R.E. Randall (Editors), The Ecology of the Forest/Savanna Boundary. *McGill Univ. Savanna Res. Ser.*, 13: 16–17.

Thompson, R.B. and Hackman, B.D., 1969. Some geological notes on areas visited by the Royal Society Expedition to the British Solomon Islands, 1965. *Philos. Trans. R. Soc. Lond.*, B 255: 189–202.

Thornthwaite, C.W., 1948. An approach towards a rational classification of climate. *Geogr. Res.*, 38: 55 — 94.

Twyford, I.T. and Wright, A.C.S., 1965. *The Soil Resources of the Fiji Islands,* I. Government of Fiji, Suva, pp. 77–86.

Udvardy, M., 1975. A classification of the biogeographical provinces of the world. *I.U.C.N. Occas. Pap.*, 8, 48 pp.

Van Balgooy, M.M.J., 1971. Plant geography of the Pacfic. *Blumea.*, (Suppl.) VI. 222 pp.

Van Royen, P., 1963a. *The Vegetation of the Island of New Guinea.* Department of Forests, Port Moresby, 70 pp. (unpublished mimeo).

Van Royen, P., 1963b. Notes on the vegetation of South New Guinea. *Sertulum Papuanum, Nova Guinea. Bot.*, 13: 195–241.

Van Royen, P., 1963c. Blanche Bay's volcanoseres. Trans. Papua New Guinea Sci. Soc., 4: 3–9. *Sertulum Papuanum* 9.

Van Steenis, C.G.G.J., 1958. Rejuvenation as a factor for judging the status of vegetation types: the biological nomad theory. In: Study of Tropical Vegetation. *Proc. Kandy Symp. UNESCO*, pp. 212–215.

Virot, R., 1956. La végétation Canaque. *Mém. Mus. Natl. Hist. Nat. N.S., B, Bot.*, 7: 1–398.

Vogel, J.C., Fuls, A. and Danin, A., 1986. Geographical and environmental distribution of C_3 and C_4 grasses in the Sinai, Negev, and Judean deserts. *Oecologia*, 70: 213–265. (Cited by Prendergast, 1989.)

Wade, L.K. and McVean, D.N., 1969. Mt Wilhelm studies, I. The alpine and sub-alpine vegetation. *Res. School Pac. Stud., Dep. Biogeogr. Geomorphol. Aust. Natl. Univ. Publ.*, B/G 1.

Walker, D., 1966. Vegetation of the Lake Ipea region, New Guinea highlands, I. Forest, grassland and 'garden'. *J. Ecol.*, 54: 454–456.

Walker, D. and Flenley, J.R., 1979. Quaternary vegetation history of the Enga Province of upland Papua New Guinea. *Philos. Trans. R. Soc. London*, B, 286: 265–344.

Whitmore, T.C., 1969. The vegetation of the Solomon Islands. *Philos. Trans. R. Soc. London*, B, 255: 259–270.

Whitmore, T.C., 1975. *Tropical Rainforests of the Far East.* Clarendon Press, Oxford, 282 pp.

Whyte, R.O., 1968. *Grasslands of the Monsoon.* Faber and Faber, London, 325 pp.

Chapter 17

REVIEW

R.T. COUPLAND

INTRODUCTION

The objective in this final chapter is to develop an overview of certain aspects of structure and function in grassland ecosystems based on the data provided by authors of the regional chapters (Sections II to VI). This supplements extensive reviews of processes that appear in the introductory chapters (Section I) of Part A and in some of the regional chapters. The reader is referred to the other parts of this volume for these comprehensive overviews; the data summaries and generalizations in this chapter are intended to assist in an understanding of the activity at the various trophic levels in grassland. The overviews in Chapters 2 to 8 of Section I (Part A) were prepared by authors who did not have access to the regional chapters. These involve the following subjects:

   Climate (pp. 7–23)
   Soils (pp. 25–52)
   Water Flow (pp. 55–71)
   Primary production (pp. 75–91)
   Micro-organisms (pp. 95–115)
   Decomposition (pp. 121–133)
   Nutrient Cycling (pp. 137–146)

Only occasional reference will be made to these introductory overviews in the following discussion. Rather, emphasis will be placed on identifying passages in Sections II to VI in which supplemental information may be found concerning the above subjects, and in which overviews occur of additional ones.

In this review, biomass values for plants are expressed as g m^{-2}, whereas those for animals and micro-organisms are in kg ha^{-1}. Conversion of the latter values to the former can be made by dividing by 10.

VEGETATION STRUCTURE AND BIOMASS

Biomass values for each of the components that comprise the total standing crop of the vegetation are summarized for various grasslands in Table 17.1. Total plant biomass for the sites considered here ranges from less than 1 kg m^{-2} to more than 8 kg m^{-2}. Care must be taken in making direct comparison of the totals between sites for the reasons stated in the footnotes to the table. Other factors that prevent direct comparison are probable differences in the effect of use — particularly for grazing — and variations — during the years of study — in weather factors from the norm.

Several references are made in this volume to the "root"/shoot ratio (really the ratio of underground to above-ground parts) in characterizing the relative distribution of biomass above and under ground. In some instances surface litter is included as part of the above-ground biomass, in others it is excluded; the values for root/shoot ratio in Table 17.1 do not include litter.

Biomass of surface litter varies widely among sites. In semiarid to dry-subhumid climates an equilibrium is attained — even in the absence of livestock — where litter biomass often does not exceed annual above-ground production. However, in some subhumid regions accumulations occur that are sufficient to impede growth (Part A, pp. 243–244). A discussion of the role of litter in annual grassland appears in Part A, pp. 331–332.

No overall simple pattern is evident from these biomass data. However, some observations in the text provide interpretations for some of the differences in the distribution between above-ground and under-ground components. The root/shoot ratio for the flooding pampa in Argentina is con-

TABLE 17.1

Maximum values (g m^{-2} of dry matter) for components of the standing crop of plant biomass in various grasslands [1]

Grassland type Location	Canopy			Litter	Under-ground parts [2]	Total plant biomass	Root/ shoot ratio	Text reference	
	Green	Dead	Total					Part	Location
Mixed grass community, Pilani, India	76	27	103	31	86	220	0.8	B	Table 4.3
Grassland on shallow loam, South Africa			208		71	279	0.3	B	p. 276
Desert grassland, New Mexico, USA			105	44	225	374	2.1	A	p. 349
Grassland on sand, Nylsvley, South Africa			171		226	397	1.3	B	p. 276
Ungrazed grassland, Khirasara, India	201	178	379	40	205	624	0.5	B	Table 4.3
Sehima community, Ratlam, India	363	316	679	275	873	1827	1.3	B	Table 4.3
Sehima community, Jhansi, India			1408	226	333	1967	0.2	B	Table 4.3
Mixed grass community, Kurukshetra, India	424	306	730	231	1040	2001	1.4	B	Table 4.3
Dichanthium grassland, Ujjain, India	457	422	879	423	925	2227	1.1	B	Table 4.3
Cottonwood, South Dakota, USA	184	210	301	452[3]	1520[3]	2273	5.0	A	Tables 10.2, 10.11
Agrostis mire, Marion Island			639		2024	2663	3.2	B	Table 15.3
Heteropogon community, Sagar, India	572	518	1090	433	1381	2904	1.3	B	Table 4.3
Flooding pampa, Argentina	222	640	862	140[3]	1956[3]	2958	2.3	A	pp. 388–391
Mixed prairie, Matador, Canada	131	504	560	268[3]	2383[3]	3211	4.8	A	Tables 10.2, 10.11
Desmostachya upland, Varanasi, India			2360	145	788	3293	0.3	B	Table 4.3
Hays, Kansas, USA	222	234	278	1251[3]	1934	3463	7.0	A	Tables 10.2, 10.11
Dickinson, North Dakota, USA	192	504	672	797[3]	2168[3]	3637	3.2	A	Tables 10.2, 10.11
Deschampsia tussock grassland, New Guinea			3431		421	3852	0.1	B	Table 15.3
Festuca community, South Georgia			2535	140	1642	4317	1.7	B	Table 15.3
Sesbania community, Kurukshetra, India	1921	1440	3361	331	900	4592	0.3	B	Table 4.3
Desmostachya community, Kurukshetra, India	838	740	1578	227	2868	4673	1.8	B	Table 4.3
Mixed grass community, Kurukshetra, India	1974	1268	3242	300	1167	4709	0.4	B	Table 4.3
Eragrostis lowland, Varanasi, India			3296	152	1282	4730	0.4	B	Table 4.3
Chionochloa tussock grassland, Campbell Island			2717		2322	5039	0.9	B	Table 15.3
Poa tussock grassland, Marion Island			2574		3236	5810	1.3	B	Table 15.3
Poa community, crest, Marion Island			4458		2001	6459	0.4	B	Table 15.3
Tussock grassland, Macquarie Island			3510	101	4800	8411	1.4	B	Table 15.3

[1] The list is in order of increasing total biomass. Where two or more values appear in the text, they have been averaged. The totals are not all comparable with one another. Some canopy totals are maximum canopy values and others are sums of green and dead components; some may include litter. "Total plant biomass" is the sum of maximum above-ground and maximum under-ground values; litter may not be included in some instances.

[2] The depth of sampling of under-ground parts is usually not indicated; the values quoted often may not include deeper layers, where roots are not abundant. Some depths of sampling are: Hays, 15 cm; Cottonwood, 60 cm; Argentina, 70 cm; Dickinson and the two South African sites, 100 cm; Matador, 150 cm.

[3] Ash-free values: ash content of surface litter and under-ground biomass is probably in the range of 20 to 25%.

sidered to be "characteristic of warmer and more humid grasslands, which generally range between 2 and 6. Cooler and drier sites have higher ratios, between 6 and 13" (Part A, p. 390). In the discussion of steppe of northern Eurasia, the biomass of under-ground parts is reported to be much greater than that of the canopy — with root/shoot ratios ranging from 8 in the cool, subhumid north to 30 or more in the desert grassland of Mongolia (Part B, p. 22). An extensive discussion of the relative distribution of biomass above and below the soil surface is presented in relation to the Indian grasslands (Part B, p. 98 and Figs. 4.4 to 4.6). The authors suggest that where growth is retarded by low annual rainfall, even a slight increase in rainfall promotes shoot growth and lateral spread of annuals. Under moderate rainfall, annuals are replaced by perennials with their storage organs under ground — resulting in an increased proportion of under-ground material. However, this proportion does not increase under higher rainfall because a more even balance is attained between growth of shoots and under-ground parts. It is suggested that the relatively very low root/shoot ratios — 0.32 to 1.39 — reported for the tussock grasslands of New Zealand result from a comparatively long growing season and mild winter.

Many references have been made to the concentration of plant biomass in the uppermost layers of soil (Table 17.2) — the zone of greatest interaction among plant species, of the deposition and translocation of reserves, of vegetative reproduction, and of the greatest activity of macro-fauna and rhizosphere micro-flora (Part B, pp. 147, 153). Usually half or more of the under-ground biomass occurs within the 10-cm layer below the soil surface. The proportion of under-ground plant biomass near the soil surface increases with increasing aridity of climate and shallowness of the soil profile (Part B, p. 18). In semi-natural meadows the roots do not penetrate as deeply as in the steppe, where the moisture supply is more limited (Part B, p. 18). In the pampa of Argentina very few roots occur at depths greater than 70 cm. In meadow steppe of the former Soviet Union "only solitary roots penetrate deeper than 75 cm".

The vertical distribution of biomass in the canopy and its effect on the radiation profile is discussed in Part B, p. 135. Although some of the grasses attained a height of 60 to 70 cm, 42% of the biomass of these species was concentrated within 10 cm of the soil surface (Table 5.5). Stands composed mainly of tall grasses display a considerable uniformity between layers in the distribution of biomass. In stands dominated by tufted grasses or herbs, most of the biomass is concentrated in the lower layers and its density diminishes rapidly towards the upper levels of the canopy.

Detailed descriptions of layering and structure of under-ground parts appear in Part B, pp. 21–22 and 136–137. The same authors have referred to the extensive surface area of the root system in relation to mass, a characteristic that is particularly important under conditions of low soil moisture which often occur in grassland.

TABLE 17.2

Percentage of the biomass of under-ground parts located in the upper layers of soil in various grasslands [1]

Grassland type / Location	Soil layer (cm)					Text reference	
	0–10	0–15	0–20	0–30	0–50	Part	Location
Flooding pampa, Argentina	65			85		A	p. 390
Short-grass steppe							
Pawnee site, Colorado, USA	55					A	p. 194
Pantex site, Texas, USA	65		75			A	p. 194
Mixed prairie							
Matador, Matador, Canada	41		53	61	74	A	Table 10.12
Dickinson, North Dakota, USA	61		71	78	86	A	Table 10.12
Cottonwood, South Dakota, USA	55		70	80	95	A	Table 10.12
Fescue prairie, Canada		73				A	p. 292
Meadow steppe, former USSR					89–94	B	p. 21
Forb–bunch-grass steppes, former USSR				76–80		B	p. 21

[1] These values are not directly comparable because of different depths of sampling in the various sites, relative to the maximum depth of rooting.

TABLE 17.3

Maximum standing crop (g m^{-2} of dry matter) of herbage in the canopy of various grasslands

Vegetation type Location	Green biomass	Dead biomass	Total biomass	Text reference Part	Location
NORTH AMERICA					
Mixed prairie	86–222	189–634	278– 672	A	Table 10.2
Short-grass steppe			47– 225	A	Table 11.5
Tall-grass prairie			258– 544	A	Table 12.11
Coastal prairie					
upland sites			500– 750	A	p. 287
lowland sites			600–1000	A	p. 287
Palouse prairie			99– 368	A	pp. 299–302, 307, 310, 311
California prairie			106–562	A	p. 327
Desert grassland			1– 120	A	p. 349
SOUTH AMERICA					
Río de la Plata grasslands	210–222	630	840	A	p. 388
EURASIA					
Meadow steppe (former USSR)			320– 420	B	p. 22
Xerophytic desert steppe (former USSR)			6– 30	B	p. 22
Kazakhstan steppe (former USSR)			62– 180	B	p. 38
Mongolian desert steppe			6– 29	B	p. 38
Meadow steppe (China)			150– 250	B	p. 65
Typical (dry) steppe (China)			80– 100	B	p. 67
Desert steppe (China)			less than 100	B	p. 67
Shrub steppe (China)			ca. 50	B	p. 68
Alpine steppe (China)			20– 35	B	p. 69
Meadows (China)			100– 370	B	Table 3.2
Northern semi-natural meadows			180–1041	B	Table 5.9
Semiarid grassland (India)	46			B	p. 99
Dry subhumid grassland (India)	1974			B	p. 99
Subhumid grassland (India)	3296			B	p. 99
Alpine grassland (India)	382–409			B	p. 99
Highland grassland (Thailand)			419	B	p. 102
AFRICA					
Saharo–Sahelian steppe			15– 60	B	p. 208
Sahelian steppe			50– 250	B	p. 208
flooded meadows			500–1500	B	p. 208
Soudano–Sahelian steppe			80– 300	B	p. 208
Eastern giant-grass region			488–1600	B	p. 229
Eastern tall-grass region			300–2000	B	p. 231
Eastern mid-grass region			117– 750	B	pp. 242, 254
Eastern annual short-grass region			30– 94	B	p. 256
Southern desert annual grasslands			0– 100	B	p. 274
Southern semiarid open grassland			37– 125	B	Table 10.1
Southern savannas			65– 264	B	Table 10.1
Southern fire-induced grasslands			50– 500	B	p. 275
Southern high-veld grassland			63– 230	B	p. 275, Table 10.1
moist vleis	450	500	950	B	p. 275
Colder regions at high elevations			100– 200	B	p. 275
OCEANIA					
Tropical tussock grassland (New Guinea)	397	2978	3375	B	Table 15.3
Hummock grassland (Australia)			94	B	p. 320
Northern tussock grassland (Australia)			125– 195	B	p. 322
Shrub–short-grass (Australia)			122– 132	B	p. 328
Tropical tall-grass (Australia)			112– 750	B	p. 338
Tussock grassland (New Zealand)			1500–8700	B	p. 387
Sub-Antarctic tussock grassland	322–899	940–2582	1258–4458	B	Table 15.3

PLANT PRODUCTION

Estimates of above-ground plant production are most frequently expressed in terms of maximum standing crop of herbage in the canopy, often referred to as "yield", "hay yield" or "standing crop at the end of the growing season". These measurements do not take into account loss to leaf senescence and fall to litter — processes that occur during the growing season. Values that are quoted throughout the text are summarized in Table 17.3. Care must be taken in making direct comparisons among these values, because of differences in sampling techniques. The height of clipping is usually not mentioned and cannot always be assumed to be at the level of the soil surface. In rangeland research, herbage yield above the height of grazing has been considered sometimes to be more relevant. Another shortcoming of most yield data is that no information is available concerning the proportion of dead material — and particularly how much of this was contributed by growth during the current growing season. Some of the estimates in Table 17.3 represent a single harvest at one site, whereas others are the range of values in one or more sites — sometimes over many years of investigation.

Canopy biomass varies widely from year to year at the same site — in response to the amount of available soil moisture in drier regions and to temperature in moister regions. Differences from site to site are related to differences in climate and to the interaction between climatic, biotic and edaphic factors. Seasonal distribution of precipitation may be more important in determining yield of herbage than total annual precipitation.

Estimates of net annual herbage production based on analyses of the dynamics of the various components of the canopy throughout the growing season provide values that are sometimes several times as great as those based on maximum standing crop — even including dead material judged to have been produced in the current growing season. The magnitude of differences between maximum green standing crop of herbage and estimated net annual above-ground production is illustrated in Table 17.4. These estimates of net production are based on several procedures and combinations of procedures for interpreting biomass data (Part A, Table 5.3). In some

TABLE 17.4

A comparison between the maximum biomass of green shoots and estimated net annual production of biomass in several grasslands (in g m^{-2})

Vegetation type Location	Maximum standing crop of green shoots	Net production of above-ground biomass	Text reference Part	Location
Mixed prairie				
Matador, Canada	86–187	255–639	A	pp. 163, 167
Dickinson, North Dakota, USA	192	580[1]	A	Tables 10.2, 10.10
Cottonwood, South Dakota, USA	164	450[1]	A	Tables 10.2, 10.10
Cottonwood, South Dakota, USA	205	398	A	Tables 10.2, 10.3
Hays, Kansas, USA	222	422[1]	A	Tables 10.2, 10.10
Flooding pampa, Argentina	222	532	A	p. 388
Mixed grass, Pilani, India	76	217	B	Table 4.3
Ungrazed grassland, Khirasara, India	201	201	B	Table 4.3
Sehima community, Ratlam, India	363	433	B	Table 4.3
Mixed grass, Kurukshetra, India	424	617	B	Table 4.3
Dichanthium community, Ujjain, India	457	520	B	Table 4.3
Heteropogon community, Sagar, India	572	914	B	Table 4.3
Desmostachya community, Kurukshetra, India	838	862	B	Table 4.3
Sesbania community, Kurukshetra, India	1921	2143	B	Table 4.3
Mixed grass, Kurukshetra, India	1974	2407	B	Table 4.3

[1] Ash-free values, which are probably about 8% below those containing ash.

instances biomass analysis was supplemented by the use of individual leaves or tillers as the unit of measurement (Part A, pp. 82–83, 167–169; Part B, pp. 390–392, 424) and the use of ash content — or a component of ash — of dead material as a marker to estimate the biomass of green material before deterioration (Part A, pp. 80, 167) — and of various other correction factors for senescence and decay. In some studies the period of measurement extended beyond the date of maximum standing crop.

Many of the values in Table 17.4 probably underestimate net production because only one or two of the above techniques were applied in analysis. Net production is more difficult to assess in grasslands with a long growing season, during which considerable senescence occurs in the canopy before maximum standing crop is attained. Therefore, the values for India in Table 17.4 probably underestimate net production to a greater extent than do those from more temperate regions.

Table 17.4 includes only instances from the text where values for net production are based on analysis beyond a single harvest and where values for maximum standing crop of green shoots are available for comparison. Several other authors refer to the relationship between these values. For example, in the Soudano–Sahelian subregion of the Sahel the sum of increments in the herbaceous layer is reported to be 25 to 36% greater than the maximum standing crop "since considerable biomass fails to accumulate due to rapid decay during the rainy season and to consumption by insects" (Part B, p. 208).

Rates of daily additions to herbage biomass are quoted for several grasslands (Part A, pp. 242, 388; Part B, pp. 102, 242, 328, 391, 424, Fig. 5.18 and Table 5.10). Average values for the growing season range from 0.25 to 3.0 g m^{-2}. Peak values for shorter periods are as high as 40 g m^{-2} day^{-1} (Part B, p. 242).

Reference is made to attempts to quantify the relationship between variations in weather and year-to-year variations in maximum standing crop and net production (Part A, pp. 194–196; Part B, pp. 391–393, 424 and Figs. 4.6 and 4.7).

Estimates of under-ground production (Table 17.5) are usually based on the difference between maximum and minimum standing crop (Part A, p. 84). They must be considered as minimum values, since losses by senescence and decomposition are not taken into account.

Values for total net primary production are calculated by summing estimates of above-ground and under-ground production, which are determined by different means. Estimates of under-ground production are compared with those of above-ground production in Table 17.5. A functional relationship has been shown between total net production and annual rainfall for certain Indian grasslands (Part B, Fig. 4.6). Net dry-matter output exponentially increases with increasing rainfall in semiarid to dry-subhumid regions, where water is the factor limiting growth. The relationships between stability, biomass production and standing crop are examined in Part B, Fig. 4.7. The proportion of produced materials that are transported to under-ground parts (Table 17.5) is a measure of stability.

Detailed discussions of photosynthesis and of carbon and energy flows through the plant components of grasslands are presented in Part A, pp. 88–91 and Part B, pp. 142–143 and 391–393.

For information on the floristic structure of grassland the reader is directed to the lengthy descriptions in each of the regional chapters.

ANIMAL BIOMASS

The values reported for biomass of native fauna are for the most part fragmentary. Except for two locations in North America, the data quoted in this volume do not represent an attempt to analyze the total animal biomass in the ecosystem. Most of the values presented are described as "live weight" or "fresh weight" and need to be converted to dry weight to be readily applicable to ecosystem analysis. The summary in Table 17.6 is intended to lead the reader to passages in the text where faunal biomass is discussed. Direct comparison from the table should not be made without reference to the text. Some entries in the table are peak seasonal values, whereas others are means of several estimates made at various times. Some are evaluations of the recent situation in the presence of domestic herbivores; others are based on estimates of probable animal populations in the past. The ranges suggest differences from season

TABLE 17.5

Estimates of annual net primary production (g m^{-2} of biomass) in various grasslands [1]

Grassland type Location	Net production			Percentage transferred under-ground	Text reference	
	Above-ground	Under-ground	Total		Part	Location
Mixed grass community, Pilani, India	217	61	278	22	B	Table 4.3
Plains community, Khirasara, India	201	155	356	44	B	Table 4.3
Apluda community, Udaipur, India	256	255	511	50	B	Table 4.3
Arrhenatherum community, Poland	253	336	589	57	B	Table 5.7
Alchemilla community, (former USSR)	263	399	662	60	B	Table 5.7
Festuca community, West Germany	316	371	687	54	B	Table 5.7
Short-grass prairie, Pawnee site, Colo., USA	137	554	691	80	A	Table 11.5
Mixed prairie, Cottonwood, S.D., USA [2]	433	269	702	38	A	Table 10.10
Mixed prairie, Hays, Kansas, USA [2]	422	288	710	41	A	Table 10.10
Mixed grass community, Jodhpur, India	164	570	734	78	B	Table 4.3
Alchemilla community, (former USSR)	347	478	825	58	B	Table 5.7
Sehima community, Ratlam, India	433	399	832	48	B	Table 4.3
Festuca community, South Georgia	492	350	842	42	B	Table 15.4
Mixed prairie, Cottonwood, S.D., USA [2]	450	478	928	52	A	Table 10.10
Short-grass prairie, Amarillo, Texas, USA	241	710	951	75	A	Table 11.5
Chionochloa community, New Zealand	550	406	956	42	B	Table 14.1
Mixed prairie, Dickinson, N.D., USA [2]	580	391	971	40	A	Table 10.10
Dichanthium community, Ujjain, India	520	464	984	47	B	Table 4.3
Mixed grass community, Ambikapur, India	436	563	999	56	B	Table 4.3
Flooding pampa, Argentina	532	496	1028	48	A	pp. 388, 390
Chionochloa community, New Zealand	628	422	1050	40	B	Table 14.1
Nardus community, Czechoslovakia	268	810	1078	75	B	Table 5.7
Mixed prairie, Matador, Canada	447	677[2]	1124	60	A	p. 167, Table 10.13 [3]
Chionochloa community, New Zealand	800	435	1235	35	B	Table 14.1
Festuca community, Czechoslovakia	957	389	1346	29	B	Table 5.7
Trisetum community, Poland	700	697	1397	50	B	Table 5.7
Mixed grass community, Kurukshetra, India	617	785	1402	56	B	Table 4.3
Alopecurus community, Czechoslovakia	1016	407	1423	29	B	Table 5.7
Sehima community, Jhansi, India	1019	497	1516	33	B	Table 4.3
Heteropogon community, Sagar, India	914	937	1851	51	B	Table 4.3
Brachypodium community, Poland	500	1367	1867	73	B	Table 5.7
Cynodon community, Behrampur, India	571	1361	1932	70	B	Table 4.3
Phalaris community, Czechoslovakia	1388	554	1942	29	B	Table 5.7
Andropogon community, Sambalpur, India	458	1972	2430	81	B	Table 4.3
Desmostachya community, Kurukshetra, India	862	1592	2454	65	B	Table 4.3
Sesbania community, Kurukshetra, India	2143	998	3141	32	B	Table 4.3
Mixed grass community, Kurukshetra, India	2407	1131	3538	32	B	Table 4.3
Desmostachya upland, Varanasi, India	2218	1377	3595	38	B	Table 4.3
Eragrostis lowland, Varanasi, India	3396	1161	4557	25	B	Table 4.3
Poa community, Macquarie Island	1911	3670	5581	66	B	Table 15.4

[1] The sites are listed according to increasing estimated net production. Where more than one estimate is available, the values are averaged.
[2] Ash-free values.
[3] Supplemented by original data.

to season, and differences in concentration as a result of habitat characteristics and movements of groups of large ungulates. Domesticated livestock are not considered in this summary.

The most detailed analysis reported of the faunal components of an ecosystem is that made at the Pawnee Site of the USIBP Grassland Biome in the short-grass steppe of Colorado (Part A, Tables

TABLE 17.6

Biomass (kg ha^{-1}) of various groups of fauna in grasslands [1]

Type of fauna and location	Biomass	Text reference Part	Location
DRY WEIGHT			
VERTEBRATES			
Rodents:			
Matador, Canada	0.005	A	p. 173
Cottonwood, S.D., USA	0.03–0.04	A	p. 173
Dickinson, N.D., USA	0.10	A	p. 173
Birds:			
Matador, Canada	0.02	A	p. 173
Cottonwood, S.D., USA	0.07	A	p. 174
INVERTEBRATES			
Invertebrates, above-ground:			
Cottonwood, S.D., USA	1.2	A	p. 174
Osage, Okla., USA	1.9	A	Fig. 12.11
Matador, Canada	3.0	A	p. 174
Invertebrates, under-ground: Matador, Canada	64	A	p. 174
Herbivorous soil macrofauna: France	1.3–35	B	p. 147
Nematodes: Matador, Canada	48	A	p. 174
Oligochaetes: Sambalpur, Orissa, India	110	B	p. 106
Enchytraeids: Matador, Canada	9.6	A	p. 174
Lumbricidae: Bródno, Poland	56–263	B	Table 5.16
Arthropds: under-ground: Matador, Canada	6.4	A	p. 174
Armadillidium vulgare: California, USA	6–490	A	p. 324
Insects:			
India	59	B	p. 106
Wisła valley, Poland	1.4–3.3	B	p. 147
Coleoptera and Lepidoptera larvae: Bródno, Poland	3–18	B	Table 5.16
Grasshoppers: Matador, Canada	0.6	B	p. 174
FRESH (LIVE, WET) WEIGHT			
VERTEBRATES			
Forest steppe, former USSR	13	B	p. 23
Desert steppe, former USSR	6.0	B	p. 23
Large herbivores:			
Leptothrium mid-grass region, E. Africa	0.28	B	p. 254
Sahel region, N. Africa	0.01–1.9	B	p. 210
Aristida annual short-grass region, Kenya	0.7–3.0	B	p. 256
Hyparrhenia tall-grass region, E. Africa	133	B	p. 234
Chrysopogon mid-grass region, E. Africa	8–180	B	p. 248
Themeda mid-grass region, Uganda	51–365	B	p. 240
Antilocapra americana, Pawnee site, Colo., USA	0.14	A	Table 11.8
Antilope cervicapra, India	60–96	B	p. 104
Axis axis, India	11	B	p. 104
Axis porcinus, India	7.2–8.0	B	p. 104
Cervus duvauceli duvauceli, India	3.6–72	B	p. 104
Mammalian carnivores: Pawnee site, Colo., USA	0.03	A	Table 11.8
Lagomorphs:			
Pantex site, Texas, USA	0.05	A	Table 11.6
Pawnee site, Colo., USA	0.25	A	Table 11.8
Rodents:			
Pawnee site, Colo., USA	0.37	A	Table 11.8
Pantex site, Texas, USA	0.60	A	Table 11.6
Arvicanthis niloticus, *Themeda* mid-grass region, E. Africa	309	B	p. 241

(*(continued)*)

TABLE 17.6 (continued)

Type of fauna and location	Biomass	Text reference	
		Part	Location
VERTEBRATES (continued)			
Birds:			
Pawnee site, Colo., USA	0.10	A	Table 11.8
Pantex site, Texas, USA	0.11	A	Table 11.6
Snakes: Pawnee site, Colo., USA	0.12	A	Table 11.8
INVERTEBRATES			
Soil invertebrates: semi-desert steppe, former USSR	0.20–0.30	B	p. 22
Nematodes:			
San Joaquin, California	0.40–1.17	A	p. 324
Cottonwood, S.D., USA	20–71	A	p. 175
Pawnee site, Colo., USA	29	A	Table 11.8
Bródno, Poland	1.6–37	B	Table 5.16
Enchytraeids:			
Behrampur (Orissa), India	2.4	B	p. 106
Bródno, Poland	4.7–5.2	B	Table 5.16
Acarina: Bródno, Poland	0.82–3.7	B	Table 5.16
Arthropods:			
Above-ground, California, USA	12	A	p. 324
Under-ground, California, USA	1260	A	p. 324
Macroarthropods: Pawnee site, Colo., USA			
Above-ground	4.0	A	Table 11.8
Under-ground	20	A	Table 11.8
Microarthropods: Pawnee site, Colo., USA	15	A	Table 11.8
Grasshoppers: *Chionochloa* communities, New Zealand	26	B	p. 396

[1] The values in this table have been converted to a uniform base from the various ones used by different authors. Hopefully, no errors have been made in conversion.

11.6 and 11.8). The biomass values quoted for that location may be summarized as follows (kg ha^{-1} live weight):

Vertebrates
    Large mammal: *Antilocapra americana* 0.14
    Mammalian carnivores 0.03
    Lagomorphs 0.25
    Rodents 0.37
    Birds 0.10
    Reptiles 0.12   1.01
Invertebrates
    Macroarthropods 24
    Microarthropods 15
    Nematodes 29   68

Details of the analysis at the Canadian IBP study site are not presented. The summary information quoted is as follows (kg ha^{-1} dry matter):

Rodents 0.005
Birds 0.02
Invertebrates, above ground 3.0
Invertebrates, under ground 64

In most grassland regions the major part of the native mammalian biomass was evidently contributed by large herbivores. Several references are made to the pronounced decline in populations of large wild ungulates after the expansion of the livestock industry. In many regions they are now rare or absent. However, they still comprise a significant portion of the total ungulate biomass in East Africa; the lower values (Table 17.6) for wild ungulates in East Africa are for areas where livestock comprise the major part of the grazing load, whereas the high values are for situations where livestock are less numerous or are absent. The highest biomass values for wild ungulates — as great as 365 kg ha^{-1} (live weight) in East Africa and a hypothetical estimate of 270 kg ha^{-1} (live weight) in the tall-grass prairie of North America — suggest that in places, at certain times, stress on the vegetation may have been similar to that at present from livestock. The capacity of grasslands to support large ungulates is much less in some regions. For example, in the short-grass steppe of north-eastern Colorado the stocking rate of cattle is reported to range from 27 to 76 kg ha^{-1} (live weight). Lagomorphs and rodents also contribute to the pressure on grassland ecosystems. *Lepus* spp. are reported to be equivalent to 2.1% of the ungulate biomass in the Serengeti plains (East Africa) during the dry season (Part B, p. 241). An extensive discussion of the effect of rodents on the grassland in the former Soviet Union is presented in Part B, pp. 42–55.

Invertebrate biomass is located mostly within the soil and is of the order of 100 to 1000 times as great as vertebrate biomass. Nematodes, enchytraeids and Acarina are the major groups under ground, whereas Orthoptera commonly predominate in biomass above the soil surface. Total faunal biomass — on a dry-weight basis — appears to be of the order of a hundredth that of plant biomass.

MICRO-ORGANISMS

The biomass of micro-organisms is of the order of 10% of that of biomass of higher plants (Table 17.7). The greatest concentrations of micro-organisms are in the soil — near the surface (Part A, p. 175, and Part B, p. 40) and in the rhizosphere (Part B, Table 2.23). Less than 1% of microbial biomass occurs above ground. Microbial biomass is greater in habitats more favourable for plant growth.

The relative abundance of the various taxa and functional groups of micro-organisms in relation to habitat, vegetation type, and soil depth is discussed in Part A, pp. 175, 212, 255–257, and Part B, pp. 24–25, 40–41, 106–107. Fungi, Bacteria and Actinomycetes contribute 99% or more of the biomass, with that of Fungi constituting 60% or more.

TABLE 17.7

Biomass (kg of dry matter ha^{-1}) of micro-organisms in various grasslands[1]

Habitat Location	Biomass	Text reference	
		Part	Location
All micro-habitats			
Protozoa, to 5 cm in soil: Pawnee site, Colo., USA	3.0	A	Table 11.7
Bacteria, to 30 cm in soil: Pawnee site, Colo., USA	267	A	Table 11.7
Fungi, to 30 cm in soil: Pawnee site, Colo., USA	380	A	Table 11.7
In soil			
All micro-organisms: 0–10 cm, Kazakhstan	390–2180	B	Table 2.22
Fungi: 0–30 cm, Matador, Canada	1770	A	p. 175
Bacteria and Actinomycetes: 0–30, Matador, Canada	460	A	p. 175
Bacteria: 0–10 cm, Czechoslovakia	30–9400	B	p. 151
Protozoa: 0–30 cm, Matador, Canada	0.69	A	p. 175
Yeasts: 0–30 cm, Matador, Canada	0.07	A	p. 175
On under-ground parts			
Bacteria and Actinomycetes: Matador, Canada	1.9	A	p. 175
On surface litter			
Fungi: Matador, Canada	6.3	A	p. 175
Bacteria and Actinomycetes: Matador, Canada	0.02	A	p. 175
On dead leaves in the canopy			
Fungi: Matador, Canada	6.6	A	p. 175
Bacteria and Actinomycetes: Matador, Canada	0.05	A	p. 175
On green leaves in the canopy			
Fungi: Matador, Canada	0.70	A	p. 175
Bacteria and Actinomycetes: Matador, Canada	0.15	A	p. 175

[1] The values in this table have been converted to a uniform base from the various ones used by different authors. Hopefully, no errors have been made in conversion. These values are not comparable between studies because of differing methods of estimation, particularly by direct counts as compared to plate counts.

ACTIVITY OF HETEROTROPHS

The roles and activities of heterotrophs in reduction and mineralization of plant materials is discussed in considerable detail in Part A, Chapters 6, 7, 8 and 11 (pp. 196–220). Many other passages in the volume deal with certain aspects of these activities in paragraphs relating to consumption, decomposition, energy flow, nitrogen fixation, nutrient cycling and turnover. Consumers apparently account for much less of the degradation of organic materials than do decomposers. No attempt is being made here to summarize passages concerned with these many complex processes. The reader may gain access to them by reference to the subject index of each of the two Parts of this volume, in which the processes are itemized.

TURNOVER OF PLANT BIOMASS

The activities of heterotrophs are the major cause of the disappearance of plant materials, which are simultaneously replaced by autotrophic processes. Turnover rates are more rapid in relatively dry regions and in grazed grasslands than in moist and ungrazed grasslands (Part B, pp. 102–103).

Residence time of shoot material in the canopy is discussed in Part A, pp. 163–165 for a semiarid grassland in which graminoid leaves die and are replaced by other leaves throughout the growing season (Part A, Table 10.14). The early leaves began to fragment and become surface litter by mid-season. The leaves that survived the winter in the canopy had disappeared from the canopy by the end of the growing season of the following year. The longevity of green leaves is much greater under more humid conditions (Part B, pp. 389–391 and 426). Dead leaves remain in the canopy in some grasslands for longer periods — in some instances resulting in the use of fire in managing grazing land (Part A, pp. 243–244 and Part B, pp. 399–404). The mass of accumulated litter on the soil surface is a function of the interaction of the rate of herbage production and the rate of decomposition. The rate of disappearance of litter in semiarid to dry-subhumid climates is such that accumulations of dead plant material on the soil surface are equivalent to no more than the yearly rate of herbage production (Part A, pp. 169, 171; Part B, p. 276). However, under more humid conditions litter may accumulate to the extent that plant growth is impeded and the floristic composition is altered — leading to the use of fire in managing grassland (Part A, pp. 243–244, 260). Turnover times of litter in the tussock grasslands of New Zealand are particularly long (Part B, p. 393).

The rate of turnover of biomass in the soil is generally much slower than above ground. Estimates of the time required for replacement of under-ground plant parts reported in this volume are mostly within a range of 2 to 4 years (Table 17.8).

TABLE 17.8

Rates of turnover of under-ground plant parts in various grasslands

Grassland type Location	Annual rate of turnover (%)	Turnover time (years)	Text reference Part	Location
Flooding pampa, Argentina	23	4.3	A	p. 390
Mixed prairie, Cottonwood, S.D., USA	14–29	3.4–7.1	A	p. 171
Tall-grass prairie, Missouri, USA [1]	26	3.9	A	p. 261
Mixed prairie, Matador, Canada	29	3.4	A	p. 171
Tall-grass prairie, Oklahoma, USA [1]	25	4.0	A	p. 261
Semi-natural meadows, Europe	25–53	1.9–4.0	B	Table 5.7
Tall-grass, Illinois, Illinois, USA	45	2.2	A	p. 261

[1] In the Oklahoma and Argentina grasslands the turnover rate of under-ground parts was faster (annual rates of 36% and 31%, respectively) in grassland grazed by livestock than in the ungrazed treatment (Part A, pp. 261, 391).

MANAGEMENT

Much information is presented concerning management of grasslands, particularly in relation to grazing and conservation. Aspects of management are discussed in all of the regional presentations — usually towards the end of the chapter.

SYSTEMATIC LIST OF GENERA [1]

MONERA

ACTINOMYCETES

CYANOBACTERIA

MYXOPHYCEAE
Nostocaceae
  Anabaena
  Nostoc
Rivulariaceae
  Calothrix
Scytonemataceae
  Scytonema
  Tolypothrix
Stigonemataceae
  Stigonema

EUBACTERIA

  Azotobacter
  Bacillus
  Clostridium
  Pseudomonas
  Rhizobium

PROTISTA

PROTOZOA

  Trypanosoma

FUNGI

ASCOMYCOTINA
Pyrenomycetes
  Sordariales
    Thielavia

DEUTEROMYCOTINA
Coelomycetes
  Sphaeropsidales
    Phoma
  Melanconiales
    Colletotrichum
Hyphomycetes
  Arthosporae
    Geotrichum
  Botryoblastosporae
    Periconia

Phialosporae
  Acrophialophora
  Aspergillus
  Fusarium
  Myrothecium
  Penicillium
Porosporae
  Curvularia
  Stenophyllum
Sympodulosporae
  Cladosporium

ZYGOMYCOTINA
Endogonales
  Endogonaceae
    Endogone
    Glomus (= *Rhizophagus*)
Mucorales
  Cunninghamellaceae
    Cunninghamella
  Mucoraceae
    Circinella
    Mucor
    Rhizopus
  Syncephalastraceae
    Syncephalastrum
  Thamnidiaceae
    Helicostylum

LICHENES

ASCOLICHENES
Cladoniaceae
  Cladonia
Parmeliaceae
  Parmelia

PLANTAE

BRYOPHYTA

HEPATICAE
Plagiochilaceae
  Plagiochila

MUSCI
Amblystegiaceae
  Drepanocladus
Dicranaceae
  Chorisodontium

Grimmiaceae
  Rhacomitrium
Polytrichaceae
  Polytrichum
Sphagnaceae
  Sphagnum
Thuidiaceae
  Thuidium

PTERIDOPHYTA

LYCOPSIDA
Lycopodiaceae
  Lycopodium

PTERIDOPSIDA
(FILICOPSIDA)
Aspidiaceae
  Aspidium
  Onoclea
  Papuapteris
Blechnaceae
  Blechnum
Cyatheaceae
  Cyathea
Gleicheniaceae
  Dicranopteris
  Gleichenia
Osmundaceae
  Osmunda
Plagiogyraceae
Pteridaceae
  Pteridium
Salviniaceae
  Salvinia
Schizaeaceae
  Lygodium
Sinopteridaceae
  Cheilanthes

SPHENOPSIDA
Equisetaceae
  Equisetum

PINOPHYTA
(GYMNOSPERMAE)

Araucariaceae
  Agathis
  Araucaria
Cupressaceae
  Callitris
  Cupressus
  Juniperus
  Tetraclinis
Cycadaceae
  Cycas
Ephedraceae
  Ephedra
Pinaceae
  Abies
  Larix
  Pinus

Podocarpaceae
  Podocarpus

MAGNOLIOPHYTA
(ANGIOSPERMAE)

LILIOPSIDA
(MONOCOTYLEDONES)
Agavaceae
  Agave
  Phormium
  Sansevieria
Alliaceae
  Allium
Amaryllidaceae
  Curculigo
Aponogetonaceae
  Aponogeton
Araceae
  Amorphophallus
  Pistia
Arecaceae (Palmae)
  Borassus
  Cocos
  Hyphaene
  Livistona
  Mauritia
  Medemia
  Sabal
Butomaceae
  Butomus
Commelinaceae
Cyperaceae
  Carex
  Carpha
  Cyperus
  Eleocharis
  Eriophorum
  Fimbristylis
  Gahnia
  Kobresia
  Machaerina
  Oreobolus
  Rhynchospora
  Schoenus
  Scirpus
  Scleria
  Thoracostachyum
  Uncinia
Dioscoreaceae
  Dioscorea
Gramineae (see Poaceae)
Iridaceae
  Bulbocodium
  Crocus
  Iris
Juncaceae
  Juncus
  Luzula
Juncaginaceae
  Triglochin
Liliaceae
  Aloë

---

[1] The taxonomic relationships of animals, plants and microorganisms are shown. Genera are listed alphabetically under families, phyla, subphyla or orders. The editor acknowledges the assistance of his colleagues at the University of Saskatchewan for assistance in classifying certain groups: C. Gillott (invertebrates), V.L. Harms (Plantae), G.A. Jones (Monera), W.J. Maher (Aves), R.A.A. Morrall (Fungi), B.R. Neal (Mammalia) and R.L. Randell (grasshoppers).

MAGNOLIOPHYTA (continued)

   Asparagus
   Asphodeline
   Asphodelus
   Astelia
   Bellevalia
   Bulbinella
   Colchicum
   Eremurus
   Erythronium
   Fritillaria
   Gagea
   Hemerocallis
   Hosta
   Lilium
   Polygonatum
   Tulipa
   Urginea
   Veratrum
  Musaceae
   Ensete
   Musa
   Scitamineae
  Orchidaceae
   Caladenia
   Prasophyllum
  Pandanaceae
   Pandanus
   Sararanga
  Poaceae (Gramineae)
   Achnatherum
   Acroceras
   Aegilops (= *Triticum*)
   Afrotrichloris
   Agropyron
   Agropyropsis
   Agrostis
   Aira
   Aleuropus
   Alloteropsis (= *Trichachne*)
   Alopecurus
   Ampelodesmos (or *Ampelodesma*)
   Andropogon
   Aneurolepidium (= *Leymus*)
   Anisantha (= *Bromus*)
   Anthephora
   Anthoxanthum
   Apluda
   Aristida
   Arrhenatherum
   Arthraxon
   Arundinaria
   Arundinella
   Arundo
   Astrebla
   Atropis
   Avena
   Avenastrum (= *Aira*, *Arrhenatherum*, *Avena*)
   Axonopus (= *Paspalum*)
   Bambusa
   Beckmannia
   Bothriochloa (= *Andropogon*, *Amphilophis*)
   Brachiaria (= *Panicum*)
   Brachyachne
   Brachypodium
   Briza
   Bromopsis (= *Bromus*)
   Bromus
   Calamagrostis
   Capillipedium
   Cenchrus
   Centosteca (= *Centotheca*)
   Chionochloa (= *Danthonia*)
   Chloris
   Chrysopogon (= *Andropogon*)
   Cleistogenes
   Coelachryum
   Coelorhachis (= *Rottboellia*)
   Coix
   Corynephorus
   Critesion (= *Hordeum*)
   Ctenium
   Cutandia
   Cymbopogon (= *Andropogon*)
   Cynodon
   Cynosurus
   Cyrtococcum
   Dactylis
   Dactyloctenium
   Danthonia
   Dendrocalamus
   Deschampsia
   Desmostachya
   Deyeuxia (= *Calamagrostis*)
   Dichanthium (= *Andropogon*)
   Dichelachne
   Digitaria
   Diheteropogon
   Dimeria
   Diplachne
   Dissochondrus
   Echinochloa
   Echinolaena (= *Panicum*)
   Eleusine
   Elionurus (or *Elyonurus*)
   Elymus
   Elytrigia (= *Agropyron*)
   Enneapogon (= *Pappophorum*)
   Enteropogon
   Eragrostis
   Eremopogon (= *Hypogynium*)
   Eriachne
   Erianthus
   Eriochloa
   Eulalia (= *Miscanthus*, *Pollinia*)
   Exotheca (= *Cymbopogon*)
   Festuca
   Garnotia
   Germainia
   Glyceria
   Harpachne
   Harpochloa (or *Harpechloa*)
   Helictotrichon (= *Avenastrum*, *Avena*)
   Heteropogon (= *Andropogon*)
   Hierochloë
   Holcus
   Hordeum
   Hymenachne (= *Panicum*)
   Hyparrhenia (= *Cymbopogon*, *Andropogon*)
   Hyperthelia (= *Hyparrhenia*)
   Imperata
   Isachne
   Ischaemum
   Iseilema (= *Anthistria*)
   Koeleria
   Lagurus
   Lasiagrostis (= *Stipa*)
   Lasiurus (= *Rottboellia*)
   Leersia
   Leptochloa
   Leptothrium
   Lepturus
   Leymus (= *Elymus*)
   Lintonia
   Lolium
   Lophatherum
   Loudetia (= *Trichopteryx*)
   Lygeum
   Manisuris
   Massia
   Megaloprotachne
   Melanocenchris
   Melica
   Melinis
   Melocanna
   Merxmuellera
   Microchloa
   Miscanthus
   Molinia
   Monachather (= *Danthonia*)
   Monocymbium
   Monostachya
   Nardus
   Ophiuros (or *Ophiurus*)
   Oplismenus
   Oropetium
   Oryza
   Oryzopsis
   Oxystenanthera
   Panicum
   Pappophorum
   Paspalidium (= *Panicum*)
   Paspalum
   Pennisetum
   Pentachistis
   Perotis
   Phalaris
   Phleum
   Phragmites
   Plectrachne
   Pleioblastus
   Poa
   Pogonarthria
   Pogonatherum
   Polypogon
   Polytoca
   Polytrias
   Psathyrostachys (= *Hordeum*)
   Pseudopogonatherum
   Pseudoraphis (= *Setaria*, *Chamaeraphis*)
   Ptilagrostis (= *Stipa*)
   Puccinellia
   Rhynchelytrum (= *Tricholaena*)
   Rottboellia
   Saccharum
   Sasa
   Schizachyrium (= *Andropogon*)
   Schmidtia
   Schoenefeldia
   Sclerodactylon
   Scolochloa
   Secale
   Sehima (= *Ischaemum*)
   Serrafalcus (= *Bromus*)
   Setaria
   Sieglingia (= *Triodia*)
   Sorghum
   Spartina
   Sphenopus
   Spinifex
   Spodiopogon (= *Erianthus*)
   Sporobolus
   Stenotaphrum
   Stipa
   Stipagrostis (= *Aristida*)
   Teinostachyum
   Tetrapogon
   Themeda
   Thuarea
   Thyridolepis
   Thysanolaena
   Trachynia (= *Brachypodium*)
   Trachypogon
   Tragus
   Tricholaena
   Triodia
   Tripogon
   Triraphis
   Triseteria
   Trisetum
   Tristachya
   Triticum
   Urochloa (= *Panicum*)
   Urochondra
   Vetiveria (= *Andropogon*)
   Viguierella
   Vulpia (= *Festuca*)
   Vulpiella
   Xerochloa
   Zea
   Zoysia
   Zygochloa
  Restionaceae
  Sparganiaceae
  Typhaceae
  Xyridaceae
   Xyris
  Zingiberaceae
   Curcuma

MAGNOLIOPSIDA (DICOTYLEDONES)

  Acanthaceae
   Blepharis
   Hypoëstes
  Aïzoaceae
   Gisekia
   Limeum
   Mollugo
   Ruschia
   Zaleya
  Amaranthaceae
   Aerva
   Amaranthus
   Psilotrichum
   Ptilotus
  Anacardiaceae
   Anacardium
   Lannea
   Mangifera
   Poupartia
   Rhus
   Sclerocarya
  Apiaceae (Umbelliferae)
   Aciphylla
   Angelica
   Anthriscus
   Archangelica
   Azorella
   Bupleurum
   Cenolophium (= *Selium*)
   Cnidium
   Crithmum
   Eryngium (= *Eringium*)
   Ferula
   Heracleum
   Hydrocotyle
   Hymenolyma
   Kyllinga
   Libanotis (= *Seseli*)
   Meum

SYSTEMATIC LIST OF GENERA

MAGNOLIOPHYTA (continued)
   Palimbia
   Peucedanum
   Pimpinella
   Pleurospermum
   Prangos
   Seseli
   Silaum
   Thapsia
  Apocynaceae
   Adenium
   Alstonia
   Carissa
  Aquifoliaceae
  Araliaceae
   Stilbocarpa
  Asclepiadaceae
   Calotropis
   Pergularia
   Vincetoxicum
  Asteraceae (Compositae)
   Achillea
   Achyrophorus
   Ajania
   Arctogeron
   Arctotheca
   Artemisia
   Aster
   Asterothamnus
   Atractylis
   Cacalia
   Calotis
   Carduncellus
   Carduus
   Carlina
   Carthamus
   Celmisia
   Centaurea
   Chromolaena
   Chrysanthemum
   Chrysocoma
   Cirsium
   Cnicus
   Cosmos
   Cousinia
   Crepis
   Crinitaria
   Cynara
   Dicoma
   Echinops
   Elytropappus
   Espeletia
   Eupatorium
   Filago
   Filifolium (= *Tanacetum*)
   Galactites
   Galatella
   Galinsoga
   Glossogyne
   Gundelia
   Helichrysum
   Helipterum
   Heteropappus
   Hieracium
   Hypochoeris
   Inula
   Ixeris
   Jurinea
   Leucanthemum
   Ligularia (= *Senecio*)
   Mikania
   Olearia
   Olgaea
   Onopordum
   Pentzia
   Pleurophyllum
   Pseudoelephantophus
   Raoulia
   Rhanterium
   Rhaponticum
   Saussurea
   Scolymus
   Scorzonera
   Senecio
   Serratula
   Silybum
   Stoebe
   Tanacetum
   Taraxacum
   Thelesperma
   Thevenotia
   Tragopogon
   Tugarinovia
  Balanitaceae
   Balanites
  Barringtoniaceae
   Careya
  Betulaceae
   Alnus
   Betula
  Bignoniaceae
   Stereospermum
  Bombacaceae
   Bombax
  Boraginaceae
   Cynoglossum
   Echiochilon
   Heliotropium
   Lappula
   Moltkia
   Myosotis
   Onosma
   Rindera
  Brassicaceae (Cruciferae)
   Alyssum
   Blennodia
   Brassica
   Capsella
   Cardamine
   Descurainia
   Diplotaxis
   Dontostemon
   Erophila
   Eruca
   Erysimum
   Farsetia
   Lepidium
   Meniocus
   Oudneya
   Pringlea
   Pseudocytisus
   Ptilotrichum
   Stenopetalum
   Zilla
  Buddlejaceae
   Buddleja
  Burseraceae
   Commiphora
  Cactaceae
   Opuntia
  Campanulaceae
   Adenophora
   Campanula
   Cyananthus
   Lobelia
  Capparidaceae
   Boscia
   Capparis
   Cleome
  Caprifoliaceae
   Lonicera
  Carpinaceae
   Ostryopsis
  Caryophyllaceae
   Arenaria
   Cerastium
   Dianthus
   Eremogone
   Gymnocarpos
   Gypsophila
   Holosteum
   Lychnis
   Sagina
   Spergularia
   Stellaria
   Thylacospermum
  Casuarinaceae
   Casuarina
   Gymnostoma
  Celastraceae
   Gymnosporia
  Chenopodiaceae
   Agathophora
   Anabasis
   Arthrocnemum
   Atriplex
   Axyris
   Bassia
   Camphorosma
   Ceratocarpus
   Chenopodium
   Climacoptera
   Corispermum
   Cornulaca
   Enchylaena
   Eurotia
   Fredolia
   Halimione
   Halocnemum
   Halopeplis
   Haloxylon
   Hammada
   Kalidium
   Kochia
   Lagenantha
   Maireana
   Nanophyton
   Noaea
   Nucularia
   Rhagodia
   Salicornia
   Salsola
   Sclerolaena
   Seidlitzia
   Suaeda
   Traganum
  Chrysobalanaceae
   Parinari (= *Parinarium*)
  Cistaceae
   Cistus
   Helianthemum
  Combretaceae
   Combretum
   Guiera
   Terminalia
  Compositae (see Asteraceae)
  Convolvulaceae
   Convolvulus
   Ipomoea
   Merremia
  Corylaceae
   Corylus
  Cruciferae (see Brassicaceae)
  Didiereaceae
   Didierea
  Dilleniaceae
   Dillenia
   Hibbertia
  Dipsacaceae
   Scabiosa
   Succisa
  Dipterocarpaceae
   Dipterocarpus
   Shorea
  Ehretiaceae
   Cordia
  Elaeocarpaceae
   Muntingia
  Elacagnaraceae
   Hippophae
  Epacridaceae
   Leucopogon
   Styphelia
  Ericaceae
   Calluna
   Dimorphanthera
   Erica
   Rhododendron
   Vaccinium
  Euphorbiaceae
   Euphorbia
   Glochidion
   Macaranga
   Manihot
   Omalanthus
   Phyllanthus
   Sapium
  Fabaceae (Leguminosae)
   (= Mimosaceae, Caesalpinaceae)
   Acacia
   Albizia
   Alysicarpus
   Anthyllis
   Arachis
   Argyrolobium
   Astragalus
   Baikiaea
   Bauhinia (= *Lysiphyllum*)
   Bonjeania
   Brachystegia
   Burkea
   Cajanus
   Calophaca
   Calopogonium
   Calycotome
   Caragana
   Cathormion
   Cassia
   Centrosema
   Ceratonia
   Colophospermum
   Colutea
   Cordeauxia
   Coronilla
   Cytisus
   Dalbergia
   Delonix
   Desmodium
   Erinacea
   Eriosema
   Erythrina
   Genista
   Glycine
   Glycyrrhiza
   Hedysarum
   Hippocrepis
   Indigofera
   Julbernardia
   Lathyrus

MAGNOLIOPHYTA (*continued*)
   Lespedeza
   Leucaena
   Lotus
   Lygos
   Lysiphyllum (= *Bauhinia*)
   Macroptilium
   Maniltoa
   Medicago
   Melilotus
   Mimosa
   Neptunia
   Onobrychis
   Orobus (= *Lathyrus*)
   Oxytropis
   Patrinia
   Phaseolus
   Piptadeniastrum
   Pithecellobium
   Prosopis
   Pterocarpus
   Pueraria
   Pycnospora
   Sarcobotrya
   Sesbania
   Stylosanthes
   Tamarindus
   Tephrosia
   Tetragonolobus (= *Lotus*)
   Thermopsis
   Trifolium
   Trigonella (see *Medicago*)
   Ulex
   Uraria
   Vicia
   Vigna
   Zornia
Fagaceae
   Nothofagus
   Quercus
Frankeniaceae
   Frankenia
Gentianaceae
   Gentiana
Geraniaceae
   Erodium
   Geranium
Globulariaceae
   Globularia
Gunneraceae
   Gunnera
Gyrocarpaceae
   Gyrocarpus
Lamiaceae (Labiatae)
   Ballota
   Betonica (= *Stachys*)
   Endostemon
   Hyptis
   Hyssopus
   Isodon
   Lagochilus
   Marrubium
   Phlomis
   Rosmarinus
   Salvia
   Schizonepeta
   Scutellaria
   Stachys
   Teucrium
   Thymus
Lauraceae
   Cassytha
Limoniaceae (see Plumbaginaceae)
Leguminosae (see Fabaceae)

Loganiaceae
   Geniostoma
Malpighiaceae
   Acridocarpus
Malvaceae
   Gossypium
   Hibiscus
   Malva
   Sida
Melastomataceae
   Melastoma
Moraceae
   Broussonetia
   Ficus
Myoporaceae
   Eremophila
   Myoporum
Myrtaceae
   Baeckea
   Decaspermum
   Eucalyptus
   Kunzea
   Leptospermum
   Melaleuca
   Myrtella
   Psidium
   Syzygium
   Tristania (= *Lophostemon*)
   Xanthomyrtus
   Xanthostemon
Nepenthaceae
   Nepenthes
Nyctaginaceae
   Boerhavia (or *Boerhaavia*)
Nymphaeaceae
   Nymphaea
Ochnaceae
   Ochna
   Schuurmansia
Oleaceae
   Fraxinus
   Olea
   Syringa
Onagraceae
   Epilobium
   Jussieua
Oxalidaceae
   Oxalis
Periplocaceae
   Cryptostegia
Pistaciaceae
   Pistacia
Pittosporaceae
Plantaginaceae
   Plantago
Plumbaginaceae
   Goniolimon
   Limoniastrum
   Limonium
Polemoniaceae
   Armeria
Polygalaceae
   Polygala
Polygonaceae
   Calligonum
   Emex
   Polygonum
   Rheum
   Rumex
Portulacaceae
   Calandrinia
   Portulaca
Primulaceae
   Androsace
   Glaux

   Primula
Proteaceae
   Banksia
   Grevillea
Ptaeroxylaceae
   Ptaeroxylon
Ranunculaceae
   Adonis
   Anemone
   Ceratocephala
   Delphinium
   Leptopyrum (= *Isopyrum*)
   Pulsatilla
   Ranunculus
   Thalictrum
   Trollius
Rhamnaceae
   Alphitonia
   Discaria
   Rhamnus
   Ventilago
   Ziziphus (or *Zizyphus*)
Rosaceae
   Acaena
   Agrimonia
   Alchemilla
   Amygdalus
   Armeniaca
   Cerasus
   Chamaerhodos
   Cliffortia
   Comarum
   Cotoneaster
   Filipendula (= *Spiraea*)
   Geum
   Hulthemia
   Leucosidea
   Pentaphylloides (= *Potentilla*)
   Potentilla
   Prinsepia
   Prunus
   Rosa
   Sanguisorba
   Sarcopoterium
   Spiraea
Rubiaceae
   Borreria
   Canthium
   Coffea
   Coprosma
   Galium
   Morinda
   Mussaenda
   Nauclea
   Spermacoce
   Timonius
   Wendlandia
Rutaceae
   Citrus
   Haplophyllum
Salicaceae
   Populus
   Salix
Salvadoraceae
   Salvadora
Sapindaceae
   Atalaya
   Dodonaea
   Guioa
   Heterodendrum
Sapotaceae
   Argania
   Butyrospermum
   Chrysophyllum
Scrophulariaceae

   Alectorolophus
   Anarrhinum
   Cymbaria
   Linaria
   Mazus
   Pedicularis
   Verbascum
   Veronica
Solanaceae
   Hyoscyamus
   Lycium
   Nicotiana
   Solanum
Sterculiaceae
   Brachychiton
   Commersonia
   Dombeya
   Theobroma
   Waltheria
Stilaginaceae
   Antidesma
Tamaricaceae
   Reaumuria
   Tamarix
Theaceae
   Camellia
   Eurya
   Wickstroemia
Thymelaeaceae
   Passerina
   Stellera
   Thymelaea
Tiliaceae
   Colona
   Grewia
Ulmaceae
   Celtis
   Ulmus
Umbelliferae (see Apiaceae)
Urticaceae
   Urtica
Vacciniaceae (see Ericaceae)
Valerianaceae
   Valeriana
   Valerianella
Verbenaceae
   Clerodendrum
   Lantana
   Pygmaeopremna
   Stachytarpheta
   Vitex
Violaceae
   Viola
Vitidaceae
   Cissus
Zygophyllaceae
   Nitraria
   Peganum
   Tribulus
   Zygophyllum

ANIMALIA

ASCHELMINTHES

NEMATODA

ANNELIDA

OLIGOCHAETA
   Enchytraeidae
   Lumbricidae
      Acanthrodilus

SYSTEMATIC LIST OF GENERA

ANNELIDA (*continued*)
  Moniligastridae
    Drawidia (or *Drawida*)
  Megascolecidae
    Lampito
    Octochaetona
  Ocnerodrilidae
    Ocnerodrilus

ARTHROPODA

  ARACHNIDA
    Acarina

  COLLEMBOLA

  CRUSTACEA
    Isopoda

  INSECTA
    Coleoptera
      Curculionidae
      Scarabaeidae
        Costelytra
        Heliocopris
      Tenebrionidae
    Diptera
      Muscidae
        Glossina
    Hemiptera
      Psyllidae
        Heteropsylla
    Hymenoptera
      Formicidae
        Paltothyreus
    Hemiptera
      Psyllidae
        Heteropsylla
    Isoptera
      Hodotermitidae
        Hodotermes
      Termididae
        Cubitermes
        Macrotermes
        Microtermes
        Odontotermes
        Trinervitermes
    Lepidoptera
      Wiseana
    Orthoptera
      Acrididae (= Locustidae)
        Angaracris
        Bryodema
        Calliptamus
        Chorthippus
        Epacromia
        Myrmeleotettix
        Locusta
        Oedaleus
    Siphonaptera
      Pulicidae
        Spilopsyllus
    Thysanoptera

  MYRIAPODA

MOLLUSCA

CHORDATA

  AMPHIBIA
    Anura
      Bufonidae
        Bufo
      Ranidae
        Rana

  REPTILIA
    Agamidae
      Phrynocephalus
    Boidae
      Liasis
      Morelia
    Colubridae
      Coluber
      Elaphe
    Lacertidae
      Eremias

  AVES
    Accipitriformes (Falconiformes)
      Accipitridae
        Accipiter
        Aquila
        Buteo
        Gypaetus
        Gyps
        Milvus
        Necrosyrtes
        Neophron
        Torgos
        Trigonoćeps
      Falconidae
        Falco
    Anseriniformes
      Anatidae
        Anas
        Cygnus
        Tadorna
    Apodiformes
      Apodidae
        Apus
    Charadriiformes
      Charadridae
        Charadrius
        Vanellus
      Laridae
        Larus
      Scolopacidae
        Philomachus
        Tringa
    Ciconiiformes
      Ardeidae
        Ardea
        Botaurus
      Threskiornithidae
        Bostrychia
    Columbiformes
      Columbidae
        Columba
    Pteroclididae
      Pterocles
      Syrrhaptes
    Galliformes
      Phasianidae
        Francolinus
        Tetraogallus
    Gruiformes
      Otididae
        Otis
      Rallidae
        Fulica
        Gallirallus
        Notornis
    Passeriformes
      Alaudidae
        Alauda
        Calandrella
        Eremophila (= *Otocoris*)
        Melanocorypha
      Corvidae
        Corvus
      Fringillidae
        Emberiza
        Serinus
      Hirundinidae
        Hirundo
      Motacillidae
        Anthus
        Motacilla
      Muscicapidae
        Oenanthe
      Ploceidae
        Quelea
        Montifringilla
        Passer
      Sylviidae
        Acrocephalus
    Pelecaniformes
      Pelecanidae
        Pelecanus
      Phalacrocoracidae
        Phalacrocorax
    Podicipediformes
      Podicipedidae
        Podiceps
    Procellariiformes
      Diomedeidae
        Diomedea
        Phoebetria
      Procellariidae
        Macronectes
    Psittaciformes
      Psittacidae
        Strigops
    Sphenisciformes
      Spheniscidae
        Aptenodytes
        Eudyptes
        Pygoscelis
    Struthioniformes
      Struthionidae
        Struthio

  MAMMALIA
    Artiodactyla
      Bovidae
        Addax
        Aepyceros
        Alcelaphus
        Ammodorcas
        Ammotragus
        Antidorcas
        Antilope
        Bubalus
        Bos
        Boselaphus
        Capra
        Cephalophus
        Connochaetes
        Damaliscus
        Dorcatragus
        Gazella
        Hemitragus
        Hippotragus
        Kobus
        Litocranius
        Nemorhaedus
        Nesotragus (= *Neotragus*)
        Oryx
        Ourebia
        Ovis
        Pantholops
        Pelea
        Procapra
        Prodorcas
        Pseudois
        Raphicerus
        Redunca
        Rhynchotragus
        Rupicapra
        Saiga
        Sylvicapra
        Syncerus (or *Synceros*)
        Taurotragus
        Tragelaphus
      Camelidae
        Camelus
      Cervidae
        Axis
        Cervus
        Rangifer
      Giraffidae
        Giraffa
      Hippopotamidae
        Hippopotamus
      Rhinocerotidae
        Ceratotherium
        Diceros
        Didermocerus
          (= *Dicerorhinus*)
        Rhinoceros
      Suideae
        Phacochoerus
        Potamochoerus
        Sus
    Carnivora
      Canidae
        Canis
        Lycaon
        Otocyon
        Vulpes
      Felidae
        Acinonyx
        Felis
        Panthera
      Herpestidae
        Helogale
        Herpestes
        Ichneumia
        Mungos
      Hyaenidae
        Crocuta
        Hyaena
      Mustelidae
        Mellivora
        Mustela
        Poecilictis
        Vormela
        Zorilla
      Protelidae
        Proteles
      Viverridae
        Genetta
        Viverricula
    Chiroptera
    Hyracoidea
      Procavidae
        Procavia
    Insectivora
      Talpidae
        Talpa
    Lagomorpha
      Leporidae
        Lepus
        Oryctolagus
      Ochotonidae
        Ochotona
    Marsupialia

CHORDATA (*continued*)
  Macropodidae
    Macropus
    Megaleia (= *Macropus*)
  Peramelidae
    Echymipera
    Isoodon
  Phalangeridae
    Trichosurus
  Perissodactyla
    Equidae
      Equus
  Pholidota
  Pinnepedia
    Otariidae
      Arctocephalus
      Phocarctos
    Phocidae
      Mirounga

  Primates
    Cercopithecidae
      Cercopithecus
      Erythrocebus
      Papio
      Theropithecus
    Lorisidae
      Galago
  Proboscidea
    Elephantidae
      Elephas
      Loxodonta
  Rodentia
    Bathyergidae
      Heliophobius
      Heterocephalus
    Cricetidae
      Desmodilliscus
      Ellobius
      Gerbillus

      Lasiopodomys
      Meriones
      Microtus
      Myospalax
      Pachyuromys
      Phodopus
      Psammomys
      Tatera
      Taterillus
    Dipodidae
      Allactaga
      Dipus
      Jaculus
    Hystricidae
      Hystrix
    Muridae
      Arvicanthis

      Mastomys
      Mus
      Rattus
      Uromys
    Rhizomyidae
      Tachyoryctes
    Sciuridae
      Euxerus (= *Xerus*)
      Citellus (= *Spermophilus*)
      Marmota
      Tamias
      Xerus
    Spalacidae
      Spalax
  Tubulidentata
    Orycteropodidae
      Orycteropus

AUTHOR INDEX [1]

Abaturov, B.D., 42, 45, 52, 54, *55*, *58*
Abdulgani, *122*
Abdulle, A.M., 253, *257*
Acocks, J.P.H., 265, 271, 272, 283, *288*
Adams, K.M., *433*
Adams, N.M., 370, *407*
Adamson, D.A., 466, *467*
Aden, A.S., 253, *257*
Aden, M.O., *259*
Agrawal, R.R., *121*
AGROTEC-CRG-/SESES, 239, 242, 245–247, *257*
Ahmad, A., 110, *117*
Ahmed, A.M., 252–254, *259*, *260*
Ahuja, L.D., 109, 113, 115, *117*, *120*
Aïdoud, A., 180, *194*
Al-Ani, T.A., 88, *117*
Al-Mufti, M.M., *117*
Alberda, Th., 125, *162*
Aleem, A., 104, *117*
Alekhin, V.V., 3, *55*
Alekseenko, L.N., 133, 135, 136, 139, 142, 144, *163*
Alexander, G., 315, *355*
Ali, A.R, 247, *257*
Allan, H.H., 361, *404*
Allen, G.H., 323, 324, 354, *359*
Allsopp, R., 234, *257*
Aly, I.M., 189, *195*, *196*
Amarsinghe, L., 88, *120*
Anderson, A.J., 393, *404*
Anderson, G.D., 238, 239, *257*
Andrew, C.S., 346, *360*
Andrew, M.H., *358*
Andrew, W.D., 87, 92, *117*
Andrews, P., 246, *257*
Andrienko, T.L., *58*
Andrushko, A.M., 49, *55*
Andrzejewska, L., 146, 147, 149, *163*
Aneja, K.R., 107, *117*, *120*
Anonymous, 270, *288*, 316, 319, *355*, 417, *431*
Appadurai, R.R., *122*
APROSC, 94, 109, *117*
Archbold, R., 461, *467*
Archer, A.C., 370, 388, *404*

Arndt, W., 339, 346, *358*
Arnol'di, L.V., 17, 18, 24, 26, *55*, *58*
Ashton, D.H., 419, 421, 422, 424, 426, *431*, *432*
Assefa, E,, 249, 250, *257*
Atabay, R., *121*
Atkinson, I.A.E., 393, *405*
Aubert de la Rue, E., 427, *431*
Aubert, M., 427, *431*
Aucamp, A.J., 276, *288*
Audley-Charles, M.G., 304, *312*
Awere-Gyeke, K., *261*

Baasher, M.M., 206, *216*, *220*
Bagine, R.K.N., 256, *257*
Bagnouls, F., 197, 200, *216*
Bailly, C., 214, *216*
Balátová-Tuláčková, E., 126, 131, 138, 158, *163*, *165*
Bannikov, A.G., 46, *55*
Bannikova, I.A., *57*
Bár, I., 153, *163*
Barbey, C., 207, 216, *216*
Barbier, J., *216*
Barker, A.P., 379, *404*
Barker, J.R, 252–254, *257*
Barker, S., 329, *355*
Barnard, A., 316, *355*
Barnes, D.L., 270, 276, 277, 281, 285, *288*
Barral, H., 207, 214, 216, *216*
Barratt, B.I.P., *407*
Barrau, J., 461, *467*
Barrie, N., *358*
Barrow, C.J., 412, 413, *431*
Barrow, P.M., 345, *355*
Barry, J.P., 207, 214, *216*
Basak, K.C., 114, *117*
Batcheler, C.L., 396, 397, *404*
Bauer, G., 85, *117*
Baylis, G.T.S., *409*
Beadle, N.C.W., 327, 329, 332, *355*
Beale, I.F., 327, 328, *355*, *356*
Beals, E.W., 225, 234, *257*
Beard, J.S., 438, *467*
Bednář, V., 154, *163*
Beese, G., 222, *261*
Behara, N., *118*, *120*
Behera, B.N., *118*
Belcher, R.A., *408*

Bell, A.T., *356*
Bell, B.D., *433*
Bellamy, J.A., 438, *467*
Belostokov, G.P., 18, *55*
Benefice, E., 1, *216*
Benjamin, R., 180, *194*, *196*
Berger-Landefeldt, U., 137, *163*
Bernus, E., 210, 214, 215, *216*
Bespalova, Z.G., 19, *55*
Beukes, B.H., 275, *289*
Bews, J.W., 167, 168, *169*, 438, *467*
Bezuidenhout, J.J., 276, 281, *288*, *289*
Bhanot, V.M., *120*
Bharucha, F.R., 91, *117*
Bhatt, S.C., *120*
Bhatti, A.G., 89, *119*
Bhimaya, C.P., 109, 113, *117*
Biddiscombe, E.F, 337, 349, 350, 354, 355, *358*
Bigalke, R.C., 280, *288*
Bille, J.C., 180, *194*, 207, 208, 213, *216*, 247, 249, 250, 257, *257*, *258*
Billings, W.D., 401, *404*
Billore, S.K., *119*
Bilyk, G.I., 3, 23, *55*, *58*
Bingham, R.E., *195*
Birand, H., 88, *117*
Birch, W.R, *263*
Bisset, W.J., 340, *359*
Blake, S.T., 323, *355*
Blakemore, L.C., 382, 397, *407*
Blanchemain, A., 191, *194*
Blankenship, L.H, 238, 239, 242, *257*, *262*
Blasco, F., 83, 84, 87, 88, 95, 96, *117*, *119*
Bleeker, P., 464, *467*, *469*
Bliss, L.C., 370, 381, 389, *404*, *407*, 417, 418, *431*
Blower, J., 226, 248, 256, *257*
Boaler, S.B., 247, *257*
Bobrovskaya, N.I., 18, *55*
Bogdan, A.V., 224, 225, 227–229, 233, 235, 238, 242, 246, 247, 256, *257*, *258*
Bogdan, V.S., 42, 54, *55*
Bolton, M., 234, *257*
Bonte, P., 215, *217*
Booysen, P. de V., 277, 285, 287, *288*, *289*
Bor, N.L., 83, 84, 112, *117*

---

[1] Page references to text are in roman type, to bibliographical entries in italics.

Borisova, I.V., 10, 13, 26, 35, *57*
Borut, A., *196*
Bosch, O.J.H., 277, *289*
Bosser, J., 300, *301*
Boswell, C.C., *407*
Botschantsev, V., *196*
Boucher, C., 269, *288*
Boudet, G., 1, 199, 206–210, *216*, *217*
Bourgeot, A., *216*, *217*
Bourlière, F., 1, 197, 207, 209, 210, *217*, 240, *257*
Bourn, D., 249, *257*
Box, E.O., 70, 71, *81*
Box, T.W., 354, *355*
Boyland, D.E., 321, 327, *355*
Bransby, D.I., *289*
Brasch, D.J., 389, *408*
Brass, L.J., 438, 461, *467*
Braun, H.M.H., 242, *257*, *263*
Braun-Blanquet, J., 18, *55*, 128, *163*
Breckle, S.W., 83, *117*
Bredon, R.M., 242, *261*
Breman, H., 208, *217*
Brewer, R, *359*
Breymeyer, A.I., 147, 148, *163*
Breytenbach, C., *289*
Bridge, J., 459, *467*, *469*
Bridger, B.A., *407*, *408*
Brooke, C., 243, *258*
Brookfield, H.C., 437, *467*
Brooks, R.R., 368, *404*
Brown, D., 63, 67, 75–77, *80*
Brown, M., 449, 463, *467*
Brownsey, P.J., 361, *404*
Brumley, C.F., 403, *404*
Brunt, M.A, *263*
Bryan, W.W., 345, 346, *355*
Buevich, Z.G., *56*
Buh, B.B., 252–254, *259*, *260*, *263*
Bulmer, R., 465, *467*
Bunderson, W.T., 248, *258*
Bunt, J.S., 426, *431*
Burbidge, N.T., 316, 319, *355*
Burns, W., 91, *117*
Burrell, J., 363, 377, *404*, *407*
Burrows, C.J., 363, 370, 374, 376–378, 389, 390, 393, *404*, *407*
Burrows, W.H., 325–328, 340, 347, 348, 355, *356*, *359*
Burstall, T., 427, *432*
Burtt, B.D., 237, 238, *258*
Buss, I.O., 228, *264*
Buxton, R.D., 248, *258*

Cairns, A., *407*, *408*
Calaby, J.H., 316, *360*
Calder, D.M., 388, *404*
Calder, J.W., 388, *404*
Callaghan, T.V., 416, 421, 423, *431*, *433*
Cameron, D.F., *357*

Canon, L., *219*
Carew, B.A.R., 191, *194*
Carnahan, J.A., 307, 310–312, 315, 338, *356*
Carrodus, B.B, 329, *356*
Cassady, J.T., 242, 247, 250, *258*
Castillo, L.C., *121*
Caughley, G., 394, 395, *404*, *405*
Celles, J.C., *216*
Chakravarty, A.K., 109, 114, 115, *117*, *118*
Champion, H.G., 83, 89, *118*
Chang, H.S. David, 69, 74, *80*
Chang Yu-ling, 72, *80*
Charley, J.L., 354, *356*
Charreau, C., 208, *217*
Chastain, A., 413, 416, 419, 421, 427–429, *431*
Chaudhri, I.I., 89, 90, *118*
Chavanes, B., 300, *301*
Cheeseman, T.F., 361, *405*
Chemin, M.C., *219*
Chen Yong-lin, *81*
Chen Zuo-zhong, 67, 75, *80*
Cheng Tse-ming, 74, *80*
Cherepanov, S.K., 4, *55*
Chernov, Yu.I., 22, 42, *55*
Cheruiyot, S.K., 247, *258*
Chin Cui-xia, 78, *81*
Choate, J.H., 337, *356*
Chosniak, I., *196*
Christian, C.S., 323, 326, 333, *356*, *359*
Christie, E.K., 322, 328, 348, *356*
Cissé, M.I., *218*
Clanet, J.C., 202, *217*
Clarke, C.M.H., 393, 395, *405*
Clarke, D.G., *408*
Clarke, G.C.S., 416, *431*
Claudin, J., *195*
Clement, J., *216*
Clements, J.F., 226, *258*
Clifford, H.T., 439, *467*
Coaldrake, J.E., 341, 345, 348, *356*
Cobb, S., 248, *258*
Cochemé, J., 197, 199, *217*
Cockayne, L., 361, 376, 377, 394, *405*, 411, 419, *431*
Coe, M.J., 234, 240, 248, 256, *258*, 209, *217*
Cole, M.M., 438, *467*
Colhoun, E.A., 412, *431*
Collett, G.I., 370, *404*
Condon, R.W., 330, 355, *356*, *358*, *359*
Connor, H.E., 361, 375–380, 382–384, *405*, *407*
Conrad, V., 393, *405*
Conte, A., 207, 216, *216*
Coode, M.J.E., 462, *467*
Cook, C.W., 180, *194*
Cooray, R., 101, 104, *120*

Cornet, A., 207–209, *217*
Corra, M., *257*
Cossins, N.J., 247, 249, 250, *258*
Costin, A.B., 345, 353, *356*, 420, 429, 430, *431*
Coulibaly, M., *216*
Coulter, J.D., 370, *405*, *408*
Cour, P., 427, *431*
Cours Darne, G., 292, *301*
Cowling, S.W., 354, *356*
Cox, J.E., *407*
Crawford, A.M., 396, *405*
Crespo, D., 193, *194*
Crocker, C.D., 87, *118*
Crocker, R.L., 321, *356*
Croker, B.H., 394, *405*
Croome, R.L., 426, *431*
Crush, J.R., 398, *405*
Cumberland, K.B., 374, *405*
Cumming, D.H., *217*, *258*
Cumpston, J.S., 427, *431*
Cusin, M., *218*
Cuthbertson, E.G., *355*
Cutler, E.J.B., 381, *406*

Dabadghao, A.K., *121*
Dabadghao, P.M., 88, 90, 112, 115, *118*, *121*, *122*
Dagar, J.C., 91, 98, *122*
Dall B.E., 74, *80*
Daly, G.T., 379, *405*
Dancette, C., 199, 202, *217*
Danckwerts, J.E., 283, *288*
Dandelot, P., 211, *217*, 226, 231, *258*
Danin, A., *470*
Darley, J., *217*
Das, R.B., 106, 115, *117*, *118*
Dash, M.C., 106, 107, *118*, *120–122*
Dasmann, R.F., 279, *288*, 437, *467*
Datta Biswas, N.R., *121*
Davies, J.G., 346, 347, *356*
Davy, E.G., 197, 199, 202, *217*
Davydov, V.A., *55*
De Haan, C., *219*
De Leeuw, P.N., *220*
De Los Llanos, C., *194*
De Rasayro, R.A., 92, *118*
De Wispelaere, G., 214, 216, *216*, *217*, *219*
De Wit, C.T., 208, *217*
Deacon, G.E.R., 416, *431*
Debroy, R., *117*, *118*
Dekeyser, P.L., 211, *217*
Delany, M.J., 211, *217*
Delhaye, R.E., 189, *194*
Delisle, J.F., 416, *431*
Delvert, J., 96, *118*
Delwaulle, J.C., 203, *217*
Deng Li-you, 69, *80*
Denis, J.P, *216*

AUTHOR INDEX

Deshler, W., 234, *258*
Deshmukh, I.K., 235, 241, 242, 246, 250, *258*
Detwyler, T.R., 427, *431*
Dewan, M.L., 86, 87, *118*
Di Castri, F, *288*, 171, *194*, *195*
Di Menna, M.E., 397, *405*
Diaité, I., *216*
Diarra, L., *218*
Diaw, O.L., *216*
Dick, R.D., 396, *405*
Dickinson, K.J.M., 401–403, *405*, *407*
Dicko, M.S., 213, *217*
Dieye, K., *216*
Digeard, J.P., *217*
Diller, H., *218*
Dinesman, L.G., 42, 51, *55*
Dirschl, H.J., 225, 237, 238, 241, 243, 248–250, 256, *258*, *259*
Djiteye, A.M., 207–209, *219*
Dmitriev, P.P., 46, 48–53, *55*–*57*
Dobrinskyi, L.N., 42, *55*
Dobson, A.T., *406*
Dokhman, G.I., 3, 23, *56*, *57*
Dokuchaev, V.V., 42, *56*
Dolan, R., 242, 256, *259*
Dolby, R., *408*
Donald, C.M., 349, 350, *356*
Donnadieu, J., *195*
Dorst, J., 211, *217*, 226, 231, *258*
Dougall, H.W, 235, 242, *258*
Doutre, M.P., *216*
Draz, O., 190, *194*
Dreux, Ph., 427, *431*
Drew, K.R., 394, *405*
Drewes, R.H., 276, 285, *288*, *289*
Drobník, J., 153, *163*
Druzina, V.D., 156, *163*
Drysdale, V.M., *258*
Dubyna, D.V., *58*
Dudal, R., 87, *118*
Dulieu, D., 207, 215, 216, *217*, *218*
Dupire, M., 215, *217*
Dwivedi, R.S., 106, *118*

Eardley, C.M., *356*
Ebersohn, J.P., 328, *356*
Eden, A.W., 427, *431*
Eden, M.J., 442, 462, *467*
Edgar, E., 361, 376, 377, *405*, *408*
Edgar, G., *355*
Edroma, E.L., 234, 242, *258*
Edwards, D.C., 221, 222, 224, 225, 227–230, 235, 238, 239, 246, 249, 255, 256, *258*
Edwards, K.A., 255, *258*
Edwards, P.J., 283, *288*
Egbunike, G.N., *194*
Ehleringer, J.R., 441, *467*
Eiten, G., 438, *467*

El Aouini, M., 191, *194*
Ellenberg, H., 129, 130, *163*
Elliott, H.F.I., 226, *258*
Eltringham, S.K., 234, *258*
Endrody-Younga, S., 281, *288*
Enting, B., 366, *405*
Epstein, H., 190, *194*, 214, *217*
Erkun, V., 110, 114, *118*
Esler, A.E., *404*
Etienne, M., 193, *194*
Eussen, J.H.H., 95, *118*
Evans, A.C., *406*
Evans, F.C., 152, *166*
Evenari, M., 171, 190, *194*, *288*
Everist, S.L, 324, 327–329, 336, *356*
Eyal, E., *194*, *196*

Fabricius, A.F., 413, 416, 417, *431*
Falls, R., *469*
Falvey, L., 102, 111–113, *118*
Famouri, J., 86, 87, *118*
FAO, 78, *80*, 168, 169, *169*, 189, *194*–*196*, 202, 210, *217*–*220*, 246, 247, 249, 250, 256, 257, *257*–*263*, 291, 300, *301*
FAO/UNESCO, 202, *217*
Fauck, R., 202, *217*
Fennessy, R., 231, 234, 241, 248, 256, *264*
Fiala, K., 138, 140, 142, 151, 152, *163*
Field, C.R., 240, 242, 256–258, *262*, *264*
Fieldes, M., 371, 372, *406*
Fiennes, R.N.T.W., 233, *258*
Findejsova, M., *166*
Fineran, B.A., 376, *405*
Fisher, F.J.F., 368, *405*
Fitzgerald, K., 325, *356*, *359*
Fitzpatrick, E.A., 319, *356*, 437, *467*
Fleming, C.A., 366–368, 393, *405*
Flenley, J.R., 442, 463, *470*
Flint, E.A., 397, *405*
Flora of the USSR, 4, *56*, 125, *163*
Floret, C., 180, 181, 192, *194*
Flux, J.E.C., 396, *405*
Fogg, G.E., 426, *431*
Foran, B.D., 278, *288*, *289*, 353, *356*
Formozov, A.N., 42, 44, 47, 49–51, 53–55, *55*, *56*
Fosberg, F.R., 459, 460, 466, *467*
Foster, J.B., 240, *258*
Fowler, N.K., 242, *259*
Fox, M.D., 466, *467*
Frame, G.W., 225, 241, *258*, *261*
Franclet, A., 189, 190, *194*
Franklin, D.A., *406*
Franquin, P., 197, 199, *217*
Fraser, K.W., 379, *405*
Friedel, M.H., 355, *356*
Frish, V.A., 52, *56*
Frith, H.J., 315, 316, 353, *356*
Frye, D., 253, 254, *259*

Fuggles-Couchman, N.R., 226, *258*
Fullard, H., 245, 255, *259*
Fuls, A., *470*
Furniss, P.R., *289*
Fursaev, A.D., 53, *56*

Gachet, J.P., 193, *194*
Gael', A.G., 25, *56*
Gage, M., 367, *405*
Gallais, J., 215, *217*
Gams, H., 4, *56*
Gandar, N.V., 281, *288*
Gaston, A., 209, 214–216, *217*, *218*
Gates, D.H., 78, *80*
Gaussen, H., 197, 200, *216*
Genov, A.P., 52, *56*, *58*
Geyger, E., 133, 137, 139–141, *163*
Gibson, T.A., 95, *118*
Gill, A.M., 353, *356*
Gillard, P., 340, 347, *356*
Gillet, H., 199, 202, 206, 210, 214, 215, *217*–*219*
Gillett, J.B., 245–247, 255–257, *259*
Gillham, M.E., 414, 427, 429, *432*
Gilliland, H.B., 222, 245–247, 255, 256, *259*
Gillison, A.N., 304–305, *312*, *358*, 435–438, 442, 449, 450, 452, 453, 455, 460, 462–464, 466, *467*, *469*
Gillon, D., 210, *218*
Gillon, Y., 210, *218*
Givnish, T.J., 437, *467*
Glassman, S., 459, *467*
Gloser, J., 142–144, 152–154, *163*, *165*, *166*
Glover, J., *259*
Glover, P.E., 225, 231, 237–239, 247, 248, *258*, *259*
Goede, A., 412, *431*
Golditch, S.S., 459, *467*
Golov, B.A., *56*
Golubintseva, V.P., 25, *58*
Good, R., 442, *467*
Goodall, D.W., 182, *194*, *288*
Gordyagin, A.Ya., 3, 25, *56*
Goudet, J.P., *216*
Gould, L.R., *433*
Gradwell, M.W., 370, *405*
Grandvaux Barbosa, L., 265, *290*
Grange, J.J. la, 427, *432*
Granier, P., 301, *301*
Grant, C.H.B., 226, *261*
Greene, D.M., 411, *432*, *433*
Greene, S.W., 411, 416, 419, 421, *431*, *432*
Greenway, P.J., 238, *259*, *262*
Greenwood, R.M., 393, *405*
Greer, D.H., 376, 381, 393, *405*
Gremmen, N.J.M., 304, *312*
Grenke, W.C., *468*

Grenot, C.J., 209, *219*
Griffiths, P.J., 222, 227, 229, 235, 240, *259*, *261*, *263*
Grigg, J.L., *410*
Grossman, D, 276, *288*
Grouzis, M., 180, *194*, 206–208, *218*
Grover, C.P., 257
Groves, R.H., *289*, 315, 327, 328, *356*
Grubb, P.J., 437, *468*
Grulich, I., 146, *163*
Grunow, J.O., 275, 277, *288*, *289*
Gugalinskaya, L.A., *55*
Guillaumet, J.L., *301*
Guillaumin, A., 455, *468*
Gupta, A., 106, *118*
Gupta, B.S., 115, *118*
Gupta, R.K., 91, 92, 109, *118*
Gupta, S.P., *121*
Gupta, S.R., 100, 102, 103, 106–108, *118*
Guricheva, N.P., 23, 42, 46, 48–53, *55–57*
Gutman, M., 180, *196*
Gwynne, M.D., 226, 227, 229–231, 234, 235, 239, 242, 243, 246, 248–250, *259*, *262*
Gyllenberg, G., 147, *163*

Haantjens, H.A., 447, *468*
Haarer, A.E., 228, *259*
Haas, H., 148, *163*
Haase, P., 363, *405*
Hackman, B.D., 454, *470*
Hagen, Y., 427, *432*
Hall, E.A., 330, *356*
Hallsworth, E.G., *359*
Haltenorth, T., 211, *218*
Halva, E., 131, 138, 140, *163*
Hamel, O., 216
Hamilton, A.C., 265, *289*
Hammond, A.L., 370, *406*
Han Rong-kuen, 78, *80*
Hancock, I.R., 441, 454, *468*
Hannibal, L.W., 95, *119*
Hanxi, Y., *122*
Happold, D.C.D., 1, 211, *217*, *218*
Hardin, G., 216, *218*
Harker, K.W., 239, *259*
Harper, J.L., 131, *163*
Harrington, G.N., 239, *258*, *259*, *262*, 353, *356*, *358*, *360*
Harris, L.D., 242, 248, *259*
Harrison, R.A., 396, *406*
Harrop, J.F., 227, *260*
Hart, D., 437, *467*
Hartley, W., 2, *2*, 168, 169, *169*, 303, *312*, 439, *468*
Harvey, J.M., *356*
Hassall, D.C., 458, *468*
Hassan, A.S., *260*

Hassanyar, A.S., 89, 114, *119*
Hatheway, W.H., *468*
Hattersley, P.W., 439, *468*
Havel, J.J., 442, *468*
Hayman, D.S., 398, *406*
Hayward, J.A., 400, *406*
Haywood, M., *195*, 214–216, *218*, 245, 246, 249, 250, *260*
Heady, H.F., 64, *80*, 221, 222, 225, 227, 230, 239, 246, *259*
Heal, O.W., 148, *164*
Heathcote, R.L., 316, *357*
Hedburg, O., 430, *432*
Hemming, C.F., 245–247, 249, 252, 253, 255, 256, *259*
Henderson, C.P., 441, 454, *468*
Hendy, C.R.C., 247, 249, 250, *259*
Hendy, K., 231, *259*
Hengmichai, P., 113, *118*
Henty, E.E., 303, *312*
Henty, E.E., 464, *468*
Herlocker, D.J., 225, 237–239, 242, 252–256, *257*, *259–261*, *263*
Heusch, B., 203, *218*
Heyligers, P.C., 438, 442, 443, 448, *468*
Hiernaux, P., 207–209, 213, *218*
Hillman, A.K.K., 242, *260*
Hillman, J.C., 242, *260*
Hirani, O.H., *120*
Hnatiuk, R.J., 363, *406*, 416, 420–424, 430, *432*, 452, *468*
Hoare, P., *118*
Hodge, C.A.H., 247, *257*
Hofmann, R.R., 280, *289*
Hogg, I.G.G., *258*
Holdgate, M.W., 411, 427, 428, *432*
Holdridge, L.R., 437, *468*
Holdsworth, D.K., 370, 400, *404*, *406*
Holgate, M.D., 342, *357*
Holland, A.A., 335, *357*
Holling, C.S., *290*
Holloway, J., 455, *468*
Holloway, J.T., 361, 374, 377–379, 388, *406*
Holmes, C.H., 83, 88, 92, *119*
Holmes, J.H.G., 111, *119*
Holmes, W.E., 325, *358*
Holt, R.M., 254, *260*
Hooker, J.D., 420, *432*
Hopcraft, D., 241, *260*
Hopcraft, D.D., *196*
Hope, G.S., 430, *432*, 438, 462, *468*
Horne, J.F.M., *257*
Hosking, W.J., 348, *358*
Hoste, C., 180, *195*, 208, *219*
Hou Hsioh-yu, 63, 73, *80*
Hounan, C.E., 316, *357*
Howard, B., *196*
Howard, W.E., 374, 379, *406*
Hubbard, C.E, 222, *260*

Hubble, G.D., 315, 319, *357*, *359*
Hugo, H.J., 214, *218*
Humbert, H., 292, *301*
Hunting Technical Services *218*
Huntley, B.J., 270, *289*, 416, 419, 427, 429, *432*
Hussain, E., 83, 88–90, 109, *119*
Hussain, T., 89, *119*
Hussein, A.J., 254, *263*
Hutasoit, J.H., *122*
Hutchings, R.J., *355*
Hutton, E.M., 346, 347, *356*

Ibrahim, K.M., 94, *119*
Ibrihim, A.M., 252–254, *259*
International Livestock Centre for Africa (ILCA), 246, *260*
Irwin, J.A.G., 347, *357*
Isachenko, T.I., 3, 20–22, *56*
Islam, M.A., 87, *119*
Ivanov, V.V., 54, *56*
Ivens, G.W., 95, *119*, 239, *260*
Ives, D.W., 381, *406*
Izmailova, N.N., *56*

Jackman, M.E., *356*
Jackson, G., 221, *260*
Jacobs, S.W.L., 304, *313*, 439, *469*
Jaffré, T., 455, *468*
Jagusch, K.T., 399, *406*
Jain, A.P., 103, *121*
Jain, H.K., 101, 107, *119*, *120*
Jain, S.K., 99, *119*
Jakrlová, J., 131, 134, 155, 156, *164*, *165*
Jakubczyk, H., 152, *164*
Jama, A.A., 252–254, *261*
Jankowska, K., 134, *164*
Jaritz, G., 193, *194*
Jarman, P.J., 279, *289*
Jarvis, J.U.M., 226, 248, *260*
Javier, E.Q., *119*
Jayan, P.K., 97, *119*
Jeannel, R.G., 427, *432*
Jenkin, J.F., 412, 413, 416, 417, 419–426, 429, 431, *432*
Jessup, R.W., 327, 329, *357*
Jiang Shu, 67, 75, *80*
Joachim, A.W.R., 87, *119*
Joffre, R., 189, 193, *194*
Johns, R.J., 444, *468*
Johnson, C.R., *468*
Johnson, P.N., 376, 393, *406*
Johnson, R.W., 306, *312*, 341, *357*, 435, 439, 440, 442, *468*
Johnston, A., 83, 88–90, 109, *119*
Johnston, G.C., *433*
Johnston, J.A., *407*
Johnstone, P.D., *407*
Jones, E., 430, *432*
Jones, R.I., *288*

Jones, R.M., 347, *359*
Joshi, M.C., 83, 98, 99, 101–103, *119, 122*
Jozwik, F.X., 325, *357*
Justice, C.O., 212, *218*

Kahurananga, J., 239, 240, *260*
Kajak, A., 147, 148, *163*
Kalenov, H., *196*
Kalkman, C., 438, 461, 462, *468*
Kalma, J.D., 445, *469*
Kalmakoff, J., 396, *405, 406, 408*
Kamau, P.N., 247, *260*
Kamenetskaya, I.V., 19, *56*
Kandiah, S., 87, *119*
Kanodia, K.C., *122*
Karamysheva, Z.V., 3, 4, 17, *56, 67, 80*
Karim, A.G.M.B., 87, *119*
Karue, C.N., 242, *260*
Kaul, R.N., 86, 87, *117, 119*
Kaushal, B.R., 106, *119*
Keig, G., *469*
Keita, M.N., 214, *218*
Keith, K., *433*
Keller, B.A., 3, *56*
Kellman, M., 461, *468*
Kelly, D., 393, *408*
Kelly, G.C., 401, 402, *406*
Kelly, R.D., 270, 276–278, 281, *289*
Kelsey, J.M., 396, *406*
Kemei, I.K., 246, *260*
Kenworthy, J., 222, 232, 235, *260*
Kerr, I.G.C., 378, *408*
Khalif, H.M., 253, 254, *259*
Khan, A., 89, 113, *119*
Khan, A.H., 89, 90, *119*
Khan, C.M.A., 113, 115, *119*
Khan, M., 90, 115, *119*
Khan, M.S., 88, 94, *119*
Khan, S.M., 113, *119*
Khattak, G.M., 84, *119*
Khatteli, M., 192, *194*
Khodasheva, K.S., 42, 45–47, 49–51, *56, 59*
Khou, S., 11, *56*
King, C., 393, *406*
Kingdon, J., 226, *260*
Kipps, E.H., *358*
Kiris-Prosvirina, I.B., 44, 50, *56*
Kirkpatrick, J.B., 458, *468*
Kirsanov, M.P., 53, *56*
Kismono, *122*
Kitamura, S., 94, *119*
Klapp, E., 127, 129, 130, 156, 157, *164*
Kleopov, Yu.D., 3, *56, 57*
Knapp, R., 1, *2,* 129, *164*
Knightly, S.P.J., 429, *432*
Knoop, W.T., 279, *289*
Kobak, K.I., *55*
Koechlin, J., 291, *301*

Kolosov, A.M., 53, *57*
Kong Zhao-chen, 62, *80*
Köppen, W., 437, *468*
Korzhinskyi, S.I., 3, *57*
Kotańska, M., 136, 138, *164*
Koteswaram, P., 86, *119*
Krajčovič, V., 140, *164*
Krasheninnikov, I.M., 3, *57*
Kreulen, D., 242, *260*
Krishnamurthy, L., 98–103, *119, 122*
Krishnaswamy, V.S., 114, *119*
Kršková, J., 134, *164*
Kruger, J.A., 277, *289*
Kruuk, H., 241, *260*
Kryazhimskyi, F.V., *55*
Kubicka, H., 152, *164*
Kubicki, A., *358*
Kuchar, P., 230, 252–254, *259, 260*
Kucherovskaya-Rozhanets, S.E., 3, *57*
Kucheruk, V.V., 42, *57*
Kulkarni, L.B., 91, *117*
Kumar, A., 101, *119*
Kumar, S., 97, *119*
Kuminova, A.V., 3, *57*
Květ, J., *165*

Lales, J.S., *121*
Lamprey, H.F., 214, 215, *218,* 239, 241, *260*
Lane-Poole, C.E., 449, 461, *468*
Lang Brown, J.R., 227, *260*
Langat, R.K., 250, *260*
Langdale-Brown, I, 221, 222, 225, 227–229, 232–234, 238, 239, *260, 263*
Lange, R.T., 177, *194,* 329, 354, *355, 357*
Langhams, P., 167, *169*
Langlands, B.W., 221, 239, *260*
Lapáček, V., 147, *164*
Lapeyronie, A., 193, *194*
Latham, M., 455, *468*
Lauenroth, W.K., *122*
Lauer, W., 430, *433*
Lavers, R.B., 394, *406, 407*
Lavrenko, E.M., 3, 4, 11, 17, 18, 23, 25, 26, 42, 44–46, 49, 51–55, *57, 59,* 64, 67, *80, 81*
Law, P.G., 427, *432*
Lawrie, J.J., 249, *260*
Laws, R.M., 240, *258*
Lawton, R.M., 202, *218, 263*
Lay, B.G., 329, *357*
Lazarides, M., 319, 327, *359*
Le Houérou, H.N., 171–173, 176, 177, 180, 181, 187–193, *194, 195,* 197–200, 202–204, 206, 208–216, *218, 219,* 245, 246, 248–250, 257, *257, 260*
Le Roux, C.J.G., 277, *289*
Leamy, M.L., 371–373, *406*
Leclercq, P., 206, 208, *217*

Lee, K.E., 396, *406,* 448, 456, *469*
Lee, W.G., 368, 370, *406–408*
Lefebure, C., *217*
Legris, P., 84, *119*
Leigh, J.H, 329, 332, 333, 342, 354, *357, 358, 360*
Lemarque, G., 209, *218*
Lemerle, C., *119*
Lenne, J.M., *357*
Leprun, J.C., 210, *216, 219*
Lesák, J., 131, 138, 140, *163*
Leuthold, B., 243, 248, *261*
Leuthold, W., 243, 248, *261*
Levakovskyi, I., 42, *57*
Levina, R.E., 32, *57*
Lewis, J.G., 256, *261*
Lewis, M.C., 416, 421, *431*
Lewis Smith, R.I., 413, 415–429, *432, 433*
Lhote, 214, *219*
Li Bo, 63, 66, *81*
Li Bo-sheng, 74, *81*
Li Hong-chang, 77, *81*
Li Jian-dong, 65, *81, 82*
Li Yu-tang, 78, *81*
Li Yue-shu, 65, *81*
Liaccos, L.G., 192, *195, 196*
Liang, T., *468*
Lieth, H., 85, *123,* 197, *220*
Lind, E.M., 224, 229, *261*
Lindsay, D.C., 429, *432*
Line, M.A., 398, *406*
Litchfield, W.H., 327, *357*
Liu Zhong-ling, *81*
Lkhagvasuren, S., *56*
Lock, J.M., 233, 242, *261*
Löffler, E., 305, 309, *312,* 442, 452, 461–463, *469*
Loh, G., *407*
Loiseau, P., 180, 187, *195*
Long, G.A., 187–191, *194–196*
Long, M.I.E, 242, *261*
Lötschert, W., 222, *261*
Lough, T.J., 388, *406*
Loutit, M.W., 398, *406*
lsmagilov, M.I., *56*
Lu Liang-shu, 79, *81*
Ludwig, D., *290*
Lusigi, W.J, *261*

Ma Shi-jun, 77, *81*
Mabbutt, J.A., 315, 316, 319, *357, 468*
MacFadyen, A., 151, *164*
Mackey, B., 325, *360*
Mackworth-Praed, C.W., 226, *261*
MacMillan, B.H., *404*
MacRae, A.H., 376, 379, *405*
Madamba, J.C., *119*
Magadan, P.B., 111, *119*
Maggs, T.M.O'C., 282, *289*

Maignan, F., 193, *196*
Main, A.R., 315, *357*
Mainguet, M., 203, *219*
Makacha, S., 225, *261*
Makarevich, V.N., 138, 140, *164*
Makepeace, W., 379, *406*
Makin, J., 255, *261*
Makulec, G., 150, *164*
Malafeev, Yu.M., *55*
Malcolm, J.R., 226, *261*
Malik, M.N., 113, *119*
Mall, L.P., 98, 100, *119*
Mallick, D.I.J., 456, *469*
Maloty, G.O.M., *196*
Malpas, R.C., 234, *258*
Maltz, E., *196*
Manière, R., *216*
Mankand, N.R., 97, 101, *120*
Mann, H.S., 113, 115, *117, 120*
Manner, H.I., 442, 464, *469*
Manning, M.S., *404*
Marbut, C.F., 167, *169*
Margalef, R., 103, *120*
Mark, A.F., 362, 364, 370, 374, 376, 377,
    379–381, 386, 388, 389, 393, 395, 400,
    401–404, *404–409*
Marsh, J., 427, *432*
Marshall, B., 242, *261*
Martynova, M.F., 139, *163*
Marwaha, S.P., *118*
Masefield, G.B., 233, 239, *261*
Masheti, S., 257, *261*
Mason, I.L., 214, *219*
Máthé, I., 143, *164*
Mathur, P.K., *120*
Mattei, F., *217*
Mauny, R., 214, *219*
MBa, A.U., *194*
Mbugua, S.W., *258*
Mbui, M.K., 247, *261*
McAlpine, J.R., 437, *469*
McCaskill, L.W., 378, 379, 399, *407*
McClure, F.A., 94, *120*
McCracken, I.J., 370, *407*
McCulloch, B., *404, 409*
McEwen, W.M., 402, *407*
McFarlane, V.M., 191, *196*
McGlone, M.S., 363, 365, 367, 374, *407,
    409*
McIlroy, R.J.M., 228, *260, 261*
McIntosh, R.J., 214, *219*
McIntosh, S.K., 214, *219*
McKay, G.H., 104, *120*
McKell, C.M., 74, *80*
McKellar, M.H., 367, *409*
McKenzie, D., 191, *196*
McKinnon, E.A., 328, *360*
McNaughton, S.J., 239, 241, 242, *261*
McNeilly, B.A., 397, *408*
McQueen, D.R., *404*

McSaveney, M.J., 361, 385, *407*
McSweeney, G.D., *407*
McVean, D.N., 430, *432, 433*, 461, 462,
    470
Mehrotra, R.S., 107, *117, 120*
Mehta, S.C., 100, *119*
Melkania, N.P., 99, *120*
Melville, G.F., 326, *357*
Menaut, J.C., 222, *261*, 269, *289*, 291,
    301
Mentis, M.T., 279, 281, *289*
Mesdaghi, M., 110, *120*
Meurk, C.D., 370, 374, 376, 389, 391–
    394, 399, *404, 407, 410*, 411, 420,
    421, 424, 429, 430, *432*
Meyer, J.F., *216*
Michel, P., 203, *219*
Michelmore, A.P.C., 221, 224, *261*
Microchnitchenko, Y., *196*
Milchunas, D.G., *122*
Miles, J.A.R., *408*
Mills, J.A., 374, 393, 395, *407*
Mills, P.F.L., 277, *289*
Milne, G., 237, *260, 261*
Milne-Redhead, E, *260*
Milogo, G., 203, *219*
Ministry of Works, 399, *407*
Miroshnichenko, E.D., 152, *164*
Mishra, R.R., 106, *120*
Misra, C.M., *119*
Misra, G., 115, *120*
Misra, R., 90, 101, *120, 122*
Moar, N.T., 367, 385, *407*
Mobayen, S., 88, *120*
Moehlman, P.D., 241, *261*
Moll, E.J., 269, *288*
Molloy, B.P.J., 365, 368, 374, 384, 385,
    388, 397, *407*
Molloy, L.F., 366, 370, 371, 382, 383,
    397, *405, 407, 408*
Monod, T., 197, 199, 203, 209, 210, 214,
    219, 222, *261*
Monteath, M.A., 399, *407*
Montrakun, S., 87, *120*
Moomaw, J.C., 230, *261*
Moore, C.W.E., 332, 335, *357*
Moore, D.M., 420, 429, 430, *431, 432*
Moore, L.B., 361, 379, 388, *408*
Moore, R.M., 315–319, 324, 326, 329,
    331, 332, 335–338, 341–344, 346,
    349–354, *357–360*
Moore, S.G., 396, *406, 408*
Moormann, F., *118*
Morat, Ph., 294, 298, *301*
Mordkovich, V.G., 22, 23, *57*
Moreau, R.E., 226, *261*
Morel, G., 209, *219*
Morgan, W.T.W., 226, 227, 229, 231,
    232, 234, 235, 241, 243, 245, 248,
    249, *261, 262*

Mori, F., 214, *219*
Morris, J.W, 281, *289*
Morris, J.Y., 378, *406, 408*
Morris, P.A., 226, *261*
Morrison, M.E.S., 224, 229, *261*
Moseley, H.N., 416, *432*
Mosi, A.K., *194*
Mossman, A.S., 279, *288*
Mostert, J.W.C., 277, *290*
Mott, J.J., 338–340, 347, 352, 358, *359*
Moulopoulos, C., 192, *195*
Mouthaan, W.L.P.J., *468*
Mueller-Dombois, D., 92, 93, 101, 104,
    107, *120*, 310, *312, 469*
Mugambi, S., 377, *410*
Mugera, J.S., 229, *261*
Mulcahy, M.J., *359*
Mulham, W.E., 332, 333, *357, 360*
Muriuki, R.M.M., 250, *261*
Muslera Pardo, E., 193, *196*
Mwova, M.B., 250, *261*
Myers, L.F., 354, *358*

Naegelé, A.F.G., 206, *219*
Nanda, P.C., 92, *118*
Nasation, A.H., *122*
Nash, R.C., *289*
Naveh, Z., 180, *196*
Naylor, J.N., 248–250, 252–254, *261*
Nemati, N., 84–86, 102, 110, 113, *120*
Nes, P., 384, 387, 393, *410*
Neubauer, H.F., 84, 88, *120*
Newman, J.C., 355, *358*
Newman, R.J., 345, 348, *358*
Nicholls, A.O., *357*
Nieuwolt, S., 222, 229, 243, 245, 255,
    261
Nikol'skyi, A.A., 45, *57*
Nix, H.A., 305, *312*, 319, 356, *359*, 435–
    437, 440, 445, *469*
Nkurunziza, E., *261*
Noble, I.R., *356*
Noble, J.C., 329, 332, 353, *357, 358*
Noel, J., 197, *216*
Norman, M.J.T., 339, 340, 346, *358, 359*
Northcote, K.H, *359*
Norton, D.A., 393, *408*
Norton-Griffiths, M., 235, 240, *261, 263*
Norwine, J.R., 199, *219*
Novopokrovskyi, I.V., 3, *57*
Nowak, E., 150, *164*
Noy-Neir, I., *194, 288*
Numata, M., 1, *2*, 63, *81*, 88, 94, *120*,
    129–131, 140, 162, *164*
Nunn, W.M., 319, *358*
Nye, G.W., 228, *261*

O'Connor, K.F., 374, 377, 378, 384, 385,
    387, 393, 399, 401, *408, 410*
O'Laughlin, C.L., 401, *408*

AUTHOR INDEX

O'Rourke, J.T, 225, 227, 242, *257*, *261*
Oberdorfer, E., 129, 130, 157, 158, *164*
Ogilvie, C.S., 96, *120*
Ogwang, B.H., 229, *261*
Olechowicz, E., 147, 148, *164*
Oliver, J., 305, *312*, *313*, 437, *469*
Ollier, C.D., 232, *262*
Omar, A.E., *260*
Orlido, N.M., *121*
Orr, D.M., 323, 325, 355, *358*, *359*
Orshan, G., 83, *120*
Orwin, D.F.G., 427, 429, *434*
Orwin, J., 378, *408*
Osborn, T.G.B, 329, *358*
Osmaston, H.A., *260*
Ostapoff, F., 411, 412, 416, *432*
Osychnyuk, V.V., 3, 23, *58*
Ouda, N.A., *117*
Ovington, J.D., 171, *196*
Owaga, M., 242, *262*
Ozanne, P.G., 344–346, 349, *359*

Pabot, H., 85, 88, *120*
Pachoskyi, I.K., 3, 25, 42, *58*
Paijmans, K., 438, 439, 444, 445, 447, 448, 452, 461–463, *469*
Paliwal, K.C., *120*
Paltridge, T.B., *358*
Pande, H., 115, *120*
Pandey, A.N., 116, *120*
Pandey, R.K., *122*
Pandeya, S.C., 91, 97, 98, 107, 108, *120*
Pankov, A.M., 42, *58*
Papanastasis, V.P., 192, *196*
Parham, J., 457, 458, *469*
Park, G.N., 401, 402, *406*
Parry, M.S., 226, *262*
Pasternak, D., *166*
Pateut, G., *216*
Pathak, P.S., 109, *120*
Pathak, S.J., *120*
Pati, D.P., 108, *118*, *120*
Patil, B.D., 109, *120*
Paton, D.F., 348, *358*
Patra, U.C., 106, *118*
Patrick, B.H., *407*
Paviard, J., 18, *55*
Payne, A.L., 319, *358*
Payton, I.J., 388, 389, *408*
Pearce, A.J., 400, 401, *408*
Pearson, F.B., 345, *355*
Peberdy, J.R., 229, 231, 234, 235, 242, 243, 249, 250, *262*
Peden, D.G., 248, 250, *262*
Pelikán, J., 146, 147, 149–151, *164*, *165*
Pelt, M., *196*
Pemadasa, M.A., 88, 93, 107, *120*
Pendleton, R.L., 87, *120*
Penman, H.L., 197, 209, *219*
Penning De Vries, F.W.T., 207–209, *219*

Pennycuick, L., 240, *261*, *262*
Penridge, L.K., *360*
Perera, K.S.O., 87, *120*
Perera, M., 92, 93, *120*
Perkins, D.F., 148, 149, 156, *164*
Perrier de la Bathie, H., 292, *301*
Perrin de Brichambaut, G., 85, *120*
Perry, R.A., 315, 316, 319, 321, 322, 326–329, 333, 334, 338–340, 354, *355*, *357–359*
Persikova, Z.I., 137, *164*
Peterman, R.M., *290*
Peterson, D.D., 237–239, *262*
Petrides, B.A., 240, *262*
Petřík, B., 134, *164*, *165*
Peyre de Fabrègues, B., 214, 216, *219*
Pfeffer, P., 106, *121*
Phillips, J., 237–239, *262*
Phillipson, D., 224, 243, *262*
Phillipson, J., *217*, *258*
Pias, J., 203, *219*
Pichi-Sermolli, R.E.G., 221, 256, *262*
Pienaar, A.J., *289*
Pigott, C.D., 242, *263*
Pilát, A., 137, *164*
Piot, J., 203, *216*, *219*, *220*
Planchenault, D., *216*
Plewczyńska-Kuraś, U., *164*
Pohlen, I.J., 371, *409*
Polhill, R.M., *260*
Pomeroy, D.E., 226, 228, 234, *262*
Pongpiachan, P., *118*
Pontanier, R., 180, 181, 192, *194*
Popov, G.F., 197–200, 209, 211, 212, *219*
Poslušná, A., 136, 137, *164*
Pouget, M., *195*
Poulet, A.R., 203, 209–212, *219*
Powell, J.M., 463, *467*
Prajapati, M.C., 109, 113, *121*
Prakash, I., 103, 104, *121*
Pratchett, D., *258*
Pratt, D.J., 226, 227, 229–231, 234, 235, 239, 242, 243, 246–250, *257*, *262*, *263*
Précsényi, I., 143, *164*
Preece, P.B., 327, *359*
Prendergast, H.D.V., 305, *313*, 439, 440, *469*, *470*
Pullen, R., 438, *468*, *469*
Purdie, A.W., 375–377, *405*
Puri, G.S., 88, *121*
Pyatin, A.M., 131, *164*

Qian Guo-zhen, 77, *81*
Quilty, P.G., 412, *432*
Qvortrup, S.A., 238, 242, *262*

Rabotnov, T.A., 20, 36, *58*, 126, 129–131, 140, 147, *164*, *165*
Rachkovskaya, E.I., 3, 20–22, *56*

Radcliffe, J.E., 399, *408*
Radde, G., 53, *55*, *58*
Radwanski, S.A., 232, *262*
Raeside, J.D., 388, 389, *408*
Rahman, K., 88, 94, *121*
Rai, P., 115, *121*
Raimondo, F.M., 253, *262*
Rajvanshi, R., *118*
Rakhmanina, A.T., 24, *58*
Rakova, M.V., *55*
Ram, J., 98, 99, 103, *117*, *121*
Rambal, S., 207–209, *217*
Rand, A.L., 461, *467*
Ratcliffe, F.N., 330, *359*
Ratera Garcia, C., 193, *196*
Rattray, J.M., 167, *169*, 221, 222, 225, 230, 234, 246, 256, *262*
Rauh, W., 291, 292, *301*
Raunkaier, C., 187, *196*
Raven, P.H., 368, *408*
Raychaudhuri, S.P., 86, *121*
Reddi, T.V., 114, *119*
Reed, C.A., 190, *196*, 214, *219*
Regál, V., 140, *164*
Reilly, P.M., 221, *262*
Reiner, E.J., 447, *469*
Ren Ji-zhou, 70, *81*
Reppe, X.N., 427, *432*
Rescikov, M.A., 3, *58*
Resource Management and Research (RMR), 221, 246, 248–250, 252–254, 256, *262*
Rethman, N.F.G., 275, *289*
Reverdatto, V.V., 3, 25, *58*
Reynders, J.J., *468*
Richards, P.W., 437, *469*
Ricou, G.A.E., 147–149, *165*
Riney, T., 378, *408*
Riou, C., 202, *219*
Riquier, J., *118*
Risley, E., 231, *262*
Rivière, R., 203, *220*
Roa, Y.N., *120*
Robbins, R.G., 438, 442, 447, *469*
Roberts, B.R., 285, *289*
Robertson, J.A., 335, *358–360*
Rodgers, W.A., 104, *121*
Rodier, J.A., 203, *220*
Rodin, L.E., 187, 189, *196*
Roe, R., 323, 324, 354, *359*
Rogers, G.M., 363, *408*
Rol, R., 88, *121*
Rollet, B., 84, *121*
Roose, E.J., 203, *220*
Ross, D.J., 397, 398, *408*, *409*
Ross, I.C., 234, 239, 259, *262*
Rossiter, R.C., 344–346, 349, 350, 354, *359*
Rothmaler, W., 125, *165*
Rotshil'd, E.V., 42, *58*

Rouveyran, J.C., 300, *301*
Roux, E., 271, 275, 277, *289*
Rovira, A.D., 426, *431*
Rowe, L.K., *408*
Rowley, J., 370, 381, 389, 400, *407, 408*
Roxburgh, S.H., 388, *409*
Rubenach, M., *432*
Rundel, P.W., 440, 441, *469*
Rusek, J., 150, *165*
Rushworth, J.E., 270, 273, 276, 277, *289*
Russell, M.J., *359*
Rutherford, M.C., 274, 276, *289*
Ruzicka, V., *166*
Rychnovská, M., 135–138, 140–145, 154, *165*
Ryecroft, H.B., 282, *289*

Sadera, P.L.K. Ole, 250, *262*
Safford, W.F., 459, *469*
Saggerson, E.P., 222, 227, 237, 255, *262*
Said, M., 109, 112, 113, *121*
Sajise, P.E., 96, 102, *121, 122*
Sale, J.B., 248, 254, *260*
Samraj, P., *117*
Sanders, N.K., 430, *433*
Sands, W.A., 241, *260*
Sanford, W., 83, *121*
Sangaré, M., 213, *217*
Santoir, C., *216*
Sarson, M., 191, *194*
Sastradipradja, D., 111, *122*
Satpathy, B., *118*
Savory, A., 285, *289*
Saxena, A.K., 84, 92, *121, 122*
Saxena, S.K., 91, *118*
Scanlon, J.C., 325, 340, *359*
Scattini, W.J., 355, *359*
Schalke, H.J.W.G., 413, *433*
Schmid, M., 95, 96, *121*, 438, 456, *469*
Schmidl, D., 241, *262*
Schmidt, W., 237–239, *262*
Schottler, J.H., *119*
Schultz, J., 238, *262*
Schulze, B.R., 413, 416, 417, *433*
Sclater, W.L., 226, *261*
Scott, D., 369, 379–381, 401, *406, 409*
Scott, J.D., 237–239, *262*
Scott, R.M., 223, 227, 230, 232, 233, 237, 245, *262*
Seavoy, R., 116, *121*
Seavoy, R.E., 461, *469*
Sebillotte, M., 180, 187, *195*
Seely, M.K., 242, *262*
Sekulic, R., 231, *262*
Selassie, A.H., 249, 250, 257, *257*
Seligman, N.G., 180, *194, 196*
Semenova-Tyan-Shanskaya, A.M., 23, *58*
Senapati, B.K., 106, *118, 121*
Senaratna, J.E., 115, 116, *121*

Seppelt, R.D., 419, *433*
Serebryakova, T.I., 133, *166*
Seredneva,T.A., 52, *55, 58*
Seth, S.K., 83, *118*
Shaabani, S.B., *263*
Shalyt, M.S., 18, 22, 25, *58*, 137, 139, *165*
Shanan, L., *194*
Shankar, K., 85, *121*
Shankar, V., 97, *119, 122*
Shankarnarayan, K.A., 88, 90, 91, 109, 112, 115, *117, 118, 121, 122*
Shantz, H.L., 167, *169*
Sharma, R., *120*
Sharma, S.C., *120*
Sharma, S.K., *118*
Sharman, M.J., *220*
Sharpe, P.J.H., *360*
Shaw, D.F., 461, *469*
Shaw, N.H., 340, 346, 353, *359*
Shear, J.A., 414, 430, *433*
Shelton, J.N., 340, *359*
Shelyag-Sosonko, Yu.R., 23, *58*
Shennikov, A.P., 125, 129, 130, 157, *165*
Shepherd, W.O., 206, *220*
Shkolnik, A., 191, *196*
Shrestha, P.B., 85, *121*
Shupranov, N.P., *58*
Silverton, J.W., 393, *409*
Simmons, D.R., 393, *409*
Simon, B.K., 304, 306, 312, *312, 313*, 439, *467, 469*
Simonetta, A.M., 248, *262*
Simpson, K.N.G., *433*
Simpson, M.J.A., *404*
Simpson, N.C., *407*
Sims, P.L., 180, *194*
Sims, R.W., 426, *433*
Sinclair, A.R.E., 240, 241, *262, 263*
Singh, J.S., 83, 84, 86, 88, 92, 95, 98–103, 106–108, 113, 115, *117, 118, 120–123*
Singh, K.P., 102, 103, 115, *120, 122*
Singh, M., 116, *122*
Singh, P., 109, *122*
Singh, S.P., *121*
Singh, V.P., 91, 97, 98, *122*
Sithamparanathan, J., *122*
Sivasupiramanian, S., 115, *122*
Škapec, L., 147, 148, *165*
Skerbek, W., *195*
Sleeman, J.R, *359*
Slemnev, N.N., *56*
Smeins, F.E., *257*
Šmíd, P., 153, *165*
Smit, I.B.J., 277, *289*
Smith, A.C., *469*
Smith, C.A., 348, *356*
Smith, J., 256, *263*
Smith, J.M.B., 308, *313*, 430, *432, 433*,
437, 438, 442, 452, 457, 462, *469, 470*
Smith, V.R., 411, 419–426, *433*
Smith-Dodsworth, J.C., 361, *404*
Sobey, W.R., 430, *433*
Sochava, V.B., 1, *2*
Soerianegara, I., 95, 113, *122*
Soerjani, M., 95, *122*
Soewardi, B., 111, *122*
Solomon, S.F., *217*
Somali National Range Agency (SNRA), 245, 248–250, *263*
Sombroek, W.G, 227, 231, 235, 242, 245, *263*
Sommerville, P., 388, *409*
Soons, J.M., 367, *409*
Sørensen, T., 93, *122*
Southcott, W.H., *359*
Specht, R.L., *194, 288*, 315, 329, *356, 359*
Speck, N.H., 333, *359*
Spegel-Roy, P., 190, *196*
Speidel, B., 129, 130, 137, 138, 140, 162, *165*
Speir, T.W., 397, *409*
Spitzer, K., 147, *165*
Springfield, H.W., 88, *122*
Stace, H.C.T., 315, *359*
Stannard, M.E., 330, *359*
Stanton, J.P., *359*
Stepanova, E.F., 3, *58*
Stephens, D., 229, *263*
Stephenson, C., 421, 424, 426, *432*
Stevens, G., 366–368, 374, *409*
Stevens, P.F., 462, *467*
Stewart, D.R.M., 256, *263*, 280, *289*
Stewart, G.A., 323, 326, 354, *356, 359*
Stewart, W.D.P., 426, *431*
Stirling, M.W., *404*
Stocker, G.C., 339, *359*
Story, R., 334, *359*
Stout, J.D., 397, *409*
Strugnell, R.G., 242, *263*
Stuart, S.N., 226, *263*
Sturtz, J.D., 339, *359*
Stuth, J,W., 247, *261*
Sudarmadi, *122*
Suganuma, T., 129, 130, 162, *165*
Suggate, R.P., 367, 385, *409*
Suijdendorp, H., 319, 320, *358, 359*
Sukhoverko, R.V., *56*
Sukopp, H., 137, *163*
Sullivan, M.E., *312*, 315, 316, *357*
Susetyo, B., 112, *122*
Sutcliffe, R.C., 85, *122*
Suwardi, I.K., *122*
Sveshnikova, V.M., 18, *58*
Swank, W.G., 240, *262*
Swartzman, G.L., 113, *122*

t'Mannetje, L., 346, *359*

AUTHOR INDEX

Tadmor, N.H., 180, *194*, *196*
Tahir, H.M., 110, *117*
Tainton, N.M, 274, 275, 284–286, *288*, *289*
Takhtajan, A., 304, *313*
Talbot, L.M., 238, 239, *257*, *263*
Talbot, M.H., 241, *263*
Taliev, V.I., 25, *58*
Tallowin, J.R.B., 421, 423, *433*
Taneja, G.C., 104, *121*
Taylor, B.W., 419, 421, 425, 426, *433*, 442, 448, 462, *470*
Taylor, H.C., 282, *289*
Taylor, N.H., 371, *409*
Taylor, R.D., 279, 281, *289*, 427, 429, *433*
Terry, P.J., *261*
Tesařová, M., 142, 151–154, *163*, *165*, *166*
Tessema, S., 242, *263*
Thalen, D.C.P., 86, 87, *117*, *119*
Thambi, A.V., 106, *118*, *122*
Theron, G.K., *288*
Thiault, M., 193, *196*
Thomas, A.S., 227, 228, 233, 234, *263*
Thomas, B.D., 239, *263*
Thomas, P.K., *121*
Thompson, K., 265, *289*
Thompson, R.B., 454, *470*
Thorbahn, P.F., 249, *263*
Thornthwaite, C.W., 437, *470*
Thornton, D.D., *261*
Thornton, R.H., 397, *409*
Thurow, T.L., 253, 254, *257*, *259*, *260*, *263*
Timmins, S.M., *407*
Tinley, K., 272, 273, *289*
Titlyanova, A.A., 145, *166*
Tkachenko, V.S., 5, 7, 8, 14, 23–25, 55, *58*
Tost, J.A., Jr., *468*
Tothill, J.C., 306, *312*, 338, 341, 347, 352, 353, *358*, *359*, 435, 439, 440, 442, *468*
Toupet, C., 214, *219*
Toutain, B., 214, 216, *217*
Traczyk, H., *166*
Traczyk, T., 138, 140, *166*
Trapnell, C.G., 221, 225, 255, 256, *259*, *263*
Tregubov, V., 88, *120*
Trilica, M.J., *257*
Trivedi, B.K., 116, *122*
Troll, C., 416, 430, *433*
Trollope, W.S.W., 283, *289*
Trotter, M.M., *404*
Trump, E.C., 237–239, *259*, *263*
Truog, E., 398, *409*
Tubiana, J., 215, *220*
Tubiana, M.J., 215, *220*

Tucker, C.J., 212, *220*
Turner, H.N., *359*
Turrill, W.G., *260*
Tussock Grassland Research Committee, 388, *409*
Twyford, I.T., 457, 458, 460, *470*
Tyulina, L.N., 50, *58*

U.S.D.A., S.C.S., Soil Survey Staff, 371–373, *409*
Udvardy, M., 437, *470*
Úlehla, J., 153, 154, *166*
Úlehlová, B., 151, 153, 154, *165*, *166*
UNESCO, 232, 243, 250, 255, 258, 259, *261*–*264*
UNESCO, 95, *117*, *119*–*123*
United Nations Sahelian Office (UNSO), 250, 252, *263*
Upadhayay, V.S., 113, *121*, *122*
Upadhayaya, S.D., 91, 98, *122*
Uranov, A.A., 131, 133, *166*

Vacher, J., *194*
Valentin, C., *216*
Valenza, J., *216*
Valter, G., 55, *55*, *58*
Valverde, J.A., 210, *220*
Van Balgooy, M.M.J., 442, *470*
Van Baren, F.A., *468*
Van der Pouw, B.J.A., *263*
Van Diepen, D., 95, *118*
Van Engelen, V.W.P., 245, 247, *263*
Van Ittersum, G., *220*
Van Rensburg, H.J., 222, 224, 225, 227, 229, *263*
Van Royen, P., 438, 448, *470*
Van Steenis, C.G.G.J., 449, *470*
Van Wijngaarden, W., 245–250, *258*, *261*, *263*
Van Wyk, J.J.P., 167, 168, *169*
Van Zinderen Bakker, E.M., Sr., 411–413, 429, 430, *433*
Vandakurova, E.V., 3, *58*
Vanpraet, C., 206, *220*
Varney, B.M., 414, *433*
Vartha, E.W., *408*
Vassiliades, G., *216*
Vats, L.K., 106, *122*
Venter, A.D., 285, *289*
Verboom, W.C., 96, *122*
Verdcourdt, B., 237, 238, *263*
Verma, C.M., 114, 115, *117*, *118*
Vernadskyi, V.I., 42, *58*
Vernon, T.R., 397, *409*
Vesey-FitzGerald, D.F., 221, 225, 238, *259*, *263*
Vicherek, J., 131, *164*, *166*
Vidal, P., 208, *217*
Vink, W., 438, 461, 462, *468*
Vinogradov, B.S., 52, *58*

Vinogradov, B.V., *196*
Virot, R., 455, *470*
Vogel, J.C., 440, *470*
Von Maydell, H.J., 213, *220*
Voronov, A.G., 42, 47, 49–51, 53, 54, 56, *58*
Vorster, L.F., 275, 277, *289*, *290*
Vos, J.G., 248, *263*
Vyas, L.N., 100, *123*
Vyas, N.L., 100, *123*
Vysotskyi, G.N., 3, 7, 25, *58*

Wace, N.M., 411, 417, 419, 420, 427, 428, *432*, *433*
Wade, L.K., 430, *433*, 461, 462, *470*
Wadia, D.N., 87, *123*
Wadsack, J., 193, *194*
Wagner, F.H., 241, *258*
Wagner, H., *169*
Walker, B.H., 202, *220*, 270, 276, 278, 279, 281, 284, 286, *289*, *290*
Walker, D., 442, 462, 463, *470*
Walker, J., 335, 340, 353, *358*–*360*
Wallen, C.C., 85, *120*
Walter, H., 61, 65, 70, 71, *81*, 85, *123*, 125, 127, 157, *166*, 197, 200, *220*, 242, *263*, 273, 274, *290*
Walton, D.W.H., 413, 416, 417, 419–429, *432*, *433*
Wang Jin-ting, 74, *81*
Wang Xian-pu, 63, *80*
Wang Yi-fung, 63, 67, 70, *81*
Wangeri, E., 83, *121*
Wardle, P., 363, 365, 367, 368, 370, 376, 388, *404*, *406*, *407*, *409*
Warfa, A.M., 253, *262*
Wasilewska, L., 150, *166*
Watson, L., 439, *468*
Webb, L.J., 324, *356*
Weigert, R.G., *166*
Weinmann, H., 275, *290*
Weiss, A., 138, 140, *165*
Wells, J.A., 376, *409*
Wendelken, W.J., *406*
Wendt, W.B., 235, *263*
Were, G., 224, 227, 230, 243, *264*
Werger, M.J.A., 269, 279, *290*
Wernstedt, F.L., 271, *290*
Weston, E.J., *358*
Wetmore, S.P., *258*
Whalley, R.D.B., 332, *358*
Whistler, W.A., 308, *313*
White, E.G., 394, 396, 397, *406*, *409*
White, F., 203, 211, *220*, 221, 224, *264*
White, R.O., 221, *264*
Whitehouse, I.E., 385, *407*
Whitmore, T.C., 437, 438, 450, 454, *470*
Whyte, R.O., 83, 84, 88, 90, 95, 114, *123*, 464, *470*
Wiedemann, A.M., 321, *360*

Wielgolaski, F.E., 149, *166*, 417, *434*
Wijesinghe, L.C.A., 115, *123*
Wilcox, D.G., 321, 327, 328, *358*, *360*
Wilcoxon, J.A., *432*
Wild, H., 265, 280, *290*
Willett, R.W., 367, *410*
Williams, C.H., 346, *360*
Williams, J., *358*
Williams, J.D., 353, *358*
Williams, J.G., 231, 234, 241, 248, 256, *264*
Williams, O.B., 315, 316, 325, 354, *355*, *356*, *360*
Williams, P.A., 370, 374–377, 381–384, 386, 387, 389–393, *410*
Williams, R.J., 315, 319, *360*
Willoughby, W.M., 354, *360*
Wilson, A.D., 191, *196*, 329, 333, 353, 354, *357*, *360*
Wilson, D., 224, 227, 230, 243, *264*
Wilson, H.D., 370, 377, *410*
Wilson, J.B., *406*, *407*, *409*
Wilson, J.G., 234, *260*, *264*
Wilson, P.R., 427, 429, *433*, *434*
Wilson, R.T., 207, 209, 210, 213, *220*, 248, *264*
Wiltshire, G.H., 275, *290*
Wing, L.D., 228, *264*
Winkworth, R.E., 319, 350, *360*

Winter, W.H., 347, *356*
Winters, R.K., 85, 86, 88, 109, *123*
Wolde-Mariam, M., 222, 232, 245, 255, *264*
Wood, F.H., 396, *410*
Wood, J.G., *358*
Wraight, M.J., 377, 379, 389, 390, 396, *406*, *410*
Wright, A.C.S., 457, 458, 460, *470*
Wu Yan, 78, *81*
Wu Zheng-yi, 62, *81*

Xi Rui-hua, *81*
Xia Wu-ping, Xia, 63, 74, *81*

Yadava, P.S., 84, 86, 98, 99, 102, 103, *122*, *123*
Yagil, R., 191, *196*
Yalden, D.W., 226, *264*
Yang Dian-chen, *82*
Yeates, G.W., 396, *410*
Yong Shi-peng, *81*
Young, R.B., 336, *356*
Young, S.A., 257
Young, S.B., 413, *434*
Young, S.R., 371, 375, *408*
Yunatov, A.A., 3, 17, 18, 26, 42, 44–46, 49, *55*, *57*, *59*

Zablan, T.A., 102, *123*
Zajonc, I., 150, *166*
Zalesskyi, K.M., 25, *59*
Zaphiro, D.R.P., 256, *263*
Zavadskaya, I.G., *55*
Zelena, V., *163*
Zeuner, F.E., 190, *196*
Zhang Rong-zhu, 76, *81*
Zhang Xiao'ai, 77, *81*
Zhang Zu-tong, 67, 79, 80, *81*
Zhao Song-qiao (Chao Sung chiao), 67, *81*
Zhen Zuo-xin, 76, 77, *81*
Zheng Hui-ying, 61, *81*
Zhu Ting-cheng, 61–63, 65, 66, 68, 71, 78, *81*, *82*
Zhu Zhi-cheng, 67, *82*
Zimina, R.P., 42, 51, 52, *59*
Zlotin, R.I., 42, 45–47, 49–52, 56, *59*, 153, *166*
Zohary, D.D., 190, *196*
Zohary, M., 86, 88, *123*
Zotov, V.D., 374, 375, 377, 379, *405*, *410*
Zou Hou-yuan, 69, *82*
Zoz, I.G., 3, *57*
Zu Yuan-gang, 62, *82*
Zulfiqar, A., 86, *123*
Zyromska-Rudzka, H., 150, *166*

SYSTEMATIC INDEX [1]

Abies pindrow Spach, 93
Acacia, 95, 96, 167, 191, 193, 204, 222, 233, 238–240, 245, 246, 252, 267, 270, 272, 278, 316, 317, 321, 326–331, 335–337, 347, 349, 350, 445, 457
   A. aneura F. Muell. (mulga), 326–328, 331, 336, 337
   A. brevispica Harma, 233
   A. bussei Sjoestedt, 246, 247
   A. calciola Ford & Ieing, 328
   A. cambagei Baker (gidgee), 327, 329, 348
   A. clavigera E. Mey., 238
   A. craspedocarpa F. Muell., 328
   A. cyanophylla Lindl., 193
   A. drepanolobium Harms ex Sjostedt, 236, 238
   A. edgeworthii T. Anders., 245, 252
   A. ehrenbergia Hayne, 199
   A. estrophiolata F. Muell. (ironwood), 328
   A. etbaica Schweinf., 246, 247
   A. excelsa Benth., 337
   A. farnesiana (L.) Willd., 298, 301, 321
   A. georginae F.M. Bail (gidgee), 327, 328
   A. gerrardii Benth., 238, 239
   A. harpophylla Benth. (brigalow), 341, 342, 345, 348
   A. hockii De Wild., 239
   A. homalophylla A. Cunn. ex Benth., 337
   A. horrida (L.) Willd., 245, 252
   A. kempeana F. Muell., 327, 328
   A. leucophloea Willd., 96
   A. linophylla W.V. Fitzg., 328
   A. mellifera (Vahl) Benth., 238, 233, 246
   A. mollissima Willd., 115
   A. nilotica (L.) Del., 233, 238, 252, 346
   A. nilotica var. *indica* (Benth.) Hill., 115
   A. nubica Bemth., 246
   A. papyrocarpa Benth. (= *A. sowdenii*), 327
   A. pendula A. Cunn. ex G. Don (Wilga), 332, 337, 344
   A. pruinocarpa Tindale, 327
   A. raddiana Savi, 178, 199
   A. reficiens Wawra, 233, 244, 246, 252, 256
   A. richii A. Gray, 458
   A. sclerosperma F. Muell., 328
   A. senegal (L.) Willd., 199, 204, 245, 252
   A. seyal Del., 199, 204, 215, 238

   A. sowdenii (= *A. papyrocarpa*), 327
   A. spirorbis Labill., 455, 457, 460, 461
   A. tetragonophylla F. Muell., 328
   A. tortilis (Forsk.) Hayne, 205, 238, 244, 245
   A. tortilis (Forsk.) Hayne subsp. *raddiana* (Salli) Brenan, 187
   A. translucens A. Cunn. ex Hook., 320
Acaena, 429
   A. adscendens Vahl (now *A. magellanica*), 421,
   A. magellanica (Lam.) Vahl, 311, 421, 426, 429
Acanthaceae, 203
Acanthrodilus (lumbricid earthworm), 426
Acarina, 150, 151, 479, 480
Accipiter gentilis (L.) (gros-hawk), 77
   A. nisus (L.) (sparrow-hawk), 77
Achillea millefolium L., 136, 139, 157, 159
   A. ptarmica, L., 158
Achnatherum splendens (Trin.) Nevski, 53, 72–74
Achyrophorus ciliatus (L.) Scop. Bip., 72
   A. maculatus Scop., 21
Acinonyx jubatus (Schreber) (cheetah), 190, 211, 226, 241
Aciphylla, 380
   A. aurea W.R.B. Oliver, 381
   A. scott-thomsonii Ckn. & Allan (speargrass), 369
Acrididae, 396, 397
Acridocarpus excelsus Juss., 298
Acrocephalus arundinaceus (L.) (great reed warbler), 77
Acroceras, 96
Acrophialophora fusispora (Saksena) Ellis, 107
Actinomycetes, 30, 40, 41, 107, 151, 480
Addax nasomaculatus De Blainville (addax antelope), 209–211
Adenium, 234
Adenophora stenophylla Hemsley, 66
Adonis vernalis L., 157, 158
   A. wolgensis Stev., 15
Aegilops, 179
Aeluropus, 89
   A. litoralis (Gouan) Parl. (= *A. laevis* Trin.), 88, 159, 161, 182
   A. mucronatus Asch., 89
   A. repens (Dasf.) Parl., 88
Aepyceros, 280
   A. melampus (Lichtenstein) (impala), 239, 280
Aerva javanica (N.L. Burman) Juss. ex Schult. (= *A. tomentosa*) (kapok bush), 321, 325
   A. tomentosa Lam., 321
Afrotrichloris hyaloptera Clayton, 253
   A. martini Chiov., 253
Agathis australis Salisb. (kauri), 361

---

[1] In this index, no attempt has been made, for larger taxonomic entities, to list all the pages where subordinate taxa are mentioned. These may be found by using the Systematic List of Genera (pp. 483–488). For some major groupings, more detailed entries will be found in the General Index.

Agathis (continued)

    A. robusta subsp. *nesophila* Whitmore, 450
Agathophora alopecuroides (Del.) Bunge, 185
Agave sisalana Perrine, 231
Agrimonia eupatoria L., 138
    A. pilosa Ledeb. (= *A. eupatoria*), 162
Agropyron, 4, 8, 14, 78, 79, 89, 94
    A. cristatum (Roem. & Schult.) Asch. & Gr. 17, 18, 48, 53, 67, 71, 159
    A. cristatum (Roem. & Schult.) Asch. & Gr. var. *pectiniforme*, 158
    A. desertorum (Fisch.) Schult., 54
    A. elongatum (Host) P. Beauv., 179, 182
    A. pectinatum (Bieb.) Beauv., 22, 52
    A. pectiniforme Roem. ex Schult., 50
    A. pseudocaesium (Pacz.) Zoz, 158, 159
    A. repens (L.) Beauv., 140, 157, 159–161
    A. scabrum (= *Elymus rectisetus*), 374
Agropyropsis lolium (Trab.) A. Camus, 182
Agrosteae, 1, 2, 168, 169, 303
Agrostideae, 188, 312
Agrostis, 308, 430, 441, 472
    A. alba L., 139
    A. borealis Hartm., 161
    A. canina L., 92, 127, 158
    A. capillaris L. (brown top), 399
    A. clavata Trin., 162
    A. magellanica Lam., 418, 419, 422, 423
    A. reinwardtii Miq., 452
    A. stolonifera L., 130, 140, 157, 159–161, 182
    A. stolonifera L. var. *gigantea* (L.) Asch. & Gr., 159
    A. syreistschikowii P. Smirn. (= *A. vinealis*), 4, 158, 159
    A. tenuis Sibth., 130, 133, 138, 140, 141, 148, 149, 152, 156–159
    A. trinii Turcz., 50
    A. vinealis, Schreb. (= *A. syreistschikowii*), 4
    A. vulgaris With., 136
Aira, 384
    A. caryophyllea L., 342
Ajania achilleoides (Turcz.) Ling, 67
    A. fruticulosa (Ledeb.) Poljak., 17, 28
Alauda arvensis (L.) (sky lark), 77
    A. gulgula (Franklin) (small sky lark), 77
Albizia anthelmintica (L.) Gamble, 245
    A. procera (Roxb.) Benth., 95, 446, 447, 450
Alcelaphus, 280
    A. buselaphus (Pallus) (hartebeest), 209, 211, 226, 234, 280
    A. buselaphus swaynei (Swayne's hartebeest), 248
    A. caama (G. Cuvier) (red hartebeest), 211
    A. lichtensteini (Peters) (Lichtenstein's hartebeest), 280
Alchemilla, 135, 158, 296, 477
    A. crinita Buser, 138
    A. jailae Juz., 130, 159
    A. leptantha Juz., 159
    A. monticola Opiz, 138, 152
    A. vulgaris L., 157
Alectorolophus major Reichenb. (= *Rhinanthus major*), 158
Allactaga sibirica Foster (Siberian jerboa), 43, 76
Allium, 4, 33, 47, 76

    A. clathratum Ledeb., 28
    A. mongolicum Regel., 28, 67
    A. pallasii Murr., 28
    A. polyrrhizum Turcz. ex Regel, 13, 17, 28, 31, 35, 39, 45, 67
    A. schoenoprasum L., 161
    A. tenuissimum L.. 45
Alloteropsis, 306
    A. semialata (R. Br.) Hitchc., 274, 278, 284
Alnus, 127, 162
Aloë, 246
Alopecurus, 477
    A. bulbosus Gouan, 182
    A. geniculatus L., 157
    A. magellanicus Lam., 419
    A. pratensis L., 125, 128, 130–132, 134–138, 140–146, 152, 154, 156–161, 182
    A. ventricosus Pers., 130, 159–161
Alphitonia zizyphyoides (Spreng.) A. Gray, 458
Alstonia muelleriana Domin, 445, 455
Alysicarpus vaginalis (L.) DC., 96
Alyssum spinosum L., 185
    A. turkestanicum Regel & Schmalh., 14, 29, 50
Amaranthus, 113
    A. albus L., 52
Ammodorcas clarkei (Thomas) (Clarke's gazelle or dibatag), 248, 253
Ammotragus lervia Pallas (barbary wild sheep), 211
Amorphophallus campanulatus Blume ex Decne., 446, 449
Ampelodesmos mauritanicum (Poir.) Dur. & Schinz, 176
Amygdalus, 8
    A. scoparia Spach, 186
Anabaena, 398
Anabasis, 7
    A. aphylla L., 54
    A. articulata (Forsk.) Moq., 185
    A. brevifolia C.A. Mey., 17, 67
    A. oropediorum Maire, 185, 190
    A. salsa (C.A.Mey.) Benth. ex Volkens, 14
    A. setifera Moq., 89, 185
Anacardium occidentale L., 231
Anarrhinum brevifolium Coss., 185
Anas clypeata (L.) (northern shoveller), 77
    A. formosa (Georgi) (Baikal teal), 77
Andropogon, 94, 95, 97, 167, 215, 225, 242, 296, 477
    A. aciculatus Retz. (= *Chrysopogon aciculatus*), 95
    A. amboinicus (L.) Merr., 95
    A. eucomus Nees, 295
    A. gayanus Kunth, 206, 273
    A. greenwayi Napper, 238
    A. ischaemum L. (= *Bothriochloa ischaemum*, 64, 69, 71, 88
    A. kelleri Hackel, 246
    A. lividus Thwaites, 93
    A. muricatus Retz., 100
    A. parviflorus Roxb., 95
    A. pseudapricus Stapf, 206, 207
    A. schirensis Hochst. ex A. Rich., 203, 206, 274
    A. tectorum Schum., 206
Andropogoneae, 1, 2, 167–169, 188, 272, 274, 294, 303, 306, 312, 440
Androsace kozo-poljanskii Ovcz., 8

Androsace (continued)

    A. maxima L., 29
    A. septentrionalis L., 50
    A. tapete Maxim., 64, 69
Anemone dichotoma L., 161, 162
Aneurolepidium chinense (Trin.) Kitag., 65
    A. dasystachys (Trin.) Nevski (= *Leymus dasystachys*), 68
Angaracris barabensis (Pallas) (grasshopper), 77
Angelica anomala Avé-Lall. (= *A. sylvestris*), 162
    A. sylvestris L., 127, 157
Angiospermae, 30
Anisantha tectorum (L.) Nevski (= *Bromus tectorm*), 14, 50
Anthephora hochstetteri Nees, 203, 206
Anthoxanthum, 296, 307
    A. alpinum Schur (= *A. odoratum*), 149
    A. angustum (Hitchc.) Owhi, 452
    A. odoratum L. (= *A. alpinum*), 135, 136, 157, 158, 182
Anthriscus sylvestris (L.) Hoffm., 135, 157
Anthus novaeseelandiae (Richards) (pipit), 77
Anthyllis henoniana Coss., 185
    A. vulneraria L., 156
Antidesma ghaesembilla Gaertn., 446, 449
Antidorcas marsupialis (Zimmerman) (springbok), 280
Antilocapra americana (pronghorn), 478, 479
Antilope cervicapra (L.) (black buck), 104, 478
Aphodelus microcarpus Salz. & Viv., 185
Aphodiinae, 281
Apiaceae, 127, 179, 369, 381
Aplophyllum davuricum (see *Haplophyllum davuricum*)
Apluda, 100, 477
    A. mutica L., 97, 449
Aponogeton, 299
Aptenodytes, 414
    A. patagonicus Miller (king penguin), 427, 428
Apterygota (subclass of insects), 150
Apus pacificus (Latham) (northern white-rumped snipe), 77
Aquila chrysaetos (L.) (golden eagle), 77
    A. verreauxi (Lesson) (Verreaux's eagle), 226
Arachis hypogaea L. (groundnut), 243, 300
Araliaceae, 363
Araucaria, 450, 462, 463, 466
    A. cunninghamii Sweet, 450
    A. hunsteinii K. Sch., 450
Araucariaceae, 361, 450, 463
Archangelica, 308
Arctogeron gramineum (L.) DC., 50
Arctotheca calendula (L.) Levyns, 344, 349
Ardea cinerea (L.) (grey heron), 77
Arenaria capillaris Poir. (see *Eremogone capillaris*)
    A. musciformis Edgew., 64, 69
Argania sideroxylon Roem. & Schult., 187
Argyrolobium linnaeanum Walp., 183
    A. uniflorum Jaub. & Spach, 185
Aristida, 89, 97, 112, 167, 215, 222, 223, 225, 238, 246, 253, 255, 256, 272, 274, 278, 294, 296, 300, 301, 317, 326, 329, 333–336, 340, 350, 478
    A. adscensionis L. (= *A. coerulescens*), 183, 237, 238, 240, 253, 256, 299
    A. arenaria (= *A. contorta*), 321

A. coerulescens Desf. (= *A. adscensionis*), 88
A. congesta Roem. & Schult., 272, 298, 299
A. contorta F. Muell. (= *A. arenaria*), 321, 328, 331, 337
A. cumingiana Trin. & Rupr., 95, 113
A. funiculata Trin. & Rupr., 204
A. heymannii Regel, 17, 28
A. hordacea Kunth, 204
A. hygrometrica R. Br., 350
A. inaeguiglumis Domin, 350
A. jerichoensis (Domin) Henr., 336, 337
A. junciformis Trin. & Rupr., 274, 275, 283
A. kelleri Hack., 251, 253
A. keniensis Henr., 240
A. latifolia Domin, 321, 322, 325
A. longiflora Schinz & Thon., 204, 206
A. meridionalis Henr., 273
A. mutabilis Trin., 199, 204–206, 256
A. pennata Trin. (= *Stipagrostis pennata*, 89, 114
A. plumosa L. (= *Stipagrostis plumosa*), 88, 89
A. pruinosa Domin, 333, 339
A. rufescens Steud., 295–298, 300
A. setacea Retz., 91, 93
A. sieberiana Trin., 253
A. similis Steud., 295, 296
A. stipitata Hack., 273
Aristideae, 312, 440
Armadillidium vulgare, 478
Armeniaca sibirica (L.) Lam., 69
    A. vulgaris Lam., 68
Armeria plantaginea All., 185
Arrhenatherum, 477
    A. elatius (L.) Mert., 130, 134, 136–138, 140, 148, 149, 152, 157, 179
Artemisia, 4, 7, 14, 22, 27–29, 33, 35–39, 45, 48, 67, 70, 73, 79, 84, 88, 161
    A. absinthium L., 50
    A. adamsii Bess., 4, 49, 50, 53
    A. anethifolia Weber, 74
    A. austriaca Jacq., 4, 22, 28, 47, 50, 54
    A. borotalensis Pall., 70
    A. campestris L. (= *A. commutata*), 172, 185
    A. capillaris Thunb., 69
    A. changaica Krasch., 48, 50, 53
    A. commutata Bess. (= *A. campestris*), 161
    A. dracunculus L., 50, 53
    A. fragrans Willd., 88
    A. frigida Willd., 17, 26, 28, 30, 31, 35, 37, 39, 45, 48, 66, 67, 71, 77
    A. giraldii Pampan., 68
    A. gmelinii Web. ex Stechm., 68
    A. gracilescens, 11, 14, 26, 28, 30, 35–40, 70
    A. herba-alba Asso (white sage), 88, 89, 172–176, 180, 185, 188, 189, 193
    A. hololeuca Bieb. ex Bess., 8
    A. japonica Thunb., 162
    A. judaica L., 185
    A. latifolia Ledeb., 4, 20, 21
    A. lerchiana Web. ex Stechm., 14, 54
    A. lessingiana Bess., 14
    A. macrocephala Jacq. ex Bess., 50

Artemisia (continued)

   A. maritima L., 47, 50, 89
   A. mesatlantica Maire, 175
   A. monosperma Del., 172, 185
   A. palustris L., 17, 53
   A. pauciflora Web., 52, 54
   A. pectinata Pall. (= *Neopaltasia pectinata*), 17, 67
   A. pontica L., 4
   A. santonica L., 14
   A. scoparia Waldst. & Kit., 15, 17, 185
   A. sericea Weber, 21
   A. sublessingiana Krasch. ex Poljak., 12, 14, 28
   A. taurica Willd., 14
   A. transiliensis Pamp., 70
   A. xerophytica Krasch., 17, 28, 67
Arthraxon, 306, 450
Arthrocnemum, 299, 345
   A. indicum (Willd.) Moq., 186
Arthropoda, 150, 151, 210
Artiodactyla, 103
Artocephalus, 414, 415, 419
Arundinaria falconeri Benth., 94
Arundineae, 307
Arundinella, 89–91, 95
   A. anomala Steud., 130, 162
   A. bengalensis (Spreng.) Druce, 91, 112
   A. furva Chase, 452
   A. hirta (Thunb.) C. Taneka, 65, 69
   A. hookeri Munro, 94
   A. nepalensis Trin., 91, 112
   A. setosa Trin., 95, 449
   A. villosa Arn., 93
Arundo, 94
   A. donax L., 186
   A. madagascariensis Kunth, 95
Arvicanthis niloticus (Desmarest) (diurnal unstriped grass mouse), 212, 241, 478
Asclepidaceae, 203
Ascomycetes, 106
Ascomycotina, 107
Asparagus gobicus Jvanova ex Gzub., 28
   A. officinalis L., 160
Aspergillus, 24, 41, 107
Asphodeline lutea (L.) Rchb., 185
Asphodelus, 172
Aspidium thelypteris (L.) Swartz (= *Dryopteris thelypteris*), 162
Astelia nervosa Banks & Sal. ex Hook. f., 386
   A. papuana Skottsb., 453
Aster altaicus Willd., 67
   A. discoideus Sond. ex Harv. & Sond., 161
   A. impatiens, 161
   A. scaber Thunb., 162
   A. tataricus L. f., 162
Asteraceae, 8, 30, 47, 67, 127, 363, 383, 386, 402
Asterothamnus, 17
Astragalus, 88, 89
   A. adsurgens Pall., 66, 67
   A. armatus Willd., 185
   A. danicus Retz., 157

   A. grubovii Sancz., 28
   A. junatovii Sancz., 28
   A. macropus Bunge, 22
   A. melilotoides Pall., 64, 67
   A. monophyllus Bunge ex Maim., 28
   A. vallestris R. Kam., 28
Astrebla, 306, 307, 310, 317, 321–326, 350, 352, 440
   A. elymoides F. Muell. ex F.M. Bailey, 321–323, 326
   A. lappacea (curly Mitchell grass) (Lindl.) Domin, 322, 323
   A. pectinata F. Muell. ex Benth., 321, 323–325
   A. squarrosa C.E. Hubb., 321, 323, 326
Atalaya hemiglauca F. Muell. ex Benth., 321
Atractylis, 179
Atriplex, 114, 191, 193, 299, 307, 317, 329–333, 337, 350, 354
   A. cana C.A. Mey., 14
   A. glauca L., 185, 193
   A. halimus L., 179, 186, 189, 193
   A. inflata F. Muell. (see *A. lindleyi* subsp. *inflata*)
   A. laciniata L. (see *A. tatarica*)
   A. leucoclada Boiss., 185
   A. lindleyi Moq. subsp. *inflata* (F. Muell.) P.G. Wilson (= *A. inflata* F. Muell.), 330
   A. malvana Aell. & Sauv., 186
   A. nummularia Lindl., 193, 331, 332
   A. rhagodioides F. Muell., 329, 331
   A. sphaeromorpha Iljin, 54
   A. stipitata Benth., 337
   A. tatarica L. (= *A. laciniata* L.), 50
   A. vesicaria Heward ex Benth., 329–333, 337
Atropis distans Griseb., 159, 160
Avena, 179
   A. barbata Potter, 183
   A. bromoides (Gouan) Trab., 182
   A. montana Vill., 182
   A. pubescens Huds., 161
Avenastrum schellianum (Hack.) Roshev., 158
Aveneae, 1, 2, 168, 188, 303, 307, 312
Axis, 104
   A. axis Exrl. (chital), 104, 105, 478
   A. porcinus Zimmerman (hog deer), 104, 106, 478
Axonopus, 96, 97
   A. affinis Chase, 346
   A. compressus (Sw.) P. Beauv., 95, 96, 116
Axyris amaranthoides L., 50, 53
   A. hybrida L., 48
Azorella, 422
   A. selago Hook. f. & Wils., 414, 416
Azotobacter, 40

Bacillus, 40
Bacteria, 24, 25, 30, 40–42, 72, 107, 108, 151, 153, 396, 397, 426, 480
Bacterium, 40
Baeckea virgata Andr., 455
Baikiaea, 266, 270, 277
   B. plurijuga Harms, 268, 273
Balanites, 204
   B. aegyptiaca (L.) Del., 199, 205
   B. orbicularis Sprague, 252
Ballota, 88

SYSTEMATIC INDEX

Bambusa latispiculata (Tramble) Holtt. [= *B. heterostachya* (Munro) Hott.], 94
Bambuseae, 96
Banksia, 445
Barbatae (section of Stipa), 14, 78
Bassia, 329
    B. dasyphylla (Fisch. & Mey.) O. Kuntze, 29
Bauhinia (= *Lysiphyllum*), 321
Beckmannia eruciformis (L.) Host, 158, 160
Bellevalia, 15
Betonica officinalis L. (= *Stachys betonica*), 159
Betula, 131, 160, 162
    B. pendula Roth, 9
    B. pubescens Ehrh., 9
Betulaceae, 68
Blechnum orientale L., 459
Blennodia lasiocarpa F. Muell., 337, 351
Blepharis, 215, 252
    B. linearifolia Pers., 204
Boerhavia diffusa L., 321
    B. repleta Hewson (= *B. diffusa*), 321, 324, 351
Bombax, 94
Bonjeania recta Reich., 185
Boophilus microplus (Canestrini) (cattle tick), 341
Boraginaceae, 127
Borassus, 307, 446
    B. flabellifer L., 96
    B. madagascariensis Boj., 297
Borreria, 215
    B. radiata DC. (= *Spermacoce radiata*), 204
Bos africanus Sanson, 214
    B. gaurus H. Smith (gaur), 104
    B. grunniens L., 70
    B. grunniens mutus (wild yak), 76
    B. ibericus Pomel, 214
    B. indicus L. (zebu), 214, 227
    B. taurus L. (domesticated cattle), 214
Boscia, 204
    B. minimifolia Chiov., 252
Boselaphus tragocamelus Pall. (nilgai), 104
Bostrychia carunculata (Ruppell) (wattled ibis), 226
Botaurus stellaris (L.) (Eurasian bittern), 77
Bothriochloa, 97, 317, 322, 333, 335, 338, 340, 341, 352, 440
    B. bladhii (Retz.) S.T. Blake, 230
    B. compressa (Hook. f.) Henr., 95
    B. decipiens (Hack.) C.E. Hubb., 334, 352
    B. erianthoides (F. Muell.) C.E. Hubb., 352
    B. ewartiana (Domin) C.E. Hubb., 340
    B. glabra (Roxb.) A. Camus, 95
    B. insculpta (A. Rich.) A. Camus, 225, 231, 239, 240, 273
    B. intermedia A. Camus, 91
    B. ischaemum (L.) Keng (see *Andropogon ischaemum*)
    B. pertusa (L.) A. Camus, 91, 94, 112, 116, 203, 206
    B. radicans (Lehm.) A. Camus, 274
Brachiaria, 95, 96, 101, 215, 306, 341
    B. brizantha (Hochst.) Stapf, 115, 273
    B. distachya (L.) Stapf, 93
    B. leucantha (K. Schum.) Stapf, 231
    B. mutica (Forsk.) Stapf (Para grass), 94, 111, 346
    B. nigropedata (Munro) Stapf, 274

    B. ovalis Stapf, 253
    B. reptans (L.) Gard. & C.E. Hubb., 112, 458
    B. serrata (Spreng.) Stapf, 272, 273
    B. subquadripara (Trin.) Hitchc., 113
Brachyachne, 112
    B. convergens (F. Muell.) Stapf, 321, 339
Brachychiton populneus R. Br. (= *Sterculia diversifolia*), 336
Brachypodium, 296, 477
    B. phoenicoides Roem. & Schult., 179, 182
    B. pinnatum Beauv., 138, 156, 159, 161
    B. ramosum (L.) Roem. & Schult. (= *B. retusum*), 179, 182
    B. retusum, 182
Brachystegia, 237, 268, 269
Brassica oleracea L., 226
Brassicaceae, 30
Briza media L., 157, 158
Bromeae, 312
Bromopsis inermis (Leyss.) Holub (= Bromus inermis), 4, 21, 50
    B. riparia (Rehm.) Holub (= *Zerna riparia*), 4, 50, 52
Bromus, 88–90, 179, 272
    B. ciliatus L., 130, 162
    B. erectus Huds., 130, 156, 158, 182
    B. hordeaceus L. (= *Serrafalcus mollis*), 349
    B. inermis Leyss. (= *Bromopsis inermis*), 4, 72, 140, 158–161
    B. racemosus L., 157
    B. rigidus Roth, 318, 344, 349
    B. riparius Rehm. (= *Bromopsis riparia*), 130, 158, 159
    B. rubens L., 183
    B. sibiricus Drob., 130, 161
    B. squarrosus L., 14, 50, 182
    B. sterilis L., 342
    B. tectorum L. (see *Anisantha tectorum*)
    B. tomentellus Boiss., 88, 182
    B. variegatus Bieb., 130, 159
Broussonetia papyrifera (L.) Vent., 445, 455
Bryodema (grasshoppers)
    B. holdereri holdereri (Krauss), 77
    B. luctuosum luctuosum (Stoll), 77
    B. tuberculatum dilutum (Stoll), 77
    B. zaisanicum fallax (Bey-Bienko), 77
Bubalus bubalis L. (water buffalo), 104, 106, 345, 353
Buddleja salvifolia Lam., 286
Bufo raddei Stranch, 77
Bulbinella, 380
Bulbocodium, 15
Bulbosae (section of Poa), 15
Bupleurum scorzonerifolium Willd., 67, 161
    B. spinosum L., 185
Burkea, 281
    B. africana Hook., 268, 274, 276, 277, 281
Buteo hemilasius (Temminck & Schelegel) (upland buzzard), 77
    B. rufofuscus augur (Ruppell) (augur buzzard), 226
Butomus umbellatus L., 158
Butyrospermum, 222, 233

Cacalia aconitifolia Bunge, 66, 72
    C. hastata L. (= *Senecio sagittatus*), 162
Cajanus, 227

Cajanus (continued)
    C. cajan (L.) Millsp., 234
Caladenia, 380
Calamagrostis, 78, 441
    C. arundinacea (L.) Roth, 159, 161
    C. epigeios (L.) Roth, 64, 66, 72, 140, 158, 162
    C. langsdorfii (Link) Trin., 130, 161, 162
    C. neglecta (Ehrh.) Gaertn., 162
Calandrella rufescens (Vieillot) (lesser sand lark), 77
Calandrinia, 328, 329, 331, 351
    C. balonensis Lindl., 321
Calligonum, 89
    C. comosum L' Hérit., 114, 186
    C. molle Litwinow, 89
    C. mongolicum Turcz., 64, 67
Calliptamus (grasshoppers)
    C. abbreviatus Ikonnikov, 77
    C. barbarus cephalotes (Fischer-Waldheim), 77
    C. italicus italicus (L.), 77
Callitris, 463
    C. columellaris F. Muell. (cypress pine), 334
Calluna vulgaris (L.) Hull, 157
Calophaca, 4, 8
    C. wolgarica (L. f.) DC., 14
Calopogonium mucunoides Desv., 346
Calothrix, 426
Calotis, 328, 337
    C. hispidula F. Muell., 351
Calotropis procera Ait., 215
Calycotome villosa (Poiret) Link, 179
Camellia sinensis (L.) O. Kuntze, 226
Camelus, 75
    C. bactrianus (L.) (Bactrian camel), 75
    C. dromedarius L. (single-humped camel), 353
Campanula glomerata L., 159, 160
    C. wolgensis P. Smirn., 21
Camphorosma monspeliaca L., 54
Canis, 226, 241
    C. adustus Sundevall (side-striped jackal), 211
    C. aureus L. (common jackal), 104, 211, 213
    C. familiaris dingo Meyer (dingo), 329
    C. simensis (Ruppell) (Simien fox), 226
Canthium oleifolium Hook., 336
Capillatae (section of Stipa), 78
Capillipedium, 306
    C. parviflorum (R. Br.) Stapf, 95, 96, 449
Capparidaceae, 203, 212
Capparis mitchellii Lindl., 337
Capra hircus L. (domestic goat), 353, 403
    C. ibex L. (ibex), 104
    C. walie (= *C. ibex*) (walia ibex), 226
Caprifoliaceae, 68
Capsella bursa-pastoris (L.) Medik., 50
Caragana, 4, 8, 17, 21, 28, 29, 71, 79
    C. balchaschensis (Kom.) Pojazk., 15
    C. frutex (L.) C. Koch., 9, 52
    C. korshinskii Kom., 68
    C. leucophloea Pojark, 26, 28, 30, 31
    C. microphylla Lam., 50, 53, 64, 67, 68

    C. pumila Pojark., 12
    C. pygmaea (l.) DC., 67
    C. stenophylla Pojark, 67
Cardamine pratensis L., 185
Carduncellus, 179
Carduus, 172, 179, 351
    C. pycnocephalus L., 349
    C. uncinatus Bieb., 50
Carex, 4, 106
    C. aquatilis Wahlenb., 161
    C. caespitosa L., 158, 161
    C. diluta Bieb., 160
    C. disticha Huds., 160
    C. duriuscula C.A. Mey., 45, 53
    C. echinata Murr., 127
    C. fusca All., 140, 182
    C. gracilis Curtis, 130, 157, 159–161
    C. hirta L., 127
    C. humilis Leyss., 130, 156, 158, 159
    C. juncella Fries (= *C. vulgaris*), 161
    C. korshinskyi Kom., 48
    C. lanceolata Boott, 162
    C. leporina L., 127
    C. lithophilla Turcz., 161
    C. moorcroftii Falc., 72
    C. nervata Franch. & Sav., 162
    C. nigra (L.) Reichard, 149, 158
    C. nubigena D. Don., 94
    C. pachystylis J. Gay (= *C. stenophylla*), 160
    C. panicea L., 127, 158
    C. pediformis C.A. Mey., 48, 53, 66, 72
    C. praecox Schreb., 21
    C. rhyzina Blytt. ex Lindbl., 161
    C. schmidtii Meinsh, 72, 73, 130, 161, 162
    C. stenophylla Wahlenb. (= *C. pachystylis*), 28, 88, 89, 158, 161
    C. supina Wahlenb., 21
    C. vulgaris Fries (see *C. juncella*)
    C. vulpina L., 159
Careya, 95
    C. herbacea Roxb., 112
Carissa lanceolata R. Br., 321
Carlina, 172, 179
Carnivora, 103
Carpha alpina R. Br., 453
Carthamus, 179
    C. lanatus L., 184
Caryophyllaceae, 8
Cassia, 215, 246, 327, 335
    C. artemisioides Gaudich. ex DC., 337
Cassytha filiformis L., 92
Casuarina, 95, 96, 331
    C. cristata Miq., 331, 336
    C. equisetifolia L., 295, 458, 459, 462
    C. junghuhniana Miq., 96
    C. oligodon L. Johnston, 451
    C. papuana S. Moore (= *Gymnostoma papuana*), 448
Cathormion umbellatum (Vahl) Kostern., 449
Cecidomyiidae, 396
Celmisia, 380

Celmisia (continued)

    C. armstrongii Petrie, 402
    C. walkeri Kirk, 402
Celtis, 228
Cenchrus, 89–91, 109, 112, 167, 215, 223, 246, 306, 316, 335, 441
    C. biflorus Roxb., 199, 204–206
    C. brownei Roem. & Schult. (= *C. inflexus*), 95
    C. ciliaris L. (buffel grass), 88, 89, 91, 94, 97, 109, 111, 114, 178, 179, 182, 188, 203, 206, 238, 246, 251, 253, 274, 299, 316, 318, 321, 325, 328, 335, 336, 347, 348
    C. inflexus R. Br. (see *C. brownei*)
    C. setigerus Vahl, 91, 97, 109, 348
Cenolophium (= *Selinum*)
    C. fischeri Koch, 160
Centaurea, 4, 157, 179
    C. jacea L., 158
Centosteca latifolia (Osb.) Trin. (= *Centotheca lappacaea*), 96
Centrosema pubescens Benth., 111, 346
Cephalophus, 226, 231
Cerastium dichotomum L., 185
Cerasus, 8
Ceratocarpus arenarius L., 29, 52
Ceratocephala, 14
    C. testiculata (Crantz) Bess., 29, 52
Ceratonia siliqua L., 187
Ceratotherium, 280
    C. simum Burchell (white rhinoceros), 280
Cercopithecus aethiops L. (vervet), 211
Cervus, 104
    C. duvauceli duvauceli Cuvier (swamp deer or "bara singha"), 104, 106, 478
    C. elaphus L. (red deer), 310, 377, 393–396, 402, 403
    C. timorensis (Blainville) (Timor deer), 310, 465
Chamaerhodos altaica (Laxm.) Bunge, 50
    C. erecta (L.) Bunge, 17, 29
Charadrius asiaticus (Pallas) (Caspian plover), 241
Cheilanthes tenuifolia (Burm.) Sw., 459
Chenopodiaceae, 14, 30, 67, 79, 177, 179, 299, 332
Chenopodium, 107, 172, 326
    C. acuminatum Willd., 29
    C. album L., 48, 50, 107
    C. aristatum L., 50, 53
    C. auricomum Lindl. (northern bluebush), 326, 329
    C. foliosum Asch., 52, 53
    C. strictum Roth, 48–50
Chionochloa, 307, 310, 365, 367–369, 372–384, 386, 388–404, 472, 477, 479
    C. acicularis Zotov (needle-leaved snow tussock), 375, 377
    C. antarctica Hook. f., 419, 422, 423
    C. archboldii (Hitchc.) Zotov, 307, 452, 454
    C. australis (Buchanan) Zotov, 375, 389, 390
    C. crassiuscula (Kirk) Zotov (curled snow tussock), 310, 372, 375, 389, 390, 395
    C. defracta Connor, 376
    C. flavescens Zotov (broad-leaved snow tussock), 363, 375, 383, 386, 389, 395
    C. lanea Connor, 376
    C. macra Zotov (slim snow tussock), 310, 369, 371, 376, 380, 382–386, 389–393, 398
    C. oreophila (Petrie) Zotov (snow patch grass), 310, 371, 375, 389, 390
    C. ovata (Buchanan) Zotov, 375
    C. pallens Zotov, 310, 375, 386, 389, 390, 393–395, 402
    C. pungens (Cheesem.) Zotov, 376
    C. rigida (Raoul) Zotov (narrow-leaved snow tussock), 310, 368, 373, 375, 376, 379, 381–386, 388–393, 397, 398, 400
    C. rubra Zotov (red tussock), 374, 383, 397
    C. teretifolia (Petrie) Zotov, 375, 395
Chiroptera, 103
Chlorideae, 188, 312, 440
Chloris, 91, 215, 223, 246, 278, 317, 329, 333, 336, 350
    C. acicularis Lindl., 334, 337, 344, 350
    C. barbata Sw., 93, 95, 112
    C. gayana Kunth, 227, 233, 238, 288, 318, 348
    C. pycnothrix Trin., 240
    C. ramosa B. Simon, 334
    C. roxburghiana Schult., 246
    C. truncata R. Br., 329, 332, 337, 344, 350, 351
    C. virgata Sw., 97, 231, 272
    C. virgata var. *ruderale*, 69
Chorisodontium aciphyllum (Hook. f. & Wils.) Broth., 418
Chorthippus (grasshoppers)
    C. brunneus (Thunberg), 77
    C. chinensis (Tarbinsky), 78
    C. dorsatus (Zetterstedt), 77
    C. dubius (Zubovsky), 77, 78
    C. fallax (Zubovsky), 78
    C. hammarstroemi hammarstroemi (Mir.), 77
Chromolaena odorata (L.) King & Robinson (= *Eupatorium odoratum*), 113
Chrysanthemum cinerariaefolium Vis., 226
    C. leucanthemum L. (= *Leucanthemum vulgare*), 136
Chrysocoma, 278
    C. ciliata, 272
Chrysophyllum, 228
Chrysopogon, 89, 92, 94–97, 109, 112, 167, 223, 231, 234, 243, 245–247, 252, 256, 257, 306, 319, 325, 350, 478
    C. aciculatus (Retz.) Trin. (= *Andropogon aciculatus*), 93–96, 112, 113, 116
    C. aucheri (Boiss.) Stapf, 89
    C. fallax S.T. Blake, 321, 339, 340, 350
    C. fulvus (Spreng.) Chiov., 89, 91, 97, 109, 116
    C. gryllus (L.) Trin., 91, 92, 182
    C. latifolius S.T. Blake, 320
    C. plumulosus Hochst., 203, 206, 222, 233, 244, 246, 247, 253
    C. serrulatus Trin., 89
    C. zeylanicus (Nees) Thw., 92, 93
Circinella, 107
    C. umbellata Van Teigh & Le Monn., 107
Cirsium oleraceum Scop., 140, 157
    C. palustre (L.) Scop., 127, 158
Cissus, 246, 449
Cistus, 172, 192
    C. albidus L., 192
    C. monspeliensis L., 192
    C. villosus L., 192
Citellus (see also *Spermophilus*), 51, 54

Citellus (continued)

    C. dauricus (Pallas) (ground squirrel), 76
    C. pygmaeus Pallas (suslik), 43, 45, 51, 53, 54
Citrus, 231
Cladonia, 419
Cladosporium, 24
Cleistogenes, 4, 17, 19, 26, 28, 29, 39
    C. songorica (Roshev.) Ohwi, 17, 26, 28, 30–33, 35–37, 39, 40
    C. squarrosa (Trin.) Keng., 14, 17, 18, 28, 44, 45, 48, 67
Cleome, 172
    C. arabica L., 184
    C. viscosa, 351
Clerodendrum serratum (L.) Moon., 95
Cliffortia, 272
Climacoptera brachiata (Pall.) Botsch., 54
Clostridium butyricum Prazmowski, 398
    C. pasteurianum Winogr., 40, 41
Cnicus, 179
Cnidium (= *Selinum*)
    C. dahuricum Fisch. & Mey., 161
Cocos nucifera L. (coconut), 231, 295, 461
Coelachyrum stoloniferum C.E. Hubb., 253
Coelorhachis glandulosa (Trin.) Stapf, 96
    C. rottboellioides (R. Br.) A. Camus, 445, 449
Coffea arabica L., 226
    C. robusta Lindau (= *C. canephora* Pierre ex Froehner), 229
Coix, 97
Colchicum autumnale L., 158
Coleoptera, 148, 150, 396, 478
Collembola, 150, 151
Colletotrichum gloeosporioides (Pensig) Pensig & Sacc, 347
Colona scabra Burret, 455
Colophospermum mopane Kirk ex Leonard, 268, 270
Coluber spinalis (Peters), 77
Columba albitorques (Ruppell) (white-collared pigeon), 226
    C. leuconta (Vigors) (snow pigeon), 77
Colutea, 180
Comarum palustre L. (= *Potentilla comarum*), 158
Combretaceae, 199, 442
Combretum, 199, 222, 230, 233, 234, 237, 239, 274
    C. ghazalense Engler and Diels, 207
    C. glutinosum Perrott. ex DC., 207
    C. goldieanum F. Muell., 442
    C. micranthum G. Don, 199
    C. molle G. Don, 237
    C. zeyheri Sond., 237
Commelinaceae, 96
Commersonia bartramia (L.) Merr., 455
Commiphora, 204, 222, 230, 233, 238, 245, 246, 252
    C. africana (A. Rich.) Engler, 199
    C. campestris Engler, 238
    C. chiovendana Gillett, 252
    C. schimperi (Berger) Engler, 238
Connochaetes gnou (Zimmermann) (black wildebeest), 280
    C. taurinus (Burchell) (wildebeest), 226, 228, 242, 279, 280
Convolvulaceae, 203
Convolvulus, 28, 29

    C. ammanii Desr., 28, 31, 35
    C. arvensis L., 50
Coprinae, 281
Coprosma, 453
Cordeauxia edulis Hemsl., 252
Cordia, 204
Corispermum mongolicum Iljin, 29
Cornulaca monacantha Del., 186
Coronilla, 180
    C. glauca L., 183
    C. minima L., 183
Corvus crassirostris (Ruppell) (thick-billed raven), 226
Corylus, 161
Corynephorus canescens (L.) Beauv., 137
Cosmos bipinnatus Cav., 94
Costelytra zealandica (White), 396
Cotoneaster, 8, 71
Cousinia, 179
Crepis biennis L., 157
Cricetidae (family of rodents), 43
Crinitaria tatarica (Less.) Czer., 10, 22, 28, 52
    C. villosa (L.) Grossh., 28
Critesion murinum (L.) A. Love subsp. *leporinum* (L.) A. Love (= *Hordeum leporinum*), 337
Crithmum maritimum L., 186
Crocus reticulatus Stev., 15
Crocuta crocuta (Erxleben) (spotted hyaena), 211, 226
Cryptostegia grandiflora, 312
Ctenium, 295
    C. elegans Kunth., 206
Cubitermes (termites), 231
Cunninghamella, 107
Cupressaceae, 463
Cupressus, 226
    C. macrocarpa Hartw., 115
Curculigo orchioides Gaertn., 449
Curculionidae, 40
Curcuma longa L., 449
Curvularia, 107
Cutandia, 179
Cyananthus lobatus Wall., 92
Cyanophyceae, 207, 215, 216
Cyathea, 307, 452–454
Cycas, 307
    C. media R. Br., 307, 446, 449
    C. rumphii Miq., 307, 456
Cygnus olor (Gmelin) (mute swan), 77
Cymbaria daurica L., 4, 45, 47, 66
Cymbopogon, 89, 91, 97, 100, 116, 273, 274, 282, 440
    C. caesius (Hook. & Arn.) Stapf, 238, 239
    C. commutatus (Steud.) Stapf, 206
    C. excavatus (Hochst.) Stapf, 273, 274
    C. giganteus (Hochst.) Chiov., 206
    C. laniger Duthie, 88
    C. nardus (L.) Rendle, 92, 93, 228, 233, 239
    C. nardus var. *confertiflorus* (Steud.) Stapf ex Bor, 92, 93, 95, 106
    C. plurinodis Stapf ex Burtt Davy, 272
    C. procerus (R. Br.) Domin, 449
    C. proximus (Hochst. ex A. Rich., 204, 206

Cymbopogon (continued)

    C. schoenanthus (L.) Spreng., 88, 89, 206
    C. validus Stapf ex Burtt Davy, 274, 278
Cynara, 179
Cynodon, 89, 97, 106, 225, 275, 288, 477
    C. arcuatus Presl, 95
    C. dactylon (L.) Pers., 90–92, 94, 96, 97, 100, 106, 160, 161, 172, 182, 188, 189, 194, 206, 225, 233, 238, 240, 253, 287
    C. plectostachyus (K. Schum.), 233, 238
Cynoglossum officinale L., 50
Cynosurus, 179
    C. coloratus Lehm., 183
    C. cristatus L., 157
Cyperaceae, 72, 96, 127, 203, 299
Cyperus, 253
    C. chordorrhizus Chiov., 253
    C. conglomeratus Rottb., 182
    C. laevigatus L., 89
    C. latifolius Poir., 299
    C. madagascariensis Roem. & Schult., 299
Cyrtococcum patens (L.) A. Camus, 96
    C. trigonum (Retz.) A. Camus, 96
Cytisus, 180

Dactylis, 287
    D. glomerata L., 127, 130, 133, 136, 142, 147–149, 152, 157, 159–162, 179, 182, 193, 287, 348
    D. hispanica Roth, 182
Dactyloctenium, 274
    D. aegyptium (L.) P. Beauv., 93, 96, 112, 238
    D. bogdanii S.M. Phillips, 256
    D. radulans (R. Br.) P. Beauv., 321, 329
    D. scindicum Boiss., 97, 246, 253
Dalbergia, 238
Damaliscus, 280
    D. dorcas dorcas, 280
    D. dorcas phillipsi Harper (blesbok), 280
    D. hunteri (Sclater) (Hunter's antelope), 248
    D. korrigum (Ogilby), 211, 231, 239
    D. lunatus (Burchell) (tsessebe), 280
Danthonia, 93, 317, 332, 333, 342, 344, 350, 351, 365, 452, 453
    D. auriculata J.M. Black, 342, 344, 351, 353
    D. bipartita (F. Muell.) Vickery (= *Monachather paradoxa*), 328
    D. cachemyriana Jaub. & Spach, 92, 98
    D. caespitosa Gaudich., 329, 332, 343, 344, 352
    D. carphoides Benth., 342, 344, 351, 353
    D. nudiflora P.F. Morris, 345, 353
    D. pallida R. Br., 341, 342
Danthonieae, 312
Decaspermum fruticosum Forsk. f., 458
    D. paniculatum Kurz, 451
Delonix elata (L.) Gamble, 245
Delphinium, 160
    D. glandiflorum L., 66, 72
Dendrocalamus longipathus Kurz, 94
Deschampsia, 441, 453, 472
    D. antarctica Desv., 419, 421

    D. caespitosa (L.) Beauv., 130, 133, 135, 138–140, 142, 152, 158
    D. klossii Ridley, 422, 423, 430, 452, 453
    D. penicillata Kirk, 419
Descurainia sophia (L.) Webb ex Prantl, 29
Desmodilliscus braueri Wettstein (pouched gerbil), 212
Desmodium, 96, 97, 101, 288, 318
    D. heterophyllum DC., 346
    D. triflorum (L.) DC., 93
    D. uncinatum DC., 229
Desmostachya, 472, 475, 477
    D. bipinnata (L.) Srapf, 90, 91, 97, 100, 106, 107, 112
Deuteromycetes, 106
Deuteromycotina, 107
Deyeuxia, 307, 453
    D. brassii (Hitchc.) Jansen, 452
Dianthus, 4, 157
    D. chinensis L., 161
Dicerorhinus sumatrensis (Fischer) (see *Didermocerus sumatrensis*)
Diceros bicornis (L.) (black rhinoceros), 226
Dichanthium, 89–91, 97, 106, 109, 112, 306, 317, 321–323, 338, 340, 341, 472, 475, 477
    D. annulatum (Forsk.) Stapf, 91, 94, 97, 100, 106, 109, 116, 203, 206
    D. caricosum (L.) A. Camus, 91, 96, 458
    D. fecundum S.T. Blake, 340, 350, 352
    D. sericeum (R. Br.) A. Camus, 323, 325, 341, 352
    D. tenuiculum (Steud.) S.T. Blake, 340
Dichelachne, 307, 317, 341, 342, 452
Dicoma, 298
Dicranopteris linearis (Burm. f.) Underwood, 448, 458–460, 463, 465
Didermocerus sumatrensis (Sumatra rhinoceros) (= *Dicerorhinus sumatrensis*), 106
Didierea trollii Capuron & Rauh, 299
Digitaria, 97, 101, 106, 112, 215, 225, 230, 306, 318
    D. abyssinica (A. Rich.) Stapf, 225, 228, 238, 239
    D. adscendens (H.B.K.) Henr., 93, 95
    D. coenicola (F. Muell.) Hughes, 337
    D. commutata Schult. subsp. *nodosa* (Parl.) Maire, 178, 182, 188
    D. decumbens Stent (= *D. eriantha* subsp. *pentzii*), 346
    D. didactyla Willd., 295
    D. eriantha Steud., 272, 274, 281, 288
    D. eriantha Steud. subsp. pentzii (Stent) Kok (= *D. decumbens* (Pangola grass), 273
    D. exilis Stapf, 209
    D. longiflora (Retz.) Pers., 93
    D. macroblephara (Hack.) Stapf, 238, 239, 246
    D. marginata Link (= *Panicum sanguinale* L.), 96
    D. milanjiana (Rendle) Stapf, 231, 238
    D. ropalotricha Buese, 95
    D. setivalva Stent, 273
    D. velutina (Forsk.) P. Beauv., 233
Diheteropogon amplectens (Nees) W. D. Clayton, 206, 274
    D. filifolius Nees, 272, 274
    D. hagerupii Hitchc., 206, 207
Dillenia ovata Wall. ex Hook. f. & Thoms., 95
Dimeria, 459

Dimeria (continued)

   D. gracilis Nees, 93
   D. lehmannii (Nees) Hack., 92
Dimorphanthera, 452
Dinornithiformes, 393
Diomedea (albatross and mollymawks)
   D. cauta Gould, 427
   D. chrysostoma Forster, 427
   D. epomophora Lesson (royal albatross), 425, 427, 429
   D. exulans L., 427
   D. melanophrys Temminck, 427
Dioscorea, 449
Diplachne fusca, 90, 115, 345
Diplotaxis harra (Forsk.) Boiss., 184
Dipodidae (family of rodents), 43
Diptera, 148, 150, 151, 396
Dipterocarpus, 96
Dipus sagitta (Pallas) (northern three-toed jerboa), 76
Discaria toumatou Raoul, 367, 374
Dissochondrus, 441
Dodonaea, 335, 337
   D. viscosa (L.) Jacq., 458
Dombeya rotundifolia Planch., 274
Dontostemon crassifolius Bunge, 28
   D. integrifolius (L.) C.A. Mey., 17
Dorcatragus megalotis (Menges) (beira), 248
Drawidia (or *Drawida*) (earthworms)
   D. calabi (Gates), 106
   D. willsii Michaelsen, 106
Drepanocladus uninatus (Hedw.) Wernst., 419
Dryopteris thelypteris (L.) Gray (see *Aspidium thelypteris*)

Echinochloa, 95, 97, 299
   E. colona (L.) Link, 112
   E. crusgalli (L.) P. Beauv., 96
   E. haploclada (Stapf) Stapf, 246, 256
   E. pyramidalis (Lam.) Hitchc. & Chase, 206, 231
   E. stagnina (Retz.) P. Beauv., 95, 206, 208
Echinops latifolius Tausch, 48, 50, 53
   E. meyeri (DC.) Iljin, 50
   E. ritro L., 50
Echiochilon fruticosum Desf., 185
Echymipera kalubu (Fischer) (mumut, bandicoot), 465
Elaphe dione (Pallas), 77
Eleocharis, 317, 345
   E. palustris (L.) R. Br., 182
Elephas maximus L. (Asian elephant), 104, 106
Eleusine, 89, 97, 225, 227
   E. compressa (Forsk.) Asch. & Schweinf., 89, 97
   E. coracana (L.) Gaertn., 226
   E. floccifolia (Forsk.) Spreng., 225
   E. indica (L.) Gaertn., 96, 106
   E. jaegeri Pilg., 222, 224, 225, 227
Elionurus (or *Elyonurus*), 295
   E. elegans Kunth, 206
   E. muticus Kunth, 273, 274
   E. tristis Hack., 296
Ellobius talpinus Pall. (mole vole), 43, 47, 49
Elymus, 89, 363

E. delilaenus Schult., 183
E. junceus Fisch. (= *Psathyrostachys juncea*), 52
E. rectisetus (Nees) Löve & Connor (= *Agropyron scabrum*), 374
Elytrigia repens (L.) Nevski (see also *Agropyron repens*), 4, 50
   E. trichophora (Link) Nevski, 4
Elytropappus, 278
Emberiza aureola (Pallas) (yellow-breasted bunting), 77
   E. cia (L.) (rock bunting), 77
   E. spodocephala (Pallas) (grey-headed black-faced bunting), 77
Emex, 172
Enchylaena, 329
   E. tomentosa R. Br., 332
Enchytraeidae, 150
Endogonaceae (endo-mycorrhizal fungi), 398
Endogone, 398
Endostemmon, 252
Enneapogon, 206, 317, 326, 328, 337, 342, 350
   E. avenaceus (Lindl.) C.E. Hubb., 329, 350
   E. brachystachyus (Jeub. & Spach) Stapf, 203
   E. cenchroides (Licht.) C.E. Hubb., 274, 298
   E. nigricans (R. Br.) P. Beauv. (= *Pappophorum nigricans* R. Br.), 342, 351
   E. persicus Boiss., 89
   E. polyphyllus (Domin) N.T. Burb., 327, 329, 334, 350
   E. scaber Lehm., 203
   E. schimperanus (A. Rich.) Renv., 253
   E. scoparius Stapf, 274
Ensete calospermum (F. Muell.) Cheesem., 446
Enteropogon acicularis Lindl., 334
Eophona (grasshopper)
Epacridaceae, 363
Epacromia coerulipes (Ivanor) (grasshopper), 77
Ephedra scoparia Lange, 89
Epilobium, 296
Equisetum arvense L., 135, 159, 161
   E. pratense Ehrh., 159
Equus, 75
   E. asinus L. (donkey), 353
   E. asinus somalicus (Fitzinger) (Somali wild ass), 248, 256
   E. burchelli Gray (Burchell's zebra), 226, 228, 234, 280
   E. caballus L. (horse), 353
   E. grevyi (Oustalet) (Grevy's zebra), 256
   E. hemionus Pall. (Asiatic wild ass), 23, 76
   E. hemonius khur Moorcroft (wild ass), 104
   E. hemonius kiang Lesson (Tibetan wild ass), 104
   E. przewalskii Poliakov (wild horse), 23, 75
   E. zebra L. (mountain zebra), 280
Eragrosteae (division of Festuceae), 1, 2, 168, 169, 303
Eragrostideae, 188, 312, 440
Eragrostis, 89, 91, 94, 96, 97, 99, 101, 106, 112, 167, 179, 215, 246, 247, 272–275, 278, 317, 319, 321, 326, 327, 350, 440, 441, 472, 477
   E. amabilis (L.) Wight & Arn., 96
   E. barteri Hubb., 206
   E. braunii Schweinf., 225
   E. capensis (Thunb.) Trin., 272
   E. chapelieri (Kunth) Nees, 298
   E. chloromelas Steud., 272

Eragrostis (continued)

    E. ciliaris (L.) R. Br., 97, 231
    E. curvula (Schrad.) Nees, 274, 275, 287, 288
    E. dielsii Pilger, 328, 329, 331, 337
    E. elongata (see *E. zeylanica*)
    E. eriopoda Benth., 320, 328, 331, 337
    E. gummiflua Nees, 274
    E. lanipes C.E. Hubb., 328
    E. lehmanniana Nees, 272, 274
    E. minor Host, 17, 28
    E. nutans (Retz.) Nees, 100
    E. obtusa Munro ex Fic. & Hiern., 272
    E. pallens Hack., 268, 273, 274, 281
    E. pilosa (L.) Beauv., 69
    E. plana Nees, 272, 274, 275
    E. rigidior Pilg., 274, 278
    E. setifolia Nees, 350
    E. tenuifolia (A. Rich.) Steud., 240
    E. tremula (Lam.) Hochst. ex Steud., 199, 206
    E. xerophila Domin, 322, 328, 350
    E. zeylanica Nees & Meyen (= *E. elongata*), 113
Eremias argus (Peters) (lizard), 77
Eremogone capillaris (Poir.) Fenzl (= *Arenaria capillaris*), 48
Eremophila (Myoporaceae), 327, 330, 331, 335
Eremophila alpestris (L.) (horned lark), 77
    E. compacta S. Moore, 328
    E. forrestii F. Muell (= *E. leucophylla*), 328
    E. fraseri F. Muell., 328
    E. gilesii F. Muell., 326, 328
    E. leucophylla Benth. (= *E. forrestii*), 328
    E. longifolia F. Muell., 337
    E. mitchellii Benth. (false sandalwood), 335, 337
Eremopogon foveolatus (Del.) Stapf, 203, 206
Eremurus, 160
Eriachne, 96, 317, 321, 326, 334, 340, 445
    E. aristidea F. Muell., 328
    E. helmsii (Domin) W. Hartley, 328
    E. obtusa R.Br., 320
Erianthus, 94
    E. ravennae (L.) P. Beauv., 94
Erica, 272
Ericaceae, 308, 462
Erinacea anthyllis Link, 185
Eriochloa, 91
    E. punctata (L.) Desv. ex Hamilt., 96
Eriophorum angustifolium Honckeny, 158
    E. latifolium Hoppe, 158
Eriosema psoraloides (Lam.) G. Don, 298
Erodium botrys (Cav.) Bertol., 344, 349
    E. cicutarium (L.) L'Her. ex Ait., 349, 351
    E. cygnorum Nees, 344, 349
Erophila verna (L.) Bess., 14
Eruca vesicaria (L.) Car., 184
Eryngium (or *Eringium*), 172
    E. campestre L., 88
    E. planum L., 21, 160
Erysimum leucanthemum (Steph.) B. Pedtsch., 28, 38
Erythrina, 442
Erythrocebus patas Schreber (red monkey), 211

Erythronium sibiricum (Fisch. & Mey.) Kryl., 161
Espeletia, 309
Eucalyptus, 95, 96, 306, 307, 316, 321, 331, 445, 456, 463
    E. alba Reinw. ex Blume, 96
    E. argillacea W.V. Fitz., 333, 340
    E. bicolor A. Cunn. ex Hook. (see *E. largiflorens*)
    E. brevifolia F. Muell. (snappy gum), 320, 333
    E. brownii, 340
    E. camaldulensis Dehnh. (river red gum), 338
    E. coccifera Hook. f., 344
    E. crebra F. Muell. (iron bark), 340
    E. dichromophloia F. Muell., 333
    E. drepanophylla F. Muell. ex Benth., 340
    E. gunnii J.D. Hook., 344
    E. largiflorens F. Muell. (= *E. bicolor* A. Cunn. ex Hook.), 338
    E. latifolia F. Muell., 338
    E. melanophloia F. Muell., 334, 335
    E. microcarpa (Maiden) Maiden, L. Johnson & Blaxell (= *E. woollsiana*), 336
    E. microneura Maiden & Blakely, 333, 340
    E. microtheca F. Muell. (coolibah), 326
    E. miniata A. Cunn. ex Schauer (woollybutt), 339
    E. niphophila (= *E. pauciflora* subsp. *niphophila*), 344
    E. oleosa (mallee) Miq., 331
    E. pauciflora Spreng. subsp. *niphophila* Maiden & Blakley (= *E. niphophila*), 344
    E. populnea F. Muell. (poplar box), 334–337
    E. pruinosa Schauer (silver-leaf box), 333
    E. saligna F. Muell., 115
    E. salmonophloia F. Muell., 331
    E. salubris F. Muell. (mallee), 338
    E. similis Maiden, 334
    E. stellulata Sieber ex DC., 344
    E. tectifica F. Muell. (McArthur River box), 338
    E. tetrodonta F. Muell. (stringybark), 339
    E. transcontinentalis Maiden, 331
    E. woollsiana R. Baker (= *E. microcarpa*), 336
Eudyptes, 414
    E. chrysocome (Forster) (rockhopper penguin), 427
    E. schlegeli Finch (royal penguin), 427
Eulalia, 97, 306, 317, 338, 340, 341
    E. aurea (Bury) Kunth [= *E. fulva* (R. Br.) O. Kuntze], 322, 325, 340, 352
    E. fulva, 322
    E. leptostachys (Pilg.) Henr., 448–450
    E. phaeothrix (Hack.) O. Kuntze, 93
    E. trispicata (Schult.) Henr., 91
Eupatorium, 113
    E. inulaefolium H.B.K., 95
    E. odoratum L. (= *Chromolaena odorata*), 95, 113
Euphorbia, 204, 234
    E. calyptrata Coss. & Dur., 184
    E. cheirolepis Fisch., 89
    E. drummondii Boiss., 351
    E. guyoniana Boiss. & Aeut., 184
    E. humifusa Schlecht., 29
    E. macroclada Boiss., 88
    E. palustris L., 160
    E. retusa Forsk., 185

Euphorbia (continued)

E. virgata Waldst. & Kit., 160
Euphorbiaceae, 179, 203
Eurotia ceratoides (L.) C.A.Mey., 28, 67
Eurya, 453, 459, 460
Euxerus erythropus (Geoffroy) Desmarest (ground squirrel), 212
Exotheca (= *Cymbopogon*), 225

Fabaceae, 30, 47, 127, 154, 159, 193, 203, 212, 442
Fagaceae, 361
Falco, 241
Farsetia aegyptiaca Turra, 185
Felis caracal Schreber (African lynx, caracal) (= *Lynx caracal*), 104, 211. 239, 241
   F. catus L. (domesticated cat), 427
   F. libyca Forster (African wild cat, desert cat) (= *F. silvestris*), 104, 211, 241
   F. margarita Loche (sand cat), 211
   F. serval Schreber (serval), 211, 226
   F. silvestris Schreber) (see *F. libyca*)
Ferula, 160, 172
Ferula, 160, 172
   F. bungeana Kitag., 29
   F. caspica Bieb., 29, 38
   F. soongarica Pall. ex Spreng., 15
Festuca, 4, 8, 14, 17, 23, 26, 28, 29, 35–37, 39, 47, 48, 52, 70, 71, 89, 90, 94, 272, 287, 308, 363, 398, 418, 425, 426, 430, 441, 472, 477
   F. arundinacea Schreb., 157, 161, 182, 188
   F. bromoides L. (see *Vulpia bromoides*)
   F. capillata Lam., 142
   F. contracta T. Kirk, 418, 421–424, 426, 429
   F. elatior L., 179, 193
   F. elatior L. subsp. *arundinacea* (Schreb.) Celak., 182, 193
   F. kryloviana Reverd., 17
   F. lenensis Drob., 17, 44, 48, 50, 53, 130, 161
   F. mairei St. Yves, 182
   F. matthewsii Cheesem. (alpine fescue tussock), 381
   F. novae-zelandiae Ckn. fescue or hard tussock), 366–368, 374, 383, 384
   F. ovina L. (see also *F. pseudovina*; *F. rupicola*), 64, 69, 70, 149, 156–159, 162, 182
   F. papuana Stapf, 453
   F. pratensis Huds., 127, 130, 133, 135, 136, 139, 140, 157–161
   F. pseudovina Hack. ex Wiesb. (= *F. ovina*), 159
   F. rubra L., 127, 128, 130, 137–141, 148, 149, 157–159, 161, 182
   F. rubra L. var. *commutata* Gaud., 157
   F. rupicola Heuff. (= *F. ovina*), 128, 130–132, 135, 137–139, 141–144, 146, 152–154, 156, 158
   F. sibirica Hack. ex Boiss., 48
   F. sulcata (Hack.) Nym.], 5, 70
   F. valesiaca Schleich. ex Gaud., (see also *F. sulcata*), 5, 10, 11, 14, 15, 20–22, 25, 26, 28, 30, 33, 35–40, 50, 52, 54, 88, 130, 131, 157, 159
   F. varia Haenke, 159
Festuceae, 1, 2, 168, 188, 303

Ficus, 94
Filago arvensis L., 29
Filifolium, 4
   F. sibiricum (L.) Kitam. (= *Tanacetum sibiricum*), 17, 64, 65, 67
Filipendula hexapetala Gilib., 158–161
   F. kamtschatica Maxim. (= *Spiraea camtschatica*), 162
   F. palmata Maxim. (= *Spiraea palmata*), 161, 162
   F. ulmaria (L.) Maxim., 130, 135, 141, 158–161
   F. vulgaris Moench, 21
Fimbristylis, 96, 317, 345, 459
   F. monostachya Link, 96
   F. pentaptera Kunth, 93
Francolinus francolinus L. (black partridge), 104
   F. pordicerianus Gmelin (grey partridge), 104
Frankenia, 330
   F. thymifolia Desf., 185
Fraxinus xanthoxyloides Wamm., 186
Fredolia aretioides Moq. & Coss., 185
Fritillaria meleagroides Patrin ex Shult. f., 158
Fulica atra (L.) (coot), 77
Fungi, 24, 30, 40, 41, 93, 107, 148, 151, 397, 398, 426, 480
Fusarium, 41, 107

Gagea bulbifera (Pall.) Salisb., 29
   G. pusilla (F.W. Schmidt) Schult. & Schult. f., 29
Gahnia, 438
   G. javanica Zoll. & Mor. ex Mor., 438
Galactites, 179
Galago senegalensis E. Geoffroy St-Hilaire (lesser galago), 211
Galatella (= *Aster*)
   G. angustissima (Tausch) Novopokz., 21
   G. divaricata (Fisch. ex M. B.) Novopokz., 22
Galinsoga parviflora Cav., 94
Galium asperifolium Wall., 94
   G. boreale L., 158–161
   G. mollugo L., 157
   G. pediformis, 48
   G. ruthenicum Willd. [= *G. verum* L. subsp. *ruthenicum* (Willd.) P. Fauzh.], 4, 21, 22
   G. verum L., 21, 47, 50, 72, 136, 139, 158–160
Gallirallus australis Sparrman (weka), 427
Garnotia, 441
Gazella bennetti Sykes (Indian gazelle), 103
   G. dama (Pallas) (gazelle, mhorr), 211
   G. dorcas (L.) (dorcas gazelle), 190, 209, 211
   G. gazella Pallas (chinkara), 104
   G. granti Brooke (Grant's gazelle), 226
   G. leptoceros F. Cuvier (rhigazelle), 209, 211
   G. pelzelni (Kohl) (Pelzeln's gazelle), 248, 256
   G. rufifrons Gray (red-fronted gazelle), 209, 211
   G. soemmeringi (Cretzschmar) (Soemmering's gazelle), 248, 253
   G. spekei Blyth (Speke's gazelle), 253, 254, 256
   G. subgutturosa (Guldenstaedt) (goitered gazelle), 75
   G. thomsoni Gunther (Thomson's gazelle), 226, 242
Genetta, 226
   G. genetta L. (common genet), 211
Geniostoma, 459

Genista, 192
   G. microcephala Coss. & Dur., 185
Gentiana algida Pall., 74
Gentiana (continued)

   G. triflora Pall., 162
Geotrichum, 24
Geranium erianthum DC., 161
   G. pratense L., 53, 135, 136, 159
   G. pseudosibiricum J. Mayer, 161
   G. sanguineum L., 159
   G. tuberosum L., 15
Gerbillus gerbillus Olivier (gerbil), 212
   G. nanus Blandford (dwarf gerbil), 212
   G. nanus indus (Thomas) (Wagner's gerbil), 103
   G. pyramidum Geoffroy (Egyptian gerbil), 212
Germainia capitata Balansa & Poitrass, 445
Geum elatum Wall., 92
   G. rivale L., 158
Giraffa camelopardalis (L.) (giraffe), 209, 211, 234, 239
Gisekia, 215
   G. pharnaceoides L., 204
Glaux maritima L., 160
Gleichenia vulcanica Bl., 453
Globularia, 172
Glochidion, 449
Glomus tenue (Greenall) Hall, 398
Glossina, 226, 229, 231, 234
Glossogyne tenuifolia (Labill.) Cass., 459
Glyceria maxima (Hartman) Holmberg, 128, 130–132, 135, 141–144, 152, 153, 157, 159, 160
Glycine, 288, 318
Glycyrrhiza glabra L., 158, 161
   G. uralensis Fisch., 73, 161
Goniolimon speciosum (L.) Boiss., 53
   G. tataricum (L.) Boiss., 50
Gossypium hirsutum L., 229
Grevillea, 445
   G. maculata, 331
Grewia, 95, 204, 230, 274
Guiera, 204
Guioa, 451
Gundelia, 179
Gunnera, 309
Gymnocarpos decander Forsk., 185
Gymnosporia linearis (L. f.) Loes., 298
Gymnostoma papuanum (S. Moore) Johnson (= *Casuarina papuana*), 448
Gypaetus barbatus (L.) (lammergeyer), 226
Gyps bengalensis (Gmelin) (white-backed vulture), 241
   G. ruppelli (Brehm) (Ruppell's vulture), 241
Gypsophila desertorum (Bunge) Fenzl, 29
Gyrocarpus hababensis Chiov., 245

Halimione portulaccoides (L.) Aellen, 186
Halocnemum strobilaceum (Pall.) M. Bib., 186
Halopeplis amplexicaulis (Vahl) Ung., 184
Haloxylon ammodendron Bunge, 90
   H. persicum Bunge ex Boiss. (= *H. ammodendron*), 89, 114
   H. salicornicum Bunge ex Boiss., 89

Hammada schmittiana (Pomel) Botsch., 181, 185
   H. scoparia (Pomel) Iljin, 185
Haplophyllum davuricum (L.) G. Don. f. (= *Aplophyllum davuricum*), 29
Harpachne schimperi A. Rich., 240
Harpochloa (or Harpechloa)
   H. falx (L.) O. Kuntze, 278, 284
Hedysarum, 180
   H. capitatum Desf., 184
   H. carnosum Desf., 184, 186
   H. coronarium L., 183, 193
   H. flexuosum L., 184
Helianthemum, 159
   H. confertum Dum., 186
   H. kahiricum Del., 185
   H. lippii (L.) Pers. subsp. *sessiliflorum* (Desf.) Murb., 185
Helichrysum, 272, 351
   H. gymnocephalum (DC.) H. Hurub., 301
Helicostylum pyriforme Bain, 107
Helictotrichon, 4, 21, 272
   H. desertorum (Less.) Nevski, 21
Heliocopris (dung rollers), 281
Heliophobius argenteocinereus Peters (solitary mole rat), 248
Heliotropium, 89, 252, 256
Helipterum, 328, 337
   H. charsleyae F. Muell., 328, 351
   H. floribundum DC., 328
Helogale parvula (Sundeval) (dwarf mongoose), 239
Hemerocallis flava L., 130, 160–162
   H. middendorfii Trautv. & Mey., 162
   H. minor Miller, 64, 66, 72, 162
Hemitragus jemlahicus (H. Smith) (Himalayan tahr), 104, 394, 395, 403
Heracleum dissectum Ledeb. (= *H. lanatum* Michx.), 130, 160, 161
   H. sibiricum L., 159, 160
   H. sphondylium L., 157
Herpestes, 226
   H. ichneumon L. (Egyptian mongoose), 211
Heterocephalus glaber Rippell (colonial mole rat), 248
Heterodendrum, 330
   H. oleifolium Desf., 336, 337
Heteropappus altaicus (Willd.) Novopokr., 29, 35
   H. hispidus (Thunb.) Less., 48, 50
Heteropogon, 94, 96, 97, 100, 105, 230, 299, 300, 306, 307, 317, 338, 340, 341, 440, 441, 472, 475, 477
   H. contortus (L.) P. Beauv. ex Roem. & Schult., 89, 91, 93, 94, 96, 105, 109, 114, 116, 203, 206, 225, 231, 239, 240, 246, 253, 272, 274, 283, 294, 296, 299–301, 310, 339, 340, 346–348, 352, 449, 458, 459
Heteropsylla cubana (Crawford), 457
Heteroptera, 148
Hibbertia lucens Sebert & Pancher., 458
Hibiscus tiliaceus L., 455
Hieracium, 379
   H. umbellatum L., 161
Hierochloë, 307, 453
   H. redolens R.Br., 452
Hippocrepis scabra DC., 185
Hippophae rhamnoides L., 68

Hippopotamus amphibius L. (hippopotamus), 211, 228, 239, 280
Hippotragus, 280
   H. equinus (Desmarest) (roan antelope), 211, 231, 234, 248, 280
   H. niger (Harris) (sable antelope), 231, 267, 280
Hirundo, 241
Hodotermes, 248
   H. mossambicus Hagen (harvester termite), 281
Holcus lanatus L., 127, 140, 141, 157, 158, 182
Holosteum, 14
Homoptera, 147, 148
Hordeum, 179, 227, 349
   H. brevisubulatum Link (= *H. secalinum*), 130, 160, 161
   H. bulbosum L., 182
   H. leporinum Link (= *Critesion murinum* subsp. *leporinum*) (barley grass), 307, 318, 337
   H. maritimum Witth., 183
   H. murinum L., 183
   H. secalinum Schreber (see *H. brevisubulatum*)
Hosta albomarginata (Hook.) Ohwi, 162
Hulthemia berberifolia Dum., 89
Hyaena hyaena L. (striped hyaena), 190, 211, 213
Hydrocotyle, 452
   H. ramiflora Maxim., 162
Hymenachne (= *Panicum*), 96
Hymenolyma trichophyllum (Schrenk) Korov., 29
Hymenoptera, 148
Hyoscyamus niger L., 52
Hyparrhenia, 167, 222, 223, 225, 229–233, 238, 239, 242, 274, 295, 300, 478
   H. cymbaria (L.) Stapf, 228, 273, 294, 298
   H. diplandra (Hack.) Stapf, 228, 233
   H. dissoluta (Nees ex Steud.) C.E. Hubb. (see *Hyperthelia dissoluta*)
   H. dregeana (Nees) Stapf, 274, 278
   H. eberhardtii (A. Camus) Hitchc., 95
   H. filipendula (Hochst.) Stapf, 222, 233–235, 240, 273
   H. hirta (L.) Stapf, 88, 182, 203, 206, 274, 275
   H. rufa (Nees) Stapf, 228, 233–235, 237, 240, 274, 278, 294–296, 298–300
Hyperthelia, 234
   H. dissoluta (Nees ex Steud.) W.D. Clayton (= *Hyparrhenia dissoluta*), 206, 233–235, 273, 274, 296, 299
   H. newtonii (Hack.) Stapf, 273
Hyphaene coriacea Gaertn., 230
   H. shatan Boj., 297–299
Hypochaeris radicata L., 342, 345
Hypoëstes, 246
Hyptis suaveolens (L.) Poit., 113
Hyssopus cretaceus Dubjan., 8
Hystrix cristata L. (North African porcupine), 212
   H. indica Kerr (porcupine), 104

Ichneumia albicauda G. Cuvier (white-tailed mongoose), 211, 226
Imperata, 90, 91, 94, 96, 102, 111–113, 230, 306, 440, 448, 451, 455
   I. cylindrica (L.) P. Beauv., 91–97, 102, 111–113, 116, 162, 228, 233, 295, 298–300, 306, 308, 310, 438, 441–443, 445–450, 454–457, 459–461, 463, 464
   I. cylindrica (L.) P. Beauv.var. africana (Anders.) C.E. Hubb., 228
Indigofera, 97, 240, 246
   I. intricata Boiss., 251, 253
   I. ruspolii Bak. f., 252
   I. spinosa Forsk., 256
Insectivora, 103
Inula, 73, 157
   I. britannica L., 158
   I. cordata Boiss., 159
   I. crithmoides L., 186
Ipomoea, 240, 246
   I. batatus (L.) Lam., 226
   I. pes-caprae (L.) Sweet, 456
   I. reptans Poir., 324
Iris, 78, 157
   I. bungei Maxim., 13, 29
   I. clarkei Baker ex Hook. f., 94
   I. ensata Thunb., 74
   I. kaempferi Sib., 72, 162
   I. lactea Pall., 45
   I. ruthenica Ker-Gawl., 160, 161
   I. scariosa Willd. ex Link, 29
   I. setosa Pall., 162
   I. songarica Schrenk, 89
   I. tenuifolia Pall., 29
Isachne, 96, 307, 441
   I. globosa (Thunb.) O. Kuntze, 96, 452
   I. myosotis Nees, 452
Isachneae, 307
Ischaemum, 96, 97, 306, 441, 446, 449, 463–465
   I. afrum (J.F. Cmel.) Dandy, 273
   I. angustifolium (Trin.) Hack., 94
   I. antephoroides (Steud.) Miq., 162
   I. barbatum Retz., 445, 447
   I. indicum (Houtt) Merrill, 92, 93
   I. muticum L., 457, 460
   I. rugosum Salisb., 91
Iseilema, 97, 317, 321, 340
   I. laxum Hack., 94, 115
   I. membranaceum (Lindl.) Domin, 321–323
   I. vaginiflorum Domin, 321
   I. wightii Anderss., 91
Isodon japonicus (Burm.) Hara (Lamiaceae), 162
Isoodon macrourus (Gould) (mumut, bandicoot), 465
Isopoda, 151
Ixeris chinensis (Thunb.) Nakai, 69
   I. dentata (Thunb.) Nakai, 162
   I. denticulata (Houtt.) Stebb., 69

Jaculus jaculus L. (jerboa), 212
Julbernardia globiflora (Benth.) Troupin, 269
Juncaceae, 127
Juncus acutiflorus Ehrh., 127
   J. atratus Krock., 158
   J. effusus L., 127, 158
   J. filiformis L., 158
   J. gerardi Loisel., 159, 160
   J. scheuchzerioides Gaud., 419

SYSTEMATIC INDEX

Juniperus phoenicea L., 172, 173, 186
Jurinea mongolica Maxim., 29
   J. multiflora (L.) B. Fedtsch., 20–22, 52
Jussieua, 299

Kalidium foliatum (Pall.) Moq., 73
Kobresia, 72, 74
   K. capillifolia (Decne.) C.B. Clarke, 74
   K. filifolia C.B. Clarke, 161
   K. graminifolia Bocklr., 74
   K. humilis (C.A. Mey.) Serg., 74
   K. pygmaea C.B. Clarke, 72, 74
   K. tibetica Maxim., 72
Kobus, 226, 228, 231
   K. defassa Ruppell (defassa waterbuck), 209, 211
   K. ellipsiprymnus Ogilby (waterbuck), 280
   K. kob Erxleben (kob waterbuck), 211
   K. leche Gray (lechwe), 280
   K. vardoni (Livingstone) (puku), 280
Kochia, 38, 79, 329
   K. prostrata (L.) Schrad., 14, 28, 37, 38, 47, 50, 52–54, 67
   K. pyramidata Benth (= *Maireana pyramidata*), 331
   K. sedifolia F. Muell. (= *Maireana sedifolia*), 329
Koeleria, 4, 8, 89, 179, 272
   K. cristata (L.) Pers. (= *K. gracilis*), 14, 21, 22, 28, 48, 53, 54, 92
   K. delavignei Czern. ex Domin, 158, 159
   K. gracilis Pers. (= *K. cristata*), 67, 71, 159, 161
   K. macrantha Schult., 17
   K. phleoides (Vill.) Pers., 183
   K. pubescens (Lamk) P. Beauv., 183
   K. pyramidata (Lam.) Beauv. (= *K. setacea* Pers.), 156
   K. splendens Presl., 182
   K. vallesiaca (Honck.) Bert., 182
Kunzea, 363
Kyllinga, 242

Lagenantha, 246
Lagochilus ilicifolius Bunge, 29
Lagomorpha, 43, 103
Lagurus lagurus Pall. (steppe lemming), 43, 47, 49–51, 54
Lamiaceae, 8, 179
Lampito mauritii Kinberg, 106
Lannea, 238
Lantana, 230
   L. camara L., 113, 301, 346
Lappula echinata Gilib (see *L. squarrosa*), 50
   L. patula (Lehm.) Menyharth, 52
   L. squarrosa (Retz.) Dumzt. (= *L. echinata*), 48, 50
Larix, 161
   L. gmelinii (Rupr.) Rupr., 10
   L. sibirica Ledeb., 10
Larus brunnicephalus (Jerdon) (brown-headed gull), 75, 77
Lasiagrostis splendens Kunth (= *Muhlenbergia alpestris*), 161
Lasiopodomys, 49
   L. brandtii Radde (Brandt's vole), 43–45, 48–50
   L. mandarinus M.-Edw. (Chinese vole), 43, 44, 46–50
Lasiurus, 89–91, 109, 112
   L. hirsutus (Forsk.) Boiss. (= *L. scindicus* = *Rottboellia hirsuta*), 88, 89, 91, 109, 114, 182, 203, 256

   L. scindicus, 182
Lathyrus alatus Sibth. & Sm., 162
   L. palustris L., 161
   L. pilosus Cham., 162
   L. pratensis L., 158–161
   L. quinquenervius (Miq.) Litv., 64, 66, 72
Lauraceae, 462
Leersia, 96, 306, 440
   L. hexandra Sw., 96, 299, 345, 445
Leiostipa (section of *Stipa*), 14, 18
Lepidium densiflorum Schrad, 53
   L. perfoliatum, 50, 54
   L. virginicum Gren. & Godr., 94
Lepidoptera, 147, 148, 150, 478
Leptochloa, 97
   L. brownii C.E. Hubb. (= *L. neesii*), 345
   L. neesii (Thwaites) Benth. (= *L. brownii*), 345
Leptopyrum fumarioides (L.) Reichenb., 53
Leptospermum, 363
   L. scoparium J.R. & G. Forst., 377
Leptothrium, 223, 250, 478
   L. senegalense (Kunth) W.D. Clayton, 222, 246, 251, 253, 256
Lepturus, 441
Lepturus incurvus (L.) Trin., 183
Lepus, 104, 241, 480
   L. capensis L. (Cape hare), 211, 239
   L. crawshayi De Winton (Crawshay's hare), 211, 239
   L. europaeus Pallas (European hare), 395, 403
   L. nigricollis dayanus Blanford (Indian desert hare), 104
   L. nigricollis F. Cuv., 104
Lespedeza, 161
   L. cyrtobotrya Miq., 162
   L. daurica (Laxm.) Schindl. ex Komarov, 66, 67, 72
   L. hedysaroides (Pall.) Kitag., 67, 72
   L. serpens Nakai, 162
Leucaena, 227, 457, 466
   L. leucocephala (Lamk.) De Wit, 311, 346, 445, 455, 456, 459, 466
Leucanthemum vulgare Lam. (= *Chrysanthemum leucanthemum*), 157–160
Leucopogon cymbulae Labill., 458
Leucosidea sericea Eckl. & Zegh., 286
Leymus, 78
   L. chinensis (Trin.) Tzvel., (= *Aneurolepidium chinense*), 4, 17, 45, 48–50, 53, 64, 65, 67, 72
   L. dasystachys (= *Aneurolepidium dasystachys*), 68
   L. ramosus (Trin.) Tzvel., 22, 50, 52, 54
Liasis (python), 465
Libanotis montana Crantz, 160
   L. sibirica Koch, 160
Ligularia, 73, 160
   L. mongolica DC., 66
   L. speciosa Fisch. & Mey., 162
Liliaceae, 4, 30, 72, 162, 179, 380, 386
Lilium amabile Seib. & Zucc., 72
   L. dauricum Ker-Gawl., 161, 162
   L. martagon L., 160, 161
   L. tenuifolium Fisch., 162
Limeum, 215

Limeum (continued)

L. viscosum (Gay) Fenzl., 204
Limoniastrum guyonianum Dur., 186
Limonium axillare (Forsk.) O. Kuntze, 246
  L. bicolor (Bunge) O. Kuntze, 73
  L. tenellum (Turcz.) Ktze., 29
Linaria, 29
Lintonia nutans Stapf, 256
Litocranius walleri (Brooke) (gerenuk), 248
Livistona, 307
  L. brassii Burret, 307, 446, 448
Lobelia, 309
Locusta migratoria L., 77, 190
Locustidae (former family name for grasshoppers), 23
Lolium, 179, 287
  L. multiflorum Lamk., 183, 193
  L. perenne L., 130, 149, 157, 182, 193, 318, 344, 345, 348, 399
  L. rigidum Gaud., 183, 318, 349
  L. temulentum L., 183
Lonicera maackii (Rupr.) Maxim., 68
Lophatherum, 96
Lophostemon, 445
Lotus, 180
  L. corniculatus L., 162, 183, 193
  L. creticus L., 183, 184, 193
  L. uliginosus Schk., 127, 184
Loudetia, 225, 274, 294, 296, 300
  L. arundinacea (A. Rich.) Steud., 233
  L. filifolia Schweick. subsp. *humbertiana* A. Camus, 298
  L. kagerensis (K. Schum.) Hutch., 225, 237
  L. simplex (Nees) C.E. Hubb., 225, 273, 274, 299
  L. simplex (Nees) C.E. Hubb. subsp. *stipoides* (Hack.) Bosser, 296
  L. togoensis (Pilger) C. E. Hubb., 206
Loxodonta, 280
  L. africana (Blumenbach) (elephant), 209, 211, 226, 228, 234, 248, 280
Lumbricidae, 150, 210
Luzula multiflora L., 158
Lycaon pictus (Temminck) (African wild dog), 211, 226
k*Lychnis flos-cuculi* L., 157, 158
k*Lycium intricatum* Boiss., 179, 186
Lycopodium cernuum L., 459
Lygeum spartum Loefl. ex L. (esparto grass), 171, 172, 177, 182
Lygodium scandens Sw., 459
Lygos raetam (Forsk.) Heywood (= *Raetama raetam* Webb), 187
Lynx caracal (Schreber) (see *Felis caracal*)
Lysiphyllum (= *Bauhinia*)
  L. cunninghamii (Benth.) De Wit, 321

Macaranga, 451
Machaerina rubiginosa (Spreng.) Koyama, 438, 452
Macronectes (petrels)
  M. giganteus Gmelin, 427
  M. halli Mathews, 427
Macropodidae, 353

Macroptilium, 318
  M. atropurpureum (DC.) Urb. (siratro), 347
Macropus agilis Gould (agile or sandy wallaby), 310, 465
  M. antilopinus (Gould) (antilopine kangaroo), 316, 353
  M. fuliginosus (Desmarest), 353
  M. giganteus Shaw (grey kangaroo), 353
  M. robustus Gould (hill kangaroo euro), 353
  M. rufus (Desmarest) (red kangaroo) (= *Megaleia rufa*), 316, 353
Macrotermes, 228, 231, 234, 241, 248
Macrotermitinae, 241, 256
Maireana, 307, 317, 329–331, 337, 350, 354
  M. aphylla (R. Br.) P.G. Wilson, 332, 333
  M. astrotricha (L. Johnson) P.G. Wilson, 329
  M. pyramidata (Benth.) P.G. Wilson (= *Kochia pyramidata*), 330, 331
  M. sedifolia (F. Muell.) P.G. Wilson (= *Kochia sedifolia*), 329–331, 337
Malva, 172
Malvaceae, 203
Mangifera indica L. (cashew), 231
Manihot, 455
  M. esculenta Crantz, 231, 300, 458
Maniltoa, 442
Manisuris clarkei (Hack.) Bor., 96
  M. granularis L., 95
Marmota, 51–53, 76
  M. bobak Müller (marmot or baibak), 43, 45, 51, 52, 76, 77
  M. brandtii, 76
  M. sibirica Radde (tarbagan), 43, 48, 51–53
Marrubium deserti De Noe, 185
  M. vulgare L., 50
Massia triseta (Nees) Balansa, 92
Mastomys erythroleucus Temminck (multimammate rat), 212
  M. huberti Wroughton (multimammate rat), 212
Mauritia, 307
Mazus delevayi Bonati, 94
Medemia nobilis Gallerand, 295, 297, 299
Medicago, 47, 66, 180, 288, 307, 315, 318, 337, 344, 345, 349
  M. arborea L., 186
  M. ciliaris Kroch, 184
  M. falcata (L.) Lamk., 158–160, 184
  M. gaetula Pomel, 184, 193
  M. hispida Gaertn., 184, 188
  M. intertexta (L.) Mill., 184
  M. laciniata Mill., 349
  M. littoralis Rhode, 184, 188, 349
  M. minima (L.) Desr., 344, 349
  M. orbicularis (L.) All., 184
  M. polymorpha L., 344, 349
  M. romanica Prod., 20–22
  M. rugosa Desr., 184, 349
  M. ruthenica (see *Trigonella ruthenica*)
  M. sativa L., 184, 188, 193, 348
  M. scutellata (L.) All., 184, 188, 349
  M. truncatula Gaertn., 184, 188, 193, 349
Megaleia rufa Desmarest (red kangaroo) (= *Macropus rufus*), 316
Megaloprotachne albescens Hubb., 273
Megascolecidae, 106

SYSTEMATIC INDEX 515

Melaleuca, 306, 445, 446, 455, 456
    M. quinquenervia (Cav.) S.T. Blake (niaouli), 306, 455, 456
Melanocenchris, 97, 112
Melanocorypha mongolica (Pallas) (Mongolian lark), 77
Melastoma, 460
    M. malabathrica L., 459
    M. offine D. Don., 95
Melica, 88
    M. ciliata L., 179, 182
    M. cupani Guss., 182
    M. minuta L., 179, 182
    M. transsilvanica Schur, 157
Melilotus, 180
    M. alba Med., 184
    M. italica Lamk., 184
Melinis, 306
    M. minutiflora P. Beauv. (molasses grass), 346, 441
Mellivora capensis (Schreber) (ratel), 239
Melocanna baccifera Skeels, 94
Meniocus linifolius (Steph.) DC., 29
Meriones hurrianae hurrianae Jerdone (desert gerbil), 103, 104
    M. libycus Lichstenstein (Libyan jird), 212
    M. unguiculatus (Milne-Edwards) (clawed gerbil), 76
Merremia, 449
Merxmuellera, 272
Meum athamanthicum, 157
Micraireae, 312
Microchloa kunthii Desv., 238, 240
Microtermes, 281
Microtermitinae, 248
Microtinae, 76
Microtus, 49, 76, 147
    M. arvalis Pallas (common vole), 43, 45–47, 50, 146, 147
    M. brandtii (Radde) (vole), 76
    M. gregalis Pallas (narrow-skulled vole), 43, 46, 48, 76
    M. oeconomus (Pallas) (vole), 76
    M. socialis Pall. (social vole), 43, 50
Mikania micrantha Kunth (= *M. cordata* var. *indica* Kitamura), 95
Milvus migrans Boddaert (black kite), 77
Mimosa asperata (see *M. pigra*)
    M. invisa Mart., 95
    M. pigra L. (= *M. asperata*), 312
Mimosaceae, 199, 204
Mirounga leonina L. (elephant seal), 414, 415, 419, 420, 423, 427
Miscanthus, 131, 162, 282, 306, 307, 451
    M. floridulus (Labill.) Warb., 446–448, 451, 455–459, 461, 464
    M. purpurascens Anderss., 162
    M. sinensis Anderss., 130, 140, 162
Molinia coerulea Moench, 130, 148, 158
Mollugo, 215
    M. nudicaulis Lam., 204, 301
Mollusca, 210
Mltkia ciliata (Forsk.) Johnst., 186
Monachather paradoxa Steudel (= *Danthonia bipartita*), 326, 328, 329, 331, 334, 350
Monocotyledoneae, 439

Monocymbium ceresiiforme (Nees) Stapf, 274
Monostachya oreoboloides (F. Muell.) Hitchc., 453
Montifringilla, 77
    M. nivalis (L.) (snow finch), 77
Morelia (python), 465
Morinda citrifolia L., 458, 459
Motacilla citreola (Tunstall) (grey wagtail), 77
Mucor genevensis Lendner, 107
    M. microsporus Namyslowski, 107
Mucorales, 24
Muhlenbergia alpestris (see *Lasiagrostis splendens*)
Mungos mungo (Gmelin) (banded mongoose), 239
Muntingia calabura L., 311, 466
Mus haussa Thomas & Hinton (haussa mouse), 212
    M. musculus L., 427
Musa (= Scitamineae) (banana), 226
Mussaenda raiateensis J.W. Moore, 458
Mustela altaica (Pallas) (weasel), 76
    M. erminea L. (stoat), 393, 403
    M. putorius (L.) (European polecat), 76
    M. sibirica (Pallas) (Siberian weasel), 76
Myoporum, 330, 335
Myosotis sylvatica Hoffm., 158
Myospalacinae (rodents), 76
Myospalax aspalax (Pallas) (zokor), 76
    M. fontanierii (Milne-Edwards) (common Chinese zokor), 76
Myriapoda, 151
Myrmeleotettix palpalis (Zubovsky) (grasshopper), 78
Myrothecium roridum Tode ex Fr., 107
Myrtaceae, 363, 462
Myrtella benningseniana (Volk.) Diels, 459
Myxoma leporipoxvirus, 430

Nanophyton erinaceum (Pall.) Bunge, 14
Nardus, 477
    N. stricta L., 128, 130, 131, 134, 136–140, 142, 147–149, 152, 154, 157–159, 182
Nauclea coadunata Roxb. ex Sm. (= *Sarcocephalus cordatus*), 447, 449
Necrosyrtes monachus (Temminck) (hooded vulture), 241
Nemorhaedus goral Hardwicke (goral), 104
Neopaltasia pectinata (see *Artemisia pectinata*)
Neophron percnopterus (L.) (Egyptian vulture), 241
Nepenthes, 460
    N. distillatoria L., 92
    N. mirabilis Lour.) Druce, 459
Neptunia monosperma F. Muell. ex Benth., 351
Nesotragus moschatus (Bon Dueben) (suni), 231
Nicotiana glauca Graham, 215
Nitraria billardieri DC., 332
    N. retusa (Forsk.) Asch., 186
Noaea mucronata (Forsk.) Asch. & Schweinf. (= *N. spinosissima* Moq.), 88, 185, 189
Nostoc, 398
    N. commune (Vauch.) Cooke, 24, 426
Nothofagus, 361, 363, 463
    N. menziesii (Hook. f.) Oerst. (silver beech), 365, 373
Notornis, 393
    N. mantelli Owen (flightless rail or "takahe"), 393–395

Nucularia perrini Batt., 186
Nymphaea, 299

Ochna pulchra Robson, 268, 274, 281
Ochotona curzoniae (Hodgson) (mouse-hare), 76
   O. daurica Pall. (Daurian pita), 43, 44, 47–50
Ochotona rutila (Severtzov) (Turkestan red pika), 76
Ochotonidae, 76
Ocnerodrilidae, 106
Ocnerodrilus occidentalis Eisen (earthworm), 106
Octochaetona surensis Michaelsen, 106, 107
Odontotermes, 231, 248
   O. gurdaspurensis Holmgren & Holmgren (termite), 106, 107
Oedaleus asiaticus (Bey-Bienko) (grasshopper), 77
   O. infernalis infernalis (Soussure) (grasshopper), 77
Oenanthe isabellina (L.) (isabelline wheatear), 77
   O. oenanthe (L.) (wheatear), 77
Olea europaea L., 187
Oleaceae, 68
Olearia colensoi Hook. f. (leatherwood), 386
Olgaea leucophylla Iljin, 67
Oligochaeta, 426
Omalanthus, 451
Onobrychis, 180
   O. arenaria (Kit.) DC., 158, 160
   O. argentea Boiss., 184
   O. gaubae Bornm., 184
   O. inermis Stev., 159
   O. sativa L., 193
   O. viciaefolia Scop., 184
Onoclea sensibilis L., 162
Onopordum, 179, 351
   O. acanthium L., 50
Onosma transrhimnensis Klok. ex M. Pop., 21
Ophiuros, 306, 440
   O. tongcalingii (Elmer) Henr., 445, 449
Oplismenus compositus Beauv., 308
Opuntia, 311
   O. ficus-indica L., 191, 193
Oreobolus pumilio R. Br., 453
Orobus lathyroides L., 161
Oropetium capense Stapf, 274
   O. thomaeum (L. f.) Trin., 97
Orthoptera, 106, 147, 148, 480
Orycteropus afer Pallas (aardvark), 211
Oryctolagus cuniculus (L.) (rabbit), 343, 353, 379, 383, 384, 420, 427
Oryx algaze (= *O. dammah*), 209
   O. beisa (Ruppell) (beisa oryx), 248, 253, 256
   O. dammah Cretzschmar (= *O. algazel*) (scimitar-horned oryx), 209–211
   O. gazella L. (gemsbok), 280
Oryza, 96, 306
   O. australiensis Domin, 345
   O. barthii A. Chev. (= *O. longistaminata* Chev. & Roehr.), 206
   O. rufipogon Griff., 94
   O. sativa L., 300, 445
Oryzeae, 440

Oryzopsis, 89
   O. aequiglumis Duthie, 89
   O. coerulescens (Desf.) Richt., 179, 182
   O. holciformis (M. Bieb.) Hack., 88, 179, 182
   O. miliacea (L.) Benth. & Hook. ex Asch. & Schweinf., 179, 182, 188, 193
   O. munroi Stapf, 89
   O. paradoxa (L.) Nutt., 179, 182
   O. thomasii (Duby) Le Houer., 179
Osmunda japonica Thunb., 162
Ostryopsis davidiana Decne., 68
Otis tarda (L.) (great bustard), 77
Otocyon megalotis (Desmarest) (bat-eared fox), 239
Oudneya africana R. Br., 186
Ourebia ourebi Zimmermann (oribi), 211, 280
Ovis ammon Blyth (argali), 104
   O. aries L. (domestic sheep), 70, 227
   O. orientalis Gmelin (urial), 104
Oxalis corniculata L., 94
Oxystenanthera auriculata Prain, 94
Oxytropis aciphylla Ledeb., 28
   O. myriophylla (Pall.) DC., 66, 67

Pachyuromys duprasi Latasste (fat-tailed mouse), 212
Palimbia salsa Bess. (= *Peucedanum redivivum*), 22
Paltothyreus (ant), 228
Pandanaceae, 307
Pandanus, 307, 439, 445, 448, 455, 459, 460, 462
   P. odoratissimus L. f., 458
   P. spiralis R. Br., 457
Paniceae, 1, 2, 96, 167–169, 188, 272, 274, 294, 303, 306, 312
Panicum, 95, 96, 112, 215, 223, 229, 230, 295, 299, 306, 318, 341, 441
   P. anabaptistum Steud., 206
   P. antidotale Retz., 89, 114
   P. coloratum L., 238
   P. decompositum R. Br., 322
   P. deustum Thunb., 225
   P. effusum R. Br., 351
   P. humidorum Buch. (see *P. perakense*)
   P. infestum Peters, 231
   P. kalaharense Mex, 273
   P. laetum Kunth, 209
   P. lutescens Weig., 95
   P. luzonense T. & C. Presl., 96
   P. maximum Jacq. (Guinea grass), 115, 228, 230, 231, 233, 237, 238, 246, 274, 311, 346, 442, 458
   P. maximum Jacq. var. *trichoglume* Eyles, 318, 348
   P. miliaceum L. var. *ruderale* Kitag., 69
   P. orientale (Rich.) Willd., 89
   P. perakense (Hook. f.) Merr. (= *P. humidorum* Buch.), 96
   P. pinnifolium Chiov., 253
   P. pseudovoeltzkowii A. Camus, 298, 299
   P. repens L., 95, 96
   P. sanguinale L. (see *Digitaria marginata*)
   P. sarmentosum Roxb., 96
   P. turgidum Forsk., 91, 182, 199, 240, 256
   P. umbellatum Trin., 295
   P. walense Mez., 113
Panthera leo (L.) (lion), 190, 209, 211, 226

SYSTEMATIC INDEX

Panthera (continued)

    P. pardus (L.) (leopard), 104, 190, 209, 211, 226
Pantholops hodgsoni Abel (chiru, Tibetan antelope), 76, 104
Papio anubis J. P. Fisher (anubis baboon), 211
Pappophoreae, 307, 312
Pappophorum nigricans R. Br. (see *Enneapogon nigricans*)
    P. persicum (Boiss.) Steud., 88
Papuapteris linearis C. Chr., 453
Parinari curatellifolia Planch. ex Benth., 269
Parmelia ryssolea (Ach.) Nyl., 30
    P. vagans Nyl., 20, 30
Paspalidium constrictum (Domin) C.E. Hubb., 334
    P. flavidum (Retz.) A. Camus, 94
    P. geminatum (Forsk.) Stapf, 94
Paspalum, 96, 97, 112, 318, 341
    P. conjugatum Berg., 95, 96
    P. dilatatum Poir., 96, 346, 348
    P. distichum L., 94
    P. notatum Fluegge, 96
    P. orbiculare Forst., 308, 458
    P. scrobiculatum Bojer, 95
    P. virgatum L., 300
Passer montanus (L.) (tree sparrow), 77
Passerina montana Thod., 272
Patrinia scabiosaefolia, 72
Pedicularis comosa L., 158
    P. dasystachis Schrenk, 15
    P. kaufmannii Pinzg., 15
Peganum harmala L., 68, 89, 185
    P. nigellastrum Bunge, 29, 31
Pelea capreolus Foster (vaal ribbok), 280
Pelecanus philippensis (Gmelin) (spot-billed pelican), 77
Penicillium, 24, 41, 107
    P. javanicum Van Beyma, 107
    P. wortmanii Klowcker, 107
Pennatae (section of Stipa), 78
Pennisetum, 95, 215, 222–225, 230, 239, 306, 318, 341
    P. asperifolium (Desf.) Kunth, 182
    P. clandestinum Hochst. ex Chiov., 93, 222, 224, 225, 227, 288, 346, 348
    P. dichotomum (Forsk.) Del. (= *P. divisum*), 89, 182
    P. divisum, 182
    P. glaucum (L.) R. Br., 214
    P. macrostachyum (Brongn.) Trin., 449
    P. mezianum Leeke, 237–240
    P. orientale L.C. Rich., 88, 89
    P. pedicellatum Trin., 206
    P. polystachyon Schult., 311, 441, 445, 454–456, 458, 460, 464, 466
    P. purpureum Schumach., 94, 102, 111, 222, 228, 229, 346
    P. ramosum (Hochst.) Schweinf., 233
    P. riparium A. Rich., 225
    P. sphacelatum (Nees) Th. Dur. & Schinz, 222, 225, 227
    P. stramineum Peter, 240
    P. unisetum (Nees) Benth., 228
Pentachistis, 272
Pentaphylloides fruticosa (L.) O. Schwarz (= *Potentilla fruticosa*), 48
Pentzia, 278

P. globosa Less., 272
Peramelidae, 465
Pergularia tomentosa L., 185
Periconia minuissima Corda, 107
Perotis, 96
    P. hildebrandtii Mez, 231
    P. indica (L.) O. Kuntze, 96
Peucedanum alsaticum L., 20, 21
    P. officinale L., 130, 160
    P. redivivum Pall. (see *Palimbia salsa*)
Phacochoerus aethiopicus (Pallas) (warthog), 211, 228, 231, 239, 253
Phalacrocorax carbo (L.) (great cormorant), 77
Phalarideae, 312
Phalaris, 179, 349, 477
    P. aquatica Desf. (= *P. tuberosa* Link), 344, 348
    P. arundinacea L., 128, 130–132, 135, 138–144, 148, 157, 159–161, 179, 183
    P. bulbosa L., 193
    P. bulbosa L. subsp. *arundinacea*, 179
    P. canariensis L., 183
    P. coerulescens Desf., 179, 183
    P. minor Retz., 94, 183
    P. truncata Guss., 179, 183, 193, 318, 344, 348
    P. tuberosa Link (see *P. aquatica*)
Phaseolus vulgaris L., 226
Philomachus pugnax (L.) (ruff), 241
Phleum, 94
    P. alpinum L., 92, 94, 419, 421
    P. phleoides (L.) Simonk., 21, 130, 158–161, 183
    P. pratense Mill., 127, 135, 136, 139, 140, 142, 147–149, 152, 157–159, 161, 183
Phlomis, 88
    P. agraria Bunge, 50, 52
    P. tuberosa L., 8, 48, 160, 161
Phocarctos hookeri (Gray) (sea lions), 414, 419
Phodopus roborooouskii (Satunin) (desert hamster), 76
    P. sungorus (Pallas) (striped hairy-footed hamster), 76
Phoebetria palpebrata Forster (light-mantled sooty albatross), 427
Pholidota, 103
Phoma, 107
Phormium, 380
Phragmites, 90, 91, 94, 112, 306
    P. communis Trin., 131, 159, 183
    P. karka (Retz.) Trin. ex Steud., 91, 94, 95, 112, 438, 446, 451, 454, 465
    P. mauritianus Kunth, 299
Phrynocephalus frontalis (Anderson) (lizard), 77
Phycomycetes, 106
Phyllanthus emblica L., 95
Pimpinella saxifraga L., 136, 158
Pinus, 84, 226
    P. halepensis L., 187
    P. merkusii Jungh., 95
    P. sylvestris L., 9
Piptadeniastrum, 228
Pistacia atlantica Desf., 187
    P. khinjuk Stocks in Hook., 187
    P. lentiscus L., 187

Pistia, 299
Pithecellobium, 442
Pittosporaceae, 462
Plagiochila crozetensis Kaal., 419
Plagiogyraceae, 439
Plantago albicans L., 185
   P. aundensis van Royen, 452
   P. coronopus L., 185
   P. crassifolia Forsk., 184
   P. erosa Wall., 94
   P. lanceolata L., 136, 157
   P. maritima L., 160
   P. minuta Pall., 29
Plectrachne, 307, 317, 319, 350, 352
   P. melvillei C.E. Hubb., 328
   P. pungens (R. Br.) C.E.Hubb., 321, 340, 350
   P. schinzii Henr., 306, 319, 321, 350
Pleioblastus chino Fr. & Sav., 162
Pleurophyllum hookeri F. Buch., 308, 419
Pleurospermum uralense Hoffm., 160, 161
Poa, 14, 15, 17, 89, 90, 94, 179, 296, 308, 317, 342, 344, 345, 353, 363, 376, 414, 419, 424, 425, 429, 441, 472, 477
   P. alpigena (Blytt) Lindm., 94
   P. alpina L., 64, 69, 149
   P. angustifolia L., 4, 158
   P. annua L., 94, 429
   P. attenuata Trin., 17, 161
   P. botryoides (Trin. ex Griseb.) Kom., 17, 53
   P. bulbosa L., 28, 37, 47, 50, 88, 89, 160, 172
   P. callosa Stapf, 453
   P. cita Edgar (formerly *P. laevis*), 367, 374
   P. colensoi Hook. f.(blue tussock), 380, 381, 392, 396, 398
   P. cookii Hook. f., 310, 419, 420, 422–425, 428, 429
   P. costiniana Vick., 345
   P. fawcettiae Vick., 345
   P. flabellata (Lam.) Hook. f., 310, 415, 419–424, 426, 428, 429
   P. foliosa Hook. f., 414, 419, 421, 423, 424, 426, 429
   P. hamiltonii T. Kirk, 425
   P. hiemata Vick, 345
   P. labillardieri Steud., 342, 343, 351, 353
   P. laevis, 374
   P. litorosa Cheesem., 419, 425
   P. nemoralis L., 161
   P. palustris L., 157, 159, 160, 162
   P. pratensis L., 92, 127, 130, 134, 135, 138, 140, 141, 149, 152, 157–159, 161, 162, 345
   P. sphondylodes Trin., 67
   P. stepposa (Kryl.) Roshev., 21
   P. subfastigiata Trin. ex Ledeb., 161
   P. trivialis L., 127, 139, 149
Poaceae, 30, 127, 193, 203, 269, 299, 300, 304, 439
Poapratensis, 148
Podiceps cristatus (L.) (great-crested grebe), 77
Podocarpaceae, 361, 363
Podocarpus, 282
Poeae, 188, 308, 312
Poecilictis libyca (Libyan striped weasel), 211
Pogonarthria squarrosa (Licht.) Pilg., 274
Pogonatherum paniceum (Lamk.) Hack., 95

Polygala japonica Houtt., 162
Polygonatum officinale All., 162
Polygonum amphibium L., 92
   P. angustifolium Pall., 48
   P. bistorta L., 130, 136, 142, 152, 154, 158, 159
   P. equisetiforme Sibth. & Sm., 186
   P. patulum Bieb., 29, 50
   P. sachalinense F. Schmidt, 162
   P. sibiricum Laxm., 65, 73
   P. sphaerostachyum Sieb. & Zucc., 74
   P. viviparum L., 72, 74
   P. maritimus Willd., 183
Polytoca bracteata R. Br., 96
Polytrias amaura (Biisc) O. Ketze, 96
Polytrichum, 419
Populus tremula L., 9
Portulaca oleracea L., 324, 351
Potamochoerus porcus (L.) (bush pig), 226, 228, 231
Potentilla, 78, 90, 452, 464
   P. bifurca L., 21, 45, 47, 48, 51, 53
   P. chinensis Ser., 67
   P. comarum (see *Comarum palustre*), 158
   P. erecta (L.) Räuschel, 139, 158
   P. freyniana Bornm., 162
   P. griffithii Hook. f., 94
   P. humifusa Willd. ex Schlecht., 21
   P. sericea L., 48
Poterium tenuifolium (see *Sanguisorba tenuifolia*)
Poupartia caffra (Sond.) H. Perr., 298, 299
Prangos, 160
   P. pabularia Lindl., 130
Prasophyllum, 380
Primates, 103
Primula elatior (L.) Hill., 136
Pringlea, 429
Pringlea antiscorbutica R. Br., 308, 428
Prinsepia uniflora Batal, 68
Procapra picticaudata Hodgson (Tibetan gazelle), 104
Procavia capensis Pallas (rock dassie), 211
Procellariidae, 425
Prodorcas gutturosa (Pallas) (Mongolian gazelle), 75
Prosopis cineraria (L.) MacBride, 115
   P. juliflora DC., 115
Proteles cristatus (Sparrman) (aardwolf), 241
Protozoa, 150, 151, 397, 480
Prunus spinosa L., 14
Psammomys obesus Cretzschmar (sand rat), 190, 212
Psathyrostachys, 52
   P. juncea (Fisch.) Nevski (= *Elymus junceus*), 52
Pseudocytisus mairei (Humb.) O. Kuntze, 186
Pseudoelephantophus spicatus (Juss.) Rohr, 113
Pseudois nayaur Hodgson (bharal), 104
Pseudomonas, 40
Pseudopogonatherum, 96
   P. irritans (Benth.) A. Camus, 449
Pseudoraphis spinescens (R. Br.) Vickery, 345, 445
Psidium guayava L., 311, 458
Psilotrichum, 252
Ptaeroxylon obliquum (Thunb.) Radlk., 282
Pteridium, 465

Pteridium (continued)

    P. aquilinum (L.) Kuhn, 94, 95, 162, 192, 228
    P. esculentum (Forst. f.) Nakai (bracken), 309, 458
Pterocarpus lucens Lepr. ex Guill. and Perrott., 215
Pterocles, 104
Ptilagrostis, 71
Ptilotrichum canescens (DC.) C.A. Mey., 28, 35
Ptilotus, 329
    P. alopecuroideus F. Muell., 331
    P. exaltatus Nees, 328, 331
    P. helipteroides F. Muell., 328
Puccinellia distans (L.) Parl., 183
    P. kashmiriana Bor, 92
    P. tenuiflora (Griseb.) Scribn. and Merr., 65, 72, 73
Pueraria phaseoloides Benth., 346
Pulsatilla, 78
    P. angustifolia Turcz., 161
    P. flavescens (Zucc.) Yuz., 21
    P. multifida (G. Pritz.) Yuz., 21
    P. patens Mill, 158
Pycnospora lutescens (Poir.) Schindl., 93
Pygmaeopremna sessilifolia (Lam.) Mold., 449
Pygoscelis, 414
    P. papua Forster (gentoo penguin), 415, 427

Quelea quelea L. (millet eater), 209
Quercus, 156, 161
    Q. coccifera L., 192
    Q. mongolica (Fisch.) ex Turcz., 66
    Q. robur L., 8

Raetama raetam Webb (see *Lygos raetam*)
Rana temporaria asiatica (David), 77
Rangifer, 427
Ranunculaceae, 127
Ranunculus, 464
    R. acer L., 127, 136, 157–159
    R. auricomus L., 158
    R. caucasicus Bieb., 159
    R. chinensis Bunge (= *R. pensylvanicus*), 162
    R. japonicus auct., 162
    R. pensylvanicus L. f. (see *R. chinensis*)
    R. repens L., 127, 157, 159
Raoulia australis Hook. f. (scabweed), 379, 383, 384
Raphicerus campestris Thunberg (steenbok), 280
Rattus, 427
Reaumuria songorica (Pall.) Maxim., 17, 64, 67
    R. stocksii Boiss., 89
Redunca arundinum Boddaert (reedbuck), 280
    R. fulvorufula Afzelus (mountain reedbuck), 280
    R. redunca (Pallas) (Bohor reedbuck), 211, 231
Restionaceae, 269
Rhacomitrium lanuginosum (Hedw.) Brid, 419
Rhagodia, 329
    R. spinescens R. Br., 332
Rhamnaceae, 68, 367
Rhamnus cathartica L., 14
    R. lycioides L., 186
Rhanterium epapposum Oliv., 186

    R. suaveolens Desf., 181, 186
Rhaponticum uniflorum DC., 66
Rheum, 53
    R. nanum Siev. (= *R. leucorrhizum* Pall.), 29
    R. rhabarbarum L., 308
    R. undulatum L., 53
Rhinanthus major (= *Alectorolophus major*), 158
Rhinoceros sondaicus Desmarest (Java rhinoceros), 106
    R. unicornis L. (Indian rhinoceros), 104, 106
Rhizobium, 348
Rhizopus nigricans Ehrenb., 107
Rhododendron, 308, 452, 454
    R. campanulatum D. Don., 93
Rhus pentaphyllum Desf., 186
    R. taitensis Guill., 451
    R. tripartitum (Ucria) DC., 179, 186
Rhynchelytrum repens (Willd.) C.E. Hubb., 240
Rhynchospora, 459
    R. rubra (Lour.) Makino, 95, 450
    R. rugosa (Vahl) Gale, 452
Rhynchotragus kirkii (Gunther) (dik dik), 231, 248
Rindera tetraspis Pall., 15
Rodentia, 43, 103
Rosa, 71
    R. hugonis Hemsl., 68
    R. xanthina Lindl., 68
Rosaceae, 47, 68, 71, 96, 127, 193
Rosmarinus, 172
    R. officinalis L., 186
Rottboellia, 306
    R. exaltata L. f., 96
    R. hirsuta Vahl) (see *Lasiurus hirsutus*)
Rottboellinae, 307, 308
Rubiaceae, 363
Rumex, 73, 172
    R. acetosa L., 136, 158
    R. acetosella L. (= *R. angiocarpa*), 345
    R. alpinus Hook. (= *R. confertus*), 160
    R. angiocarpa Murb. (= *R. acetosella*), 345
    R. confertus Willd. (= *R. alpinus*), 160
    R. crispus L., 160
    R. haplorhizus Czern.ex Turcz., 159–161
    R. nepalensis Spreng., 94
Rupicapra rupicapra (L.), 394, 396, 403
Ruschia, 272
Rutaceae, 462

Sabal, 307, 446
Saccharum, 89–91, 94, 112, 229, 306, 464
    S. arundinaceum Retz., 91
    S. bengalense Retz., 90
    S. edule Hassk. (= *S. officinarum*), 465
    S. narenga Wall., 95
    S. officinarum L. (see *S. edule*)
    S. robustum Brandes & Jesswiet ex Grassl, 438, 445
    S. spontaneum L., 90, 91, 94–96, 100, 106, 438, 447, 449, 451, 454, 460, 464, 465
Sagina japonica (Sw. ex Steud.) Ohwi, 94
Saiga tatarica L. (saiga), 23
Salicornia, 299

Salicornia (continued)

   S. arabica L., 186
   S. fruticosa L., 186
   S. herbacea L., 186, 160
Salix, 131
Salsola, 7, 76
   S. arbuscula (Pall.) Reise, 89
   S. australis R. Br. (= *S. pestifer*), 15, 29, 49, 50
   S. baryosma (Schult.) Dandy, 186
   S. collina Pall., 29, 53, 69
   S. kali L. (see also *S. pestifer*), 321, 324, 351
   S. laricina Pall., 54
   S. passerina Bunge, 64, 67
   S. pestifer Nels. (see *S. australis*)
   S. tamariscina Pall., 15
   S. tetrandra Forsk., 186
   S. vermiculata L., 186
   S. vermiculata L. var. *villosa* Del., 174, 190
Salvadora persica L., 205
Salvia, 157
   S. dumetorum Andrs., 158
   S. nutans L., 50
   S. stepposa Shost., 21
Salvinia, 299
Sanguisorba (= *Poterium*), 73
   S. minor Scop., 156, 185, 193
   S. officinalis L., 44, 48, 53, 66, 72, 130, 142, 148, 149, 152, 157, 158, 160–162
   S. parviflora (Maxim.) Takeda, 72, 162
   S. tenuifolia Fisch. ex Link (= *Poterium tenuifolium*), 162
Sansevieria, 246
Sapium, 233
Sararanga, 307
Sarcobotrya strigosa (Benth.) H. Vig., 301
Sarcocephalus cordatus (see *Nauclea coadunata*)
Sarcopoterium spinosum L. Zoh., 179, 186
Sasa, 162
   S. chokaiensis Makino ex Koidz., 162
   S. nipponica Makino & Shibata, 130, 162
   S. veitchii (Carrière) Rehder, 162
Saussurea salicifolia (L.) DC., 53
   S. superba DC., 74
Scabiosa, 157
Schizachyrium brevifolium Stapf, 95
   S. fragile (R. Br.) A. Camus, 334, 340
   S. sanguineum (Retz.) Alst., 273, 274
Schizonepeta multifida (L.) Briq., 50
Schmidtia pappophoroides Steud., 274
Schoenefeldia gracilis Kunth, 199, 204, 206
   S. transiens (Pilg.) Chiov., 246
Schoenus, 445
Schoenus nigricanus L., 183
Schurumansia henningsii K. Sch., 451
Scirpus holoschoenus L., 183
   S. sylvaticus L., 127
Sciuridae (family of rodents), 43, 54
Scleria, 459
   S. laevis Willd., 97
Sclerocarya birrea (A. Rich.) Hochst., 199

Sclerodactylon macrostachyum (Benth.) A. Camus, 300
Sclerolaena, 329, 331, 337, 338, 349, 350
   S. convexula R.H. Anders, 337
   S. divaricata (R. Br.) Domin, 332
   S. obliquecuspis (R.H. Anders) Ulbrich, 332
   S. paradoxa R. Br., 332
   S. uniflora R. Br., 337
   S. ventricosa (J.M. Black) A.J. Scott, 332
Scolochloa festucacea (Willd.) Link, 161
Scolymus, 179
Scorzonera capito Maxim., 29
   S. divaricata Turcz., 29
   S. tuberosa Pall., 29
Scrophulariaceae, 8
Scutellaria scordifolia Fisch. ex Schrank, 48, 53
Scytonema, 215, 216
Secale montanum Guss., 88, 183
Sehima, 90, 91, 97, 100, 105, 112, 472, 475, 477
   S. galpinii Stent, 273
   S. nervosum (Rottl.) Stapf, 91, 93, 100, 105, 114, 116, 339, 340, 350
Seidlitzia rosmarinus Bunge, 89
Selinum (see *Cenolophium; Cnidium*)
Senecio, 160, 272
   S. cannabifolius Less. (= *S. palmatus*), 162
   S. chrysanthemoides DC., 94
   S. gregorii F. Muell., 351
   S. integrifolius (L.) Cloizv., 21
   S. jacobaea L., 161
   S. palmatus (*S. cannabifolius*)
   S. sagittatus (see *Cacalia hastata*)
Serinus pusillus (Pallas) (gold-footed serin), 77
Seriphidium (subgenus of Artemisia), 4, 14
Serrafalcus mollis Parl. (= *Bromus hordeaceus*), 349
Serratula centauroides L., 53
   S. coronata L., 160, 161
   S. tinctoria L., 158
Sesbania, 472, 475, 477
   S. bispinosa (Jacq.) Wight, 100
Seseli ledebourii G. Don, 20–22
Setaria, 96, 179, 215, 233, 306, 318
   S. anceps Stapf ex Massey, 111
   S. flabellata Stapf, 272
   S. geniculata (Lamk.) P. Beauv., 96
   S. glauca (L.) P. Beauv., 94
   S. incrassata (Hochst.) Hack., 233, 242
   S. pallidefusca (Schumach.) Stapf & C.E. Hubb., 94
   S. palmifolia (Koem.), 465
   S. sphacelata (Schumach.) Stapf & C.E. Hubb. ex M.B. Moss, 115, 225, 227, 231, 233, 273
   S. verticillata (L.) P. Beauv., 96, 240
   S. viridis (L.) Beauv., 28, 69
   S. woodii Hack., 273
Shorea, 94, 96
Sida, 321, 351
   S. acuta Burm. f., 113
   S. corrugata Lindl., 324
   S. rhombifolia L., 457
Sieglingia decumbens (L.) Bernh., 157
Silaum silaus (L.) Schinz & Thell., 158

SYSTEMATIC INDEX   521

Silybum, 172, 179, 351
Smirnovia (section of Stipa), 17
Solanaceae, 203
Solanum, 240, 246
Sorghum, 95, 215, 300, 306, 310, 317, 334, 338–341
    S. aethiopicum (Hackel) Rupr. ex Stepf., 204
    S. almum Parodi, 318, 348
    S. intrans F. Muell. ex Benth., 339
    S. nitidum (Vahl) Pers., 93, 96
    S. plumosum (R. Br.) P. Beauv., 96, 338, 350
    S. purpureo-sericeum (A. Rich.) Asch. & Schweinf., 247
    S. serratum (Thunb.) O. Kuntze, 95
    S. stipoideum (Ewart & J.R. White) C.E. Hubb., 339, 350
    S. vulgare Pers., 231
Spalacidae (family of rodents), 43
Spalax microphtalmus Guld. (mole rat), 43, 45, 47, 49, 50
Sparganiaceae, 439
Spartina arundinacea Carmich., 419, 420
Spergularia marginata Kittel, 184
    S. salina Presl., 184
Spermacoce radiata (= *Borreria radiata*), 204
Spermophilus, (see *Citellus*)
Sphagnum, 411
Sphenopus divarticatus (Gouan) Rchb., 183
Spilopsyllus cuniculi (Dale) (European rabbit flea), 430
Spinifex, 96
Spiraea, 4, 8, 21, 71
    S. camtschatica (see *Filipendula kamtschatica*)
    S. hypericifolia L., 15
    S. palmata Pall. (see *Filipendula palmata*)
Spodiopogon sibiricus Trin., 65
Sporoboleae, 188, 312
Sporobolus, 91, 96, 97, 223, 225, 230, 238, 242, 246, 274, 278, 295, 345, 441
    S. agrostoides Chiov., 225
    S. australasicus Domin, 333
    S. berterianus Hitchc. & Chase, 95
    S. capensis (Willd.) Kunth, 272, 274
    S. caroli Mez, 337, 344
    S. consimilis Fresen., 238
    S. diander (Retz.) P. Beauv., 95
    S. elongatus R. Br., 458
    S. festivus, 240, 301
    S. helvolus (Trin.) Th. Dur. & Schinz, 206, 238, 246, 247, 256
    S. indicus (L.) R. Br., 91, 112, 458
    S. ioclados (Trin.) Nees, 178, 183, 188, 203, 206, 238
    S. maderaspatanus Bor, 101
    S. marginatus Hochst., 89, 91, 97
    S. pyramidalis P. Beauv., 233, 272–275
    S. ruspolianus Chiov., 246, 253
    S. somalensis Chiov., 253
    S. spicatus (Vahl) Kunth, 238, 253
    S. variegatus Stapf, 246
    S. virginicus (L.) Kunth, 96, 300, 317, 345, 458
Stachys betonica Benth. (see *Betonica officinalis*)
    S. recta L, 50
Stachytarpheta urticaefolia (Salisb.) Sims, 457
Stellaria graminea L., 21
    S. radians L., 162

Stellera, 89
    S. chamaejasme L., 44, 46, 48, 51
Stenopetalum, 328
Stenophyllum, 24
Stenotaphrum dimidiatum (L.) Brongn., 295
Sterculia diversifolia (see *Brachychiton populneus*)
Stereospermum euphorioides DC., 298
    S. variabile H. Perr., 294, 298, 299, 301
Stigonema ocellatum (Dillw.) Thuret., 426
Stilbocarpa polaris A. Gray, 308, 419, 421, 426
Stipa, 4, 8, 14, 17–21, 23, 26–29, 32, 35–37, 39, 47, 48, 50, 52–54, 67, 70, 71, 78, 79, 88, 89, 171, 317, 329, 332, 333, 336, 342, 344, 350
    S. aliena Keng, 71, 74
    S. aristiglumis F. Muell. (plains grass), 337, 343, 344, 352
    S. baicalensis Roshev., 17, 44, 53, 63–65, 78
    S. barbata Desf., 88, 89, 174, 183
    S. bigeniculata Hughes, 343, 351, 353
    S. borysthenica Klok., 17
    S. breviflora Griseb., 64, 67
    S. bungeana Trin., 63, 64, 68, 69, 78
    S. capensis Thunb., 183
    S. capillata L., 5, 8, 11, 14, 17, 50, 70, 71, 157
    S. dasyphylla (Lindem.) Trautv., 14
    S. effusa Nakai ex Honda, 66
    S. falcata Hughes, 337, 342, 344, 353
    S. glareosa P. Smirn., 13, 17, 28, 39, 64, 67, 78
    S. gobica Roshev., 13, 17–19, 26, 28, 30, 31, 35, 37, 39, 63, 64, 67, 68, 78
    S. grandis P. Smirn., 17, 63, 64, 66, 67, 77
    S. joannis Roshev., 78
    S. kirghisorum P. Smirn., 17, 28, 70
    S. klemenzii Roshev., 17, 64, 67
    S. korshinskyi Roshev., 14, 22
    S. krylovii Roshev., 5, 17, 45, 48–50, 53, 64, 67, 78
    S. lagascae Roem. & Schult., 174, 181, 183
    S. laxiflora Keng, 70
    S. lessingiana Trin. & Rupr., 5, 7, 8, 10–12, 14, 15, 22, 26, 28, 30, 35–40, 50, 52, 54, 78
    S. orientalis Trin., 13, 14, 70
    S. parviflora Desf., 174, 183
    S. pennata L., 5, 14, 18
    S. pulcherrima C. Koch, 7, 14
    S. purpurea Griseb., 63, 64, 69, 71
    S. rubens P. Smirn. (see *S. zalesskii*), 14
    S. sareptana A. Beck., 14, 15, 17, 18, 28, 30, 37, 50, 52, 54
    S. setacea R. Br., 337, 344
    S. sibirica (L.) Lam., 50, 53, 66
    S. szowitsiana Trin., 89
    S. tenacissima L. (alfa grass), 171, 172, 175, 176, 183
    S. tirsa Stev., 14, 24
    S. trichotoma Trin. & Rupr., 287
    S. ucrainica P. Smirn., 14, 25
    S. variabilis Hughes, 329, 332, 337, 343, 344, 352
    S. zalesskii Wilensky (= *S. rubens*), 7, 9, 14, 17, 20–22
Stipagrostis, 171, 215, 267, 272, 274
    S. ciliata (Desf.) De Winter, 183, 272
    S. gonadostachys De Winter, 267, 272
    S. obtusa (Del.) Nees, 183, 272
    S. pennata (Trin.) De Winter (= *Aristida pennata*), 183

Stipagrostis (*continued*)

    S. plumosa (L.) Monro ex T. Anders (= *Aristida plumosa*), 178, 183

    S. pungens (Desf.) De Winter, 171, 183

Stipeae, 307, 312

Stoebe vulgaris Leuyns, 272

Strigops habroptilus Gray (flightless parrot or "kakapo"), 393

Struthio camelus L. (ostrich), 209, 226, 248, 253, 254

Stylosanthes, 235, 345

    S. hamata (L.) Taub, 111, 347

    S. humilis Humb. & Bonp. ex Kunth (Townsville stylo), 315, 318, 340, 341, 346, 347

    S. scabra Vog., 347, 348

Styphelia, 452, 453

Suaeda, 73, 246

    S. australis Moq., 345

    S. corniculata Bunge, 65, 72

    S. fruticosa L., 186

Succisa pratensis Moench, 158

Sus scrofa cristatus Wagn. (wild boar), 104

Sylvicapra grimmia (L.) (duiker), 211, 226, 279

Syncephalastrum racemosum (Cohn.) Schroet, 107

Syncerus caffer caffer Sparrman (buffalo, savanna subsp.), 279, 280

Syncerus caffer Sparrman (African buffalo), 211, 225, 226, 228, 234, 248

Syringa oblata Lindl., 68

Syrrhaptes paradoxus (Pallas) (sand grouse), 77

Syzygium richii (A. Gray) Merr. & Perry, 458

Tachyoryctes, 226

Tadorna tadorna (L.) (shel-duck), 77

Talpa europaea (European mole), 146

Tamaricaceae, 67

Tamarindus indica L., 298

Tamarix chinensis Lour., 73

    T. pallassii Desv., 114

    T. stricta Boiss., 114

Tamias sibiricus (Laxmann) (Siberian chipmunk), 76

Tanacetum, 54

    T. achilleifolium (Bieb.) Sch. Bip., 50, 54

    T. santolina C. Winkl., 28,

    T. sibiricum Fisch. (= *Filifolium sibiricum*), 65

Taraxacum ceratophorum (Ledeb.) DC., 161

    T. officinale Weber, 127, 157, 345

Tatera indica indica Hardwicke (Indian gerbil), 103

Taterillus arenarius F. Petter (gerbil), 212

    T. gracilis Thomas (gerbil), 212

    T. pygargus Cuvier (gerbil), 212

Taurotragus, 280

    T. derbyanus Gray (giant eland), 211

    T. oryx (Pallas) (eland), 226, 234, 280

Teinostachyum dullooa Gramble, 94

Tenebrionidae (darkling beetles), 23, 40

Tephrosia, 252

Terminalia, 222, 233, 234, 442

    T. arostrata Ewart & O. Davies, 321

    T. sericea Burch., 267, 269, 270, 274

    T. volucris Benth., 321

Tetraclinis articulata Benth., 187

Tetragonolobus siliquosus L., 184

Tetraogallus tibetanus (Gould) (Tibetan snowcock), 77

Tetrapogon bidentatus Pilg., 246

    T. cenchriformis (A. Rich.) W.D. Clayton, 246

    T. villosus Desf., 89, 246

Teucrium, 159, 172

Thalictrum, 160

    T. minus L., 125, 159–161

    T. simplex L., 66, 72, 160, 161

Thapsia, 172

    T. garganica L., 185

Themeda, 89–91, 94–97, 102, 112, 116, 167, 223, 226, 231, 234, 235, 237, 239, 241, 242, 306, 307, 310, 317, 338–342, 344, 350–352, 440, 448, 478

    T. anathera (Nees) Hack., 89, 91, 92, 112

    T. arguens (L.) Hack., 93

    T. arundinaceae (Roxb.) Ridley, 95

    T. australis (R. Br.) Stapf (= *T. triandra*), 306, 321, 338–340, 342, 343, 345, 351, 353, 438

    T. avenacea (F. Muell.) Maiden & Betche, 343, 352

    T. gigantea (Nees) Hack., 95, 96

    T. hookeri (Griseb.) A. Camus, 94

    T. japonica (Willd.) Max., 69

    T. quadrivalvis (L.) O. Kuntze, 96, 97, 225, 298

    T. tremula (Nees) Hack., 92, 93

    T. triandra Forsk. (= *T. australis*), 64, 91, 96, 203, 206, 222, 225, 227, 231, 233, 236–240, 242, 243, 272–274, 283, 286, 306, 308, 310, 321, 438, 443, 446–448, 450, 454, 455

Theobroma cacao L. (cacao), 461

Thermopsis lanceolata R. Br., 47, 50, 53

Theropithecus gelada (Ruppell) (gelada baboon), 226

Thevenotia, 179

Thielavia, 107

Thielavia terricola Gilman & Abbott, 107

Thoracostachyum sumatranum Kurz, 438

Thuarea, 96

    T. involuta (G. Forster) Roem. & Schult., 458

Thuidium abietinum (Hedw.) B.S.G., 20

Thylacospermum caespitosum (Camb.) Schinz, 64, 69

Thymelaceae, 68

Thymelaea hirsuta Endl., 186

Thymus marschallianus Willd., 21

    T. serphyllum L., 68

Thyridolepis mitchelliana (Nees) S.T. Blake, 326, 328, 334, 337, 350

Thysanolaena maxima (Roxb.) O.K., 448

Thysanoptera, 148

Timonius timon (Spreng.) Merr., 445, 449, 451

Tolypothrix, 398, 426

Torgos tracheliotus (Forster) (nubian vulture), 241

Trachynia (= *Brachypodium*), 179,

Trachypogon spicatus (L. f.) O. Kuntze, 272, 274, 296, 298

Traganum nudatum Del., 186

Tragelaphus, 248

    T. buxtoni (Lydekker) (mountain nyala), 226

    T. imberbis Blyth (lesser kudu), 253

    T. scriptus (Pallas) (bushbuck), 209, 211, 226, 228, 231

    T. strepsiceros (Pallus) (greater kudu), 209, 211, 279

Tragopogon brevirostris DC., 160

Tragus, 246
    T. berteronianus Schult., 240
Tribulus, 215, 246, 256, 321
    T. terrestris L., 240, 274
Tricholaena monachne (Trin.) Stapf & C.E. Hubb., 273
    T. teneriffae (L. f.) Link, 88
Trichosurus vulpecula (Kerr.) (Australian brush-tailed possum), 403
Trifolium, 180, 288, 296, 315, 345, 399
    T. ambiguum Bieb., 159
    T. angustifolium L., 342
    T. arvense L., 342
    T. fragiferum L., 184, 193
    T. glomeratum L., 349, 351
    T. lupinaster L., 72, 160–162
    T. medium L., 135
    T. montanum L., 125, 139, 158, 159
    T. pratense L., 157–161, 184, 288, 348
    T. repens L. (white clover), 136, 157–159, 161, 162, 184, 227, 288, 318, 345, 346, 348, 399
    T. resupinatum L., 184
    T. semipilosum Fresen, 225
    T. squarrosum L., 184
    T. stellatum L., 184
    T. subterranean L., 184, 193, 310, 318, 344, 348–351
    T. tomentosum L., 184
Triglochin palustris L., 160
Trigonella, 66, 89, 180
    T. arabica Delile, 184
    T. cachemiriana Cambers., 89
    T. ruthenica L. (= *Medicago ruthenica*), 64, 66, 67
Trigonoćeps occipitalis (Burchell) (white-headed vulture), 241
Trinervitermes bettonianus (Sjöstedt), 241
    T. rhodesiensis (Sjostedf.), 281
Tringa totanus (L.) (redshank), 77
Triodia, 307, 317, 319, 350, 352, 440
    T. basedowii E. Pritzel (hard spinifex), 306, 319, 321, 328, 350
    T. intermedia Cheel, 320
    T. irritans R. Br., 337, 338
    T. lanigera Domin, 321
    T. longiceps J.M. Black, 319, 321
    T. mitchellii Benth., 333, 334
    T. pungens R. Br., 310, 319–321, 333, 350
    T. wiseana C.A. Gardner, 321
Tripogon, 97
Triraphis schinzii Hack., 273
Triseteria flavescens Baumg. (= *Trisetum flavescens*), 179, 183
Trisetum, 441, 477
    T. flavescens (L.) Beauv. (= *Triseteria flavescens*), 130, 136, 138, 157
    T. sibiricum Rupr., 162
Tristachya leucothrix Nees, 274, 278, 284
Tristania (= *Lophostemon*), 445
Triticeae, 312
Triticum, 226
Trollius asiaticus L., 160, 161
    T. caucasicus Stev., 159
    T. chinensis Bunge, 162
    T. europaeus L., 158

Trypanosoma, 235
Tugarinovia mongolica Iljin, 29
Tulipa biebersteiniana Schult. & Schult. f., 15
    T. biflora Pall., 15
    T. patens Agardh ex Schult. f., 29, 38, 52
    T. schrenkii Regel, 15
    T. uniflora (L.) Bess. ex Baker, 17
Typhaceae, 439

Ulex parvifolius Pourret, 192
Ulmus, 66
Uncinia dikei Nelmes, 419
Ungulata (hoofed mammals), 23, 40
Uraria, 97
Urginea maritima (L.) Baker, 185
Urochloa, 215
    U. mosambicensis (Hack.) Dandy, 237, 238, 274
    U. panicoides P. Beauv., 233, 272
Urochondra setulosa (Trin.) C.E. Hubb., 246, 253
Uromys (rats), 465
Urtica cannabina L., 53

Vaccinium, 452
Valeriana officinalis L., 53, 162
    V. tuberosa L., 29
Valerianella, 14
Vanellus vanellus (L.) (lapwing), 77
Ventilago viminalis Hook., 321, 336
Veratrum nigrum L., 72
Verbascum, 73
Veronica incana L., 4
    V. longifolia L., 159
    V. sibirica L. (= *V. virginica*), 162
    V. spicata L., 21
    V. spuria L., 21
    V. supina, 21
    V. virginica L. (see *V. sibirica*)
Vetiveria, 95
    V. nigritana (Benth.) Stepf., 206
    V. zizanioides (L.) Nash, 91, 106, 112
Vicia, 180
    V. amoena Fisch., 72, 160–162
    V. cracca L., 139, 159–162
    V. megalotropis Ledeb., 160, 161
    V. pseudorobus Fisch. & Mey., 162
    V. sepium L., 161
    V. unijuga A. Br., 162
Vigna sinensis (L.) Hassk. (cow peas), 249
Viguierella madagascariensis A. Camus, 298
Vincetoxicum sibiricum Decne., 29
Viola oreades Bieb. (= *V. altaica* Ker-Gawl.), 159
Vitex negundo L., 64, 68, 69
Viverricula indica Desmarest (small Indian civet), 104
Vormela peregusna (Güldenstradt) (marbled polecat), 76
Vulpes, 77
    V. bengalensis Shaw (Indian fox), 104
    V. corsac (L.) (corsax fox), 76
    V. pallida Cretzschmar (sand fox, pale fox), 211
    V. ruppelii Schinz (Ruppell's fox), 211

Vulpia, 179
   V. bromoides (L.) S.F. Gray (= *Festuca bromoides*), 349, 351
   V. dertonensis Volk., 384
   V. myuros (L.) C.C. Gmelin, 344, 349
Vulpiella stipoides (L.) Maire, 183

Waltheria indica L., 298
Wendlandlia paniculata (Roxb.) DC., 451
Wickstroemia chamaedaphne Meisn., 68
Wiseana (porina caterpillars), 396

Xanthomyrtus, 452
Xanthostemon, 453
Xerochloa barbata R. Br., 345
   X. imberbis R. Br., 345
Xerus erythropus Desmarest (striped ground squirrel), 211
Xyris papuana Royen, 452

Zaleya pentandra (L.) Jeffrey, 240
Zea mays L., 226, 282, 300
Zerna riparia (Rehm.) Nevski (= *Bromopsis riparia*), 4, 50, 52
Zilla spinosa (L.) Prantl., 186
Ziziphus (or *Zizyphus*), 204
   Z. lotus (L.) Desf., 179, 186
   Z. mauritiana Lam., 298, 301
   Z. rugosa Lamk., 95
   Z. spinosa (Bge.) Schneid., 64, 68
Zorilla striatus Perry (zorilla, striped polecat), 211
Zornia, 101, 215
   Z. glochidiata Reichenb. ex DC., 204
Zoysia, 96, 131
   Z. japonica Steud. (= *Z. pungens* Willd.), 130, 140, 162
Zygochloa paradoxa (R. Br.) S.T. Blake, 321
Zygophyllaceae, 30, 67, 203
Zygophyllum, 272
   Z. album L., 186
   Z. dumosum Boiss., 186
   Z. fabago L., 89
   Z. rosovii Bunge, 29
   Z. xanthoxylon (Bunge) Maxim., 28

//go directly to index

GENERAL INDEX

aardvark (*Orycteropus afer*), 211
abandoned cropland, 83, 116, 228, 253, 274, 282
above-ground plant parts (*see also* biomass; decomposition; nutrients; primary production)
–, anatomy of, 133
acid decarboxylation, 305, 439
–, C_3 and C_4 photosynthetic pathways, 188, 305, 306, 351, 435, 439
–, C_3 grasses, 305–308, 451–454
–, C_4/C_3 metabolism in grasses, effect of temperature, 440
–, –, effect of elevation, 441
–, C_4/C_3 syndrome, distribution among grass genera, 441
–, C_4 grasses, 306, 307, 438–440, 446, 450, 451, 455–457, 459, 462–465
–, C_4 megatherm types, 308
acid soils and soil acidity, 87, 96, 157, 158, 175, 202, 227, 245, 291, 337, 344, 348, 349, 351, 460
acidophilous soils, 158
actinomycetes, 281
adaptation of animals (*see also particular groups*), 23, 75, 76
–, to adverse environments, 25, 75
–, to drought, 210
adaptation of plants, 18–20, 129, 136, 348, 440, 421
–, mechanisms of, 145
–, morphological, 129, 136
–, to adverse environments, 18, 19, 32, 64, 136
–, to drought, 19, 26, 207, 451
–, to fire, 446, 452
–, to fluctuating environments, 18, 19, 32, 64, 98, 136
–, to grazing, 18, 98
–, to high altitudes, 179
–, to low supply of soil nutrients, 25
–, to low temperature, 452
–, to moisture supply, 18, 126, 142
–, to temperature, 19, 68
–, to trampling, 18
Adelaide region (South Australia, Australia), 342
Aegean coastal region (Turkey), 85
Afghanistan (southern Asia), 84–89, 108, 109, 114
–, grasslands of, 88, 89, 114
Africa (*see also* East Africa; North Africa; Sahel; southern Africa; *and particular territories, regions and localities*), 167–301, 308, 309, 366, 441, 452
African grasslands, (*see also particular territories and regions*), 167–301
Agadès region (Niger), 200
age composition of plant populations, 19, 36, 37, 46, 131–133, 137
agricultural ecosystems, 83, 151, 153
agriculture

–, sedentary, 78
–, slash and burn (*see* shifting cultivation; slash and burn agriculture)
agrocoenosis, 151, 153
Ahmedabad region (Gujarat, India), 97
Ahuriri Valley (South Island, New Zealand), 366
Airai region (Babelthuap island, Belau, Micronesia), 460
Aktyubinsk region (Kazakhstan), 47, 49–51, 54
Albania (southern Europe), 171
albatross, 427, 429
–, light-mantled sooty (*Phoebetria palpebrata*), 427
–, royal (*Diomedea epomophora*), 425
alcohol hydrolysates in soil, 42
Aldan river (Yakutsk region, Siberia, Russia), 161
alder (*Alnus*), 162
alfa grass (*Stipa tenacissima*), 171
algae, 24, 30, 40, 41, 216, 422
–, blue green, as nitrogen-fixers, 398
– in soil, 30, 148
–, populations of, 24
Algeria (North Africa) (*see also particular regions and localities*), 169, 171, 174, 175, 189, 190
Alice Springs region (Northern Territory, Australia), 315, 326, 328
alkaline soils, 63, 91, 202, 237, 238, 246, 349
alluvial plains, 256
alluvial soils, 63, 86, 87, 86–89, 91, 95, 104, 137, 141, 145, 152–154, 156, 157, 159, 161, 178, 179, 203, 227, 232, 238, 246, 291, 294, 337, 371
alpine fault, 370
alpine plant species, 306, 375
Alps [mountains] (Europe), 1
Altai [mountains, region] (Central Asia), 11, 12, 15, 16, 43, 53, 70, 76, 161
altitudinal differentiation of vegetation, 12, 15, 66, 68, 70, 71, 74, 78
aluminium, 154, 155
Ambikapur region (Madhya Pradesh, India), 86, 100
Amga river (Yakutsk region, Siberia, Russia), 161, 162
amino acids in soil, 42
amphibians, 75, 77
Amsterdam Island (sub-Antarctic), 427
Anatolian plateau (Turkey), 87, 88, 110
Andes mountains (South America) (*see also Part A*), 452
Andringitra mountains (Madagascar), 298
andropogonoid species, 341, 442, 451
Aneityum [island] (Vanuatu), 456
anemochoric grasses, 19
anemophilous grasses, 19
angiosperms, 304, 456

Angola (southern Africa), 265, 270
animal diets (*see also particular groups and species*), 47
animal/vegetation interactions, 42–55
animals, effects on vegetation (*see also* browsing; grazing; trampling), 23
Anse Longue region (New Caledonia), 455
Antarctic Peninsula, 416
Antarctica, 366
antelope, addax(*Addax nasomaculatus*), 209, 211
–, Hunter's (*Damaliscus hunteri*), 248
–, roan (*Hippotragus equinus*), 211, 231, 280
–, sable (*Hippotragus niger*), 231, 267, 280
–, Tibetan (*Pantholops hodgsoni*), 76
ants, 208
– (*Paltothyreus* spp.), 228
–, consumption by, 209
Arabia (Asia Minor), 2, 83, 87
Arabian Gulf states, 254
Arabian Peninsula, 83
Arabian plateau, 87
arable (seeded, exotic) grassland, 79, 102, 109, 111, 114, 115, 142, 143, 147, 192, 193, 216, 234, 235, 249, 282, 301, 310, 311, 325, 339, 345, 346, 399
arable land, 150, 151, 155, 156, 159, 168
Aravalli ranges (Rajasthan, India), 91, 97
Arctic region, 418
argali (*Ovis ammon*), 104
aridity, 69, 78, 85, 99, 250, 255, 281, 270, 316, 340, 355
–, effect on vegetation, 5, 7, 22
–, – – –, species diversity, 4
aridity index, 61, 65–67
arthropods, 209
–, biomass of, review of Parts A and B, 478, 479
Arthur's Pass (Southern Alps, South Island, New Zealand), 374
ash content of plants, 39–40
Ashburton catchment (South Island, New Zealand), 391
Asia (*see also* Central Asia; Eurasia; South-east Asia; southern Asia; *and particular territories, regions and localities*), 1, 2, 61–80, 83–117
Askaniya Nova reservation, Kherson region, Ukraine, 25, 42
aspartate-forming plants, 440
ass(es) (*Equus* spp.) (*see also* donkey), 104, 108
–, Asiatic wild, (*Equus hemionus*), 23
–, Somali wild (*Equus asinus somalicus*), 248
Assam hills (India), 85
Assam [state] (India), 85, 91, 106, 112
assectator, definition of, 39
assimilation surface, 133, 139
Atherton tablelands (Australia), 442
Atjeh region (Indonesia), 95
Atlantic coastal region (Africa), 177, 199
Atlantic Ocean, 125
Atlantic/sub-Atlantic region (Europe), 158
Auckland Islands (sub-Antarctic), 419, 427
Augara-Sayan (former Soviet Union), 2
Australasia, 439
Australia (*see also particular states, regions and localities*), 303–308, 310–312, 315–355, 366, 435, 440, 441, 445, 456, 463
–, exotic grasslands (pastures) of, 345–350

–, – – –, map of, 318
–, semiarid crescent of, 307
Australian grassland regions
–, Acacia shrub–short-grass (*Eragrostis–Aristida–Eriachne–Enneapogon*), 317, 326, 327–329, 336, 337, 347, 349, 350
–, aquatic and saline (*Eleocharis–Fimbristylis–Sporobolus*), 317, 345
–, mesophytic subhumid tropical grasslands, 352
–, saltbush xerophytic mid-grass (*Atriplex–Maireana–Chloris–Stipa*), 317, 329–333, 337, 349
–, saltbush–short-grass communities, 354
–, subalpine sod tussock-grass (*Poa–Themeda–Danthonia*), 317, 344, 345, 353
–, subtropical tall-grass, 340
–, temperate short-grass (*Danthonia–Stipa–Enneapogon*), 317, 342–344, 346, 348, 349, 254
–, temperate tall-grass (*Themeda–Poa–Dichelachne–Danthonia*, 317, 341, 342, 348
–, tropical tall-grass, 317, 333, 338-341, 346, 347
–, true grasslands, 317, 325, 344, 352, 353,
–, xerophytic hummock-grass (*Triodia–Plectrachne*), 317, 319-321, 333, 337, 340, 347, 348, 350, 352
–, xerophytic mid-grass, 317, 333–338, 340, 344, 347, 349, 350, 354
–, xerophytic tussock-grass (*Astrebla–Iseilema*), 317, 321–325, 346, 350, 352, 354
Australian grasslands, 315–355
–, areal extent of, 319, 323, 326
–, climate of, 316, 319
–, extent of, 310, 311
–, fauna in, 353
–, livestock in, 324, 330, 333, 335, 340, 342–346, 349, 351, 353, 354
–, map of, 317
Australian vegetation
–, eucalypt shrublands, 316, 317, 319, 329, 331, 337, 345, 349
–, eucalypt woodlands, 319, 334–342, 344
–, forests, 315, 316, 341, 342, 344–346, 348, 353
–, frontage woodlands, 333, 337
–, heathlands, 317, 342, 345, 346, 348
–, low woodlands, 316, 321, 329, 333
–, mallee communities, 331, 337, 338, 345
–, Mediterranean annual pastures, 318, 345, 349
–, monsoon grasslands, 338, 340
–, mulga communities, 326, 327, 331, 336
–, savannas, 435
–, seasonally-flooded lands, 326, 345
–, sedgelands, 317, 345
–, semiarid crescent, 316, 333, 338, 355
–, shrub woodlands, 316, 329, 330, 331, 333–337, 344, 348, 349, 354, 355
–, shrublands, 316, 326, 327, 329, 352, 354, 355
–, subtropical woodlands, 316, 340, 341
–, successional (seral) communities of, 352
–, temperate perennial pastures, 318, 348, 349
–, temperate shrublands, 353
–, temperate woodlands, 316, 336, 342–345, 349, 353
–, tropical annual pastures, 318, 346, 347
–, tropical mesophytic communities, 322, 323
–, tropical perennial pastures, 318, 346

Australian vegetation (*continued*)

–, tropical woodlands, 338, 339–341
–, vegetation, disclimax, 337, 343
–, woodlands, 333, 335, 336, 340, 352, 353
–, xerophytic perennial pastures, 318, 347
autotrophs in soil, 150
Azerbaydzhan (*see also* former Soviet Union), 158, 159
Azov, Sea of (Azovskoye More) (former Soviet Union), 12, 52

baboon, anubis (*Papio anubis*), 211
–, gelada (*Theropithecus gelada*), 226
bacteria, 24, 25, 30, 40–42, 107, 151, 281, 398
–, anaerobic, 72
–, biomass of, 24
–, – –, review of Parts A and B, 480
–, denitrifying, 40–41
– in plant litter, population of, 426
– in soil, biomass of, 151
– – –, populations of, 426
– – –, secondary production by, 151
– in tussock root masses, populations of, 426
–, nitrifying, 40–42
–, nitrogen-fixing, 40
–, non-sporulating, 25
–, oily-acidic (butyric acid), 40–41
–, populations of, 24, 107
baibakoviny (*Marmota* hills), 51, 52
Bali (island) (Indonesia), 95
Baluchistan area (Pakistan), 89
Baluran region (Indonesia), 95
bamboos, 94, 96, 439, 466
Ban Me Thuot region (Vietnam), 96
bananas (*Musa* spp.), 226, 227, 229
bandicoot (mumut) (*Echymipera kalubu*), 465
– (*Isoodon macrourus*), 465
Bangladesh (southern Asia) (*see also particular regions and localities*), 84–88, 94, 108, 109
–, grasslands of, 94, 109
Barkly Tableland (Northern Territory, Australia), 321, 326
barley (*Hordeum* sp.), 227
beans (*Phaseolus vulgaris*), 226
beech, silver (*Nothofagus menziesii*), 365
beetles, consumption by, 151
Behrampur region (Orissa, India), 86, 100, 106
beira, (*Dorcatragus megalotis*), 248
Belaya river (former Soviet Union), 160
Belgium (Europe), 157
Belorussiya (*see also* former Soviet Union), 158
Bena region (Timor, Indonesia), 112
Bengal Basin (India), 87
bharal (*Pseudois nayaur*), 104
Bhartum region (Sudan), 200
Bhutan (southern Asia) (*see also particular localities*), 84, 86, 94, 108
–, grasslands of, 94
Bihar [state] (India), 90
bioclimates, classification of, 305
biomass (*see also* biomass *of particular groups of organisms*)

–, of animals (*see also* carrying capacity; stocking rate; *and particular groups and species*), 22, 256
–, – –, review of Parts A and B, 476–480
–, of macroarthropods (*see also particular groups*)
–, of micro-organisms (*see also particular groups*), 154
–, of microarthropods (*see also particular groups*)
–, of plants (*see also* primary production), 4, 18, 21–24, 31, 32, 34, 36–41, 44–47, 51, 54, 421, 423, 424, 430
–, – –, above-ground parts (*see also* forage yield; hay yield herbage yield), 22, 38, 99–101, 133, 134, 136, 137, 277, 338, 339, 389, 422
–, – –, above-ground parts, seasonal dynamics of, 37–39, 139
–, – –, biomass/rainfall relationships, 242
–, – –, consumable by livestock, 187, 213
–, – –, litter, 276
–, – –, proportional representation by species, 38–39
–, – –, ratio of green to dead biomass in canopy, 132, 134, 145, 387
–, – –, review of Parts A and B, 471–474
–, – –, under-ground parts, 39, 99–101, 137, 139, 154, 276, 389, 422
–, – –, – –, diversity of, 151
–, – –, – –, dynamics of, 139
–, – –, – –, rate of decomposition of, 151
–, – –, – –, reserve substances in, 153
–, – –, – –, vertical distribution of, 18, 21, 22, 37, 99, 137
–, – –, vertical distribution in canopy, 134–137
–, – –, vertical distribution of, 391
birch (*Betula* sp.), 160, 162
birds (avifauna) (*see also regions and particular groups and species*) 75-77, 208, 209
–, activity in distributing diaspores, 146
–, adaptations of, 75
–, biomass of, review of Parts A and B, 478, 479
–, browsing, 393
–, consumption by, 209
–, effect on energy balance of, 146
–, extinct species of, 393
–, fertilization by excreta of, 427
–, flightless, 427, 428
–, food habits of, 146
–, food preference of, 395
–, game, 465
–, granivorous, 146
–, ground-nesting, 419
–, –, effect on plants by, 427
–, herbivorous, 146
–, introduced, 427
–, migration of, 146
– of prey, 75, 77
–, omnivorous, 77
–, predators, 146
bittern, Eurasian (*Botaurus stellaris*), 77
Black Sea region (former Soviet Union), 12, 19, 20, 22, 85, 158
blesbok (*Damaliscus dorcas phillipsi*), 280
Bloemfontein (South Africa), 271
blue grass (*Dichanthium sericeum*), 341
blue grasses (*Bothriochloa* spp., 341
blue tussock (*Poa colensoi*), 380, 381, 392

blue-green algae, addition to soil nutrients by, 309
–, rate of nitrogen fixation by, 426
bluebushes (*Maireana* spp.), 307, 330, 331, 333
Bluegrass Downs (Queensland, Australia), 352
Bluegrass–Browntop Downs (Queensland, Australia), 352
boar, wild (*Sus scrofa cristatus*), 104
Bohemia (Czechoslovakia), 147
Bohemian–Moravian uplands (Czechoslovakia), 128, 134, 136, 140, 147, 154
Bollon district (Queensland, Australia), 334
Bombay Deccan (India), 2
bontebok (*Damaliscus dorcas dorcas*), 280
Botswana, 265, 268, 269, 270, 280, 270
Bougainville [island (Papua New Guinea), 435, 448, 454
Bouvetøya (sub-Antarctic islands), 427
bracken (*Pteridium esculentum*), 458
Brahmaputra river (Pakistan/India), 87, 91
Brazil (South America) (*see also* Part A), 199
brigalow (*Acacia harpophylla*), 317, 341, 345, 348
Bródno (Poland), 148–150, 152
brown top (*Agrostis capillaris*), 399
browse, 216
–, effect of lopping woody plants for, 192
–, nutritive value of, 213, 216, 286
–, root/shoot ratio of, 213
–, yield of, 213
browsing, 172, 192, 248, 279, 288
– by birds, 393
brussels sprouts (*Brassica oleracea* var. *gemmifera*), 226
bryophytes, 78, 419, 421, 422, 453
Bua [province] (Vanua Levu, Fiji), 458
buck, black, *Antilope cervicapra*, 104
buffalo, 190
–, African (savanna subspecies) (*Syncerus caffer caffer*), 279
–, African (*Syncerus caffer*), 211, 345, 353
–, wild water (*Bubalus bubalis*), 104–106, 108, 110
buffel grass (*Cenchrus ciliaris*), 321, 348
Bukidnon [province] (Philippines), 102
Bulawayo region (Zimbabwe), 271
Bulgaria (southern Europe), 1
Bulnai mountains (Central Asia), 15
Bulolo region (Papua New Guinea), 466
Bulolo–Manki region (Papua New Guinea), 462
Bulolo–Watut divide (Papua New Guinea), 448, 450
Bulolo–Wau region (Papua New Guinea), 442
bunch (tussock) growth form, 363
bunch grasses, 50, 65, 225
–, effect of grazing on, 23
–, microtherm, 4
–, sclerophyllous, 4
–, xerophilous, 4
bunting, grey-headed black-faced (*Emberiza spodocephala*), 77
–, rock (*Emberiza cia*), 77
–, yellow-breasted (*Emberiza aureola*), 77
Burkina-Faso (formerly Upper Volta, West Africa), 198, 204, 205
Burma (southern Asia), 84, 86, 108
burning of vegetation (*see also* fire; *and particular regions*), 86, 89, 116, 162, 167, 221, 222, 227, 228, 230, 233, 235, 239, 241, 242, 247, 248, 250, 273, 278, 282–284, 292, 300, 301, 351, 374, 378, 379, 382
–, effect on floristic composition of, 320
–, effect on flowering of grasses of, 389
–, effect on herbage palatability of, 389
–, effect on herbage yield of, 284
–, effect on nutrient transfers of, 309
–, effect on root reserves of, 284
–, effect on seed supply of, 319
–, effect on vegetation of, 84, 189, 192, 380, 325
–, fuel load required for, 116, 206, 273, 284, 340, 355
–, objectives in, 284
–, techniques used in, 216, 284, 379
–, to control weedy grasses, 339
–, to control woody plants, 300, 301, 335, 348, 355
–, to remove accumulated plant litter, 300
–, to remove old growth, 341
–, to stimulate grasses, 341, 345
burr medics, 337, 349
bushbuck (*Tragelaphus scriptus*), 209, 211, 226
bustard, great (*Otis tarda*), 77
butany, (*Marmota* hills), 48, 51, 53
buzzard, augur (*Buteo rufofuscus augur*), 226
–, upland (*Buteo hemilasius*), 77

C_4/C_3 metabolism in grasses (*see* acid decarboxylation)
cabbage (*Brassica oleracea* var. *capitata*), 226
cabbage, Macquarie Island (*Stilbocarpa polaris*), 308
cacao (*Theobroma cacao*) plantations, 461
cactus, spineless (*Opuntia ficus-indica* var. *inermis*), 193
caespitose grasses, 64, 67, 74, 78
calcarious soils and calcic soils (calcium and calcium carbonate), 86, 94, 97, 154, 237–239, 245, 246, 248, 292, 328, 330
calcium, export of, 155
–, mobility in plants, 155
–, pool of, 384
–, uptake by plants, 385
California [state] (U.S.A.) (*see also* Part A), 351
CAM-based vegetation, 306
Cambodia, grasslands of, *veal* types, 87, 96
camel(s), 84, 108, 171, 188, 190, 213, 248, 249, 252, 254, 256
– (*Camelus dromedarius*), 353
– Bactrian (*Camelus bactrianus*), 75
–, biomass of, 254
–, grazing area of, 191
Cameroons (Central Africa), 198
Campbell Island (sub-Antarctic), 411, 414, 419–425, 427, 429, 430
Canaries Current (Atlantic Ocean), 199
Canberra region (A.C.T., Australia), 344
canopies, open, 112
canopy (*see also* above-ground plant parts; biomass; herbage, etc.)
– architecture (*see also* layering; structure; biomass), 98, 132, 134–36
–, nutrient flow through, 156
Canterbury Plains (South Island, New Zealand), 361
Cape Province (South Africa), 269, 270, 280, 280
Cape Verde [island] (Atlantic Ocean), 198

GENERAL INDEX 529

Cape York Peninsula (Queensland, Australia), 306, 334, 340
caracal (*Felis caracal*), 239
carbohydrate reserves in plants
–, above-ground, 143
–, under-ground, 143
carbon
–, content in soil, 97
– cycle, 137, 153, 154
carbon/nitrogen ratio, effect on decomposition rate, 154
carbon-dioxide evolution in soil, 42, 103, 107, 108, 152, 153
carnivores, 40, 146, 226, 241
– of livestock, 256
carnivorous mammals, 75, 76
carnivorous vertebrates, populations of, 76
carnivory, 23
Caroline Islands (Micronesia) (*see also particular regions and localities*), 435
carotene, 202, 213, 216
Carpathian mountains (Czechoslovakia), 140
carrying (grazing) capacity for livestock (*see also* stocking rate), 64, 78, 83, 108, 109, 112–114, 116, 189, 208, 210, 212, 229, 231, 235, 242, 250, 256, 284, 321, 323, 326, 328, 331, 333, 334, 337–342, 346, 348, 349, 399
cashew (*Anacardium occidentale*), 231
Caspian Sea region (Central Asia), 12, 25, 45, 53, 85, 87, 158, 159
cassava (*Manihot* spp.), 231, 300, 455
cat(s), 311
– (*Felis catus*), 427
– (*Felis* spp.), 104, 211
–, African wild (*Felis libyca*), 241
–, effect on bird population by, 430
caterpillars, 396
cattle, 108–110, 180, 188–192, 210, 212–214, 226–229, 231, 233–235, 241–243, 248–250, 254, 256, 273, 281, 282, 300, 311, 322, 323, 325, 326, 328, 333, 334, 336, 339–341, 345, 346, 348, 352, 353, 354, 427, 428, 446, 457, 460
–, biomass of, 235, 254
–, gains in weight of, 322, 329, 340, 346, 348
–, populations of, 242
–, species of, 214
–, zebu (*Bos indicus*), 227
Caucasus region (former Soviet Union), 139, 159
cellulolytic activity, 151
Central America, 168 (*see also Part A*)
–, savannas of, 461
Central Asia (*see also particular territories, regions and localities*), 3, 4, 10, 17, 23, 43, 74–76, 125, 160, 161
Central Chernozem Reservation (Kursk region, Russia), 45–47, 49, 52
Chad (Central Africa), 198, 210
Chagai–Kharan region (Pakistan), 90
chamaephytes, 127, 128, 176
chamois (*Rupicapra rupicapra*), 394
Changai mountains (Mongolia), 15
Changbai Shan (mountains) (China), 72
Chari river (Cameroons/Chad), 202
Charleville region (Queensland, Australia), 328
cheetah (*Acinonyx jubatus*), 190, 211, 226
Chellala region (Algeria), 187

chenopodiaceous shrubs, 327, 329–332, 336, 350
Chester Burn (Fiordland, South Island, New Zealand), 376
China (eastern Asia) (*see also particular provinces, regions and localities*), 2, 4, 10, 11, 61–80, 125
–, map of geographical features of, 62
Chinese grasslands, 61–80
–, alpine meadow, 61, 72, 73, 75, 78
–, alpine steppe, 61, 63, 64, 69–71, 76, 78
–, area occupied by, 61
–, azonal types of, 72–74
–, classification of, 64, 71
–, climate of, 61–64
–, desert steppe, 61–64, 67, 68, 70, 75–77, 79, 80
–, desertified steppe, 76, 77
–, fauna in, 74–78
–, forest-steppe, 66
–, halophytic (saline) meadow, 72–74
–, livestock in, 61, 64, 71, 76, 78–80
–, map of, 61
–, marshy meadow, 72
–, meadow steppe, 61–67, 76–80
–, meadows, types of, 72–74
–, mountain steppes, 71
–, salinized steppe, 76
–, steppe meadow, 66
–, typical (dry) steppe, 61–67, 70, 75–77, 79, 80
–, typical meadow, 72, 73
–, zonal grassland, 64–71
Chinese vegetation, altitudinal zonation of, 66, 68, 70, 71, 74, 78
–, desert, 67, 70
–, desert zones, 73
–, forest, 61, 63, 66, 68–70, 72, 74, 75, 80
–, forest zone, 72
–, shrub forest, 68
–, shrub steppe, 61, 63, 64, 67–69, 77
–, successional (seral) communities of, 69
–, woodland, 66
Chingiz-Tau mountains (Kazakhstan), 12, 13
chipmunk, Siberian (*Tamias sibiricus*), 76
chiru (*Pantholops hodgsoni*), 104
chital (*Axis axis*), 104, 105
Chittagong hills (Bangladesh), 94
chlorophyll content, 98
Cholistan desert (Pakistan), 89, 104, 113
Chuyskaya steppe (Altai region, former Soviet Union), 43, 53
Ciscaucasia (former Soviet Union), 2
civet, small Indian (*Viverricula indica*), 104
classification of grasslands, 1, 2, 5, 64, 71, 88, 128–130, 265, 305, 306, 417, 435–440
climate (*see also particular regions*)
–, change of, 374, 388, 397
– diagrams, 127, 270, 271
–, mediterranean type, 171
–, seasonality of, 202
–, stability of, 309
clover, white (*Trifolium repens*), 348
co-edificator, 35, 39
coconut (*Cocos nucifera*), 92, 231, 295, 461

coenopopulation, definition of, 20
coenopopulations, 20, 36–37, 46
coffee (*Coffea* spp.), 226, 229
Colombo region (Sri Lanka), 85
comparison between grassland regions, 26–42, 78
competition
–, among animals, 45, 310
–, among plants, 100, 253
–, between birds and ungulates, 393
–, between deer and birds, 394
–, between grasses and woody plants, 279
–, between native herbivores and livestock, 241, 353
–, between shrubs and grasses, 328
–, between trees and grasses, 340
competitive ability of grasses, 64
condensation, 370, 400
conservation (protection) of grasslands, 78, 401, 403
consumer functions, 147
consumers, 150
–, flow of materials through, 148
consumption (*see also particular consumer groups and species*), 146
– by ants, 209
– by beetles, 151
– by birds, 209
– by earthworms, 151
– by goats, 191
– by grasshoppers, 281, 394
– by invertebrates, 149, 151
– by large mammals, 241
– by livestock, 147, 149, 180, 181, 191, 213, 394
– by sheep, 394
– by soil fauna, 151
– by termites, 281
– by wild herbivores, 45
– of herbage, 188
– of litter, 241
control of mammalian pests, 403
control of weeds, 113, 116, 287
control of woody plants (*see also* burning; fire), 335
–, by browsing, 287
–, by mowing, 335
–, by ring barking (girdling), 335
–, by sheep, 354
convergence zone, inter-tropical, 197, 221
–, sub-Antarctic, 305, 306, 309, 411, 412, 416, 418, 421
cool-season plant species, 272, 274, 343, 344, 350, 351
coot, (*Fulica atra*), 77
copper, 154, 155
– in soil, 346, 348
–, mobility in plants of, 155
cormorant, great (*Phalacrocorax carbo*), 77
Corsica [island] (Mediterranean Sea), 189
cotton (*Gossypium hirsutum*), 229, 234
cow-peas (*Vigna sinensis*), 116, 249, 254
Craigieburn Range (South Island, New Zealand), 369, 378
Crete [island] (Mediterranean Sea), 189
Crimea mountains (previously Jaila mountains) (Ukraine), 159
cropland (*see also* agricultural ecosystems; arable land; slash and burn agriculture; shifting cultivation), 25, 50, 61, 69, 88, 110, 113, 117, 143, 188, 190–193, 228–230, 233, 234, 242, 243, 249, 254, 273, 282, 292, 300, 310, 316, 341, 349, 355, 379, 451
–, abandoned (*see* abandoned cropland)
–, invasions of pests from grassland to, 83
–, length of the fallow period in, 214, 231, 254, 274
–, rate of expansion of, 214
Crozet archipelago (sub-Antarctic), 413, 419, 427
cryptogams, 41, 418
cryptophytes, 71, 98, 446, 449
cryptophytic lianes, 449
Cunnamulla region (Queensland, Australia), 322–324
cutting (clearing) of woody plants (*see also* deforestation), 221, 224, 225, 228–230, 233, 273, 277, 279, 292, 328, 342, 343, 345, 346, 348, 458
–, for fuel, 249
–, – –, effect of, 192
Cyprus [island] (Mediterranean Sea), 171
Czechoslovakia (central Europe) (*see also particular regions and localities*), 127, 128, 131, 132, 134–136, 138–141, 143–147, 149, 152

Dacca–Mymensingh region (Bangladesh), 88
Dahra region (Senegal), 206
Dakar (Senegal), 202
Daraundi river (Nepal), 94
darkling beetles (Tenebrionidae), 23, 40
Darlac plateau (Vietnam), 96
Darling Downs (Queensland, Australia), 344
Darwin region (Northern Territory, Australia), 345
dassie, rock (*Procavia capensis*), 211
Daurian region (former Soviet Union), 4
Dda Hinggan Ling (*see* Greater Khingan mountains)
De Grey (Western Australia, Australia), 303
Deccan plateau (India), 86, 87, 91
decomposer system, energy flow through, 106
decomposers, 148, 150–153, 148
–, nutrient flow through, 156
decomposition, 42, 46, 101, 103, 127, 132, 140, 146, 148, 151, 152, 154, 208, 281
– chain, 146
– of grass leaves, rate of, 426
– of litter, rate of, 234, 397, 398
– of under-ground plant parts, 107, 108
– processes, 151
–, rate of, 72, 151, 107, 108, 281
–, – –, differences between species, 426
deer, 377, 393–395, 402–404
–, axis (*Axis* spp.), 104
–, diet of, 394
–, effect of grazing by, 393
–, efficiency of energy use by, 394
– farming, 404
–, hog (*Axis porcinus*), 104, 106
–, populations of, 403
–, red (*Cervus elaphus*), 310, 393, 402
–, rusa (Timor) (*Cervus timorensis*), 465
–, – –, populations of, 310
–, swamp (*Cervus duvauceli duvauceli*), 104, 106

deforestation, (see also cutting of woody plants), 2, 80, 125, 129, 367, 374
degradation (degeneration, deterioration) of grassland (grazing land), 76, 78, 83, 113–115, 175, 176, 189, 191, 192, 206, 215, 235, 247, 249, 292, 275, 278, 295, 301, 325, 328, 346, 378, 379, 400, 402, 429, 430, 455, 458, 466
–, by introduced mammals, 403
–, by reindeer, 429
–, effects of fuel-gathering in, 113
–, rate of, 278
–, through mismanagement, 272
degradation of soil, 156, 191, 192, 245, 292, 294, 295, 300
–, Manioc (*Manihot esculenta*) stage, 458
degradation of woodland (forests), 171, 176
Delhi region (Orissa, India), 86, 91, 100
demographic structure of plant populations, 19, 36, 37, 46, 131–133, 137
Deniliquin region (New South Wales, Australia), 343
Denmark (Europe), 157
denudation–revegetation, effect of rodents on, 54
Dera Ghazi Khan region (Pakistan), 89
desalinization, 54
desert animals, 74
desert plant species, 31
desert soils, 86
desert steppe, phenology of, 20
–, plant species of, 31
desertification, 64, 76, 203, 215
deserts, various types of (*see particular territories and regions*)
desiccation, plant, 19
detritivores, 146
detritophages, metabolism of, 151
detritus, 151
– food chain, 127, 146, 151, 153, 155, 309, 426
diaspore types, 34
Diego Ramirez (Chile, sub-Antarctic)), 419
digestibility of herbage by livestock, 111, 242, 286, 394
dik dik (*Rhynchotragus kirkii*), 231, 248
disclimax communities (*see also* fire disclimaxes *and particular regions*), 389, 458
disclimax species, 344
diseases of animals, 76
dissemination of diaspores, 32, 34
disseminule distribution in grasses, 19
distribution of grass tribes, 168, 169, 303, 304
distribution of plants
–, types of, 31
–, within communities, 35
disturbance of vegetation, 18, 25, 36, 45, 47–51, 54, 55, 76, 312
diversity (floristic), 3, 7, 20, 33, 64, 65, 67–69, 72, 78, 129, 131, 132, 134, 145, 151, 155–57, 159, 160, 203, 297, 350, 351
–, effects of grazing on, 148
– of morphological plant types, 131
– of plants, morphological, 132
–, species, of soil organisms, 151
Djibouti (East Africa), 255
Dnieper (Dnepr) river (former Soviet Union), 16, 43, 159
dog, African wild, (*Lycaon pictus*), 211, 226

Don river (Volgograd region, former Soviet Union), 16, 53, 160
Donets river (former Soviet Union), 7, 19
Donetsk region (Ukraine), 7, 8, 14
donkey (*see also* ass), 84, 190, 256, 257
–, wild (*Equus asinus*), 353
–, wild Tibetan (*Equus hemionus*), 76
Drakensberg mountains (South Africa), 269
Drakensberg region (South Africa), 280
drought, 5, 15, 20, 33, 47, 181, 202, 203, 206, 207, 210, 215, 233, 241, 249, 256, 270, 272, 279, 281, 284, 304, 325, 326, 328, 330, 336, 352–354, 417
–, definition of, 270
–, effects on animals of, 75
–, effects on vegetation of, 5, 62, 63
–, seasonal, 167, 171, 370
drought-evading strategy, 193
drought-resistant plants, 67, 115, 233, 326
duiker, Grimm's (*Sylvicapra grimmia*), 211, 226, 279
duikers (*Cephalophus* spp.), 226, 231
dung rollers (large *Heliocopris* sp.), activity of, 281
dung-beetles, activity of, 281
Dunstan Mountains (South Island, New Zealand), 381
dust-storms, 78
Dvina river region (northern Caucasus, former Soviet Union), 139
dwarf half-shrubs, 66, 67, 79
dwarf shrubs, 180, 246, 252, 253, 256
Dyje river (Czechoslovakia), 128, 131, 142, 146
dynamics of grasslands, 17–26
Dzhanibek research station (Ural region, former Soviet Union), 53

eagle, golden (*Aquila chrysaetos*), 77
–, Verreaux's *Aquila verreauxi*), 226
earthworms, 106, 107, 426
–, beneficial effects of, 396
–, comparison of untilled and tilled sites, 150
–, consumption by, 151
–, effect on soil-nutrient mobility of, 309
–, energetics of, 106
–, populations of, 396
–, role of, 23
East Africa, (*see also particular territories, regions and localities*), 169, 221–257
–, *Aristida* annual short-grass region, geology and soils of, 255–256
–, *Chrysopogon* mid-grass region, geology and soils of, 245
–, *Hyparrhenia* tall-grass region, geology and soils of, 232, 233
–, *Leptothrium* mid-grass region, geology and soils of, 252
–, *Panicum–Hyparrhenia* tall-grass region, geology and soils of, 229, 230
–, *Pennisetum* giant-grass region, geology and soils of, 227, 228
–, *Pennisetum* mid-grass region, geology and soils of, 222, 223
–, *Themeda* mid-grass region, geology and soils of, 235, 237
East African grasslands, 221–257
–, alpine meadow zone, 223
–, annual communities, 222, 247, 255

East African grasslands (*continued*)

–, *Aristida* annual short-grass communities, 255, 256
–, *Aristida* annual short-grass region, 255–257
–, avifauna of, 226, 231, 234, 241, 256
–, *Cenchrus–Chloris* subregion, 246
–, *Chrysopogon* mid-grass communities, 244–247
–, *Chrysopogon* mid-grass region, 243–250
–, elevation and climate of, 222, 227, 229, 232, 235, 243, 245, 250, 255
–, fauna of, 225, 226, 228, 231, 234, 239–241, 248, 253, 254, 256
–, *Hyparrhenia* tall-grass communities, 233, 234
–, *Hyparrhenia* tall-grass region, 232–235
–, land use in, 226, 227, 229, 231, 234, 235, 241–243, 248–250, 254–257
–, *Leptothrium* mid-grass communities, 251–253
–, *Leptothrium* mid-grass region, 250–255
–, livestock in, 225, 229, 231, 234, 235, 240–243, 246–250, 252, 254, 256
–, map of, 223
–, *Panicum–Hyparrhenia* tall-grass communities, 230, 231
–, *Panicum–Hyparrhenia* tall-grass region, 229–231
–, *Pennisetum* giant-grass communities, 228
–, *Pennisetum* giant-grass region, 227–229
–, *Pennisetum* mid-grass communities, 224, 225
–, *Pennisetum* mid-grass region, 222–227
–, *Sporobolus* subregion, vegetation of, 246, 247
–, successional (seral) plant communities of, 222, 224, 225, 228, 231, 233, 237–239, 242, 245, 247, 252, 253
–, swamps, 232
–, *Themeda* mid-grass communities, 236–239
–, *Themeda* mid-grass region, 235–243
East African vegetation
–, bushland, 222, 224, 226, 229–231, 233–235, 237–239, 243, 245–250, 253, 254
–, dwarf shrubland, 222, 252
–, forest, 221, 222, 224–230, 233, 237, 239, 246
–, savanna, 222, 228, 230, 241
–, shrubland, 222, 238, 248, 252–254
–, wooded grasslands, 233, 234
–, woodland, 221, 222, 233, 234, 237, 238, 240, 246
–, woodland savanna, 222
Eastern Goldfields (Western Australia, Australia), 331
ecological types of plant species, 31
ecosystem dynamics, 17–26
ecosystems, agricultural, 83, 151, 153
edaphic factors (*see* soils)
edificator, 23, 35, 39
–, definition of, 18
Efogi region (Papua New Guinea), 463
Egypt (North Africa), 171, 189
El Fasher (Sudan), 200
eland (*Taurotragus oryx*), 226, 280
–, giant (*Taurotragus derbyanus*), 211
Elbe river (Germany), 139, 140
elephant grass (*Pennisetum purpureum*), 94
elephant(s), 104, 106
–, activity of, 268
– (*Loxodonta africana*), 209, 211, 224, 226, 232, 248, 280

–, populations of, 106, 234
Emerald region (Queensland, Australia), 325, 352
enchytraeids
–, biomass of, 106
–, – –, review of Parts A and B, 478, 479
–, populations of, 106
endangered species, 155
endemic animals, 226, 280, 300
endemic birds, 393
endemic plants, 203, 297, 298, 377, 425, 428
energy flow (energetics) (*see also particular groups and species*), 41, 44
–, rates of, 153
–, role of phytophagous insects in, 147
–, through consumers, 149
–, through plants, 392
–, through ungulates, efficiency of, 310
energy relationships, animal/plant, 44–45
enzymes in soil, 42
enzymes, role in decomposition of, 42, 151
enzymic pathways, 305
ephemerals, 14, 15, 17, 21, 30, 32, 33, 36, 70, 159, 160, 204, 274, 321, 452
ephemeroids, 14, 15, 17, 30, 32, 33, 159
ericaceous species, 454
Eritrea (East Africa), 225
Erromango island (Vanuatu), 459, 456, 457, 460
esparto grass (*Lygeum spartum*), 171
Espiritu Santo [island] (Vanuatu), 456
Ethiopia (East Africa), 198, 221, 222, 224–226, 232, 234, 240, 243, 245, 246, 248, 255, 257
Ethiopian highlands, 222, 225, 226, 243, 246, 255
eucalypts, 443
Eungella tableland (Australia), 441
Euphrates river (Iraq), 87
Eurasia (*see also particular territories, regions and localities*), 1–55, 125–162
Eurasian grasslands (*see also particular territories and regions*), 3–55, 85, 88, 125, 129, 130, 158–161
–, alpine (mountain) steppe, 1, 17, 53
–, alpine meadows, 1
–, arid types, 1
–, biocoenoses, 24
–, Black Sea–Kazakhstan steppes, 4, 10
–, bunch-grass steppes, 1, 5, 7–9, 14, 18, 20, 22, 45, 48–50, 54
–, bunch-herb type of steppe, 17
–, Central-Asian (North-Gobian) desert-steppe, 29
–, classification of, 1, 2, 5, 128–130
–, climate of, 26–27
–, climate of, degree of continentality, 10
–, community types of, 30–31
–, Daurian steppes, 4
–, desert steppes, 1, 2, 5, 7, 11, 13, 15, 17–21, 26–30, 29, 31, 37, 42
–, desertified communities, 3, 5
–, distribution of, 1
–, dry steppe, 20–22, 24, 43–45, 49, 50, 52, 53
–, fauna of, 22–24, 40, 42–55
–, forb–bunch-grass steppes, 19, 21, 22, 45, 49, 51, 52
–, forb–grass steppes, 1

Eurasian grasslands (*continued*)

–, forb steppes, 1
–, forest steppe, 1, 5, 8–11, 16, 17, 19, 22, 23, 43, 46, 47, 54
–, forest-steppe zone, 8
–, geographical types of, 30–31
–, hemi-psammophytic and psammophytic subtypes of steppe, 8
–, herbaceous swamps, 1
–, Khomutovskaya steppe (Sea of Azov region), 7, 8, 14, 52
–, lithophytic plant communities, 71
–, livestock in, 3, 39, 45, 128, 148
–, maps of, 31
–, meadow steppe, 5, 7, 8, 10, 18–22, 44–47, 49, 50, 52, 53
–, monsoonal types, 1
–, nemoral grasslands, 1
–, northern steppe, 1
–, phyto-geographical characteristics of, 4–17
–, phytocoenoses, 20, 127–129
–, pseudo-savannas, 1
–, semi-desert (desertified) steppe, 1, 3–5, 7, 11–15, 17, 18, 20–30, 40, 47, 49–51, 53, 125, 159
–, steppe meadows (hydrophilous steppes), 158
–, steppified meadows of, 19, 53
–, Streletskaya steppe, 45–47, 49, 52
–, sub-Arctic grasslands, 1
–, subalpine meadows, 1
–, subtropical steppes, 1
–, temperate mountain bunch-grass steppes, 11
–, Transbaykal steppes, 4
–, Transvolga–Kazakhstan steppes, 18
–, true (typical) bunch-grass steppe, 5, 7–11, 20, 23, 26, 30
–, vegetation of, ecosystem organization, animals, 22–23
–, vegetation of, ecosystem organization, micro-organisms, 24–25
–, vegetation of, ecosystem organization, plants, 17–22
–, xerophytic steppes, 19
–, zonal types of steppe, 4
Eurasian semi-natural grasslands, 1, 2, 125–162
–, alluvial communities, 159
–, anthropogenic meadows, 159, 161
–, climate of, 126
–, consumers of, 146–148
–, damp meadow, 144
–, distribution of, 126
–, dry meadow, 144
–, ecological types of, 156–162
–, forb communities, 158
–, forest-steppe, 125, 130, 157–161
–, hay meadows, 125, 127–129, 131–135, 141, 143, 144, 146–149, 156–158, 162
–, hygrophytic meadows, 131, 143, 151, 157, 159
–, meadows, 125, 127–129, 131, 132, 134–139, 143–149, 152–155, 157, 158, 160
–, meso-xerophytic communities, 131, 156, 158, 159
–, mesohalophytic communities, 161
–, mesohygrophytic communities, 157–159, 161
–, mesophytic communities, 125, 133, 136, 137, 142–144, 151, 153, 154, 157–162
–, montane communities, 158, 159
–, mossy communities, 158
–, pastures, 125, 127–129, 131, 136, 137, 146–149, 157, 159–161
–, peat-bog grasslands, 125, 158, 162
–, psammophytic grassland, 136, 137
–, short-forb mixed communities, 159
–, short-forb–sedge communities, 159
–, short-grass communities, 157, 162
–, short-grass–forb communities, 159, 161
–, soils of, 126
–, sub-xerophytic communities, 157
–, subcontinental type, 158
–, submontane communities, 158, 159
–, swamps, 125
–, synanthropic communities, 161
–, tall-forb communities, 158, 160, 162
–, tall-forb–grass meadows, 161
–, tall-grass communities, 157, 158–162
–, tall-grass–forb communities, 157–159, 161
–, types of, 126
–, waterlogged communities, 158, 160
–, xero-mesophytic communities, 128, 136, 137, 161
Eurasian steppe regions
–, Asian (Dauro–Mongolian) subregion, 11
–, Azov–Black Sea sub-province, 16
–, Balkano–Mesian forest-steppe province, 11, 16
–, Big lakes pan sub-province, 17
–, Black Sea (Pontic) province, 5, 7, 8, 11, 16, 24, 25
–, Black Sea–Kazakhstan subregion, 5, 7–17, 20, 21, 22, 24–26
–, Central Asian subregion, 10, 13, 15–17, 26
–, Central Kazakhstan sub-province, 16, 24
–, East European forest-steppe province, 11, 16
–, Eastern Kazakhstan sub-province, 16
–, Eastern Mongolian sub-province, 17
–, Hangay–Daurian mountain forest-steppe province, 11, 17
–, Lower Selenga sub-province, 17
–, Manchurian forest-steppe (meadow-steppe) province, 11, 17
–, Middle Dnieper sub-province, 16
–, Middle Khalkha sub-province, 17
–, Middle Russian (Middle Don) sub-province, 16
–, Mongolian province, 11, 17
–, Mongolo–Altaian mountain-steppe sub-province, 17
–, Nerchinsk–Onon sub-province, 17
–, North-eastern Gobian sub-province, 17
–, North-Gobi desert-steppe province, 11, 13, 17
–, Saur–Tarbagatai (Tacheng) sub-province, 16
–, Shanxi–Gansuian forest-steppe and steppe province, 11, 17
–, Songhuian sub-province, 17
–, South Khinganian mountain forest-steppe sub-province, 17
–, Southern Altai province, 16
–, Transkama–Transvolga sub-province, 16
–, Transvolga–Kazakhstan province, 9–13, 16, 20–22, 26
–, West Siberian forest-steppe province, 11, 16
–, Western Hangay sub-province, 17
–, Western Kazakhstan sub-province, 16
–, Yergeni–Transvolga sub-province, 16
Eurasian steppe zone, sub-division of, 11, 18

Eurasian vegetation
–, altitudinal zonation of, 12, 15
–, desert regions, 129, 158, 159, 160, 161
–, forest, 8, 43, 125–127, 129, 130, 150, 151, 153, 157–162, 167, 168
–, shrub communities, 8
–, steppe forest, 8
–, steppified forests, 9
–, successional (seral) communities of, 1, 19, 23, 25, 49, 54, 55, 125, 127, 129, 131, 133, 145, 161
–, woodlands, 125
Europe (*see also particular territories, regions and localities*), 1, 2, 12, 43, 125, 127–131, 139, 145, 148, 155–159, 161, 241
European semi-natural grassland types, *Agropyro–Rumicion* Nordh 1940, 157
–, *Arrhenatheretum elatioris* Braun 1915, 157
–, *Arrhenatherion* W. Koch 1926, 157
–, *Calthion* Tx. 1936, 157.
–, *Cynosurion* Tx. 1947, 157
–, *Eu-Nardion* Luquet. 1926, 157
–, *Festucetea rubrae*, 161
–, *Festucion vaginatae* Soo 1929, 157
–, *Festucion valesiacae* Klika 1931, 157
–, *Festuco–Cynosuretum* Tx. 1940, 157
–, *Junco–Molinietum* Preisg. 1951, 158
–, *Lolio–Cynosuretum* Tx. 1937, 157
–, *Magnocaricion* W. Koch 1926, 157
–, *Mesobromion* W. Koch 1926, 156
–, *Molinion* W. Koch 1926, 158
–, *Phragmition* W. Koch 1926, 257
–, *Polygono–Trisetion* Br.-Bl. et Tx. ex Marschall 1947, 157
–, *Violion caninae* Schwick. 1944, 157
–, *Xerobromion* Br.-Bl. et Moor 1938, 157
European steppe, 78
euryxerophilous forbs, 7
eutrophication, 129, 154
euxerophytes, 159
evapo-transpiration, 139, 188, 197, 202, 209, 224, 370, 392, 399
–, rates of, 275
evaporation, 62, 202, 208
–, rate of, 143
evaporative stress, 252
excrements, 151
exotic (introduced) plant species (*see also* naturalized plant species), 114, 115, 162, 291, 300, 301, 304, 309, 310, 312, 315, 316, 321, 326, 335, 336, 339, 342, 344–347, 349–351, 353, 378–380, 428, 441, 442, 445, 455–458, 460
expansion of grassland, 455
–, effect of hurricanes, 454
–, effect of tectonic disturbances, 454

Falkland Islands (South Atlantic), 419
famine conditions, 109
fauna (*see also particular regions, groups and species*), 77
– in soil (*see also* macrofauna; mesofauna; microfauna; omnivores), 153, 281, 426
–, – –, activity of, 107
–, – –, assimilation efficiency of, 151
–, – –, biomass of, 146

–, – –, comparison of untilled and tilled sites, 150
–, – –, consumption by, 151
–, – –, effect of soil-forming processes on, 150
–, polyphagous, 150
–, native, impact on vegetation, 427
–, –, role in supplying nutrients for plants, 427
–, re-introduction of native species, 104
–, seasonal migrations of, 209
faunas, Central-Asian, 77
Feldberg (Germany), 141
Fennoscandia (sub-Arctic), 308
feral animals, 353, 354
Ferlo region (Senegal), 210
fern(s), 127, 129, 448, 453, 459
–, finger (*Papuapteris linearis*), 453
–, fire (*Dicranopteris linearis*, 458
fertilization of soil, 79, 102, 115, 127–129, 131, 133, 136–140, 145–149, 152–154, 157, 158, 161, 188, 193, 214, 216, 225, 227, 229, 235, 275, 277, 278, 287, 288, 344, 346, 348, 349, 380, 399, 455
–, effect on floristic composition of, 350, 351
–, effect on plant growth of, 425, 427
–, effect on plant succession of, 275
fescue tussock, alpine (*Festuca matthewsii*), 381
–, hard (*Festuca novae-zelandiae*, 366, 368, 383
Fété Olé (Ferlo region, Senegal), 209
figs, 190
Fiji (South-west Pacific) (*see also particular islands, provinces and localities*), 435, 456, 457, 460, 466
–, vegetation of, 457–459, 462
–, – –, fernland, 458
–, – –, Megatherm-seasonal grassland, 462
–, – –, open grassland, 458
–, – –, Sigatoka sand dunes, 458
–, – –, swamp, 457
–, – –, *talasiga* (sun-burnt) grassland, 446, 455, 458, 460, 462
finches (*Montifringilla* spp.), 77
finger millet (*Eleusine coracana*), 226, 234
Finland (northern Europe), 147
Finschhafen/Saruwaket (Huon) peninsula (Papua New Guinea), 450
Fiordland (South Island, New Zealand), 375–377
Fiordland mountains (South Island, New Zealand), 375, 394
Fiordland National Park (South Island, New Zealand), 393
fire (*see also* burning of vegetation), 25, 26, 64, 116, 266, 269, 271–279, 282–284, 287, 292, 295–297, 299–301, 202, 363, 365, 368, 374, 377, 388, 389, 400
– climaxes, disclimaxes and subclimaxes, 92, 224, 246, 269, 273, 274, 276–279, 283, 283, 284, 460
–, effect on vegetation of, 25, 167, 176, 206, 324, 388, 400, 442, 446, 447, 449, 450, 452, 454, 457, 458, 463
–, following hurricanes, 461
–, from lightning strikes, 461
– patterns in vegetation, 452
–, subsequent to frost damage, 463
–, use of, 189, 192, 266, 283, 284, 292, 297, 305, 306, 309, 320, 353, 377, 429, 443, 445, 458
–, use in gardening of, 461
–, use in hunting of, 282, 304, 447, 465
–, use in New Zealand by Polynesians, 363

GENERAL INDEX

fire-adaptive mechanisms of plants, 447
fire-derived grassland, 453
fire-maintained grassland, 239, 453
fire-prone vegetation, 461
–, related to nutrient content of herbage, 461
–, related to oil content of plants, 462
–, relationship to soil fertility, 461
fire-resistant vegetation, 292
fire-tolerant species, 224
fleas, 430
flood-plains, 87, 125, 129, 154, 159, 160, 203, 256, 333
flooding, 2, 90, 129, 130, 132, 139, 152, 155, 157, 159–62, 190, 203, 206, 299, 300, 321, 447
floras, Austro–New Guinean, 442
–, Caucasian, 159
–, Central-Asian, 31
–, holarctic, 31, 296
–, Mediterranean, 30, 31
–, palaeo-tropical, 203
Flores (island) (Indonesia), 95
floristic composition of vegetation, 8, 10, 18, 26, 27–30, 40, 83, 92, 94, 95, 102, 134, 159, 160, 177, 207, 270, 272, 278, 285, 327, 350, 355, 377
–, effect of grazing on, 148
–, fluctuations in, 206
–, relationship to soil fertility, 352
floristic discontinuity, 419
flower development, 18
flower opening in grasses, 19
flowering, 138
– period, 19, 20, 78
– types, 34
Fly river system (Papua New Guinea), 443
Fly–Digul region (Papua New Guinea), 445
fodder shrubs, 193
food chains, 127, 146, 153, 155
–, grazing, 127, 146, 151, 155
forest animals, 74, 75
forests, various types of (*see particular territories and regions*)
formation of grassland, 224
fox (*Vulpes* spp.), 211
–, bat-eared (*Otocyon megalotis*), 239
–, corsax (*Vulpes corsac*), 76
–, Indian (*Vulpes bengalensis*), 104
–, Simien (*Canis simensis*), 226
France (Europe), 149, 157, 171, 193
Frankenwald research station (Johannesburg region, South Africa), 275
frost, 74, 222, 270–272, 276, 319, 396, 416, 429, 453
–, effect on vegetation, 463, 464
–, in relation to fire, 463
frost-free period, 5
Fulda Valley (Germany), 141
functional groups of grasses, 305
fungal/bacterial biomass ratio, 108
fungi, 24, 30, 40, 41, 92, 97, 151, 281, 426
–, biomass of, review of Parts A and B, 480
– in soil, 93, 148
– – –, populations of, 106
–, populations of, 107

–, successions of, 107
fungivores in soil, biomass of, 150
–, population of, 150
Futuna Island, Vanuatu, 461

Gal Oya region (Sri Lanka), 106
galago, lesser (*Galago senegalensis*), 211
Gambia (West Africa), 198
game birds, 465
game cropping, 241
game reptiles, 465
Ganga river (India), 87
Gangetic plains (India), 86, 91
Gansu [province] (China), 61, 67
gardening activities, effect on vegetation, 454, 455–458, 460
Gascoyne River catchment (Western Australia, Australia), 328
gaur (*Bos gaurus*), 104
gazelle(s) (*Gazella* spp.), 103, 104, 209, 211
–, Clarke's (*Ammodorcas clarkei*), 248
–, dorcas (*Gazella dorcas*), 190
–, goitered (*Gazella subgutturosa*), 75
–, Grant's (*Gazella granti*), 226
–, Mongolian (*Prodorcas gutturosa*), 75
–, Pelzeln's (*Gazella pelzelni*), 248, 256
–, Soemmering's (*Gazella soemmeringi*), 248, 253
–, Speke's (*Gazella spekei*), 253
–, Thomson's (*Gazella thomsoni*), 226
gemsbok (*Oryx gazella*), 280
genet (*Genetta* spp.), 226
–, common (*Genetta genetta*), 211
geobioelements, 154, 155
geomorphology, 203
geophytes, 127, 128, 380
George region (South Africa), 168
gerbil(s) (*Gerbillus* spp.), 212
– (*Gerbillus nanus indus*), 103
– (*Taterillus* spp.), 212
–, clawed (*Meriones unguiculatus*), 76
–, desert (*Meriones hurrianae hurrianae*), 103
–, Indian (*Tatera indica indica*), 103
–, pouched (*Desmodilliscus braueri*), 212
gerenuk, (*Litocranius walleri*), 248
Germany (Europe) (*see also particular localities*), 138–141, 148, 157
germination, 19, 131
gidgee (*Acacia cambagei*), 329, 348
giraffe (*Giraffa camelopardalis*), 209, 211, 234
glacial movement, effect on grassland development, 442
glacial refugia, 413
glaciation, 366–368, 375
–, effects of, 368
–, effect on vegetation of, 374
glycophytic species, 177
goat(s) (*Capra hircus*), 84, 108–110, 188–192, 213, 248, 250, 254, 256, 287, 353, 403
–, adaptation to drought, 191
–, anatomy and physiology, 191
–, biomass of, 254
–, consumption by, 191

goats (*continued*)

–, feeding behaviour, 191
–, grazing area of, 191
Gobi Altai mountains (Central Asia), 15
Gobi region (Central Asia), 11, 13, 15, 17, 26, 31
Gobi-Mongolian plant species, 31
Gondwanaland, 366
goral, (*Nemorhaedus goral*), 104
Gorkiy region (former Soviet Union), 158
Gose-Elbe lowland (Germany), 141
Gough Island (South Atlantic), 419, 427
Grande Terre [island] (Kerguelen archipelago), 411, 421, 428
granivores, seed consumption by, 207
–, seed storage by, 207
grapes, 190
grass leaves, morphology of, 451
grass tillers, structure of, 390
grass tussocks, as nesting sites by birds, 427
grasses, autecology of, 421
–, longevity of leaves of, 421
–, sclerophyllous, 295
–, stoloniferous, 224, 238
grasshoppers, 77, 190, 208
–, activity in relation to rangeland condition, 396, 397
–, biomass of, 396
–, consumption by, 281, 394
–, consumption by, compared to cattle, 281
–, effect of climate on, 397
–, effect on vegetation, 396, 397
–, energetics of, 396, 397
–, feeding behaviour of, 397
–, population dynamics of, 396, 397
–, populations of, 78
–, production/ingestion efficiency of, 396
–, proportion of herbage consumed by, 396, 397
–, secondary production by, 397
grazing
–, absence of, by indigenous mammals, 377, 393
– by birds, 393
– by livestock (*see also* overgrazing), 2, 3, 18, 23, 25, 26, 39, 44–47, 61, 63, 64, 70, 71, 73, 74, 78, 83, 89, 91, 96–101, 103, 104, 106–116, 125, 127–129, 131, 136, 146–148, 151, 155–157, 159, 161, 162, 171, 178, 188, 206, 210, 212–216, 221, 222, 224, 225, 227–231, 233–235, 239–243, 246–250, 253–257, 269, 273, 274, 278, 279, 282, 300, 301, 315, 316, 320–326, 328–330, 332–335, 337, 338, 340–346, 348–354, 460
– – –, effect on floristic composition of, 323, 350, 428, 429
– – –, effect on herbage yield, 113
– – –, effect on land management, effect on seed supply, 131
– – –, effect on lichens, 429
– – –, effect on shrubs, 329
– – –, effect on structure of vegetation, 112
– – –, effect on water yields, 345
– – –, effects on vegetation of (*see also* succession of vegetation), 39, 104, 112, 113, 148, 167, 171, 174, 215, 216, 225, 240, 242, 246, 253, 276, 278, 279, 288, 300, 310, 325, 328, 337, 340, 343, 344, 354, 374, 379, 382, 388, 389, 400, 401, 428, 449, 457

– – –, pattern of, 109
– – –, proportion of herbage consumed, 147
– – –, return of urine and faeces, 147
– by reindeer, effects on vegetation of, 428
– by sheep, effects on vegetation, 367
– capacity (*see* carrying capacity)
– food chain, 127, 146, 151, 155
– habits of livestock, 354
–, intensity of, 178, 193, 240, 242
– land, management of (*see* grazing systems; land management; land use; range management, etc.)
–, proportion of herbage removed in, 281
– regimes, 139, 189
–, resistance of plants to, 247, 326
–, season of, 113, 284–286
–, selective, 192, 282, 283, 285, 300, 354, 429
–, sensitivity of plants to, 329, 330, 428
– systems, 110, 113, 116, 188–193, 216, 219, 250, 285, 301, 324, 337, 351, 354
– –, camel-based nomadism, 256
– –, nomadic, 78
Great Dividing Range (Western Australia, Australia), 307
Greater Khingan mountains (Da Hinggan Ling) (Mongolia/China), 15, 63, 72, 76
grebe, great crested (*Podiceps cristatus*), 77
Greece (southern Europe), areal extent of grassland in, 171
gros-hawk (*Accipiter gentilus*), 77
groundnut (*Arachis hypogaea*), 243, 300, 341
grouse, sand (*Pterocles* spp.), 104
–, – (*Syrrhaptes paradoxus*), 77
growth- (life-) forms of plants (*see also particular growth- (life-) forms*), 4, 18, 28, 32–34, 127, 182–187, 308
–, adaptability to grazing, 18
–, adaptability to moisture supply, 18
–, adaptability to trampling, 18
–, spectra, 128
growth of plants, daily rate of, 328
growth strategies, 145
Guadalcanal [island] (Solomon Islands), 454, 455, 457
Guadalcanal Plain, Solomon Islands, 445, 454
Guam [island] (South-west Pacific), 466
Guinea grass (*Panicum maximum*), 458
Guinean, ecoclimatic zone, areal extent of, 198
Gujarat [state] (India) (*see also particular regions and localities*), 2, 97
Gulf of Carpentaria region (Australia), 334, 340, 345, 352
Gulf of Finland (northern Europe), 139
Gulf of Guinea (off Central Africa), 197, 202
gull, brown-headed (*Larus brunnicephalus*), 75, 77

Haast Pass (Southern Alps, (South Island, New Zealand), 365, 372
half-shrubs, 4, 5, 7, 8, 14, 17, 20, 21, 26, 30, 32–36, 54, 66, 67, 79, 89, 127, 129
halo-mesophytic herbs, 73
halomorphic soils (*see* alkaline soils; saline soils)
halophytes, 159, 160, 177, 188, 246, 299, 307, 374
Hamersley Range (Western Australia, Australia), 321
hamster, desert (*Phodopus roborouskii*), 76
–, striped hairy-footed (*Phodopus sungorus*), 76

Hangay region (Mongolia), 11, 17, 40, 44, 46–53
Hangayn Nuruu (Khangai or Changai mountains) (Mongolia), 15
Harare region (Zimbabwe), 271
hare(s) (*Lepus* spp.), 211
– – –, biomass of, 241
– (*Lepus nigricollis dayanus*), 104
–, Cape *Lepus capensis*, 239
–, Crawshay's *Lepus crawshayi*, 239
–, European (*Lepus europaeus*), 395
–, –, population density of, 396
hartebeest(s) (*Alcelaphus* spp.), 211
–, Lichtenstein's (*Alcelaphus lichtensteini*), 280
–, red (Bubal) (*Alcelaphus buselaphus*), 209, 226, 280
–, Swayne's (*Alcelaphus buselaphus swaynei*), 248
Haryana [state] (India), 86, 99, 100, 106–109
Hawaii (Polynesia), 303, 304, 435, 440, 441, 466
–, coastal grassland in, 310
Hawaii Volcanoes National Park, 441
hay meadows, export of forage from, 147
hay yield, 72, 286
haylands, 2, 79, 80
Hazara region (Pakistan), 89
Hazarganji region (Pakistan), 89
hazel (*Corylus*), 161
Heard Island (sub-Antarctic), 411, 413, 416, 427
heat flux, 153
Heilongjiang [province] (China), 61
heliophilous species, 128
hemi-ephemeroids, 15, 30, 32
hemi-cryptophytes, 127, 128, 452
Hentey region (Mongolia), 50
Hentiyn Nuruu (mountains) (Mongolia), 4, 10
herbage
– available to livestock, 328
– biomass, methods of estimating, 212
– –, variability in, 270
– intake by livestock, 242
–, nutritive value of, 74, 79, 80, 111, 294, 300, 301
–, proportion consumable, 213
– yield (maximum standing crop), 64, 65, 67–69, 71–73, 78, 79, 102, 109, 112, 113, 115, 116, 132, 138, 139, 140, 144, 146, 148, 149, 154, 160, 174, 180, 187, 193, 205, 207–209, 227, 256, 274, 277, 284, 229, 231, 235, 242, 250, 254, 287, 310, 320, 322, 323, 328, 340, 348, 349
– –, as affected by system of grazing, 189
– –, effect of cool climate on, 275
– –, effect of grazing on, 275
– –, effect of shrub density on, 328, 335
– –, effect of slope aspect on, 399
– –, effect of soil moisture on, 275
– –, effect of soil type, 180
– –, fluctuations in, 207
– –, related to bioclimatic groups of species, 309
– –, related to moisture supply, 180, 187
herbivores
–, diet of, 202
–, wild (*see also particular groups and species*), 23, 42, 77, 146, 149, 202, 240, 242, 243, 248, 254, 256, 353, 420
–, –, activities of, 266

–, –, biomass (standing crop) of, 45, 234, 242, 243, 248, 250, 256, 280, 281
–, –, consumption by, 45
–, –, feeding habits of, 280
–, –, fluctuations in biomass of, 248
–, –, nutrient flow through, 156
–, –, populations of, 279, 280
–, –, seasonal movements of, 202
herbivory, 23
heron, grey (*Ardea cinerea*), 77
heteropsyllid insect, 457
heterotrophs, activity of, review of Parts A and B, 481
hibernating types of plants, 33
hibernation of animals, 76
hill soils, 87
Himalaya mountains (southern Asia), 70, 308
Himalaya region, 84, 85, 87, 88, 92, 93, 98, 99, 103, 109, 110
hippopotamus (*Hippopotamus amphibius*), 211, 228, 280
Hisiu coastal region (Papua New Guinea), 437, 445
Hodna basin (Algeria), 187, 189
Hokkaido island (Japan), 162
Holland (Netherlands) (Europe), 157
homoeostatic systems, 131
Homoptera, populations of, 147
Honiara region (Guadalcanal, Solomon Islands), 455, 456, 458
Honshu island (Japan), 140, 162
horse, 73, 108, 189, 190, 311, 428
–, Mongolian wild (*Equus przewalskii*), 23, 75
–, wild (*Equus caballus*), 353
Hulunbeier plateau (Inner Mongolia, China), 67
human impact, by trampling and waste disposal, 430
–, effect on spread of grasslands, 442
–, nature of, 190–192, 282, 283
–, on vegetation, 129, 161, 214–216, 243, 292, 297, 309, 427, 449, 454, 455, 457
–, – –, by gathering fuel, 176, 178, 189, 249
human migration because of drought, 109
human population, 188, 191, 210, 215, 229, 282, 315, 316
–, density of, 234, 303
–, rates of increase, 214, 212
humans, dependence of, on grasslands, 465
Humboldt Mountains (South Island, New Zealand), 402, 403
humidity co-efficient, 80
humification, 25, 151
hummock grasses, 311, 328, 340, 350
Hungary, grasslands of, *puszta* type, 1
hunting (over-hunting), 210, 231, 241, 249, 304, 394, 404
hunting, use of fire in, 282, 304, 447, 465
hurricanes, effect on vegetation, 456
hyaena, 213
–, spotted (*Crocuta crocuta*), 211, 226
–, striped (*Hyaena hyaena*), 190, 211
hydro-mesophytes, 128
hydromorphic soils (*see also* waterlogged soils), 95, 299
hydrophytes, 72, 128, 145, 299
hydrosere, 131, 139, 143, 144
hygromesophytes, 160
hygrophilous grasses, 95

ibex (*Capra ibex*), 104
–, walia (*Capra walie*), 226
ibis, wattled (*Bostrychia carunculata*), 226
Iceland (North Atlantic), 1, 308
Île de la Possession (Crozet archipelago), 427
impact of rain-drops on bare soil, 278, 279
impala (*Aepyceros melampus*), 239, 280
importance of grasslands, 155
Inaccessible Island (Tristan da Cunha), 427
increaser species, in response to grazing, 253, 278
India (southern Asia) (*see also particular states, regions and localities*), 83–88, 90–92, 97–101, 103, 105–109, 111–116, 214, 243, 366, 442
–, grasslands of, 90–92, 97, 99–101, 105, 107, 109–116
–, – –, alpine communities, 103
–, – –, *Dichanthium–Cenchrus–Lasiurus* region, 91
–, – –, distribution map of, 90
–, – –, invertebrates in, 106
–, – –, *Phragmites–Saccharum–Imperata* region, 91
–, – –, *Sehima–Dichanthium* region, 90
–, – –, temperate–alpine region, 91, 93
–, – –, *Themeda–Arundinella* region, 91
–, – –, ungulates, distribution of, 104
–, – –, vertebrates in, 104
Indian Ocean, 221, 229
Indo–Gangetic plain (India/Bangladesh), 85, 87, 94
Indo-China (South-east Asia), 96
Indo-malesia, 307, 446, 450
Indo-malesian affinities, 306
Indonesia (southern Asia) (*see also particular islands and localities*), 83, 84, 86, 95, 108, 111–113, 116
–, grasslands of, 95, 111–113
Indus plain (Pakistan), 86, 87, 89
infiltration (percolation) of water into soil, 181, 256, 278
–, rate of, 181, 278
Inner Mongolia (Nei Mongol Zizhiqu) [autonomous region] (China) (*see also particular localities*), 11, 61, 65–68, 72, 76, 79
Inner Mongolian plateau, 63, 65, 66, 73
insect(s) (*see also particular groups and species*), 40, 77, 103, 190
–, affecting livestock, 341
–, biomass of, 23, 106, 147
–, – –, underground/aboveground ratio of, 147
–, consumption by, 146, 147, 208, 209
–, destructive effects of, 396
–, effects on livestock, 213
–, effect on primary production, 146
–, herbivorous, 106
–, –, populations of, 147
–, –, role in energy flow of, 147
–, populations of, 106, 140
–, predaceous, 106
–, root-feeders, 396
–, secondary production by, 148
introduced animals, 310, 311, 345, 377, 379, 393, 394, 402, 427, 430, 465
–, competition with livestock, 353
–, effects on vegetation, 310, 420
invasion (*see also* competition)

– by dwarf shrubs, 283, 287
– by eucalypts, 449
– by exotic plants, 307, 310, 311, 441, 454, 456, 466
– – – –, related to disturbance, 466
– – – –, related to grasslands, 466
– – – –, related to grazing by livestock, 466
– – – –, related to logging, 466
– –, susceptibility to, 466
– by graminoids, 253
– by grasses, 224, 251, 283, 287, 311, 337, 450, 454, 460, 464
– by naturalized plants, 428
– by plants, 113, 179, 215, 351, 384
– by shrubs, 192, 253, 272, 273, 283
– by trees, 116, 129, 228, 249, 272, 292, 450
– by weeds, 69, 112, 148, 288, 310, 312, 342, 351
– by woody plants, 278, 279, 292, 301, 311
– of animals, 76
– – –, seasonal, 419
– of exotic animals, 311
invertebrates, 210, 396, 397
–, biomass of, 22, 23, 106
–, – –, review of Parts A and B, 478. 479
–, consumption by, 149, 151
–, impact on vegetation of, 420
– in soil, 147
– – –, biomass of, 147
– – –, populations of, 147
–, number of species of, 146
–, populations of, 106
Inyanga region (Zimbabwe), 271
ion concentration in plants, 73
Iran (southern Asia) (*see also particular localities*), 84–88, 102, 108, 110, 113, 171, 190
–, grasslands of, 88, 89, 102, 110, 113
Iraq (southern Asia) (*see also particular localities*), 84–88, 108, 171
–, grasslands of, 88, 89
Irimu valley (Papua New Guinea), 450
iron, 154, 155
–, mobility in plants of, 155
irrigation, 79, 129, 188, 399
Irtysh (Siberia, Russia), 161
Isplinji valley (Pakistan), 89
Israel (Middle East), 171
Italy (southern Europe), 171

jackal(s) (*Canis* spp.), 211, 226
– (*Canis aureus*), 104
Jaldapara region (West Bengal, India), 106
Jammu and Kashmir [state] (southern Asia), 2, 85, 91, 109, 110
Japan (eastern Asia) (*see also particular islands*), 2, 125, 128–131, 140, 162
–, grasslands of, 162
Jawa (Java) (island) (Indonesia), 95
–, savannas in, 95
jerboa (*Jaculus jaculus*), 212
jerboa, Mongolian (Siberian) five-toed (*Allactaga sibirica*), 43, 76
–, northern three-toed (*Dipus sagitta*), 76

Jhansi region (Uttar Pradesh, India), 86, 100, 105, 113, 114
Jilin [province] (China), 61
jird, Libyan (*Meriones libycus*), 212
Jodhpur region (Rajasthan, India), 86, 100, 103
Jordan (Middle East), 171
Juba river (Somalia), 246
Judean desert, 440

Kalahari desert, 268, 270, 276, 277
–, aeolian deposits of, 270
–, sand deposits of, 273
Kalat region (Pakistan), 109
Kalbinskiy mountains (Kazakhstan), 12
Kama river (former Soviet Union), 159, 160
Kamchatka peninsula (Japan), 162
Kameničky (Czechoslovakia), 128, 134, 147–149, 152, 154
Kamysh-Samarsk lakes (former Soviet Union), 159
kangaroo, 304, 353
–, antilopine kangaroo (*Macropus antilopinus*), 316, 353
–, grey (*Macropus fuliginosus*), 353
–, grey (*Macropus giganteus*), 353
–, hill (euro) (*Macropus robustus*), 353
–, *maganir* (*Macropus agilis*), 465
–, red (*Macropus rufus*, 316, 353
Kanha National Park (Madhya Pradesh, India), 105
kapok bush (*Aerva javanica*), 321
Karaganda region (Kazakhstan), 9, 11, 15
Karan Phuli river (Bangladesh), 94
Karkloof (Natal, South Africa), forests of, 282
Kashmir valley (Jammu and Kashmir, southern Asia), 110
Kasserine-Ferina (Tunisia), 176
Katherine region (Northern Territory, Australia), 339, 352
Kathmandu valley (Nepal), 85
kauri (*Agathis australis*), 361
Kavo range (Guadalcanal, Solomon Islands), 454
Kazakhstan (*see also* former Soviet Union; *and particular regions and localities*), 2, 3–5, 7–18, 20–22, 24–43, 47, 49–52, 54, 159, 161
–, grasslands of, 20, 78
Kaziranga (Assam, India), 106
Keetmanshoop (Namibia), 271
Kenya (East Africa) (*see also particular regions and localities*), 222, 224, 229–234, 236, 237, 240–246, 248–250, 255, 256
Kenyan highlands, 255
Kerguelen archipelago (*see also particular islands*), 305, 308, 412, 413, 416, 419, 427, 429, 430
kestrels, (*Falco* spp.), 241
Khangai mountains (Mongolia), 15
Kherson region (Ukraine), 25, 42, 158
Khirasara (Gujarat, India), 86, 99–101, 108
Khomutovskaya steppe (Sea of Azov region, former Soviet Union), 7, 8, 14, 52
Khuzestān [province] (Iran), 87, 88
Kimberley Range (Western Australia, Australia), 319–321, 325
Kimberley region (Western Australia, Australia), 303
King Edward Cove (South Georgia, sub-Antarctica), 418
kite, black (*Milvus migrans*), 77
Kojonup region (Western Australia, Australia), 349
Kokchetav region (Kazakhstan), 20, 21

Kokchetavskaya hills region (Kazakhstan), 21
Koksengir mountains (Kazakhstan), 11, 15
Kongwa region (Tanzania), 243
Konotoef region (Timor, Indonesia), 112
Koonamore region (South Australia, Australia), 330
Kopet-Dag (former Soviet Union), 2
kudu (*Tragelaphus* spp.), 248
kudu, greater (*Tragelaphus strepticeros*), 209, 211, 279
Kulekhani river (Nepal), 94
Kumait (Middle East), 171
Kunlun Shan (mountains) (China), 70
Kurdistan, 153
Kursk region (Russia), 45, 52, 158
Kurukshetra region (Haryana, India), 86, 99, 100, 106–108
Kuruktag mountains (Xinjiang Uygur autonomous region, China), 76
Kushum river (Kazakhstan), 159
Kyushu island (Japan), 162

Lae region (Papua New Guinea), 446
lagomorphs, 43
–, feeding habits of, 44
Lake Baykal (Siberia, Russia), 4
Lake Eyre region (South Australia, Australia), 316
Lake Ladozhskoye district (Russia), 136, 139, 140
Lake Mobutu Sese Seko (Lake Albert) (East Africa), 234
Lake Myola (Owen Stanley ranges, Papua New Guinea), 463
Lake Victoria (East Africa), 221, 227, 229, 239
Lalmai elevation (Bangladesh), 88
lammergeyer (*Gypaetus barbatus*), 226
Lammerlaw Range (South Island, New Zealand), 400
land
–, abuse of, 203
–, carrying capacity of humans of, 214
– management, 78–80, 108–117, 155
– –, role of fire in, 429, 461–463, 461
– –, use of fire in (*see* fire; burning)
– mismanagement, 301
–, ownership of, 234, 243, 249, 254
– use (*see also* cropland; *and particular grassland regions*), 78–80, 108, 155, 210–212, 282, 464, 465
– –, competition between pastoralism and farming, 214
– –, itinerant agriculture, 300
– –, patterns of, 197, 199
– –, proportion in each category of, 168
– –, semi-nomad farming systems, 215
– –, socio-cultural factors in, 188
– –, subsistence agriculture, 229, 249
– –, transhumant pastoralism, 190, 199, 210, 214, 215
– –, tree farming, 188, 190, 226, 229, 231, 234
Lanžhot (Czechoslovakia), 128, 132, 141, 144–146, 150, 152–154
Laos (South-east Asia), 84, 86, 96, 108
–, grasslands of, 96
lapwing (*Vanellus vanellus*), 77
larch (*Larix* sp.), 161
lark, horned (*Eremophila alpestris*), 77
–, lesser sand (*Calandrella rufescens*), 77
–, Mongolian (*Melanocorypha mongolica*), 77
–, sky (*Alauda arvensis*), 77

Lawes region (Queensland, Australia), 344
layering of vegetation (*see also* biomass; canopy; structure), 20–22, 35, 36, 98, 204, 206, 208, 273, 274, 270, 452
leaf
– area, 133–36, 136, 141
– development, 139
– elongation and die-back, 386
– growth, rate of, 388, 390
– morphology, 308
leaf-area index, 98, 136, 137, 139, 141, 422
leatherwood (*Olearia colensoi*), 386
leaves, longevity of, 426
Lebanon (Middle East), 171
lechwe (*Kobus leche*), 280
legumes, 64, 66, 67, 91, 92, 97, 110, 113, 114, 116, 117, 180, 188, 227, 229, 231, 249, 286, 318, 326, 339, 342, 344–349, 351, 452
lemming, steppe (*Lagurus lagurus*), 43
Lena river south of Aldan (near Yakutsk), 161
leopard (*Panthera pardus*), 104, 190, 209, 211, 226
Lepidoptera, number of species of, 147
–, relationship to ecosystem stability, 147
Lesotho (southern Africa), 168
Lesser khingan mountains (Xiao Hinggan Ling) (Mongolia), 4, 72
Lesser Sunda Islands (Indonesia), 95
Liaoning [province] (China), 61
Libya (North Africa), 171, 190, 193
lichens, 20, 30, 78, 422, 429
life-cycles of plants, 15, 17, 132
life-forms of plants (*see* growth-forms)
lightning strikes, frequency of, 461
Linguere region (Senegal), 201
lion (*Panthera leo*), 190, 209, 211, 226
litter
–, animal, 45, 51
–, plant (*see also* biomass; decomposition), 23, 38, 42, 46, 100, 132
–, –, accumulation of, 241, 242, 279, 284, 285, 426
–, –, accumulation under protection of, 103, 133
–, –, biomass of, 275, 277
–, –, C/N ratio above ground, 352
–, –, formation of, 153, 277, 384, 387
–, –, nutrient flow through, 156
–, –, of under-ground plant parts, 127
–, –, rate of decomposition of, 46, 107, 108, 151, 152, 154, 276, 281, 393, 397
–, –, resulting from animal activity, 146
Liverpool Plains (New South Wales, Australia), 344, 352
Liverpool Range (New South Wales, Australia), 344
livestock (*see also* carrying capacity; grazing; overgrazing; range management; stocking rate *and particular species and regions*)
–, adaptations of, 353, 354
–, biomass of, 231, 235, 241, 250, 254–256
–, – –, seasonal fluctuation in, 254
–, consumption by, 147, 149, 180, 181, 191, 213, 394
–, dietary preference of, 333
–, diets of, 189
–, diseases of, 190, 191, 213, 235

–, draught animals, 213
–, gain in weight of, 346, 348
–, impact of, 190–192, 214
–, management systems for, 210
–, meat production by, 213
–, milk production by, 213
–, mortality, 213
–, pack animals, 213
–, parasites of, 213
–, populations of, 109, 110, 188, 191, 210, 213, 291
–, – –, increases in, 78, 212
–, –, species composition of, 190
–, predators of, 189, 190, 213
–, reproductive rates of, 213
–, seasonal movements of, 109, 189, 210, 231, 254, 256
–, weight gains, 213
livestock/rodent interactions, 45
lizard (*Eremias argus*), 77
– (*Phrynocephalus frontali*), 77
locusts, 23, 190, 208
–, consumption by, 190
–, effects of, 190
loess plains, 160
Loess plateau (Huangtu Gaoyuan) (China), 61, 66–69
logging, 450, 455, 466
Lombok (island) (Indonesia), 95
loss of soil by erosion, rate of, 203, 278
Louga region (Senegal), 206
lumbricids, 150
–, biomass of, review of Parts A and B, 478
Luzon (Philippines), 2, 303
lynx, African (*Felis caracal*), 211

Macquarie Island (sub-Antarctic), 305, 308, 311, 411–414, 416, 417, 419–427, 429, 430
Macquarie region (New South Wales, Australia), 344
macrofauna
–, biomass of, review of Parts A and B, 478, 479
– in soil, 148
– – –, biomass of, 150
– – –, populations of, 150
Macuata [province] (Vanua Levu, Fiji), 458
Madagascar (Indian Ocean) (*see also particular regions and localities*), 169, 229, 291–301
–, avifauna of, 300
–, climate of, 392
–, fauna of, 300
–, livestock in, 291, 300, 301
–, successional (seral) plant communities of, 295, 301
–, vegetation of, azonal (edaphic) vegetation, 299
–, – –, Central Domain, 292, 294–298, 300, 301
–, –, – –, gramineous formations of, 292
–, – –, deciduous forest of the Western Domain, 292
–, – –, Eastern Domain, 292, 294, 296, 297
–, – –, – –, gramineous formations of, 292
–, – –, fire-resistant vegetation, 292
–, – –, forest, 292, 295–301
–, – –, grasslands of, 291–301
–, – –, – –, areal extent of, 291, 297
–, – –, high mountain vegetation, 292

GENERAL INDEX 541

Madagascar, vegetation of (*continued*)

–, – –, map of, 292
–, – –, marshy grasslands, 299
–, – –, regions, 292
–, – –, Sambirano domain, 292
–, – –, savannas, 291, 292, 294–301
–, – –, shrubland, 292
–, – –, Southern Domain, 292, 294, 299–301
–, – –, – –, gramineous formations of, 292
–, – –, thicket vegetation, 292
–, – –, Western Domain, 292, 294, 295–297, 298, 300, 301
–, – –, – –, gramineous formations of, 292
–, – –, xerophilous thicket, 292
Madhya Pradesh [state] (India) (*see also particular regions and localities*), 109
magnesium, 154
–, export of, 155
–, mobility in plants of, 155
–, pool of, 384
Mahaweli Ganga river (Sri Lanka), 94
maize (*Zea mays*), 116, 226, 229, 231, 264, 249, 287
Makran coastal region (Pakistan), 90
malacophyllous plants, 216
Malaita [island] (Solomon Islands), 455
Malawi (Central Africa), 265
Malaysia (southern Asia), 2, 84, 86, 95, 96, 108
–, grasslands of, 95, 96
Malekula [island] (Vanuatu), 456
Mali (West Africa) (*see also particular regions and localities*), 168, 198, 200, 201, 206–208, 213, 215
malic enzyme, 440
mallee-form *Eucalyptus* spp, 307
malvaceous species, 457
mammals, 23, 40, 210, 225, 226, 234, 256, 420
–, carnivorous, 75, 76
manganese, 154, 155
–, mobility in plants of, 155
mango (*Mangifera indica*), 231
Manipur [state] (india), 91
Manki region (near Bulolo, Papua New Guinea), 450
manuring, 419, 422, 423, 425
– by animals, effects on soils, 425
– by birds, effects on soils, 425
– by sea-birds, 424, 425
– by seals, 424, 425
–, destructive effects of, 414
map
– of Australian exotic grasslands (pastures), 318
– of Australian grasslands, 317
– of Chinese grasslands, 61
– of Eurasian grasslands, 31
– of geographical features of China, 62
– of grassland regions in East African grasslands, 223
– of grasslands regions of Pakistan, 90
– of Indian grasslands, 90
– of physical features of New Zealand, 362
– of Sahelian region, 198
– of South-west Pacific, 436
– of southern cool-temperate islands, 412

– of sub-Antarctic islands, 412
– of vegetation in Madagascar, 292
– of vegetation in New Zealand, 364
– of vegetation in southern African grasslands, 266
Marion Island (sub-Antarctic), 310, 411, 413, 416–427, 429, 430
Markham people (Papua New Guinea), 447
Markham Valley (Papua New Guinea), 443, 446, 448–450, 465
marmot (or baibak) (*Marmota bobak*), 43, 76
marsupials, 351, 353
Masai Rangeland Development Commission, 243
Massenerhebung effect, 308, 437,
Maulasar region (Rajasthan, India), 104
Maungatua Range (South Island, New Zealand), 373
Mauritania (West Africa), 198
meadow soils, 63, 72
Mediterranean region (*see also particular regions, localities and territories*), 171–194
–, areal extent of grasslands, 171
–, arid savanna, 178
–, arid steppe, 182–187
–, arid steppic zone, 180, 199
–, chamaephytic steppic communities, 174, 176, 177
–, climate of, 172, 173
–, climax grassland, 171
–, crassulescent steppic communities, 176, 177
–, distribution of plant species in various vegetation types, 182–187
–, dwarf-shrub steppes, 171
–, fauna of, 190
–, garrigues, 192
–, gramineous steppic communities, 176
–, grazing lands of, 171–194
–, halophytic steppe, 179
–, high mountain pastures, 177
–, livestock in, 171, 173, 176, 177, 188–193
–, maquis shrubland, 175, 193
–, marshes, 177
–, meadows, 177, 182–187, 193
–, open woodlands, 176
–, shrubland (tall shrubs without trees), 171, 173, 175, 176, 180, 182–189, 191–193
–, steppe, 173, 175
–, steppic types, 176
–, successional (seral) plant communities of, 171, 172, 174, 176, 188
–, swamps, 177
–, tragacanthic steppic communities, 176, 177
–, white sage (*Artemisia herba-alba*) steppes, 180
–, woodland, 171, 173, 176, 182–187
Mediterranean Sea, 1
Meghna river
– (Pakistan), 87
– (Laos), 87
Melanesia (South-west Pacific) (*see also particular islands*), 304, 435
Melbourne region (Victoria, Australia), 344, 353
melkosopochnik, 8, 12
Menyamya region (Papua New Guinea), 464
Menzenschwand (Germany), 141

mesofauna in soil, 148
–, biomass of, 150
–, populations of, 150
meso-microtherms, 64
mesophytes, 23, 60, 65, 37, 68, 72, 73, 126, 128, 129, 145, 158, 160
mesotherm plant species, 308, 309
mesoxerophilous grasses, 5
meso-xerophytes, 65–67, 128
Messina region (South Africa), 271
metabolic activity in soil, 108
methods of study, trough–peak analysis, 101
Mezen' region (former Soviet Union), 159
micro-elements, 154
micro-organism/plant interactions, 154
micro-organisms (see also particular groups), 24–25, 40–42, 106, 107, 151, 397–398
–, activity of, 107, 281
–, biomass of, 40, 41, 107
–, cellulose-decomposing, 40–41
–, energy content of, 41
– in soil, 148, 153, 154
– – –, activity of, 42, 107, 153, 425
– – –, populations of, 151
– – –, vertical distribution of, 41
–, nitrogen-fixing non-symbiotic, 398
–, – symbiotic bacteria, 153
–, populations of, 41, 281
–, rate of decomposition by, 46
–, review of Parts A and B, 480
–, species composition of, 40–41
–, species diversity of, 40
–, synusiae, species composition of, 41
microbiocoenoses (microbial communities), 40
microbiovers in soil
–, biomass of, 150
–, populations of, 150
microclimate, 44, 47, 64, 416
microcoenoses (sub-communities), 30
microfauna in soil, 30, 426
microfungi, 30
Micronesia (South-west Pacific) (see also particular islands), 435, 441, 460
–, vegetation of, 459, 460, 463–465
–, – –, disclimax mosaics, 459
–, – –, forest plantations, 460
–, – –, Megatherm man-induced grassland, 464
–, – –, palm grassland, 460
–, – –, short fern and megatherm grassland, 463, 465
–, – –, swamps, 460
–, – –, tedh, 460
microphyllous species, 206, 216, 363, 367
microrelief, 42, 44, 53
microtherm bioclimate, definition of, 437
microtherm species, 69, 308
migration
–, of animals, 76
–, – –, because of drought, 109
–, of birds, 77
–, of insects, 77

–, of plants, 157, 442
Mikhaylovskaya virgin-land (Ukraine), 5
millet, 199
– (Pennisetum glaucum), 214
– (Sorghum vulgare), 231
– eater (Quelea quelea), 209
mineralization
–, of organic substances, 25, 151
–, –, rate of, 46
minerals (see nutrients and particular elements)
mission grass (Pennisetum polystachyon), 458
Mitchell grass (Astrebla spp.), 306, 310
Mitchell Grass Downs (Australia), 352
moas (flightless birds), 393
modification of vegetation, 310, 346
Moen Island (Truk group, Eastern Caroline Islands, Micronesia), 460, 464
Mokhotlong (Lesotho, South-east Africa), 271
mole, European, (Talpa europaea), 146
mole rat (Spalax microphtalmus), 43
– – (Tachyoryctes spp.), 226
mollymawks, 427
molybdenum in soil, 346, 348, 349
Monaro Highlands (Australia), 344
Mongolia (see also particular regions and localities), 2, 3, 4, 8, 10, 11, 15, 17–20, 22, 24–26, 30, 32, 33, 36, 40, 43–46, 48–50, 53, 61, 67, 78
–, steppes of, 3–55
Mongolian atmospheric high-pressure system, 62, 63
mongoose (Herpestes spp.), 226
– (Ichneumia albicauda), 226
–, banded (Mungos mungo), 239
–, dwarf (Helogale parvula), 239
–, Egyptian (Herpestes ichneumon), 211
–, white-tailed (Ichneumia albicauda), 211
monkey, red (Erythrocebus patas), 211
monocoenoses, 145
monospecific grassland, 102
monsoon(s), 61, 202, 229, 243, 255, 291
–, influences of, 78
–, Pacific, 11
–, regions, 86, 88, 98, 99, 101, 103, 104, 108
–, season, 109, 114
Mopti (Mali), 201
Morava river (Czechoslovakia), 128, 131, 135, 140, 144, 154, 155, 158
Morehead river region (Papua New Guinea), 439, 445
Morehead–Bensbach region (Papua New Guinea), 445
Morocco (North Africa), 168, 169, 171, 173, 175, 180, 190, 193
mortality, of plants, 36
mosaics, 457
Moscow region (Russia), 158
mosses, 30, 127, 420, 429
Mount Andringitra (Madagascar), 296
Mount Ankaratra (Madagascar), 296
Mount Aspiring National Park (South Island), 402, 403
Mount Bagana (Kieta region, Papua New Guinea), 448
Mount Bangeta region (Saruwaket plateau, Papua New Guinea), 454

Mount Brown (near Owen Stanley range, Papua New Guinea), 449
Mount Cardrona (South Island, New Zealand), 385
Mount Cook (South Island, New Zealand), 363
Mount Dayman (Papua New Guinea), 464
Mount Digimi (Kubor range. New Guinea), 442
Mount Egmont (North Island, New Zealand), 366
Mount Elgon (Kenya), 222
Mount Kenya (Kenya), 222
Mount Khogis (New Caledonia), 455
Mount Kilimanjaro (Tanzania), 222
Mount Lamington (Papua New Guinea), 448
Mount Ngauruhoe (North Island, New Zealand), 366
Mount Ruapehu (North Island, New Zealand), 366
Mount Suckling (Papua New Guinea), 309
Mount Tabwemasana (Espiritu Santo Vanuatu, Vanuatu), 308, 446, 456, 457
Mount Tongariro (North Island, New Zealand), 366
Mount Wilhelm (Papua New Guinea), 422, 423, 430, 453
mouse (*Mus musculus*), 311, 427, 428
mouse, diurnal unstriped grass (*Arvicanthis niloticus*), 241
–, fat-tailed (*Pachyuromys duprasi*), 212
–, Hausa (*Mus haussa*), 212
mouse-like rodents, 44, 46
mouse-hare (*Ochotona curzoniae*), 76
mowing, to control woody plants, 335
Mozambique (southern Africa), 265, 269, 270
Mozambique Belt, 237
mule, 108, 190, 311, 428
Murchison Mountains (Fiordland, South Island), 375, 376, 394, 395
Murree hills (Pakistan), 89
Mustang river (Nepal), 94
Myagdi river (Nepal), 94
mycorrhiza in grasses, 398
myxomatosis (*Myxoma leporipoxvirus*), 430

Nadi region (Viti levu, Fiji), 447
Nadzab region (Papua New Guinea), 446
Nairobi National Park (Kenya), 240, 242
Namara coastal region (Turkey), 85
Namib desert (Namibia), 265, 267, 269, 272
Namibia (southern Africa), 168, 265, 269, 270
Natal [province] (South Africa), 269, 272, 274, 276, 282, 283, 287
–, mid-elevation grasslands of, 272
–, mesic grasslands of, 274
–, tall-grass veld of, 276
native-herbivore/livestock biomass, 248
naturalized plant species (*see also* exotic plant species), 311, 312, 315, 316, 318, 337, 345–349, 429
Nauru [island] (South-west Pacific), 303
Near East, 192, 193
Negev desert (Israel), 190, 440
nematodes, 150
–, biomass of, review of Parts A and B, 478, 479
–, distribution of, 396
–, feeding habits of, 396, 426
– in soil, biomass of, 150
– – –, populations of, 150

Nen Jiang plain (China), 61
Neon basin (near Mount Albert Edward, Papua New Guinea), 463
Nepal (southern Asia) (*see also particular regions and localities*), 84, 85, 87, 88, 94, 106, 108, 109
–, grasslands of, 94, 109
–, *terai* plain in, 85, 87, 94
Nepati valley (Nepal), 106
Nercha river (Mongolia/China), 4
New Amsterdam island, 419
New Britain (South-west Pacific), 435, 448, 461
New Caledonia (South-west Pacific), 306, 435, 456, 457, 466
–, megatherm savannas in, 455, 456
–, *niaouli* savannas in, 306, 455, 456
–, sedgeland in, 455
–, woodland savanna in, 455
New Georgia [island] (Solomon Islands), 455
New Guinea (South-west Pacific) (*see also and particular regions and localities*), 304, 305, 307–309, 366, 422, 423, 430, 435, 437–439, 441–443, 456, 459, 461, 463, 466
–, *kunai* grasslands of, 438
–, microtherm grassland of, 309
New Ireland Province (Papua New Guinea), 307, 435
New South Wales [state] (Australia) (*see also particular regions and localities*), 329–334, 336, 337, 341–344, 346, 348, 352, 353
New Zealand (*see also and particular regions and localities*), 303–305, 308–311, 361–404, 428
–, biogeography and geology, 365, 366
–, biotas of, 304
–, climate of, 370
–, fauna of, 393–397
–, grasslands of, 361–404
–, – –, alpine communities, 370, 395, 396, 401, 404
–, – –, *Chionochloa* communities, 363, 365, 368, 369, 371–373, 375, 378–382, 386, 398, 402, 403
–, – –, herbfields, 392
–, – –, high-alpine zone, 376, 389, 392
–, – –, low-alpine communities, 365, 369–372, 374–376, 392, 375, 379–382, 385, 386, 392, 398, 402, 403
–, – –, Lowland tussock communities, 365, 374, 379, 396, 398, 399
–, – –, maritime communities, 376
–, – –, montane tussock communities, 366–368, 373
–, – –, moorland and moorland-type communities, 375–377
–, – –, relict stands, 389
–, – –, semi-aquatic stands, 392
–, – –, semi-desert communities, 379, 383, 384
–, – –, short-tussock communities, 380, 399
–, – –, soils of, 371–373
–, – –, subalpine communities, 368, 373
–, – –, successional (seral) plant communities of, 367, 379, 403, 404
–, – –, tall-tussock communities, 365, 374, 387
–, livestock in, 377
–, map of physical features of, 362
–, vegetation of, 373–393
–, – –, alpine regions, 383
–, – –, alpine zone, 378
–, – –, altitudinal distribution of vegetation, 383

New Zealand, vegetation of (*continued*)

–, – –, distribution of vegetation, 373–380
–, – –, forest, 361, 363, 365, 366, 368, 370, 372–374, 378, 388, 393, 395, 397, 398, 401–404
–, – –, grass-line, 371, 372
–, – –, map of, 364
–, – –, oceanic influence on vegetation, 393
–, – –, savanna, 363
–, – –, savannas, tropical, 392
–, – –, shrubland, 361, 363, 367, 395
–, – –, snow-line, 368
–, – –, subalpine scrub, 363
–, – –, subalpine zones, 376
–, – –, tall scrub, 363
–, – –, tree line, 363, 365, 367, 370, 374, 375
–, – –, tundra, 389
–, – –, volcanic regions, 363, 374, 401
–, – –, wetlands, 363
Niger (West Africa), 198, 203, 214
Niger river (West Africa), 202
Nigeria (West Africa), 198
Nikkachu region (Bhutan), 94
nilgai (*Boselaphus tragocamelus*), 104
Nilgiri hills (Kerala/Tamil Nadu, India), 92
Ningxia Huizu Zizhiqu [autonomous region] (China), 61, 67
Niono region (Mali), 206
nitrification, 351
nitrogen, 102, 103, 112, 202, 213, 216
–, budget of, 154
–, competition for, 154
–, content of herbage, 235, 242, 273
–, cycle of, 153, 154
–, deficiencies of, 209
–, export of, 155
– fixation by legumes, 346, 351
– –, non-symbiotic, rate of, 398
– –, rate of, by blue-green algae, 426
–, flow of, from birds to grassland, 425
–, –, rate of, 154
– in soil, 231, 239, 245, 351, 352, 384
– – –, content of, 25, 40, 42, 46, 51, 65, 86, 87, 97, 115
– – –, deficiency of, 425
– – –, effect of rodents on, 51
–, mobility in plants of, 155
–, pool of, 384
– uptake by plants, 385
nitrogen-fixation by soil micro-organisms, 153
nitrophilous species, 350
nitrophily, 351
nomadic and semi-nomadic pastoralism, 243, 248, 254
nomadic grazers, 109
nomadic pastoral communities, 109
nomadic pastoralists, 214
nomadism, 188–190, 199, 209, 215, 250, 253, 355
North Africa (*see also and particular territories and localities*), 168, 171, 180, 187–193
North America (*see also Part A*), 1, 160, 199, 241, 441
north-Gobian plant species, 31
North-West Frontier Province (Pakistan), 89

Northern Mariana Islands (South-west Pacific), 303
Northern Territory (Australia) (*see also particular regions and localities*), 315, 319, 324, 327, 328, 330, 333, 338–340, 345, 346, 352, 353
Norway (northern Europe), 149
nutrient needs of livestock, 189
nutrient system, grassland–marine, 427
nutrient(s) (mineral) (*see also particular elements*), 129, 157, 158
–, above-ground pools of, 384
–, atmospheric deposition of, 309
–, budgets of, 156
–, content in plant biomass, 79, 80, 156, 202, 225, 233, 384
–, cycles (cycling) of, 72, 137, 150, 153
–, deficiency of, 209, 216
–, flow, 425, 426
–, – from precipitation, 425
–, – through under-ground plant parts, 156
–, inputs by seals, 425, 427
–, – through birds, 309, 423, 425
–, – through marine animals, 309, 407
–, loss during decomposition of, 398
–, losses of, during burning, 309
–, mineralization of, 351, 352, 189
–, pathways of, 309
–, plant, 425, 426
–, pools and fluxes of, 384
–, rate of turnover of, 46
–, restitution of, 103
–, transfer pathways of, 387
–, transfers of from sea to land of, 309
–, under-ground pools of, 384
–, uptake by plants of, 103, 384, 385
–, utilization by plants, efficiency of, 426
nutrients
–, atmospheric, transfer from the sea, 425
– in soil (*see also particular elements*), 86, 136, 177, 227, 228, 253, 425
– – –, additions by blue-green algae, 309
– – –, deficiencies of, 112, 245, 348, 421, 425
– – –, effect of earthworms on mobility of, 309
–, returned to soil by livestock, 147
–, transfers from animals to soil, 425
nutritive value
– – of browse, 213
– – of herbage, 213, 239, 241, 242, 283, 319, 320, 322–324, 326, 333, 337, 339, 340, 345, 346, 354, 393–395
nyala, mountain (*Tragelaphus buxtoni*), 226
Nyanza Shield (East Africa), 237
Nylsvley (Transvaal, South Africa), savanna of, 281

oak (*Quercus*), 156, 161
Oceania (*see also particular territories, regions and localities*), 168, 303–467
–, anthropogenic grasslands of, 309
–, – – –, hekistotherm type, 305
–, – – –, megatherm seasonal type, 306, 307, 309, 310
–, – – –, megatherm type, 305–307, 309, 310, 311
–, – – –, mesotherm type, 304–307, 310, 311
–, – – –, microtherm type, 304–306, 308–310

Oceania (*continued*)

–, comparison of different grassland regions in, 308, 430
–, grasslands of, 303–312
–, – –, hekistotherm sub-antarctic communities, 308
–, – –, hummock communities, 311
–, – –, megatherm communities, 306, 307
–, – –, mesotherm communities, 307–309
–, – –, microtherm communities, 308
–, – –, microtherm/hekistotherm communities, 308
–, – –, montane grasslands, 305, 307, 308
–, – –, origin, 304, 305
–, – –, palm communities, 307
–, – –, seasonally flooded communities, 309
–, – –, sedgeland, 311
–, – –, subalpine communities, 305, 308
–, – –, tussock communities, 306, 308, 311
–, – –, xerophytic communities, 307
–, livestock in, 315, 316, 319, 321
–, vegetation of, bioclimate classification, 305
–, – –, forests, 309, 463
–, – –, savannas, 304, 306, 307
–, – –, woodland savannas, 304, 306
–, – –, woodlands, 307
Odra river (Poland/Czechoslovakia), 157
Oka river (Russia), 139, 140
Old Man Range (South Island, New Zealand), 367, 379, 380, 383, 386, 388
oligochaetes
–, biomass of, 106
–, – –, review of Parts A and B, 478
–, secondary production of, 106
oligonitrophiles, 25
olives, 188, 190
omnivores, 77
– in soil, biomass of, 150
– – –, populations of, 150
Onega (former Soviet Union), 159
Opava river (Czechoslovakia), 138
Orange Free State [province] (South Africa), 274
–, mesic grasslands of, 274
orchid(s), 156
– (*Caladenia* spp.), 380
– (*Prasophyllum* spp.), 380
Ord River region (Western Australia, Australia), 325
Ordos plateau (northern part of the loess plateau) (China), 67
Orël (Russia), 158
organic matter in soil, content of, 63, 79, 86, 87, 153, 154
organization of grasslands, 17–26
oribi (*Ourebia ourebi*), 211, 280
Orissa [state] (India) (*see also particular localities*), 107, 108
Orthoptera, biomass of, 106
–, populations of, 106
oryx (*Oryx* spp.), 209, 211
–, beisa (*Oryx beisa*), 248
oryzoid grasses, 440
osmotic pressure, 18
ostrich (*Struthio camelus*), 209, 226, 248
Otago region (South Island, New Zealand), 365

Otradnoj (former Soviet Union), 152
overbrowsing by goats, 191, 192
overgrazing (overstocking) (*see also* grazing), 76, 78, 79, 84, 89, 94, 96, 110, 112, 113, 169, 176, 179, 188, 189, 191, 192, 206, 210, 215, 216, 235, 240, 241, 246, 247, 249, 273, 274, 278, 279, 282, 297, 299–301, 311, 319, 321, 325, 328, 332, 350, 354, 380, 402, 429
–, effects of, 88, 192
–, resistance of vegetation to, 278
overuse of land, 245, 249
Owen Stanley range (Northern Province, Papua new Guinea), 443, 448

Pacific grasslands
–, bioclimatic framework for, 435–437
–, origins of, 441, 442, 447, 460
Pacific Ocean (*see also particular territories and islands*), 125, 304
Pago volcano (New Britain, South-west Pacific), 448
Pahang [state] (Malaysia), 96
Pakistan (southern Asia) (*see also particular regions and localities*), 83–90, 93, 102, 104, 108, 109, 112, 113, 115, 117
–, grasslands of, 89, 90, 109, 112, 113, 115
–, – –, area of, 109
–, – –, *Chrysopogon* region, 89
–, – –, *Dichanthium–Cenchrus–Lasiurus* region, 89
–, – –, distribution map of, 90
–, – –, *Saccharum* azonal community, 89
–, – –, *Themeda–Arundinella* region, 89
–, map, of grasslands regions of, 90
palatability of herbage, 72, 79, 80, 88–90, 112, 113, 171, 174, 177, 192, 212, 213, 225, 226, 228, 229, 231, 235, 239, 240, 242, 246, 247, 253, 273, 276, 275, 278, 283, 285, 300, 319, 320, 321, 325, 326, 328, 330, 335, 350, 377, 381, 389
Palestine (Middle East), 2
palmoid life forms, 439
palms, 446, 448
Pamir–Altai mountains (Central Asia), 160
pandanoid life form, 439
panicoid grasses, 440
panicoid species, 341
Paparoa National Park (New Zealand), 403
paperbark tree (*Melaleuca quinquenervia*), 455
Papua New Guinea (*see also particular regions and localities*), 303–310, 437–439, 442–454, 462–466
–, bioclimatic regions, 444
–, megatherm bioclimate, 449
–, vegetation of, 443–454, 466
–, – –, alpine tussock grassland, 454
–, – –, bog, 443, 453
–, – –, fire-refugic thicket, 454
–, – –, forest, 447, 449
–, – –, heath, 452, 453
–, – –, megatherm seasonal grassland, 438, 444–449
–, – –, mesotherm grassland, 446
–, – –, montane megatherm-mesotherm grassland, 448–450
–, – –, montane mesotherm grassland, 450, 451
–, – –, montane mesotherm-microtherm grassland, 451–453
–, – –, open grassland, 443, 444, 448, 449, 451–453
–, – –, seasonally inundated grassland, 445

Papua New Guinea, vegetation of (*continued*)

–, – –, sedgeland, 445, 450
–, – –, short grassland, 452
–, – –, subalpine tussock grassland, 452–454
–, – –, swamp, 446–448
–, – –, tree-fern savanna, 453
–, – –, vine forest, 466
–, –, woodland, 445
–, vegetation zones of, 444
parasites of plants, 92
–, biomass in soil of, 150
–, populations in soil of, 150
parasitic (smut and rust) micro-fungi, 30
parrot, flightless (kakapo) (*Strigops habroptilus*), 393
partridges (*Francolinus*) spp.), 104
pastoral afar tribe, 257
pastoralism, 189
pastures (*see also* arable grassland), 79, 168, 179, 182–187, 189, 190, 225, 227–229, 231, 234, 235, 249
pea, pigeon (*Cajanus cajan*), 234
peat, peaty and organic soils, 72, 86, 296, 299, 372, 374, 414, 426, 429
peaty soils, nature of, 417
Pechora region (former Soviet Union), 159
pelican, spot-billed (*Pelecanus philippensis*), 77
penguin(s) (*Aptenodytes* spp.), 414
– (*Pygoscelis* spp.), 414
–, effect on grassland, 311
–, gentoo (*Pygoscelis papua*), 415, 427
–, king (*Aptenodytes patagonicus*), 427, 428
–, rockhopper (*Eudyptes chrysocome*), 427
–, royal (*Eudyptes schlegeli*), 427
periodicity of plant growth, 4, 101
–, seasonal dormancy, 4, 20, 32, 33, 78, 188
–, seasonal variation in rate of growth, 3, 26, 421
permafrost below ground, 11, 69, 72, 74
petrel(s), 425
–, giant (*Macronectes* spp.), 427
phanerophytes, 127, 128
phenological types of plants, 33
phenology of plants, 19, 20, 26, 27, 32–34, 36, 78, 97–99, 101, 138, 202, 206, 274–276, 333, 380, 381, 421
phenotypes, 33
Philippines (South-west Pacific), 83, 84, 86, 96, 102, 108, 111, 113, 303, 304, 307
–, grasslands of, 96, 102, 111, 113
phosphorus, 102, 103
–, content of herbage, 202
–, deficiencies of, 209
–, export of, 155
– in soil, 227, 245, 288, 347–349, 351, 384, 425
– – –, content of, 86, 87, 97, 115
–, mobility in plants, 155
–, pool of, 384
–, uptake by plants, 385
phosphorus-fixing soils, 288
photoperiodic amplitude, 309
photosynthesis
–, in grasses, 19

–, rate of, 142, 143, 154
photosynthetic C_3 and C_4 pathways (*see* acid decarboxylation)
photosynthetic efficiency, 207, 208
photosynthetically active radiation (PhAR), 134, 135, 208, 392
phreatophytes, 204
physiognomic structure of vegetation, 270
physiognomy of grasslands, 128, 270
phytocoenotypes, 31, 32, 39
phytogeographic elements, 31
phytomass, (*see* biomass *of particular plant parts*)
pig, 311, 430
–, bush (*Potamochoerus porcus*), 226
pigeon, snow (*Columba leuconta*), 77
–, white-collared (*Columba albitorques*), 226
pika, Turkestan red (*Ochotona rutila*), 76
Pilani region (Rajasthan, India), 86, 100
Pilbara Range (Western Australia, Australia), 319, 321
Pin-au-Haras (France), 149
pine (*Pinus* spp.), 84
piosphere, 354
pipit (*Anthus novaeseelandiae*), 77
pita, Daurian (*Ochotona daurica*), 43
Pitcairn Island (Southern Pacific), 303
plakor, 7, 31
plant/soil relationships, 381–385
Pleiku region (Vietnam), 96
plover, Caspian (*Charadrius asiaticus*), 241
poaching, 210, 241, 248–250, 254
Podor region (Senegal), 200
poisonous plant species, 80, 192, 213, 328, 335, 446
Poland (Central Europe)(*see also particular localities*), 134, 138, 140, 146–150, 152
polecat, European (*Mustela putorius*), 76
–, marbled (*Vormela peregusna*), 76
–, striped (*Zorilla striatus*), 211
pollination, 19, 32, 34, 446
–, method of, 34
Polynesia (Pacific Ocean), 304, 435
Polynesians, 305
Ponape region (Eastern Caroline Islands, Micronesia), 459, 465
ponding of cold air, effect of, 463, 464
pooid grasses, 453
porcupine (*Hystrix indica*), 104
–, North African (*Hystrix cristata*), 212
Port Hedland region (Western Australia, Australia), 321, 348
Port Moresby region (Papua New Guinea), 442, 443, 445, 463
Portugal (southern Europe), 171, 173
possum, Australian brush-tailed (*Trichosurus vulpecula*, 403
potassium
–, export of, 155
–, in soils, 97, 115
–, mobility in plants, 155
–, pool of, 384
–, uptake by plants, 385
precipitation (*see also* climate *of particular regions*)
–, effect on herbage yield, 276
–, effects of fluctuations in, 4
–, gradient of, 199, 203

precipitation (continued)

–, variability of, 199, 207
precipitation /temperature relationships, 63
predation of birds
–, by mammals, 428
–, by mustelids, 393
predator satiation, 393
predators, 146
– of livestock, 189, 190, 213
preservation of grasslands, 401, 403, 429
prickly pear (*Opuntia* spp.), 311
primary production (*see also* biomass; growth; photosynthesis), 180, 181, 187, 188, 207–209, 242, 274–278, 389–393, 399, 411, 421
–, above-ground, energy content of, 144
–, daily rate of, 102, 139, 208, 391, 424
–, effect of burning on, 392
–, effect of elevation on, 392
–, effect of insects on, 146
–, effect of soil moisture content, 102
–, energy fixed in, 147, 208
–, in relation to plant biomass, 103, 208
–, methods of study of, 389–392
–, of above-ground parts, 100, 101, 138, 141, 147, 153, 154, 423, 430
–, of under-ground parts, 100, 101, 138–40, 142, 153, 154, 423
–, review of Parts A and B, 475–477
–, stability of, 145
–, total biomass production, 102, 207
Primorskiy region (Siberia, Russia), 161, 162
Prince Edward Island (sub-Antarctic), 411, 419, 429
prions, burrow-nesting, 425
productivity, of ecosystems, 45
–, primary (*see* primary production)
–, secondary (*see* secondary production)
protein (*see* nitrogen)
protozoa, 150, 151
–, biomass of, review of Parts A and B, 480
psammophytes, 78
Psël river (Ukraine), 5
puku (*Kobus vardoni*), 280
pulse crops, 190
Puluthi region (Timor, Indonesia), 112
pumice areas, 363
Punjab [state] (India), 86, 89, 91, 107, 109
pyrethrum (*Chrysanthemum cinerariaefolium*), 226
pyrophytes, 192

Qilian Shan (mountains) (China), 71
Qinghai [province] (China), 75
Qinghai plateau (China), 71
Qinghai–Tibetan plateau (China), 72, 77
Queen Elizabeth National Park (Uganda), 242
Queensland [state] (Australia) (*see also particular regions and localities*), 303, 309, 316, 319, 321–323, 325–329, 333–336, 339–348, 352, 459
Quetta region (Pakistan), 89, 109

rabbit(s), 311
– (*Oryctolagus cuniculus*), control of, 383, 384
–, control of, 311, 430
–, effect on vegetation of grazing by, 311, 420, 428, 429
–, European (*Oryctolagus cuniculus*), 343, 344, 351, 353, 379, 383, 384, 427
radiation (*see* solar radiation)
rail, flightless (*takahe*) (*Notornis mantelli*), 393, 394
–, –, diet of, variation in, 395
–, –, effect on plants, 393
–, –, feeding habit of, 394
–, flightless (wekas), 311
rain-shadow
–, effect of, 229, 255, 367
–, grasslands of, 374
– regions, 307, 308, 363, 367, 370, 376, 378, 384, 390, 391, 394, 401–403, 456, 457
Raj Mahal hills (Nepal), 85
Rajkot region (Punjab, India), 107
Rakaia Valley (South Island, New Zealand), 382
Ramu rift valley system (Papua New Guinea), 443, 446, 448, 450, 465
range (grazing land) management, 188–190, 192–194, 210–212, 281, 283, 284, 287, 353–355, 429, 431
–, objectives of, 354
–, techniques of, 229
range management, socio-political factors affecting, 216
rangeland
– condition, categories of, 109, 113, 114, 176, 187, 244, 254
–, exploitation of, 190
–, improvements of (*see also* renovation), 192–194, 216, 254, 399
rat(s), 311
– (*Rattus* spp.), 427
– (*Uromys* sp.), 465
–, colonial mole (*Heterocephalus glaber*), 248
–, multimammate (*Mastomys*), 212
–, Nile (*Arvicanthis niloticus*), 212
–, sand (*Psammomys obesus*), 212
–, solitary mole (*Heliophobius argenteocinereus*), 248
ratel (*Mellivora capensis*), 239
Ratlam region (Madhya Pradesh, India), 86, 100
raven, thick-billed (*Corvus crassirostris*), 226
Red Sea coastal plain (East Africa), 255, 256
redshank, (*Tringa totanus*), 77
reducer organisms, 281
reduction
– of herbivore dung, 281
– of plant organic matter, 281
reedbuck (*Redunca arundinum*), 280
–, Bohor (*Redunca redunca*), 211, 231
–, mountain (*Redunca fulvorufula*), 280
refugia (relicts) of vegetation, 69, 443
Registan desert (Afghanistan), 89
reindeer (*Rangifer* sp.), 311, 427
Remarkables (South Island, New Zealand), 371
renovation (regeneration, rejuvenation) of degraded grassland, 108, 113, 116, 193, 194, 225, 286, 288
–, by artificial seeding, 286, 287
–, by fertilization, 286, 287

renovation (regeneration, rejuvenation) of degraded grassland (*continued*)

–, by re-development of the climax, 286
–, by weed control, 286
reptiles, 40, 75, 75, 420, 465
–, biomass of, review of Parts A and B, 478
respiration in soil (*see* carbon dioxide; root respiration)
– of plants, 142, 146, 153, 154
review of Parts A and B
–, activity of heterotrophs, 481
–, animal biomass, 476–480
–, micro-organisms, 480
–, primary production, 475–477
–, structure and biomass of vegetation, 471–474
–, turnover of plant biomass, 481
rhinoceros (*Rhinoceros* spp.), 104, 106
–, black (*Diceros bicornis*), 226, 248
–, white (*Ceratotherium simum*), 280
rhizomatous grasses and sedges, 4, 17, 22, 23, 32, 33, 47, 49, 50, 52, 64, 65, 74, 78, 224, 238, 253
rhizomatous grasses, effect of grazing on, 23
rhizomatous species, adaptations of, 129
rhizosphere, 24, 40–42, 147
–, micro-flora in, 137, 153
rhubarb, (domestic (*Rheum rhabarbarum*), 308
ribbok, vaal (*Pelea capreolus*), 280
rice (*Oryza sativa*), 300
–, wild, 306
Rift Valley region (East Africa), 236, 243
ring barking (girdling), to control woody plants, 282, 335
River Nile (North-east Africa), 232, 233
river terraces, 129, 130, 157, 159–61
Riverina region (New South Wales, Australia), 331–333
Rockhampton region (Queensland, Australia), 339
rodent/livestock interactions, 45
rodent mounds, vegetation on, 49–55
rodent/soil-fauna relationships, 146
rodent(s), 2, 42–55, 75, 76, 103, 190, 208, 210, 212, 248, 279, 300
–, adaptation to dry habitats of, 76
–, amount of excrement by, 45, 51
–, amount of soil excavated by, 51, 54
–, areas occupied by, 49
–, biomass of, 147, 241
–, – –, review of Parts A and B, 478
–, catalytic role of, 44–45
–, consumption by, 44, 45, 47, 146, 147, 149, 190, 209
–, degree of disturbance by, 48
–, density of burrows (mounds), 44, 49, 51–54, 76, 147
–, density of colonies, 49
–, diet of, 47
–, effect of burrowing by, 47–54
–, effect of trampling by, 47
–, effect on chemical composition of soil, 51–52, 54
–, effect on degradation of grasslands by, 76
–, effect on demographic structure of plant populations of, 46
–, effect on denudation/revegetation by, 54
–, effect on desalinization of, 54
–, effect on formation and decay of plant litter of, 46

–, effect on grassland of, 42–54, 190
–, effect on humus content of soil, 51
–, effect on length of vegetative period of, 46
–, effect on microbial decomposition of, 46
–, effect on microclimate of, 47
–, effect on plant succession of, 76
–, effect on radiation, 47
–, effect on soil moisture, 47, 54
–, effect on soil of, 44, 51–52
–, effect on structure of vegetation of, 46–47
–, effect on vegetative period, 47
–, effect on yield of phytomass, 45–47
–, energy relationships of, 44–45
–, energy values of faeces of, 147
–, energy values of food consumed by, 147
–, energy values of urine of, 147
–, feeding habits of, 44–47, 103, 104
–, feeding selectivity of, 46
–, in solonetz soils, 54
–, populations of, 45, 48, 52–54, 76, 146
–, role in desalinization of, 54
–, role in steppification of, 54
–, size of burrows (mounds) of, 49, 51–54
–, size of colonies of, 47
–, structure of burrows (mounds) of, 49, 51, 52
–, types of burrows of, 48
–, types of disturbance by, 48–54
–, vegetation on mounds, 49, 52–55
–, winter reserves for, 46, 47
Romania (southern Europe), 1, 10
root(s) (*see also* biomass; primary production; under-ground plant parts)
–, depth of penetration of, 136
root respiration, 108, 142
–, rate of, 152, 154
root/shoot ratio, 22, 99–03, 137, 138, 273, 276, 308, 389, 391, 392, 430, 452
–, effect of age of stand on, 137
–, review of Parts A and B, 477
root systems
–, fibrous, 18, 155
–, types of, 4
rubber vine, Madagascan (*Cryptostegia grandiflora*), 312
ruff (*Philomachus pugnax*), 241
Ruhuna National Park (Sri Lanka), 104
run-off water, 155, 158, 159, 188–190, 203, 216, 247, 286, 309, 316, 324, 327, 336
rushes, 419
Russia (*see also* former Soviet Union; Siberia; *and particular regions and localities*), 45
Ryazan' (former Soviet Union), 158
rye-grass, perennial (*Lolium perenne*), 348

Sagar region (Madhya Pradesh, India), 86, 97, 100
Sahara desert, 169, 197, 203, 214
–, shifting margins of, 215
Sahel region (*see also particular territories, regions and localities*), 197–216
Sahelian region, climate of, 197, 199–202
–, ecoclimatic subzones of, 197

GENERAL INDEX

Sahelian region (*continued*)

–, fauna of, 209–212
–, grasslands of, annual communities, 204, 205, 207, 215
–, livestock in, 197, 202, 203, 206, 208–210, 212–216
–, map of, 198
–, Saharan ecoclimatic zone, areal extent of, 198
–, Saharo–Sahelian subzone, 197, 199–203, 206, 208–215
–, – –, vegetation of, 203, 204
–, Sahelian ecoclimatic zone, areal extent of, 198
–, Sahelian subzone *sensu stricto*, 197, 199–212, 214
–, – –, vegetation of, 204–206
–, Soudanian ecoclimatic zone, 210, 215
–, – –, areal extent of, 198
–, Soudano–Sahelian subzone, 197, 199–202, 206–215
–, – –, vegetation of, 206
–, vegetation of, 169, 197–216
–, – –, bushland, 208
–, – –, savanna, 209
–, – –, successional (seral) communities of, 215, 216
–, – –, thornscrub, 204
–, – –, woodland, 197, 206, 214, 216
Sahelian steppes, 294
Sahelian zone, 222
saiga (*Saiga tatarica*), 23
Sakhalin island (Japan), 162
saline depressions, 374
saline soils and soil salinity, 8, 12, 63, 73, 77, 87, 89–91, 159, 161, 168, 177, 179, 202, 237, 238, 246, 252, 256, 299, 331, 414
–, response of plants to 73
salinization of soil, 3, 54, 76, 78, 159
salinization/desalinization of soil, effect of rodents on, 54
Salisbury (Zimbabwe), 269
saltbushes (*Atriplex* spp.), 307, 317, 329–333, 337, 349, 354
salts from the sea, 425
Sambalpur region (Orissa, India), 86, 100, 106
sand dunes, 89, 91, 96, 114, 177, 188, 202, 251, 252, 294
–, formation of, 112
sandy soil, 4, 7, 8, 14, 15, 20, 27, 30, 31, 41, 63, 78, 86, 91, 114, 160, 173, 177, 178, 180, 202–204, 225, 227, 229, 231, 232, 237, 238, 245, 246, 252, 253, 267, 268, 270, 273–278, 319, 325, 337, 339, 350
–, effect on herbage yield of, 181
saprophages, 23
saprophagy, 23
saprophytic micro-organisms, 46
Sardinia [island] (Mediterranean Sea), 189
Saudi Arabia (Middle East), 171
Saur ridges (Kazakhstan), 12, 16
Savaii region (Western Samoa), 308
savanna, definition of, 435–437
savannas, various types of (*see particular territories and regions*)
savannization, 460
scabweed (*Raoulia australis*), 383, 384
sclerophyllous shrubs, 175
sclerophyllous xerophytes, 216
sea lions (*Phocarctos hookeri*), 414
seal(s), 425, 429
–, disturbance of grassland by, 427

–, effect on grassland of, 311
–, elephant (*Mirounga leonina*), 414, 415
–, fur (*Arctocephalus* spp.) 414, 415
seasonal aspects, 138, 380, 381
seasonal dynamics of vegetation, 206
seasonal rhythm of activity of animals, 23
secondary production (*see also particular heterotrophic groups*), 394
– by deer, 394
– by grasshoppers, 397
– by insects, 148
– by livestock, 111
– by oligochaetes, 106
– by soil bacteria, 151
– by ungulates, 109
– by wild ungulates, 280, 281
– – – –, compared with livestock, 281
– efficiency of energy use, by deer, 394
– – – – –, by sheep, 394
sedge-like plants, 438
sedges, 4, 5, 8, 28, 45, 419, 438, 446, 448, 453, 459
–, bunch, mesoxerophilous, 5
seed
– dormancy, 131
– production (yield), 36, 38, 131, 207, 209, 393
– reserve in the soil, 36, 207
– supply, 131
seedling
– establishment, 36
– survival in grasses, 19
seedlings, populations of, 38
Selenga river (former Soviet Union), 4, 17
Semipalatinsk region (Kazakhstan), 12, 13
Senegal (West Africa) (*see also particular regions and localities*), 198, 203, 205, 206, 208
sensible heat flux, 153
Sepik Plains (Papua New Guinea), 309
Sepik river flood-plains (Papua New Guinea), 443, 447
seral species, 275
Serengeti plains (Tanzania), 239–241
serin, gold-footed (*Serinus pusillus*), 77
serval (*Felis serval*), 211, 226
Shaanxi [province] (China), 61
Shanxi [province] (China), 11, 17, 61
she-oak (*Casuarina equisetifolia*), 295
Shebelli river valley (Somalia), 250, 254
sheep (*Ovis aries*), 64, 65, 67–70, 74, 78, 94, 108–110, 147, 149, 159, 180, 188–192, 213, 214, 227, 243, 248, 250, 254, 256, 282, 311, 315, 319–321, 323, 324, 328, 329, 331–334, 336, 337, 342, 344, 345, 348, 349, 351–354, 367, 378–380, 394, 399, 400, 427–429
–, barbary wild (*Ammotragus lervia*), 211
–, biomass of, 254
–, consumption by, 394
–, diet of, 394
–, dietary preferences of, 332
–, efficiency of energy use by, 394
–, populations of, 315
–, Tibetan (a subspecies of *Ovis aries*), 70
–, use in control of woody plants, 354

shel-duck (*Tadorna tadorna*), 77
shifting cultivation (*see also* slash and burn agriculture), 95, 96, 113, 234, 254, 282, 300
–, effect of, 192
Shilka river (Siberia, Russia), 162
Shillong plateau (Nepal), 85
Shimba Hills area (Kenya), 231
Shiwalik hills (Jammu and Kashmir, southern Asia), 110
shoveller, northern (*Anas clypeata*), 77
shrubs
–, density of, 329
–, – –, effect on livestock, 333
–, lopped branches as feed for livestock, 336
–, nutrient content of, 336
–, thorny, 363, 374
Siberia (*see also* former Soviet Union; *and particular regions and localities*), 4, 9, 10, 12, 14, 78, 130, 160, 161
Sicily [island] (Mediterranean Sea), 189
Signy Island] (South Atlantic), 426
silicon, 154
Simpson Desert (Australia), 321
Sinai desert (Egypt), 440
Sind region (Pakistan), 89
sisal (*Agave sisalana*), 231
sky lark, small (*Alauda gulgula*), 77
slash and burn agriculture (*see also* shifting cultivation), 116, 167, 228, 231, 304, 306, 310, 443, 466
snails, 151
snake(s), 77, 300
–, python (*Liasis* spp.), 465
–, – (*Morelia* spp.), 465
sneezewood (*Ptaeroxylon obliquum*), 282
snipe, northern white-rumped (*Apus pacificus*), 77
snow, 67, 75, 77, 125, 159, 161, 162, 367, 371, 399, 416, 417, 419, 421, 453
snow grasses (*Poa* spp.), 344
snow patch grass (*Chionochloa oreophila*), 371
snow tussock(s) (*Chionochloa* spp.), 365, 374, 393
–, distribution of, 375, 376, 382
–, broad-leaved (*Chionochloa flavescens*), 363
–, curled (*Chionochloa crassiuscula*), 372, 375
–, mid-ribbed (*Chionochloa pallens*), 375, 402, 403
–, narrow-leaved (*Chionochloa rigida*), 368, 373, 379, 381, 382, 385, 388
–, needle-leaved (*Chionochloa acicularis*), 377
–, slim (*Chionochloa macra*), 369, 371, 380
snowcock, Tibetan (*Tetraogallus tibetanus*), 77
Snowdonia (Wales, U.K.), 149
social values, 249
sodium
–, in soil, 97, 384
–, mobility in plants of, 155
–, pool of, 384
Sogeri plateau (Papua New Guinea), 449
soil(s) (*see also particular soil types and particular grassland regions*), 281
– bacteria (*see* bacteria in soil)
– catenas, 227, 228, 233, 237, 238
–, chemical composition of, 47, 51, 155
–, colour of, black, 86, 233, 237, 238, 247, 273, 277, 341, 343

–, – –, brown–grey and grey–brown, 245, 371, 382
–, – –, brown–yellow–red, 232
–, – –, chestnut, brown and dark brown, 4, 13, 15, 22, 26, 27, 40, 52, 54, 63, 65, 66, 86, 87, 202, 223, 227, 230, 321, 327, 332, 337, 343
–, – –, dark-red, 227
–, – –, grey, 86, 227, 230, 233, 237, 238, 321, 332, 340, 343
–, – –, red, dark-red and reddish, 86, 223, 227, 230, 232, 237, 245, 246, 252, 277, 319, 321, 327, 328, 333, 338
–, – –, reddish-brown, 87, 327, 332, 337, 343, 344
–, – –, yellow, 87, 230, 319, 327, 333, 338
–, – –, yellow–brown, 372, 381, 382, 399
–, – –, yellow–grey, 371, 382
–, – –, yellow–red, 227, 230, 232
–, duricrust in, 291
–, effect of rodents on, 44, 51–52
– erosion (by wind and water), 63, 78, 112, 113, 136, 187, 192, 215, 233, 243, 245, 247, 250, 253, 273, 278, 279, 283, 284, 286, 291, 292, 294–296, 300, 301, 325, 328, 330, 332, 354, 366, 368, 369, 372, 378, 383, 399, 429
– –, extent of losses by, 203, 278
– –, protection from, 80
–, exchange capacity of, 245
– fauna (*see* fauna in soil)
– ferruginous crusts in, 299
– fertility, 87, 92, 115, 129, 151, 155, 156, 160, 214, 215, 224, 225, 227, 228, 230, 231, 235, 239, 241, 242, 245, 252, 253, 256, 275, 277, 287, 291, 294, 297, 309, 346, 349, 351, 371, 372, 424, 442, 448, 462, 464, 466
– –, relationship to fire-prone vegetation, 461
– –, relationship to floristic composition, 352
– formation, 156
– fungi (*see* fungi in soil)
– fungivores (see fungivores in soil)
–, hard-pan in, 238, 269, 327, 328
– humus, 18, 25, 51, 72, 151
– –, effect of rodents on content of, 51
– invertebrates (see invertebrates in soil), 147
–, leached, 238, 291, 344, 346, 283, 414
– macrofauna (*see* macrofauna in soil)
– mesofauna (*see* mesofauna in soil)
– microbivores (*see* microbiovers in soil)
– micro-organisms (*see* micro-organisms in soil)
– microfauna (*see* microfauna in soil)
– moisture content, 136, 145, 177, 247, 270, 340
– – –, effect of C_4/C_3 metabolism on, 440
– – –, effect of rodents on, 47, 54
– – –, effect on foliage cover of, 139
– – –, effect on herbage yield, 275
– – –, effect on primary production, 102
– – –, effect on root/shoot ratio of, 99
– nematodes (*see* nematodes in soil)
– nitrogen (*see* nitrogen in soils)
– nutrients (*see* nutrients in soil)
– –, response of plants to low supply of, 25
– omnivores (*see* omnivores in soil)
– organic matter, 63, 79, 86, 87, 153, 154, 245, 252, 253
– organisms, activity in relation to plant biomass, 151
– –, activity of, 155
– –, proportional distribution of, 150

GENERAL INDEX

soil(s), organismis (*continued*)

– –, species diversity of, 151
– respiration (*see* carbon-dioxide evolution from soil)
– structure, 23, 31, 63, 155, 156
– –, influence of plants on, 156
–, surface crust of, 278
–, surface-sealing of, 247
– type, effect on herbage yield, 180
–, types of (*see also* alkaline soils; alluvial soils; calcareous soils; saline soils; peaty soils, etc.)
–, – –, alfisols, 327
–, – –, altitudinal zonation of, 223
–, – –, andepts (andisols), 373
–, – –, aridisols, 202, 371, 373
–, – –, cambic and luvic arenosols, 202
–, – –, chernozems, 4, 12, 14, 17, 19, 22, 45–47, 49, 52, 65, 126, 153, 156, 158, 159
–, – –, colluvial, 294, 371
–, – –, cracking clay, 306, 321, 340, 343, 350, 352
–, – –, dystrochrepts, 372
–, – –, dystrophic, 277, 282
–, – –, eldefulvic, 372, 381
–, – –, eluvial, 245
–, – –, entisols, 373
–, – –, ferralitic and ferruginous soils, 202, 291, 296, 298, 299
–, – –, ferric luvisols, 202
–, – –, fulviform, 372
–, – –, gleyed, 72, 372, 383
–, – –, grumosols, 233
–, – –, gypsum and gypsic, 245, 246, 252, 253
–, – –, histosols, 373
–, – –, inceptisols, 371, 373
–, – –, laterites and lateritic, 86, 87, 97, 227, 230, 232
–, – –, latosols and latosolic, 223, 227, 232
–, – –, lithosols, 87
–, – –, loess and loess-like soils, 68, 86, 159, 371, 372
–, – –, palliform, 371
–, – –, podic, podzolized and podzols, 87, 158, 272, 327, 338, 343, 372, 374, 382
–, – –, psamments, 202
–, – –, regosols, 87
–, – –, sierozems, 67, 86, 87
–, – –, sitiform, 371
–, – –, skeleton soils, 87
–, – –, sodic, 245
–, – –, solods, solodic and solodized solonetz, 327, 343, 349
–, – –, solonchak, 12, 54, 86
–, – –, solonetz, 12, 14, 27, 53, 54
–, – –, spodosols, 373
–, – –, texture-contrast, 327, 328, 330, 338, 343, 352, 353
– water, rate of infiltration, 181, 278
soil-fauna/rodent relationships, 146
soil-forming processes, effect of soil fauna on, 150
soil-moisture stress, 142, 341, 399
soil/plant relationships, 97, 238, 239, 271, 272, 274, 299, 321, 440, 447, 448, 451, 460
soil-water potential, 142
solar energy
–, efficiency of fixation by plants of, 391

–, input of, 144, 153
–, reflection of, 153
solar radiation, 47, 62, 74, 153
–, effect of rodents on, 47
–, efficiency of conversion by plants of, 63, 102, 134, 143, 144
–, intensity of, 309
–, profile in the canopy of, 134, 135
Solling (Germany), 141, 148
Solling hills (Germany), 140
Solomon Islands (South-west Pacific), 307, 435, 441, 454, 455
–, vegetation of, 454–458
–, – –, megatherm seasonal grasslands, 455, 457
–, – –, short grassland, 455
solonetz soils, effect of rodents on, 54
Somalia (East Africa) (*see also particular regions and localities*), 168, 229, 231, 243, 245, 246, 248–251, 254–256
Sonbhadra region (Uttar Pradesh, India), 116
sorghum, 199, 234, 249, 254
souslikoviny (rodent mounds), 53, 54
South Africa (*see also* southern Africa *and particular states, regions and localities*), 169, 265, 267–272, 277, 283, 429
South African Savanna Ecosystem Study (Nylsvley, Transvaal), 276
South African vegetation
–, high veld, 269, 270, 272, 275, 280
–, *karoo* type, 269
–, mountain grasslands, 270
South America (*see also Part A*), 168, 169, 304, 309, 366, 416
South Australia [state] (Australia) (*see also particular regions and localities*), 316, 321, 327, 329, 330, 342
South-east Asia (*see also* southern Asia *and particular territories, regions and localities*), 84, 86–88, 95, 106, 111, 304, 442
South Georgia [island] (sub-Antarctic), 310, 411, 413, 415–421, 423–430
South Johnstone (Queensland, Australia), 316
South-east Asia (*see also particular territories and localities*)
–, grasslands of, 95–97
–, – –, climax communities, 95
–, – –, mammals of, 106
–, vegetation of, open woodlands, 95
–, – –, savannas, 95
South-west Pacific (*see also particular territories, regions and localities*), 303–309, 311, 435–467
–, bioclimates of, classification, 435, 440
–, – –, megatherm seasonal type, 442, 446, 466
–, – –, megatherm types, 437, 444, 454, 457, 459
–, – –, mesotherm bioclimate 437, 444, 459
–, – –, microtherm type, 444
–, geographical map of, 436
–, grasslands of, 303, 306, 435–466
–, – –, anthropogenic grasslands, 442
–, – –, classification, 305, 306, 435–440
–, – –, fauna of, 465
–, – –, lowland communities, 443
–, – –, megatherm seasonal communities, 443, 439
–, – –, short-grass communities, 464
–, – –, swamp, 438, 440, 443, 463, 464
–, – –, tall-grass communities, 438, 448
–, vegetation of, forest, 444

South-west Pacific, vegetation of (*continued*)

–, – –, *kunai* species of, 448
–, – –, low scrub, 459
–, – –, megatherm seasonal eucalypt woodland savanna, 444
–, – –, savannas, 435, 443, 465
–, – –, successional (seral) communities of, 441–443, 445, 446, 448, 449, 455, 456, 458, 459, 464
–, – –, tree-line boundaries, 442, 451
–, volcanic areas, 447, 448
southern Africa (*see also particular territories, regions and localities*), 265–288
–, geology and soils of, 270
–, grasslands of, 265–288
–, – –, annual communities, 274
–, – –, avifauna of, 279
–, – –, classification, 265
–, – –, climate of, 270
–, – –, climatic-climax communities, 272, 275, 278, 283, 287
–, – –, climax communities, 265, 266, 269, 272–274, 275, 277–279, 283
–, – –, desert grassland, 265, 266, 267, 269, 272, 274, 279
–, – –, disclimax communities, 279
–, – –, edaphic-climax grasslands, 267, 278
–, – –, fauna of, 279–281, 279
–, – –, fire-climax communities, 269, 273, 275–278, 287
–, – –, high veld, 265, 266, 270, 272, 274, 275, 277, 280
–, – –, high-elevation grasslands, 272
–, – –, hydromorphic communities, 267, 270, 273, 280
–, – –, induced (biotic) communities, 265, 269, 269, 273, 275–278, 287, 282
–, – –, livestock in, 278, 280–282, 285, 288
–, – –, low veld communities, 267
–, – –, mesic grasslands, 282
–, – –, mountain communities, 265, 266, 269, 270, 280
–, – –, savanna grasslands, 265, 266, 269, 275, 281
–, – –, seasonally flooded or saturated communities, 265, 266, 269, 272, 275, 280, 283
–, – –, semiarid, open grassland, 277
–, – –, sour veld communities, 270, 273, 283, 285, 286
–, – –, subclimax communities, 277
–, – –, successional (seral) plant communities of, 272–275, 277, 278
–, – –, swamp, 266
–, – –, sweet veld communities, 273, 274, 283–285
–, – –, veld, definition of, 265
–, vegetation of, *Acacia* associations, 270
–, – –, arid savannas, 273, 276
–, – –, *Baikiaea* associations, 268, 270, 273, 277
–, – –, *Burkea* savanna at Nylsvley, 281
–, – –, bush, 269, 273, 283
–, – –, desert, 272
–, – –, desert scrub, 269
–, – –, forest, 266, 269, 271–273, 279, 282, 283, 286
–, – –, *karoo* (dwarf shrubland), 266, 269, 272, 283
–, – –, *macchia* communities, 266, 283
–, – –, map of, 266
–, – –, mesic savannas, 270, 268, 277
–, – –, *miombo* communities, 266, 268–270, 273, 280
–, – –, mixed (mesic) savanna, 268, 277

–, – –, mixed savannas, 266, 269, 270, 274, 276
–, – –, moist savannas, 269, 270, 280
–, – –, savannas (wooded grasslands), 265, 266–274, 276–284, 287, 288
–, – –, sclerophyll scrub, 269
–, – –, scrub savanna, 272
–, – –, semi-desert scrub, 269
–, – –, semiarid savannas, 266–270, 276, 277, 281, 288
–, – –, temperate sclerophyllous *fynbos* (*macchia*) invade, 272
–, – –, thorn savannas, 267
–, – –, wet savannas, 269
–, – –, woodland, 265, 267–269, 272, 273, 277, 280, 282
–, – –, xeric savannas, 269, 271
Southern Alps (South Island, New Zealand), 308, 361, 363, 365, 366, 368, 370, 372, 374, 375, 389
southern Asia (*see also particular territories, regions and localities*), 02, 83–117
–, climate of, 85, 127
–, grasslands of, 83–117
–, – –, alluvial grassland, 104
–, – –, alpine communities, 88, 90, 92, 99
–, – –, areal extent of, 83
–, – –, classification of, 88
–, – –, climate of, 84–86
–, – –, desert types, 103
–, – –, disclimax and subclimax communities, 84, 91
–, – –, fauna of, 106–106, 103
–, – –, hydrophilous steppe, 88
–, – –, marshland, 88, 94, 112
–, – –, meadows, 108, 110
–, – –, salt marshes, 87
–, – –, soils of, 86, 87
–, – –, successional (seral) communities of, 83, 96, 103, 112, 113, 116
–, – –, swamps, 91, 94, 105
–, – –, types of, 88–99
–, livestock in, 83, 88, 89, 108–110, 112–114
–, pastures of, 83, 84, 93, 102, 108, 109, 111, 113, 115–117
–, vegetation of, forests, 83–85, 87–97, 103, 106, 113
–, – –, open forests, 83, 84, 95, 96,
–, – –, savanna woodlands, 84
–, – –, savanna-like vegetation, 88
–, – –, savannas, 83, 87, 88, 90, 92, 93, 95, 96, 104
southern cool-temperate islands, map of, 412
Southern Hemisphere, 366, 414
Southern Highlands (Tanzania), 225, 235
Soviet Union (former) (*see also* Azerbaydzhan; Belorussiya; Kazakhstan; Russia; Siberia; Tadzhikistan; Ukraine; Uzbekistan; *and particular regions and localities*), 1, 3, 4, 8, 10, 19, 25, 51, 53, 54, 61, 67, 76, 78, 83, 129, 136,138–140, 152, 158, 159
–, grasslands of, 3–55
Spain (southern Europe), 190, 193
–, areal extent of grassland in, 171
sparrow, tree (*Passer montanus*), 77
sparrow-hawk (*Accipiter nisus*), 77
spear grass, black (*Heteropogon contortus*), 449
speargrass (*Aciphylla scott-thomsonii*), 369, 381
species composition, effect of fertilization on, 275
spinifex (*Plectrachne* spp. and *Triodia* spp.), 319–321, 328

GENERAL INDEX

spinifex hummock grasses, 306
springbok (*Antidorcas marsupialis*), 280
squirrel, ground (*Citellus dauricus*), 76
-, - (*Euxerus erythropus*), 212
-, striped ground (*Xerus erythropus*), 211
Sri Lanka (southern Asia) (*see also particular regions and localities*), 2, 84, 85, 87, 88, 92, 99, 104, 107, 108, 110, 112, 115, 116
-, grasslands of, 92–94, 107, 112, 115, 116
-, - -, *damana* (lowland) types, 87, 92, 93, 99, 101
-, - -, *patana* (montane) types, 87, 92, 93
-, - -, *talawa* type, 92
-, - -, vertebrates of, 104
-, - -, wet *villu* type, 92, 94
-, vegetation of, savannas, 93
St. Bathans Range (South Island, New Zealand), 382
St. Helena [island] (Atlantic Ocean), 428
St. Paul [island] (Indian Ocean), 419
St. Petersburg (formerly Leningrad) district (Russia), 139
St-Louis (Senegal), 202
stability of vegetation, 10, 18, 20, 25, 34–36, 73, 78, 103, 126, 155, 162, 278, 279, 288, 325, 351, 352, 354, 375, 377, 388, 389
standing crop, (*see* biomass)
steenbok (*Raphicerus campestris*), 280
steppe/desert plant species, 31
steppes, various types of (*see particular territories and regions*)
steppification of semi-deserts, role of rodents on, 54
Stewart Island (New Zealand), 376
stoat (*Mustela erminea*), 393
stocking rate of livestock (*see also* carrying capacity), 65, 67–69, 108–112, 116, 180, 189, 193, 212, 227, 231, 242, 243, 250, 254, 256, 322–324, 328, 300, 333, 337, 344, 348, 349, 394
-, control of, 379
stony (gravelly) soils, 8, 25, 48, 67, 71, 86, 204, 227, 233, 267, 383
Streletskaya steppe (Kursk region, Russia), 45–47, 49, 52
Streltsovskaya Steppe Reservation (Voroshilovgrad region, Ukraine), 7, 24, 45, 52
strontium, 154, 155
-, mobility in plants of, 155
structure of vegetation, 419, 420
-, effect of burning on, 84
-, effect of rodents on, 46–47
-, review of Parts A and B, 471–474
sub-Antarctic islands (*see also particular islands*), 411–431
-, climate of, 309, 413, 414, 416, 417, 430
-, geographic data of, 413
-, livestock in, 427
-, map of, 412
-, vegetation of, 303–306, 308, 309, 311, 411–431
-, - -, avifauna of, 420, 425, 429
-, - -, bryophyte communities, 416
-, - -, classification of grasslands, 417
-, - -, cool-temperate grassland, 419
-, - -, distribution of species, 419
-, - -, graminoid tundra, 417
-, - -, herbfield vegetation, 411
-, - -, low forest, 411

-, - -, low sub-Antarctic tundra grassland, 417, 418
-, - -, medium tall grassland, 417, 419
-, - -, mire grassland, 414, 418, 419, 421–423, 430
-, - -, peat bogs, 418
-, - -, refugial sites, 430
-, - -, short grassland, 417, 423
-, - -, successional (seral) plant communities of, 420, 427, 429–431
-, - -, tall scrub, 411
-, - -, tundra-like feldmark, 414
-, - -, tussock grassland, 414, 415, 417, 419, 420, 422–425, 427–430
sub-Antarctic microtherm grasses, 308
sublethal period in plants, 19
succession of plant species
-, cyclic, between closed forest and grassland, 462
-, -, human role in, 466
-, following drought, 207
- on rodent mounds, 49–50, 52, 54
-, rate of, 388
successional (seral) plant communities (*see particular regions*)
succulent species, 246, 292
Sudan (North Africa), 169, 198, 206, 210, 215
sugar cane (*Saccharum* spp.), 229, 310
Sujfun river (Siberia, Russia), 162
sulphur
-, in soil, 348
- pool, 384
Sumatra (island) (Indonesia), 86, 95, 106
Sumba (island) (Indonesia), 95
Sumbawa (island) (Indonesia), 95
Sumy region (Ukraine), 5
Sundarbans tidal forest (India/Bangladesh), 91
suni (*Nesotragus moschatus*), 231
sunshine, 63
surchiny, (*Marmota* hills), 51, 52
suslik (*Citellus pygmaeus*), 43
suslikoviny (*Citellus* hills), 51
Swahili people (East Africa), 230
swallows (*Hirundo* spp.), 241
swan, mute (*Cygnus olor*), 77
Swat region (Pakistan), 89
sweet potatoes (*Ipomoea batatus*), 226
synusiae, 4, 18, 32, 35, 36, 78
Syria (Middle East), 2, 171
Sysola region (former Soviet Union), 159

Tabora (Tanzania), 271
Tadzhikistan (*see also* former Soviet Union), 160
Tahoua region (Niger), 200
tahr (*Hemitragus jemlahicus*), 104
Tamakaduwa region (Sri Lanka), 94
Tambov region (former Soviet Union), 158
Tamworth region (New South Wales, Australia), 344
Tanna [island] (Vanuatu), 456
Tanzania (East Africa) (*see also particular regions and localities*), 221, 222, 225, 226, 229, 235, 237, 239–243, 248, 265, 270
-, *miombo* woodland of, 221
Tararua Range (North Island, New Zealand), 386

tarbagan (*Marmota sibirica*), 43
Tarbagatai mountains (China), 12, 15, 16
Tarim river (China), 76
Tasmania [state] (Australia), 303, 304, 342–344, 348, 353, 429
tea (*Camellia sinensis*), 226
teal, Baikal (*Anas formosa*), 77
tectonism, effect on vegetation, 456
Temir Sonal Experimental Station (Aktyubinsk region, Kazakhstan), 54
temperature accumulations, 63–69
termite(s), 106, 107, 208, 226, 228, 231, 234, 238, 241, 248, 256, 281
–, activity of, 281
–, biomass of, 256
–, consumption by, 241, 256
–, effects on micro-organisms, 107
–, feeding habits of, 106
–, harvester (*Hodotermes mossambicus*), 281
–, –, activity of, 281
–, –, consumption compared to ungulates, 281
–, –, pest on rangeland, 281
– (*Macrotermes* spp.), 228
–, mounds of, 445
–, – –, density of 228, 234
–, secondary production by, 241
Thailand (southern Asia), 84, 86, 87, 95, 102, 108, 111, 113
–, grasslands of, 95, 102, 111, 113
–, open forests in, 87
thar, Himalayan (*Hemitragus jemlahicus*), 394, 403
thermophilous grasses, 67, 69
therophytes, 127, 128
therophytism, 351
Thie valley (New Caledonia), 456
Thimbu region (Bhutan), 94
thorn forests, 83, 97
thorn scrub, 84
thorn-bushes, 68
three-awn, spiny (*Stipagrostis pungens*), 171
Tibetan plateau (Xizang Gaoyuan) (China), 61, 69, 70, 72, 75, 77
tick, cattle (*Boophilus microplus*), 341
Tien Shan (mountains) (China), 70, 71, 76, 77, 160
Tien Shan region, 2
Tierra del Fuego (South America) (*see also Part A*), 419
Tigris river (Iraq/Turkey), 87
tillering, 18, 26, 283, 284, 325
Timor (island) (Indonesia), 95, 112
Tombouctou (Timbucktoo) (Mali), 200
topi (*Damaliscus korrigum*), 211, 231
topographic features, *melkosopochnik*, 8
–, *sopki* (small hills), 12, 15, 53
Torres Strait (Papua New Guinea/Australia), 445
trace elements, 349
–, deficiencies of, 346
trade winds, 199, 202, 291
trampling by animals (*see also* grazing), 64, 112, 162, 177, 207, 208, 329
–, effects on vegetation, 18, 47, 414, 429
trampling by reindeer, effect on grassland, 429
Trangie region (New South Wales, Australia), 337

Transbaykal region (former Soviet Union), 4, 10, 17, 43, 49, 51, 53
Trans-Fly region (Papua New Guinea), 445, 465
translocation of assimilates, 137, 142, 157, 202
–, model of, 143
transpiration, 208
–, loss of energy in, 153
–, rate of, 18, 139, 143–48, 400
Transvaal [province] (South Africa), 268–270
Transvolga region (former Soviet Union), 2, 9–13, 16, 18, 20–22, 26
tree, densities of, 340
tree-ferns, 307, 452–454
treelessness, reasons for, 272, 273
trees within grassland regions, 8
tribal areas, 249
tribal law, 243
Tristan da Cunha (South Atlantic), 419
trunk-forming grasses, 451, 452
trypanosomiasis (sleeping sickness), 235
Tsavo National Park (Kenya), 248
Tselinograd region (Kazakhstan), 10, 22
tsessebe (*Damaliscus lunatus*), 280
tsetse flies (*Glossina* spp.), 226, 229, 231, 233–235, 243, 249, 254
Tula (former Soviet Union), 158
Tunisia (North Africa), 169, 171, 176–181, 190, 193
Turkey (Middle East) (*see also particular regions and localities*), 2, 83–88, 108, 110, 114, 171
–, grasslands of, 88, 89, 110, 114
turnover
–, of invertebrate biomass, 106
–, of nutrients, 208
–, – –, rate of, 46
–, of organic material, 45, 46, 281, 389, 393
–, – – –, rate of, 46, 102, 129, 138, 151, 276, 426
–, review of Parts A and B, 481
tussock (bunch) grasses, 93, 112, 176, 307, 311, 320, 321, 340, 342, 363, 376, 415, 419, 426, 438, 453
–, adaptations of, 431
tussock peat, 419
–, earthworms in, 426
tussocks, regeneration of, 421
Tuvalu (islands) (South-west Pacific), 303
types of grasslands (*see* classification *and particular territories and regions*)

Udaipur ergion (Rajasthan, India), 86, 100
Uganda (East Africa), 222, 224, 225, 227, 228, 232–235, 239, 240, 242, 243
Ujjain region (Madhya Pradesh, India), 86, 97, 100
Ukraine (*see also* former Soviet Union; *and particular regions and localities*), 1, 5, 7, 8, 12, 14, 24, 25, 42, 43, 45, 50, 52
Ulaanbaatar (Mongolia), 43, 44, 53
under-ground plant parts (*see also* biomass; decomposition; nutrients; primary production)
–, age of, 137
–, structure of, 136, 137
–, surface area of, 18, 137, 139
–, translocation to, 143

GENERAL INDEX
555

under-ground plant parts (*continued*)

–, vertical distribution of, review of Parts A and B, 473
ungulate communities, estimating populations of large, 279
ungulates (*see also* herbivores *and particular groups and species*)
– (hoofed animals), 23, 40, 73, 75, 226, 231, 300
–, biomass of, 109, 209, 231, 240, 241
–, – –, review of Parts A and B, 478
–, dispersal rates on introduction of, 394, 395
–, effects on vegetation of, 23
–, efficiency in energy conversion, 310
–, large wild, 239, 243
–, migration of, 75
–, naturalized, population dynamics of, 395
–, populations of, 104, 240, 241
–, role in favouring bunch grasses, 23
–, seasonal migrations of, 240, 248
–, wild, 250, 253
–, –, competition with livestock, 256
United Kingdom (Western Europe), 157, 429
Ural mountains (former Soviet Union), 12, 53, 54, 129, 158
Ural river (former Soviet Union), 12, 43, 158, 160
urial (*Ovis orientalis*), 104
uses of grasslands
–, as a source of salt, 465
–, as arrow shafts, 464
–, as food, 465
–, as gene pools, 83
–, as roofing thatch for dwellings, 464
–, for grazing for draught animals, 83, 300
–, for medicine, 80, 155
–, in making ropes and mats, 80, 171
–, in paper making, 80, 171
uses of livestock
–, dung, 214
–, hair and wool, 213, 214
–, hides and skins, 213
Ussuri region (Siberia, Russia), 162
Ussuri river (Siberia, Russia), 162
Uttar Pradesh [state] (india) (*see also particular localities*), 90, 91, 109, 115
Uzbekistan (*see also* former Soviet Union), 160
Uzen' river (former Soviet Union), 159

Vanua Levu [island] (Fiji), 457
Vanuatu (formerly New Hebrides, South-west Pacific), 435, 442, 454–457, 466
–, vegetation of, 456, 457, 459–461
–, – –, cleared forest, 460
–, – –, forest–grassland mosaics, 457, 460
–, – –, lowland grassland, 457
–, – –, megatherm grassland, 456, 459, 461
–, – –, mesotherm grasslands, 456, 457
–, – –, secondary forest, 461
–, – –, tall grassland, 457, 461
vapour pressure deficit, 202
Varanasi region (Uttar Pradesh, India), 86, 99, 100, 106
vegetative growing season, 5
vegetative reproduction, 137
verbenaceous species, 457

vertebrate herbivores, 426
vertebrate predators, 104, 146
vertebrates, 45, 103–106, 210, 393–396
–, biomass of, 23
–, feeding habits of, 104
vervet, (*Cercopithecus aethiops*), 211
Victoria River region (Western Australia, Australia), 321
Victoria [state] (Australia) (*see also particular regions and localities*), 321, 343, 344, 348, 352, 353
Vietnam (South-east Asia) (*see also particular regions and localities*) (South-east Asia), 84, 86, 87, 96, 108
–, grasslands of, 96
vitamins, 202
Viti Levu [island] (Fiji, South-west Pacific), 457, 458, 462
volcanic activity, 222, 227
volcanic ash, 373
volcanic mountains, 366
volcanic soils ,340
volcanism, effect on vegetation, 448
vole (*Microtus brandtii*), 76
– (*Microtus oeconomus*), 76
–, Brandt's (*Lasiopodomys brandtii*), 43
–, Chinese (*Lasiopodomys mandarinus*), 43
–, common (*Microtus arvalis*), 43
–, mole (*Ellobius talpinus*), 43
–, narrow-skulled (*Microtus gregalis*), 43, 76
–, social (*Microtus socialis*), 43
Volga river (former Soviet Union), 43, 51, 53, 53, 54
Vologda (former Soviet Union), 159
Voroshilovgrad region (Ukraine), 7, 24, 45, 52
vulture, Egyptian (*Neophron percnopterus*), 241
–, hooded (*Necrosyrtes monachus*), 241
–, nubian (*Torgos tracheliotus*), 241
–, Ruppell's (*Gyps ruppelli*), 241
–, white-backed (*Gyps bengalensis*), 241
–, white-headed (*Trigonoćeṗs occipitalis*), 241

wagtail, grey (*Motacilla citreola*), 77
wallabies, 304, 353, 465
wallaby (*Macropus agilis*), 310
Wanderrie banks (Australia), 328
warbler, great reed warbler (*Acrocephalus arundinaceus*), 77
warm-season plant species, 272, 319, 343, 344, 350, 351
warthog (*Phacochoerus aethiopicus*), 211, 228
Wassi Kussa river region (Papua New Guinea), 445
water
– balance, ombrothermic diagrams of, 200
– erosion (*see* soil erosion)
– regime, 145
– relations in plants, 18, 34, 143, 144
– saturation deficit, 18, 145
– stress, 129, 157, 157, 275, 440, 451
– use, efficiency of, 180, 187, 208, 209, 391
– yield, effect of vegetation on, 400
water-table, 91, 129, 157, 179, 189, 203, 204, 238, 278, 269, 420
–, in relation to fire, 462
waterbuck(s) (*Kobus* spp.), 209, 211, 226
– (*Kobus ellipsiprymnus*), 280
waterfowl, 77

waterlogged soils, (see also hydromorphic soils), 90, 91, 225, 247, 272, 273, 353
watermelons, 254
weasel (*Mustela altaica*), 76
–, Libyan striped (*Poecilictis libyca*), 211
–, Siberian (*Mustela sibirica*), 76
weka (*Gallirallus australis*), 427
West Africa (see also particular territories, regions and localities), 221
West Bengal [state] (India), 90, 91, 106
West Cape (Fiordland, New Zealand), 377
Western Australia [state] (Australia) (see also particular regions and localities), 307, 315, 320, 321, 327–331, 333, 337, 338, 342–344, 349–351, 353
Western Ghats (mountains) (India), 91
Western Hemisphere (see also Part A), 1, 169
Western Samoa (Polynesia), 435
Whanganui National Park (New Zealand), 403
wheat (*Triticum* spp.), 145, 226
wheatear (*Oenanthe oenanthe*), 77
–, isabelline (*Oenanthe isabellina*), 77
wild browsers, 280
wild fires, 374
wild grazers, 280
wild ungulates, stocking rate of, 210
wildebeest (*Connochaetes taurinus*), 236, 242, 279
–, black (*Connochaetes gnou*), 280
wildfires, 189, 191, 192, 203, 353
–, effect on rate of mineralization of nutrients, 352
wildlife (see also particular groups and species of animals), 197, 202, 415
– biomass, 254
– –, fluctuation in, 254
–, economic value of, 241, 256
–, overgrazing by, 242
Wiluna region (Western Australia, Australia), 331
Wimmera region (Victoria, Australia), 352
wind, 62, 63, 74
– erosion (see soil erosion)
wind-breaks, 115
wire grass (*Sporobolus indicus*), 458
Wisła river (Poland), 140, 147
wood
– production, 214
–, scarcity of, 78
woodlands, various types of (see particular territories and regions)

woody *miombo* species, 273

xeromesophilous forbs, 5
xeromesophytes, 72, 160
xeromorphic plant species, 67, 203, 306
xeromorphous steppe, 88
xeromorphy, plant, 18
xerophilous plant species, 5, 7, 64, 66, 67, 71, 299
xerophily, gradations of, 5
xerophytes, 64, 67, 161, 177, 179, 216, 307, 340
xerophytism, 351
Xilin Gol river (China), 67
Xinjiang Uygur Zizhiqu [autonomous region] (China), 71, 75, 76

yak, wild (*Bos grunniens mutus*), 74, 76
Yakutsk region (former Soviet Union), 161
Yap Island (Western Caroline Islands, Micronesia), 460, 463
yeasts, 151
yeasts, biomass of, review of Parts A and B, 480
Yellow river (Huang He) (China), 69
yellow-woods (*Podocarpus* spp., 282
Yenisey river (former Soviet Union), 161
Yergeni (Volgograd region, former Soviet Union), 12, 16, 54
Yergeni heights, 12
Yugoslavia (Eastern Europe), 1, 171

Zabaykalye (Transbaykal, former Soviet Union), 4
Zagros mountains (Iran), 87, 88
Zaïre river, 221
Zambia, 265, 268, 270, 270, 280
Zavolzhye (see Transvolga region), 12
zebra (*Equus burchelli*), 226, 228, 280
–, Grevy's (*Equus grevyi*), 256
–, mountain (*Equus zebra*), 280
Zimbabwe, 265, 267–271, 277, 281
zinc, 154, 155
– in soil, 346, 348
–, mobility in plants of, 155
Zinder (Niger), 201
Znojmo (Czechoslovakia), 127
zokor (*Myospalax aspalax*), 76
–, common Chinese (*Myospalax fontanierii*), 76
zoogenic regulation, 46
Zwartkop, (Natal, South Africa), forest of, 283